T0292558

Studies in Computational Intelligence

Volume 621

Series editor

Janusz Kacprzyk, Polish Academy of Sciences, Warsaw, Poland
e-mail: kacprzyk@ibspan.waw.pl

About this Series

The series "Studies in Computational Intelligence" (SCI) publishes new developments and advances in the various areas of computational intelligence—quickly and with a high quality. The intent is to cover the theory, applications, and design methods of computational intelligence, as embedded in the fields of engineering, computer science, physics and life sciences, as well as the methodologies behind them. The series contains monographs, lecture notes and edited volumes in computational intelligence spanning the areas of neural networks, connectionist systems, genetic algorithms, evolutionary computation, artificial intelligence, cellular automata, self-organizing systems, soft computing, fuzzy systems, and hybrid intelligent systems. Of particular value to both the contributors and the readership are the short publication timeframe and the worldwide distribution, which enable both wide and rapid dissemination of research output.

More information about this series at http://www.springer.com/series/7092

Rami Abielmona · Rafael Falcon
Nur Zincir-Heywood · Hussein A. Abbass
Editors

Recent Advances in Computational Intelligence in Defense and Security

Springer

Editors
Rami Abielmona
Larus Technologies Corporation
Ottawa, ON
Canada

Rafael Falcon
Larus Technologies Corporation
Ottawa, ON
Canada

Nur Zincir-Heywood
Dalhousie University
Halifax
Canada

Hussein A. Abbass
University of New South Wales
Canberra, ACT
Australia

ISSN 1860-949X   ISSN 1860-9503   (electronic)
Studies in Computational Intelligence
ISBN 978-3-319-26448-6   ISBN 978-3-319-26450-9   (eBook)
DOI 10.1007/978-3-319-26450-9

Library of Congress Control Number: 2015955362

Springer Cham Heidelberg New York Dordrecht London

Printed on acid-free paper

Springer International Publishing AG Switzerland is part of Springer Science+Business Media
(www.springer.com)

Foreword

Complex, volatile, and uncertain—this is the new reality of our world today that now causes us to rethink how we anticipate, manage, and shape desired outcomes as, and preferably before, they happen. Where behaviors were once determined by fixed and known outcomes or end-states, set far downstream, today's behaviors struggle to keep up with an ever-changing environment where the end-state or destination is as obscured as the road and the journey ahead of it. The signs are less clear and the choice of paths more than we can manage or possibly know. By all accounts, we no longer enjoy the sanctity of what former U.S. Secretary of Defense Donald Rumsfeld often referred to as the *Known–Knowns*, and have skipped clear past the *Known–Unknown* domain, and landed squarely in *Unknown–Unknown* territory, but not for the reasons one might think.

It is not that we lack the understanding or even the capacity to rationalize the complex nature of the changing world around us. Rather much the opposite. We've become very aware that the speed and the nature of change have now outstripped our ability to manage it deliberately using the same orthodox methods. Take for instance today's highly charged information-driven environment. While it may have brought the world closer together, and this is a good thing, the downside is that we now live in a world where data overload makes the simplest bits of information difficult to see even when placed directly in front of us. And when we do see, we don't truly know. The challenge we face in this new era will be in methodically transforming the *Unknown–Unknown* space back to its more manageable semi-state—the *Known–Unknown*—a state albeit fluid, but with more predictable outcomes.

So how do we stay ahead of a constantly evolving landscape where not only will we face a mountain of data coming at us from every source imaginable (and some not so obvious), but we aren't even sure where to begin, or how to interpret what we find actually has relative value, or how quickly we should act before the value is lost? And when we do act, how confident are we with the quality and reliability of what governments and defense and security agencies often refer to as *actionable intelligence*? The reality is that as much as there are noteworthy examples where actionable intelligence saved the day, there are just as many instances, if not more, where it all went horribly wrong because we simply trusted that it was good enough. *Trust, but verify*, as former US President Ronald Reagan used to say; and in today's fast-paced chaotic environment, a robust and highly intuitive verification process is fundamental.

In much the same way a pilot flying today's extraordinarily sophisticated aircraft, who when encountering a complex flight dynamics problem, is trained to trust his or her instruments when the natural sensory cues suggest he or she do otherwise, dealing with today's complex problems requires an equally robust and highly federated system of inputs and validations to reduce the risks of making irreversible or fatal mistakes. This isn't a criticism on present-day piloting skills, but rather an acknowledgment that flying these incredibly sophisticated machines is far easier to do with the best decision-support tools at the pilot's disposal. Our security environment is no different.

Today's security environments are so multifaceted that the simplest of errors in judgment or a missing piece or an action out of sequence can have severe consequences. Moreover, where there was once a clear demarcation both in time and space between, for example, military and civilian objectives, present-day joint and combined operations now include a variety of state-sponsored and independent actors, each with specific requirements that further complicate the decision-making process. The demands on near flawless evidence-based decision making are so extraordinarily high that tolerance for getting it wrong is virtually zero. Even if *To err is human, forgiveness will surely not be divine* when the stakes are so high. Possessing the right intelligence tools at the right time and for the right circumstances is paramount.

Recent Advances in Computational Intelligence in Defense and Security offers a very practical and intuitive glimpse into the leading-edge science of predicative analysis in complex problem sets. Using the most advanced *Computational Intelligence* (CI) tools and techniques ranging from game theory—to fuzzy logic— to swarm intelligence (and much more), CI provides both the discipline and depth to help us foresee and more effectively deal with many of today's and tomorrow's seemingly intractable problems. It is the perfect marriage of art and science, much in the same way early Artificial Intelligence was envisioned to be—a combination of both the human experience and machine logic with highly intuitive and multi-layered rule-based precision.

And finally, while the case studies used in this book may be focused, for the most part, on Defense and Security, I encourage you to think of CI in a much broader context. And don't be fooled by the book's technical flavor either. While it may appear to be written by scientists for scientists, it is very much highly recommended reading for the person(s) who seek to better understand, manage, and shape the complex environments that surround them with the help of some of the most powerful decision-support tools around today. So enjoy the book and see firsthand the power of Computational Intelligence and begin to imagine the applications and potential it has to offer in your world.

Rick Pitre
Brigadier-General (Retired)
Royal Canadian Air Force, CD

Contents

Recent Advances in Computational Intelligence in Defense and Security

Rami Abielmona, Rafael Falcon, Nur Zincir-Heywood and Hussein Abbass

1 Introduction

Given the rapidly changing and increasingly complex nature of *global security*, we continue to witness a remarkable interest within the defense and security communities in novel, adaptive and resilient techniques that can cope with the challenging problems arising in this domain. These challenges are brought forth not only by the overwhelming amount of data reported by a plethora of sensing and tracking modalities, but also by the emergence of innovative classes of decentralized, mass-scale communication protocols and connectivity frameworks such as *cloud computing* [5], *sensor and actuator networks* [7], *intelligent transportation systems* [1], *wearable computing* [2] and *the Internet of Things* [6]. Realizing that traditional techniques have left many important problems unsolved, and in some cases, not adequately addressed, further efforts have to be undertaken in the quest for algorithms and methodologies that can accurately detect and easily adapt to emerging threats.

Computational Intelligence (CI) [4] lies at the forefront of many algorithmic breakthroughs that we are witnessing nowadays. This vibrant research discipline offers a broad set of tools that can deal with the imprecision and uncertainty prevalent in the real world and can effectively tackle ill-posed problems for which traditional (i.e., hard computing) schemes do not provide either a feasible or an efficient solution. The term CI is not exclusive to a single methodology; rather, it acts as a large umbrella under which several biologically and linguistically motivated techniques have been developed [3]—some of them enjoying unprecedented popularity these days [4]. CI has expanded its traditional foundation (pillared on *artificial neural networks*, *fuzzy systems* and *evolutionary computation*) to accommodate other related

R. Abielmona · R. Falcon (✉)
Larus Technologies Corporation, 170 Laurier Ave West - Suite 310,
Ottawa, ON K1P 5V5, Canada
e-mail: rafael.falcon@larus.com

N. Zincir-Heywood
Dalhousie University, Halifax, Canada

H. Abbass
University of New South Wales, Canberra, Australia

© Springer International Publishing Switzerland 2016
R. Abielmona et al. (eds.), *Recent Advances in Computational Intelligence in Defense and Security*, Studies in Computational Intelligence 621,
DOI 10.1007/978-3-319-26450-9_1

problem-solving approaches that have recently emerged and also functionally pursue the same goals of tractability, robustness and low solution cost [3, 4], including but not withstanding: *rough sets, multi-valued logic, connectionist systems, swarm intelligence, artificial immune systems, granular computing, game theory, deep learning* and the *hybridization* of the aforementioned systems.

As a recognition of the influence CI algorithms are increasingly having upon the security and defense realm, the IEEE Computational Intelligence Society (CIS) created a *Task Force on Security, Surveillance and Defense*[1] (SSD) in February 2010 to showcase recent and ongoing efforts in the application of CI methods to the SSD domain. The flagship event organized by the Task Force, as a forum to exchange ideas and contributions in these topics, is the *IEEE Symposium on Computational Intelligence for Security and Defense Applications* (CISDA), which originated in 2007 and has been annually held since 2009. Other related initiatives are the *Computational Intelligence for Security, Surveillance and Defense* (CISSD) Special Session held at WCCI 2010/2014 and at SSCI 2011/2013; the *Soft Computing applied to Security and Defense* (SoCoSaD) Special Session organized under ECTA 2014; the *Workshop on Genetic and Evolutionary Computation in Defense, Security and Risk Management* held during GECCO 2014 and 2015; and the Canadian Tracking and Fusion Group (CTFG) annual workshops since 2011.

This volume is another endeavour undertaken by the IEEE CIS SSD Task Force and a step in the right direction of consolidating and disseminating the role of CI techniques in the design, development and deployment of security and defense solutions. The book serves as an excellent guide for surveying the state of the art in CI employed within SSD projects or programs. The reader will find in its pages how CI has contributed to solve a wide range of challenging problems, ranging from the detection of buried explosive hazards in a battlefield to the control of unmanned underwater vehicles, the delivery of superior video analytics for protecting critical infrastructures or the development of stronger intrusion detection systems and the design of military surveillance networks, just to name a few. Defense scientists, industry experts, academicians and practitioners alike (mostly in computer science, computer engineering, applied mathematics or management information systems) will all benefit from the wide spectrum of successful application domains compiled in this volume. Senior undergraduate or graduate students may also discover in this volume uncharted territory for their own research endeavors.

We received 53 initial submissions in November 2014 as a response to the Call for Book Chapters, out of which 25 were accepted following the recommendations emanating from the peer-review process conducted by the Technical Program Committee composed of 74 experts and researchers in the field from 22 countries. The 25 accepted chapters were co-authored by 75 contributors from the following countries: Australia (2), Belgium (1), Canada (24), China (1), Cuba (3), India (5), Italy (9), Saudi Arabia (1), Singapore (3), Spain (7), Thailand (3), Tunisia (1), UK (2) and USA (13). It is important to note that 73 % of the contributors are affiliated with academic institutions, 17 % with industry and the remaining 10 % with government.

[1]http://www.ieeeottawa.ca/ci/ssdtf/.

1.1 Volume Organization

The book is structured into five major parts corresponding to the themes that naturally emerged out of the accepted contributions, i.e., physical, cyber and biometric security, situational/threat assessment and mission planning/resource optimization. They are representative of five strategic areas within defense and security that evidence the burgeoning interest of the CI community in developing cutting-edge solutions to entangled problems therein.

Part I: Physical Security and Surveillance [4 chapters]

The problem of detecting buried explosive hazards using forward-looking infrared and ground-penetrating radar sensors is described in Chap. 2 "*Computational intelligence methods in forward-looking explosive hazard detection*". The authors elaborate on the prescreening phase (detection of candidate points in the image) and then on the classification phase. They report the performance of different approaches in the latter phase, ranging from kernel methods to more advanced algorithms like deep belief and convolutional networks to learn new image space features and descriptors.

In the Chap. 3 entitled "*Classification-driven video analytics for critical infrastructure protection*", the authors are concerned with alleviating the burden of an operator that constantly monitors several video feeds to detect suspicious activities around a secured critical infrastructure. The automated solution proposed in this chapter extracts the objects of interest (i.e., car, person, bird, ship) from the image using an iteratively updated background subtraction method, then the object is classified by an artificial neural network (ANN) coupled to a temporal Bayesian filter. The next step is determining the behavior of the object, e.g., entering a restricted zone or stopping and dropping an object. Relevant alerts are issued to the operator should a suspicious event be identified. The authors tried their approach in the automated monitoring of a dumpster, a doorway and a port.

A model-based event correlation framework for critical infrastructure surveillance is put forward in Chap. 4 "*Fuzzy decision fusion and multiformalism modeling in physical security monitoring*". The framework named DETECT (DEcision Triggering Event Composer & Tracker) stores detected threat scenarios using event trees and then recognizes those scenarios in real time. A multiformalism approach for the evaluation of fuzzy detection probabilities using fuzzy operators upon Bayesian Networks and Generalized Stochastic Petri Nets is presented. The authors considered a threat scenario of a terrorist attack in a metro railway station to illustrate the applicability of their methodology.

Chapter 5 "*Intelligent radar signal recognition and classification*" investigates a classification problem for timely and reliable identification of radar signal emitters by implementing and following an ANN-based approach. The idea is to determine the type of radar given certain characteristics of its signal described by a group of attributes (some of them having missing values). Two separate approaches were considered. In the first one, missing values are removed using listwise deletion and then a feedforward neural network is used for classification. The other approach leans on a multiple-imputation method to produce unbiased estimates of the missing data before it is passed to the ANN. In both cases, competitive classification accuracies were obtained.

Part II: Cyber Security and Intrusion Detection Systems [5 chapters]
Chapter 6 *"An improved decision system for URL accesses based on a rough feature selection technique"* addresses corporate security; in particular, internal security breaches caused by employees accessing dangerous Internet locations. The authors propose a classification system that detects anomalous and potentially insecure situations by learning from existing white (allowed) and black (forbidden) URL lists. It then decides whether an unseen new URL should be allowed or denied. The system's performance is boosted by the removal of irrelevant features (guided by rough set theory) and handling class imbalances, with a reported classification accuracy reaching about 97 %.

Chapter 7 *"A granular intrusion detection system using rough cognitive networks"*, the authors designed an intrusion detection system from a Granular Computing angle to classify network traffic as either normal or abnormal. The proposed methodology relies on rough cognitive networks (RCNs), a recently introduced granular system that combines the causal representation inherent to fuzzy cognitive maps with the imprecision-handling abilities provided by rough set theory. The RCN parameters are learned from data using Harmony Search as the underlying optimization engine. RCNs were evaluated against seven other traditional classifiers and were found to be a competitive model that produces high detection rates and low false alarm rates.

Chapter 8 *"NNCS: randomization and informed search for novel naval cyber strategies"* argues that software security can be improved by providing adequate degrees of redundancy and diversity to counter both hardware and software faults. The proposed scheme relies on component rule bases written in a schema-based Very High Level Language. Deviations from the constructed model are likely indicators of a cyber attack. The authors illustrate the benefits of their proposal with a battle management example.

Developing classifiers that can identify sophisticated types of cyber attacks is the main goal of Chap. 9 *"Semi-supervised classification system for the detection of Advanced Persistent Threats"*. The authors define an anomaly score metric to detect the most anomalous subsets of traffic data. The human expert is then required to label the instances within this set, after which a classifier is built based on both labeled and unlabeled data. Genetic programming, decision trees and support vector machines were independently used to construct the classifier.

Chapter 10 *"A benchmarking study on stream network traffic analysis using active learning"* aims at comparing the performance of previously existing active learning and query budgeting strategies as well as an adaptive ANN approach on streaming network traffic to detect malicious network activity such as botnets. The analysis revolves around two new metrics that account for class imbalance as well as the traditional accuracy and detection rate measures. Results are quite encouraging and confirm that the Hoeffding Tree classifier behaves particularly well on the data sets under consideration.

Part III: Biometric Security and Authentication Systems [5 chapters]
Handwritten signatures have long been used as an authentication system given that they are intrinsically endowed with specificity related to an individual. In Chap. 11 *"Visualization of handwritten signatures based on haptic information"*, the authors discuss how to integrate haptic technologies to capture other aspects like kinesthetic and tactile feedback from the user. The study is centered around visualizing and understanding the internal structure of the haptic data (position, force, torque and orientation) in an unsupervised fashion. Special emphasis is made on several dimensionality reduction methods, including CI-based ISOMAP and Genetic Programming.

Reducing the number of false positives in a biometric identification system is at the heart of Chap. 12 *"Extended metacognitive neuro-fuzzy inference system for biometric identification"*. The authors introduce a neurofuzzy inference system along with a sequential evolving learning algorithm as a cognitive component of an architecture that also features a metacognitive component. The latter is responsible for actively regulating the learning of the cognitive component by deciding what, when and how to learn from the available data. The proposed architecture is first benchmarked on a set of medical datasets and then on two real-world biometric security applications, namely signature verification and fingerprint recognition. The comparison with four other authentication systems confirms that the proposed architecture yields a superior performance.

Travel documentation at this time relies either on paper documents or on electronic systems requiring connectivity to core servers and databases for verification purposes. Chapter 13 *"Privacy, security and convenience: biometric encryption for smartphone-based electronic travel documents"* proposes a new paradigm for issuing, storing and verifying travel documents. This smartphone-based approach enables a new kind of biometric checkpoint to be placed at key points throughout the international voyage that does not require storage of biometric information, which simplifies things from a policy and privacy perspective. The authors expect their architecture to enhance system security as well as the privacy and convenience of international travelers.

Digital watermarking allows enforcing authenticity and integrity of an image, which is a major security concern for many industries. The optimization of the embedding parameters for a bi-tonal watermarking system is pursued in Chap. 14 *"A dual-purpose memory approach for dynamic particle swarm optimization of recurrent problems"*. The authors propose a memory-based Dynamic Particle Swarm Optimization method. This memory can operate in either generative or regression mode and is implemented via a Gaussian Mixture Model of candidate solutions estimated in the optimization space, which provides a compact representation of previously found PSO solutions. Results indicate that the computational burden of this watermarking problem is reduced by up to 90.4 % with negligible impact on accuracy.

Chapter 15 *"Risk assessment in authentication machines"* presents an approach for building a risk profiler for use in authentication machines. The proposed risk profiler provides a risk assessment at all phases of the authentication machine life-cycle. The key idea is to utilize the advantages of belief networks to solve large-scale multi-source fusion problems. The authors have extended the abilities of belief networks by incorporating Dempster-Shafer Theory measures. The main goal is to increase the reliability of security risk assessment for authentication machines using the computational-intelligence-based fusion of results from different models, metrics, and philosophies of decision-making under uncertainty.

Part IV: Situational Awareness and Threat Assessment [5 chapters]
To counter piracy attempts, maritime operators need to quickly and effectively allocate some mobile resources (defender units) to assist a target given the available information about the attackers. In Chap. 16 *"Game theoretical approach for dynamic active patrolling in a counter-piracy framework"*, the authors introduce a decision support system (DSS) to that end. The DSS has been designed using Game Theory in order to handle the attractiveness of targets and model strategies for attackers and defenders. Game Theory has proved to be a robust tool to identify the best strategy for the defenders given the information and capabilities of opponents. In the proposed framework, the optimal strategy is modeled as the equilibrium of a time-varying Bayesian-Stackelberg game.

A naval mine is an underwater explosive device meant to damage or destroy surface ships or submarines. An influence mine is a type of naval mine that is triggered by the influence of a vessel or submarine rather than requiring direct contact with it. The ship classification unit (SCU) of an influence mine determines whether the sensed vessel is a target or not, which will cause it to detonate accordingly. In Chap. 17 *"mspMEA: the microcones separation parallel multiobjective evolutionary algorithm and its application to fuzzy rule-based ship classification"*, the author uses a parallel multiobjective evolutionary algorithm (MOEA) based on the concept of microcones to speed up the optimization of the fuzzy rule-based classifiers used to emulate the SCU contained in modern influence mines. A speedup factor of 16.58 % was achieved over a cone-based MOEA algorithm.

Detecting a target in a Synthetic Aperture Radar (SAR) image is a challenging issue since SAR images do not look similar to optical images at all. In Chap. 18 *"Synthetic aperture radar (SAR) automatic target recognition (ATR) using fuzzy co-occurrence matrix texture features"*, the authors put forward a methodology for detecting three types of military vehicles from SAR images without using any pre-processing methods. The texture features generated from the fuzzy co-occurrence matrix are passed on to a multi-class SVM and to a radial basis function (RBF) neural network. The ensemble average is utilized as an information fusion tool. The classification results are superior to those obtained via gray level co-occurrence matrices.

Text mining techniques are important for security and defense applications since they allow detecting possible threats to security and public safety (such as mentions of terrorist activities or extremist/radical texts). Chapter 19 *"Text mining in social media for security threats"* discusses information extraction techniques from social media texts (Twitter in particular) and showcases two applications that make use of these techniques: (1) extracting the locations mentioned in tweets and (2) inferring the users' location based on all the tweets generated by each user. The former task is accomplished via a sequence-based classifier followed by disambiguation rules whereas the latter is tackled through deep neural networks.

The increasing worldwide use of mobile devices has also sparked a growing number of malware apps that should be automatically flagged and vetted by security researchers. Chapter 20 *"DroidAnalyst: synergic Android framework for static and dynamic app analysis"* features an automated web-based app vetting and malware analysis framework for Android devices that integrates the synergy of static and dynamic analysis to improve the accuracy and efficiency of the identification process. DroidAnalyst generates a unified analysis model that combines the strengths of the complementary approaches with multiple detection methods to boost the app code analysis. Machine learning methods such as random forests are employed to generate a set of features with multiple detection methods based on the static and dynamic module analysis.

Part V: Strategic/Mission Planning and Resource Management
[6 chapters]

Chapter 21 *"Design and development of intelligent military training systems and wargames"* elaborates on an architectural approach for designing composable, multi-service and joint wargames that can meet the requirements of several military establishments. This architecture is realized by the design and development of common components that are reused across applications and variable components that are customizable to different training establishments' training simulators. Some of the important CI techniques (such as fuzzy cognitive maps, game trees, case-based reasoning, genetic algorithms and fuzzy rule-based systems) that are used to design these wargame components are explained with suitable examples, followed by their applications to two specific cases of Joint Warfare Simulation System and an Integrated Air Defence Simulation System for air-land battles.

Due to operational requirements, helicopters are now being frequently used for missions beyond what their original design permits. There is thus the need to monitor their usage and more accurately determine the life of its critical components. The methodology outlined in Chap. 22 *"Improving load signal and fatigue life estimation for helicopter components using computational intelligence techniques"* enables the prediction of the load signals (i.e., the time-varying measurement of the load) on the helicopter components using existing flight data and avoiding the installation of additional sensors. The prediction is performed by means of CI techniques (e.g., fuzzy sets, neural networks, evolutionary algorithms) and statistical techniques (e.g., residual variance analysis). The predicted load signals then form the basis for estimating the fatigue life of the component, i.e., the length of time that the component can be safely operated with minimal or acceptable risk of failure. The presented techniques certainly attained a more accurate prediction of the peak values in the load signal.

Defense and security organizations rely on the use of scenarios for a wide range of activities. Scenarios normally take the form of linguistic stories, whereby a picture of a context is painted using storytelling principles. In Chap. 23 *"Evolving narrations of strategic defense and security scenarios for computational scenario planning"*, the authors illustrate how evolutionary computation techniques can be used to evolve different narrations of a strategic story. A representation of a story is put forth that allows evolution to operate on it in a simple manner. Through a set of linguistic constraints and transformations, it is guaranteed that any random chromosome gets transformed into a unique coherent and causally consistent story. The same representation could be used to design simulation models that evaluate these stories. The proposed approach paves the way for automating the evaluation process of defense and security scenarios on multiple levels of resolution, starting from a grand strategic level down to a tactical level.

Chapter 24 "*A review of the use of computational intelligence in the design of military surveillance networks*" surveys the state of the art in the application of CI methods to design various types of sensor networks, including wireless/fixed sensor, mobile ad hoc and cellular networks, as these constitute the backbone for realizing Intelligence, Surveillance and Reconnaissance (ISR) military operations. The authors also list important defense and security applications of these networked systems, review the CI methods and their usage and outline a number of research challenges and future directions.

Given the prolific number of sensing modalities available nowadays, determining on which platform a sensor should be mounted to collect measurements during the next observation period is far from being a trivial task. Chapter 25 "*Sensor resource management: intelligent multi-objective modularized optimization methodology and models*" proposes a new sensor tasking framework named OPTIMA that aims at solving this problem. OPTIMA features a Sensor Resource Analyzer module and a Sensor Tasking Algorithm (Tasker) module. The latter leans on multiobjective evolutionary optimization methods to consider timing constraints, resolution and geometric differences among the sensors with the goal of fulfilling some tasking requirements related to maximizing the available sensor resources for search, optimizing sensor resources for tracking and better defending the high-priority assets.

Chapter 26 entitled "*Bio-inspired topology control mechanism for unmanned underwater vehicles*" addresses the problem of having a group of unmanned underwater vehicles (UUVs) cooperatively self-organize in order to protect valued assets in unknown 3D underwater spaces. The topology control mechanism is rooted in particle swarm optimization and employs Yao-graph-inspired metrics to craft its fitness function. The self-organization protocol only requires neigborhood-limited UUV information to collectively guide the UUVs to make movement decisions in these unknown 3D spaces. The algorithm is able to provide a user-defined level of protection for different maritime vessel applications. The proposed methodology is illustrated with three examples: (1) uniform coverage of the underside of a maritime vessel; (2) plane formation to cover a given dimension in the 3D space and (3) forming a sphere around a given asset such as a fully submerged submarine while maintaining connectivity.

Our hope is that the wealth of technical contributions gathered in this book helps create further momentum and drive forward many other theoretical and practical aspects of the fascinating synergy between CI methods and the defense and security problem spaces. Enjoy the reading!

References

1. Barfield, W., Dingus, T.A.: Human Factors in Intelligent Transportation Systems. Psychology Press, New York (2014)
2. Hong, J., Baker, M.: Wearable computing. IEEE Pervasive Comput. **2**, 7–9 (2014)
3. Kacprzyk, J., Pedrycz, W.: Springer Handbook of Computational Intelligence. Springer, New York (2015)
4. Kruse, R., Borgelt, C., Klawonn, F., Moewes, C., Steinbrecher, M., Held, P.: Computational Intelligence: a Methodological Introduction. Springer Science & Business Media, Berlin (2013)
5. Lu, G., Zeng, W.H.: Cloud computing survey. In: Applied Mechanics and Materials, vol. 530, pp. 650–661. Trans Tech Publ (2014)
6. Perera, C., Zaslavsky, A., Christen, P., Georgakopoulos, D.: Context aware computing for the internet of things: a survey. Commun. Surv. Tutor. IEEE **16**(1), 414–454 (2014)
7. Verdone, R., Dardari, D., Mazzini, G., Conti, A.: Wireless Sensor and Actuator Networks: Technologies, Analysis and Design. Academic Press, London (2010)

Part I
Physical Security and Surveillance

Computational Intelligence Methods in Forward-Looking Explosive Hazard Detection

Timothy C. Havens, Derek T. Anderson, Kevin Stone, John Becker
and Anthony J. Pinar

Abstract This chapter discusses several methods for *forward-looking* (FL) *explosive hazard detection* (EHD) using FL *infrared* (FLIR) and FL *ground penetrating radar* (FLGPR). The challenge in detecting explosive hazards with FL sensors is that there are multiple types of targets buried at different depths in a highly-cluttered environment. A wide array of target and clutter signatures exist, which makes detection algorithm design difficult. Recent work in this application has focused on fusion methods, including fusion of multiple modalities of sensors (e.g., GPR and IR), fusion of multiple frequency sub-band images in FLGPR, and feature-level fusion using multiple kernel and iECO learning. For this chapter, we will demonstrate several types of EHD techniques, including kernel methods such as *support vector machines* (SVMs), *multiple kernel learning* MKL, and feature learning methods, including deep learners and iECO learning. We demonstrate the performance of several algorithms using FLGPR and FLIR data collected at a US Army test site. The summary of this work is that deep belief networks and evolutionary approaches to feature learning were shown to be very effective both for FLGPR and FLIR based EHD.

Keywords Sensor fusion · Explosive hazard detection · Aggregation · Multiple kernel learning · Deep learning · Fuzzy integral

T.C. Havens (✉) · J. Becker · A.J. Pinar
Department of Electrical and Computer Engineering/Department of Computer Science,
Michigan Technological University, Houghton, MI 49931, USA
e-mail: thavens@mtu.edu

J. Becker
e-mail: jtbecker@mtu.edu

D.T. Anderson
Department of Electrical and Computer Engineering, Mississippi State University,
Mississippi State, MS, USA
e-mail: anderson@ece.msstate.edu

K. Stone
Department of Electrical and Computer Engineering, University of Missouri,
Columbia, MO, USA
e-mail: kes25c@mail.missouri.edu

© Springer International Publishing Switzerland 2016
R. Abielmona et al. (eds.), *Recent Advances in Computational Intelligence
in Defense and Security*, Studies in Computational Intelligence 621,
DOI 10.1007/978-3-319-26450-9_2

1 Introduction

An important goal for the U.S. Army is remediating the threats of explosive hazards as these devices cause uncountable deaths and injuries to both Civilians and Soldiers throughout the world. Since 2008, explosive hazard attacks in Afghanistan have wounded or killed nearly 10,000 U.S. Soldiers; worldwide, explosive devices on average cause 310 deaths and 833 wounded per month [25]. Systems that detect these threats have included *ground-penetrating-radar* (GPR), *infrared* (IR) and visible-spectrum cameras, and acoustic technologies [9, 10, 37]. Past research has examined both handheld and vehicle-mounted systems and much progress has been made in increasing detection capabilities [7, 14]. *Forward-looking* (FL) systems are an especially attractive technology because of their ability to detect hazards before they are encountered; standoff distances can range from a few to tens of meters. A drawback of forward-looking systems is that they are not only sensitive to explosive devices, *unexploded ordnance* (UXO), and landmines, but also to other objects, both above and below the ground. Because these sensors are standoff sensors, the area being examined for targets is much larger than with downward-looking systems. Thus, clutter is a serious concern. Furthermore, the explosive hazard threat is very diverse—they are made from many different materials, including wood, plastic, and metal, and come in many different shapes and sizes—and this threat continues to evolve. This means that it is nearly impossible to detect explosive hazards solely by a modeling-based approach, and, hence, *computational intelligence* (CI) methods are very appropriate. Previous work has shown that if *forward-looking infrared* (FLIR) or visible-spectrum imagery is combined with L-band FLGPR, *false alarm* (FA) rates can be reduced significantly [2, 16, 18, 19, 44, 45]. Hence, we focus on CI methods for sensor-fused forward-looking detection of explosive threats, comparing CI to other machine learning approaches.

The structure of the remainder of this study is as follows. Section 2.2 describes the preprocessing of the sensor data into a format that is ready for prescreening and feature extraction. The prescreener algorithms are described in Sect. 2.3, and the feature extraction is detailed in Sect. 2.4. In Sect. 3 we describe kernel learning methods, including *support vector machine* (SVM)-based methods, *multiple kernel* (MK) methods, and a fuzzy integral-based MK learner. Methods that learn the features implicitly, such as *deep belief networks* (DBNs), *convolutional neural networks* (CNNs), and iECO feature learning, are described in Sect. 4. Results for the various learning algorithms will be presented in the respective parts of Sects. 3 and 4. We summarize in Sect. 5. Table 1 contains the acronyms used in this chapter. Next, we describe the sensing technologies used to demonstrate the various EHD algorithms in this chapter.

Table 1 Acronyms

UXO	Unexploded ordnance	EHD	Explosive hazard detection
GPR	Ground-penetrating radar	IR	Infrared
FL	forward looking	DL	Downward looking
LW	long-wave	MW	Mid-wave
UTM	Universal traverse mercator	CI	Computational intelligence
FA	False alarm	ROC	Receiver operating characteristic
MK	Multiple kernel	SK	Single kernel
MKLGL	MK learning-group lasso	SVM	Support vector machine
FIMKL	Fuzzy integral MKL	CNN	Convolutional neural network
RBM	Restricted Boltzmann machine	DBN	Deep belief network
CFAR	Constant false-alarm rate	NAUC	Normalized area under the curve
iECO	Improved evolution constructed	CLAHE	Contrast-limited adaptive histogram equalization
HOG	Histogram of oriented gradients	LBP	Local binary patterns
MSER	Maximally stable extramal regions	GMM	Gaussian mixture models
SIFT	Scale-invariant feature transform	AOI	Area of interest

2 Explosive Hazard Detection: Background Knowledge

2.1 Sensing Technologies for FLEHD

FLGPR GPR has long been an interest to the U.S. Army for EHD, and *downward-looking* (DL) systems have been shown to be very effective in operational scenarios. However, DL systems fail to provide a standoff range from the threat; the array is located directly above the threat upon detection. Hence, there has been much focus on improving standoff distances by using FL systems. FLGPR aims to improve standoff by aiming the GPR array forward, often with the center of the beam aimed 10–15 m in front of the vehicle. Since the angle of incidence at which the beam hits the ground surface is important for penetration—the more orthogonal the beam is to the surface, the better the ground penetration—the arrays are usually built on some type of boom above the vehicle. Still, due to the geometry of the FL problem, much array energy is lost to specular reflection from the ground surface. Hence, FLGPR *signal-to-noise ratios* (SNRs) are not nearly as good as with DLGPR systems. Furthermore, the index of refraction of the soil is significantly different than that of the air, which causes a refraction—or bending—of the radar beam at the ground surface, further complicating image formation. These, and other challenges, mean that FLGPR-based EHD is not as simple as looking for local regions of high intensity; more complex EHD strategies are necessary. We talk about several approaches in this chapter.

(a)   (b)

Fig. 1 FLGPRs under research and development for use in EHD. **a** ALARIC L-B and FLGPR. **b** L/X-Band FLGPR

Many FLGPR systems have been designed specifically for EHD, including the two shown in Fig. 1. View (a) shows the ALARIC system, which combines an L-band FLGPR and a visible spectrum imaging system, while (b) shows an FLGPR that combines L- and X-band radar arrays. The FLGPR results shown in this chapter will focus on data recorded with the L/X-band system shown in Fig. 1b. The government-furnished FLGPR data is composed of complex radar data as well as motion data of the vehicle from several lanes at an arid U.S. Army test site.

FLIR While numerous frequency ranges in the infrared portion of the electromagnetic spectrum have been investigated for EHD, e.g., *mid-wave* IR (MWIR) and combinations of IR bands for "disturbed earth" detection, we focus on recent advancements in anomaly detection in *long-wave* IR (LWIR). However, without loss of generality the vast majority of mathematics and algorithms discussed herein are naturally applicable to both MWIR and LWIR imagery with little-to-no change. LWIR or thermal imagers are passive (i.e., they do not require illuminators) and detect infrared radiation in approximately the 8–14 μm wavelength. Objects with a temperature above absolute zero emit infrared radiation in this range at their surface. The amount of emitted thermal radiation increases with temperature. The exact relationship between an object's temperature and the amount of emitted thermal radiation depends on the emissivity, a quantity representing a material's ability to emit thermal radiation that varies with wavelength. A thermal imager actually sees not only the emitted radiation of the object, but also transmitted radiation, i.e., radiation from an external source which passes through the object toward the imager, and/or reflected radiation, i.e., radiation from an external source which reflects off the object toward the imager. These factors complicate assigning absolute temperature values to objects. However, in EHD we can exploit the fact that buried objects will likely possess a different thermal conductivity, thermal capacity, or density than the surrounding soil, resulting in either a cooling or warming of the soil immediately surrounding the object. This most often leads to a change in temperature at the surface above the object and results in a measurable change in the amount of emitted

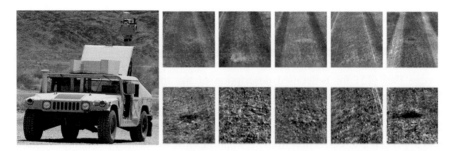

Fig. 2 Example of *thermal scarring* in FLIR with targets of varying difficulty at a fixed vehicle stand off distance. (*left*) NVESD FLEHD multi-sensor ground vehicle platform, (*top row*) LWIR and (*bottom row*) MWIR imagery. Columns are different (center aligned) targets co-registered in MWIR and LWIR. Note, the MWIR camera has a higher resolution (more pixels on target)

thermal radiation compared to areas of the ground free of such objects. Figure 2 shows this phenomenon, referred to in many circles as *thermal scarring*.

However, FLIR is not without flaw. One challenge is diurnal cross-over, the time-period during which the buried object comes to near thermal equilibrium with its surroundings making targets, for all intents, unidentifiable. Another factor is the difference in emitted radiance seen at the soil surface (even for the same soil composition and object) varies based on factors such as the amount of incident thermal radiation, which is dependent on time of day, time of year, and current weather conditions. These are just some of the factors that emphasize the need to include and fuse different sensing technologies to solve this extremely challenging real-world problem.

The FLIR data used in our experiments was collected from two cameras. The first camera, called DVE, was uncooled and used the DRS Infrared Technologies U6000 microbolometer detector which has a spectral response of $8-14 \mu m$. The DVE camera captured 8-bit single channel imagery with a resolution of 640×480, and horizontal and vertical fields of view of 40 and $30°$, respectively. The second camera was a SELEX L20, which produces a 16 bit single channel image with resolution 640×512. The SELEX camera had a spectral response of $8-10 \mu m$, and horizontal and vertical fields of view of 15 and $12°$, respectively. Both cameras were mounted on a mast at the back of the vehicle as shown in Fig. 2. The mast height was approximately 3.35 m and had a downward look angle of $6.3°$. An inertial navigation system was mounted next to the cameras, and the time at which each image was captured was recorded. This allowed precise georeferencing using the dense 3D scene reconstruction technique described in [46].

The government-furnished data consists of numerous runs from three lanes at an arid U.S. Army test site. The number of targets per lane varied from 44 to 79, and the area of the lanes ranged from 3,600–4,200 square meters. Emplaced targets were buried between 1–6 in. deep, and varied in metal content (some had no metal).

2.2 Sensor Processing

FLGPR Preprocessing We use a backpropagation procedure to form the radar images (see [15] for a detailed description of the imaging algorithm). In brief, the radar images are formed by coherently summing successive backpropagation images, accounting for platform motion effects on phase and beam pattern effects. The images are formed on a 2.5 cm-spaced grid for each antenna polarization. We also apply a phase correction to the L-band FLGPR to account for vehicle motion during the swept-frequency transmission [4]. The end results of the FLGPR imaging and preprocessing are complex images for each of the L- and X-band polarizations on a rectangular grid coordinate system. In Sect. 2.3, we discuss how we take each FLGPR image $I_p(u, v)$ and indicate candidate detections.

FLIR Preprocessing Numerous algorithms have been applied to the government-furnished FLIR data for preprocessing. However, these algorithms are not the subject of investigation in this chapter as they are not focused on CI. The reader can refer to [3, 42, 43, 46] for more details. In general, these preprocessing algorithms are focused on deinterlacing, denoising, and global or local contrast enhancement. For the DVE images, preprocessing typically consists of deinterlacing, denoising, and *contrast limited adaptive histogram equalization* (CLAHE) [3]. For the SELEX, the 16-bit data was converted to 8-bit by contrast stretching, with saturation limits at 0.05 and 99.95 percent of the original pixel values, so the resulting values filled the entire 16-bit range. After contrast stretching the pixel values were divided by 256 and CLAHE was run. Next we describe how the initial hit locations are determined.

2.3 Prescreeners

Prescreener is a term used for a weak detection scheme by which candidate detections are found and passed on to stronger classification algorithms. The main ideas are to (i) reduce the computational load of the classificaiton algorithms, and (ii) improve classification accuracy by only training on target-like candidate detections.

FLGPR Prescreener The result of the radar preprocessing method described in the Sect. 2.2 is a coherently integrated image $I_p(u, v)$, where (u, v) are the image coordinates: one image for each polarization of the L-band FLGPR (HH and VV polarizations) and one image of the X-band FLGPR (VV polarization). It is well known that penetration depth increases with wavelength; hence, the L-band will have a deeper penetration than the X-band radar. Thus, we use the L-band radar as the detection radar for the method proposed here; although, we will show results for X-band detection and classification too.

The prescreening detector is the first algorithm that indicates candidate detection locations—a block diagram is shown in Fig. 3a. In [15], we proposed two methods to indicate the presence of a target, both of which could be considered to be

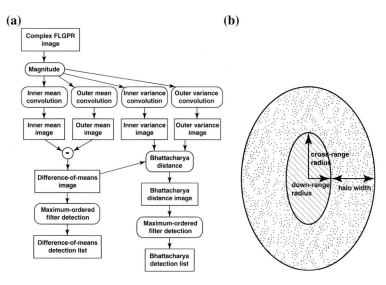

Fig. 3 **a** Block diagram of prescreener detection algorithm. **b** Elliptical convolution kernels used in prescreener. Detection is indicated by comparing the distribution of pixel intensities in inner ellipse to the distribution of pixel intensities in outer halo [15]

a *constant FA rate* (CFAR) detector. The first prescreener indicates a hit by taking the mean of the pixels in the inner ellipse and comparing that to the mean of the pixels in the outer halo (as shown in Fig. 3b. Essentially, the prescreener identifies regions that have values that are higher than the surrounding regions. The second prescreener uses a signed Bhattacharyya distance between the distributions of the pixel values in the center region and outer halo to indicate a hit. For a more detailed description of these prescreeners, see [15]. In our experiments, we have determined the following prescreener parameters to be good choices for this system: down-range radius = 0.25 m; cross-range radius = 0.5 m; and halo width = 0.75 m. These values are related to the impulse-response of the FLGPR system and to expected target sizes. Furthermore, for this chapter we will only present results for the difference-of-means prescreener, which has been shown to be more effective than the Bhattacharyya prescreener for FLGPR data [15].

One could simply threshold the output of the prescreener to indicate a detection; however, this can result in many detections in one local region. Hence, we use a maximum order-filter with a 3 m (cross-range) by 1m (down-range) rectangular kernel to reduce the presence of closely grouped hits. The prescreeners are rough first-look algorithms for indicating candidate detections—they merely indicate if a region of pixels is different in intensity than the surrounding pixels. They do not, however, consider higher-level features, such as texture or shape, that might indicate better the difference between clutter and true detections. Hence, at each detection location, we then extract a set of shape- or texture-based features, described in Sect. 2.4.

FLIR Prescreener In [2], we outlined a FLIR prescreener for EHD which was later extended to FLGPR in [46]. This prescreener consists of an ensemble of trainable size-contrast (CFAR) filters, i.e., local dual sliding window detectors. Each size-contrast filter has seven parameters: the inner window height and width, the pad height and width (which determine the size of the outer window), a Bhattacharyya distance threshold, a squared difference between the mean values threshold, and three state parameters, referred to as DType (which determines whether the detector will trigger only on bright on dark regions, dark on bright regions, or both). At each pixel, the mean and variance of the inner and outer windows are computed, the Bhattacharyya distance and squared difference between the mean values is calculated and these two values are compared against their corresponding threshold. If both values are greater than their threshold, and the DType condition is met, then the corresponding detector fires. When a detector fires, it projects the inner window center pixel coordinate into UTM coordinates. Next, a clustering algorithm is run on all UTM coordinates generated from individual frames. Specifically, mean-shift, a mode seeking clustering algorithm, with an Epanechnikov kernel is used. Mean-shift was chosen as the application requires a fast clustering algorithm (in the offline training phase, the algorithm has to run hundreds of thousands of times on potentially large data sets: 10,000+ points) that also does not require the user to set the number of clusters. We have compare mean-shift results to the *basic sequential algorithmic scheme* and did not see a significant different in performance. Herein, this clustering step is referred to as spatial mean-shift, and it results in candidate hit locations. Next, mean-shift is run a second time on the hit locations from the combination of multiple frames (this is referred to as temporal mean-shift). Each mean-shift step requires two parameters: the kernel bandwidth and the minimum number of points around a peak in order to keep that cluster. Mean-shift works by performing gradient ascent on the kernel density estimator,

$$\hat{f}(x) = \sum_{i=1}^{N} K\left(x_i - x\right), \quad K\left(x_i - x\right) = k\left(\|x_i - x\|^2\right), \tag{1}$$

where K is the kernel function, N is the number of data points, and normalizing constants have been omitted for brevity. Taking the gradient of this function with respect to x and setting it to zero results in the following (well known) iterative update equation:

$$x_{t+1} = \frac{\sum_{i=1}^{N} k'\left(\|x_i - x_t\|^2\right) x_i}{\sum_{i=1}^{N} k'\left(\|x_i - x_t\|^2\right)}, \tag{2}$$

where, $k'(x)$ denotes the derivative of $k(x)$ with respect to x, and t denotes the iteration. For the Epanechnikov kernel with bandwidth parameter h, the update equation reduces to:

$$x_{t+1} = \frac{\sum_{x_i \in L} x_i}{|L|}, \quad k_{epn}(v) = \begin{cases} 1 - \frac{v}{h} & 0 \le v \le h \\ 0 & \text{else} \end{cases} \tag{3}$$

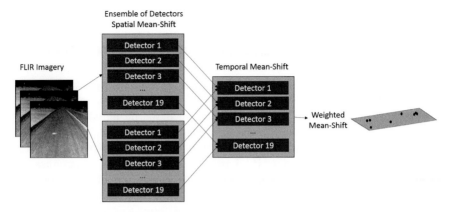

Fig. 4 Illustration of FLIR prescreener, which uses an ensemble of detectors (trained under different criteria) and spatial and temporal weighted mean-shift

where L is the set of all points for which k_{epn} is non-zero and $|\cdot|$ is cardinality. Mean-shift is initialized at every hit location, and the update procedure is run until convergence. For this application, convergence is defined as a change of less than 1 cm between updates (remember that the points are in UTM coordinates). Refer to [46] for additional algorithm speedups. Figure 4 illustrates the proposed FLIR prescreener.

A *genetic algorithm* (GA) is used to learn the detector parameters. To this end, we explored two methodologies. The first, referred to as *one-per-rate*, trains a single detector for each desired *detection rate*. The primary objective of the GA is to achieve the desired detection rate with the secondary objective of minimizing the *false alarm rate* (FAR). In [2], 19 detectors were trained at desired detection rates ranging from 0.05 to 0.95 in step sizes of 0.05. The idea behind training many detectors is that the resulting ROC curve after fusion should be better than if a single detector were trained and only its thresholds allowed to vary. The second method, referred to as *one-per-target*, trains a single detector for each ground truth encounter in the training data. The primary objective of this GA is to detect the specific target with the secondary objective of minimizing the FAR. For both cases, weighted mean-shift is used to fuse the detectors (each trained with a different objective function). A weight is learned for each detector using separable *covariance matrix adaptation evolution strategy* such that the *normalized area under the curve* (NAUC) is maximized on the training data. Reference [2] reports the learned detector parameters and aggregation weights for a prior experiment.

In [46], a few improvements to the above FLIR prescreener were outlined. The first improvement was allowing confidence information to be passed from the size-contrast filter to the spatial mean-shift step and from the spatial mean-shift step to the temporal mean-shift step. Previously, UTM coordinates resulting from a size-contrast filter triggering were treated identically during spatial mean-shift. However, this discards the Bhattacharyya distance and mean difference information which is useful for locating the strongest response, which generally corresponds to the center of the object. Likewise, information about the peaks found during spatial mean-shift, such as the number of points surrounding each peak, could be useful for the temporal mean-shift step. To remedy this, mean-shift was replaced with weighted mean-shift in both steps, and two new parameters were added to each detector to control whether confidence information is passed on. This leaves it up to the GA to decide if the confidence information is useful. The second improvement was the introduction of a different grouping algorithm as an alternative to weighted mean-shift. The alternative method, also proposed in [46], is an ordered filter approach inspired by the MUFL FLGPR prescreener introduced in [16]. Lastly, the separable CMA-ES optimization for finding weights for the weighted mean-shift step which combines detectors was eliminated as it tended to overfit the training data. Instead, three heuristics were used to generate weights, and the set of weights which performs best in terms of NAUC on the training data was chosen. The first method assigns equal weight to all detectors; the second method assigns weights based on detection rate and the third method assigns weights based on FAR.

2.4 Feature Extraction

While our FLIR and FLGPR prescreeners achieve relatively high positive detection rates, meaning they often do better than what an expert can identify visually, they still suffer from an unacceptable FAR (relative to U.S. Army requirements). In order to address this deficiency, we have explored, extended and created a number of new image space features and descriptors, including *convolutional neural networks* (CNNs) [43], *improved Evolution COnstructed* (iECO) features [38], "soft" (importance map weighted) features [42], histogram of cell-structured Gabor energy filter and Shearlet filter bank responses [38, 46], *histogram of gradients* (HOG) [32] and *local binary pattern* (LBP) [15, 17, 35] and "soft" edge histogram descriptor features [2, 46]. In [2], additional anomaly evidence map features in FLIR were proposed, which include features from *maximally stable extremal regions* (MSERs) [33] and *Gaussian mixture models* (GMMs) [41] for change detection. Unlike a CFAR (or size-contrast) filter, which is often utilized as a local contrast feature, the above image space features focus on texture and shape. In addition, we do not use features "directly", e.g., a single image gradient. Instead, high(er)-level image space descriptors are formed by "pooling" features within a given spatial *area of interest* (AOI), e.g., HOGs, LBPs, or edge descriptors. Furthermore, it is important to not just simply extract features and pool their values over a large spatial AOI as that often leads to ambiguous configurations of patterns. Instead, we preserve the spatial properties of image patterns by using a cell-structured (partially overlapping to allow patterns to drift some in translation across detections) grid for a given AOI. It is usually of great benefit to extract features at different scales in a given AOI, e.g., multi-scale HOG. Convention is to concatenate these multi-scale and multi-cell features together into a single long feature vector of high dimensionality and let a classifier (or fea-

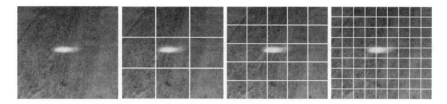

Fig. 5 Multiple cell-structured configurations for feature extraction at a single scale to preserve the spatial context of features. Note that cells are not shown as overlapping for visual simplicity

ture selection algorithm) learn which are most relevant to a particular task at hand. Figure 5 shows the use of multiple cells at a single scale.

The first feature introduced is the LBP. The LBP is a sort of texture or pattern feature and it is calculated at each pixel according to

$$LBP_n = \sum_{k=0}^{n} s\left(i_k - i_c\right) 2^k,$$

where LBP_n is the LBP code, i_c is the window center value, i_k is the value of the kth neighbor and function $s(x)$ is 1 if $x \geq 0$ and 0 otherwise. Ojala extended the LBP for neighborhoods of different shapes and sizes [35]. The circular (radius r) neighborhood version, $LBP_{n,r}$ includes bilinearly interpolating values at non-integer image coordinates. Ojala also observed that there is a limited number of transitions or discontinuities in the circular presentation of 3×3 texture patterns and that these uniform patterns, $LBP_{n,r}^u$, are fundamental properties of local image texture, meaning they provide the vast majority of all patterns (accounting for 90 % at $u = 2$). The u stands for no more than u 0–1 or 1–0 transitions, e.g., 00011110 has 2 transitions and 00101001 has 5 transitions. Last, the LBP is turned into a descriptor by binning the patterns into a histogram over an AOI. For example, for $u = 2$ there are only 59 patterns (thus histogram bins) for a neighborhood of size 8. In addition, Ojala put forth a rotation robust version that consists of shifting the binary patterns until there is a 1 in the first digit [35]. This reduces the number of patterns for a neighborhood size of 8 to only 9. Last, most normalize the resultant histogram by its ℓ_1 or ℓ_2-norm.

Another feature is the famous HOG, popularized by David Lowe in the *scale invariant feature transform* (SIFT); however it was first explored by Edelman in the context of wet science and later popularized by Dalal-Triggs for HOG-based person detection [11]. It is important to note that SIFT technically consists of keypoint detection, a feature descriptor and detection. The HOG (the feature descriptor in SIFT) involves the extraction of a gradient vector per pixel in an image. For a given AOI, one computes the magnitude of each gradient, $\|(\partial I(x, y)/\partial x, \partial I(x, y)/\partial y)\|$, and its respective orientation. A histogram of B bins (a user defined or learned parameter) is specified and each pixel's gradient magnitude, per cell, is added to the bin with respect to its orientation. For example, for 360° and 8 bins each bin spans 45° and for a cell structured configuration of 4×4 we obtain a 128-length feature vector. Note, convention involves bilinearly interpolating each gradient magnitude for the closest and next closest bin. Also, while SIFT identifies and then rotates the descriptor with respect to its major orientation bin(s), this is an optional step that the user must determine relative to the given detection task at hand. In our FLEHD investigations, we do not perform the rotation step.

In [46], we proposed a "soft" edge histogram descriptor feature. The edge histogram descriptor is inspired by the MPEG-7 edge histogram descriptor, which has five simple convolution operators that represent vertical, horizontal, diagonal, anti-diagonal and non-directional edge classes. The operators for the first four classes closely resemble the standard Sobel and Prewitt edge operators. At each pixel, the five operators are applied and the absolute value of the response to each is computed. The pixel is assigned to the class of the operator generating the largest response. In [46], we extended this feature to make it less sensitive to noise. We allow a pixel to contribute to all classes by creating a histogram at each pixel location and we accumulate the individual pixel histograms inside a window to form the final descriptor. A pixel's histogram is constructed by computing the absolute value of the response to each of the edge convolution operators and then dividing each of those values by the sum, i.e. taking the l_1 norm. Linear interpolation is performed to distribute the pixel's contribution between the edge classes and the non-edge class by comparing the sum of the absolute values of the operator responses to the edge threshold. If the sum is greater than or equal to the edge threshold then the non-edge class is assigned zero. Otherwise, the non-edge class is assigned one minus the fractional value of the sum divided by the edge threshold, and that fractional value is multiplied to the value of each of the edge classes in the histogram. We introduced a further change, the addition of two new edge masks; making the total descriptor length seven. We extract two edge histogram descriptors per cell using edge thresholds of 15 and 35. Therefore, edge histogram descriptor gives $7 \times 2 = 14$ features per cell.

In [40, 42], we created a softened version of the HOG, LBP, and edge histogram descriptor based on the extraction and utilization of an *importance map*. An importance map, one per each image, is simply a [0, 1]-valued image that is the same size as the original image. Each pixel in an importance map informs us about the relevance or significance of that pixel for a given task at hand. The importance map is used to weight features, such as HOGs and LBPs, as they are added to a descriptor like a histogram. The motivation for importance maps is that current image space descriptors unfortunately extract both background (e.g., clutter, tire tracks, foliage, etc.) and foreground (target) information. In many cases, the number of encountered foreground features are extremely few relative to the background information and their presence in the descriptor can be dwarfed. Most researchers ignore this fact and pass the problem down the processing pipeline. That is, most extract all features in an AOI and leave it up to the classifier or feature selection to determine what is important. Instead, our goal is to extract feature-rich information in target areas and more-or-less ignore extraneous information in other parts of an AOI. In [40], Scott and Anderson used this philosophy and showed improvement in aircraft detection in remote satellite imagery across different parts of the world and times of the year based on importance-weighted multi-scale texture and shape descriptors. Their importance maps were based on fuzzy integral-based fusion of differential morpho-logical map profiles for soft object extraction. In [42], we extend this technique to FLEHD, introducing a new way to derive an importance map for FLIR. In FLIR, we are interested in detecting circular or elliptical (due to perspective deformation in FL imagery) shapes for anomaly detection. Hence, we exploited this information and created a frequency and orientation selective bank of Gabor energy filters, which we later reduced down to a single Shearlet filter, to build an importance map. The real-valued Gabor or Shearlet image is normalized between min and max across an AOI. It is then blurred with a Gaussian kernel to spread out the filter response, as many features reside at or around the edges of an object. The result is then re-normalized, according to its min and max, back into [0, 1] (values that represent the relative worth of different pixels in the AOI relative to the task at hand). The soft HOG, LBP, and edge histogram descriptor features are calculated as before, however as these fea-tures are being added to their respective bins in the histogram they are multiplied by their corresponding per-pixel importance map weights $E(x, y)$. The features that we describe in this section can now be used to further reduce the number of FAs by training classifiers to indicate prescreener hits as either FAs or true-positives. Next we discuss kernel methods that can accomplish this task.

3 Kernel Methods for EHD

Consider some non-linear mapping function $\phi : \mathbf{x} \to \phi(\mathbf{x}) \in \mathbb{R}^{D_K}$, where D_K is the dimensionality of the transformed feature vector $\mathbf{x}$. With kernel clustering, we do not need to explicitly transform $\mathbf{x}$, we simply need to represent the dot product $\phi(\mathbf{x}) \cdot \phi(\mathbf{x}) = \kappa(\mathbf{x}, \mathbf{x})$. The kernel function κ can take many forms, with the polynomial $\kappa(\mathbf{x}, \mathbf{y}) = (\mathbf{x}^T \mathbf{y} + 1)^p$ and *radial-basis-function* (RBF) $\kappa(\mathbf{x}, \mathbf{y}) = \exp(\sigma \|\mathbf{x} - \mathbf{y}\|^2)$ being two of the most well known. Given a set of n objects X, we can thus construct an $n \times n$ kernel matrix $K = [K_{ij} = \kappa(\mathbf{x}_i, \mathbf{x}_j)]^{n \times n}$. This kernel matrix K represents all pairwise dot products of the feature vectors associated with n objects in the transformed high-dimensional space—called the *Reproducing Kernel Hilbert Space*.

The main goal of kernel methods is to transform the feature vectors $\mathbf{x}$ such that the new representations, $\phi(\mathbf{x})$, are advantageous to the classification problem. We present three methods for learning classifiers in kernel spaces, SVM, MKLGL, and FIMKL, which we now describe.

3.1 Single Kernel

One of the most popular kernel methods for classification is the SVM. The SVM attempts to find an optimal separating hyperplane between two classes of training data; for the case of EHD, we use it to find a hyperplane between features that describe FAs and those of true positives. For a detailed description of the SVM, see [8]. The *single-kernel* SVM (SKSVM) is defined as

$$\max_{\alpha} \left\{ \mathbf{1}^T \alpha - \frac{1}{2}(\alpha \circ \mathbf{y})^T K (\alpha \circ \mathbf{y}) \right\}, \tag{4a}$$

subject to

$$0 \le \alpha_i \le C, \ i = 1, \dots, n; \ \alpha^T \mathbf{y} = 0, \tag{4b}$$

where $\mathbf{y}$ is the vector of class labels, $\mathbf{1}$ is the n-length vectors of 1s, $K = [\kappa(\mathbf{x}_i, \mathbf{x}_j)] \in \mathbb{R}^{n \times n}$ is the kernel matrix, and $\circ$ indicates the Hadamard product [5]. The value C determines how many errors are allowed in the training process [8]. Note that SKSVM reduces to the linear SVM for the kernel $\kappa(\mathbf{x}_i, \mathbf{x}_j) = \mathbf{x}_i^T \mathbf{x}_j$ (which is simply the Euclidean dot product).

One of the drawbacks of using the above SVM formulation is that it treats each datum equally; hence, when there is an imbalance between the number of datum in each class, then the SVM decision boundary is driven primarily by the data from the class with more data points. This is a problem in explosive hazards detection as there are typically many more FA detections than there are true positives—the true positives only comprise a small overall area of the lane. To attack this issue, we use a formulation of the SVM for imbalanced data which uses a different error cost for positive (C^+) and negative (C^-) classes. Specifically, we change the constraints of the kernel SVM formulation at (4) to

$$0 \le \alpha_i \le C^+, \forall i | y_i = +1; \ 0 \le \alpha_i \le C^-, \forall i | y_i = -1; \ \alpha^T \mathbf{y} = 0; \tag{5}$$

where C^+ is the error constant applied to the positive class and C^- is the error constant applied to the negative class. In our application, the positive class is true positives and the negative class is FAs. We set $C^+ = n^-/n^+$ and $C^- = 1$, where n^- is the number of objects in the negative class and n^+ is the number of objects in the positive class. This essentially allows for fewer errors in the true positive class.

We use LIBSVM to efficiently solve the SKSVM problem [6]. The output of LIB-SVM is a classifier model that contains the vector α and the bias b. A measured feature vector $\mathbf{x}$ can be classified by computing

$$y = \text{sgn} \left[\sum_{i=1}^{n} \alpha_i y_i \kappa(\mathbf{x}_i, \mathbf{x}) - b \right], \tag{6}$$

where sgn is the signum function. We now show the application of SKSVM to our FLEHD problem.

Application of SKSVM to FLGPR EHD Figure 6 shows selected results of training the SKSVM on FLGPR lanes A, B, and D, and testing on Lane C. The results are compared to random performance, which is the ROC achieved by uniform random selection of hit locations at given FA rates. View (a) shows the prescreener ROC

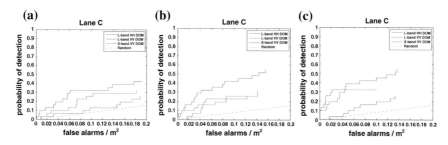

Fig. 6 ROC curves showing testing performance of (**a**) FLGPR prescreener, and SKSVM classifier with RBF kernel for (**b**) single HOG feature and (**c**) combination of HOG and LBP features. Percent NAUC improvements are shown for each of the L-band (HH and VV polarizations) and X-band FLGPRs. The performance of a uniform random detector is shown by the *dotted line*. **a** Prescreener. **b** L-HH: 21 %; L-VV: 32 %; X: 64 %. **c** L-HH: 20 %; L-VV: 36 %; X: 7 %

curve for Lane C for the three FLGPR sensors, while views (b) and (c) show the results of using the SKSVM classifier to reject FAs. The kernel used for this experiment is the RBF kernel, which is well-known to be effective for most data. View (b) shows the ROC curve using only the HOG feature, while view (c) shows the results when combining the HOG and LBP features. As the figure illustrates, the SKSVM is able to reduce the number of FAs significantly. Interestingly, the combination of features is detrimental to SKSVM performance for the X-band FLGPR. This is because the addition of the LBP feature to the SKSVM for the X-band radar results in over-training (the training or resubstitution results are nearly perfect), which negatively affects the test lane performance.

3.2 Multiple Kernel

MKL extends the idea of kernel classification by allowing the use of combinations of multiple kernels. The kernel combination can be computed in many ways, as long as the combination is a Mercer kernel [34]. In this chapter we assume that the kernel K is composed of a weighted combination of pre-computed kernel matrices, i.e.,

$$K = \sum_{k=1}^{m} \sigma_k K_k, \tag{7}$$

where there are m kernels and σ_k is the weight applied to the kth kernel. The composite kernel can then be used in the chosen classifier model; we will use the SVM. Thus, MKL SVM extends the SKSVM optimization at (4) by also optimizing over the weights σ_k,

$$\min_{\sigma \in \Delta} \max_{\alpha} \left\{ \mathbf{1}^T \alpha - \frac{1}{2} (\alpha \circ \mathbf{y})^T \left(\sum_{k=1}^{m} \sigma_k K_k \right) (\alpha \circ \mathbf{y}) \right\}, \tag{8a}$$

subject to (typically)

$$0 \le \alpha_i \le C, \; i = 1, \dots, n; \; \alpha^T \mathbf{y} = 0, \tag{8b}$$

where Δ is the domain of σ. Note that this is the *same* problem as SKSVM if the kernel weights are assumed constant [28]. This property has been used by many researchers to propose *alternating optimization* procedures for solving the min-max optimization problem. That is, solve the inner maximization for a constant kernel K, and then update the weights σ_k to solve the outer minimization, and repeat until convergence. We use the optimization procedure proposed by Xu et al. called MKL *group lasso* (MKLGL) [47]. This method is efficient as it uses a closed-form (i.e., non-iterative) solution for solving the outer minimization in (8a);

$$\sigma_k = \frac{f_k^{2/(1+p)}}{\left(\sum_{k=1}^{m} f_k^{2p/(1+p)} \right)^{1/p}}, \; k = 1, \dots, m, \tag{9a}$$

$$f_k = \sigma_k^2 (\alpha \cdot \mathbf{y})^T K_k (\alpha \cdot \mathbf{y}), \tag{9b}$$

where p is the norm on the domain constraint, $\|\sigma\|_p = 1, p > 1$.

We further modify the MKLGL algorithm, as we did for SKSVM, to allow for unbalanced classes—i.e., we apply the constraints C^+ and C^- as shown at (5). The MKLGL training algorithm is outlined in Algorithm 1. The MKLGL is simple to implement and is efficient as the update equations for σ_k are closed-form. MKL can be thought of as a classifier fusion algorithm. It can find the optimal kernel among a set of candidates by automatically learning the weights on each kernel. The individual kernels can be computed in many ways—see our previous papers on this topic for more discussion on the formation of the kernel matrices [15, 17].

Algorithm 1: MKLGL Classifier Training [47]

Data: $(\mathbf{x}_i, y_i)$ - feature vector and label pairs; K_k - kernel matrices
Result: α, σ_k - MKLGL classifier solution
Initialize $\sigma_k = 1/m, k = 1, \dots, m$ (equal kernel weights)
while *not converged* **do**
  | Solve unbalanced SKSVM for kernel matrix $K = \sum_{k=1}^{m} \sigma_k K_k$
  |_ Update kernel weights by Eq. (9)

Application of MKLGL to FLGPR EHD The MKLGL algorithm is applied in the same way as the SKSVM—it acts to classify prescreener hits as either FAs or true positives. Figure 7 shows results of the MKLGL classifier using an ensemble of RBF

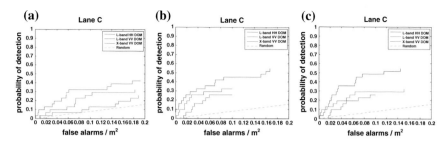

Fig. 7 ROC curves showing testing performance of (**a**) FLGPR prescreener, and MKLGL classifier for (**b**) single HOG feature and (**c**) combination of HOG and LBP features. Percent NAUC improvements are shown for each of the L-band (HH and VV polarizations) and X-band FLGPRs. The performance of a uniform random detector is shown by the *dotted line*. **a** Prescreener. **b** L-HH: 21 %; L-VV: 46 %; X: 67 %. **c** L-HH: 21 %; L-VV: 47 %; X: 67 %

kernels on the same training and testing lanes as shown for SKSVM in Fig. 6. The NAUC results show that the MKLGL is able to match and sometimes improve upon the results obtained using the SKSVM. The MKLGL improvement of the L-band VV NAUC is especially noteworthy.

3.3 Fuzzy Integral-Based Multiple Kernel (FIMKL)

The *Fuzzy Integral-based* MK (FIMKL) [22, 23] extends MKL by using a non-linear aggregation operator, the *fuzzy integral* (FI). The fusion of information using the Sugeno or Choquet FI has a rich history; for a recent review, see [1]. Depending on the problem domain, the input to the FI can be experts, sensors, features, similarities, pattern recognition algorithms, etc. The FI is defined with respect to the *fuzzy measure* (FM), a monotone and often normal capacity. With respect to a set of m information sources, $X = \{x_1, \ldots, x_m\}$, the FM encodes the (often subjective) *worth* of each subset in 2^X. For a finite set of sources, X, the FM is a set-valued function $g : 2^X \rightarrow [0, 1]$ with the following conditions:

1. (Boundary condition) $g(\phi) = 0$,
2. (Monotonicity) If A, B $\subseteq X$ with A $\subseteq$ B, then $g(A) \leq g(B)$.

Note, if X is an infinite set, there is a third condition guaranteeing continuity and we often assume $g(X) = 1$ (although it is not necessary in general). Numerous FI formulations have been proposed to date for generalizability, differentiability, and to address different types of uncertain data [1]. In [22, 23], we investigated the Sugeno and Choquet FIs for MKL. We proposed a solution based on sorting at the *matrix level*. Assume each kernel matrix K_k has a numeric "quality." This can be computed, for example, by computing the classification accuracy of a base-learner that uses kernel K_k (or by a learning algorithm like a GA). Let $v_k \in [0, 1]$ be the kth kernel's *quality*. These qualities can be sorted, $v_{(1)} \geq v_{(2)} \geq \ldots \geq v_{(m)}$. Given m base Mercer kernels, $\{\kappa_1, \dots, \kappa_m\}$, FM g, and a sorting $v_{(1)} \geq v_{(2)} \geq \dots \geq v_{(m)}$, the difference-in-measure Choquet FI is computed by

$$\mathcal{K}_{ij} = \sum_{k=1}^{m}(G_{\pi(k)} - G_{\pi(k-1)})(K_{\pi(k)})_{ij} = \sum_{k=1}^{m} \omega_k (K_{\pi(k)})_{ij}, \; i,j \in \{1, \dots, n\}, \qquad (10)$$

where $\omega_i = \left(G_{\pi(i)} - G_{\pi(i-1)}\right)$, $G_{\pi(i)} = g\left(\{x_{\pi(1)}, \dots, x_{\pi(i)}\}\right)$, $G_{\pi(0)} = 0$, and $\pi(i)$ is a sorting on X such that $h(x_{\pi(1)}) \geq \dots \geq h(x_{\pi(m)})$. The MK formulation at (10) produces a Mercer kernel as multiplication by positive scalar and addition are *positive semidefinite* (PSD) preserving operations. Since (10) involves per-matrix sorting, it can be compactly written in a simpler (linear algebra) form, i.e., $\mathcal{K} = \sum_{k=1}^{m} \omega_k K_{\pi(k)}$.

Prior works in MKL rely on the relatively *linear convex sum* (LCS) formulation. It is often desired due to its advantage in optimization, e.g., MKLGL. Both FIMK and LCS MK are of type convex sum, i.e., $w_k \in \mathfrak{R}_+^m$ and $\sum_{k=1}^{m} w_k = 1$. However, one is linear, the other is not, and the weights are derived from the FM. The Choquet FI is capable of representing a much larger class of aggregation operators. For example, it is well known that the Choquet FI can produce, based on the selection of FM, the maximum, minimum, *ordered weighted average* (OWA), order statistics, etc. However, the machine learning LCS form is simply m weights anchored to the individual inputs. The LCS is a subset (one of the aggregation operators) of the FI.

In [22, 23], we reported improved SVM accuracies and lower standard deviations over the state-of-the-art MKLGL on publicly available benchmark data. We proposed a GA, called FIGA, based on learning the densities for the Sugeno λ-FM. In that work we demonstrated that the GA approach is more effective than MKLGL, even in light of the fact that our GA approach used far fewer component kernels. In particular, the FIGA approach achieved a mean improvement of nearly 10 % over MKLGL on the Sonar data set. The performance of FIGA comes at a cost though, as MKLGL is much faster in terms of actual running time than FIGA. We also saw that FIGA using a combination of FM/FIs is somewhat more effective than the FIGA LCS form. These findings are not surprising as our intuition tells us that the nonlinear aggregation allowed by the FM/FI formulation is more flexible than just the LCS aggregation; hence, these results reinforce our expectation. Overall, these results are not surprising as different data sets require different solutions, and while an LCS may be sufficient for a given problem, it may not be appropriate for a different problem. Also, it should be noted that the FM/FI formulation includes LCS aggregation as a subset of its possible solutions; hence, when LCS is appropriate the FM/FI aggregation can mimic the LCS. In summary, the learner (GA vs GL) appears to be the most important improvement factor, followed by a slight improvement by using the nonlinear FM/FI aggregation versus LCS. While FIMKL has not been applied to date for EHD, this computational intelligence method is reviewed as it is an improvement to classical MKL and stands to be of relevance and benefit to EHD.

4   Deep Learners and Feature Learning for EHD

Deep learning architectures were initially designed to mimic the human brain, more specifically, the neocortex [36]. This part of the brain has been shown to have six layers and a forward-backward structure to classify image data collected by the eye [26]. In brief, deep learning architectures extend "shallow" neural networks by adding multiple hidden layers—these additional layers act as generalized feature detectors. Deep learning algorithms have been shown to perform very well on a variety of classification tasks, such as facial recognition [29], document classification [30], and speech recognition [39]. We will present results for two types of deep learning architectures: *deep belief networks* (DBNs) and *convolutional neural networks* (CNNs).

4.1   Deep Belief Networks

DBNs are a type of deep learning network formed by stacking *Restricted Boltzmann Machines* (RBMs) in successive layers to reduce dimensionality, making a compressed representation of the input. DBNs are trained layer by layer using greedy algorithms and information from the previous layer. In this subsection, we will first discuss RBMs and how to train them, then move on to training DBNs.

RBMs are simple binary learners that consist of two layers: one visible and one hidden. The visible layer is the input layer and typically consists of an n-length vector of normalized values. The hidden layer is the feature representation layer. The defining equation of the RBMs is the energy equation,

$$E(\mathbf{v}, \mathbf{h}) = -\mathbf{b}^T \mathbf{v} - \mathbf{c}^T \mathbf{h} - \mathbf{v}^T W \mathbf{h}, \tag{11}$$

where $\mathbf{v}$ is the input vector, $\mathbf{h}$ is the hidden feature vector, $\mathbf{b}$ and $\mathbf{c}$ are the visible and hidden layer biases, respectively, and W is the weight matrix that connects the layers. It should be noted that weights only exist between the hidden and visible layers, that is to say, that the nodes in either layer are not interconnected. $\mathbf{v}$ is the input and used to train hidden layer $\mathbf{h}$ as

$$\mathbf{h} = \sigma(\mathbf{c} + W^T \mathbf{v}). \tag{12}$$

The hidden layer is then used to reconstruct the visible layer in the same manner,

$$\mathbf{v}_{recon} = \sigma(\mathbf{b} + W \mathbf{h}). \tag{13}$$

The reconstruction of the visible layer $\mathbf{v}_{recon}$ is then used in (12) to form $\mathbf{h}_{recon}$ and then the weight update is calculated as

Fig. 8 Illustration of DBN
training: numbers in
rectangles indicate the
number of neurons in each
layer. **a** Pretraining.
b Unrolling. **c** Fine tuning

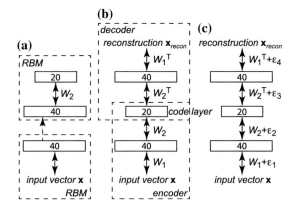

$$\Delta W = \epsilon \left(\left[\mathbf{vh}^T \right]_{data} - \left[\mathbf{vh}^T \right]_{recon} \right), \tag{14}$$

where ϵ is the learning rate. Iterated over several epochs, this weight update performs
a type of gradient descent called contrastive divergence [36].

To form a DBN, layers of RBMs are stacked as shown in Fig. 8a, where the hidden
layer of the lower RBM becomes the input/visible layer of the next RBM. Once the
input RBM is trained, its reconstructed hidden layer $\mathbf{h}_{recon}$ is used to create the visible
layer of the next RBM by

$$\mathbf{v}_{n+1} = \sigma(\mathbf{c}_n + W_n^T \mathbf{h}_{recon,n}) \tag{15}$$

where n denotes the layer number. The $(n+1)$th RBM is now trained and this cycle is
repeated for the number of layers desired. After all layers have been trained, the DBN
is typically then mirrored to make an encoder-decoder as shown in Fig. 8b [21]. An
input to the encoder-decoder thus produces a reconstruction of itself, where

$$\text{encoder: } \mathbf{x}_{n+1} = W_n \mathbf{x}_n; \tag{16a}$$

$$\text{decoder: } \mathbf{x}_{recon,n-1} = W_{n-1}^T \mathbf{x}_{recon,n}; \tag{16b}$$

and $x_1 \in \mathbb{R}^d$ is the input vector and $x_{recon,1} = x_{recon} \in \mathbb{R}^d$ is the reconstruction.
Note that the final hidden layer in the encoder is the first layer in the decoder, $\mathbf{x}_{n+1} = \mathbf{x}_{recon,n+1}$, where n is the number of RBMs in the DBN. Fine-tuning of the weight
matrices can be performed as shown in Fig. 8c. This fine-tuning is often done using
stochastic gradient descent (backpropagation) or Hinton's *up-down* algorithm [20].
Note that this gives the DBN more flexibility as the weight matrices are adjusted for
each of the encoder and decoder.

Application of DBNs to FLGPR EHD To apply DBNs to the FLEHD problem, we take the extracted features from each prescreener hit location in the training data and apply the DBN to learn the representation of the FAs; this is due to the imbalance between the number of FA and target examples in the training data. The reconstruction *root mean-square error* (RMSE),

$$RMSE = \sqrt{\sum_{i=1}^{d} \left(x_i - x_{recon,i} \right)^2}, \tag{17}$$

of the DBN is thus a measure of how well an input feature vector matches to the learned representation of the FAs—true positives ideally have high RMSE and false positives ideally have low RMSE. Hence, the RMSE can be directly used as the confidence of a true positive in the ROC curve. The DBNs for the results here are trained on three lanes of data and then tested on a separate lane (in essence, 4-fold cross-validation).

Since DBNs are flexible in their construction, we tested many different architectures, learning rates, and epoch limits. The best DBN we found for overall EHD performance was a network that uses two hidden layers of sizes 40 and 20, giving a full encode-decode stack architecture of [$\mathbf{x}$ 40 20 40 $\mathbf{x}_{recon}$], where $\mathbf{x}$ is the $d \times 1$ input feature vector and $\mathbf{x}_{recon}$ is the $d \times 1$ reconstruction (see Fig. 8). The learning rate is 0.9, and 30 epochs of contrastive divergence was used for RBM training.

Several combinations of features were tested with the DBN classifier. Figure 9 illustrates selected results from our comprehensive evaluation of DBNs for FLGPR EHD. These ROC curves show the performance of the DBN classifier on Lane C (training on Lanes A, B, and D). The percent NAUC improvements clearly show that the DBN significantly improves NAUC, by up to 85 % for the case of the X-band FLGPR using HOG & LBP features (note that the X-band FLGPR also has the most room for improvement in this case).

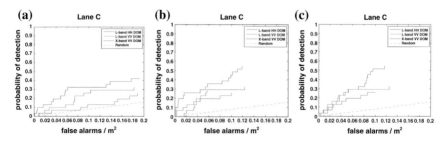

Fig. 9 ROC curves showing testing performance of **a** FLGPR prescreener, and DBN classifiers for **b** single HOG feature and **c** combination of HOG and LBP features. Percent NAUC improvements are shown for each of the L-band (HH and VV polarizations) and X-band FLGPRs. The performance of a uniform random detector is shown by the *dotted line*. **a** Prescreener. **b** L-HH: 28 %; L-VV: 52 %; X: 53 %. **c** L-HH: 23 %; L-VV: 52 %; X: 85 %

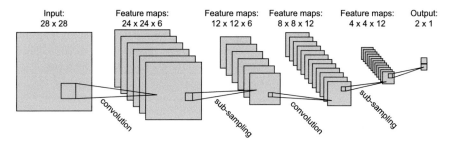

Fig. 10 Illustration of a convolutional neural network [31, 36]

4.2   Feature Learning

Convolutional Neural Networks CNNs are a type of neural network with a unique architecture. Inspired by the visual system, these networks consist of alternating convolutional and sub-sampling layers. The convolutional layers generate feature maps by convolving kernels over the data from the previous layers and then the sub-sampling layers downsample the feature maps [36]. CNNs work directly on the 2D data as opposed to most other forms of deep networks which reorganize the data into 1D feature vectors. Figure 10 illustrates a convolutional neural network.

The *l*th convolutional layer is generated from a *j*th feature map by

$$a_j^l = \sigma \left(b_j^l + \sum_{i \in M_j^l} a_j^{l-1} * k_{ij}^l \right), \tag{18}$$

where σ is the activation function, usually hyperbolic tangent or sigmoid, b_j^l is a scalar bias, M_j^l is an index vector of feature maps i in layer $l-1$, $*$ is the 2D convolution operator and k_{ij}^l is the kernel used on map i in layer $l-1$. A sub-sample layer l is generated from a feature map j by

$$a_j^l = \text{down}(a_j^{l-1}, N^l), \tag{19}$$

where down is a down-sampling function, such as mean-sampling, that down-samples by factor N^l [36]. The output layer is then generated by

$$o = f(\mathbf{b}^o + W^o \mathbf{x}_v), \tag{20}$$

where $\mathbf{x}_v$ denotes a feature vector concatenated from the feature maps of the previous layer, $\mathbf{b}^o$ is a bias vector, and W^o is a weight matrix. The parameters to be learned are thus k_{ij}^l, b_j^l, $\mathbf{b}^o$ and W^o. Gradient descent is used to learn these parameters and this can be efficiently performed through the use of convolutional backpropagation [36].

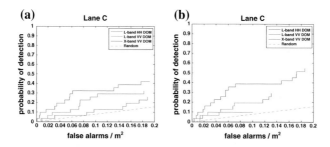

Fig. 11   ROC curves showing testing performance of **a** FLGPR prescreener, and **b** CNN classifier using the HOG feature. Percent NAUC improvements are shown for each of the L-band (HH and VV polarizations) and X-band FLGPRs. The performance of a uniform random detector is shown by the *dotted line*. **a** Prescreener. **b** L-HH: 13 %; L-VV: 1.4 %; X: 28 %

Application of CNNs to FLGPR EHD Unlike the SVM, MKLGL, and DBN, the CNN operates on 2D feature maps. Fortunately, the HOG, LBP, and FFST are all 2D features and thus can be used as input for the CNN; we also used the raw image data (imagelet) surrounding each prescreener hit as input to the CNN. The output of the CNN is a 2-element vector—one element to indicate FA and one to indicate true positive. As shown in Fig. 10, we use two convolutional layers and two subsampling layers. The learning rate was 0.9 and 350 epochs were used for training, which were shown to be good choice in a more comprehensive parameter study we performed. Figure 11 shows selected results from our comprehensive evaluation of CNNs for FLGPR EHD. These ROC curves show the performance of the CNN classifier on Lane C (training Lanes A, B, and D). As is evidenced by the percent NAUC improvement values, the CNN is the least effective of the classifiers that we have applied to the FLGPR EHD. Furthermore, many of the results (which we do not show) that we compiled using the CNN were very poor. Hence, we do not recommend the CNN at this time for FLGPR EHD.

Application of CNNs to FLIR EHD In [43], CNNs were evaluated for EHD in FLIR imagery. Image chips extracted at prescreener alarm locations were fed directly as input to a CNN. CNN classification results were compared to a baseline algorithm which extracted five hand-engineered feature sets and performed classification using a SVM. Due to the lack of training data for training a conventional deep CNN model, two alternative CNN approaches were explored.

The first approach used a deep CNN model pre-trained on the ImageNet dataset [12]. This model is referred to herein as DPT-CNN, and was made available through the open source python package DeCAF [13]. DPT-CNN uses the architecture proposed by Krizhevsky et al. in [27], which won the *ImageNet Large Scale Visual Recognition Challenge 2012* (ILSVRC2012). The architecture consists of five convolutional layers, some followed by *Rectified Linear Unit* (ReLU) activation, response normalization, and max pooling, followed by three fully connected layers. The last fully connected layer is fed to a 1000-way softmax function.

DPT-CNN was trained on the ILSVRC2012 training data, which consisted of more than 1.2 million training images from 1000 object classes. It was shown in [13] that values from intermediate layers of this pre-trained network work well as features for new vision tasks. Specifically, tasks with small amounts of training data, where a deep CNN trained directly performed poorly. For our tests, the alarm image chips were input to DPT-CNN, and intermediate values were saved at six stages. These intermediate values were used to train a SVM which was evaluated the same way as the baseline algorithm. The first two sets of intermediate values came from the second and first fully connected layers after ReLU activation. These are referred to as FC7-ReLU and FC6-ReLU, respectively. The ReLU activation takes the form $\phi(v) = \max(0, v)$. The next two intermediate values, POOL5 and CONV5, came from the last convolutional layer. CONV5 is before max-pooling, and POOL5 is after pooling. The last two sets, RNORM1 and POOL1, came from the first convolutional layer. RNORM1 is after pooling and response normalization. POOL1 is after pooling but before response normalization.

The fully connected layer outputs, which no longer convey spatial position, were not expected to be useful for this EHD task since position in the image chip is extremely important. The POOL5 and CONV5 features do retain some spatial information. The RNORM1 and POOL1 features retain more, but since they are not deep features they may not be as descriptive. Table 2 shows the NAUC results for the DPT-CNN features, as well as for the baseline algorithm and the best individual baseline feature, for a three lane leave-one-lane-out cross validation test using two FLIR cameras.

As expected, the fully connected layer features did not perform well. Performance improved significantly when moving to the POOL5 features, and again when moving to the CONV5 features. The CONV5 features compared well with the top performing hand-engineered image feature, multi-scale HOG, even outperforming it on Lane A. Surprisingly, the POOL1 features scored better overall than the CONV5 features on the DVE camera image chips, but show a pronounced drop on the SELEX image chips.

Table 2 DPT-CNN: DVE/Selex cameras: NAUC at 0.01 FA/m^2

Feature	All lanes	Lane A	Lane B	Lane C
FC7-ReLU	0.435/0.451	0.321/0.355	0.420/0.418	0.556/0.573
FC6-ReLU	0.479/0.469	0.353/0.365	0.480/0.454	0.598/0.582
POOL5	0.557/0.501	0.404/0.386	0.600/0.500	0.658/0.609
CONV5	0.615/0.566	0.471/0.423	0.655/0.604	0.712/0.662
RNORM1	0.623/0.525	0.454/0.389	0.709/0.553	0.699/0.624
POOL1	0.624/0.519	0.458/0.386	0.710/0.545	0.695/0.617
BASE: HOG	0.645/0.584	0.453/0.421	0.722/0.615	0.753/0.708
BASE: ALL	**0.677/0.610**	0.496/0.445	**0.766/0.652**	0.762/**0.727**
CONV5 + BASE	0.676/0.607	**0.508/0.449**	0.748/0.649	**0.764**/0.714

*Bold indicates best result for each camera and lane

The DPT-CNN results indicated that a deep CNN model was not necessary to achieve good performance on this FLIR EHD task. This was not particularly surprising since the task requires little in the way of translation, scale, or orientation invariance. The primary difficulty is intra-class variation. Thus, a second CNN approach using a shallow CNN trained directly using the image chips was pursued. The shallow architecture consists of a single convolutional layer followed by an output layer containing a single neuron followed by the sigmoid activation function. The output of the sigmoid was used as the alarm confidence value. For all experiments, weights were learned using *stochastic gradient descent* (SGD) with momentum and the cross entropy error function. To address class imbalance, for each training pattern presentation an example was chosen randomly from either the true target class or the false alarm class with equal probability. Evaluation was performed using the same methodology as for DPT-CNN.

In [24], Jarrett et al. found that the single most important factor for recognition accuracy in a CNN model, considering architecture choices such as activation function, sub-sampling type, and response normalization, was the use of a rectifying non-linearity. While they used the *absolute value function* (AVF), the ReLU in Krizhevskys architecture performs a similar operation. Therefore, the first experiment evaluated performance when using either no non-linearity, ReLU, or AVF following the convolutional layer. These results are presented in Table 3. Both activations improved performance. AVF performed better than ReLU, and was chosen for further experiments.

We next investigated forcing the convolutional filters to have zero-mean and zero-phase. The intuition being that only the non-dc frequency characteristics are important, and that shifting of the output is meaningless for classification. To enforce these characteristics, transformation functions were inserted before the variables in question were used. The transformation functions modify their inputs to enforce the desired constraint. During SGD learning, derivatives are propagated through the transformations. For example, if the original convolutional layer is OUTPUT = CONV(INPUT,X), to enforce zero-mean for the kernel X the expression becomes OUTPUT = CONV(INPUT,$G(X)$), where $G(X)$ modifies X to have the zero-mean characteristic. No significant performance improvement was seen from enforcing either constraint.

Table 3 Rectifying Nonlinearity: DVE Camera: NAUC at 0.01 FA/m^2

Convolution filter radius	None, # filters			ReLU, # filters			AVF # filters		
	4	8	16	4	8	16	4	8	16
3	0.492	0.482	0.503	0.519	0.517	0.499	0.555	**0.573**	0.566
5	0.508	0.510	0.509	0.550	0.552	0.565	0.600	0.602	**0.603**
7	0.487	0.483	0.475	0.555	0.545	0.537	**0.596**	0.580	0.592

*Bold indicates best result for each filter radius

Table 4 Learning in freq domain: DVE camera: NAUC at 0.01 FA/m^2

Convolution filter radius	Spatial—# of filters			Frequency—# of filters		
	4	8	16	4	8	16
3	0.555	0.573	0.566	0.636	0.636	**0.640**
5	0.600	0.602	0.603	0.617	**0.619**	0.618
7	0.596	0.580	0.592	0.614	0.613	**0.619**

*Bold indicates best result for each filter radius

We then experimented with learning the convolutional filters' frequency domain representations instead of their spatial domain representations. This was done by using the inverse FFT as a transformation function. Table 4 shows the results for learning the convolutional filters in the frequency domain versus the spatial domain. Zero-mean and zero-phase were enforced in the frequency domain. A slight performance improvement was seen across all combinations.

Based on these results, shallow CNN networks with eight zero-mean, zero-phase filters learned in the frequency domain were scored on the DVE and SELEX data. Table 5 shows the per lane results for various kernel radii, as well as the DPT-CNN and baseline results for comparison. Overall, the shallow CNN results were very similar to those of DPT-CNN. The shallow CNN achieved a slightly better overall result on DVE, and a slightly worse overall result on SELEX when comparing to the CONV5 features of DPT-CNN. When comparing to the POOL1 and RNORM1 features, the shallow CNN SELEX result is much better. The baseline algorithm, which includes features that cannot be expressed via convolution, outperforms both CNN approaches.

iECO Feature Learning In [38], the algorithm *improved Evolutionary COnstructed* (iECO) feature descriptors (referred to hereafter as simply iECO) was put forth for FLIR-based EHD. The iECO algorithm is a feature learning technique that looks to

Table 5 Shallow CNN: NAUC at 0.01 FA/m^2

	DVE camera				SELEX camera			
	All lanes	Lane A	Lane B	Lane C	All lanes	Lane A	Lane B	Lane C
CNN radius 3	0.635	0.464	0.734	0.700	0.562	0.397	0.626	0.656
CNN radius 5	0.616	0.478	0.694	0.670	0.559	0.413	0.611	0.645
CNN radius 7	0.612	0.460	0.697	0.673	0.557	0.409	0.628	0.626
Pre-trained CNN	0.624	0.458	0.710	0.695	0.566	0.423	0.604	0.662
Baseline	**0.677**	**0.496**	**0.766**	**0.762**	**0.610**	**0.445**	**0.652**	**0.727**

*Bold indicates best result

exploit important cues in data that often elude non-learned (often referred to as "hand crafted") features such as HOGs, LBPs and edge histogram descriptors. Each hand crafted feature is ultimately an attempt to more-or-less sculpt (force) a signal/image into some predisposed mathematical framework which may or may not reveal the information that a user/system needs. Instead of coming to the table with a limited set of tools and trying to make everything look like a nail, iECO learns the tool based on the task at hand.

While the field of deep learning has demonstrated state-of-the-art performance, the ECO (and iECO respectively) work of Lillywhite et al. has the advantage over CNNs of interpretability (it is not a black box) and it does not predispose the solution to that of convolution. At its core, ECO is the GA-based learning and (ensemble-based) use of a population of chromosomes that are compositions of functions (image processing transformations). Each chromosome is of variable length and the goal is to learn the image transformations and respective parameters relative to some task. An advantage of this approach, versus CNNs, is that it makes use of a relatively wide set of different heterogeneous image transformations to seek a new tailored solution. In [38] we used 19 different image transformations which range from a Harris corner detector to a square root, Hough circle, median blur, rank transform, LoG, mathematical morphology and Shearlet and Gabor spatial frequency domain filtering, just to name a few. In many cases, emergent behaviour arises and the chromosomes can be manually examined and studied, potentially revealing additional domain information such as what features or physics in IR or GPR are most important for a task like EHD. Figure 12 shows the iECO process (not learning, but application of iECO to a given AOI).

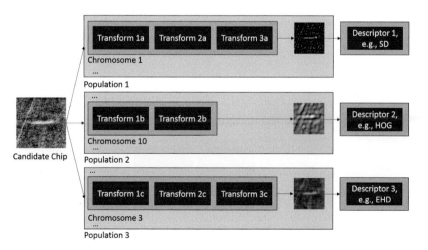

Fig. 12 iECO applied to a prescreener hit in FLIR. Learned iECO chromosomes, in different populations, are applied to the input image. Finally, different descriptors are extracted relative to each transformed image

In iECO, we address a shortcoming of the ECO features– the so-called "features" which are the unrolling of image pixels into a single vector. ECO suffers from the curse of dimensionality and the naive unrolling does not intelligently take into account various spatial and scale cues. In [38], we extract ECO features relative to different high-level descriptors and cell-structured configurations. Specifically, we explored the HOG, EHD and statistical features; which include the local mean, standard deviation, kurtosis, l_2-norm, and the difference between the local values and their corresponding global values. A separate GA population is maintained and a separate search is conducted relative to each high-level descriptor. iECO, like CNN learning is not a computationally trivial task. As a result, we have not yet attempted to learn the different populations in a single simultaneous algorithm. Furthermore, in [38] we showed that each descriptor learns/prefers different chromosomes that have varying fitness values. With respect to classification, we experimented with taking a single best chromosome per descriptor (highest fitness), taking the top 50 % of chromosomes relative to each descriptor, and the identification of the top 5 most diverse chromosomes (which is currently a manual process). Our results indicate that the concatenation of multiple chromosome features leads to improved performance. Furthermore, we showed that if one pipeline is applied to a different descriptor than it was learned for, then the result is a significant drop in performance (fitness). This is interesting as it tells us that iECO appears to learn a tailored pre-processing of imagery relative to each descriptor in order to better highlight salient information. Figure 13 shows different learned iECO pipelines.

In [38], we introduced constraints on each individual's chromosome to help promote population diversity and prevent infeasible solutions. This allows us to search for quality solutions faster and it typically results in shorter length chromosomes that are computationally simpler to realize (which is important for a real-time causal EHD system). In ECO, there are no direct mechanisms incorporated into the GA,

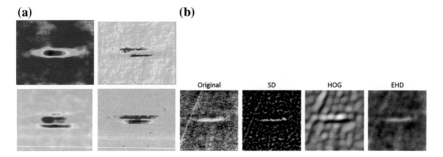

Fig. 13 iECO on FL LWIR. **a** Average iECO output of four chromosomes across 50 different buried targets. Each image is scaled to [0, 1] for visual display and they are shown in Matlab jetmap color coding, where *blue* is 0 and *red* is 1. These images show that diversity exists across chromosomes and different aspects of targets are learned, e.g., local contrast, orientation specific edge information, etc. **b** Output of highest fitness chromosones for each descriptor for a single target. These images show that each descriptor prefers a different iECO pipeline. **a** Average iECO output. **b** Different iECO populations

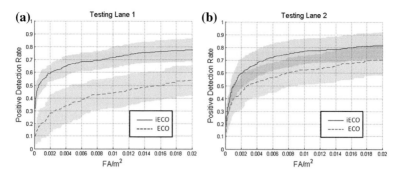

Fig. 14 Vertically averaged ROC curves with 95 % confidence intervals. **a** Testing lane 1. **b** Testing lane 2

outside of mutation, to promote diversity in the population. In [38], we introduced diversity promoting constraints that consider the uniqueness and complexity of the ECO's search space. We designed a set of diversity promoting constraints that define what percentage of the population is allowed overlapping genes at each layer of the individual's gene segment. Next, we addressed the issue of the occurrence of the same gene back-to-back. Such a scenario is undesirable, e.g., it does not typically make sense to perform a rank transform back-to-back. In addition, this increases the computational complexity of the system as a consequence of the unnecessary image transforms. We combat this by collapsing consecutive uses of the same gene type, i.e., if any gene occurs more than once consecutively then only the first occurrence is retained. Elitism is used in iECO.

In summary, in [38] we showed that the above diversity promoting constraints and the combination of high-level image descriptors leads to the discovery of significantly higher quality solutions for EHD. We showed that iECO continuously identifies higher performance solutions, i.e., an impressive drop in the FAR for a given PDR, populations are more diverse, which was verified manually, and the resultant chromosomes are significantly shorter and thus give rise to a simpler system (computationally and memory utilization-wise) to realize. Figure 14 is ROC results for iECO versus ECO features. iECO clearly outperforms ECO.

5 Conclusions

This chapter described the EHD problem and various methods for preprocessing, prescreening, and false rejection for FLGPR and FLIR. The methods discussed for FLGPR-based EHD were SKSVM, MKLGL, DBNs, and CNNs. The best overall detection and classification method for the FLGPR was the DBN using a combination of HOG and LBP features, showing up to 85 % improvement in NAUC. The weakest FLGPR-based method was the CNN. In the future, we are going to investigate more advanced CNN architectures and training methods for CNNs as applied to FLGPR

EHD. Several EHD methods were discussed for FLIR-based detection, including baseline SVM detection, CNNs, and iECO feature learning. Several CNN architectures were tested. While the CNN architectures showed promise, especially those that use frequency-domain AVF filters, the baseline SVM-based feature-fusion approach outperformed the CNN. Lastly, iECO feature learning was demonstrated for FLIR-based EHD. In the future, we aim to further apply our fuzzy integral-based multiple kernels methods for EHD as FIMKL has been shown to be superior to MKLGL for benchmark data sets. We also aim to extend the deep learning approaches for online and active learning for EHD.

Acknowledgments This work is funded in part by a National Institute of Justice grant (2011-DN-BX-K838), U.S. Army (W909MY-13-C0013, W909MY-13-C0029) and Army Research Office (W911NF-14-1-0114 and 57940-EV) in support of the U.S. Army RDECOM CERDEC NVESD. Superior, a high performance computing cluster at Michigan Technological University, was used in obtaining results presented in this work.

References

1. Anderson, D.T., Havens, T.C., Wagner, C., Keller, J., Anderson, M.F., Wescott, D.J.: Extension of the fuzzy integral for general fuzzy set-valued information. IEEE Trans. Fuzzy Syst. **22**(6), 1625–1639 (2014)
2. Anderson, D.T., Stone, K., Keller, J.M., Spain, C.: Combination of anomaly algorithms and image features for explosive hazard detection in forward looking infrared imagery. IEEE J. Sel. Topics Appl. Earth Obs. Remote Sens. **5**(1), 313–323 (2012)
3. Anderson, D.T., Stone, K., Keller, J.M., Rose, J.: Anomaly detection ensemble fusion for buried explosive material detection in forward looking infrared imaging for addressing diurnal temperature variation. In: Proceedings of the SPIE, vol. 8357, p. 83570T (2012)
4. Becker, J., Havens, T.C., Pinar, A., Schulz, T.J.: Deep belief networks for false alarm rejection in forward-looking ground-penetrating radar. In: Proceedings of the SPIE (2015)
5. Boser, B.E., Guyon, I.M., Vapnik, V.N.: A training algorithm for optimal margin classifiers. In: ACM Workshop on COLT, pp. 144–152 (1992)
6. Chang, C.C., Lin, C.J.: LIBSVM: a library for support vector machines. ACM Trans. Intell. Sys. Tech. **2**(3), 1–27 (2011)
7. Collins, L.M., Torrione, P.A., Throckmorton, C.S., Liao, X., Zhu, Q.E., Liu, Q., Carin, L., Clodfelter, F., Frasier, S.: Algorithms for landmine discrimination using the NIITEK ground penetrating radar. Proc. SPIE. **4742**, 709–718 (2002)
8. Cortes, C., Vapnik, V.N.: Support-vector networks. Mach. Learn. **20**(3), 273–297 (1995)
9. Costley, R.D., Sabatier, J.M., Xiang, N.: Forward-looking acoustic mine detection system. Proc. SPIE. **4394**, 617–626 (2001)
10. Cremer, F., Chavemaker, J.G., deJong, W., Schutte, K.: Comparison of vehicle-mounted forward-looking polarimetric infrared and downward-looking infrared sensors for landmine detection. Proc. SPIE. **5089**, 517–526 (2003)
11. Dalal, N., Triggs, B.: Histograms of oriented gradients for human detection. In: IEEE Computer Society Conference on Computer Vision and Pattern Recognition, 2005. CVPR 2005. vol. 1, pp. 886–893 (2005)
12. Deng, J., Dong, W., Socher, R., Li, L.J., Li, K., Fei-Fei, L.: Imagenet: a large-scale hierarchical image database. In: IEEE Conference on Computer Vision and Pattern Recognition, 2009. CVPR 2009, pp. 248–255 (2009)

13. Donahue, J., Jia, Y., Vinyals, O., Hoffman, J., Zhang, N., Tzeng, E., Darrell, T.: Decaf: A deep convolutional activation feature for generic visual recognition. CoRR abs/1310.1531 (2013), http://arxiv.org/abs/1310.1531

14. Gader, P.D., Grandhi, R., Lee, W.H., Wilson, J.N., Ho, K.C.: Feature analysis for the NIITEK ground penetrating radar using order weighted averaging operators for landmine detection. In: Proceedings of the SPIE, vol. 5415, pp. 953–962 (2004)

15. Havens, T.C., Becker, J.T., Pinar, A.J., Schulz, T.J.: Multi-band sensor-fused explosive hazards detection in forward-looking ground-penetrating radar. In: Proceedings SPIE, vol. 9072, p. 90720T (2014)

16. Havens, T.C., Ho, K.C., Farrell, J., Keller, J.M., Popescu, M., Ton, T.T., Wong, D.C., Soumekh, M.: Locally adaptive detection algorithm for forward-looking ground-penetrating radar. In: Proceedings of the SPIE, vol. 7664, p. 76442E (2010)

17. Havens, T.C., Keller, J.M., Stone, K., Ho, K.C., Ton, T.T., Wong, D.C., Soumekh, M.: Multiple kernel learning for explosive hazards detection in FLGPR. In: Proceedings of the SPIE, vol. 8357, p. 83571D (2012)

18. Havens, T.C., Spain, C.J., Ho, K.C., Keller, J.M., Ton, T.T., Wong, D.C., Soumekh, M.: Improved detection and false alarm rejection using ground-penetrating radar and color imagery in a forward-looking system. In: Proceedings of the SPIE, vol. 7664, p. 76441U (2010)

19. Havens, T.C., Stone, K., Keller, J.M., Ho, K.C.: Sensor-fused detection of explosive hazards. In: Proceedings of the SPIE, vol. 7303, p. 73032A (2009)

20. Hinton, G., Osindero, S., Teh, Y.W.: A fast learning algorithm for deep belief nets. Neural computation **18**(7), 1527–1554 (2006)

21. Hinton, G.E., Salakhutdinov, R.R.: Reducing the dimensionality of data with neural networks. Science **313**(5786), 504–507 (2006)

22. Hu, L., Anderson, D.T., Havens, T.C.: Multiple kernel aggregation using fuzzy integrals. In: IEEE International Conference Fuzzy Systems, pp. 1–7 (2013)

23. Hu, L., Anderson, D.T., Havens, T.C., Keller, J.M.: Efficient and scalable nonlinear multiple kernel aggregation using the choquet integral. In: Laurent, A., Strauss, O., Bouchon-Meunier, B., Yager, R. (eds.) Information Processing and Management of Uncertainty in Knowledge-Based Systems. Communications in Computer and Information Science, vol. 442, pp. 206–215. Springer (2014)

24. Jarrett, K., Kavukcuoglu, K., Ranzato, M., LeCun, Y.: What is the best multi-stage architecture for object recognition? In: 2009 IEEE 12th International Conference on Computer Vision, pp. 2146–2153 (2009)

25. JIEDDO COIC MID: Global IED monthly summary report (2012)

26. Kandel, E.R., Schwartz, J.H., Jessell, T.M.: Principles of Neural Science, 4 edn. McGraw-Hill, New York (2000)

27. Krizhevsky, A., Sutskever, I., Hinton, G.E.: Imagenet classification with deep convolutional neural networks. In: Pereira, F., Burges, C.J.C., Bottou, L., Weinberger, K.Q. (eds.) Advances in Neural Information Processing Systems 25, pp. 1097–1105. Curran Associates, Inc. (2012)

28. Lanckriet, G.R.G., Cristianini, N., Bartlett, P., Ghaoui, L.E., Jordan, M.I.: Learning the kernel matrix with semidefinite programming. J. Mach. Learn. Res. **5**, 27–72 (2004)

29. Lawrence, S., Giles, C.L., Tsoi, A.C., Back, A.D.: Face recognition: a convolutional neural-network approach. IEEE Trans. Neural Netw. **8**(1), 98–113 (1997)

30. LeCun, Y., Bottou, L., Bengio, Y., Haffner, P.: Gradient-based learning applied to document recognition. Proc. IEEE **86**(11), 2278–2324 (1998)

31. LeCun, Y., Jackel, L.D., Bottou, L., Brunot, A., Cortes, C., Denker, J.S., Drucker, H., Guyon, I., Müller, U., Säckinger, E., Simard, P., Vapnik, V.: Comparison of learning algorithms for handwritten digit recognition. In: International conference on artificial neural networks, vol. 60 (1995)

32. Lowe, D.G.: Object recognition from local scale-invariant features. In: International Conference Computer Vision, pp. 1150–1157 (1999)

33. Matas, J., Chum, O., Urban, M., Pajdla, T.: Robust wide-baseline stereo from maximally stable extremal regions. Image Vis. Comput. **22**(10), 761–767 (2004)

34. Mercer, J.: Functions of positive and negative type and their connection with the theory of integral equations. Philos. Trans. R. Soc. A **209**, 441–458 (1909)
35. Ojala, T., Pietikainen, M., Harwood, D.: A comparative study of texture measures with classification based on featured distributions. Patt. Recognit. **29**(1), 51–59 (1996)
36. Palm, R.B.: Prediction as a candidate for learning deep hierarchical models of data. Ph.D. thesis, Technical University of Denmark (2012)
37. Playle, N., Port, D.M., Rutherford, R., Burch, I.A., Almond, R.: Infrared polarization sensor for forward-looking mine detection. Proc. SPIE. **4742**, 11–18 (2002)
38. Price, S.R., Anderson, D.T., Luke, R.H.: An improved evolution-constructed (iECO) features framework. In: IEEE Symposium Series on Computational Intelligence (2014)
39. Sarikaya, R., Hinton, G.E., Ramabhadran, B.: Deep belief nets for natural language call-routing. In: IEEE International Conference Acoustics, Speech and Signal Processing, pp. 5680–5683 (2011)
40. Scott, G.J., Anderson, D.T.: Importance-weighted multi-scale texture and shape descriptor for object recognition in satellite imagery. In: IEEE International Geoscience and Remote Sensing Symposium, pp. 79–82 (2012)
41. Stauffer, C., Grimson, W.E.L.: Adaptive background mixture models for real-time tracking. In: IEEE Computer Society Conference on Computer Vision and Pattern Recognition, 1999, vol. 2 (1999)
42. Stone, K., Keller, J.M.: Clutter rejection by cluster analysis in an automatic detection system for buried explosive hazards in forward looking imagery. In: Proceedings of the SPIE (2013)
43. Stone, K., Keller, J.M.: Convolutional neural network approach for buried target recognition in FL-LWIR imagery. In: Proceedings of the SPIE (2014)
44. Stone, K., Keller, J.M., Ho, K.C., Gader, P.D.: On the registration of FLGPR and IR data for the forward-looking landmine detection system and its use in eliminating FLGPR false alarms. In: Proceedings of the SPIE, vol. 6953 (2008)
45. Stone, K., Keller, J.M., Popescu, M., Havens, T.C., Ho, K.C.: Forward-looking anomaly detection via fusion of infrared and color imagery. In: Proceedings of the SPIE, vol. 7664, p. 766425 (2010)
46. Stone, K.E., Keller, J.M., Anderson, D.T., Barclay, D.B.: An automatic detection system for buried explosive hazards in fl-lwir and FL-GPR data. In: Proceedings of the SPIE Conference Detection and Sensing of Mines, Explosive Objects, and Obscured Targets (2012)
47. Xu, Z., Jin, R., Yang, H., King, I., Lyu, M.R.: Simple and efficient multiple kernel learning by group lasso. In: Proceedings of the Interence Conference Machine Learning, pp. 1175–1182 (2010)

Classification-Driven Video Analytics for Critical Infrastructure Protection

Phillip Curtis, Moufid Harb, Rami Abielmona and Emil Petriu

Abstract At critical infrastructure sites, either large number of onsite personnel, or many cameras are needed to keep all key access points under continuous observation. With the proliferation of inexpensive high quality video imaging devices, and improving internet bandwidth, the deployment of large numbers of cameras monitored from a central location have become a practical solution. Monitoring a high number of critical infrastructure sites may cause the operator of the surveillance system to become distracted from the many video feeds, possibly missing key events, such as suspicious individuals approaching a door or leaving an object behind. An automated monitoring system for these types of events within a video feed alleviates some of the burden placed on the operator, thereby increasing the overall reliability and performance of the system, as well as providing archival capability for future investigations. In this work, a solution that uses a background subtraction-based segmentation method to determine objects within the scene is proposed. An artificial neural network classifier is then employed to determine the class of each object detected in every frame. This classification is then temporally filtered using Bayesian inference in order to minimize the effect of occasional misclassifications. Based on the object's classification and spatio-temporal properties, the behavior is then determined. If the object is considered of interest, feedback is provided to the background subtraction segmentation technique for background fading prevention reasons. Furthermore any undesirable behavior will generate an alert, to spur operator action.

P. Curtis (✉) · M. Harb · R. Abielmona
Research & Engineering Division, Larus Technologies, Ottawa, Canada
e-mail: pcurtis@eecs.uottawa.ca

M. Harb
e-mail: moufid.harb@larus.com

R. Abielmona
e-mail: rami.abielmona@larus.com

P. Curtis · R. Abielmona · E. Petriu
School of Electrical Engineering and Computer Science, University of Ottawa,
Ottawa, Canada
e-mail: petriu@eecs.uottawa.ca

© Springer International Publishing Switzerland 2016
R. Abielmona et al. (eds.), *Recent Advances in Computational Intelligence
in Defense and Security*, Studies in Computational Intelligence 621,
DOI 10.1007/978-3-319-26450-9_3

Keywords Artificial neural networks · Bayesian inference · Video analytics · Classification · Surveillance

1 Introduction

The use of video feeds within surveillance applications is becoming quite popular due to the increased demand for ensuring the security of buildings and other infrastructures, as well as the declining cost and increased precision of digital video cameras. The increased usage of multiple video sources to cover a large perimeter surrounding a critical infrastructure imposes a large burden on the system operators, who cannot physically concentrate on simultaneously observing all the remotely distributed video feeds. This leads to fatigued, stressed, and overworked operators who end up possibly missing important events [1].

The increased processing power, and reduced costs, of current computing technologies can be used to help solve this problem, mainly through the application of computer vision (CV) and computational intelligence (CI) techniques. Video analytics pairs CV with CI in order to understand the activities occurring and behaviors exhibited by the various actors within video feeds. Using CV techniques, objects can be detected and extracted from the video stream. These objects can then be classified based on supervised learning techniques, and their behavior monitored for undesirable events. Beyond this work, semantic analysis of the segmented objects [2] can be applied in order to improve the prediction of intent and threat that is imparted to the infrastructure.

The operator can then be alerted when these undesirable events occur through the annotation of the video stream, as well as other alert mediums, to indicate this fact so that a decision on the potential response can be made. By providing these alerts to the operator, attention can be directed to specific events from a single video feed among many, thereby improving operator response to undesirable events that may be otherwise delayed or missed due to distractions or fatigue.

The solution proposed in this work uses a background subtraction method of extracting objects of interest, which is updated adaptively based on the classes detected and observed behavior. After the objects have been extracted from the scene, an artificial neural network (ANN) classifier combined with a temporal Bayesian filter is used to classify the object. The behavior of the classified object, such as entering a restricted zone, stopping, and abandoning another object, is determined. Based on these behaviors, alerts and annotations to the video are enacted (if necessary), and the information is fed back into the background subtraction model. This feedback of information is used to keep objects belonging to classes of interest in the foreground model, even when the object becomes stationary.

The proposed solution is capable of detecting several behaviors of interest in surveillance activities, including restricted zone intrusion by objects of select classes (e.g. *car, person, bird,* or *maritime vessel*), abandoned object detection and stopped object detection, while handling the issues of background fading inherent in most background subtraction techniques. It is implemented using C++, in-part using the open source OpenCV library [3].

The rest of this work is structured as follows. Section 2 briefly reviews relevant works. Section 3 unveils the proposed behavior-driven classification methodology and Sect. 4 illustrates its application within critical infrastructure protection. Section 5 sheds light on the empirical evaluation before some final conclusions and future directions are elaborated upon in Sect. 6.

2 Literature Review

The first subsection reviews a subset of classification techniques found in the literature, while relevant computer vision techniques are discussed in the second subsection.

2.1 Classification Techniques

Classifiers exist in two different flavors: unsupervised and supervised. Unsupervised techniques extract knowledge from a scene without a priori knowledge, and are typically used for clustering data and discovering interesting properties of the input data. Supervised classifiers, however, involve training the classifier, through a reinforcement machine learning technique, by introducing many labelled samples of each class that is needed to be identified. For each class, the data is typically processed by extracting a feature vector that is then fed into the classifier. There are several supervised classification techniques that are commonly used for image and video processing, with the most popular being the support vector machine (SVM), boosted classifiers, k-nearest neighbor (kNN), and the artificial neural network (ANN).

The SVM is a binary classifier that maps the feature vector into a multi-dimensional vector space and defines a partition (the classification threshold) such that the margin of classification between each class within the vector space is maximal [4, 5]. By ensuring the distance between the feature vectors representing classes is maximal, discrimination of features representing each class is made easier, and determining which class an object belongs to becomes the detection of which side of the hyper-planar class partition the feature vector lies. To form a multiclass classifier using SVMs, several strategies are employed, such as using an ensemble of binary SVMs in a one versus one or one versus all methods. In a one versus one strategy, an SVM is trained to discriminate between each pair of classes, and a voting strategy is used to decide on the outcome. In a one versus all strategy, there is an SVM for each class used to determine membership to the class or not.

The resulting class is the result with the dominant outcome. Finally, [6] introduces a technique for optimizing the direct multiclass SVM, instead of having to decompose the problem into many binary SVMs.

Boosted classifiers use many weak classifiers, such as a decision tree with only a few branches that are only slightly better than random chance. These weak classifiers are then combined to produce a stronger result [7, 8]. These classifiers are typically fast and simple to use, allowing for much parallelization, but at the cost of longer training periods. An example of boosted classifiers within the field of computer vision is the Voila Jones object detector [9]. This detector uses Haar wavelet based features within the Adaboost framework to successfully detect faces, and other objects, within images.

The kNN algorithm is perhaps one of the simplest to implement, as it classifies a new data sample by assigning it the class of the most common class among its k nearest neighbors. Some strategies may enforce weighting the contribution of each neighbor according to its distance to the new data sample. With large high dimensional datasets however, determining the kNN becomes computationally expensive, and so techniques to approximate the kNN have been developed. FLANN [10, 11] is one such technique that is commonly used to approximate the kNN algorithm in computer vision problems.

ANNs are inspired from neural biology, and are quite flexible in modelling any desired system [12, 13]. They consist of several inter-networked neurons. An individual neuron accepts a weighted combination of input values that get processed by a typically non-linear activation function to generate an output value; it is the weights and biases for all the neurons in the network that get adapted during training based on the desired output. The multi-layer perceptron (MLP) is a feed forward type ANN that can be trained by back-propagation techniques, and it is widely used for classification tasks due to its simple structure, computational efficiency, and ability to approximate any function to within defined error bounds [14].

2.2 Computer Vision Techniques

Determining image content can occur in several different ways. One way is to perform segmentation that selects regions of the image based on spatio-temporal properties, followed by classification of the segments. Another way is to use a windowing methodology with a detector in order to locate and classify objects that are of interest.

Segmentation is the clustering of regions sharing similar spatio-temporal properties, such as color, texture, location, and motion that may be performed by supervised, or unsupervised, methods. Image segmentation techniques, such as the watershed algorithms [15] and k-means clustering [16, 17], only use spatial properties of a single image to perform the segmentation. While producing good results, they tend to take an extensive amount of computational time, and are not directly suitable for segmenting video streams. Moreover, they may result in an over segmentation (i.e. each object having one or more segments), or an under segmentation (i.e. fewer segments than objects) depending on scene complexity and parameters employed. On the other hand, video streams can take temporal properties into consideration, thereby using the additional information to minimize the computational resources required for segmentation, while at the same time providing more information to mitigate over, and under, segmentation.

Some video segmentation techniques [18–20] rely on performing an accurate segmentation based on the first frame using slower image-based techniques, and then track the intra-frame changes, refining the segmentation in each subsequent frame. These techniques work well in situations with minimal amounts of object motion between frames, however, when there are significant changes, such as the introduction of new objects, a reinitialization of the segmentation may be necessary, due to a breakdown in the corresponding regions between frames. In situations where this occurs frequently, the goal of reducing segmentation-related computational resources by tracking changes between frames is prevented.

Other methods, such as the GrabCut [21], CamShift [22], and MILTrack [23] algorithms require a region to be selected that initializes the video segmenter model to track this selected region within each frame of the video. The selected region is identified in subsequent frames by applying the learned region model, which creates a heat map of possible locations that the region may lay in the new frame. The region is then extracted either through an application of maximum a posteriori estimation or through traditional thresholding. These techniques work well at tracking specific individual objects, but do not fare well at detecting, extracting and tracking generalized classes of objects.

Feature point based techniques; such as SIFT [24], SURF [25], and ORB [26], identify interesting features within a scene, and characterize them through their local spatial properties. By comparing the properties of these features between frames, it is possible to determine where each feature point has moved. In order to detect objects, along with their classes and localizations, these feature points must be combined with other techniques, such as a bag of visual words, a classification algorithm, and a windowing technique [27]. These techniques require much computational time to determine the vectors for each key point, and obtaining a precise bounding box of the objects of interest may prove difficult, with the advantage being that they simplify the feature correspondence problem between frames while reducing the volume of data processed in subsequent analysis.

Other techniques model the scene stochastically, taking advantage of the time dimension, such that when a new object is introduced, it can more easily be detected. Mixture of Gaussians (MoG) belongs to such a class of techniques [28–30]. The MoG algorithm works by modelling each pixel using a mixture of Gaussian distributions, such that when a value for a particular pixel is observed that does not match any of the existing Gaussians for that pixel; it is flagged as containing something new, and belonging to the foreground. After a certain amount of time, these foreground values are modelled as Gaussians in the background model. This allows the background model to adapt to changing illumination or a dynamic scene. While this integration of foreground objects into the background model may be beneficial in some situations, it becomes a problem when the object of interest becomes part of the background when stationary, as this object may exhibit behaviors that are being monitored. Additionally, these techniques tend to require a relatively static scene, with any illumination variation occurring slowly, and using a stationary camera. Some research has been made in providing methods to model dynamic backgrounds [31] with illumination invariance [32, 33] characteristics.

Figure 1 demonstrates the fade to background problem. In Fig. 1a, the scene is static containing no objects of interest. This becomes integrated into the background model. Sometime later, a person enters the scene, shown in Fig. 1b. In Fig. 1c, the person moves a little bit more and then stops; after some time, the person is fully integrated into the background model, precluding their representation in the foreground model in Fig. 1d. If persons are the class of interest in this scenario, then this

Fig. 1 Demonstrating fade to background problem

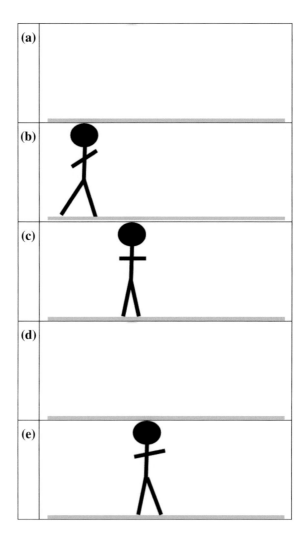

is undesirable behavior, as once an object is integrated into the background model, it is not extracted, and therefore is effectively invisible to the computer vision algorithm. For objects belonging to classes of interest it is desirable to prevent this. Figure 1e shows the person moving again, and hence the person appears in the foreground model at locations where the person has not been previously observed.

3 Proposed Solution

The proposed solution is based on three interconnected modules (see Fig. 2) which include an object extraction module, a classification module, and a behavior engine that generates feedback to the object extraction module, as well as the annotated output frame and any necessary system alerts. Sections 3.1, 3.2, and 3.3 discuss the three respective modules of proposed solution.

3.1 Object Extraction

The object extraction technique that has been employed (see Fig. 3) uses a background subtraction based approach. This is followed by a dilating morphological operation to fill in possible gaps, and then a standard 8-wise connected components labelling algorithm is applied to the foreground model. A Kalman tracker combined with a nearest neighbor matching technique is utilized to perform correspondence of detected objects between frames.

The background subtraction based segmenter is the MoG technique [30] previously described in Sect. 2.2 that models the variations of each pixel in the scene over time by a mixture of Gaussian distributions. Any measurement that does not fit into these distributions is considered as an anomaly, and labelled as belonging to a foreground object. As there may be holes in the foreground model,

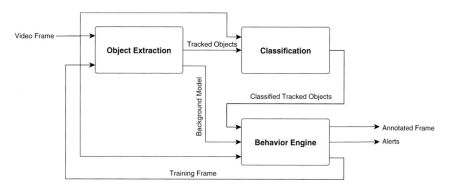

Fig. 2 Block diagram illustrating the proposed solution

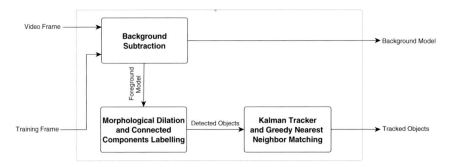

Fig. 3 Block diagram illustrating the object extraction process

due to the moving object being similar in color to the background model in some locations, a dilation morphological operator is applied that fills in these smalls gaps, following which each pixel in the resulting model is grouped together using an 8-wise connect components labelling algorithm, and a rectilinear bounding box is then fitted for each labelled object. In order to prevent the background model from incorporating objects that are of interest, the learning parameter is set to zero when the video frame is first introduced, and then set back to its regular value when the training image for that particular frame has been decided by the behavior engine, as detailed in Sect. 3.3. The Kalman tracker is then used to predict the locations of bounding boxes that previously detected objects will have in the current frame. Corresponding matches between frames is performed by a greedy nearest neighbor algorithm using the Euclidean distance of bounding boxes obtained from objects extracted in the current frame and those predicted by the Kalman tracker.

3.2 Classification

The classifier architecture, as shown in Fig. 4, contains a feature extractor that produces a feature vector based on the current appearance for each tracked object. These feature vectors are then fed into a parallel bank of Multi-Layer Perceptron (MLP) ANN binary classifiers. The final classification is chosen using a one versus all strategy, where the highest activation level among the active classifiers is chosen as the winner. If no classifiers reach the threshold of classification, then the object is deemed as belonging to an *unknown* class. Using a parallel bank of MLPs allows for the easy addition/removal/retraining of any particular class, at the expense of a higher computational burden.

The output from the ANN classifier is then fed into a temporal Bayesian classification filter. As the Bayesian filter requires many positive classifications to reinforce the belief in the resulting classification, this minimizes the impact of temporary misclassifications; effectively minimizing the false positive and negative classification rates produced by the overall system.

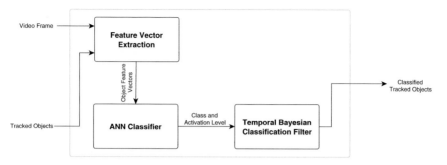

Fig. 4 Block diagram illustrating the classification process

The features that are provided to the ANN classifier are extracted from the contents of a subimage defined by the object's bounding box. The first feature is the mean color corrected red, green and blue (RGB) values of the subimage. To generate the color corrected image, two factors are initially calculated using the mean greyscale value, $\overline{grey}$, the mean red channel value, $\overline{red}$, and the mean green channel value, $\overline{green}$, as shown in (1). These factors are then applied to each r row and c column pixel location in the red (I_{red}), green (I_{green}), and blue (I_{blue}) channels of the image to create the color corrected image, I_{cor}, as shown in (2). The second feature is a greyscale version of the subimage that has been rescaled to 4×4 pixels in size, and the final feature is a black and white thresheld version of the greyscale image using the OTSU algorithm [34], which is then resampled, with each pixel representing the corresponding ratio of the positively thresheld pixels against the total number of pixel represented by the new subsampled pixel. An example of the extraction of these features is shown in Fig. 5. This results in a total feature vector length of 35 elements. The latter are then normalized to be between -1 and 1, with the normalization limits chosen based on the range observed for each parameter in the dataset used for training.

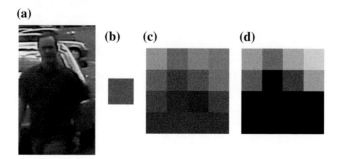

Fig. 5 Demonstrating the feature vectors extracted from a scene: **a** the subimage defined by a detected objects bounding box, **b** the first feature, **c** the second feature, and **d** the third feature

$$(f_1 \quad ,f_2) = \overline{grey} \cdot \left(1/\overline{red},\ 1/\overline{green}\right) \tag{1}$$

$$I_{cor}(r,c) = \left(I_{red}(r,c)*f_1 \quad I_{green}(r,c)*f_2 \quad I_{blue}(r,c)*f_1\right) \tag{2}$$

The ANN is a simple feed forward (FF) type MLP that has input, hidden, and output layers. The output layer has 2 neurons with binary output values in the range of 0–1 to indicate that the object belongs to the class for the first output neuron, and similarly to indicate that the object does not belong to the class for the second output neuron. Each classifier has a different number of hidden layer neurons that was found via the training process. By using a short feature vector, the speed of the classification is improved, at the cost of potentially higher misclassification rates.

The classes that are currently classified by the MLP classifier's binary output are *bird, person, car*, and *maritime vessel*. These classes were trained using a combination of the Visual Object Classes Challenge−Pattern Analysis, Statistical Modelling, and Computational Learning (PASCAL) [35] dataset of 2007 and 2008, and an internally maintained dataset of images for the targeted categories. While the PASCAL dataset contains annotations for *person, bird, cat, cow, dog, horse, sheep, aeroplane, bicycle, boat, bus, car, motorbike, train, bottle, chair, dining table, potted plant, sofa*, and *monitor*, the four aforementioned classes were chosen since *persons* and *cars* are typically objects that are of interest for critical infrastructure protection applications, while *birds* and *maritime vessels* are of concern within maritime situational awareness applications.

Matlab's scaled conjugate gradient back-propagation method [36] was used for training since the dataset is quite large and this method can tackle such data with low memory consumption. The trained configuration was then implemented in a custom optimized C++ module. Table 1 shows the training and testing results of

Table 1 Results of training and testing for each of the NN classifiers

Class	Result	Data Size	CC=AC (%)	TP (%)	FN (%)	TN (%)	FP (%)
Person	Tr	12684	96.2	46.3	3.7	49.2	0.83
	Ts	14976	59.8	21.9	28.1	34.0	16.0
Maritime Vessel	Tr	12889	98.8	38.2	11.8	50.0	0.01
	Ts	14976	96.5	11.8	38.2	49.2	0.76
Car	Tr	6904	98.3	38.5	11.5	50.0	0.01
	Ts	6347	89.7	8.6	41.4	47.7	2.37
Bird	Tr	6904	98.6	44.3	5.7	50.0	0.03
	Ts	6347	89.5	13.2	36.8	46.3	3.7

(*Tr* Training; *Ts* Testing; *AC* Accuracy; *CC* Correct classification; *TP* True positive; *FN* False negative; *TN* True negative; *FP* False negative)

Fig. 6 TP, TN, FP, FN
Values of testing on unseen
images

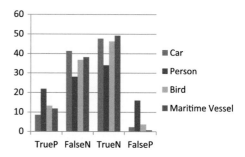

the designed classifiers. The classifiers were capable of classifying a significant
number of images that contain an object of a targeted category. As shown in Fig. 6,
the person classifier had the highest false positive rate $X_{k,n}$, for the kth object, $O_{k,n}$,
at frame n being chosen based on the highest activation output of all the classifiers
being used, while ensuring that there is a sufficient delta, α, between the activation
levels. If there is no single class dominant or if the dominant class has an activation
level below a threshold, β, then the classification of the object is considered to be
unknown, as in (3).

$$
X_{k,n} =
\begin{cases}
\arg \max_{i} \{ANN_i(O_{k,n})\}, & iff \quad \begin{matrix} ANN_i(O_{k,n}) > \beta, \\ |ANN_i(O_{k,n}) - ANN_j(O_{k,n})| > \alpha, i \neq j \end{matrix} \\
unknown, & otherwise
\end{cases} \quad (3)
$$

To prevent the effect of temporary misclassifications in the form of false positives and false negatives, a Bayesian inference predictor, (4), has been implemented
to perform temporal filtering of the ANN classifier output, where $P(X_n|O_n)$ is the
probability that the object belongs to class X at time n given the current observation
O_n, $L(X_n|O_n)$ is the likelihood that the observation O results in the classification X
for the current observation at time n (which is determined by the normalized output
of the ANN classifier) and $P(X_{n-1}|O_{n-1})$ is the probability that the object belongs to
class X observation O at the previous instant in time, $n-1$.

$$
P(X_n|O_n) = \frac{P(X_{n-1}|O_{n-1}) \cdot L(X_n|O_n)}{P(X_{n-1}|O_{n-1}) \cdot L(X_n|O_n) + (1 - P(X_{n-1}|O_{n-1})) \cdot (1 - L(X_n|O_n))} \quad (4)
$$

The object's current class is decided by the dominant probability out of all
classes, including the unknown object class. Note that when the output of the MLP
is unknown, the Bayesian temporal filter is not updated in order to prevent situations that the classifier does not recognize, such as uneven lighting or occlusion,
from suppressing the current classification of the object.

#	MLP classification output		Bayesian filter output	
	Class	Prob.	Class	Prob.
1	Unknown	0.50	Unknown	0.50
2	Person	0.60	Person	0.60
3	Car	0.75	Car	0.67
4	Person	0.68	Person	0.52
5	Person	0.75	Person	0.76
6	Person	0.69	Person	0.88
7	Bird	0.6	Person	0.83

Table 2 An example demonstrating the effectiveness of Bayesian filtering on classification

An example demonstrating the effectiveness that Bayesian filtering has on the classification of an object is shown in Table 2. This scenario catalogs the classification of a *person* object for 7 successive instances in time, with the MLP classification output and the Bayesian classification filter output. As can be observed, the object is first detected with an *unknown* classification. In the second frame, the correct classification of *person* with a probability of 0.6 is produced from the MLP, and similarly for the Bayesian filter. At frame 3, the *person* is misclassified as a *car* with a probability of 0.75, the Bayesian filter output is *car*, but with a lower probability due to dissimilar classes reducing the confidence of classification. At the 4th, 5th, and 6th instances of time, the MLP output class is *person* with a varying probabilities, these repeated observations increase the confidence in the objects class being a *person* within the Bayesian classification filter, resulting in an output classification of *person* at 0.88 probability at time 6. At time 7, the MLP yields a misclassification with the output being *bird* with probability of 0.6. This only slightly lowers the Bayesian output of *person* slightly to 0.83. This example shows that temporary misclassification from the MLP do not affect the long term classification of an object thanks to the Bayesian classification filter, as long as there have been repeated measurements from the correct classification to reinforce the class belief.

3.3 Behavior Engine

The behavior engine module, as shown in Fig. 7, consists of a behavior analysis unit that looks for specific behavior from certain classes of tracked objects, followed by an annotation unit that generates an operator output in the form of an annotated video frame and alerts, as well as a unit that generates a training frame for feedback into the object extractor module.

The behavior analyzers that are currently implemented include intrusion detection, abandoned object, and counting object analyzers. The intrusion detection analyzer monitors for the intrusion of a restricted zone (e.g. a preselected subregion

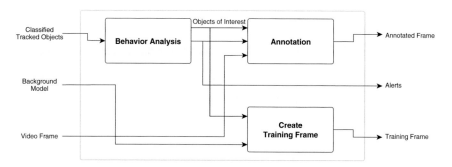

Fig. 7 Block diagram illustrating the behavior engine

of the image bounded by a polygon) by an object from a selected class, or classes. When objects of the selected class are detected within this zone, it triggers an alert. The abandoned object analyzer monitors for the separation of a smaller object from a larger parent object of a specific class, or classes, within a particular predefined subregion of the image space that is defined by a polygon. When this smaller object remains stationary for a period of time within the monitoring zone, and it is separated by the parent object of a particular class, an alert is triggered. The counting object analyzer counts the number of objects from a particular class or classes that has crossed a predefined subregion of the image space indicated by a bounding polygon.

The annotation unit marks up the video stream, highlighting objects and classes of interest, as well as providing alerts based on the behavior analysis. The training frame creation unit generates the training frame based on the background model, the current frame, and the objects of interest produced from the behavioral analysis unit, such that the background subtraction will not integrate objects of interest into the background module. For the case of *maritime vessels*, as the wake generated in the water is also considered foreground and part of the *maritime vessel* object segment by the object extraction module. For objects of this class, the wake is filtered using a color based filter (as a wake is generally light gray to white), thereby allowing the integration of the wake into the background model, limiting future detections to only the *maritime vessel* itself, and not the wake, while still preventing the *vessel* from fading into the background. The produced training frame is then fed back into the object extraction module, and the background model is adjusted appropriately.

4 Case Studies: Critical Infrastructure Protection

In this work, three scenarios are considered. The first scenario, shown in Fig. 8a, consists of monitoring a pair of dumpsters for their unauthorized usage. The second scenario, shown in Fig. 8b, consists of monitoring a doorway for unauthorized

(a)   **(b)**   **(c)**

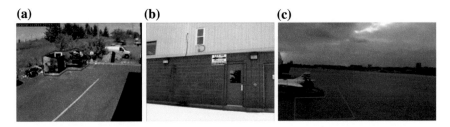

Fig. 8 Demonstrating the three considered scenarios of **a** monitoring of a dumpster, **b** monitoring of a doorway for suspicious activities, and **c** monitoring of a port for maritime vessels

access. The final scenario, shown in Fig. 8c, monitors the entrance way of a port for *maritime vessels*. In each scenario, the region of interest for monitoring intrusion is a polygon that is drawn on the images in blue.

4.1   Scenario 1: Monitoring of a Dumpster

In this scenario, the dumpster is located at the rear of a building that will have infrequent vehicular traffic. The region around the dumpster will be monitored for objects of type *person* intruding within that zone, as when this occurs there may be somebody putting garbage in the dumpster. When the operator receives the alert, they must make a decision as to whether the *person* is authorized to use the dumpster. Furthermore, as there is vehicular traffic expected, *cars* should not cause alerts, but they should be prevented from fading into background, as cars should not be permanent residents of the scene.

This scenario commences when a *car* enters from the bottom left corner and drives up to the dumpster. A *person* then exits the car, grabs a bag of garbage, tosses it into the garbage bin, and then drives away to the right. The *car* should still be detected and classified, but it should not cause an alert, nor should it fade into the background model. Furthermore, a *person* by the dumpster indicates a condition that should be handled by an operator to ensure the *person* is authorized to use that resource, and as a result an alert will be sent under this situation. The alert will cease once the condition is alleviated, which occurs when the *person* leaves the intrusion zone; when he enters the *car* and drives away.

4.2   Scenario 2: Monitoring of a Doorway

In this scenario, a doorway is monitored for intrusion by objects of type *person*. When an object of type *person* is detected within this zone, an alert will be sent to the operator, as this may indicate a possible attempt to access the building by an unauthorized individual. Additionally, objects of class *person* will have their recent tracking history kept, indicated by the path followed by the bounding box centroid, annotated for the operator to view the direction that the *person* came from where they have visited. Furthermore, the abandonment of suspicious objects by a *person* within this zone will be monitored, as this may indicate a potential danger.

This scenario commences when a *person* walks in from the right. This *person* stops at the door, drops a bag, and then proceeds to walks away towards the left of the scene. An alert should be triggered when the *person* intrudes upon the monitored zone, which will cease when the he leaves it. Furthermore, after a period of time, the abandoned bag left by the door by the *person* should trigger an abandoned object alert. Throughout the whole scenario, the tracking history of the *person* should be displayed indicating the path followed, and as the bag does not belong to a known class, its classification should remain *unknown*. Both the abandoned object and the *person* should not be integrated into the background model.

4.3   Scenario 3: Monitoring of a Port

In this final scenario, a port is monitored for intrusion of *maritime vessels*. Maritime *vessels* are the object of interest, and as a result they should not fade into the background model. When *maritime vessels* enter the port region, this will trigger an alert to the operator, who should follow up by making sure that they are authorized to access this particular zone. Furthermore, as it may be of interest to the operator to determine the path that *maritime vessels* follow and as such the recent tracking history indicating the centroid of the bounding box is shown for each maritime vessel detected. Finally, any wake included with the segmented *maritime vessel* should be integrated into the background model.

This scenario commences with a *maritime vessel* coming from the right hand side, and entering the port. When the vessel crosses into the intrusion zone, an alert is generated. Furthermore, a recent track history is displayed for this maritime vessel, indicating where it has recently been located within the video feed.

5   Experimental Results

The scenarios described in Sect. 4 were captured using three different cameras and frame rates. The first camera is a Vivotek security camera, capturing at a variable frame rate at a resolution of 640×480, which was used to acquire the video of Scenario 1. The second camera is a Logitech webcam, capturing at 30 fps at a resolution of 640×480, which was used to acquire the video of Scenario 2. The video in Scenario 3 was acquired from archival footage, and has a resolution of 848×480 at 30 fps. Each video was saved in the MP4 video format, and subsequently processed offline to allow for the repeatability and thorough analysis of the results. This is not to state that the system can only operate offline; in fact the proposed system is capable of operating online in real-time in a wide range of situations. The video processing algorithm was implemented in optimized C++, using a combination of OpenCV and in-house libraries. The first three subsections detail the results obtained from each of the three scenarios.

5.1 Scenario 1

The key moments of the first scenario are shown in Fig. 9, where the first column contains the frame number for the corresponding row, the second column contains the annotated output video frames, the third column contains the detected objects, and the fourth column contains the feedback training frames. The *car* comes into view and is classified as *unknown* in frame 246. In frame 249, it is correctly classified as a *car*; notice that it has not been introduced into the background model. The *car* continues moving until frame 297, where it stops. Notice that the *car* crossing into the intrusion polygon does not trigger an alert, hence highlighting that the behavioral module correctly distinguishes between classes when processing behaviors. In frame 387, an object that has been correctly classified as a *person* has exited the *car* with a garbage bag in hand and is about to toss it into the dumpster. In frame 390, an alert is generated as an intrusion has been detected by an object classified as a *person*, which causes the intrusion polygon to alternate between blue, green, and red. In frame 447, the *person* has reentered the *car*, and drives away in frame 479. Notice that in the training images, the regions of the image that correspond to *unknown* objects, *cars*, and *persons* have not been introduced into the feedback training image, thereby preventing objects of potential interest from being incorporated into the background model, even with the *car* being stationary for over 42 s.

Table 3 provides key moments of the classifier performance of the *car* object in Scenario 1. The first column is the frame number that the classification took place in, the second column indicates the classification of the object from the previous frame (*unknown* with a probability of 0.5 by default for new objects), the third column indicates the output classification and probability of the MLP ANN classifier, the fourth column is the current classification after the temporal Bayesian inference filter has been integrated with the MLP ANN observation. The object is first detected at frame 245, when it is classified as unknown with probability of 1.0000 by the MLP ANN classifier. As previously mentioned, since the Bayesian temporal filter is not updated upon an unknown classification, the resulting classification is still *unknown* with a probability of 0.5000. This situation remains unchanged until frame 249, when the MLP ANN finally recognizes the object as a *car* with a probability of 0.9997. The output to the Bayesian filter becomes *car* with a probability of 0.9997. In the following frame (#250), the MLP ANN classifier produces another classification of *car* with a probability of 0.9997. This results in the reinforcement of the Bayesian belief that the object is a *car*, but now with a probability of 1.0000. In frame 285, the MLP ANN classifier produces a

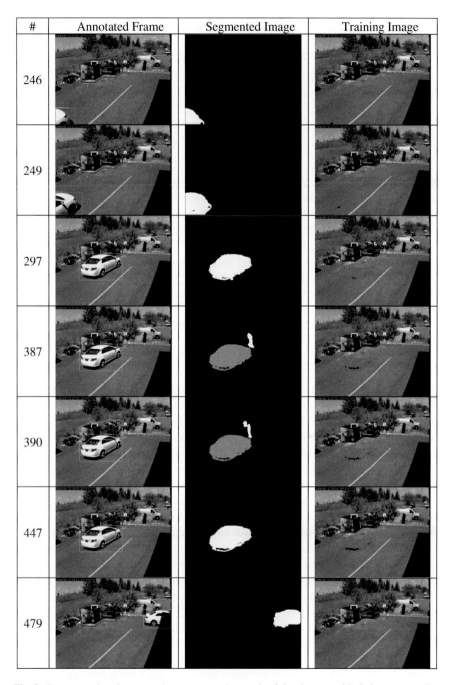

#	Annotated Frame	Segmented Image	Training Image
246			
249			
297			
387			
390			
447			
479			

Fig. 9 Demonstrating the annotation, segmentation, and training images with their corresponding video frame # of key moments that occurred during Scenario 1

Table 3 Demonstrating key instances in the classification of the *car* object in Scenario 1

Frame #	Previous classification		MLP ANN classifier classification		Current classification	
	Class	Prob.	Class	Prob.	Class	Prob.
245	Unknown	0.5000	Unknown	1.0000	Unknown	0.5000
249	Unknown	0.5000	Car	0.9997	Car	0.9997
250	Car	0.9997	Car	0.9997	Car	1.0000
285	Car	1.0000	Person	0.9997	Car	1.0000

misclassification with the class being *person* with a probability of 0.9997. Due to the high belief that the Bayesian temporal classification filter currently has, the resulting probability is still *car* with probability of 1.0000, thereby preventing the misclassification from affecting the culminating classification, and any potential action based on that classification.

These are demonstrably the expected results based upon the description made of the scenario in Sect. 4.1, as well as the expected system behavior, illustrates the correct operation of the classification-driven video analytics system.

5.2   Scenario 2

The key moments of the second scenario are shown in Fig. 10, using the same column order as previously defined for Fig. 9. In frame 239, a *person* enters the frame from the right and is initially misclassified as a *bird*. By frame 244, this individual is now mostly in the scene and is correctly classified as a *person*. In frame 276, he enters the region by the door, triggering an intrusion alert, causing the outlining polygon to alternate between blue, green, and red. In frame 326, the individual stops for a bit, and drops a bag. By frame 401, he has walked away from the door, but the bag has been identified as an *unknown* object, still triggering the intrusion alert. In frame 459, this unknown object has been determined to be an abandoned object, which has created yet another alert, indicated by the thicker red boundary around the object with an 'A' drawn in the interior. At frame 459, the *person* has completely left the scene and the abandoned object is still triggering both the intrusion alert, as well as the abandoned object alert. Furthermore, a track in green indicating the individual's center of gravity over time has been traced through the scene. Finally, all objects corresponding to the *person* and *unknown* classes have not been fed back into the training image, while other classes have, such as when the individual was misclassified as a bird in frame 239, due to the interest being on persons and unknown objects. This keeps both of the monitored objects, person and unknown, from being integrated into the background model, thereby allowing the detection, tracking, and behavior analysis to take place for objects of these classes in subsequent frames.

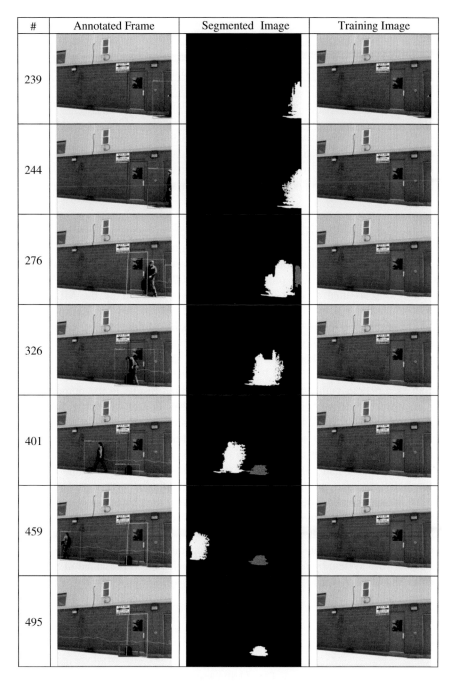

#	Annotated Frame	Segmented Image	Training Image
239			
244			
276			
326			
401			
459			
495			

Fig. 10 Demonstrating the annotation, segmentation, and training images with their corresponding video frame # of key moments that occurred during Scenario 2

Table 4 Demonstrating key instances in the classification of the *person* object in Scenario 2

Frame #	Previous classification		MLP ANN classifier classification		Current classification	
	Class	Prob.	Class	Prob.	Class	Prob.
217	Unknown	0.5000	Unknown	1.0000	Unknown	0.5000
231	Unknown	0.5000	Bird	0.7470	Bird	0.7470
232	Bird	0.7470	Unknown	1.0000	Bird	0.7470
244	Bird	0.7470	Person	0.9310	Person	0.8206
245	Person	0.8206	Person	0.9271	Person	0.9831
246	Person	0.9831	Person	0.9995	Person	1.0000
271	Person	1.0000	Bird	0.8311	Person	1.0000

Table 4 provides key moments of the classifier performance of the *person* object in Scenario 2. The organization is identical to that previously described for Table 3. In this scenario, the *person* object first enters the scene at frame 217, where it is classified as *unknown* with a probability of 1.0000. However in frame 231, this object is misclassified as *bird* with a probability of 0.7470 by the MLP ANN classifier, resulting in the output classification of *bird* by the Bayesian temporal filter. The following frame, the MLP ANN resumes its classification of the object as *unknown* with a probability of 1.0000, but as previously discussed, the Bayesian temporal classification filter is not updated when the MLP ANN classification is unknown. In frame 244, the MLP ANN classifier finally correctly classifies the output as a *person* with probability of 0.9310, which results in the output of the Bayesian filter of *person* with a probability of 0.8206. In each of the two following frames, number 245 and 246, the MLP ANN classifier produces a classification of *person* with probabilities of 0.9271 and 0.9995 respectively. This reinforces the Bayesian belief that the correct classification is *person* with the probabilities evolving to 0.9831 and 1.0000 in those two successive frames. In frame 271, the MLP ANN classifier produces a misclassification of *bird* with a probability of 0.8311, which does not affect the Bayesian belief that the object is a *person* with a probability of 1.0000, thereby further demonstrating the benefit of the temporal Bayesian classification filter.

Again, these results demonstrate the correct operation of the classification-driven video analytics system by following the expected behavior as presented in Scenario 2 described in Sect. 4.2.

5.3 Scenario 3

As with Fig. 9 from Scenario 1, Fig. 11 illustrates the key moments in Scenario 3. This scenario begins with a *maritime vessel* approaching from the right hand side of the scene. The object extraction algorithm does not initially detect the object when

#	Annotated Frame	Segmented Image	Training Image
355			
496			
1052			
1281			
1418			
1621			

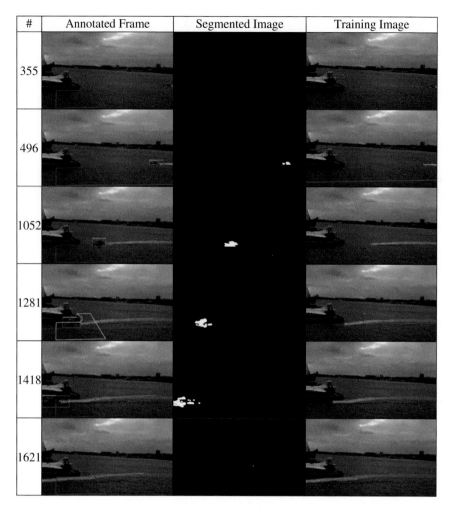

Fig. 11 Demonstrating the annotation, segmentation, and training images with their corresponding video frame # of key moments that occurred during Scenario 3

it is introduced in frame 355, due to the object having much similarity with the background, and what is not similar is small, and hence discarded as noise. However on frame 496, the object is finally detected, and successfully classified as a *maritime vessel*, where it is prevented from being integrated into the background model, as shown in the corresponding training image. The vessel travels for a while, and enters the intrusion zone about the port in frame 1052, thereby triggering an alert. Furthermore, the tracking history is annotated, as indicated by the line that follows the *maritime vessel*'s bounding box centroid. In frame 1281, the vessel is part way through the intrusion zone, where the alert is still being raised, and the tracking history is still being shown. In frame 1418, the vessel is almost out of the

Table 5 Demonstrating key instances in the classification of the *maritime vessel* object in Scenario 3

Frame #	Previous classification		MLP ANN classifier classification		Current classification	
	Class	Prob.	Class	Prob.	Class	Prob.
496	Unknown	0.5000	Maritime vessel	0.5671	Maritime vessel	0.5671
506	Maritime vessel	0.6872	Unknown	0.5269	Maritime vessel	0.6872
612	Maritime vessel	0.9997	Maritime vessel	0.8485	Maritime vessel	1.0000

intrusion zone, with the alert still being raised. By frame 1621, the vessel has completely left the scene, and the alert has ceased. Furthermore the recent tracking history for the vessel is still being displayed. Notice that the track does not intersect with the edge of the scene, as the object becomes too small, and is dropped, before it completely leaves the scene. Finally, during each instance shown, when the object is classified as *maritime vessel*, the vessel is not integrated into the background model, but the wake is introduced to the training frame, allowing for the possibility of still monitoring it if it ceases motion and stops, while ignoring the wake.

Table 5 provides key moments of the classifier performance of the *maritime vessel* object in Scenario 3, with the organization being identical to that previously described for Table 3 from Scenario 1. As previously shown in Fig. 11, the object is first detect in frame 496, and is assigned the *maritime vessel* class with probability 0.5671. Until frame 506, the belief in the *maritime vessel* class has been reinforced to a level of 0.6872, when the MLP returns a classification of unknown with a level of 0.5269. As previously described, *unknown* classification does not change the values of the Bayesian classification filter; hence the classification is still *maritime vessel* with the previous probability. By frame 612, the object has been completely reinforced through repeated measurements to be of class *maritime vessel* with probability of 1.0000, with no misclassifications, aside from the occasional unknown, from occurring.

The results presented in this subsection indicate that the classification-driven video analytics system is able to operate in both land-based and maritime-based environments. Additionally, the expected outcome of Scenario 3 described is Sect. 4.3 was met, further verifying the correct operation of the system.

6 Conclusion

The proposed classification-driven video analytics system correctly extracts interesting objects from the scene. It then tracks, classifies, determines the behavior of these objects. Finally, the system provides relevant alerts to the operator so that a potential action can be determined and eventually enacted. The performance of the proposed system was demonstrated in the three scenarios presented in Sect. 4 of this work. Furthermore, the system was shown to operate effectively in different types of operational environments, namely land based (Scenarios 1 and 2) and maritime based (Scenario 3).

In the proposed object extraction module, objects of interest are extracted based on the MoG background subtraction technique. A combination of Kalman tracking and greedy nearest neighbor matching is then used to track these objects throughout an image sequence, which is clearly demonstrated in the results where a track is drawn that follows the motion of the *person* through the scene in Scenario 2, and a track is drawn that follows the motion of the *maritime vessel* in Scenario 3. These tracked objects are then classified. The proposed classification module contains four accurate parallel ANN classifiers for the following classes: *car, person, bird,* and *maritime vessel* in a one versus all configuration. In order mitigate the inevitable small amounts of misclassification, and improve both the reliability and stability of the end classification; a temporal Bayesian filter is applied to reduce the effect of these occasional misclassifications. This was demonstrated in Scenario 1 where the *car* was misclassified as a *person* in frame 285, but the output from the Bayesian filter was still a car, and additionally demonstrated in Scenario 2 where the *person* was initially considered a *bird* in frame 231, but later confirmed to be a *person* in frame 244.

After the objects have been classified, they are processed by the proposed behavior analysis module. In Scenario 1 that monitors the dumpster; *cars* do not trigger an intrusion alert, but are still monitored. On the other hand, the *person* does trigger the intrusion alert, thereby demonstrating that behavior can be determined for each object on a per class basis. In Scenario 2, the *person* crosses into the intrusion region, which triggers an alert. Furthermore, the *person* leaves a bag behind, which continues the intrusion alert, while also producing an abandoned object alert, further demonstrating the capability of the system to simultaneously monitoring for different types of behaviors. In Scenario 3, the *maritime vessel* enters the intrusion zone triggering an alert.

The knowledge of behaviors and object classes is fed back into the MoG segmenter by adjusting the training image to not contain objects that are of interest for the particular monitoring application, and hence reducing the chance of forgetting, or not observing, interesting objects that could be of highest importance for critical infrastructure protection. Furthermore, in Scenario 3, the wake is introduced into the background model, while preventing the vessel from being integrated, demonstrating the efficacy of a color based filter for handling features with uniform color profiles such as wake.

Future enhancements are being planned for the current system. Firstly, the segmenter will be enhanced to handle more dynamic scenes with camera movement, which will permit a greater range of applicable scenarios. Secondly, the computer vision techniques will be made more illumination-invariant such that they can handle greater light variation across the scene, which will improve performance within a wider variety of uncontrolled environments. The classification module will be enhanced to handle more classes of objects, as well as determine additional properties of classes. Finally, the addition of further in-depth behavior analysis capability will be developed, such as vandalism and fire detection.

References

1. Weiss, L.G.: Autonomous robots in the fog of war. IEEE Spectr. **48**(8), 30–34, 56–57 (2011)
2. Amato, A., Di Lecce, V., Piuri, V.: Semantic Analysis and Understanding of Human Behavior in Video Streaming. Springer, New York (2013)
3. OpenCV dev team: OpenCV 3.0.0-dev documentation. WWW: http://docs.opencv.org/master/index.html, (2014). Accessed June 2014
4. Anthony, G., Gregg, H., Tshilidzi, M.: Image classification using SVMs: one-against-one versus one-against-all. In: Proceedings of the Asian Conference on Remote Sensing (2007)
5. Banerjee, B., Bhattacharjee, T., Chowdhury, N.: Image object classification using scale invariant feature transform descriptor with support vector machine classifier with histogram intersection kernel. In: Information and Communication Technologies: Communications in Computer and Information Science, pp. 443–448. Springer, Berlin (2010)
6. Cramer, K., Singer, Y.: On the algorithmic implementation of multiclass kernel-based vector machines. J. Mach. Learn. Res. **2**, 254–292 (2001)
7. Shotton, J., Winn, J., Rother, C., Criminisi, A.: TextonBoost for image understanding: multi-class object recognition and segmentation by jointly modeling texture, layout, and context. Int. J. Comput. Vis. **81**(1), 2–23 (2009)
8. Lienhart, R., Kuranov, A., Pisarevsky, V.: Empirical Analysis of Detection Cascades of Boosted Classifiers for Rapid Object Detection. Intel Corporation, Santa Clara (2002)
9. Voila, P., Jones, M.: Rapid object detection using a boosted cascade of simple features. In: Proceedings of the Converence on Computer Vision and Pattern Recognition, pp. 511–518 (2001)
10. Muja, M., Lowe, D.G.: Fast approximate nearest neighbors with automatic algorithm configuration. In: International Conference on Computer Vision Theory and Applications (VISAPP'09) (2009)
11. Muja, M., Lowe, D.G.: Scalable nearest neighbor algorithms for high dimensional data. In: IEEE Transactions on Pattern Analysis and Machine Intelligence (PAMI), vol. 36 (2014)
12. Zhang, G.P.: Neural networks for classification: a survey. IEEE Trans. Syst. Man Cybern. Part C Appl. Rev. **30**(4), 451–462 (2000)
13. Goerick, C., Noll, D., Werner, M.: Artificial neural networks in real time car detection and tracking applications. Pattern Recogn. Lett. Neural Netw. Comput. Vis. Appl. **17**(4), 335–343 (1996)
14. Haykin, S.: Neural Neworks: A Comprehensive Foundation, 2nd edn. Prentice-Hall Inc., Upper-Saddle River (1999)
15. Roerdink, J.B., Meijster, A.: The watershed transform: definitions, algorithms and parallelization strategies. Fundamenta Informaticae **41**, 187–228 (2001)
16. Alsabti, K., Ranka, S., Singh, V.: An Efficient k-means Clustering Algorithm. Syracuse University SURFACE Electrical Engineering and Computer Science, Syracuse (1997)
17. Ding, C., He, X.: K-means clustering via principal component analysis. In: Proceedings of the International Conference on Machine Learning, Banff (2004)
18. Ryan, K., Amer, A., Gagnon, L.: Video object segmentation based on object enhancement and region merging. In: IEEE Internation Conference on Multimedia and Expo, Toronto (2006)
19. EL Hassani, M., Jehan-Besson, S., Brun, L., Revenu, M., Duranton, M., Tschumperlé, D., Rivasseau, D.: A time-consistent video segmentation algorithm designed for real-time implementation. In: VLSI Design, vol. 2008, p. 12 (2008)
20. Hu, W.-C.: Real-time on-line video object segmentation based on motion detection without background construction. Int. J. Innov. Comput. Inf. Control **7**(4), 1845–1860 (2011)
21. Rother, C., Kolmogorov, V., Blake, A.: GrabCut: Interactive foreground extraction using iterated graph cuts. ACM Trans. Graphics (SIGGRAPH) **23**, 309–314 (2004)
22. Bradski, G.R.: Computer Vision Face Tracking for Use in a Perceptual User Interface. Intel Corporation (1998)

23. Babenko, B., Yang, M.-H., Belongie, S.: Visual tracking with online multiple instance learning. In: IEEE Transactions on Pattern Analysis and Machine Intelligence (PAMI), August (2011)
24. Lowe, D.G.: Object recognition from local scale-invariant features. Proc. Int. Conf. Comput. Vis. **2**, 1150–1157 (1999)
25. Bay, H., Ess, A., Tuytelaars, T., Van Gool, L.: SURF: speeded up robust features. Comput. Vis. Image Underst. (CVIU) **110**(3), 346–359 (2008)
26. Rublee, E., Rabaud, V., Konolige, K., Bradski, G.: ORB: and efficient alternative to SIFT or SURF. In: IEEE Internation Conference on Computer Vision (ICCV) (2011)
27. Csurka, G., Dance, C.R., Fan, L., Willamowski, J., Bray, C.: Visual categorization with bags of keypoints. In: ECCV International Workshop on Statistical Learning in Computer Vision (2004)
28. Friedman, N., Russell, S.: Image segmentation in video sequences: a probabilistic approach. In: Proceedings of the Conference on Uncertainty in Artificial Intelligence (UAI'97) (1997)
29. KaewTraKulPong, P., Bowden, R.: An improved adaptive background mixture model for real-time tracking with shadow detection. In: Proceedings of the European Workshop on Advanced Video Based Surveillance Systems (2001)
30. Zivkovic, Z.: Improved adaptive Gaussian mixture model for background subtraction. In: Proceedings of the International Conference of Pattern Recognition (2004)
31. Sheikh, Y., Shah, M.: Bayesian modeling of dynamic scenes for object detection. IEEE Trans. Pattern Anal. Mach. Intell. **27**(11), 1778–1792 (2005)
32. Avgerinakis, K., Briassouli, A., Kompatsiaris, I.: Real Time Illumination Invariant Motion Change Detection. In: ACM-MM ARTEMIS International Workshop. Firenze, Italy (2010)
33. Drew, M.S., Wei, J., Li, Z.-N.: Illumination-invariant color object recognition via compressed chromaticity histograms of color-channel-normalized images. In: IEEE International Conference on Computer Vision, Bombay, India (1998)
34. Otsu, N.: A threshold selection method from gray-level histograms. IEEE Trans. Syst. Man Cybern. **9**(1), 62–66 (1979)
35. The PASCAL Visual Object Classes Homepage. WWW: http://pascallin.ecs.soton.ac.uk/challenges/VOC/ (2014). Accessed 7 July 2014
36. Banerjee, B., Bhattacharjee, T., Chowdhury, N.: Image object classification using scale invariant feature transform descriptor with support vector machine classifier with histogram intersection kernel. In: Information and Communication Technologies: Communications in Computer and Information Science, pp. 443–448. Springer, Berlin (2010)

Fuzzy Decision Fusion and Multiformalism Modelling in Physical Security Monitoring

Francesco Flammini, Stefano Marrone, Nicola Mazzocca
and Valeria Vittorini

Abstract Modern smart-surveillance applications are based on an increasingly large number of heterogeneous sensors that greatly differ in size, cost and reliability. System complexity poses issues in its design, operation and maintenance since a large number of events needs to be managed by a limited number of operators. However, it is rather intuitive that redundancy and diversity of sensors may be advantageously leveraged to improve threat recognition and situation awareness. That can be achieved by adopting appropriate model-based decision-fusion approaches on sensor-generated events. In such a context, the challenges to be addressed are the optimal correlation of sensor events, taking into account all the sources of uncertainty, and how to measure situation recognition trustworthiness. The aim of this chapter is twofold: it deals with uncertainty by enriching existing model-based event recognition approaches with imperfect threat modelling and with the use of different formalisms improving detection performance. To that aim, fuzzy operators are defined using the probabilistic formalisms of Bayesian Networks and Generalized Stochastic Petri Nets. The main original contributions span from support physical security system design choices to the demonstration of a multiformalism approach for event correlation. The applicability of the approach is demonstrated on the case-study of a railway physical protection system.

F. Flammini
Ansaldo STS, Innovation Unit, Via Argine 425, Naples, Italy
e-mail: francesco.flammini@ansaldo-sts.com

S. Marrone (✉)
Department of Mathematics and Physics, Second University of Naples,
Viale Lincoln 5, Caserta, Italy
e-mail: stefano.marrone@unina2.it

N. Mazzocca · V. Vittorini
Department of Electrical Engineering and Information Technologies,
University of Naples Federico II, Via Claudio 21, Naples, Italy
e-mail: nicola.mazzocca@unina.it

V. Vittorini
e-mail: valeria.vittorini@unina.it

© Springer International Publishing Switzerland 2016
R. Abielmona et al. (eds.), *Recent Advances in Computational Intelligence in Defense and Security*, Studies in Computational Intelligence 621,
DOI 10.1007/978-3-319-26450-9_4

Keywords Physical security · Surveillance · Multiformalism modelling · Decision fusion · Performance evaluation · Fuzzy correlation

1 Introduction

In modern society, the assurance of a secure environment is of paramount importance due to the growing risk factors threatening critical infrastructures. For that reason, the number and the diversity of sensors used in modern wide-area surveillance are continuously increasing [1]. The types of sensors include environmental probes (e.g., measuring temperature, humidity, light, smoke, pressure and acceleration), intrusion sensors (e.g., magnetic contacts, infrared/microwave/ultrasound motion detectors) Radio-Frequency Identifiers (RFID), geographical position detectors, Smart-cameras and microphones with advanced audio-video analytics capabilities and Chemical Biological Radiological Nuclear and explosive (CBRNe) detectors.

Different types of sensing units are often integrated in smart-sensors that can be part of a Wireless Sensor Networks (WSN); these sensors can also feature on-board 'intelligence' through programmable embedded devices with dedicated operating systems, processors and memory [2, 3].

Such a wide range of sensors provides a large quantity of heterogeneous information. The information supports Physical Security Information Management (PSIM) operators by using pre-processing, integration and reasoning techniques. In a contrary case, there is a serious risk of overwhelming the operators with unnecessary warnings or alarms: they would not be able anymore to perform their task and they could underestimate critical situations [4, 5].

In such a context, (semi-)automatic situation recognition becomes essential. The reliability assurance of sensor networks has been dealt with several approaches, first of them the sensor information fusion approach. There are many scientific works in this field [6, 7, 8]; this notwithstanding, this chapter focuses on multiformalism technique that, at the best of our knowledge, has received little attention by the scientific community.

However, not much work has been done in the research literature to develop frameworks and tools aiding surveillance operators to take advantage of recent developments in sensor technology. In other words, most researchers seem to ignore the apparent paradox according to which more complex is the sensing system, more complex are the tasks required to manage and verify their alarms by the operators.

The challenges to be addressed are the optimal correlation of sensor events, taking into account all the sources of uncertainty (i.e. imperfect threat modelling, sensor false alarm probability, etc.), and how to measure situation recognition trustworthiness. Those challenges can only be addressed using fuzzy probabilistic modelling approaches since 'exact' modelling do not allow to represent those uncertainties.

We have addressed the issue of automatic situation recognition by developing a framework for model-based event correlation in infrastructure surveillance. The framework—named DETECT (DEcision Triggering Event Composer & Tracker)— is able to store in its knowledge base any number of threat scenarios described in the form of Event Trees, and then recognize those scenarios in real-time, providing early warnings to PSIM users [9, 10].

In this paper, we adopt a model-based evaluation approach supporting the quantitative assessment of DETECT effectiveness in reducing the number of false alarms and in increasing the overall trustworthiness of the surveillance system. The evaluation is dependent on sensor technologies and scenario descriptions, and it is based on stochastic modelling techniques. Some mappings are performed from Event Trees to other formalisms like Fault Trees (FT), Bayesian Networks (BN), Petri Nets (PN) and their extensions. Those formalisms are widespread in dependability modelling and allow engineers to perform several useful analyses, including 'what if' and 'sensitivity', accounting for false alarms and even sensor hardware faults.

The choice of these formalisms relies on a comparison of their modelling power and efficiency: a complete report about this comparison is in [11]. In brief, FTs are very easy to build and analyse, but they have a limited modelling power. On the other hand, PNs feature a great expressive power but they are limited by the well-known state-space explosion problem. BNs represent a good trade-off between those two extremes. The practice of multiformalism modelling, i.e. the integration of different formal modelling languages into a single composed model, has proven to be effective in several applications of dependability [12] and safety [13] evaluation. This paper describes a multiformalism approach for the evaluation of detection probabilities using Bayesian Networks and Generalized Stochastic Petri Nets (GSPN).

Generally speaking, the method used for the analysis, which is the main original contribution of this paper, allows to:

- Support design choices in terms of type and reliability of detectors, redundancy configurations, scenario descriptions.
- Demonstrate the effectiveness of the overall approach in practical surveillance scenarios, in terms of improved trustworthiness in threat detection with respect to single sensors;
- Define fuzzy event-correlation operators able to detect threat events in noisy environments;
- Define multiformalism approaches for event correlation allowing the combined exploitation of modelling power and solution efficiency of different formalisms.

A threat scenario of a terrorist attack in a metro railway station is considered, to show the practical application of the methodology.

The rest of this paper is structured as follows. Section 2 provides an overview of the related literature on DETECT and trustworthiness evaluation of surveillance systems. Section 3 describes the general fuzzy process used for the analysis. Section 4 customizes the general process to the Bayesian Networks modelling formalism as well as to Petri Nets (Sect. 5). Section 6 presents a multiformalism approach exploiting the power of both BNs and GSPNs. Section 7 introduces the case-study application using a metro-railway threat scenario. Section 8 summarizes the results of the analyses and discusses the achievements. Finally, Sect. 9 provides conclusions and hints about future developments.

2   Background

2.1   Related Works

The first concept of DETECT has been described in [9], where the overall architecture of the framework is presented: it includes the composite Event Description Language (EDL), the modules for the management of detection models and the scenario repository. In [10], an overall system including a middleware for the integration of heterogeneous sensor networks is described and applied to railway surveillance case-studies. Reference [14] discusses the integration of DETECT in a PSIM system [15], presenting the reference scenario that will be also used in this paper. To detect redundancies while updating the scenario repository (off-line issue) and to increase the robustness of DETECT with respect to imperfect modelling and/or missed detections (on-line issue), distance metrics between Event Trees are introduced in [16].

A survey of state-of-the-art in physical security technologies and advanced surveillance paradigms, including a section on PSIM systems, is provided in [17]. Contemporary remote surveillance systems for public safety are also discussed in [18]. Technology and market-oriented considerations on PSIM can be also found in [19, 20].

In [21] the authors address the issue of providing fault-tolerant solutions for WSN, using event specification languages and voting schemes; however, no model-based performance evaluation approach is provided. A similar issue is addressed in [22], where the discussion focuses on different levels of information/decision fusion on WSN event detection using appropriate classifiers and reaching a consensus among them in order to enhance trustworthiness. Reference [23] describes a method for evaluating the reliability of WSN using the FT modelling formalism, but the analysis is limited to hardware faults (quantified by the Mean Time Between Failures, MTBF) and homogenous devices (i.e. the WSN motes). Performance evaluation aspects of distributed heterogeneous surveillance systems are instead addressed in [24], which only lists the general issues and some pointers to the related literature. Reference [25], about the trustworthiness analysis of sensor networks in cyber-physical system, is apparently one of the most related to the topics of this paper, since it focuses on the reduction of false alarms by clustering sensors according to their locations and by building appropriate object-alarm graphs; however, the approach is quite different from the one of DETECT and furthermore it applies to homogeneous detectors. Another general discussion on the importance of the evaluation of performance metrics and human factors in distributed surveillance systems can be found in [26]; however, no hints are provided in that paper about how to perform such an evaluation on real systems.

Regarding the dependability modelling approach used in this paper, it is based on the results of the comparison among formalisms (i.e. Fault Trees, Bayesian Networks and Stochastic Petri Nets) in terms of modelling power and solving efficiency that has been reported in [11], and also applied in [13] to a different case-study using an approach known as 'multiformalism'. Some applications to physical security are in [27, 28] respectively exploiting GSPNs and BNs.

In recent years, the scientific community has also attempted to raise the abstraction level of modelling approaches and to combine high-level modelling with formal methods using model-driven techniques. These approaches rely on meta-modelling and model-transformation techniques: UML-CI is proposed to model critical infrastructures focusing on management aspects [29]; CIP_VAM is a recent UML profile addressing the physical protection of critical infrastructures and providing tool support for the automatic generation of vulnerability models based on Bayesian Networks [30]; SecAM extends MARTE and MARTE-DAM in order to allow security specification and modelling of critical infrastructures and to enable survivability analysis [31].

2.2 Event Description Language

Threats scenarios are described in DETECT using a specific Event Description Language (EDL) and stored in a Scenario Repository. In such a way, it is possible to store permanently all scenarios using an interoperable format (i.e. XML). A high-level architecture of the framework is depicted in Fig. 1.

A threat scenario expressed by EDL consists of a set of basic events detected by the sensing devices. An event is a happening that occurs at some locations and at some points in time. Events are of course related to sensor data (i.e. temperature

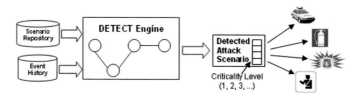

Fig. 1 The DETECT framework

higher than a pre-set threshold). Events are classified as *primitive events* and *composite events.*

A primitive event is a condition on a specific sensor that is associated with some parameters (e.g., event identifier, time of occurrence). A composite event is a combination of primitive events using specific operators. Each event is denoted by an *event expression*, whose complexity grows with the number of involved events. Given the expressions $E_1, E_2, \ldots, E_n$, each application on them through any operator is still an expression. Event expressions are represented by *Event Trees*, where primitive events are the leaves, and internal nodes represent EDL operators.

DETECT can support the composition of complex events in EDL through a *Scenario GUI* (Graphical User Interface), used to draw threat scenarios using a user-friendly interface. Furthermore, in the operational phase, a model manager macro-module has the responsibility of performing queries on the Event History database for the real-time feeding of detection models corresponding to threat scenarios, according to predetermined policies. Those policies, namely *parameter contexts*, are used to set a specific consumption mode of the occurrences of the events collected in the database.

The EDL is based on the Snoop event algebra [32], considering the following operators: OR, AND, ANY, SEQ. As an example, Fig. 2 shows a simple event tree representing the scenario (E_1 AND E_2) OR E_3.

The semantics of the Snoop operators are as follows:

- **OR**. Disjunction of two events E_1 and E_2, denoted (E_1 *OR* E_2). It occurs when at least one of its components occurs.
- **AND**. Conjunction of two events E_1 and E_2, denoted (E_1 *AND* E_2). It occurs when both events occur (the temporal sequence is ignored).
- **ANY**. A composite event, denoted $ANY(m, E_1, E_2, \ldots, E_n)$, where $m \leq n$. It occurs when m out of n distinct events specified in the expression occur (the temporal sequence is ignored).
- **SEQ**. Sequence of two events E_1 and E_2, denoted (E_1 *SEQ* E_2). It occurs when E_2 occurs provided that E_1 has already occurred. This means that the time of occurrence of E_1 has to be less than the time of occurrence of E_2.

Fig. 2 A simple event tree

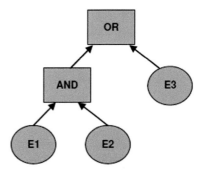

Furthermore, *temporal constraints* can be specified on operators, to restrict the time validity of logic correlations. In order to take into account appropriate event consumption modes and to set how the occurrences of primitive events are processed, four parameter contexts are defined. Given the concepts of *initiator* (the first constituent event whose occurrence starts the composite event detection) and *terminator* (the constituent event that is responsible for terminating the composite event detection), the four different contexts are described as follows: (1) *Recent,* only the most recent occurrence of the initiator is considered; (2) *Chronicle,* the (initiator, terminator) pair is unique. The oldest initiator is paired with the oldest terminator; (3) *Continuous,* each initiator, starts the detection of the event; and (4) *Cumulative,* all occurrences of primitive events are accumulated until the composite event is detected.

The effect of the operators is then conditioned by the specific context in which they are placed. When a composite event is recognized, the output of DETECT consists of:

- the identifier(s) of the detected/suspected scenario(s)[1];
- the temporal value related to the occurrence of the composite event (corresponding to the event occurrence time of the last component primitive event, given by the sensor timestamp);
- an alarm level (optional), associated with scenario evolution (used as a progress indicator and set at design time);
- other information depending on the detection model (e.g., 'likelihood' or 'distance', in case of heuristic detection).

3   Fuzzy Decision Modelling Process

The advantage of the modelling and analysis activity is twofold. On one hand, it can be used during the design phase since it allows to evaluate quantitatively different design options for sensing and decision mechanisms allowing cost/effectiveness trade-off in protection systems design. In fact, the sensing strategies can differ in the number of sensors, in their reliability and/or their event detection performance; decision options are related to the logics that can be applied for correlating primitive events. On the other hand, the model can be used at run-time due to the possibility of tuning the models using data collected in the operational phase (i.e. event history log files merged with operator feedback about false negative/positive), allowing incremental refinement of detection models.

---

[1]The difference between detected and suspected scenario depends on the partial or total matching between the real-time event tree and the stored threat pattern.

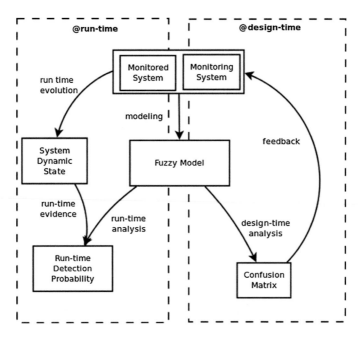

Fig. 3 The modelling and analysis process

Figure 3 shows how the aforementioned objectives can be achieved in an integrated process, in which both the monitored and monitoring systems are represented using probabilistic modelling formalisms.

Fuzzy model evaluation enables two possibilities:

- When used at design-time, the analyses can be used to compute the probability of having an alarm and its confusion matrix (i.e. the false positive and false negative probabilities). Such information can be used to improve the system by using more accurate or redundant sensors.
- When used at run-time, the detected events can be used as the evidence in the models. In such a way, the probability that the configuration of the primitive events is representative of the composite event (i.e. the threat scenario) can be dynamically adapted. Consequently, alarms can be generated only when the confidence in the detection is greater than a certain threshold.

It is essential to develop an appropriate modelling methodology supporting the design phase. In the context of surveillance systems trustworthiness evaluation, models of interest can be structured in layers as depicted in Fig. 4.

The three layers of fuzzy models are:

- *Event layer*: this layer is devoted to modelling the actual cause-consequence relations in real environments. It determines how complex situations can be broken down into basic events. It is usually the output of physical security

Fig. 4 Fuzzy model layered
structure

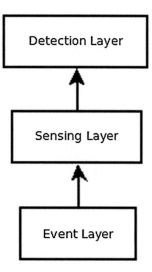

Fig. 4 Fuzzy model layered
structure

surveys, vulnerability analysis and risk assessment. In its most trivial form, it is
constituted by the sequence of basic events associated with a threat scenario.

- *Sensing layer*: this layer models the sensors as objects with their characteristics
  (e.g., event detection capabilities, hardware reliability, detection performance)
  and the basic sensing actions with respect to the events identified in the lower
  layer.
- *Decision layer*: this layer addresses the (probabilistic) combination of simple
  events using EDL operators. It is important to notice that this layer is built on
  top of the Sensing layer, since it does not deal with events occurring in the
  reality but with the ones generated by the sensing system. Those events can be
  different according to sensor types, deployment granularity, and detection
  performance.

In this context, some general concepts can be refined. Let us define: A, the set of
the alarms associated to threat scenarios; S, the set of the sensors; E, the set of the
events that can occur in the real environment; $a \in A$, $e \in E$, $s \in S$.

4 Instantiating the Process with Bayesian Networks

In this Section, we instantiate the process schema shown in Fig. 5 using the BN
formalism, which is especially suited to be adopted in situation recognition scenarios like the one we are addressing in this paper.

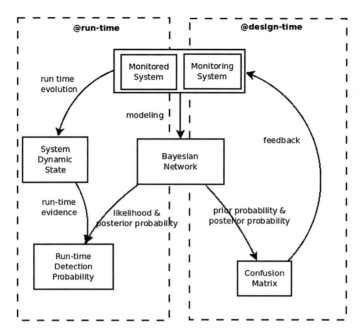

Fig. 5 Customization of the process to the BN formalism

4.1 Process Customization

The customization of the process presented in Sect. 3 to the BN formalism is provided in Fig. 5.

Three different indexes can be computed by solving the BN model [33]:

- *Prior probability*, $P(a)$, that is the likelihood of occurrence of an alarm before any evidence relevant to the alarm has been observed. This index is the probability that an alarm is raised and it may be used at the design time of a PSIM system to predict the expected alarm rate, provided that the rate of primitive events is known a priori (or somehow predictable).
- *Posterior probability*, $P(a \mid e, s)$, that is the conditional probability that an alarm is raised after some evidence is given. This index represents the probability of having an alarm in specific conditions, e.g., when some events happen (e.g., intrusion) and some others are generated by the surveillance system (e.g., sensor failure). It is useful at both design- and run- times. When used at design time it can be used to evaluate the performance of the detection system (i.e. the confusion matrix[2]). Also, the Posterior probability may be used to perform a

---

[2]In this case of event detection, the confusion matrix accounts for binary events which can be *true* (i.e. occurred) or *false*. In DETECT, the positive false probability is given by $P(a = \text{true} \mid e = \text{false})$ while the negative false probability is $P(a = \text{false} \mid e = \text{true})$.

'what-if' analysis to evaluate the performance degradation in case of sensor failures. When used at run-time, a posterior analysis on the model fed with real evidence of events and/or sensor failures may provide a surveillance operator with alerts if probabilities are higher than a certain threshold.

- *Likelihood*, $P(e \mid a, s)$, that is the probability of observing an element of E(real threat scenario) given evidence in A and S. In practice, it can be used to determine the probability that the alarm is trustworthy given that it has been generated. This kind of analysis is useful at run-time since it can support the decision making of the operators.

In the customization of the general process, prior-probability is used for design-time analysis while likelihood and posterior analysis are used in the context of run-time analysis.

4.2 Fuzzy BN Model Structure

The layered model presented in Sect. 3 is implemented by a Bayesian Network where the BN nodes modelling the elements of the Event Layer are at the bottom, the ones representing the Sensing Layer are in the middle, the ones translating EDL operators in the Detection Layer are on the top.

The mapping between EDL and BN is based on the following rules.

BN_R1: for each event e ∈ E, a Boolean BN node N(e) is created: the variable is 'true' when the related attack event e occurs, 'false' otherwise. Let pr(e) be the probability of occurrence of the event (computed by the ratio of the occurrence period T(e) and the reference time unit), the Conditional Probability Table (CPT) of this kind of node is represented in Table 1.

BN_R2: for each sensor s ∈ S, a ternary *{true, false, unknown}* BN node N(s) is created: the variable is 'true' when the sensor s is working properly, 'false' otherwise. Assuming *pr(e)* is the probability that the sensor is failed, the CPT of this kind of node is represented in Table 2. The value of *pr(e)* is computed by the formula *MTBF(s)/(MTBF(s) + MTTR(s))*

Table 1 CPT of the node translating event

N(e)	
True	False
pr(e)	*1-pr(e)*

Table 2 CPT of the node translating sensors

N(s)	
True	False
1-pr(s)	*pr(s)*

Fig. 6 BN pattern for the
sensing layer

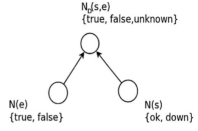

$N_D(s,e)$
{true, false, unknown}

$N(e)$
{true, false}

$N(s)$
{ok, down}

Table 3 CPT of $N_D(s,e)$ BN
node

N(e)	N(s)	$N_D(s,e)$		
		True	False	Unknown
False	Down	0	0	1
False	Ok	fpp(s)	1-fpp(s)	0
True	Down	0	0	1
True	Ok	1-fnp(s)	fnp(s)	0

> where *MTBF(s)* is the Mean Time Between Failures and *MTTR(s)* is the
> Mean Time To Repair of the sensor.

BN_R3: When a sensor *s* ∈ *S* is in charge of detecting an event *e* ∈ *E*, a ternary
{true, false, unknown} BN node $N_D(s, e)$ is created having as parents
both *N(s)* and *N(e)* as in Fig. 6. The variable is 'true' when the sensor
s detects the event, 'false' if it is detecting an event not occurring,
unknown if no information comes from the sensor. Assuming *fnp(s)* and
fpp(s) respectively the false negative and the false positive probabilities,
the CPT of this kind of node is represented in Table 3.

BN_R4: for each *a* ∈ *A*, a ternary *{true, false, unknown}*BN node is generated,
namely *N(a)*. The Parents and the CPT of these nodes are determined
according to the position of the operator inside the detection tree and to
the nature of the operator:

 a. let $P_E(a)$ and $P_A(a)$ be the sets of the events and of the alarms on the
   event tree preceding *a*. The set of the parents of *N(a)* is $N_D(s, x)$ ∪ *N(y)*
   where *x* ∈ $P_E(a)$, *s* is in charge of detecting *x*, and *y* ∈ $P_A(a)$;
 b. the CPTs are built according to the nature of the operator. Bayesian
   Networks allow the definition of operators implementing either
   'sharp' or 'fuzzy' logics, including AND (Table 4), ANY (Table 5),
   noisy-AND (Table 6).

The DETECT framework allows associating a certain amount of uncertainty to
operators, implementing a sort of fuzzy event composition. This is easily modelled
with BN using the so-called 'noisy logic gates', in which the correlation of events is
affected by a 'modelling confidence' error:$0 < k < 1$. In case of error (with probability *k*), we suppose an equal probability distribution for all the other cases. As an
example, Table 6 describes the CPT for the noisy-AND operator.

Table 4 CPT of the AND operator

E1	E2	AND		
		True	False	Unknown
False	False	0	1	0
False	True	0	1	0
False	Unknown	0	1	0
True	False	0	1	0
True	True	1	0	0
True	Unknown	0	0	1
Unknown	False	0	1	0
Unknown	True	0	0	1
Unknown	Unknown	0	0	1

Table 5 CPT of the ANY operator

E1	E2	E3	ANY		
			True	False	Unknown
True	True	True	1	0	0
True	True	False	1	0	0
True	True	Unknown	1	0	0
True	False	True	1	0	0
True	False	False	0	1	0
True	False	Unknown	0	0	1
True	Unknown	True	1	0	0
True	Unknown	False	0	0	1
True	Unknown	Unknown	0	0	1
False	True	True	1	0	0
False	True	False	0	1	0
False	True	Unknown	0	0	1
False	False	True	0	1	0
False	False	False	0	1	0
False	False	Unknown	0	1	0
False	Unknown	True	0	0	1
False	Unknown	False	0	1	0
False	Unknown	Unknown	0	0	1
Unknown	True	True	1	0	0
Unknown	True	False	0	0	1
Unknown	True	Unknown	0	0	1
Unknown	False	True	0	0	1
Unknown	False	False	0	1	0
Unknown	False	Unknown	0	0	1
Unknown	Unknown	True	0	0	1
Unknown	Unknown	False	0	0	1
Unknown	Unknown	Unknown	0	0	1

Table 6 CPT of the
noisy-AND operator

E1	E2	Noisy-AND		
		True	False	Unknown
False	False	k/2	1−k	k/2
False	True	k/2	1−k	k/2
False	Unknown	k/2	1−k	k/2
True	False	k/2	1−k	k/2
True	True	1−k	k/2	k/2
True	Unknown	k/2	k/2	1−k
Unknown	False	k/2	1−k	k/2
Unknown	True	k/2	k/2	1−k
Unknown	Unknown	k/2	k/2	1−k

Please note that, as combinatorial formalisms, Fault Trees and Bayesian Networks cannot precisely model the SEQ operator since they do not allow taking into account state and time dependent properties. To overcome such a limitation, more powerful formalisms are needed, like Dynamic Bayesian Networks or Petri Nets. However, it is possible to approximate an SEQ operator by an AND. In fact, since the SEQ requires the occurrence of events in a certain order, the set of cases in which e.g., *SEQ(E1, E2)* is true is a subset of the set in which *AND(E1, E2)* is true. Thus, by substituting the SEQ with AND in the trustworthiness model, we are overestimating the false positive rate for the specific scenario.

5   Instantiating the Process with Petri Nets

In this Section we instantiate the process schema shown in Fig. 3 using the GSPN formalism [34], which introduces a higher level of complexity but it is able to cope with all cases of situation recognition, including the ones that can only be approximated by BNs.

5.1   Process Customization

The customization of the process presented in Sect. 3 to the GSPN formalism is provided in Fig. 7.

In the context of this work, analysing a GSPN means evaluating the steady-state probability of threat detection. In particular, two different steady-state measures are relevant and widespread in most GSPN applications: mean number of tokens in places and throughput of transitions. These measures can be used for the scope of this work in two different contexts:

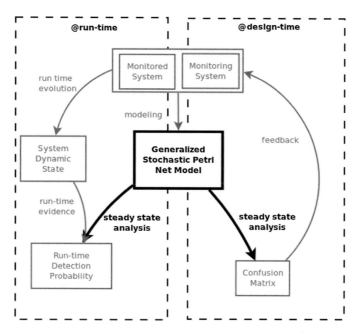

Fig. 7 Customization of the process using the GSPN formalism

- *@design-time*, there is no observation of occurred events, the GSPN measures computing P(a) are evaluated without an initial marking.
- *@run-time*, by setting the correct initial marking both P(a | e, s) and P(e| a, s) probabilities can be computed.

5.2 Fuzzy GSPN Model Structure

The layered model presented in Sect. 3 is replaced by a GSPN model where the GSPN subnets translating elements of the Event Layer are at the bottom, the ones representing the Sensing Layer are in the middle, and the ones translating EDL operators in the Detection Layer are on the top. The model structure is a simplification of the model introduced in [27].

The mapping between EDL and GSPN is based on the following rules.

GSPN_R1: for each event *e* *E*, a GSPN pattern is generated as shown in Fig. 8. The pattern is constituted by a place *Pl(e)* and an exponentially distributed timed transition *Tr(e)* connected by two arcs (ordinary and inhibitor). The rate of the transition is set to *1/T(e)* (the inverse of the occurrence period of *e*).

Fig. 8 Event GSPN pattern

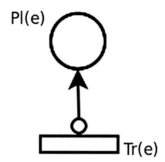

GSPN_R2: for each sensor *s* *S,* a GSPN pattern is generated as shown in Fig. 9. The rule generates the *Net(s)* subnet constituted by two places and two timed stochastic transitions representing the classical up-down model for components subject to failure and repair.

GSPN_R3: when a sensor *s* *S* is in charge of detecting an event *e* *E*, the subnet *Net(s,e)* of Fig. 9 is used, that represents the appropriate pattern. This subnet is connected to the one translating the Event layer using the place *Pl(e)*. In *Net(s,e)*, only the false negative probability is shown to keep the model simple. It is important to underline that the model needs the specification of the time $T_D(s,e)$ for completing the detection of the event *e* by sensor *s*: this value is the inverse of the rate of the *Detecting* transition. The detection network ends with a place *D(s,e)*, containing a token if the event *e* is detected by the sensor *s*.

GSPN_R4: for each a ∈ A, a GSPN subnet is generated:

    a. let $P_E(a)$ and $P_A(a)$ respectively be the sets of the events and of the alarms on the event tree that precede *a*. The GSPN subnet translating *a* is connected to the *D(s,e)* places of the nets translating elements of $P_E(a)$ and to the *D(a)* places of the nets translating elements of $P_A(a)$;

Fig. 9 Sensor related GSPN patterns

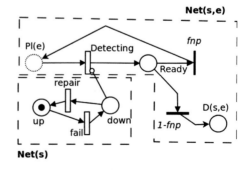

Fig. 10 GSPN pattern for the
AND operator

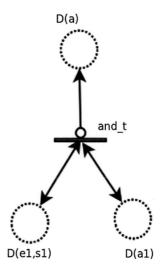

Fig. 11 GSPN pattern for the
SEQ operator

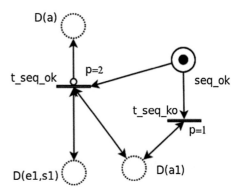

b. the specific subnets implementing the EDL detecting operators in GSPN are built according to the nature of the operators. Some examples are: ordinary AND (Fig. 10), SEQ operator (Fig. 11), and noisy-AND operator (Fig. 12). All the nets are built under the hypothesis that the operator works on the event $e1$ and the result of $a1$ (another detection operator).

To fully translate EDL into a GSPN models, further subnets are required to reset properly the network. The details of such 'control networks' as well as the criteria of analysis are discussed in reference [27].

Fig. 12 GSPN pattern of the
noisy AND

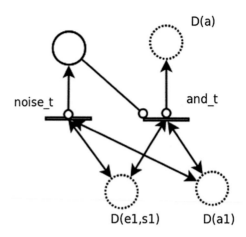

6   A Multiformalism Unifying Process

This section provides a possible approach based on a multiformalism paradigm. Multiformalism is a technique in which more formal languages are jointly employed to model a (typically complex) system. In the context of this work, two different ways to translate EDL into BN and GSPN models have been presented, each one with specific strengths and limitations.

By exploiting the advantages of both languages, it is possible to define a solution process for EDL associating the most suitable formalism for each sub-tree. In particular, we focus our attention on the following points: (1) BNs are easier to analyze, scale better with respect to model size and are easier to analyze during run-time; moreover, BN models allow easier modeling of false-positive detection behaviors; (2) GSPNs are able to deal with time both in explicit (e.g., duration of detection activities) and implicit (e.g., precise sequence of events) forms.

A simple exploitation of these criteria in the construction of a multiformalism approach is based on the following solution steps:

1. Each minimal sub-tree *st* of the EDL tree model containing a SEQ operator, and that is not contained in another sub-tree, is translated into a GSPN mode *Net(st)*.
2. Each *Net(st)* is solved and the probability of detecting the sub-tree *pr(st)* is evaluated.
3. The overall EDL model is translated into a BN where each *st* is translated into a BN node *N(st)* having the CPT in the form described in Table 7.
4. The BN model is solved.

The translation and solution process described above is applied to the case-study scenario in next Section.

Table 7 CPT of the
SEQ-subtree

N(st)	
True	False
pr(st)	*1-pr(st)*

7 Modelling Trustworthiness in a Specific Scenario

The effectiveness of the modelling approach described in the previous section is demonstrated in this section using a case-study in the mass-transit domain. Mass transit systems are vulnerable to many threats, including terrorist attacks. For this reason, surveillance systems for mass-transit feature a growing number of heterogeneous sensing devices. In such a context, the quantitative evaluation of model trustworthiness and sensitivity to sensor faults is very important. In fact, such a model-based design allows to improve the robustness of surveillance systems and to reduce the number of unnecessary alerts. In particular, at design time, the results of the model analysis provide valuable information to assess the level of redundancy and diversity required to the sensors. That allows designers to find the most appropriate configuration complying with performance targets specified in client requirements. In fact, feedbacks from the model evaluation suggest changes about sensor dislocation and sensing technologies. An estimation of detection model trustworthiness is also important during operation to define confidence thresholds for triggering high-level warnings and even automatic response actions.

Let us consider a threat scenario similar to the chemical attack with Sarin agent occurred in the Tokyo subway on March 20, 1995, which caused 12 fatalities and 5500 injured [35]. The technologies available to early detect and assess the threat include intelligent cameras, audio sensors and specific standoff CWA (Chemical Warfare Agents) detectors. For several reasons (inner technology, installation environment, etc.), the single events reported by these sensing devices can feature non-negligible levels of false alarms, and hence cannot be simply trusted. Therefore, events detected by the sensors are correlated in a threat scenario representation, which has been already introduced in reference [14]. The current CWA detection technologies include Ion Mobility Spectroscopy (IMS), Surface Acoustic Wave (SAW), Infrared Radiation (IR), etc. They are employed in ad hoc standoff detectors, characterized by different performances. One of the most accurate devices, the automatic scanning, passive, and IR sensor can recognize a vapour cloud from several kilometres with an 87 % detection rate. As already mentioned, it is possible to combine heterogeneous sensors (e.g., IMS/SAW and IR) to detect the same event, and to correlate their detections according to appropriate criteria, either logical, temporal, and/or spatial.

The threat scenario is a CWA attack on a subway platform. Let us assume the following set of events that are very likely to occur in such a scenario:

1. Attackers drop the CWA that spreads in the surrounding environment.
2. Contaminated passengers start to fall-down.
3. Around the contaminated area, other people scream and run away.

It is assumed that the subway station is equipped with smart-cameras (i.e. 'intelligent' cameras with video-analytics), microphones (i.e. audio sensors with audio pattern recognition), plus both IMS/SAW and IR CWA detectors. The scenario can be formally described by means of the notation "sensor description (sensor ID) :: event description (event ID)":

- *Intelligent Camera (S1) :: Fall of person (E1)*
- *Intelligent Camera (S1) :: Abnormal running (E2)*
- *Intelligent Camera (S2) :: Fall of person (E1)*
- *Intelligent Camera (S2) :: Abnormal running (E2)*
- *Audio sensor (S3) :: Scream (E3)*
- *IMS/SAW detector (S4) :: CWA detection (E4)*
- *IR detector (S5) :: CWA detection (E4)*

The Event Tree model of the CWA threat scenario is depicted in Fig. 13.

The OR operators correlate the events "person falling" and "person running", detectable by two redundant intelligent cameras monitoring the platform. The other child node (E3-S3)of the ANY operator represents the event "person screaming" detectable by the intelligent microphone. When 2 out of these 3 events are detected in a certain (limited) time frame, the situation can be reliably considered abnormal so that a warning to the operator can be issued. The SEQ operator represents the upward CWA spread detectable by two redundant CWA sensors, installed at different levels. Finally, the AND operator at the top of the tree represents the composite event associated with the complete CWA threat scenario. In the following, two different models are proposed and compared: a BN model generated according to the rules described in Sect. 4 and a multiformalism model generated using the algorithm presented in Sect. 6 as well as the transformation rules in Sects. 4 and 5.

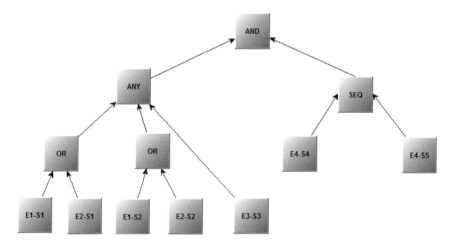

Fig. 13 Event tree associated to the CWA threat scenario

7.1 The BN Model

As described in the previous section, an event occurrence can be 'true' with a probability p, or false with a probability $1-p$. Each sensor can be available (i.e. working properly) with a probability q, or unavailable (i.e. not working properly) with a probability $1-q$. Each event detected by a sensor can be 'true', 'false', or 'unknown' according to event occurrence and the availability of the sensor at the time the event is occurring. Moreover each sensor, for each detected event, has a couple of values *fnp* and *fpp* which are the sensor false-negative and false-positive probabilities. For the sake of brevity, only the BN model is built and analysed according to the modelling methodology described in the previous sections. The three parts of the model are depicted in Figs. 14, 15 and 16.

The Event Layer (Fig. 14) is constituted by a node E that represents the actual CWA attack, while E1, E2, E3 and E4 are the primitive events that can be detected by the sensors.

The interface between the Event Layer and the Sensor Layer is the set of E1, E2, E3 and E4 nodes. In the Sensor Layer (Fig. 15), there are five nodes (S1, S2, S3, S4 and S5) representing sensors, and seven nodes (E1_S1, E2_S1, E1_S2, E2_S2, E3_S3, E4_S4 and E4_S5) representing the sensed events.

Finally, the overall BN model is represented in Fig. 17: as already stated, the SEQ operator has been implemented by an AND operator, introducing a modeling error.

Fig. 14 Event layer of the CWA Bayesian network

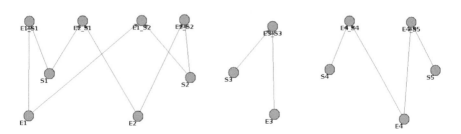

Fig. 15 Sensing layer of the CWA Bayesian network

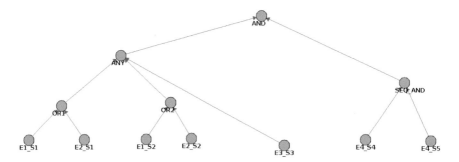

Fig. 16 Detecting layer of the CWA Bayesian network

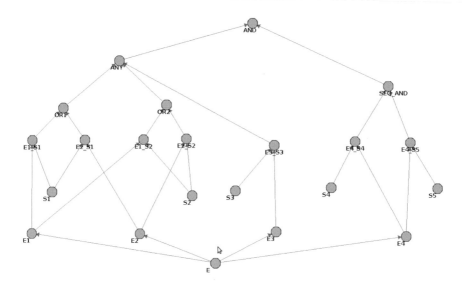

Fig. 17 BN model of the CWA threat scenario

7.2 The Multiformalism Model

Since there is a single SEQ sub-tree, related to the SEQ operator and the E4-S4 and E4-S5 nodes, the multiformalism model is obtained by replacing this sub-tree with the GSPN depicted in Fig. 18.

A slightly different GSPN model is depicted with respect to the proposed patterns, to make the model analysable by a steady-state construction of the tangible state-space in case of an observed attack.

This model is solved by computing the throughputs of the transitions *det* and *nondet*, representing the rates of detection and non-detection of the attack event respectively. Let *th(det)* and *th(nondet)* be the values of these transitions. We are interested in evaluating *PT* and *PF* that are respectively the posterior probability of

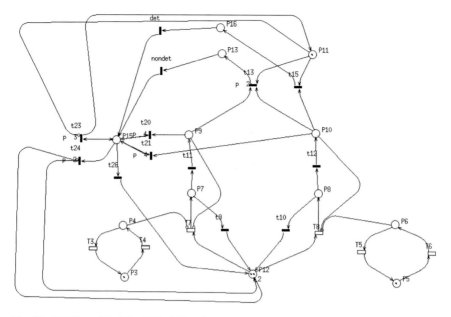

Fig. 18 GSPN model of the SEQ-AND sub-tree

detecting and not detecting by the SEQ operator. These indices are computed as follows:

$$PT = th(det)/(th(nondet) + th(det))$$
$$PF = th(nondet)/(th(nondet) + th(det))$$

Once the GSPN model is solved, the entire sub-tree may be collapsed into a BN node. In other words, the *st* sub-model in the original BN network (Fig. 19) becomes a single BN node as in Fig. 20.

In this last model, the SEQ_AND node is associated the CPT in Table 8, while the AND node on top of the BN replaces—with respect to the full BN model—its CPT with the one in Table 9 (GSPN do not easily support multi-value models).

8 Evaluation and Discussion of the Results

The model has been evaluated using the parameters summarized in Table 10, where (non-conditional) probabilities refer to a standard time frame of 1 h. Parameters have been assigned realistic pseudo-data, since exact values depend on risk assessment results, specific sensor technology as well as operational reports from the real environment. Some parameters apply only to the BN model, some only to the multiformalism model, while others to both.

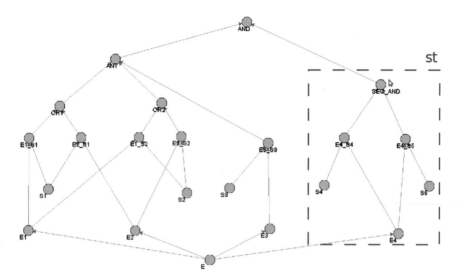

Fig. 19 Original CWA BN model

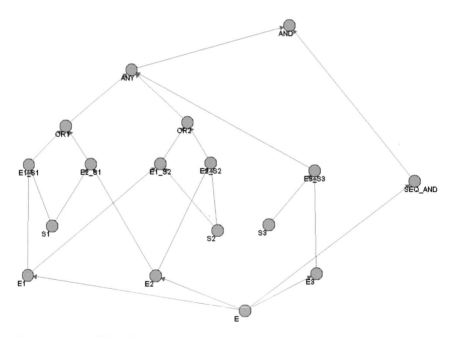

Fig. 20 Reduced CWA BN model

Table 8 Reduced SEQ-AND node CPT

E	SEQ_AND	
	True	False
True	PT	PF
False	0	1

Table 9 CPT of the top event in the reduced model

SEQ_AND	ANY	AND		
		True	False	Unknown
False	False	0	1	0
False	True	0	1	0
False	Unknown	0	1	0
True	False	0	1	0
True	True	1	0	0
True	Unknown	0	0	1

Table 10 CWA model parameters

Name	Description	Model	Node	Value
attackProb	Probability of having a CWA attack	Both	BN/E	10^{-6}
running	Probability of a running man in normal conditions (not related to an attack)	Both	BN/E1	4×10^{-1}
falling	Probability of a falling man in normal conditions (not related to an attack)	Both	BN/E2	10^{-3}
screaming	Probability of a scream in normal conditions (not related to an attack)	Both	BN/E3	5×10^{-3}
U_1	Unavailability of sensor 1	Both	BN/S1	2×10^{-4}
U_2	Unavailability of sensor 2	Both	BN/S2	2×10^{-4}
U_3	Unavailability of sensor 3	Both	BN/S3	10^{-4}
U_4	Unavailability of sensor 4	BN	BN/S4	2×10^{-5}
U_5	Unavailability of sensor 5	BN	BN/S5	10^{-5}
$MTBF_4$	Mean time between failures of sensor 4	Multi	GSPN/T4	199996 h
$MTBF_5$	Mean time between failures of sensor 5	Multi	GSPN/T6	399996 h
$MTTR_{4,5}$	Mean time to repair of sensors 4 and 5	Multi	GSPN/T3 GSPN/T5	4 h
TD_4	Time of detection of sensor 4	Multi	GSPN/T4	0.083 h
TD_5	Time of detection of sensor 5	Multi	GSPN/T4	0.014 h
Sfp_{11} Sfp_{12}	Sensor false positive probability of sensor 1 (resp. 2) when sensing event 1	Both	BN/E1_S1 BN/E1_S2	3×10^{-2}
Sfn_{11} Sfn_{12}	Sensor false negative probability of sensor 1 (resp. 2) when sensing event 1	Both		2×10^{-2}

(continued)

Table 10 (continued)

Name	Description	Model	Node	Value
Sfp_{21} Sfp_{22}	Sensor false positive probability of sensor 1 (resp. 2) when sensing event 2	Both		2×10^{-2}
Sfn_{21} Sfn_{22}	Sensor false negative probability of sensor 1 (resp. 2) when sensing event 2	Both	BN/E2_S1 BN/E2_S2	3×10^{-2}
Sfp_{33}	Sensor false positive probability of sensor 3 when sensing event 3	Both	BN/E3_S3	2×10^{-2}
Sfn_{33}	Sensor false negative probability of sensor 3 when sensing event 3	Both		1.2×10^{-2}
Sfp_{44}	Sensor false positive probability of sensor 4 when sensing event 4	BN	BN/E4_S4 GSPN/t11 GSPN/t9	0.8×10^{-2}
Sfn_{44}	Sensor false negative probability of sensor 4 when sensing event 4	Both		0.2×10^{-2}
Sfp_{55}	Sensor false positive probability of sensor 5 when sensing event 5	BN	BN/E5_S5 GSPN/t12 GSPN/t10	0.7×10^{-2}
Sfn_{55}	Sensor false negative probability of sensor 5 when sensing event 5	Both		0.3×10^{-2}

8.1 CWA Scenario Analysis Using the BN Model

At design-time, the evaluation addresses both prior and posterior probabilities. The distribution of prior probability of the model is reported in Table 11, that also highlights the related meaning at the PSIM system level, regarding the specific scenario. Posterior probability analysis has been performed to evaluate the confusion matrix (see Table 12). The left column represents the evidence, that can be *true* (CWA threat happening) or *false*. The other columns represent the probability of CWA alarm being generated:

- 'Alarm on', which can be a true positive, *tp*, or false positive, *fp*, depending whether the evidence is 'true' or 'false', respectively
- 'Alarm off', which can be a *tp* or a *fp*, depending whether the evidence is 'false' or 'true', respectively)
- Inactive, due to the unavailability of essential sensors.

The results show that the rate of alarms, in general, and the probability of *fp* and *fn* in particular, are largely acceptable, according to recent ergonomics studies [5].

Table 11 Prior probability distribution of the CWA scenario

Value	Meaning	Probability
True	Alarm on	2.273×10^{-5}
False	Alarm off	0.999977
Unknown	Alarm inactive	2.7×10^{-7}

Table 12 Confusion matrix of the CWA scenario

Evidence	Alarm on	Alarm off
True	0.995 (tp)	0.22×10^{-4} (fn)
False	0.5×10^{-2} (fp)	0.999978 (tp)

Table 13 Robustness with respect to sensor availability

Sens.	Event (E)	Alarm on	Alarm off
S1	True	0.982	0.62×10^{-5}
	False	0.005	0.99997
	Unknown	0.013	2.38×10^{-5}
S2	True	0.982	0.62×10^{-5}
	False	0.005	0.99997
	Unknown	0.013	2.38×10^{-5}
S3	True	0.994	0.257×10^{-5}
	False	0.004	0.99994
	Unknown	0.002	3.43×10^{-5}
S4	True	0	0
	False	0.003	0.9972
	Unknown	0.997	0.0028
S5	True	0	0
	False	0.002	0.9969
	Unknown	0.998	0.0031

In particular, *fp* are much less than the ones generated by single sensors. The evaluation of those parameters is essential to ensure system effectiveness and usability in real environments. More sophisticated analyses can be performed on the model to evaluate the robustness of the design. The first set of posterior probability evaluations aims at computing the confusion matrices in presence of single sensor failures; this is accomplished by setting *S1-S5* evidences one by one in the BN model to 'false' (i.e. sensor off). Table 13 summarizes the results of such analysis.

The second robustness analysis aims at validating the event tree with respect to variations in threat patterns: specifically, we consider the case when some features (*E1, E2, E3*) are not present and therefore we calculate *fn* when *E1, E2, E3* events are false. The results of such posterior probability analysis are reported in Table 14.

Table 14 Robustness with respect to scenario variations

Unobserved event	Alarm on	Alarm off
E1	0.9934	0.0066
E2	0.9941	0.0059
E3	0.9938	0.0062

Table 15 Confusion matrix of the multiformalism model

Evidence	Alarm on	Alarm off
True	0.8589 (tp)	0.141 (fn)
False	0 (fp)	1 (tn)

8.2   *CWA Scenario Analysis Using the Multiformalism Model*

First, the PT and PF values of the GSPN model related to the *st* sub tree have been computed: PT equals to 0.859, while PF is 0.141. These values are used in the CPT of the substituting BN node. Table 15 reports the confusion matrix of the whole BN model where the SEQ_AND sub-tree has collapsed into a single BN node.

The differences between the confusion matrix of Table 15 and the one reported in Table 12 are due to the absence of false-positive effects of the GSPN, as well as to the more realistic evaluation with the introduction of detection times and of the exact SEQ operator model in the GSPN.

9   Conclusions and Future Work

In this paper, we have provided a structured trustworthiness modelling approach especially suited to surveillance systems featuring situation recognition capabilities based on Event Trees, which is the threat specification formalism used in the DETECT framework.

The effectiveness of the approach described in this paper is twofold. At design time, the results of the analysis provide a guide to support the choice and deployment of sensors with respect to risk assessment results. At run-time, trustworthiness indices can be associated to detection models and hence to alarms reported to the operators, accounting for sensor performance and reliability. Furthermore, at run-time: sensor status (e.g., events detected, hardware failures, etc.) can be used for the on-line updating of performance and reliability indices; the feedback of the operators over a significant time period can be used to fine-tune trustworthiness parameters: e.g., the *fp* probability can be estimated by counting the average number of false alerts generated by single sensors or by DETECT and by normalizing that number according to the reference time frame. Fuzzy correlation operator and multiformalism technique address a greater modelling power and solving efficiency.

Future developments will address the following issues: evaluation results are going to be extended using further models and simulation campaigns, data coming from on-the-field experimentations and long-term observations is going to be integrated in the models and used to validate them.

References

1. Garcia, M.L.: The Design and Evaluation of Physical Protection Systems. Butterworth-Heinemann, Boston (2001)
2. Flammini, F., Gaglione, A., Mazzocca, N., Moscato, V., Pragliola, C.: Wireless sensor data fusion for critical infrastructure security. Adv. Intell. Soft Comput. **53**, 92–99 (2009)
3. Flammini, F., Gaglione, A., Ottello, F., Pappalardo, A., Pragliola, C., Tedesco, A.: Towards wireless sensor networks for railway infrastructure monitoring. In: Proceeding ESARS 2010, pp. 1–6, Bologna, Italy (2010)
4. Zhu, Z., Huang, T.S.: Multimodal Surveillance: Sensors, Algorithms and Systems. Artech House Publisher, Boston (2007)
5. Wickens, C., Dixon, S.: The benefits of imperfect diagnostic automation: a synthesis of the literature. Theor. Issues Ergon. Sci. **8**(3), 201–212 (2007)
6. Dong, M., He, D.: Hidden semi-Markov model-based methodology for multi-sensor equipment health diagnosis and prognosis Eur. J. Oper. Res. **178**(3), 858–878 (2007)
7. Guo, H., Shi, W., Deng, Y.: Evaluating sensor reliability in classification problems based on evidence theory. IEEE Trans. Syst. Man Cybern. Part B: Cybern. **36**(5), 970–981 (2006). doi:10.1109/TSMCB.2006.872269
8. Luo, H., Tao, H., Ma, H., Das, S.K.: Data fusion with desired reliability in wireless sensor networks. IEEE Trans. Parallel Distrib. Syst. **23**(3), 501–513 (2012)
9. Flammini, F., Gaglione, A., Mazzocca, N., Pragliola, C.: DETECT: a novel framework for the detection of attacks to critical infrastructures. In: Martorell, et al. (eds.) Safety, Reliability and Risk Analysis: Theory, Methods and Applications, Proceedings of ESREL'08, pp. 105–112 (2008)
10. Flammini, F., Gaglione, A., Mazzocca, N., Moscato, V., Pragliola, C.: On-line integration and reasoning of multi-sensor data to enhance infrastructure surveillance. J. Inf. Assur. Secur. (JIAS) **4**(2), 183–191 (2009)
11. Bobbio, A., Ciancamerla, E., Franceschinis, G., Gaeta, R., Minichino, M., Portinale, L.: Sequential application of heterogeneous models for the safety analysis of a control system: a case study. RESS **81**(3), 269–280 (2003)
12. Flammini, F., Marrone, S., Iacono, M., Mazzocca, N., Vittorini, V.: A Multiformalism Modular Approach to ERTMS/ETCS Failure Modelling. Int. J. Reliab. Qual. Saf. Eng. Vol. 21(1) 450001 World Scientific Publishing Company (2014). doi:10.1142/S0218539314500016
13. Flammini, F., Marrone, S., Mazzocca, N., Vittorini, V.: A new modelling approach to the safety evaluation of N-modular redundant computer systems in presence of imperfect maintenance. RESS **94**(9), 1422–1432 (2009)
14. Flammini, F., Mazzocca, N., Pappalardo, A., Pragliola, C., Vittorini, V.: Augmenting surveillance system capabilities by exploiting event correlation and distributed attack detection. In: Proceeding 2011 International Workshop on Security and Cognitive Informatics for Homeland Defence (SeCIHD'11), LNCS 6908, pp. 191–204 (2011)
15. Bocchetti, G., Flammini, F., Pragliola, C., Pappalardo, A.: Dependable integrated surveillance systems for the physical security of metro railways. In: IEEE Proceeding of 3rd ACM/IEEE International Conference on Distributed Smart Cameras (ICDSC 2009), pp. 1–7 (2009)
16. Flammini, F., Pappalardo, A., Pragliola, C., Vittorini, V.: A robust approach for on-line and off-line threat detection based on event tree similarity analysis. In: Proceeding of Workshop on Multimedia Systems for Surveillance (MMSS), pp. 414–419 (2011)
17. Flammini, F., Pappalardo, A., Vittorini, V.: Challenges and emerging paradigms for augmented surveillance. In: Effective Surveillance for Homeland Security: Combining Technology and Social Issues. Taylor & Francis/CRC Press, Boca Raton (2013) To appear
18. Räty, T.D.: Survey on contemporary remote surveillance systems for public safety. IEEE Trans. Sys. Man Cyber Part C, No. 40, **5**, 493–515 (2010)

19. Hunt, S.: Physical security information management (PSIM): The basics. http://www.csoonline.com/article/622321/physical-security-information-management-psim-the-basics (2011)
20. Frost & Sullivan: Analysis of the Worldwide Physical Security Information Management Market. http://www.cnlsoftware.com/media/reports/Analysis_Worldwide_Physical_Security_Information_Management_Market.pdf (2012)
21. Ortmann, S., Langendoerfer, P: Enhancing reliability of sensor networks by fine tuning their event observation behavior. In: Proceeding 2008 International Symposium on a World of Wireless, Mobile and Multimedia Networks (WOWMOM '08). IEEE (2008)
22. Bahrepour, M., Meratnia, N., Havinga, P.J.M.: Sensor fusion-based event detection in wireless sensor networks. In: 6th Annual International Conference MobiQuitous 2009, 13–16 July 2009
23. Silva, I., Guedes, L.A., Portugal, P., Vasques, F.: Reliability and availability evaluation of wireless sensor networks for industrial applications. Sensors **12**(1), 806–838 (2012)
24. Legg, J.A.: Distributed multisensor fusion system specification and evaluation issues. Defence Science and Technology Organisation, Edinburgh, South Australia 5111, Australia (2005)
25. Tang, L.-A., Yu, X., Kim, S., Han, J., Hung, C.-C., Peng, W.-C.: Tru-Alarm: Trustworthiness analysis of sensor networks in cyber-physical systems. In: ICDM '10, IEEE Computer Society, Washington, USA (2010)
26. Karimaa, A.: Efficient video surveillance: performance evaluation in distributed video surveillance systems. In: Lin, W., (ed.) Video Surveillance, ISBN: 978-953-307-436-8, InTech (2011)
27. Flammini, F., Gentile, U., Marrone, S., Nardone, R., Vittorini, V.: A petri net pattern-oriented approach for the design of physical protection systems. In: proceedings of Computer Safety, Reliability, and Security; LNCS **8666**, 230–245 (2014)
28. Drago, A., Marrone, S., Mazzocca, N., Tedesco, A., Vittorini, V.: Model-Driven Estimation of Distributed Vulnerability in Complex Railway Networks. In: Ubiquitous Intelligence and Computing, 2013 IEEE 10th International Conference on Autonomic and Trusted Computing (UIC/ATC), pp.380–387, 18–21 Dec. 2013 doi:10.1109/UIC-ATC.2013.78
29. Bagheri, E., Ghorbani, A.A.: UML-CI: a reference model for profiling critical infrastructure systems. Inf. Syst. Frontiers **12**(2), 115–139 (2010)
30. Marrone, S., Nardone, R., Tedesco, A., D'Amore, P., Vittorini, V., Setola, R., Cillis, F.D., Mazzocca, N.: Vulnerability modeling and analysis for critical infrastructure protection applications. Int. J. Crit. Infrastruct. Prot. **6**(34), 217–227 (2013). doi:http://dx.doi.org/
31. Rodrìguez, R.J., Merseguer, J., Bernardi, S.: Modelling security of critical infrastructures: a survivability assessment. Comput. J. (2014). doi:10.1093/comjnl/BXU096
32. Chakravarthy, S., Mishra, D.: Snoop, an expressive event specification language for active databases. Data Knowl. Eng. **14**(1), 1–26 (1994)
33. Charniak, E.: Bayesian Networks without Tears, AI Magazine, 1991
34. Ajmone-Marsan, M., Balbo, G., Conte, G., Donatelli S., Franceschinis, G.: Modelling with Generalized Stochastic Petri Nets; Wiley Series in Parallel Computing. John Wiley and Sons, New York ISBN: 0–471-93059-8 (1995)
35. National Consortium for the Study of Terrorism and Responses to Terrorism (START).: Global Terrorism Database [199503200014] (2012). http://www.start.umd.edu/gtd

Intelligent Radar Signal Recognition and Classification

Ivan Jordanov and Nedyalko Petrov

Abstract This chapter investigates a classification problem for timely and reliable identification of radar signal emitters by implementing and following a neural network (NN) based approach. A large data set of intercepted generic radar signals, containing records of their pulse train characteristics (such as operational frequencies, modulation types, pulse repetition intervals, scanning period, etc.), is used for this research. Due to the nature of the available signals, the data entries consist of a mixture of continuous, discrete and categorical data, with a considerable number of records containing missing values. To solve the classification problem, two separate approaches are investigated, implemented, tested and validated on a number of case studies. In the first approach, a listwise deletion is used to clean the data of samples containing missing values and then feed-forward neural networks are employed for the classification task. In the second one, a multiple imputation (MI) model-based method for dealing with missing data (by producing confidence intervals for unbiased estimates without loss of statistical power, i.e. by using all the available samples) is investigated. Afterwards, a feedforward backpropagation neural network is trained to solve the signal classification problem. Each of the approaches is tested and validated on a number of case studies and the results are evaluated and critically compared. The rest of the chapter is organised as follows: the next section (*Introduction and Background*) presents a review of related literature and relevant background knowledge on the investigated topic. In Sect. 2 (*Data Analysis*), a broader formulation of the problem is provided and a deeper analysis of the available data set is made. Different statistical transformation techniques are discussed and a multiple imputation method for dealing with missing data is introduced in Sect. 3 (*Data Pre-Processing*). Several NN topologies, training parameters, input and output coding, and data transformation techniques for facilitating the learning process are tested and evaluated on a set of case studies in Sect. 4 (*Results and Discussion*). Finally, Sect. 5 (*Conclusion*) summarises the results and provides ideas for further extension of this research.

I. Jordanov (✉) · N. Petrov
School of Computing, University of Portsmouth, Portsmouth PO1 3HE, UK
e-mail: ivan.jordanov@port.ac.uk

N. Petrov
e-mail: nedyalko.petrov@port.ac.uk

© Springer International Publishing Switzerland 2016
R. Abielmona et al. (eds.), *Recent Advances in Computational Intelligence in Defense and Security*, Studies in Computational Intelligence 621,
DOI 10.1007/978-3-319-26450-9_5

101

Keywords Radar signal recognition and classification · Surveillance · Neural networks · Data analysis · Multiple imputation · Missing data

1 Introduction and Background

What an irony of fate when Robert Watson-Watt was pulled over in a RADAR (**RA**dar **D**etection **A**nd **R**anging) speed trap during his visit in Canada in the late 1950s. He joked that had he known radar would be used for speed traps, he would never have invented it. Nowadays, this is what most people associate the radar with, but when Watson-Watt invented his primitive radar system in the mid 1930s, it was secretly developed for military purposes. Later, in 1940, it played a vital role in the Battle of Britain, providing early warning of incoming Luftwaffe bombers. During the World War II, USA scientists made the Watson-Watt's radar a lot smaller, more efficient and reliable. This made possible a compact radar unit to be used for warning fighter pilots of enemy aircraft approaching from behind. Also, four of these units were carried on each of the nuclear bombs dropped over Hiroshima and Nagasaki to monitor the bomb distance to the ground, so that detonation could be triggered at a pre-set altitude for maximum destruction. Vigorous development of radar technology after the war led to a wide range of military applications for detecting, locating, tracking, and identifying objects, for surveillance, navigation and weapon guidance purposes for terrestrial, maritime, and airborne systems at small to medium and large distances (from ballistic missile defence systems to fist sized tactical missile seekers) [1].

Later, civilian applications emerged and became wide-spread. This began in air traffic control systems to guide commercial aircrafts in the vicinity of the airports and during their flight and in the sea navigation, used by ships in maritime collision avoidance systems. Nowadays, radars are beginning to serve the same role for the automobile and trucking industries in self-braking systems in cars, crash avoidance and parking assist [2, 3].

Police traffic radar are used for enforcing speed limits; airborne radars are used not only for weather forecast, large-scale weather monitoring, prediction and atmospheric research, but also for environmental monitoring of forestry conditions and land usage, water and ice conditions, pollution control, etc.; space-born (both satellite and space shuttle) serve for space surveillance and planetary observation; in sport they are used for measuring the speed of tennis and baseball serves [1].

A basic block-scheme of a radar system is shown in Fig. 1. Radars are considered to be "active" sensors, as they use their own source of illumination (a transmitter) for locating targets. They transmit energy towards a target and then catch the reflected signal to identify the target. The problem is that (especially for a long range radars) a powerful transmitter and very sensitive receiver are needed

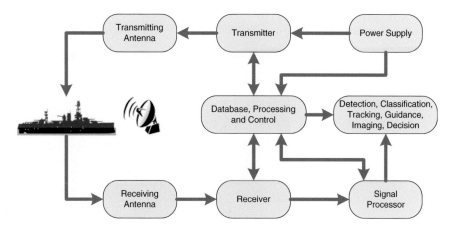

Fig. 1 A block diagram of a basic radar system. Radars operate by transmitting electromagnetic energy toward targets and processing the observed echoes

because the energy spreads out on its way to the target, scatters on reflection and further spreads out on its way back (in general, the decrease of the received signal is proportional to the fourth power of the target distance). The radars range, resolution and sensitivity are generally determined by their transmitter and waveform generator. Although the typical radar systems operate in the microwave region of the electromagnetic spectrum with frequency range of about 200 MHz to about 95 GHz (with corresponding wavelengths of 0.67 m to 3.16 mm), there are also radars that function at frequencies as low as 2 MHz and as high as 300 GHz [4].

The application of the Doppler effect revolutionized the cosmology enabling Doppler spectroscopy to become a powerful tool for finding extrasolar planets and proving the expansion of the universe (the light spectrum of stars (or galaxies) receding from us exhibits *redshift* (increased bandwidth and reduced frequencies), and *blueshift* (higher frequencies and lower bandwidth) if they are moving towards us), but also expanded dramatically the use of radiolocation radars. For the Doppler radars, the reflection from an approaching target electromagnetic wave exhibits higher frequency than the transmitted one and vice versa, a moving away target returns lower frequency wave. The difference between the sent and received frequencies can then be used to estimate the target speed. The problem is that this difference is a very small one, e.g., an incoming target with a 100 km/h increases the received frequency by less than 1e-6, which needs very precise circuits to measure.

A Doppler weather radar with a parabolic antenna situated within a large tiled dome is shown in Fig. 2 [5]. A system with such a radar can measure the distance and lateral speed of falling rain drops, hail particles, or snowflakes, allowing forecasters to predict storms' evolving locations. The presence of debris in the air is

Fig. 2 A Doppler weather radar (*Photo* Brownie Harris/Corbis)

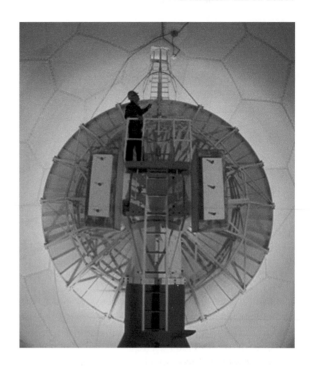

used in similar radar systems to detect tornadoes and define their location, velocity and direction, allowing projections of their movement in real time.

The classical radar imaging uses its antenna to focus a radio frequency beam on a target and capture its reflection to create the image. To work over a long-range it requires powerful transmitters and sensitive receivers because of the way the transmitted energy spreads out on its way to the target and then scatters on reflection. Also, to achieve higher resolution of the image, it needs narrower beams which means that the airborne or space-born platform will need much larger antenna than it could carry. The application of a synthetic aperture technique solves this problem by enabling the use of a smaller antenna through simulating a virtual one with aperture defined by the travel distance of the physical antenna.

The use of the Doppler effect further enhanced the angular resolution in synthetic-aperture radars (SAR) [6] enabling them to acquire surprisingly clear and crisp images [7]. The SAR have been long used on planes and satellites (Fig. 3) for military reconnaissance, mapping ground terrain with intelligence imagery, revealing enemy facilities for enhancing situational awareness and all this in any type of weather, in total darkness and through cloud cover and foliage [8, 69]. They also proved to be very useful in diverse range of civil applications, e.g., in earthquake damage assessment [9], ice [10] and snow monitoring [11], oceanography, polar ice caps and coastal regions imagery, oil pollution monitoring, solid earth science, hydrology, ecology and planetary science [12, 13].

Fig. 3 JAXA's ALOS-2
Earth-observation radar sat
may help the Japanese navy
keep track of ship movements
in the region. *Photo* JAXA
Concept

Another type developed especially to look underground and through walls is the Ground-penetrating radar (GPR), also known as surface-penetrating radar (SPR) [14]. GPR has recently proved to be efficient non-invasive technology with applications in archaeology [15, 16], mining–for both identifying underground rock strata and monitoring instabilities [14, 17], and for optimal irrigation and pollution monitoring [18, 19]. It has been also used for helping police, emergency response and firefighters 'to see' through building walls to locate hostages or help people trapped by fire or under a rubble of a collapsed building [20]. Its ability to see under surface metallic and non-metallic objects makes it useful mapping tool for detection and localisation of underground cables and pipes [21], and buried objects of historical and archaeological importance [22].

The IEEE standard letter nomenclature for the common nominal radar bands is given in Table 1, [23]. The millimetre wave band is sometimes further decomposed into approximate sub-bands of 36–46 GHz (Q band), 46–56 GHz (V band), and 56–100 GHz (W band). The lower frequency bands are usually preferred for longer range surveillance applications due to the low atmospheric attenuation and high available power, and vice versa the higher frequencies tend to be used for shorter range applications and higher resolution, due to the smaller achievable antenna beam widths for a given antenna size, higher attenuation, and lower available power [1]. The radars from the first category (considered a form of radar radiolocation) are capable of covering distances of up to hundreds of kilometres (using high-power transmitters concentrated in a relatively narrow radio bandwidth) and the second group covers radar systems that operate at low power levels, over much smaller distances.

Based on their characteristics, features and application areas, radars can be classified in terms of the following criteria [24]:

- **purpose and function:** surveillance, tracking, guidance, reconnaissance, imaging, data link;
- **frequency band:** radar systems have been operating at frequencies as low as 2 MHz and as high as 300 GHz (see Table 1). Criteria for frequency selection for surveillance radar can be found in [4, 25];
- **waveform:** continuous wave, pulsed wave, digital synthesis;
- **beam scanning:** fixed beam, mechanical scan (rotating, oscillating), mechanical scan in azimuth, electronic scan (phase control, frequency control and mixed in azimuth/elevation), mixed (electronic-mechanical) scan, multi-beam configuration;
- **location:** terrestrial (stable, mobile), marine-borne, air-borne, space-borne;
- **spectrum of collected data:** range (delay time of echo), azimuth (antennae beam pointing, amplitude of echoes), elevation (3D—radar, multifunctional, tracking), height (derived by range and elevation), intensity (echo power), radar cross section (RCS)—(derived by echo intensity and range), radial speed (measurement of differential phase along the time on target due to the Doppler effect—it requires a coherent radar), polarimetry (phase and amplitude of echo in the polarisation channels: horizontally transmitted—HH, horizontally received—HV, VH, VV), RCS profiles along range and azimuth (high resolution along range, imaging radar);
- **configuration:** monostatic (same antenna with co-located transmitter and receiver), bi-static (two antennas), multistatic (one or more spatially dispersed transmitters and receivers). Further detail on variety of radar configurations can be found in [26];
- **signal processing:** coherent (Moving Target Detector/Pulse-Doppler/Super-resolution Signal Processor/Synthetic Aperture Radars (SAR)), non-coherent (integration of envelope signals, moving window, adaptive threshold (Constant False Alarm Rate (CFAR)) and mixed [6];

Table 1 Letter nomenclature for nominal radar frequency bands (IEEE, 2003)

Band	Frequencies	Wavelengths
HF	3–30 MHz	100–10 m
VHF	30–300 MHz	10–1 m
UHF	300 MHz–1 GHz	1–0.3 m
L	1–2 GHz	0.3–0.15 m
S	2–4 GHz	15–7.5 cm
C	4–8 GHz	7.5–3.75 cm
X	8–12 GHz	3.75–2.5 cm
Ku	12–18 GHz	2.5–1.67 cm
K	18–27 GHz	16.7–11.1 mm
Ka	27–36 GHz	11.1–7.5 mm
Q, V, W	36–300 GHz	7.5–1 mm

- **transmitter and receiver technologies:** antenna—reflector plus feed, array (planar, conformal), corporate feed; transmitter—magnetron, klystron, wideband amplifiers (high-power travelling wave tubes (TWT)), solid state; and receiver—analogue and digital technologies, base band, intermediate frequency sampling, low-power TWT;
- **area of application:** large-scale weather forecast and monitoring, air traffic control and guidance (terminal area, en route, collision avoidance, airport apron); police traffic radar used for enforcing speed limits; air defence; anti-theatre ballistic missile defence; vessel traffic surveillance; remote sensing (application to crop evaluation, geodesy, astronomy, defence); environmental monitoring of forestry conditions and land usage; pollution control; geology and archaeology (ground penetrating radar); meteorology (hydrology, rain/hail measurement); study of atmosphere (detection of micro-burst and gust, wind profilers); space-born altimetry for measurement of sea surface height; acquisition and tracking of satellites; monitoring of space debris; marine—navigation and ship collision avoidance; others [5, 12–15].

Radar detection, classification and tracking of targets against a background of clutter and interference are considered as "the general radar problem". For military purposes, the general radar problem includes searching for, interception, localisation, analysis and identification of radiated electromagnetic energy, which is commonly known as radar Electronic Support Measures (ESM). They are considered to be a reliable source of valuable information regarding threat detection, threat avoidance, and, in general, situation awareness for timely deployment of counter-measures [27, 28]. A list of ESM abbreviations is given in Table 2.

A real-time identification of the radar emitter associated with each intercepted pulse train is a very important function of the radar ESM. Typical approaches include sorting incoming radar pulses into individual pulse trains [29], then comparing their characteristics with a library of parametric descriptions, in order to get list of likely radar types. This can be very difficult task as there may be radar modes

Table 2 Commonly adopted ESM abbreviations

Abbreviation	Meaning
EW	Electronic warfare
MOP	Modulation on pulse
PA	Pulse amplitude
PDW	Pulse descriptor word
PPI	Pulse-to-pulse interval
PRI	Pulse repetition interval
PD	Pulse duration
PW	Pulse width
RF	Radio frequency
TOA	Time of arrival
ST	Scanning type
SP	Scan period

for which there is no record in ESM library; overlaps of different radar type parameters; increases in environment density (e.g., Doppler spectrum radars, transmitting hundreds of thousands of pulses per second); agility of radar features, such as radio frequency and scan, pulse repetition interval, etc.; multiplication and dispersion of the modes for military radars; noise and propagation distortion that lead to incomplete or erroneous signals [30].

1.1 Neural Networks in Radar Recognition Systems

There are wide variety of approaches and methods used for radar emitter recognition and identification. For example, [31] investigate a specific emitter identification technique applied to ESM data and by analysing the radar pulses try to extract unique features for each radar, which can be later used for identification. A wavelet transform is employed in [32] for the feature extraction phase in radar signal recognition, as in [33], where they use it before employing probabilistic support vector machines SVMs for the radar emitter recognition task. SVMs are also used in [8, 34] for solving a similar problem. In [35] the authors focus their research on the estimation of a common modulation from a group of intercepted radar pulses and use it as a basis for specific emitter identification. A variety of novel radar emitter recognition algorithms, incorporating clustering and competitive learning, and investigating their advantages over the traditional methods are proposed in [32, 36–42, 70–73].

Among those approaches, a considerable part of the research in the area incorporates NN, due to their parallel architecture, fault tolerance and ability to handle incomplete radar type descriptions and inconsistent and noisy data [43]. NN techniques have previously been applied to several aspects of radar ESM processing [28], including Pulse Descriptor Word (PDW) sorting [44, 45] and radar type recognition [46]. More recently, many new radar recognition systems include NNs as part of a clutter reduction system to improve the information managed by automatic identification systems, such as the detection, positioning, and tracking of surrounding ships [47], or as a key classifier [48–52]. Some examples of NN architectures and topologies used for radar identification recognition and classification based on ESM data include Multilayer Perceptron (MLP) [43], Radial Basis Function (RBF) neural networks as a signal detector [46, 53], a vector neural network [54], and a single parameter dynamic search neural network [50].

In many cases, the NNs are hybridised with other techniques, including fuzzy systems [55], clustering algorithms [29, 56], wavelet packets [32, 57], or Kalman filters [30]. When implementing their "What-and-Where fusion strategy" [30] use an initial clustering algorithm to separate pulses from different emitters according to position-specific parameters of the input pulse stream, and then apply fuzzy ARTMAP (based on Adaptive Resonance Theory (ART) neural network) to classify streams of pulses according to radar type, using their functional parameters. They also complete simulations with a data set that has missing input pattern components and missing training classes and then incorporate a bank of Kalman filters to demonstrate high-level performance of their system on incomplete, overlapping and complex radar data. In [48] higher order spectral analysis (HOSA) techniques are used to extract information from low probability of intercept (LPI) radar signals to produce 2D signatures, which are then fed to a NN classifier for detecting and identifying the LPI radar signal. The work presented in [49] investigates the potential of NNs (MLPs) when used in Forward Scattering Radar (FSR) applications for target classification. The authors analyse collected radar signal data and extract features, which are then used to train NN for target classification. They also apply K-Nearest Neighbour classifier to compare the results from the two approaches and conclude that the NN solution is superior. In [58] an approach combining rough sets (for data reduction) and NN as a classifier is proposed for radar emitter recognition problem, while [59] combines wavelet packets and neural networks for target classification.

The common denominator of all referenced approaches is that they use predominantly supervised NN learning. This means that there is an available data set (or it is on-line collected), on which the NN can be trained and later used to determine the type of the radar emitters detected in the environment. During the training, the NN is presented with labelled samples from the available dataset and the NN weights are adjusted in order to minimise the difference between the NN output and the available target (supervised learning). This difference is expressed by an error function that is minimised by adjusting the NN weights. One of the most popular methods for training is backpropagation (BP), but, as it uses Newton and quasy-Newton deterministic minimisation methods, it could become trapped in a local minimum and in this way to converge to a suboptimal training. Another drawback of the BP algorithm is that it can, sometimes, be slow and unstable. After training, the NN is tested for its ability to generalise, in other words, its ability to correctly classify samples that have not been shown during the learning process.

Among other considerations, the complexity of the training includes selecting the way of showing the samples to the network (i.e. how the training data set is organised and presented to the NN—'batch mode', 'on-line mode', etc.). Another important question is when to stop the training—achieving a zero error function does not always lead to an optimal training. The reality shows that at some point of the learning process, the NN starts to memorise rather than to generalise—this happens when the NN starts to overfit. In order to avoid the overfitting, an additional data subset (called validation subset), is used in parallel with the training set. Initially, the errors on both sets will decrease, but at some point the validation error will start to rise, while the training error will continue to decrease. This point is an indication of overfitting and the training should be stopped, with the current weights assumed to be optimal. This training approach is known as split sample training, where the available dataset is split in training, validation and testing subsets. There are also other training approaches, such as k-fold crossover, or bootstrapping, each with their own specific advantages and drawbacks [43]. One advantage of the k-fold crossover, for example, is that it can be applied when limited number of samples is available for training.

In addition, often before approaching training, the available data set needs to be pre-processed, e.g., [60] use feature vector fusion before feeding the NN classifier. Radar signal processing has specific features that differentiate it from most other signal processing fields. Many modern radars are coherent, meaning that the received signal, once demodulated to baseband, is complex-valued rather than real-valued and as it can be seen from Table 2, many of the collected data is categorical. Another specificity of the radar data sets is that there are usually many missing or incomplete data. Therefore, the problems of representation and statistical pre-processing of the available dataset are very important steps that need to be considered, before starting the actual training. This may also include transformation techniques, such as linear discriminant analysis and principal component analysis, in order to reduce the dimensionality of the problem and dispose of redundant information in the dataset.

1.2 Dealing with Missing Data

According to statistical analysis, the nature of missing data can be classified into three main groups [61–63]: missing completely at random (MCAR), where the probability that an observation is missing is unrelated to its value or to the value of any other variables; missing at random (MAR)—that missingness does not depend on the value of the observed variable, but on the extent of the missingness correlation with other variables that are included in the analysis (in other words, the cause of missingness is considered); and missing not at random (MNAR)—when the data are not MCAR or MAR (missingness still depends on unobserved data). The problem associated with MNAR is that it yields biased parameter estimates, while MCAR and MAR analysis yield unbiased ones (at the same time the main consequence of using MCAR is loss of statistical power), [63].

Dealing with missingness requires an analysis strategy leading to least biased estimates, while not losing statistical power. The problem is these criteria are contradictory and in order to use the information from the partial data in samples with missing data (keeping up the statistical power), and substituting the missing data samples with estimates, inevitably brings bias.

The most popular approaches in dealing with missing data generally fall in three groups: Deletion methods; Single imputation methods; and Model-based methods [62, 64, 65].

Deletion methods include pairwise and listwise deletion. The pairwise deletion (also called "unwise" deletion) keeps as many samples as possible for each analysis (and in this way uses all available information for it), resulting in incomparable analysis, as each is based on different subsets of data, with different sample sizes and different standard errors. The listwise deletion (also known as complete case analysis) is a simple approach, in which all cases with missing data are omitted. The advantages of this technique include comparability across the analyses and it leads to unbiased parameter estimates (assuming the data is MCAR), while its main disadvantage is that there may be substantial loss of statistical power (because not all information is used in the analysis, especially if a large number of cases is excluded).

The single imputation methods include mean/mode substitution, dummy variable method, and single regression. Mean/mode substitution is an old procedure, currently rejected due to of its intrinsic problems, e.g., it adds no new information (the overall mean stays the same), reduces the variability, and weakens the covariance and correlation estimates (it ignores relationship between variables). The dummy variable technique uses all available information about missing observation, but produces biased estimates. In the regression approach, linear regression is used to predict what the missing value should be (based of the available other variables) and then uses it as an actual value. The advantage of this technique is that it uses information from the observed data, but overestimates the model fit and the correlation estimates, and weakens the variance [62].

Most popular, "modern" model-based approaches, fall into two categories: multiple imputation (MI) and maximum likelihood (ML) methods (often referred to as full-information maximum likelihood), [63]. Their advantage is that they model the missingness and give confidence intervals for estimates, rather than relying on a single imputation. If the assumption for MAR holds, both groups of methods result in unbiased estimates (i.e., tend to "preserve" means, variances, co-variances, correlations and linear regression coefficients) without loss of statistical power.

ML identifies a set of parameter values that produces the highest (log) likelihood and estimates the most likely value that would result in the observed data. It has the advantage that both complete and incomplete cases are used, in other words, it utilises all of the information and produces unbiased parameter estimates (with MCAR/MAR data). The MI approach involves three distinct steps: first, sets of plausible data for the missing observations are created and these sets are filled in separately to create many 'completed' datasets; second, each of these datasets is analysed using standard procedures for complete datasets; and thirdly, the results from previous step are combined and pooled into one estimate for the inference. The aim of the MI process is not just to fill in the missing values with plausible estimates, but also to plug in multiple times these values by preserving important characteristics of the whole dataset. As with most multiple regression prediction models, the danger of overfitting the data is real and can lead to less generalisable results than would have been possible with the original data [66].

The advantage of the MI technique is that it provides more accurate variability by making multiple imputations for each missing value (it considers both variability due to sampling and variability due to imputation) and its disadvantage is that it depends on the correctly specified model. Also, it requires cumbersome coding, but the latter is not an issue due to the existence of easy to use off-shelf software packages. For the purpose of this investigation, a free, open source R statistical software is used.

2 Data Analysis

For the purpose of this research, a data set composed of 29,094 intercepted generic
data samples is used. Each of the captured signals is pre-classified by experts in one
of 26 categories, in regards to the platform that can carry the radar emitter (aircraft,
ship, missile, etc.) and in one of 142 categories, based on the functions it can
perform (3D surveillance, weather tracking, air traffic control, etc.).

Each data entry represents a list of 12 recorded pulse train characteristics (signal
frequencies, modulation type, pulse repetition intervals, etc. that will be considered
as input parameters), a category label (specifying the radar function and being
treated as system output) and a data entry identifier (for reference purposes only)
(Table 3).

A more comprehensive summary of the data distribution is presented in Table 4,
where an overview of the type, range and percentage of missing values for the
recorded signal characteristics is given. The collected data consists of both
numerical (integer and float) and categorical values, therefore coding of the categorical fields to numerical representations will be required during the data
pre-processing stage. Also, due to the large number of missing values for some of
the parameters, approaches for handling of missing data will be considered.

3 Data Pre-processing

The pre-processing of the available data is of a great importance for the subsequent
machine learning stage and usually can significantly affect the overall success or
failure of the application of a given classification algorithm. In this context, the
main objective of this stage is to analyse the available data for inconsistencies,
outliers and irrelevant entries and to transform it in a form that could facilitate the
underlying mathematical apparatus of the machine learning algorithm and lead to
an overall improvement of the classifier's performance.

3.1 Data Cleaning and Imputation

Data cleaning (also known as data cleansing or scrubbing) deals with detecting and
removing errors and inconsistencies from data, in order to improve its quality [67].
The most important tasks carried out on this stage would include identification of
outliers (entries that are significantly different from the rest and could be a result of
an error), resolving of data inconsistencies (values that are not consistent with the
specifications or contradict expert knowledge), dealing with missing data (removing
the missing values, assigning those values to the attributes' mean, using statistical

Table 3 Sample radar data subset

ID	FN	RFC	RFmin	RFmax	PRC	PRImin	PRImax	PDC	PDmin	PDmax	ST	SPmin	SPmax
84	SS	B	5300	5800	K	–	–	S	–	–	A	5.9	6.1
4354	AT	F	2700	2900	F	1351.3	1428.6	S	–	–	A	9.5	10.5
7488	3D	B	8800	9300	K	100	125	S	13	21	B	1.4	1.6
9632	WT	F	137	139	T	–	–	V	–	–	D	–	–
9839	3D	S	2900	3100	J	–	–	V	99	101	A	9.5	10.5

Missing values (i.e. values that could not have been intercepted or recognised) are denoted by '–'. The rest of the acronyms are defined in Table 4

Table 4 Data description and percentage of missing values

Field	Field description	Type	Categories	Missing (%)
ID	Reference for the line of data	I	–	–
FN	Function performed by the radar ('3D'—3D surveillance, 'AT'—airtraffic control, 'SS'—surface search, 'WT'—weather tracker, etc.)	C	142	1.35
RFC	Type of modulation used by the radar to change the frequency of the radar from pulse to pulse ('A'—agile, 'F'—fixed, etc.)	C	12	20.75
RFmin	Min frequency that can be used by the radar	R	–	11.15
RFmax	Max frequency that can be used by the radar	R	–	11.15
PRC	Type of modulation used by the radar to change the Pulse Repetition Interval (PRI) of the radar from pulse to pulse ('F'—fixed, etc.)	C	15	15
PRImin	Min PRI that can be used by the radar	R	–	46.70
PRImax	Max PRI that can be used by the radar	R	–	46.70
PDC	Type of modulation used by the radar to change the pulse duration of the radar from pulse to pulse ('S'—stable)	C	5	12.92
PDmin	Min pulse duration that can be used by the radar	R	–	46.05
PDmax	Max pulse duration that can be used by the radar	R	–	46.05
ST	Scanning type—method that the radar uses to move the antenna beam ('A'—circular, 'B'—bidirectional, 'W'—electronically scanned, etc.)	C	28	11.33
SPmin	Min scan period that can be used by the radar	R	–	59.35
SPmax	Max scan period that can be used by the radar	R	–	59.35

In column "Type": I—integer; C—categorical; R—real values

algorithms to predict the missing values) or removing redundant data in different representations.

At this stage of the pre-processing phase, two data sets are prepared. For the purposes of the first two case studies (presented later in this chapter), a data set only containing samples with complete data values is extracted, with the data that could not have been fully intercepted and recognised removed by applying listwise deletion. The second data set (used for the final case study) is received after applying multiple imputation, performed as described below.

3.2 Dealing with Missing Data—Data Imputation

To estimate the values of the missing multivariate data, a sequential imputation algorithm, presented in [68] is used. According to it, if the available data set is denoted with Y and the complete subset with Y_c, the procedure starts from the complete subset to estimate sequentially the missing values of an incomplete observation Y^*, by minimizing the covariance of the augmented data matrix $Y^* = [Y_c, x^*]$. Subsequently the data sample x^* is added to the complete data subset and the algorithm continues with the estimate of next data sample with missing values.

Implementations in R of the original algorithm (available under the function name "*impSeq*") and two modifications of it (namely "*impSeqRob*" and "*impNorm*") are considered and tested. As the original algorithm uses the sample mean and covariance matrix, it is vulnerable to the presence of outliers, but this can be enhanced by including robust estimators of location and scatter (which is rea- lised in the "*impSeqRob*" function). However, the outlyingness metric can be computed for a complete dataset only, therefore the sequential imputation of the missing data is done first and then the outlyingness measure is computed and used to define whether the observation is an outlier or not. If the measure does not exceed a predefined threshold, the observation is included in the next stage of the algo- rithm. In our investigation, however, the use of modified "*impSeqRob*" and "*impNorm*" versions did not produce better results when tested on complete dataset (which may be simply due to the lack of outliers), so the "*impSeq*" function was adopted.

After employing MI on the data samples with missing continuous values, a second dataset of 15656 observations is received, which is more than double the size of the first dataset. Table 5 shows the inputted values produced by the MI algorithm for the sample subset, presented previously in Table 3.

3.3 Data Coding and Transformation

This stage of the pre-processing aims to transform the data into a form that is appropriate for feeding to the selected classifier and would facilitate faster and more accurate machine learning.

In particular, a transformation known as coding is applied to convert the cate- gorical values presented in the data set into numerical ones. Three of the most broadly applied coding techniques are investigated and evaluated—continuous, binary and introduction of dummy variables.

For the first type of coding, each of the categorical values is substituted by a natural number, e.g., the 12 categories for the RFC input are encoded with 12 ordinal numbers, the 15 PRC categories—with 15 ordinal numbers, etc. A sample of data subset coded with continuous values is given in Table 6.

Table 5 Sample radar data subset with imputed values for the missing data entries

ID	FN	RFC	RFmin	RFmax	PRC	PRImin	PRImax	PDC	PDmin	PDmax	ST	SPmin	SPmax
84	SS	B	5300	5800	K	963.2	5625	S	5.8	17	A	5.9	6.1
4354	AT	F	2700	2900	F	1351	1428	S	4	6.3	A	9.5	10.5
7488	3D	B	8800	9300	K	100	125	S	13	21	B	1.4	1.6
9632	WT	F	137	139	T	622.6	31,312	V	61.1	93.1	D	12	47.8
9839	3D	S	2900	3100	J	2058	48,128	V	99	101	A	9.5	10.5

Table 6 Sample subset with imputed radar data and natural number coding for the 'RFC', 'PRC', 'PDC', and 'ST' signal characteristics

ID	RFC	RFmin	RFmax	PRC	PRImin	PRImax	PDC	PDmin	PDmax	ST	SPmin	SPmax
84	2	5300	5800	7	963.2	5625	1	5.8	17	1	5.9	6.1
4354	4	2700	2900	4	1351	1428	1	4	6.3	1	9.5	10.5
7488	2	8800	9300	7	100	125	1	13	21	2	1.4	1.6
9632	4	137	139	11	622.6	31,312	2	61.1	93.1	4	12	47.8
9839	9	2900	3100	6	2058	48,128	2	99	101	1	9.5	10.5

Binary coding, wherein each non-numerical value is substituted by $log_2 N$ (where N is the number of categories taken by that variable) new binary variables (i.e. taking value of either 0 or 1), is illustrated in Table 7 for 32 categories.

Finally, the non-numerical attributes are coded using dummy variables. In particular, every N levels of a categorical variable are represented by introducing N dummy variables. An example of dummy coding for 32 categorical levels is shown in Table 8.

Taking into account the large number of categories presented for the categorical attributes in the input data set (Table 4), continuous and binary codings are considered for transforming the input variables. On the other hand, binary and dummy variable codings are chosen for representing the output parameters.

Finally, in order to balance the impact of the different input parameters on the training algorithm, data scaling is used. Correspondingly, each of the conducted experiments in this chapter is evaluated using 3 forms of the input data set: the original data (with no scaling); normalised data (scaled attribute values within [0, 1] interval); and standardised data (i.e. scaling the attribute values to a zero mean and unit variance). A sample binary coded and standardised data subset is given in Table 9.

Table 7 Example of binary coding for 32-level categorical variable

Original category		Encoded variables				
Index	Label	B1	B2	B3	B4	B5
1	'2D'	0	0	0	0	0
2	'3D'	0	0	0	0	1
3	'AA'	0	0	0	1	0
...						
16	'CS'	0	1	1	1	1
...						
32	'ME'	1	1	1	1	1

Table 8 Example of dummy coding for 32-level categorical variable

Original category		Encoded variables								
Index	Label	D1	D2	D3	D4	D5	...	D16	...	D32
1	'2D'	1	0	0	0	0	...	0	...	0
2	'3D'	0	1	0	0	0	...	0	...	0
3	'AA'	0	0	1	0	0	...	0	...	0
...										
16	'CS'	0	0	0	0	0	...	1	...	0
...										
32	'ME'	0	0	0	0	0	...	0	...	1

Table 9 Sample subset with imputed radar data and binary coding

ID	RFC	RFmin	RFmax	PRC	PRImin	PRImax	PDC	PDmin	PDmax	ST	SPmin	SPmax
84	0001	5300	5800	0110	963	5625	0	5.8	17	00000	5.9	6.1
4354	0011	2700	2900	0011	1351	1428	0	4	6.3	00000	9.5	10.5
7488	0001	8800	9300	0110	100	125	0	13	21	00001	1.4	1.6
9632	0011	137	139	1010	622	31,312	1	61.1	93.1	00011	12	47.8
9839	1000	2900	3100	0101	2058	48,128	1	99	101	00000	9.5	10.5

3.4 System Training

The investigated neural network topologies include one hidden layer, with fully connected neurons in the adjacent layers and batch-mode training. For a given experiment with P learning samples, the error function is presented as:

$$E_P = \frac{1}{2} \sum_{p=1}^{P} \sum_{i=1}^{L} (x_i^p - t_i^p)^2, \qquad (1)$$

where for each sample $p = 1, ..., P$ and each neuron of the output layer $i = 1, ..., L$, a pair (x_i, t_i) of NN output and target values, respectively, is defined.

4 Results and Discussion

A number of experiments are designed, implemented, executed and evaluated to test and validate the performance of the proposed intelligent system for identification and classification of radar signals. Two separate approaches are considered and the related results are grouped and presented in the following two case studies. MATLAB® and its *Statistics*, *Neural Networks* and *Global Optimisation* toolboxes are used for coding and running of all the experiments.

4.1 Case Study 1—Listwise Deletion and Feedforward Neural Networks

For the purposes of the first case study, samples that contain incomplete data (i.e. data that was not fully intercepted or recorded) are removed from the considered data set, resulting in a subset of 7693 complete data samples of radar signal values.

Subsequently, depending on the experiment to be performed, the samples are sorted by experts in several groups of major interest according to their application. In two classes for the first two experiments ("*Civil*" and "*Military*"), and in 11 classes for the purpose of the final one (4 from the "*Civil*" and 7 from the "*Military*" application areas).

A randomly selected, no missing data sample subset (after listwise deletion) is presented in Table 10. Its first column (the ID attribute) is retained for referencing purposes only and it is not used during the classifier's training.

Next, a coding transformation (as described in Sect. 3.2) is applied to convert the categorical values in the data set to numerical ones. Taking into account the large number of categories in the inputs (Table 4), continuous and binary codings are

Table 10 Sample radar data subset with no missing values, received after listwise deletion

ID	FN	RFC	RFmin	RFmax	PRC	PRImin	PRImax	PDC	PDmin	PDmax	ST	SPmin	SPmax
983	AT	F	15,700	17,700	F	100	142.9	S	0.03	0.05	A	0.9	1.1
1286	SS	A	5500	5800	K	909.1	1111.1	S	0.6	0.8	A	1.9	2.1
4846	SS	F	172	180	F	2439	2564.1	S	1.6	1.8	G	28	32
12,097	3D	D	5250	5850	F	2703	2777.8	S	3	3.3	A	5.8	6.2
28,059	WT	F	5300	5700	F	1127	1132.5	S	0.75	0.85	C	12	60

considered for transforming the input variables. On the other hand, binary and dummy variable representations are used for transforming the output parameters.

In order to balance the impact of the different input parameters on the training algorithm, data scaling is applied. Respectively, each of the experiments conducted for the purposes of this case study is evaluated using three forms of the input data set—the data itself (with no scaling), after normalisation (i.e., scaling the attribute values to fall within a specific range, for example [0 1]), and after standardisation (i.e. scaling the attribute values to a zero mean and unit variance). A sample binary encoded and normalised data subset is given in Table 11.

The investigated NN topologies include one hidden layer, with fully connected neurons in the adjacent layers and batch-mode training. For a given experiment with P learning samples, the error function is given with Eq. 1. Supervised NN learning with Levenberg-Marquardt algorithm and tangent sigmoid transfer function is used. A split-sample technique using randomly selected 70 % of the available data for training, 15 % for validation and 15 % for testing, and mean squared error (MSE) is adopted for evaluating the learning performance. The stopping criteria is set to 500 training epochs, gradient reaching less than 1.0e-06 or if 6 consequent validation checks fail, whichever occurs first.

For the purposes of the first experiment, the categorical attributes of the input data are coded with consecutive integers. In this way a total of 12 input variables are received (Table 6). Two neural network topologies are examined—12-10-1 (12 neurons in the input, 10 neurons in the hidden and 1 neuron in the output layers) and 12-10-2, where the output parameter is coded as one binary neuron taking values 0 ("*Civil*") and 1 ("*Military*") for the first topology and 2 binary neurons, taking values 10 ("*Civil*") and 01 ("*Military*") for the second topology (Fig. 4). The performance of each of the topologies is investigated, evaluated and compared after training with the original, normalised and standardised data. The results are summarised in Table 12 and Fig. 5.

The second experiment investigates two additional NN topologies: 22-22-1 and 22-22-2, where the output parameter is again coded by one binary neuron (*0* for "*Civil*" and *1* for "*Military*") for the first topology and by two binary neurons for the second one (*10* for "*Civil*" and *01* for "*Military*"). Again, the performance of each of the topologies is investigated, evaluated and compared using the original data, after normalisation and after standardisation. The results are summarised in Table 13.

Similarly to the first experiment, sample confusion matrices are presented in Fig. 6 for a 22-22-2 NN classifier trained with standardised input data. A very high accuracy of 84.3 % on the testing data set is achieved after 114 epochs and activation of the validation check stopping criteria (unsatisfactory performance on the validation data set in six successive iterations).

The final experiment in this case study investigates a broader output space of 11 classes (4 from the "*Civil*" and 7 from the "*Military*" domain) and evaluates a 22-22-11 NN classifier with unscaled, normalised and standardised training data using dummy variable coded outputs. Summary of the obtained results is presented in Table 14 and a sample confusion matrix for the investigated classifier with

Table 11 Sample radar data subset with no missing values (listwise deletion), after binary encoding and normalisation

ID	RFC enc.				RFmin	RFmax	PRC enc.				PRImin	PRImax	PDC enc.	PDmin	PDmax	ST enc.				SPmin	SPmax
983	0	0	1	1	0.228	0.249	0	0	1	1	0.0006	0.0003	0	0.000010	0.000014	0	0	0	0	0.00025	0.00030
1286	0	0	0	0	0.080	0.082	0	1	1	0	0.0056	0.0022	0	0.000303	0.000363	0	0	0	0	0.00054	0.00058
4846	0	0	1	1	0.002	0.003	0	0	1	1	0.0151	0.0051	0	0.000815	0.000828	0	0	1	1	0.00789	0.00877
12,097	0	0	0	0	0.076	0.082	0	0	1	1	0.0167	0.0055	0	0.001533	0.001526	0	0	0	0	0.00163	0.00170
28,059	0	0	1	1	0.077	0.080	0	0	1	1	0.0070	0.0023	0	0.000379	0.000386	0	0	1	0	0.00338	0.01644

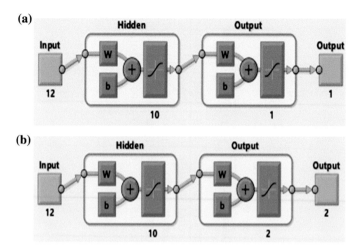

Fig. 4 Investigated NN topologies for case study 1: 12 neurons in the input layer; 10 in the hidden; and 1 (**a**), or 2 (**b**) neurons in the output layer

Table 12 Classification performance (over the testing set) for continuous input coding and 12-10-N topologies with no data scaling, after normalisation and after standardisation

NN topology	Inputs scaling	Classification accuracy (%)
12-10-1	No scaling	78.12
	Normalisation	80.82
	Standardisation	80.76
12-10-2	No scaling	80.14
	Normalisation	81.60
	Standardisation	82.18

Table 13 Classification performance (over the testing set) for binary input coding and 22-22-N topologies with no data scaling, after normalisation and after standardisation

NN topology	Inputs scaling	Classification accuracy (%)
22-22-1	No scaling	81.90
	Normalisation	83.34
	Standardisation	83.01
22-22-2	No scaling	81.77
	Normalisation	83.90
	Standardisation	84.30

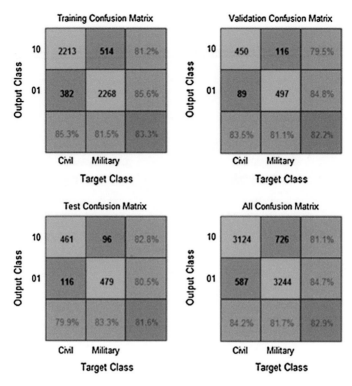

Fig. 5 Classification results for 12-10-2 NN classifier with normalised input data and a validation stop after 118 epochs. The values in *green* specify the correctly classified samples for each class (10—Civil, 01—Military)

standardised input training data is given in Fig. 7, where a good recognition rate of 67.49 % can be observed.

Although a straightforward comparison with radar classification studies reported by other authors might be misleading, due to the different data sets, model parameters and training methods used, the achieved results appeared to be strongly competitive when compared to the ones reported in [30, 32, 48, 49, 60]. Furthermore, additional improvement is expected, if further statistical pre-processing techniques, missing data handling routines, NN topologies or training algorithm parameters are investigated (as shown in the next two case studies).

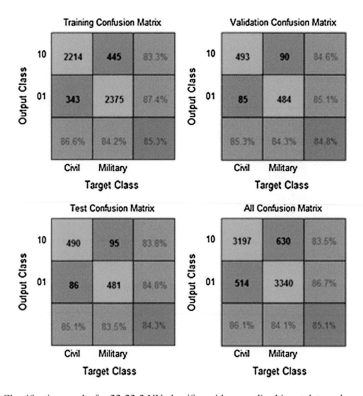

Fig. 6 Classification results for 22-22-2 NN classifier with normalised input data and a validation stop after 114 epochs. The values in *green* specify the correctly classified samples for each class (*10—"Civil"*, *01—"Military"*)

Table 14 Classification performance (over the testing set) for binary input coding and 22-22-11 topology with no data scaling, after normalisation and after standardisation

NN topology	Inputs scaling	Classification accuracy (%)
22-22-11	No scaling	61.94
	Normalisation	66.70
	Standardisation	67.49

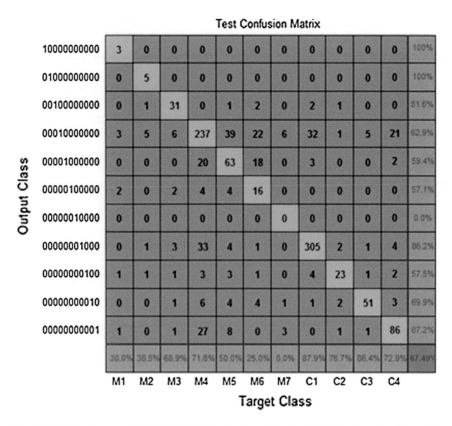

Fig. 7 Classification results for 22-22-11 NN classifier with standardised data on 7 military (*M1* —"*Multi-function*", *M2*—"*Battlefield*", *M3*—"*Aircraft*", *M4*—"*Search*", *M5*—"*Air Defense*", *M6*—"*Weapon*" and *M7*—"*Information*") and 4 civil classes (*C1*—"*Maritime*", *C2*—"*Airborne Navigation*", *C3*—"*Meteorological*" and *C4*—"*Air Traffic Control*")

4.2 Case Study 2—Multiple Imputation and Feedforward Neural Networks

The second case study follows the same sequence of experiments and NN topologies, as introduced in the first study, however, this time an extended dataset, received after multiple imputation of the missing data values (as described in Sect. 3) is used.

For the purposes of the first experiment in this study, the categorical attributes of the input data are coded with consecutive integers. Two NN topologies are examined—12-10-1 and 12-10-2, where the output parameter is coded as one binary neuron taking values 0 ("*Civil*") and 1 ("*Military*") for the first topology and 2 neurons, taking binary values 10 ("*Civil*") and 01 ("*Military*") for the second one.

The performance of each of the topologies is investigated, evaluated and compared using training with the original data (no pre-processing), and after normalisation and standardisation. The results are summarised in Table 15 showing up to 5 % accuracy improvement for the case introducing imputation.

Sample confusion matrices for a 12-10-2 NN classifier trained with normalised input data and a validation stop activated after 106 epochs are given in Fig. 8. They demonstrate improved accuracy rates (especially for the *"Military"* class) when compared to the case studies using listwise deletion to cope with the incomplete data samples (Fig. 5).

The second experiment in this study investigates two additional NN topologies —22-22-1 and 22-22-2, where the output is again coded by one binary neuron (*0* for *"Civil"* and *1* for *"Military"*) for the first topology and by two binary neurons for the second one (*10* for *"Civil"* and *01* for *"Military"*).

The NN performance for each of the topologies is investigated, evaluated and compared using the original, normalised and standardised data for both the cases— with and without imputed values. The performance results are summarised in Table 16, again showing improved NN performances for the cases with imputed data.

The final experiment investigates a broader output space of 11 classes (4 *"Civil"* and 7 *"Military"*) and evaluates 22-22-11 NN classifiers with the original, normalised and standardised training data, and with dummy variable coded outputs. Summary of the obtained results when training on data subsets with and without imputation is presented in Table 17.

Sample confusion matrices for the imputed 22-22-11 NN case, trained with standardised input data and a validation stop activated after 98 epochs are presented in Fig. 9. Although the results seem slightly inferior to the listwise deletion case (Fig. 7), they give higher statistical confidence because of the increased number of samples.

It can also be seen from Fig. 9 that although the accuracy of the NN classifier is relatively the same (compared to the NN trained after listwise deletion (Fig. 7)), the

Table 15 Classification performance (over the testing set) for continuous input coding and 12-10-N topologies with no data scaling, after normalisation and after standardisation

Topology	Input data	% Accuracy	
		No imputation	With imputation
12-10-1	No scaling	78.1	83.3
	Normalised	80.8	84.5
	Standardised	80.8	85.2
12-10-2	No scaling	80.1	82.1
	Normalised	81.6	83.6
	Standardised	82.1	84.5

Comparison between NN training with data received after listwise deletion and after multiple imputation

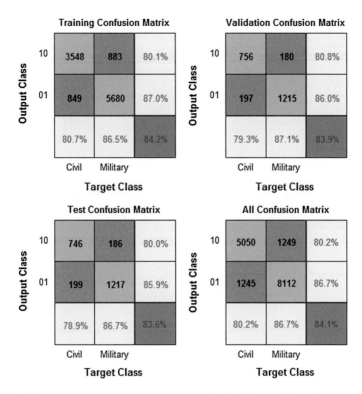

Fig. 8 Classification results for imputed data case for 12-10-2 NN classifier with normalised input data and a validation stop after 106 epochs. The values in green specify the correctly classified samples for each class (*10—"Civil"*, *01—"Military"*)

number of hits is largely increased and with a better distribution. This is especially evident for the *'M7'* class, for which there were no hits in the case without imputation. The best accuracy is again achieved for the *'M4'* and *'C1'* classes, but the more important achievement as a result of the imputation is the uniform

Table 16 Classification performance (over the testing set) for binary input coding and 22-22-N topologies with no data scaling, after normalisation and after standardisation

Topology	Input data	% Accuracy	
		No imputation	With imputation
22-22-1	No scaling	81.9	85.6
	Normalised	83.3	87.3
	Standardised	83.1	87.2
22-22-2	No scaling	81.8	84.8
	Normalised	83.9	85.0
	Standardised	84.3	86.8

Comparison between NN training with data received after listwise deletion and after multiple imputation

Table 17 Classification performance (over the testing set) for binary input coding and 22-22-11 topology with no data scaling, after normalisation and after standardisation

Topology	Input data	% Accuracy	
		No imputation	With imputation
22-22-11	No scaling	61.9	66.1
	Normalised	66.7	66.4
	Standardised	67.5	66.7

Comparison between NN training with data received after listwise deletion and after multiple imputation

distribution of correctly classified samples. As illustrated in Fig. 7, the class accuracy variance for the classification with no missing data is very high, from 0 to 87.9 %, whereas in the case using imputed data (Fig. 9), it is between 22.6 and 87.4 %. In other words, while keeping the best accuracy almost the same, the

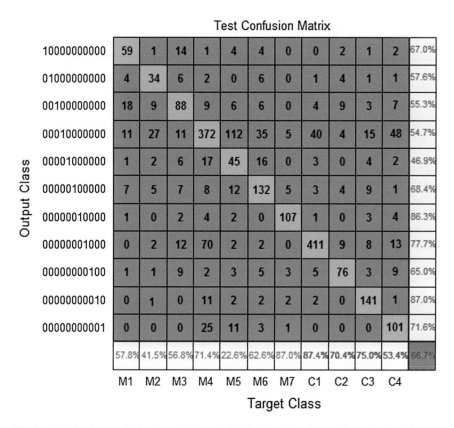

Fig. 9 Classification results for inputed data and 22-22-11 NN classifier with standardised data on 7 military (*M1*—"*Multi-function*", *M2*—"*Battlefield*", *M3*—"*Aircraft*", *M4*—"*Search*", *M5*—"*Air Defense*", *M6*—"*Weapon*" and *M7*—"*Information*") and 4 civil classes (*C1*—"*Maritime*", *C2*—"*Airborne Navigation*", *C3*—"*Meteorological*" and *C4*—"*Air Traffic Control*")

minimum accuracy is improved by more than 22 %. This should be attributed to the greater number of available training and testing samples as a result of the imputation, which increases the statistical power of the dataset and subsequently improves the classification performance of the NN.

5 Conclusion

Reliable and real-time identification of radar signals is of crucial importance for timely threat detection, threat avoidance, general situation awareness and timely deployment of counter-measures. In this context, this chapter investigates the potential application of NN-based approaches for timely and trustworthy identification of radar types, associated with intercepted pulse trains.

A number of experiments are designed, implemented, executed and evaluated for testing and validating the performance of the proposed intelligent systems for solving the investigated classification tasks. The different experiments study a variety of NN topologies, data transformation techniques and missing data handling approaches.

The simulations are divided in two broad case studies, each of which conducts several sub-experiments. In the first one, all the signals are pre-classified by experts into between 2 and 11 classes, depending on the experiment, and then a listwise deletion is used to clean the data from incomplete samples. As a result, very competitive classification accuracy of about 81, 84 and 67 % is received for the different recognition tasks.

In the second one, a study applying a multiple imputation model-based approach for dealing with the large number of missing data (contained in the available radar signals data set) is investigated. The experiments conducted for the purposes of the first case study are repeated, but this time using the imputed data set for training of the classifiers. An improved accuracy of up to 87.3 % is achieved. The results are compared and critically analysed, showing overall improved accuracy when the NN are trained on the larger subset with imputed values.

Although a straightforward comparison to radar classification studies, reported by other authors might be misleading, due to the different data sets, model parameters, data transformations, training and optimisation methods used, the achieved results are strongly competitive to the ones reported in [30, 42, 48, 49, 52, 60].

Potential areas for further extension of this research include investigation of additional statistical transformation techniques, such as Principal Component Analysis (PCA), Non-Linear Principal Component Analysis (NLPCA), and Linear Discriminant Analysis, for decreasing the dimensionality of the problem and increasing the separability between the classes. In terms of classifiers, we presented supervised learning and classification, but unsupervised learning techniques (such as self-organising maps (SOM)) can also be considered, as well as varying other training parameters and exploring additional NN topologies. Finally, additional classes can be introduced, in order to achieve more specific classification of the intercepted radar data.

References

1. Richards, M.A.: Fundamentals of Radar Signal Processing. Tata McGraw-Hill Education, New Delhi (2005)
2. Dickmann, J., Appenrodt N., Brenk, C.: Making Bertha see. IEEE Spectrum, 41–45 (2014)
3. Mure-Dubois, J., Vincent, F., Bonacci, D.: Sonar and radar SAR processing for parking lot detection. In: Proceedings of the IEEE International Radar Symposium (IRS), pp. 471–476 (2011)
4. Skolnik, M.I.: Introduction to Radar Systems. McGraw Hill, Boston (2001)
5. Lazarus, M.: Radar everywhere. IEEE Spectr. 52(2), 52–59 (2015)
6. Cumming, I.G., Wong, F.H.: Digital Processing of Synthetic Aperture Radar Data: Algorithms and Implementation. Artech House, Norwood (2005)
7. Krieger, G., Younis, M., Gebert, N., Huber, S., Bordoni, F., Patyuchenko, A., Moreira, A.: Advanced concepts for high-resolution wide-swath SAR imaging. In: 8th European Conference on Synthetic Aperture Radar (EUSAR), pp. 1–4. VDE (2010)
8. Zhai, S., Jiang, T.: A new sense-through-foliage target recognition method based on hybrid differential evolution and self-adaptive particle swarm optimization-based support vector machine. Neurocomput. 149(1), 573–584 (2015)
9. Brunner, D., Lemoine, G., Bruzzone, L.: Earthquake damage assessment of buildings using VHR optical and SAR imagery. IEEE Trans. Geosci. Remote Sens. 48(5), 2403–2420 (2010)
10. Arcone, S.A., Spikes, V.B., Hamilton, G.S.: Stratigraphic variation within polar firn caused by differential accumulation and ice flow: interpretation of a 400 MHz short-pulse radar profile from West Antarctica. J. Glaciol. 51(174), 407–422 (2005)
11. Storvold, R., Malnes, E., Larsen, Y., Høgda, K., Hamran, S., Mueller, K., Langley, K.: SAR remote sensing of snow parameters in Norwegian areas—Current status and future perspective. J. Electromagn. Waves Appl. 20(13), 1751–1759 (2006)
12. Evans, D.L., Alpers, W., Cazenave, A., Elachi, C., Farr, T., Glackin, D., Holt, B., Jones, L., Liu, W.T., McCandless, W.: Seasat—a 25-year legacy of success. Remote Sens. Environ. 94(3), 384–404 (2005)
13. Secades, C., O'Connor, B., Brown, C., Walpole, M.: Earth observation for biodiversity monitoring: a review of current approaches and future opportunities for tracking progress towards the Aichi Biodiversity Targets, CBD Technical Series 72, (2014)
14. Persico, R.: Introduction to Ground Penetrating Radar: Inverse Scattering and Data Processing. Wiley, New York (2014)
15. Goodman, D., Piro, S.: GPR Remote Sensing in Archaeology. Springer, New York (2013)
16. Sambuelli, L., Bohm, G., Capizzi, P., Cardarelli, E., Cosentino, P.: Comparison between GPR measurements and ultrasonic tomography with different inversion algorithms: an application to the base of an ancient Egyptian sculpture. J. Geophys. Eng. 8(3), 106–116 (2011)
17. Francke, J.: Applications of GPR in mineral resource evaluations. In: Proceedings of the 13th IEEE International Conference on Ground Penetrating Radar (GPR), pp. 1–5 (2010)
18. Lambot, S., Slob, E.C., van den Bosch, I., Stockbroeckx, B., Vanclooster, M.: Modeling of ground-penetrating radar for accurate characterization of subsurface electric properties. IEEE Trans. Geosci. Remote Sens. 42(11), 2555–2568 (2004)
19. Soldovieri, F., Prisco, G., Persico, R.: A strategy for the determination of the dielectric permittivity of a lossy soil exploiting GPR surface measurements and a cooperative target. J. Appl. Geophys. 67(4), 288–295 (2009)

20. Peabody, J.E., Charvat, G.L., Goodwin, J., Tobias, M.: Through-wall imaging radar. Lincoln Lab. J. **19**(1), 62–72 (2012)
21. Pettinelli, E., Di Matteo, A., Mattei, E., Crocco, L., Soldovieri, F., Redman, J.D., Annan, A.P.: GPR response from buried pipes: Measurement on field site and tomographic reconstructions. IEEE Trans. Geosci. Remote Sens. **47**(8), 2639–2645 (2009)
22. Slob, E., Sato, M., Olhoeft, G.: Surface and borehole ground-penetrating-radar developments. Geophysics **75**(5), 75A103–175A120 (2010)
23. IEEE Standard Letter Designations for Radar-Frequency Bands. Standard 521–2002 (2003). doi:10.1109/IEEESTD.2003.94224
24. Capraro, G.T., Farina, A., Griffiths, H., Wicks, M.C.: Knowledge-based radar signal and data processing: a tutorial review. IEEE Sig. Process. Mag. **23**(1), 18–29 (2006)
25. Galejs, R.J.: Volume surveillance radar frequency selection. In: Proceedings of the IEEE International Radar Conference, pp. 187–192 (2000)
26. Johnsen, T., Olsen, K.E.: Bi-and multistatic radar. DTIC Document (2006)
27. Schleher, D.C.: Electronic warfare in the information age. Artech House, Boston, London (1999)
28. Sciortino, J.C.: Autonomous ESM systems. Naval Eng. J. **109**(6), 73–84 (1997)
29. Wang, Z., Zhang, D., Bi, D., Wang, S.: Multiple-parameter radar signal sorting using support vector clustering and similitude entropy index. Circ. Syst. Sig. Process. **33**(6), 1985–1996 (2014)
30. Granger, E., Rubin, M.A., Grossberg, S., Lavoie, P.: A what-and-where fusion neural network for recognition and tracking of multiple radar emitters. Neural Netw. **14**(3), 325–344 (2001)
31. D'Agostino, S., Foglia, G., Pistoia, D.: Specific emitter identification: Analysis on real radar signal data. In: Proceedings of the IEEE European Radar Conference (EuRAD), pp. 242–245 (2009)
32. Feng, B., Lin, Y.: Radar signal recognition based on manifold learning method. Int. J. Control Autom. **7**(12), 399–440 (2014)
33. Li, J., Ying, Y.: Radar signal recognition algorithm based on entropy theory. In: Proceedings of the 2nd IEEE International Conference on Systems and Informatics (ICSAI 2014), pp. 718–723 (2014)
34. Zhang, G., Hu, L., Jin, W.: A novel approach for radar emitter signal recognition. In: Proceedings of the IEEE Asia-Pacific Conference on Circuits and Systems, pp. 817–820 (2004)
35. Lunden, J., Koivunen, V.: Automatic radar waveform recognition. IEEE J. Sel. Top. Sig. Process. **1**(1), 124–136 (2007)
36. Hassan, H.: A new algorithm for radar emitter recognition. In: Proceedings of the 3rd IEEE International Symposium on Image and Signal Processing and Analysis (ISPA), pp. 1097–1101 (2003)
37. Liu, H.-J., Liu, Z., Jiang, W.-L., Zhou, Y.-Y.: Approach based on combination of vector neural networks for emitter identification. IET Sig. Proc. **4**(2), 137–148 (2010)
38. Ting, C., Wei, G., Bing, S.: A new radar emitter recognition method based on pulse sample figure. In: Proceedings of the 8th IEEE International Conference on Fuzzy Systems and Knowledge Discovery (FSKD), pp. 1902–1905 (2011)
39. Wang, L., Ji, H., Shi, Y.: Feature extraction and optimization of representative-slice in ambiguity function for moving radar emitter recognition. In: Proceedings of the IEEE International Conference on Acoustics Speech and Signal Processing (ICASSP), pp. 2246–2249 (2010)
40. Zeng, Y., Li, O.: A new algorithm for signal emitter recognition. In: Proceedings of the IEEE International Conference on Image Analysis and Signal Processing (IASP), pp. 446–449 (2010)
41. Pang, J., Lin, Y., Xu, X.: The improved radial source recognition algorithm based on fractal theory and neural network theory. Int. J. Hybrid Inf. Technol. **7**(2), 397–402 (2014)

42. Li, J., Ying, Y.: Radar signal recognition algorithm based on entropy theory. In: Proceedings of the 2nd IEEE International Conference on Systems and Informatics (ICSAI), pp. 718–723 (2014)
43. Ripley, B.D.: Pattern Recognition and Neural Networks (2008)
44. Kamgar-Parsi, B., Kamgar-Parsi, B., Sciortino Jr., J.C.: Automatic Data Sorting Using Neural Network Techniques. DTIC Document (1996)
45. Pape, D.R., Anderson, J.A., Carter, J.A., Wasilousky, P.A.: Advanced signal waveform classifier. In: Proceedings of the 8th Optical Technology for Microwave Applications, pp. 162–169 (1997)
46. Ince, T.: Polarimetric SAR image classification using a radial basis function neural network. PIERS **41**(4), 636–646 (2010)
47. Vicen-Bueno, R., Carrasco-Álvarez, R., Rosa-Zurera, M., Nieto-Borge, J.C., Jarabo-Amores, M.-P.: Artificial neural network-based clutter reduction systems for ship size estimation in maritime radars. EURASIP J. Adv. Sig. Process. **9**(1), 1–15 (2010)
48. Anjaneyulu, L., Sarma, N., Murthy, N.: Identification of LPI radar signals by higher order spectra and neural network techniques. Int. J. Inf. Commun. Technol. **2**(1), 142–155 (2009)
49. Ibrahim, N., Abdullah, R.R., Saripan, M.: Artificial neural network approach in radar target classification. J. Comput. Sci. **5**(1), 23 (2009)
50. Yin, Z., Yang, W., Yang, Z., Zuo, L., Gao, H.: A study on radar emitter recognition based on SPDS neural network. Inf. Technol. J. **10**(4), 883–888 (2011)
51. Zhang, Z.-C., Guan, X., He, Y.: Study on radar emitter recognition signal based on rough sets and RBF neural network. In: Proceedings of the IEEE International Conference on Machine Learning and Cybernetics, pp. 1225–1230 (2009)
52. Peipei, D., Hui, L.: The radar target recognition research based on improved neural network algorithm. In: Proceedings of the Fifth International Conference on Intelligent Systems Design and Engineering Applications (ISDEA), pp. 1074–1077 (2014)
53. Chen, S., Cowan, C.F., Grant, P.M.: Orthogonal least squares learning algorithm for radial basis function networks. IEEE Trans. Neural Netw. **2**(2), 302–309 (1991)
54. Shieh, C.-S., Lin, C.-T.: A vector neural network for emitter identification. IEEE Trans. Antennas Propag. **50**(8), 1120–1127 (2002)
55. Lin, C.-M., Chen, Y.-M., Hsueh, C.-S.: A self-organizing interval type-2 fuzzy neural network for radar emitter identification. Int. J. Fuzzy Syst. **16**(1), 20 (2014)
56. Liu, J., Lee, J.P., Li, L., Luo, Z.-Q., Wong, K.M.: Online clustering algorithms for radar emitter classification. IEEE Trans. Pattern Anal. Mach. Intell. **27**(8), 1185–1196 (2005)
57. Li, L., Ji, H., Wang, L.: Specific radar emitter recognition based on wavelet packet transform and probabilistic SVM. In: Proceedings of the IEEE International Conference on Information and Automation (ICIA), pp. 1308–1313 (2009)
58. Ting, C., Jingqing, L., Bing, S.: Research on rough set-neural network and its application in radar signal recognition. In: Proceedings of the 8th International Conference on Electronic Measurement and Instruments (ICEMI), pp. 764–768 (2007)
59. Azimi-Sadjadi, M.R., Yao, D., Huang, Q., Dobeck, G.J.: Underwater target classification using wavelet packets and neural networks. IEEE Trans. Neural Netw. **11**(3), 784–794 (2000)
60. Lee, S.-J., Choi, I.-S., Cho, B., Rothwell, E.J., Temme, A.K.: Performance enhancement of target recognition using feature vector fusion of monostatic and bistatic radar. Prog. Electromagnet. Res. **144**, 291–302 (2014)
61. Baraldi, A.N., Enders, C.K.: An introduction to modern missing data analyses. J. Sch. Psychol. **48**(1), 5–37 (2010)
62. Enders, C.K.: Applied Missing Data Analysis. Guilford Press, New York (2010)
63. Graham, J.W.: Missing data analysis: making it work in the real world. Annu. Rev. Psychol. **60**(1), 549–576 (2009)
64. Horton, N.J., Lipsitz, S.R.: Multiple imputation in practice: comparison of software packages for regression models with missing variables. Am. Stat. **55**(3), 244–254 (2001)
65. Little, R.J., Rubin, D.B.: Statistical Analysis with Missing Data. Wiley, New York (2002)
66. Osborne, J.W., Overbay, A.: Best practices in data cleaning. Sage, Thousand Oaks (2012)

67. Rahm, E., Do, H.H.: Data cleaning: Problems and current approaches. IEEE Data Eng. Bull. **23**(4), 3–13 (2000)
68. Verboven, S., Branden, K.V., Goos, P.: Sequential imputation for missing values. Comput. Biol. Chem. **31**(5), 320–327 (2007)
69. Morring Jr., F., Perrett, B.: L-band SAR Satellite May Help JAXA's New Military Job, Aviation Week & Space Technology (2014)
70. O'Reilly, D., Bowring, N., Harmer, S.: Signal processing techniques for concealed weapon detection by use of neural networks. In: IEEE 27th Convention of Electrical & Electronics Engineers in Israel (IEEEI), pp. 1–4 (2012)
71. Lee, J.-H., Choi, I.-S., Kim, H.-T.: Natural frequency-based neural network approach to radar target recognition. IEEE Trans. Sig. Process. **51**(12), 3191–3197 (2003)
72. Khairnar, D., Merchant, S., Desai, U.: Radar signal detection in non-gaussian noise using RBF neural network. J. Comput. **3**(1), 32–39 (2008)
73. Briones, J., Flores, B., Cruz-Cano, R.: Multi-mode radar target detection and recognition using neural networks. Int. J. Adv. Rob. Syst. **9**(177), 1–9 (2012)

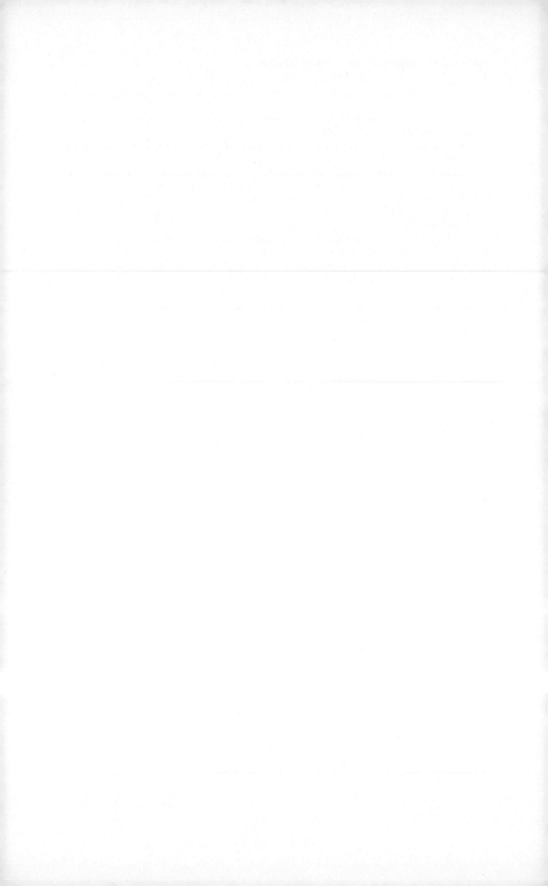

Part II
Cyber Security and Intrusion Detection Systems

An Improved Decision System for URL Accesses Based on a Rough Feature Selection Technique

P. de las Cuevas, Z. Chelly, A.M. Mora, J.J. Merelo
and A.I. Esparcia-Alcázar

Abstract Corporate security is usually one of the matters in which companies invest more resources, since the loss of information directly translates into monetary losses. Security issues might have an origin in external attacks or internal security failures, but an important part of the security breaches is related to the lack of awareness that the employees have with regard to the use of the Web. In this work we have focused on the latter problem, describing the improvements to a system able to detect anomalous and potentially insecure situations that could be dangerous for a company. This system was initially conceived as a better alternative to what are known as black/white lists. These lists contain URLs whose access is banned or dangerous (black list), or URLs to which the access is permitted or allowed (white list). In this chapter, we propose a system that can initially learn from existing black/white lists and then classify a new, unknown, URL request either as "should be allowed" or "should be denied". This system is described, as well as its results and the improvements made by means of an initial data pre-processing step based on applying Rough Set Theory for feature selection. We prove that high accuracies can be obtained even without including a pre-processing step, reaching between 96 and 97 % of correctly classified patterns. Furthermore, we also prove that including the use of Computa-

P. de las Cuevas (✉) · A.M. Mora · J.J. Merelo
Department of Computer Architecture and Computer Technology,
University of Granada, Granada, Spain
e-mail: paloma@geneura.ugr.es

A.M. Mora
e-mail: amorag@geneura.ugr.es

J.J. Merelo
e-mail: jmerelo@geneura.ugr.es

Z. Chelly
Laboratoire de Recherche Opérationelle de Décision Et de Contrôle de Processus,
Institut Supérieur de Gestion, Tunis, Tunisia
e-mail: zeinebchelly@yahoo.fr

A.I. Esparcia-Alcázar
University of Valencia,Valencia, Spain
e-mail: aesparcia@s2grupo.es

© Springer International Publishing Switzerland 2016
R. Abielmona et al. (eds.), *Recent Advances in Computational Intelligence
in Defense and Security*, Studies in Computational Intelligence 621,
DOI 10.1007/978-3-319-26450-9_6

tional Intelligence techniques for pre-processing the data enhances the system performance, in terms of running time, while the accuracies remain close to 97 %. Indeed, among the obtained results, we demonstrate that it is possible to obtain interesting rules which are not based only on the URL string feature, for classifying new unknown URLs access requests as allowed or as denied.

Keywords Computational intelligence · Rough sets · Feature selection · Corporate security policies · Internet access control · Data mining · Blacklists and whitelists

1 Introduction

Security is an inclusive term that refers to a diversity of steps taken by individuals, and companies, in order to protect computers or computer networks that are connected to the Internet. The Internet was initially conceived as an open network facilitating the free exchange of information. However, data which is sent/received over the Internet travel through a dynamic chain of computers and network links and, as a consequence, the risk of intercepting and changing the data is high. In fact, it would be virtually impossible to secure every computer connected to the Internet around the world. So, there will likely always be weak links in the chain of data exchange [7]. Yet, companies have to find out a way for their employees to safely interact with customers, clients, and anyone who uses the Internet while protecting internal confidential information. Companies have, also, to alert the employees from the Internet misuse while doing their job.

Most of the time, employees have a misguided sense of security and believe that it is an IT problem, a purely technical issue, and they naively believe that an incident may never happen to them [36]. Actually, the employees' web misuse is one of the main causes of security breaches [3], so that making them security-conscious has become a security challenge.

The reality is that every department must be involved in readiness planning and establishing security policies and procedures to minimize their risks. Such strategies are mainly handled by means of *Corporate Security Policies* (CSPs) which basically are a set of security rules aiming at protecting company assets by defining permissions to be considered for every different action to be performed inside the security system [19].

The basic idea behind these CSPs is usually to include rules to either allow or deny employees' access to non-confident or non-certified websites, which are referenced by their URLs in this chapter. Moreover, several web pages might be also controlled for productivity or suitability reasons, given the fact that the employees who connect to these might have working purposes or not. In fact, some of the CSPs usually define sets of allowed or denied web pages or websites that could be accessed by the company employees. These sets are usually included in two main lists; a white list (referring to "permitted") and a black list (referring to "non-permitted"). Both lists, the white and the black, act as a good and useful control tools for those URLs included in them, as well as for the complementary. For instance, the URLs which are not included in a white list will automatically have a denial of access [24].

The aim of this paper is going beyond this traditional and simple decision making process. By using black and/or white lists, we either allow or deny users' requests/connection based, only, on the URLs provided in the lists. Yet, updating these lists is a never ending task, as numerous malicious websites appear every day. For instance, Netcraft reports from November of 2014 [30] showed that there are about 950 million active websites. But McAfee reported [27] that, at the end of the first quarter of 2014, there were more than 18 million new suspect URLs (2 million associated domains), and also more than 250 thousand new phishing URLs (almost 150 thousand associated domains).

With this situation in mind, in this chapter, our aim is to define a tool for automatically making allow or deny decisions with respect to URLs that are not included in the aforementioned lists. This decision would be based on that made for similar URL accesses (those with similar features), but instead of using only the URL strings included in the lists, we will consider other parameters of the request/connection.

For this reason, the problem has been mapped to a *classification* problem in which we start from a set of unlabelled patterns that model the connection properties from a huge amount of *actual*[1] URL accesses, known as sessions. After that, we assign a label to many of them, considering a set of *actual*[2] security rules (CSPs) defined by the Chief Security Officer (CSO) in the company. This was the approach followed in [28], and which we extend in this chapter.

In order to extract conclusions from the resulting studied dataset and to properly apply a classification algorithm, a pre-processing step is needed. In fact, to obtain an accurately trained classifier, there is a need to extract as much information as possible from the connections that the employees normally make throughout the workday. This translates into high computational requirements, which is why we introduce in this paper techniques for data reduction. More precisely, we aim to apply a feature selection technique to extract the most important features from the data at hand. Among the well known feature selection techniques proposed in literature, we propose to use a Computational Intelligence method: the Rough Set Theory (RST) [31]. RST has been experimentally evaluated with other leading feature selection techniques, such as Relif-F and entropy-based approaches in [18], and has been shown to outperform these in terms of resulting classification performance.

After pre-processing and based on the reduced dataset, we will apply several classification algorithms, testing them and selecting the most appropriate one for this problem. The selected classifier should be capable of dealing with our data while producing high accuracies and being lightweight in terms of running time. Moreover, as we want to further test the reliability of the results, in this work we propose different experimental setups based on different data partitions. These partitions are formed either by preserving the order of the data or by taking the patterns in a random

---

[1] Taken from a log file released to us by a Spanish company.

[2] The set of rules has been written by the same company, with respect to its employees.

way. Finally, given that the used data presents unbalance, we aim to apply balancing techniques [16] to further guarantee the fairness of our obtained results.

In this chapter, we want to improve the accuracies obtained in our previous work [28], as well as see if the new incorporated method (namely Rough Sets) for feature selection yields to better rules. This is meant to be done not only by applying RST for feature selection, but also by improving the quality of the original data set, by means of erasing information that may be redundant.

The rest of the paper is structured as follows. Next section describes the state of the art related to Data Mining (DM), Machine Learning (ML), and Computational Intelligence (CI) techniques applied to corporate security. Also, related works about URL filtering will be reviewed. Data description is detailed in Sect. 3. Then, Sect. 4 describes the basic concepts of Rough Set Theory for feature selection which we have used for data pre-processing. Section 5 gives an overview of the followed methodology, as well as the improvements done after our first results obtained in [28]. Then Sect. 6 depicts the results, and discusses the obtained rules which are different for every used classifier. Finally, conclusions and future trends are given in Sect. 7.

2   State of the Art

Our work tries to obtain a URL classification tool for enhancing the security in the client side, as at the end we want to get if a certain URL is secure or not, having as reference a set of rules (derived from a CSP) which allow or deny a set of known *HTTP* requests. For this, different techniques belonging to Data Mining (DM), Machine Learning (ML), and Computational Intelligence, have been applied. This section gives an overview in a number of solutions given to protect the user, or the company, against insecure situations.

First, as we want to add a good pre-processing phase to our system, in order to improve it, Sect. 2.1 gives an overview of the state of the art related to data analysis and pre-processing. Then, in Sect. 2.2 we try to analyse similar systems, as well as define which advantages our system provides.

2.1   Data Analysis and Pre-processing

Performing DM means analyzing the database we have [12] which in our case is a log of HTTP requests. The work discussed in [45] presents an exhaustive review of works which study database cleaning and their conclusion is that a database with good quality is decisive when trying to obtain good accuracies; a fact which was also demonstrated in [6]. To analyse the data that we have at hand, we have based our work on two main processes: data pre-processing on the URL dataset and the application of balancing techniques depending on the data.

While performing data pre-processing, we have focused first on the kind/type of data included in the HTTP requests in the log file that is used as input file. We realised that many URL strings are redundant in the dataset and thus we aimed to eliminate them which is seen a cleaning approach.

Many cleaning techniques have been proposed in literature [45] in order to guarantee the good quality of a given dataset. Most of these techniques are based on updating a database by adding or deleting instances to optimize and reduce the initial database. These policies include different operations such as deleting the outdated, redundant, or inconsistent instances; merging groups of objects to eliminate redundancy and improve reasoning power; re-describe objects to repair incoherencies; check for signs of corruption in the database and controlling any abnormalities in the database which might signal a problem. Working with a database which is not cleaned can become sluggish and without accurate data users will make uninformed decisions. In this work, we have maintained the our HTTP request dataset by focusing on a specific kind of data that should be eliminated: redundant URL strings. Section 5.3.1 explains in detail the process that we have adopted to eliminate these redundant data.

Still with the data pre-processing task, we have focused as a second step on checking the importance of the set of features presented in the HTTP requests log file. Thus we tried to select the most informative features from the initial feature set. At this point, we have introduced an extra technique, a data reduction technique, that was not included in our first work presented in [28]. Feature reduction is a main point of interest across a wide variety of fields and focusing on this step is crucial as it often presents a source of significant information loss. Many techniques were proposed in literature to achieve the task of feature reduction and they can be categorized into two main heads; techniques that transform the original meaning of the features, called the "transformation-based approaches", and the second category is a set of semantic-preserving techniques known as the "selection-based approaches".

Transformation based approaches, also called "feature extraction approaches", involve simplifying the amount of resources required to accurately describe a large set of data. Feature extraction is a general term for methods that construct combinations of variables to represent the original set of features but with new variables while still describing the data with sufficient accuracy. The transformation based techniques are employed in situations where the semantics of the original database will not be needed by any future process. In contrast to the semantics-destroying dimensionality reduction techniques, the semantics-preserving techniques, also called "feature selection techniques", attempt to retain the meaning of the original feature set. The main aim of this kind of techniques is to determine a minimal feature subset from a problem domain while retaining a suitably high accuracy in representing the original features [23]. In this work, we mainly focus on the use of a feature selection technique, instead of a feature extraction technique, as it is crucial to preserve the semantics of the features in the URL data that we dispose at hand, and among them, select the most important/informative ones which nearly preserve the same performance as the initial feature set.

Yet it is important to mention that most feature selection techniques proposed in the literature suffer from some limitations. Most of these techniques involve the user for the task of the algorithms parameterization and this is seen as a significant drawback. Some feature selectors require noise levels to be specified by the user beforehand, some simply rank features leaving the user to choose their own subset. There are those that require the user to state how many features are to be chosen, or they must supply a threshold that determines when the algorithm should terminate. All of these require the user to make a decision based on its own (possibly faulty) judgment [17]. To overcome the shortcomings of the existing methods, it would be interesting to look for a method that does not require any external or additional information to function appropriately. Rough Set Theory (RST) [31], which will be deeply explained in Sect. 4, can be used as such tool.

As previously stated and apart from applying a data pre-processing process, we aim to apply balancing techniques, depending on the distribution of patterns per class, in order to ensure the fairness of our results. This is due to the fact that using "real data"[3] may yield to highly unbalanced data sets [5]. This is our case, as the log file includes a set of URL accesses performed by humans, and indeed we obtained an unbalanced dataset. In order to deal with this kind of data there exist several methods in literature known as balancing techniques [5]. These methods can be categorized into three main groups [16]:

- *Undersampling the over-sized classes*: This category aims at reducing the considered number of patterns for the classes with the majority.
- *Oversampling the small classes*: This category aims at introducing additional (normally synthetic) patterns in the classes with the minority.
- *Modifying the cost associated to misclassifying the positive and the negative class*: This category aims at compensating the unbalance in the ratio of the two classes. For example, if the imbalance ratio is 1:10 in favour of the negative class, the penalty of misclassifying a positive example should be 10 times greater.

Techniques belonging to the first group have been applied to some works, following a random undersampling approach [14]. However, those techniques have the problem of the loss of valuable information.

Techniques belonging to the second group have been so far the most widely used, following different approaches, such as SMOTE (Synthetic Minority Oversampling Technique) which is a method proposed in [4] for creating 'artificial' samples for the minority class in order to balance the amount of them, with respect to the amount of samples in the majority class. However this technique is based on numerical computations, considering different distance measures, in order to generate useful patterns (i.e., realistic or similar to the existing ones).

The third group implies using a method in which a cost can be associated to the classifier accuracy at every step. This was done for instance in [1], where a Genetic Programming (GP) approach was used in which the fitness function was modified in order to consider a penalty when the classifier makes a false negative (an element from the minority class was classified as belonging to the majority class). However almost all the approaches deal with numerical (real, integer) data.

---

[3]Data which was gathered from the real world, and was not artificially generated.

For our purposes, we will focus on techniques of the first and second group, as we will use state-of-the-art classifiers. Details about the balancing techniques used in our work will be explained in Sect. 5.2.

2.2 Related Work and Contribution

The works that we are interested in are those which scope is related with the users' information and behaviour, and the management (and adaptation) of Information or Corporate Security Policies (ISPs).

In this line, in [13] a combined biometrics signals with ML methods in order to get a reliable user authentication in a computer system was proposed. In [20] a method was presented named *user-controllable policy learning* in which the user gives feedback to the system every time that a security policy is applied, so these policies can be refined according to that feedback to be more accurate with respect to the user's needs.

On the other hand, policies could be created for enhancing user's privacy, as proposed in [9], where a system able to infer privacy-related restrictions by means of a ML method applied in a social network environment was defined. The idea of inferring policies can be also considered after our results, given the fact that we are able to obtain new rules from the output of the classifiers, but in the scope of the company, and focused on ISPs.

In the same line, in [21, 22] a system was proposed which evolves a set of computer security policies by means of GP, taking again into account the user's feedback. Furthermore, the work presented in [37] took the same approach as the latter mentioned work, but also bringing event correlation into it. The two latter works are interesting in our case, though they are not focused on company ISPs; for instance, our case with the allowed or denied HTTP requests.

Furthermore, it is worth mentioning a tool developed in [15], taking the approach of "greylisting", and which *temporarily* rejects messages that come from senders who are not in the black list or in the white list, so that the system does not know if it is a spam message or not. And, like in our approach, it works trying to have a minimal impact on the users.

Finally, a system named MUSES (from Multiplatform Usable Endpoint Security System) [29] is being developed under the European Seventh Framework programme (FP7). This system will include event treatment on the user actions inside a company, DM techniques for applying the set of policies from the company ISP to the actions, allowing or denying them, CI techniques for enhancing the system performance, and ML techniques for improving the set of rules derived from these policies, according to user's feedback and behaviour after the system decisions [34]. The results of this work could be applied in this system, by changing the pre-processing step, due to the fact that the database is different. But overall, our conclusions can be escalated to be included in such a system.

In the next Section, we will describe the problem we aim to solve, in addition to the data from which the data sets are composed.

3 Problem and Data Description

The problem to solve is related with the application of corporate security policies in order to deal with potential URL accesses inside an enterprise. To this end a dataset of URL sessions (requests and accesses) is analysed. These data are labelled with the corresponding permission or denial for that access, following a set of rules. The rules themselves act as a mix between a black list and a white list. The problem is then transformed into a classification one, in which every new URL request will be classified, and thus, a grant or deny action will be assigned to that pattern.

The analysed data come from an `access.log` of the Squid proxy application [38], in a real Spanish company. This open source tool works as a proxy, but with the advantage of storing a cache of recent transactions so future requests may be answered without asking the origin server again [43].

Every pattern, namely an URL request, has ten associated variables. These patterns are described in Table 1 in which we have indicated the type of each variable; either if it is numeric or nominal/categorical. The table has, however, not only ten but eleven described variables. This is due to the fact that we decided to consider the 'Content Type' of the requested web page as a whole, but also its Main Content Type (MCT) separately. By adding more information through a new feature, we intended to see if more general rules could be obtained by the classifiers, given that there are less possible values for an MCT than for a whole 'Content Type'.

Table 1 Independent variables corresponding to a URL request through *HTTP*

Variable name	Description	Type	Rank
`http_reply_code`	Status of the server response	Categorical	20 values
`http_method`	Desired action to be performed	Categorical	6 values
`duration_milliseconds`	Session duration	Numerical	integer in [0,357170]
`content_type`	Media type of the entity-body sent to the recipient	Categorical	85 values
`content_type_MCT`	**Main Content Type** of the media type	Categorical	11 values
`server_or_cache_address`	IP address	Categorical	2343 values
`time`	connection hour (in the day)	Date	00:00:00 to 23:59:59
`squid_hierarchy`	It indicates how the next-hop cache was selected	Categorical	3 values
`bytes`	Number of transferred bytes during the session	Numerical	integer in [0,85135242]
`client_address`	IP address	Categorical	105 values
`URL`	Core domain of the URL, not taking TLD into account	Categorical	976 values

The URLs are parsed as detailed in Sect. 5.1

The dependent variable or class is a label which inherently assigns a decision (and so the following action) to every request. This can be: *ALLOW* if the access is permitted according to the CSPs, or can be *DENY*, if the connection is not permitted. These patterns are labelled using an 'engine' based in a set of security rules, that specify the decision to make. This process is described in Sect. 5.1.

These data were gathered along a period of two hours, from 8.30 to 10.30 am (30 min after the work started), monitoring the activity of all the employees in a medium-size Spanish company (80–100 people), obtaining 100,000 patterns. We consider this dataset as quite complete because it contains a very diverse amount of connection patterns, going from personal to professional issues. Moreover, results derived from the experiments and which are described in Sect. 6 show that this quantity of data might be big enough, but a more accurate outcome would be given with, for instance, a 24 h long log.

Later on, Sect. 5 will describe how the data coming from the proxy log is labelled due to the application of the aforementioned rules, and the result will be an initial URL dataset with 12 features. Then, at this stage and after describing the data, it seems necessary to describe the technique that we have used for the URL data pre-processing. Rough Set Theory for feature selection is depicted in the next Section.

4 Rough Set Based Approach for Feature Selection

As previously mentioned, it is important to perform data pre-processing on the initial URL dataset. To do so, it seems necessary to think about a technique that can, on the one hand, reduce data dimensionality using information contained within the dataset and, on the other hand, be capable of preserving the meaning of the features. Rough Set Theory (RST) [31] can be used as such a tool to discover data dependencies and to reduce the number of attributes contained in the URL dataset using the data alone, requiring no additional information [17]. In this Section, the basic concepts of RST for feature selection are highlighted.

4.1 Preliminaries of Rough Set Theory

Data are represented as a table where each row represents an object and where each column represents an attribute that can be measured for each object. Such table is called an "Information System" (IS). Formally, an IS can be defined as a pair $IS = (U, A)$ where $U = \{x_1, x_2, \ldots, x_n\}$ is a non-empty, finite set of objects called the *universe* and $A = \{a_1, a_2, \ldots, a_k\}$ is a non-empty, finite set of attributes. Each attribute or feature $a \in A$ is associated with a set V_a of its value, called the *domain* of a. We may partition the attribute set A into two subsets C and D, called *condition* and *decision* attributes, respectively [31].

Let $P \subset A$ be a subset of attributes. The indiscernibility relation, denoted by $IND(P)$, is an equivalence relation defined as: $IND(P) = \{(x, y) \in U \times U : \forall a \in P, a(x) = a(y)\}$, where $a(x)$ denotes the value of feature a of object x. If $(x, y) \in IND(P)$, x and y are said to be *indiscernible* with respect to P. The family of all equivalence classes of $IND(P)$ (Partition of U determined by P) is denoted by $U/IND(P)$. Each element in $U/IND(P)$ is a set of indiscernible objects with respect to P. Equivalence classes $U/IND(C)$ and $U/IND(D)$ are called *condition* and *decision* classes.

For any concept $X \subseteq U$ and attribute subset $R \subseteq A$, X could be approximated by the R-*lower* approximation and R-*upper* approximation using the knowledge of R. The lower approximation of X is the set of objects of U that are surely in X, defined as: $\underline{R}(X) = \bigcup \{E \in U/IND(R) : E \subseteq X\}$. The upper approximation of X is the set of objects of U that are possibly in X, defined as: $\overline{R}(X) = \bigcup \{E \in U/IND(R) : E \cap X \neq \emptyset\}$. The boundary region is defined as:

$$BND_R(X) = \overline{R}(X) - \underline{R}(X)$$

If the boundary region is empty, that is, $\overline{R}(X) = \underline{R}(X)$, concept X is said to be R-*definable*. Otherwise X is a rough set with respect to R.

The positive region of decision classes $U/IND(D)$ with respect to condition attributes C is denoted by $POS_c(D)$ where:

$$POS_c(D) = \bigcup \overline{R}(X)$$

The positive region $POS_c(D)$ is a set of objects of U that can be classified with certainty to classes $U/IND(D)$ employing attributes of C. In other words, the positive region $POS_c(D)$ indicates the union of all the equivalence classes defined by $IND(P)$ that each for sure can induce the decision class D.

4.2 Reduction Process

The aim of feature selection is to remove unnecessary features to the target concept. It is the process of finding a smaller set of attributes, than the original one, with the same or close classification power as the original set. Unnecessary features, in an information system, can be classified into irrelevant features that do not affect the target concept in any way and redundant (superfluous) features that do not add anything new to the target concept.

RST for feature selection is based on the concept of discovering dependencies between attributes. Intuitively, a set of attributes Q depends totally on a set of attributes P, denoted $P \rightarrow Q$, if all attribute values from Q can be uniquely determined by values of attributes from P. In particular, if there exists a functional dependency between values of Q and P, then Q depends totally on P. Dependency can be defined in the following way: For $P, Q \subset A$, Q depends on P in a degree k ($0 \leq k \leq 1$), denoted $P \rightarrow_k Q$, if $k = \gamma_P(Q) = |POS_P(Q)|/|U|$; If k = 1 Q depends totally on P, if $k < 1$ Q depends partially (in a degree k) on P, and if k = 0 Q does not depend on P.

RST performs the reduction of attributes by comparing equivalence relations generated by sets of attributes. Attributes are removed so that the reduced set, termed "Reduct", provides the same quality of classification as the original. A *reduct* is defined as a subset R of the conditional attribute set C such that $\gamma_R(D) = \gamma_C(D)$. Thus, a given data set may have many attribute reduct sets. In RST, a *reduct* with minimum cardinality is searched for; in other words an attempt is made to locate a single element of the minimal reduct set. A basic way of achieving this is to generate all possible subsets and retrieve those with a maximum rough set dependency degree. However, this is an expensive solution to the problem and is only practical for very simple data sets. Most of the time, only one reduct is required as, typically, only one subset of features is used to reduce a data set, so all the calculations involved in discovering the rest are pointless. Another shortcoming of finding all possible reducts using rough sets is to inquire about which is the best reduct for the classification process. The solution to these issues is to apply a *heuristic attribute selection* method [46].

Among the most interesting heuristic methods proposed in literature, we mention the *QuickReduct* algorithm [35] presented by Algorithm 1.

Algorithm 1 The QuickReduct Algorithm

1: C: the set of all conditional features;
2: D: the set of decision features;
3: $R \leftarrow \{\}$;
4: **do**
5:     $T \leftarrow R$
6:     $\forall x \in (C-R)$;
7:     **if** $\gamma_{R \cup \{x\}}(D) > \gamma_T(D)$;
8:         $T \leftarrow R \cup \{x\}$;
9:     **end if**
10:    $R \leftarrow T$;
11: **until** $\gamma_R(D) == \gamma_C(D)$
12: **return** R

The *QuickReduct* algorithm attempts to calculate a reduct without exhaustively generating all possible subsets. It starts off with an empty set and adds in turn, one at a time, those attributes that result in the greatest increase in the rough set dependency metric. According to the QuickReduct algorithm, the dependency of each attribute is calculated and the best candidate is chosen. This process continues until the dependency of the reduct equals the consistency of the data set. For further details about how to compute a reduct using the QuickReduct algorithm, we kindly invite the reader to refer to [35].

5 Followed Methodology

Before classification techniques are applied, a data pre-processing step has been performed. First, the raw dataset is labelled according a set of *initial corporate security rules*, i.e., every pattern is assigned to a label indication if the corresponding URL request/access would be ALLOWED or DENIED considering these rules. This step is necessary in order to transform the problem into a classification one. However, in order to apply the rules they must be transformed from their initial format into another one that can be applied in our programs, a hash or map. This is described in Sect. 5.1. This Subsection also details how the patterns of the navigation data log (URL sessions) are parsed, in order to build a hash to perform the matting/labelling process.

At the end of the 'parsing' phase, the two hashes are compared in order to obtain which entries of the log should be ALLOW or DENY, known as the *labelling* step. This is similar to perform a decision process in a security system. This step results in that there are 38972 pattern belonging to class ALLOW (positive class) and 18530 of class DENY (negative class), so just a 67.78 % of the samples belong to the majority class. This represents a very important problem, since a classifier that is trained considering these proportions is supposed to classify all the samples as ALLOW, getting a theoretically quite good classification accuracy equal or greater than 68 %. However, in Sect. 6 we will see that, despite the fact that some denied patterns are classified as allow, the overall performance of the classifiers is better than expected.

It is worth to mention that there is not the same amount of patterns in the two classes. This means that the dataset is unbalanced, and therefore Sect. 5.2 describes the balancing techniques used for dealing with this situation. Finally, in Sect. 5.3 we explain the applied methods in the pre-processing phase. What we want to prove is that by adding this phase, it enhances the results of our previous work presented in [28].

Based on the generated pre-processed and balanced dataset and as a final step, a supervised classification process [25] has been conducted. For this step, Weka Data Mining Software [42] has been used, in order to select the best set of classifiers in order to deal with these data. These classifiers will be further tested in Sect. 6.

5.1 Building the Dataset

In previous sections, it was stated that the data to work with was not originally presented in the form of a dataset. Instead, 'raw' data was gathered. In order to have the data in the form of a dataset, ready to be pre-processed, as well as being adequate to act as an input for the classifiers, a parsing process must be performed.

First, in this work we have considered Drools [41] as the tool to create and manage rules in a business environment. This so called Business Rule Management System (BRMS) has been developed by the JBoss community under an Apache License and it is written in Java. Though this platform consists of many components; here

(a)

```
rule "name"
    attributes
    when
        /* Left Side of the Rule
        */
    then
        /* Right Side of the
        Rule */
end
```

(b)

```
%rules = (
    rule =>{
        field => xxx
        relation => xxx
        value => xxx
        action => [allow, deny]
    },
);
```

Fig. 1 **a** Structure of a rule in Drools Expert. **b** Resulting rule, after the parsing, in a global hash of rules

we focus on Drools Expert and the Drools Rule Language (DRL, [40]). Then, the defined rules for a certain company are included in a file with a `.drl` extension; the file that needs to be parsed to obtain the final set of rules. To obtain the needed knowledge from the rules file, it is necessary to know the format of this type of language, because it is essential for the parsing process.

In Fig. 1a, we display the typical rule syntax in DRL. Two main parts should be obtained from the parsing method that will be applied: both left and right sides of the rule, taking into account that the left side is where the company specifies the conditions required to apply the action indicated in the right side. Also, for describing the conditions, Squid syntax is used (see Sect. 3), having thus the following structure: `squid:Squid(conditions)`. Finally, from the right side of the rule, the *ALLOW* or *DENY* label to apply on the data which matches with the conditions, will be extracted. The parser that we have implemented applies two regular expressions, one for each side of the rule, and returns a hash with all the rules with the conditions and actions defined. The 'before and after' performing the parsing over the `.drl` file is presented in Fig. 1.

Then, the log file is analysed. Usually, the instances of a log file have a number of fields (which will be later referred as features/attributes of a connection pattern), in order to have a registration of the client who asks for a resource, the time of the day when the request is made, and so on. In this case, we have worked with an *access.log* (see Sect. 3) file, converted into a CSV format file so it could be parsed and transformed in another hash of data. All ten fields of the Squid log yield a hash like the one depicted in Fig. 2. Once the two hashes of data were created, they were compared in such a way that for each rule in the hash of rules, it was determined how many entries in the data log hash are covered by the rule, and so they were applied the label that appears as 'action' in the rule.

Among the tasks to be performed, is the one to extract from a whole URL the part that was more interesting for our defined purposes. It is important to point out that in a log with thousands of entries, an enormous variety of URLs can be found, since some can belong to advertisements, images, videos, or even some others does not have a domain name but are given directly by an IP address. For this reason, we have taken into account that for a domain name, many subdomains (separated by dots) could be considered, and their hierarchy grows from the right towards

Fig. 2 Hash in Perl with an
example entry. The actual
hash used for this work has a
total of 100,000 entries, with
more than a half labelled as
ALLOW or *DENY* after the
comparing process

```
%logdata = (
  entry =>{
    http_reply_code => xxx
    http_method => xxx
    duration_miliseconds => xxx
    content_type_MCT => xxx
    content_type => xxx
    server_or_cache_address => xxx
    time => xxx
    squid_hierarchy => xxx
    bytes => xxx
    url => xxx
    client_address => xxx
  },
);
```

the left. The highest level of the domain name space is the Top-Level Domain
(TLD) at the right-most part of the domain name, divided itself in country code
TLDs and generic TLDs. Then, a domain and a number of subdomains follow the
TLD (again, from right to left). In this way, the URLs in the used log are such as
http://subdomain...subdomain.domain.TLD/ other_subdirectories. However, for the
ARFF[4] file to be created, only the domain (without the subdomains and the TLD)
should be considered, because there are too many different URLs to take into consideration. Hence, applying another regular expression, the data parser obtains all
the core domains of the URLs, which makes 976 domains in total.

5.2   Balancing the Dataset

While analysing the data, we observed that more than half of the initial amount of
patterns are labelled, and that the ratio is 2:1 in allows to denies. The 2:1 ratio means
that the data is unbalanced, and therefore we have performed different approaches
from the first and second groups of data balancing techniques, which were introduced
in Sect. 2.1:

- Undersampling: we will randomly remove samples of the majority class until the
  amount in both classes are similar. In other words, we will reduce the amount of
  'denied' patterns by a half.
- Oversampling: we will introduce more samples in the minority class, in order
  to get a closer number of patterns in both classes. This has to be done due to
  the impossibility of creating synthetic data when dealing with categorical values,
  given that there is not a proper distance measure between two values in a category.
  Actually, since the number of samples in the majority class is almost twice the
  minority one, we have just duplicated all of those belonging to the minority class.

---

[4]Format of Weka files.

5.3 Pre-processing the Data

Section 2.1 explains that having a good quality database is crucial when good accuracy values are required. For this reason, we have maintained the dataset by performing a removal of patterns in the log which we found as redundant and applied a feature selection technique, based on Rough Sets.

5.3.1 Erasing Redundant Information

The first thing to do is studying the data in order to look for the patterns that are repeated. Hence, after having analysed the log of connecting patterns, we studied the field *squid_hierarchy* and saw that had two possible values: DIRECT or DEFAULT_PARENT. The Squid FAQ reference [39], and the Squid wiki [44] explain that, as a proxy, the connections are made, firstly to the Squid proxy, and then, if appropriate, the request continues to another server. These connections are registered in Squid in the same way, with the same fields, with the exception of the client and server IP addresses. From the point of view of classification, if one of these two entries happens to be in the training file, and the other in the testing file, it would mean that the second would be correctly classified because of all the attribute values that both have in common. However, this also means that the good percentages that we obtained may not be real, but biased. That is why the second step is about removing entries that we called "repeated" (in the explained sense). This step is performed over the original, unbalanced, dataset. After the removal, a new file was created.

5.3.2 Performing Feature Selection

For pattern classification, our learning problem has to select high discriminating features from the input database which corresponds to the URL information dataset. To perform this task, we apply rough set theory.

Technically, we may formalize our problem as an information system where universe $U = \{x_1, x_2, \ldots, x_N\}$ is a set of pattern identifiers, the conditional attribute set $C = \{c_1, c_2, \ldots, c_N\}$ contains each feature of the information table to select and the decision attribute D of our learning problem corresponds to the class label of each pattern. The input database has a single binary decision attribute. Hence, the decision attribute D has binary values d: either the HTTP request is allowed or denied. The condition attribute feature D is defined as follows:

$$D = \{Allow, Deny\}$$

For feature selection, we apply the rough *QuickReduct* algorithm which was previously explained in Sect. 4.2. First of all, the dependency of the entire database $\gamma_C(D)$ is calculated. To do so, the algorithm has to calculate the positive region for the whole attribute set C: $POS_C(D)$. Once the consistency of the database is measured, the feature selection process starts off with an empty set and moves to calculate the dependency of each attribute c apart: $\gamma_c(D)$. The attribute c having the greatest value of dependency is added to the empty set. Once the first attribute c is selected, the algorithm adds, in turn, one attribute to the selected first attribute and computes the dependency of each obtained attributes couple $\gamma_{\{c,c_i\}}(D)$. The algorithm chooses the couple having the greatest dependency degree. The process of adding each time one attribute to the subset of the selected features continues until the dependency of the obtained subset equals the consistency of the entire database already calculated; i.e., $\gamma_C(D)$.

From the initial dataset containing 12 features and after applying the rough feature selection technique, we obtained a list of 9 features. The features kept after the feature selection process are the following:

- http_reply_code
- duration_miliseconds
- content_type
- server_or_cache_address
- time
- bytes
- url
- client_address

On the contrary, the following features were erased by applying Rough Set for feature selection:

- http_method
- content_type_MCT
- squid_hierarchy

In Sect. 6.2, we will show that by applying rough set theory for selecting the most important features is a good way of maintaining the good quality of the database, and the system performance will improve significantly.

5.4 Classification Methods

The choice of the classifiers to apply and would be admitted before we make a test selection phase – known as a 'pre-selection phase' – is based on two main requirements. First and as our goal consists of obtaining a set of rules able to classify unknown URL connection requests, we need classifiers based on decision trees or rules, so that we can study their output in addition to their accuracy. Second and as mentioned in Sect. 3, the features of the data used for this work are mainly categorical, but also numerical. Thus, among the classifiers based on trees or rules, we need classifiers that are able to handle these types of features. Consequently, we made a first selection over all the classifiers in Weka which complied with these requirements.

Yet, it is important to mention that previously in [28], we studied a set of classifiers that may be applied to the nature of our dataset; classifiers that fit our requirements. The selected classifiers are indeed based on the obtained good classification accuracy. In fact, we demonstrated that the classifiers that lead to better classification results were:

J48 This classifier generates a pruned or unpruned C4.5 decision tree. Described for the first time in 1993 by [33], this machine learning method builds a decision tree selecting, for each node, the best attribute for splitting and create the next nodes. An attribute is selected as 'the best' by evaluating the difference in entropy (information gain) resulting from choosing that attribute for splitting the data. In this way, the tree continues to grow till there are not attributes anymore for further splitting, meaning that the resulting nodes are instances of single classes.

Random Forest This manner of building a decision tree can be seen as a randomization of the previous C4.5 process. It was stated by [2] and consist of, instead of choosing 'the best' attribute, the algorithm randomly chooses one between a group of attributes from the top ones. The size of this group is customizable in Weka.

REP Tree Is another kind of decision tree, it means Reduced Error Pruning Tree. Originally stated by [32], this method builds a decision tree using information gain, like C4.5, and then prunes it using reduced-error pruning. That means that the training dataset is divided into two parts: one devoted to make the tree grow and another for pruning. For every subtree (not a class/leaf) in the tree, it is replaced by the best possible leaf in the pruning three and then it is tested with the test dataset if the made prune has improved the results. A deep analysis about this technique and its variants can be found in [10].

NNge Nearest-Neighbor machine learning method of generating rules using non-nested generalised exemplars, i.e., the so called 'hyperrectangles' for being multidimensional rectangular regions of attribute space [26]. The NNge algorithm builds a ruleset from the creation of this hyperrectangles. They are non-nested (overlapping is not permitted), which means that the algorithm checks, when a proposed new hyperrectangle created from a new generalisation, if it has conflicts with any region of the attribute space. This is done in order to avoid that an example is covered by more than one rule (two or more).

PART It comes from 'partial' decision trees, for it builds its rule set from them [11]. The way of generating a partial decision tree is a combination of the two aforementioned strategies "divide-and-conquer" and "separate-and-conquer", gaining then flexibility and speed. When a tree begins to grow, the node with lowest information gain is the chosen one for starting to expand. When a subtree is complete (it has reached its leaves), its substitution by a single leaf is considered. At the end the algorithm obtains a partial decision tree instead of a fully explored one, because the leafs with largest coverage become rules and some subtrees are thus discarded.

These methods will be deeply tested on the dataset (balanced and unbalanced) in the following section.

6 Results

This section presents the obtained results for the different configurations of the experimental setup. First, Sect. 6.1 depicts the first results, summarising what was proved in [28]. This means that, for the chosen classifiers, described in Sect. 5.4, results are displayed for the dataset when it is in its initial—unbalanced—form, as well as after the balancing process. At the end of this Subsection, we introduce the results obtained once the dataset is released from the redundant patterns. Then, Sect. 6.2 presented the results obtained when applying rough set theory as a feature selection technique to the balanced generated dataset. This subsection justifies that the use of rough sets enhances the system performance in terms of both execution/running time and classification accuracy. Finally, examples of the obtained rules which were taken from the classifiers' output, are discussed in Sect. 6.3.

6.1 Results About Classification

Several experiments have been conducted, once a subset of classification methods has been chosen (see Sect. 5.4). In order to better test the methods, two different divisions (training-test) have been done; namely 90–10 % and 80–20 %. Also, it is worth mentioning that we have included Naïve Bayes in the result tables, as it is normally used as a reference classifier in classification problems [12].

Moreover, the way in which those divisions were built has been considered as: randomly built, or sequentially built. We say that the training and test files were randomly built when the patterns are taken from the original dataset and, by generating a random number, they have a certain probability to belong to the training file, and another to belong to the test file. On the contrary, the training and test files are built sequentially when the patterns inside them strictly follow the same order in time as the original dataset, before being divided. The aim of the sequential division is to compare if the online activity of the employees, considering URL sessions, could be somehow "predicted", just using data from previous minutes or hours. In the case of the random distribution of patterns, we have done three different pairs of training-test files. These files have been built considering that similar patterns (in the whole dataset) are placed in the same file, in order to avoid biasing the classification.

Table 2 Percentage of correctly classified patterns for the unbalanced dataset with 12 features

	80 % Training–20 % Test		90 % Training–10 % Test	
	Random (mean)	Sequential	Random (mean)	Sequential
Naïve Bayes	91.60 ± 1.25	85.53	92.89 ± 0.12	83.84
J48	97.56 ± 0.20	88.48	97.70 ± 0.15	82.28
Random Forest	97.68 ± 0.20	89.77	97.63 ± 0.13	82.59
REP Tree	97.47 ± 0.11	88.34	97.57 ± 0.01	83.20
NNge	97.23 ± 0.10	84.41	97.38 ± 0.36	80.34
PART	97.06 ± 0.19	89.11	97.40 ± 0.16	84.17

As stated in Sect. 5.2, the dataset presents unbalance in the data due to the fact that there are more patterns classified as 'allow' than 'deny'. Therefore, two data balancing methods have been applied to all the files to get similar pattern amounts in both classes: undersampling (random removal of ALLOW patterns) and oversampling (duplication of DENY patterns).

Classification results for the unbalanced data are presented in Table 2. Mean and standard deviation are shown for the three different tests done in the random pattern distribution approach.

As it can be seen from Table 2, all five classifiers achieved a high performance classifying in the right way the test dataset. Also, having low values of standard deviation means that the obtained accuracies are stable; and this can be seen from the obtained results as well.

For a 80–20 % division, results based on the sequential data have lower values than those obtained from the random data, but still they are considered as good (> 85 %). This is due to the occurrence of new patterns from a certain time. Some requests may happen just at one specific time of the day, or in settled days. Then, the classifier may not find enough similarity in the patterns to correctly classify the entries in the test file. On the other hand, the loss of 5 to 6 points in the results of the 90–10 % division is somehow expected as it reinforces the previous mentioned hypothesis.

The classifier that lightly stands out over the others is *Random Forest*, being the best in almost every case for randomly made divisions, and it also has good results for sequentially made divisions. However, if we focus on the standard deviation, *REP Tree* is the chosen one, as its results present robustness.

Once balancing is performed, resulting datasets were used as inputs for the same classifiers, and results are shown in Tables 3 and 4. Table 3 shows the classifiers' accuracy for the balanced dataset with 12 features, applying undersampling technique, and Table 4, with the application of oversampling technique. For each one, the 90–10 % and 80–20 % divisions were also made.

Applying Undersampling In comparison with those results from Table 2, these go down one point (in the case of randomly made divisions) to six points (sequential divisions). The reason why this happens is that when randomly removing

Table 3 Percentage of correctly classified patterns for the balanced dataset with 12 features, applying undersampling technique

	80 % Training–20 % Test		90 % Training–10 % Test	
	Random (mean)	Sequential	Random (mean)	Sequential
Naïve Bayes	91.30 ± 0.20	84.94	91.74 ± 0.13	85.43
J48	97.05 ± 0.25	84.29	96.85 ± 0.35	76.44
Random Forest	96.61 ± 0.17	88.59	96.99 ± 0.13	79.98
REP Tree	96.52 ± 0.13	85.54	96.55 ± 0.10	77.65
NNge	96.56 ± 0.42	85.28	96.33 ± 0.05	81.93
PART	96.19 ± 0.14	85.16	96.09 ± 0.10	79.70

Table 4 Percentage of correctly classified patterns for the balanced dataset with 12 features, applying oversampling technique

	80 % Training–20 % Test		90 % Training–10 % Test	
	Random (mean)	Sequential	Random (mean)	Sequential
Naïve Bayes	91.18 ± 0.16	82.35	91.77 ± 0.28	81.81
J48	97.40 ± 0.03	85.66	97.37 ± 0.06	74.24
Random Forest	97.16 ± 0.19	89.03	97.25 ± 0.33	81.33
REP Tree	97.13 ± 0.25	85.41	97.14 ± 0.09	76.81
NNge	96.90 ± 0.28	83.46	96.91 ± 0.06	78.73
PART	96.82 ± 0.09	84.50	96.68 ± 0.11	78.16

ALLOW patterns, we really are losing information, i.e., key patterns that could be decisive in a good classification of a certain set of test patterns.

Applying Oversampling Here we have duplicated the DENY patterns so their number could be up to that of the ALLOW patterns. However, it does not work as well as in other approaches which uses numerical computations for creating the new patterns to include in the minority class. Consequently, the results have been decreased.

In both cases, it is noticeable that if we take the data in a sequential way, instead of randomly, results will decrease. Also, it is clear that due to the fact that performing undersampling some patterns are lost while in the case of oversampling they all remain, and this leads to have better results with the *oversampling* balancing technique. Then, in this case the algorithm with best performance is *J48*, though *Random Forest* follows its results very closely in random datasets processing, and *REP Tree*, which is better than the rest when working with sequential data. Nevertheless, generally speaking and given the aforementioned reasons, performing data balancing methods decreases the results.

Once this first study is finished, the next step is to erase the duplicated requests. And then, we test the obtained reduced dataset to see if it has some influence on the results. As it seems that the best results are obtained for an unbalanced dataset,

Table 5 Percentage of correctly classified patterns for unbalanced data, after the removal of entries that could lead to misclassification

	80 % Training–20 % Test		90 % Training–10 % Test	
	Random (mean)	Sequential	Random (mean)	Sequential
Naïve Bayes	93.01 ± 0.32	82.61	93.09 ± 0.91	83.04
Random Forest	96.97 ± 0.47	91.03	96.79 ± 0.97	80.60
J48	96.90 ± 0.26	87.78	96.50 ± 1.00	84.49
NNge	96.21 ± 0.28	81.17	96.11 ± 1.13	81.92
REP Tree	96.97 ± 0.40	87.75	96.62 ± 0.87	85.57
PART	96.84 ± 0.18	86.68	96.55 ± 0.87	83.61

and also for a training-test random division, we choose this configuration for the following experiments.

The results are displayed in Table 5. We can see that the results slightly decrease in comparison to the ones obtained originally, but they are still good, and definitely better than Naïve Bayes.

The way it happened for the original datasets, results for files with the patterns taken consecutively lower significantly. And as previously explained, this happens due to the possible loss of information. Best results are obtained by both *Random Forest* and *REP Tree* classifiers, with a 96 % of accuracy.

First, we concluded that not balancing the dataset was better for obtaining good results, also that taking the samples randomly instead of a sequential way is more adequate. Finally, we noticed that we have successfully reduced the dataset and did not lose good accuracies. For this reason, the dataset with the redundant patterns erased is the chosen one to perform the feature selection. Results are described in next subsection.

6.2 Results About Feature Selection

In this section, our aim is to prove two hypotheses. First, we want to prove that applying rough set theory for feature selection reduces the running time when testing the classifiers. Second and based on the reduced feature set of data, we want to prove that the accuracies remain the same, or even improve in comparison to the original set of features.

The resulting reduced dataset, from the previous subsection, was used to test the same chosen classifiers, plus JRip. This is a classifier which consists of a propositional rule learner, the so-called Repeated Incremental Pruning to Produce Error Reduction (RIPPER) algorithm. It was proposed in [8] as an improved version of the Incremental Reduced Error Pruning (IREP) algorithm. The reason why this JRip classifier was added to the list, is because we cannot compare the size of the trees for the Random Forest classifier, as the size of the forest is chosen when running it. Then, we added JRip for making the comparison more complete.

All the experiments were made with the same computer, in the following conditions: Toshiba Laptop with Intel™Core i7-3630QM, CPU@2.40 GHz× 8; RAM 3.8 GB; operating system 64 bit Ubuntu Linux 14.04 LTS; and Weka version 3.6.10, with 3 GB assigned for memory usage. Table 6 shows the results of the comparison between performance before and after applying feature selection. Though the complexity of the trees generated by the classifiers grows after applying fea ture selection, even obtaining more rules for PART classifiers, the running time lowers by an average of 40 %.

Now, if we focus on the results of the accuracies which are summarised in Table 7, we notice that the classification accuracies are nearly the same. This comparison was made in the same conditions which are unbalanced datasets, and with a 10 fold cross-validation technique for training-test.

Moreover, results in Table 8 show the same behaviour as for previous experiments. Results from this table are compared to those obtained from Table 2. This is because both tables are sharing nearly the same conditions, except that each one has a specific number of features.

We can see that, for example, for a division of 80 % training–20 % test, the results after feature selection are better for the *Random Forest* and *PART* classifiers. Only

Table 6 Comparison between rule/tree complexity and running times (in seconds) for the initial data set, which had 12 features, and the resulting one, having 9 features, after applying Rough Set Theory for feature selection

	12 features	9 features
J48	Size of the tree 8113, 1.7 ± 0.41 (s)	Size of the tree 10191, 1.17 ± 0.17 (s)
Random Forest	10 trees, 3.32 ± 0.61 (s)	10 trees, 2.28 ± 0.19 (s)
REP Tree	Size of the tree 8317, 1.40 ± 0.31 (s)	Size of the tree 8817, 0.87 ± 0.10 (s)
NNge	1341 exemplars, 66.65 ± 4.04 (s)	1294 exemplars, 64.18 ± 3.76 (s)
PART	966 rules, 40.28 ± 2.12 (s)	998 rules, 37.34 ± 1.67 (s)
JRip	87 rules, 164.99 ± 72.29 (s)	64 rules, 115.48 ± 60.88 (s)

Table 7 Comparison between the obtained accuracies for the initial data set, which had 12 features, and the resulting one, having 9 features, after applying Rough Set Theory for feature selection

	12 features	9 features
Naïve Bayes	92.30 ± 0.15	92.19 ± 0.09
J48	97.37 ± 0.29	97.36 ± 0.30
Random Forest	97.61 ± 0.24	97.62 ± 0.25
REP Tree	97.34 ± 0.25	97.35 ± 0.25
NNge	97.15 ± 0.25	97.13 ± 0.25
PART	97.34 ± 0.26	97.26 ± 0.25
JRip	92.84 ± 0.91	91.97 ± 1.25

Table 8 Percentage of correctly classified patterns for the unbalanced dataset with 9 features

	80 % Training–20 % Test		90 % Training–10 % Test	
	Random (mean)	Sequential	Random (mean)	Sequential
J48	97.10 ± 0.23	87.83	97.33 ± 0.80	84.51
Random Forest	97.61 ± 0.49	88.06	97.76 ± 0.83	83.71
REP Tree	97.17 ± 0.15	87.79	97.39 ± 0.58	85.73
NNge	96.63 ± 0.50	82.18	97.27 ± 1.12	80.94
PART	97.24 ± 0.12	87.88	97.29 ± 0.86	85.11

NNge seems to generate lower classification results, but still considered as interesting as it is higher than 96 %.

Finally, we have proved our first hypothesis: applying Rough Set Theory for feature selection significantly improves the computational cost of the system. Also, we proved that our second hypothesis is also true, because the obtained accuracies after applying rough set theory for feature selection are the same, even slightly better, than the ones obtained before the pre-processing phase.

6.3 Discussion About the Obtained Rules

One of the main objectives of this chapter is to find a method (classifier) that can build rules not dependent on the URL, in order to get a behaviour quite different from the classical black and white lists. Thus, it could made a decision about new connection requests based on other, more general, features.

In the performed experiments, the majority of the obtained rules/trees are based on the URL in order to discriminate between the two classes. However, we also found several ones which consider other variables/features rather than the URL itself to make the decision. For instance:

```
IF server_or_cache_address = "173.194.34.225"
AND http_method = "GET"
AND duration_milliseconds > 52
THEN ALLOW


IF server_or_cache_address = "173.194.78.103"
THEN ALLOW


IF content_type = "application/octet-stream"
AND server_or_cache_address = 192.168.4.4
AND client_address = 10.159.86.22
THEN ALLOW
```

```
IF server_or_cache_address = "173.194.78.94"
AND content_type_MCT = "text"
AND content_type = "text/html"
AND http_reply_code = "200"
AND bytes > 772
THEN ALLOW
```

In their presented format, these rules are considered adequate to fulfill our purposes, since they are somehow independent of the URL to which the client requests to access. Thus, it would be potentially possible to allow or deny the access to unknown URLs just taking into account some parameters of the request.

When the features considered in the rule can be known in advance, such as http_method, or server_or_cache_address, for instance, the decision could be made in real-time, and thus, a granted URL (whitelisted) could be DENIED, or the other way round.

Tree-based classifiers also yield to several useful branches in this sense, but they have not been plotted here because of the difficulty for showing/visualizing them properly.

Focusing on the presented rules, it can be noticed that almost all of them also depend on very determining features/values, such as server_or_cache_ address, or even on the client_address, what we have called 'critical features'. These features create several non-useful rules, mainly in the case of the client IP address, because it will not be correct to settle that a specific IP can or cannot access to some URLs.

Thus, we have conducted two additional experiments in this line by removing, first, the url feature in a new dataset, and second, erasing the three critical features: url, server_or_cache_address and client_address from the dataset. Then we have trained again the classifiers. These experiments have been performed over the unbalanced data, considering a 10-fold cross validation test.

The results of classification accuracies in each of the two tests are shown in Table 9.

As expected, and as it can be seen in Table 9, the percentages of accuracy have been decreased. Results are more influenced and decrease in the case where three features have been discarded. However, the results are still quite good, having in mind that the remaining features are more general than those removed.

Table 9 Percentage of correctly classified patterns for the unbalanced dataset without the set of critical features, namely URL, server_or_cache_address, and client_address

	Without URL feature	Without URL and IP addresses features
J48	93.62	90.53
Random Forest	94.42	91.75
REP Tree	92.58	89.61
PART	93.40	88.25
JRip	87.45	85.60

In addition, it is worth to analyse the set of rules that the classifiers[5] have generated as models. Thus, having a look at these other rules, in the case without `url` (i.e., 11 features), the rules are pretty similar to those presented before. Thus, we can find, among the most important rules (in the sense that the classification accuracy depends in a big part on them) the following ones:

```
IF bytes >= 1075
AND time >= 29633000
AND time <= 30031000
AND client_address = "10.159.52.182"
AND content_type_MCT = "image"
AND content_type = image/jpeg
THEN DENY

IF server_or_cache_address = "173.194.66.121"
AND client_address = "192.168.4.4"
AND time <= 33603000
THEN ALLOW

IF client_address = "10.159.188.11"
AND bytes <= 2166
AND content_type_MCT = "text"
THEN ALLOW
```

These rules, actually most of them in the generated model, still depend on the rest of critical features (server and client IP addresses). Due to this reason, we conducted the second experiment omitting these variables. In this case the generated rules by all the classifiers are closer to what we aimed to obtain. Some examples of relevant rules (those with high influence in the obtained accuracy) are the following:

```
IF http_reply_code = "200"
AND content_type = "application/json"
AND time <= 33635000
AND bytes <= 3921
THEN ALLOW

IF content_type = "text/plain"
AND duration_milliseconds >= 7233.5
THEN DENY

IF content_type = "application/octet-stream"
AND bytes <= 803
THEN ALLOW
```

---

[5]Trees can be deployed as rules.

```
IF bytes <= 1220
AND time <= 33841000
AND http_reply_code = "404"
AND squid_hierarchy = DEFAULT_PARENT
AND duration_milliseconds <= 233
AND bytes <= 722
THEN ALLOW
```

As it can be seen these are more general rules which could be much more useful for classifying new, previously unknown, URL access requests in a company. These rules could be taken as a reference to build a decision system. However, there are still some considerations that should be taken into account since all the rules have been created using a very specific data. Thus, there are several rules that cannot be used as they are specific for some other companies, and should be supervised somehow (maybe by an expert).

Moreover, some of these features depend on the session itself, i.e., they will be computed after the session is over, but the idea in that case would be 'to refine' somehow the existing set of URLs in the white list. Thus, when a client requests access to a whitelisted URL, this will be allowed, but after the session is over and depending on the obtained values, one of these classifiers could label the URL as DENIED for further requests. This could be a useful decision-aid tool for the Chief Security Officer (CSO) inside a company, for instance.

7 Conclusions and Future Work

In this paper various classification methods have been applied in order to perform a decision process inside a company, according to some predefined corporate security policies. This decision is focused on allowing or denying URL access requests by considering previous decisions on similar requests, and not having specific rules in an already defined white/black list for those URLs. Thus, the proposed method would allow or deny an access to a URL based on additional features rather than the specific URL string, only. This could be very useful since new URLs could be automatically 'whitelisted' or 'blacklisted' depending on some of the connection parameters, such as the content_type of the access or the IP of the client which makes the request.

To this aim, we have started from a big dataset (100,000 patterns) with employees' URL requests information, and by considering a set of URL access permissions, we have composed a labelled dataset (57,000 patterns). Over that set of data, we have tested several classification methods, after some data balancing techniques have been applied. Then, the best five classifiers have been deeply proved over several training and test divisions, and with two methods: by leaving the order in time when the URL were requested, and by taking them in a random way.

The results show that classification accuracies are between 95 and 97 %, even when using the unbalanced datasets. However, they have been diminished because of the possible loss of data that comes from performing an undersampling (removing patterns) method; or taking the training and the datasets in a sequential way from the main log file, due to the fact that certain URL requests can be made only at a certain time.

After that, we have shown that maintaining the dataset is crucial in order to improve the performance of the system, mainly, in terms of classification accuracy. We have shown that by erasing the duplicated data, the accuracies remain inside the range of 96–97 %, which means that indeed there was redundant information in the dataset.

The resulting dataset was the one over which we have performed feature selection by means of rough sets, and we have proved that by selecting the most interesting features we could improve the classification accuracy of the system while being lightweight in terms of running time. In this way, we can conclude that our proposed approach has been successful and it would be a useful tool in an enterprise.

Future lines of work include conducting a deeper set of experiments trying to test the generalisation power of the method, maybe by considering bigger data divisions, bigger data sets (from a whole day, or a week), or by adding some kind of 'noise' to the dataset. Moreover, considering the good classification results obtained in this work, the next step could be the application of our methodology in the real system from which data was gathered, counting with the opinion of expert CSOs, in order to know the real value of the proposal.

The study of other classification methods could be another research branch, along with the implementation of a Genetic Programming approach, which could deal with the unbalance problem. This can be done by using a modification of the cost associated to misclassifying patterns as done in [1].

Finally, we also aim at extracting additional information from the URL string. This information could be transformed into additional features that could be more discriminative than the current set of obtained rules. Moreover, a data process involving grouping data into sessions (such as number of requests per client, or average time connection) will be also considered.

Acknowledgments The authors would like to thank GENIL-SSV'2015 for ensuring the visit of Dr. Zeineb Chelly to be part of this project. We thank Dr. Zeineb Chelly from Institut Supérieur de Gestion, Tunisia for her technical insight, recommendations and suggestions and for her assistance during the practical experiments. This paper has been funded in part by European project MUSES (FP7-318508), along with Spanish National project TIN2011-28627-C04-02 (ANYSELF), project P08-TIC-03903 (EVORQ) awarded by the Andalusian Regional Government, and projects 83 (CANUBE), and GENIL PYR-2014-17, both awarded by the CEI-BioTIC UGR.

References

1. Alfaro-Cid, E., Sharman, K., Esparcia-Alcázar, A.: A genetic programming approach for bankruptcy prediction using a highly unbalanced database. In: Giacobini, M. (ed.) Applications of Evolutionary Computing. Lecture Notes in Computer Science, vol. 4448, pp. 169–178. Springer, Heidelberg (2007). http://dx.doi.org/10.1007/978-3-540-71805-5_19
2. Breiman, L.: Random forests. Mach. Learn. **45**(1), 5–32 (2001)
3. Breivik, G.: Abstract misuse patterns—a new approach to security requirements. Master thesis. Department of Information Science. University of Bergen, Bergen, N-5020 NORWAY (2002)
4. Chawla, N.V., Bowyer, K.W., Hall, L.O., Kegelmeyer, W.P.: Smote: synthetic minority oversampling technique. J. Artif. Int. Res. **16**(1), 321–357 (2002). http://dl.acm.org/citation.cfm?id=1622407.1622416
5. Chawla, N.: Data mining for imbalanced datasets: an overview. In: Maimon, O., Rokach, L. (eds.) Data Mining and Knowledge Discovery Handbook, pp. 853–867. Springer, USA (2005). http://dx.doi.org/10.1007/0-387-25465-X_40
6. Chelly, Z.: New danger classification methods in an imprecise framework. Ph.D. thesis. Laboratoire de Recherche Opérationelle de Décision et de Contrôle de Processus, Institut Supérieur de Gestion, Tunisia (2014)
7. Cheswick, W.R., Bellovin, S.M., Rubin, A.D.: Firewalls and Internet Security: Repelling the Wily Hacker. Addison-Wesley Longman Publishing Co., Inc., Boston (2003)
8. Cohen, W.W.: Fast effective rule induction. In: Proceedings of the Twelfth International Conference on Machine Learning, pp. 115–123 (1995)
9. Danezis, G.: Inferring privacy policies for social networking services. In: Proceedings of the 2nd ACM Workshop on Security and Artificial Intelligence. AISec 2009, pp. 5–10. ACM, New York (2009). http://doi.acm.org/10.1145/1654988.1654991
10. Elomaa, T., Kaariainen, M.: An analysis of reduced error pruning. Artif. Intell. Res. **15**, 163–187 (2001)
11. Frank, E., Witten, I.H.: Generating accurate rule sets without global optimization. In: Shavlik, J. (ed.) Fifteenth International Conference on Machine Learning, pp. 144–151. Morgan Kaufmann, San Francisco (1998)
12. Frank, E., Witten, I.H.: Data Mining: Practical Machine Learning Tools and Techniques, 3rd edn. Morgan Kaufmann Publishers, San Francisco (2011)
13. Greenstadt, R., Beal, J.: Cognitive security for personal devices. In: Proceedings of the 1st ACM Workshop on Workshop on AISec. AISec 2008, pp. 27–30. ACM, New York (2008). http://doi.acm.org/10.1145/1456377.1456383
14. Guo, X., Yin, Y., Dong, C., Yang, G., Zhou, G.: On the class imbalance problem. In: Fourth International Conference on Natural Computation. ICNC 2008, vol. 4, pp. 192–201, October 2008
15. Harris, E.: The Next Step in the Spam Control War: Greylisting (2003)
16. Japkowicz, N., Stephen, S.: The class imbalance problem: a systematic study. Intell. Data Anal. 6(5), 429–449, October 2002. http://dl.acm.org/citation.cfm?id=1293951.1293954
17. Jensen, R., Shen, Q.: Semantics-preserving dimensionality reduction: Rough and fuzzy-rough-based approaches. IEEE Trans. Knowl. Data Eng. **17**(1), 1 (2005)
18. Jensen, R., Shen, Q.: Fuzzy-rough sets assisted attribute selection. IEEE Trans. Fuzzy Syst. **15**(1), 73–89 (2007)
19. Kaeo, M.: Designing Network Security. Cisco Press, Indianapolis (2003)
20. Kelley, P.G., Hankes Drielsma, P., Sadeh, N., Cranor, L.F.: User-controllable learning of security and privacy policies. In: Proceedings of the 1st ACM Workshop on Workshop on AISec. AISec 2008, pp. 11–18. ACM, New York (2008). http://doi.acm.org/10.1145/1456377.1456380
21. Lim, Y.T., Cheng, P.C., Clark, J., Rohatgi, P.: Policy evolution with genetic programming: a comparison of three approaches. In: IEEE Congress on Evolutionary Computation. CEC 2008. (IEEE World Congress on Computational Intelligence), pp. 1792–1800, June 2008

22. Lim, Y.T., Cheng, P.C., Rohatgi, P., Clark, J.A.: Mls security policy evolution with genetic programming. In: Proceedings of the 10th Annual Conference on Genetic and Evolutionary Computation. GECCO 2008, pp. 1571–1578. ACM, New York (2008). http://doi.acm.org/10.1145/1389095.1389395
23. Liu, H., Motoda, H.: Feature Extraction, Construction and Selection: A Data Mining Perspective. Springer, USA (1998)
24. Ludl, C., McAllister, S., Kirda, E., Kruegel, C.: On the effectiveness of techniques to detect phishing sites. In: Hämmerli, B.M., Sommer, R. (eds.) Detection of Intrusions and Malware, and Vulnerability Assessment, pp. 20–39. Springer, Heidelberg (2007)
25. MacQueen, J., et al.: Some methods for classification and analysis of multivariate observations. In: Proceedings of the fifth Berkeley Symposium on Mathematical Statistics and Probability, pp. 281–297, no. 14. California, USA (1967)
26. Martin, B.: Instance-based learning: nearest neighbor with generalization. Master's thesis, University of Waikato, Hamilton, New Zealand (1995)
27. McAfee: Mcafee labs threats report, June 2014 . http://www.mcafee.com/uk/about/newsroom/research-reports.aspx
28. Mora, A., De las Cuevas, P., Merelo, J.: Going a step beyond the black and white lists for url accesses in the enterprise by means of categorical classifiers. In: Proceedings of the International Conference on Evolutionary Computation Theory and Applications (ECTA). SCITEPRESS, pp. 125–134 (2014)
29. Mora, A., De las Cuevas, P., Merelo, J., Zamarripa, S., Juan, M., Esparcia-Alcázar, A., Burvall, M., Arfwedson, H., Hodaie, Z.: MUSES: a corporate user-centric system which applies computational intelligence methods. In: Shin, D. et al., (ed.) 29th Symposium On Applied Computing, pp. 1719–1723 (2014)
30. Netcraft: November 2014 web server survey (2014). http://news.netcraft.com/archives/category/web-server-survey/
31. Pawlak, Z., Polkowski, L., Skowron, A.: Rough set theory. In: Wah, B.W. (ed.) Wiley Encyclopedia of Computer Science and Engineering. Wiley, Hoboken (2008)
32. Quinlan, J.R.: Simplifying decision trees. Man Mach. Stud. **27**(3), 221–234 (1987)
33. Quinlan, J.R.: Programs for Machine Learning. Morgan Kaufmann, San Mateo (1993)
34. Seigneur, J.M., Kölndorfer, P., Busch, M., Hochleitner, C.: A survey of trust and risk metrics for a BYOD mobile working world. In: Third International Conference on Social Eco-Informatics (2013)
35. Shen, Q., Jensen, R.: Rough sets, their extensions and applications. Int. J. Autom. Comput. **4**(3), 217–228 (2007)
36. Stanton, J.M., Stam, K.R., Mastrangelo, P., Jolton, J.: Analysis of end user security behaviors. Comput. Secur. **24**(2), 124–133 (2005)
37. Suarez-Tangil, G., Palomar, E., Fuentes, J., Blasco, J., Ribagorda, A.: Automatic rule generation based on genetic programming for event correlation. In: Herrero, A., Gastaldo, P., Zunino, R., Corchado, E. (eds.) Computational Intelligence in Security for Information Systems, Advances in Intelligent and Soft Computing, vol. 63, pp. 127–134. Springer, Heidelberg (2009). http://dx.doi.org/10.1007/978-3-642-04091-7_16
38. Team, S.: Squid website (2013). http://www.squid-cache.org/
39. Team, S.: Squid faq—squid log files (2014)
40. Team, T.J.D.: Drools documentation. version 6.0.1.final (2013). http://docs.jboss.org/drools/release/6.0.1.Final/drools-docs/html/index.html
41. Team, T.J.D.: Drools website (2013). http://www.jboss.org/drools.html
42. Waikato, U.: Weka (1993), University of Waikato, September 2014, http://www.cs.waikato.ac.nz/ml/weka/
43. Wessels, D.: Squid: The Definitive Guide, 1st edn. O'Reilly Media Inc., Sebastopol (2004)
44. Wiki, S.: Squid hierarchy (2014)
45. Wilson, D.C., Leake, D.B.: Maintaining case-based reasoners: dimensions and directions. Comput. Intell. **17**(2), 196–213 (2001)
46. Zhong, N., Dong, J., Ohsuga, S.: Using rough sets with heuristics for feature selection. J. Intell. Inf. Syst. **16**(3), 199–214 (2001)

A Granular Intrusion Detection System Using Rough Cognitive Networks

Gonzalo Nápoles, Isel Grau, Rafael Falcon, Rafael Bello
and Koen Vanhoof

Abstract Security in computer networks is an active research field since traditional approaches (e.g., access control, encryption, firewalls, etc.) are unable to completely protect networks from attacks and malwares. That is why Intrusion Detection Systems (IDS) have become an essential component of security infrastructure to detect these threats before they inflict widespread damage. Concisely, network intrusion detection is essentially a pattern recognition problem in which network traffic patterns are classified as either normal or abnormal. Several Computational Intelligence (CI) methods have been proposed to solve this challenging problem, including fuzzy sets, swarm intelligence, artificial neural networks and evolutionary computation. Despite the relative success of such methods, the complexity of the classification task associated with intrusion detection demands more effective models. On the other hand, there are scenarios where identifying abnormal patterns could be a challenge

G. Nápoles · I. Grau · R. Bello
Computer Science Department, Central University of Las Villas, Carretera
Camajuaní Km 5.5, 54830 Santa Clara, Cuba
e-mail: gnapoles@uclv.edu.cu

I. Grau
e-mail: igrau@uclv.edu.cu

R. Bello
e-mail: rbellop@uclv.edu.cu

R. Falcon (✉)
Larus Technologies Corporation, 170 Laurier Ave. West - Suite 310,
Ottawa, ON K1P 5V5, Canada
e-mail: rafael.falcon@larus.com; rfalcon@uottawa.ca

R. Falcon
Electrical Engineering & Computer Science, University of Ottawa,
800 King Edward Ave., Ottawa, ON K1N 6N5, Canada

G. Nápoles · K. Vanhoof
Hasselt Universiteit Campus Diepenbeek, Agoralaan Gebouw D,
BE3590 Diepenbeek, Belgium
e-mail: gonzalo.napoles@student.uhasselt.be

K. Vanhoof
e-mail: koen.vanhoof@uhasselt.be

© Springer International Publishing Switzerland 2016
R. Abielmona et al. (eds.), *Recent Advances in Computational Intelligence
in Defense and Security*, Studies in Computational Intelligence 621,
DOI 10.1007/978-3-319-26450-9_7

as the collected data is still permeated with uncertainty. In this chapter, we tackle the network intrusion detection problem from a classification angle by using a recently proposed granular model named Rough Cognitive Networks (RCN). An RCN is a fuzzy cognitive map that leans upon rough set theory to define its topological constructs. An optimization-based learning mechanism for RCNs is also introduced. The empirical evidence indicates that the RCN is a suitable approach for detecting abnormal traffic patterns in computer networks.

Keywords Intrusion detection system · Computational intelligence · Granular computing · Rough set theory · Fuzzy cognitive maps · Rough cognitive networks · Harmony search

1 Introduction

The 21st century has brought forth a digital age in which we are all immersed. Up-and-coming information communication and processing paradigms such as the Internet of Things (IoT) [4], Cloud Computing [47], Software-Defined Networks [32] and Wearable Computing [25] are increasingly gaining momentum and rapidly permeating every facet of mankind. These new architectural frameworks bring a unique set of challenges with them, among which cybersecurity is one of paramount importance. The computer systems that constitute the backbone of critical infrastructure behind a plethora of industrial and societal processes often become prey to sophisticated malicious attacks that originate at any node in the entangled World Wide Web. As a result, governments and businesses are adapting their legislative bodies to account for the prevention, detection and mitigation of the risks and threats associated with these potentially devastating attacks [39].

Intrusion Detection Systems (IDS) [43] have become an essential component of security infrastructure to detect these threats before they inflict widespread damage, since traditional approaches (e.g., access control, encryption, firewalls, etc.) are unable to completely protect networks from attacks and malwares. The purpose of an IDS is to analyze the network traffic, either the incoming one or existing logs of past traffic activities, and identify anomalous behaviours that could reasonably be taken as cues of the presence of an intruder in the system. Concisely described, network intrusion detection is essentially a pattern recognition problem in which network traffic patterns are classified as either normal or abnormal.

Although traditional statistical techniques have enjoyed success in analyzing traffic flows as part of an IDS operation, the network security research community is increasingly leaning on Computational Intelligence (CI) solutions due to their ability to adapt to complex environments, handle noise and uncertainty and remain computationally tractable and robust.

More recently, the advent of Granular Computing (GrC) [6, 26, 52] as an innovative information representation and processing framework has largely influenced the way CI systems are being conceived nowadays. This is due to the fact that

GrC provides reasoning constructs at higher levels of abstraction that better capture human understanding of the real world. From classification [55] to clustering [51], time-series prediction [72] and decision making [50], granular models are becoming prominent tools for the analysis of large volumes of data as they operate upon information granules (i.e., constructs of order higher than plain numeric or symbolic atoms) and can better represent and manifest the dynamics of human-centric world modeling.

In this chapter, we tackle network intrusion detection via a GrC model and demonstrate its advantages over several traditional classification schemes. Our study makes the following contributions: (1) we model network intrusion detection as a classification problem and apply a recently introduced granular model, named "Rough Cognitive Network"(RCN), to the analysis of archived traffic data in computer networks for intrusion detection purposes; (2) we put forth a learning mechanism for RCNs that is based on self-adaptive Harmony Search [44]; (3) we empirically evaluate the RCN performance in conjunction with that of seven well-established classifiers in the literature. The experimental evidence confirms that RCNs are a plausible model to discriminate between normal and abnormal traffic patterns in network data as it attains high detection rates (i.e., successfully identified abnormalities) and low false negative rates (misidentified anomalies).

The rest of this chapter is structured as follows. Section 2 briefly surveys relevant works in intrusion detection systems, with special emphasis on CI-based solutions. Section 3 elaborates on the two precursor formalisms leading up to RCNs: rough set theory (RST) and fuzzy cognitive maps (FCMs). Then, the RCN topology learning and classification inference process are dissected in Sect. 4 while Sect. 5 describes the proposed optimization-based RCN parameter learning method. The experimental analysis is unveiled in Sect. 6 before conclusions and future work directions are outlined in Sect. 7.

2 Related Work

In this section, we briefly review several published works that are relevant to our study. They provide the necessary background to understand the contents of this chapter.

2.1 Intrusion Detection Systems

The literature in the IDS arena is quite vast. This field appears often interwoven with other similar terms such as "network anomaly detection" or "network intrusion detection" and the common underlying problem has been addressed through a myriad of techniques. In a recent and comprehensive survey [8] covering publications in this field from 2000 to 2012, 28 % of the papers surveyed approached IDS from a supervised learning angle (i.e., classification), as we do in this chapter. However, unsupervised learning (via clustering) was the preferred choice of 21 % of the papers given that labeled data could be scarce and/or difficult to access in certain cases where privacy concerns impede the sharing of such information.

The statistical methods and systems applied to intrusion detection [45, 61, 66, 79] first construct a general statistic model of the observed traffic data, either via parametric techniques (which assume the knowledge of the type of probability distribution is available and then try to learn their parameters) or by means of non-parametric techniques, which do not lay any assumption on the type of the data distribution. Once this model has been fitted to the data, any point (traffic pattern) with low probability of having been generated by the underlying data model is labeled as an outlier and hence flagged as suspicious.

The use of computational intelligence methods in the IDS realm has been well documented in the 2010 survey compiled by Wu and Banzhaf [73]. Artificial neural networks (ANNs) [11, 40, 67, 78, 81], fuzzy sets [16, 21, 29, 68], evolutionary computation [5, 18, 24, 31, 38, 57–59], artificial immune systems (AIS) [70, 75], fuzzy cognitive maps [62–64, 74, 83], rough sets [2, 13, 14] and swarm intelligence (SI) [19, 20, 29] techniques, all representative methods of the wider CI/Soft Computing (SC) family, and their hybrids [15, 22, 63, 64, 74] have all been wielded against complex network traffic datasets to identify attack vectors or suspicious activities either in a supervised or unsupervised fashion.

2.2 Rough Set Theory in Network Security

Rough sets and fuzzy cognitive maps have been independently applied to network intrusion detection [8, 73], although the number of reported works thus far is not significant compared to the volume of documented applications of other CI techniques.

Chen et al. [13] employ rough set theory in the preprocessing stage of their proposed network intrusion detection scheme in order to remove irrelevant attributes prior to the operation of the Support Vector Machine (SVM)-based classifier. A similar use (attribute dimensionality reduction) is evoked by Li and Zhao with their Fuzzy SVM [41] and by Zhang et al. in the context of their Artificial Immune System (AIS)-based technique [82], where the number of attributes that describe an antibody is shortened using the lower and upper approximations of each rough concept. Shrivastava and Jain [60] also boost the network traffic classification power of their SVM via rough-set-based feature selection by dropping 35 irrelevant attributes out of 41 initially gathered to describe the traffic flows in their system. An analogous rationale is pursued by Sivaranjanadevi et al. in their work [65] and by Poongothai and Duraiswamy in [53].

Fuzzy and rough sets are integrated into a partitive clustering engine in [14] to address network intrusion detection from an unsupervised perspective; the proposed clustering method yielded superior results compared to other classical unsupervised techniques.

Finally, rough sets are used in [2] to induce classification rules via the LEM2 algorithm so as to create a potent classifier capable of detecting network intrusions with high detection rate and low false alarm rate. The classification results of LEM2 are found to be more interpretable and can be obtained in a shorter time than those of the K-nearest neighbor classifier, which are more accurate yet more resource-demanding.

2.3 Fuzzy Cognitive Maps in Network Security

Xin et al. [74] derive fuzzy features from the network data and pass them on to a fuzzy cognitive map (FCM) in order to model more complex attack vectors.

Siraj et al. [63] used FCM and fuzzy rule bases to model causal knowledge among different intrusion variables in an interpretable fashion. Suspicious events are mapped to nodes in FCM, which function as neurons that trigger alerts with different weights depicting on the causal relations between them. So, an alert value for a particular machine or a user is calculated as a function of all the activated suspicious events at a given time. This value reflects the safety level of that machine or user at that time.

Siraj et al. [64] chose FCMs and fuzzy rule bases as the vehicles for causal knowledge acquisition within the decision engine of an intelligent IDS deployed at the Mississipi State University. The system fuses information from a variety of intrusion detection sensors. In particular, the FCMs are used at two levels: (i) to model individual suspicious events such as 'high login failure' or 'SYN flood' and (ii) to ascertain the overall impact of various suspicious events (input concepts) for each host computer and system user (output concepts).

Afterwards, Siraj and Vaughn [62] also leaned upon FCMs to cluster network intrusion alerts based on discovered similarities among the raw features extracted from sensor data. The FCM is thus acting as a fusion machine where intrusion evidence for a particular network resource that originates at different clusters is amalgamated.

Zhong et al. [83] consider a distributed attack scenario and resort to an FCM to describe the entities that are part of it as well as their relationships.

The study authored by Jazzar and Bin Jantan [27] focuses on IDS designed around the Self Organizing Map (SOM) neural network given its ability to process large volumes of data with low computational overhead. Having realized that these systems still exhibit a high false alarm rate, they coupled the SOM with an FCM in order to refine the clustering performed by the former approached. The FCM's role is to calculate the relevance of odd concepts (neurons) to a network attack. By doing so, irrelevant concepts can be left out and other concepts may come to the forefront of the intrusion analysis.

Krichene and Boudriga [37] devised a methodology to automatically determine responses to security incidents. The underlying formalism that allows attack identification, complexity reduction and response elicitation is termed an *incident response probabilistic cognitive map*. These maps differ from traditional FCMs in that they are capable of modeling different relationships between symptoms, actions and unauthorized results as pertaining to a network attack. A function that enables the identification of those concepts that are tied to a set of events is also part of the proposed scheme. The authors illustrate their proposal on a real-world denial of service (DoS) attack against a web server.

Zaghdoud and Al-Kahtani [80] bring forth a multi-layered architecture for intrusion detection and response. They employ an FCM to gauge the impact of a confirmed intrusion event belonging to a known class upon the compromised system. The FCM nodes represent components of the computer network system or security concepts whereas the edges symbolize the influence exercised by one component upon another; these influences must be carefully taken into consideration now that a network intrusion has been confirmed.

2.4 Discussion

Our proposed granular classifier, the Rough Cognitive Network, borrows from both aforementioned techniques: RST and FCM; however, their synergy is dictated by a topological arrangement of the FCM nodes into symbolic and higher-order information granules, the latter of which correspond to the three RST-based regions (positive, boundary, negative) of the decision concepts (classes) induced by a similarity relationship over the set of input attributes in the data set under consideration. To the best of our knowledge, this hybridization scheme is completely different from previous efforts to combine both methodologies, and so is certainly the RCN application to the IDS domain.

3 The Forerunners of Rough Cognitive Networks

As mentioned before, in this paper we design an IDS which uses an RCN for detecting potentially atypical (and likely dangerous) patterns. One could briefly define an RCN as a Sigmoid Fuzzy Cognitive Map where concepts represent granules of information. In this section, we summarize the mathematical underpinnings behind Rough Set Theory and Fuzzy Cognitive Maps, which are the two core building blocks of the granular model proposed in this chapter.

3.1 Rough Set Theory

Rough Set Theory (RST) is a robust and mature theory for handling uncertainty in the form of inconsistency in the data [1, 49]. The RST framework employs two exact set approximations to describe a generic or real-world concept. Let us assume a decision system $S = (U, A \cup d)$, where U is a non-empty finite set of objects called the universe, A is a non-empty finite set of attributes, while $d \notin A$ denotes the decision attribute. Any subset $X \subseteq U$ can be approximated by two crisp sets: the lower and upper approximations. These sets are defined as $B_*X = x \in U : [x]_B \subseteq X$ and $B^*X = x \in U : [x]_B \cap X \neq \emptyset$ where the equivalence class $[x]_B$ comprises the set of inseparable objects associated to the target instance x that are described using $B \subseteq A$.

Based on the lower and upper approximations, we can compute the positive, negative and boundary regions of any concept X. The positive region $POS(X) = B_*X$ includes those objects that are certainly contained in X; the negative region $NEG(X) = U - B^*X$ involves those objects that are certainly not contained in X, whereas the boundary region $BND(X) = B^*X - B_*X$ represents the objects whose membership to the set X is uncertain, i.e., they might be members of X. These regions are in fact information granules and provide a valuable knowledge when facing decision-making or pattern classification problems.

Based on the positive, negative and boundary regions, Yao [76] defined two types of rules: *deterministic* decision rules for the positive region and *undeterministic* decision rules for the boundary region. More recently Yao [77] introduced the three-way decisions model. Rules constructed from the three regions are associated with different actions [23]. A positive rule suggests a decision of *acceptance*, a negative rule makes a decision of *rejection* and a boundary rule implies a decision of *abstaining*. The three-way decisions play an important role in decision-making problems [42].

In the classical RST formulation, the indiscernibility relation is defined as an equivalence relation; hence, two objects will be inseparable if they are identical with respect to a set of attributes $B \subseteq A$. The equivalence relation R induces a partition of the universe U on the basis of the attributes in B. However, this definition is extremely strict. For example, a decision system with millions of objects will be categorized as inconsistent if two objects are equivalent but they have different decision classes (i.e., two experts might have different perceptions about the same observation). But are two objects really significant in a universe comprised of millions of objects?

To counter the above stringent definition, the equivalence requirement on R is relaxed. In fact, if we adopt a "weaker" inseparability relation then we could tackle problems having numerical (or mixed) attributes. Two inseparable objects, according to some similarity relationship R, will be tossed together in the same set of not identical (but reasonably similar) instances. Equation 1 shows the indiscernibility relation adopted in this paper, where $0 \leq \varphi(x, y) \leq 1$ is a similarity function. This binary relation determines whether two objects x and y are inseparable or not (i.e., as long as their similarity degree $\varphi(x, y)$ is greater than or equal to a user-specified threshold ξ).

Despite the clear advantages of using this approach to cope with problems having numerical features, selecting the correct value for the similarity threshold ξ could be a challenge.

$$R : yRx \Leftrightarrow \varphi(x, y) \geq \xi. \tag{1}$$

If the threshold $\xi = 1$ then the similarity relation R will be reflexive, transitive and symmetric, leading to Pawlak's model for discrete (nominal) domains. If $\xi < 1$ then the similarity relation will be reflexive and symmetric but not transitive.

Another aspect to be considered when designing a similarity relation is the adequate selection of the similarity function. Equation 2 shows a variant which combines both numerical and categorical attributes. It provides a more general formulation for addressing decision-making problems having different features.

$$\varphi(x, y) = \frac{1}{|A|} \sum_{i=1}^{|A|} \omega_i \delta(x(i), y(i)). \tag{2}$$

In the above equation, A is the set of features describing the problem, $0 \leq \omega_i \leq 1$ represents the relative importance of the ith attribute, $x(i)$ and $y(i)$ denote the values of the ith attribute associated with the objects x and y respectively, and δ is the attribute-wise similarity function. The greater $0 \leq \varphi(x, y) \leq 1$, the more similar the objects x and y. Equations 3 and 4 display the attribute-wise similarity functions adopted in this research study. The function δ_1 is used when we want to compare two values of a discrete attribute, whereas δ_2 is used for comparing two values of a numerical attribute (L_i and H_i denote the lowest and highest value of the ith attribute, respectively).

$$\delta_1(x(i), y(i)) = \begin{cases} 1, \ x(i) = y(i) \\ 0, \ x(i) \neq y(i) \end{cases}. \tag{3}$$

$$\delta_2(x(i), y(i)) = 1 - \frac{|x(i) - y(i)|}{H_i - L_i}. \tag{4}$$

Equations 5 and 6 respectively formalize how to compute the lower and upper approximations of a concept X, where $R(x)$ denotes the similarity class of the object x. These exact sets are the basis for granulating the available information about the concept using RST, and they become the core of Granular Fuzzy Cognitive Maps [48].

$$B_*X = \{x \in U : R(x) \subseteq X\}. \tag{5}$$

$$B^*X = \bigcup_{x \in X} R(x). \tag{6}$$

As a result, an object can simultaneously belong to multiple similarity classes, so the covering induced by the similarity relation R over the universe U is not necessarily a partition [7]. Therefore, similarity relations do not induce a partition of the universe, but rather generate similarity classes. It suggests that an object could simultaneously belong to different similarity classes, and consequently the instance x could activate several granular regions. In such cases, the decision-making stage becomes really difficult for the expert, since it has to consider non-trivial decision patterns.

3.2 Fuzzy Cognitive Maps

Fuzzy Cognitive Maps (FCM) are recurrent neural networks for modeling and simulation [34] consisting of concepts and their causal relations. Concepts are equivalent to neurons denoting objects, variables, or entities related to the system under investigation whereas the weights associated with the connections among neurons denote the strength of the *causality* among such nodes. It should be highlighted that causal relations are quantified in the range $[-1; 1]$. This value is the result of the numerical evaluation of a fuzzy linguistic variable, which is usually assigned by experts during the modeling phase [36]. The activation value of the neurons is also fuzzy in nature and regularly takes values in the range $[0; 1]$ although the interval $[-1; 1]$ is used too. The magnitude of the activation is also meaningful for the model: the higher the activation value of a map concept, the stronger its influence over the system under consideration.

Equation 7 mathematically formalizes the rule for updating the activation value of concepts in an FCM, assuming A^0 is the initial configuration. This rule is iteratively repeated until a fixed point attractor or a maximum number of iterations T is reached. At each step t a new state vector is produced, and after a large enough number of iterations, the map will arrive at one of the following states: (i) fixed equilibrium point, (ii) limited cycle or (iii) chaotic behavior [35]. If the FCM reaches a fixed-point attractor, then we can conclude that the map has converged. In such cases, the final output corresponds to the desired state (i.e., the system response for the activation vector).

$$A_i^{t+1} = f(\sum_{j=1}^{M} w_{ji}A_j^t + w_{ii}A_i^t), i \neq j. \tag{7}$$

In the above equation $f(.)$ represents a monotonically non-decreasing nonlinear function which is used for transforming the activation value of each concept (the weighted combination of the activation levels). The most used functions are: the bivalent function, the trivalent function, and sigmoid variants [10]. In this paper we will focus on sigmoid functions since it has been shown that they exhibit superior prediction capabilities [10].

4 Rough Cognitive Networks

Rough Cognitive Networks (RCNs) [48] are an extension of three-way decision rules introduced by Yao [76]. In a nutshell, we can define an RCN as a sigmoid FCM where concepts denote information granules, namely, the RST-derived positive, boundary and negative regions of the original problem as well as the set of decision classes in the problem at hand.

The RCN methodology not just allows solving mixed-attribute problems, but also provides accurate inferences since it uses a recurrent inferential process to converge to a stable attractor, which comprises the most fitting decision class. It should be pointed out that the complexity of this model does not depend on the number of attributes in the decision system, but on the number of decision classes. In this section, we explain how to learn an RCN from data. Furthermore, we introduce a supervised learning algorithm for computing the required RCN parameters, which enhances the value of our proposal.

4.1 Information Granulation and Network Design

As mentioned before, a central aspect when designing an RCN is the process related to the construction of positive, negative and boundary regions. Let us assume a pattern classification problem and a partition $X = X_1, \ldots, X_k, \ldots, X_N$ of the universe U according to the decision attribute, where each subset X_k denotes a decision class and comprises all instances labeled as d_k. These information granules will be expressed as map concepts. More precisely, input concepts denote positive, negative and boundary regions associated with each subset X_k; they are subsequently used for activating the network.

In the RCN model, the output neurons do not influence other neurons since they are target concepts. Once the FCM inference process is done (this point will be clarified next), the activation degree of each output concept (decision class) will be gauged. After the map concepts are defined, we establish causal connections among such neurons, where the direction and intensity of the causal weights are computed according to the set of rules below:

- R_1 : IF C_i is P_k AND C_j is d_k THEN $w_{ij} = 1.0$.

- R_2 : IF C_i is P_k AND C_j is $d_{(v \neq k)}$ THEN $w_{ij} = -1.0$.

- R_3 : IF C_i is P_k AND C_j is $P_{(v \neq k)}$ THEN $w_{ij} = -1.0$.

- R_4 : IF C_i is N_k AND C_j is d_k THEN $w_{ij} = -1.0$.

In the above rules, C_i and C_j denote two map concepts, P_k and N_k are the positive and negative region for the kth decision respectively, whereas $-1 \leq w_{ij} \leq 1$ is the causal weight between the cause C_i and the effect C_j. More precisely, rules R_1 and R_2 define the relation between positive regions and decision neurons. If the positive region P_k is activated (rule 1), then the decision d_k must be stimulated as well, since we confidently know that objects belonging to the positive region P_k will be categorically members of the concept X_k. Accordingly, decisions $d_{(v \neq k)}$ must be inhibited (rule 2) because an object cannot simultaneously belong to different positive regions.

The third rule follows an analogous reasoning: if a positive region P_k is activated then positive regions unrelated to the decision d_k (i.e., $P_{(v \neq k)}$) will be inhibited. If the negative region N_k is activated (rule 4), then the map will inhibit the decision, but we cannot conclude anything about other decisions. Moreover, we incorporated an additional rule for handling the intrinsic knowledge concerning the RST boundary regions:

- R_5 : IF C_i is B_k AND C_j is d_v AND $(BND(X_k) \cap BND(X_v) \neq \emptyset)$ THEN $w_{ij} = 0.5$.

Observe that not all boundary regions are included in the RCN's topology. This is dictated by the learning procedure on the training data: if a boundary region is empty $(BND(X_k) = \emptyset)$ then the neuron B_k will be removed from the modeling in order to simplify the network topology. On the other hand, we need to establish causal links between each boundary neuron and decision classes involving some degree of uncertainty; otherwise the causal connection will be removed from the map as well.

The above topology construction scheme implies that an RCN for a problem with $|D|$ decision classes will have at most $3|D|$ input neurons (assuming all boundary regions are in), $|D|$ decision (output) neurons and $3|D|(1 + |D|)$ causal relations. Additionally, for each neuron we add a self-reinforcement connection with causality $w_{ii} = 1$ which partially preserves the initial excitation.

4.2 Inference Using Rough Cognitive Networks

The final phase concerns the network exploitation, where the activation value of input and decision concepts play a pivotal role. In this scheme, to classify a test instance O_i, first the excitation vector A_i will be calculated using the similarity class $R(O_i)$ and its relation to each RST-based region. For instance, let us assume that $|POS(X_1)| = 20$, $|R(O_i)| = 10$, whereas the number of objects that belong to the positive region is given by the expression: $|R(O_i) \cap POS(X_k)| = 7$. This implies that the activation degree of the neuron P_1 is $7/20 = 0.35$. It denotes the conditional probability of accepting d_1 given the similarity class $R(O_i)$, that is $Pr(d_k|R(O_i))$. Analogously, we can compute the activation degree of other input concepts related to each decision class. Rules $R_6 - R_8$ formalize this procedure as follows:

- R_6 : IF C_i is P_k THEN $A_i^0 = \frac{|R(O_i) \cap POS(X_k)|}{|POS(X_k)|}$.
- R_7 : IF C_i is N_k THEN $A_i^0 = \frac{|R(O_i) \cap NEG(X_k)|}{|NEG(X_k)|}$.
- R_8 : IF C_i is B_k THEN $A_i^0 = \frac{|R(O_i) \cap BND(X_k)|}{|BND(X_k)|}$.

Once the activation vector A^0 has been computed, we trigger the FCM inference rule until a fixed point attractor, or a maximal number of iterations T is reached. This process will stress a pattern using the similarity class of the instance O_i to do that, which is desirable in problems with insufficient positive evidence where selecting the proper class could be difficult. Afterward one can use the output vector for making a decision (e.g., we can sort the alternatives according to the preference degrees calculated by the map inference process). When dealing with pattern classification problems, the final output will be the concept having the highest activation, or alternatively it could be a random class if the input similarity class only activates negative and/or boundary regions.

5   Learning Methodology for Rough Cognitive Networks

As mentioned before, the basis for computing the set of positive, negative and boundary regions is the proper estimation of the similarity threshold ξ in Eq. 1. If this value is too small then positive regions will be small as well, leading to poor excitation of neurons. This step is quite important when selecting the most adequate decision: the higher the activation of the positive region, the more desirable the decision (although the model will compute the final decision taking into account all the evidence). If this threshold ξ is excessively large then boundary regions will be large, thus increasing the uncertainty.

In this section, we present a learning algorithm for tuning the model parameters, which is based on the Harmony Search (HS) metaheuristic [44]. The method needs to adjust two kinds of parameters: the weight ω_i of each attribute and the similarity threshold ξ. This approach leads to a numerical optimization problem with $|A| + 1$ variables and will be solved using an adaptive variant of the HS procedure.

The HS metaheuristic is a simple-trajectory search method, which only evaluates one potential solution at a time, instead of evaluating a set of potential solutions (as it occurs with population-based metaheuristics). This HS design choice is relevant for our learning methodology since evaluating a solution means computing the set of lower and upper approximations, which could be computationally expensive as the number of objects in the training data set increase.

During the optimization phase, the algorithm randomly creates a harmony memory with size HMS and iteratively improves a new harmony from the HM. If the improved harmony is better than the worst harmony in the HM, then the new solution replaces the worst harmony. Despite its algorithmic simplicity, HS suffers from a serious problem common to other metaheuristics: its search capabilities are quite sensitive to the specified parameter vector.

For this reason in this paper we adopt an improved variant, called Self-adaptive Harmony Search (SHS), which is capable of adjusting its own parameters [71]. The SHS method not only alleviates the parametric sensitivity issue, but also significantly enhances the accuracy of the solutions. Algorithm 1 shows the pseudocode of this metaheuristic, where N is the maximal number of iterations, HMCR

(Harmony Memory Consideration Rate) is a parameter that controls the balance between exploitations and exploration, while $R_1 = U - x$ and $R_2 = x - L$, assuming that L and U respectively denote the lowest and the highest values for each problem variable in the harmony memory.

On the other hand, PAR is the pitch adjustment rate and determines whether further adjustment is required to a harmony drawn from the harmony memory. In this variant, the PAR factor is linearly decreased over time. Experiments reported by the authors [71] suggested that moderate size of the harmony memory (e.g., 50) and large values of HMCR (e.g., 0.9) are adequate choices for these parameters. Based on these considerations, we used these values during the experiments and simulations performed in the next section. The *rand()* function draws a random number uniformly distributed in the unit interval.

Algorithm 1. Self-adaptive Harmony Search

```
Initialize the memory
FOR i = 1 TO N DO
    IF rand() < HMCR THEN
        Select a random pitch x from the memory
        IF rand() < PAR THEN
            x = x + rand(R₁, R₂)
        END
    ELSE
        x = x + rand(a, b)
    END
    Select the worst harmony y from the memory
    IF (y is worse than x) THEN
        Replace the worst harmony y with the new pitch x
    END
END
Select the best solution S from the memory
RETURN S
```

The other component of the optimization problem to be specified is the objective function. Equation 8 shows the function $G(.)$ used in this study, where the parameters denote the set of weights W, the similarity threshold ξ and the set of instances ϕ to be used for training the model, respectively. On the other hand, $\aleph_{R(W,\xi)}(x)$ is the output vector computed by the RCN which is obtained from the similarity threshold defined by the function $R(W, \xi)$, whereas the function $Y(x)$ is the known class vector associated with the instance x and D is the set of decision classes in the problem. It should be also mentioned that $\|.\|_L$ refers to a norm (e.g., the L_1-norm, L_2-norm or L_∞-norm) that is used to calculate the error.

$$\text{minimize } G(W, \xi, \phi) = \sum_{x \in \phi} \frac{\|\aleph_{R(W,\xi)}(x) - Y(x)\|_L}{|\phi||D|}. \tag{8}$$

If $G(W, \xi, \phi) = 0$ then the RCN, using the similarity relation R, is capable of recognizing all patterns stored in the training set; otherwise the value $1 - G(W, \xi, \phi)$ stands for the model accuracy. The proposed parameter tuning method not only estimates the introduced parameters, but also allows determining the relevance of each attribute, which contributes to elicit further knowledge about the problem.

6 Detecting Intrusion in Computer Networks

In this section we study the performance of the proposed granular cognitive network for detecting abnormal traffic behavior in computer networks. As mentioned before, this problem can be envisioned as a challenging pattern classification task having two decision classes: either 'normal' or 'abnormal'. In order to perform our simulations, we used an improved variant of the NSL-KDD dataset [17] which is a widely used benchmark when testing IDS [19, 22, 23]. In the following section, we summarize the most important features of both training and testing NSL-KDD datasets.

6.1   Description of the NSL-KDD Dataset

Perhaps the most popular dataset for evaluating the performance of anomaly detection models is KDD'99 [30]. The KDD training dataset consists of 4,900,000 network connection vectors, each of which contains 41 features. Such features could be gathered in three groups: (i) basic features, (ii) traffic features and (iii) content features.

The first group comprises attributes extracted from a TCP/IP connection, whereas the second one includes time-based features computed in a window interval (e.g., connections in the past 2 s having the same destination host or the same service as the current connection). It should be stated that there are several slow-probing attacks that scan the ports using a much larger time interval than 2 s and accordingly these attacks will not produce any intrusion patterns. Finally, the third group contains features related to attacks having a single connection, which do not have intrusion frequent sequential patterns. In such cases, attacks are embedded in the data portions of packets, hence forcing the Intrusion Detection System to catch suspicious behavior in the data portion (e.g., number of failed login attempts) instead of in the connections.

On the other hand, in the training set each record is labeled as either "normal" or "abnormal" with exactly one specific attack type (i.e., Probing Attack, Denial of Service Attack, User to Root Attack and Remote to Local Attack).

It is essential to mention that the KDD'99 dataset was built based on the data captured in DARPA'98 which has been criticized by McHugh [46]. It suggests that some of the existing problems in the dataset DARPA'98 remain in KDD'99. More recently, Tavallaee and collaborators [69] conducted a statistical analysis where two important issues were detected. The first important deficiency in the KDD'99 dataset is the huge number of redundant records (78 and 75 % of records are duplicated in the train and test set, respectively). Consequently, this will cause learning algorithms to be biased towards the more frequent records. As a second issue they noticed that this dataset has poor difficulty level: about 98 % of the records in the train set and 86 % of the records in the test set were correctly classified with 21 learned machines (7 learners, each trained 3 times with different training sets).

To solve the aforementioned issues, Tavallaee et al. [69] removed all the redundant records in both train and test sets. Moreover, they randomly sampled correctly classified records in such a way that the number of selected instances from each difficulty level group is inversely proportional to the percentage of records in the original dataset. This refinement process gave rise to two improved datasets called KDDTrain+ and KDDTest+ which include 125,973 and 22,544 records, respectively. As well, they created another test set called KDDTest-21 by removing the records that were correctly classified by all 21 learners. This dataset contains 11,850 records, which are more difficult to classify. Because of its increasing popularity and sound verification procedure, we adopted Tavallaee et al's data sets for our experimentation.

6.2 Numerical Simulations

Next we study the behavior of RCN across the selected dataset. Figure 1 displays the network topology that allows solving the prediction problem (i.e., where each

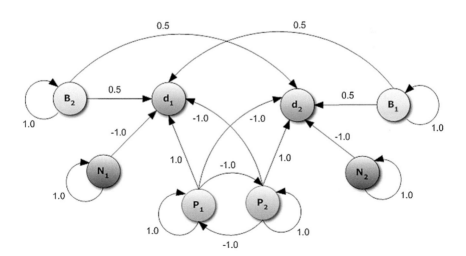

Fig. 1 The proposed Rough Cognitive Network for intrusion detection. The d_1 concept corresponds to the normal traffic class and the d_2 concept represents the abnormal traffic class. The P_i, B_i and N_i nodes denote the positive, boundary and negative regions for these two classes, $i \in \{1, 2\}$

instance is classified as either "normal" or "abnormal"). More exactly, d_1 = "normal", d_2 = "abnormal", P_i denotes the positive region associated to the ith class, N_i is the negative region related to the ith class while B_i is the ith boundary region. Note that boundary concepts are allowed regardless of the inconsistency of the features in the target problem because only two decision classes are possible. More explicitly, if the problem has inconsistent instances, then both classes will be equally affected; otherwise, the activation value of the (empty) boundary regions will remain inactive during the inference process.

6.2.1 Comparison with Traditional Classifiers Over KDDTest+

The first experiment consists of studying the prediction ability of our model regarding the following set of traditional classifiers: J48 decision tree [54], NBTree [33], Random Forest [9], Random Tree [3], Multilayer Perceptron [56], Naive Bayes [28], and Support Vector Machine [12]. For experimental purposes, we adopted the first 20 % of the records in KDDTrain+ for training all models. Figure 2a summarizes the accuracy achieved for each learner, whereas Fig. 2b displays some representative samples of the solution space associated with the similarity threshold to be explored by the learning algorithm. In other words, Fig. 2b illustrates the performance of our granular network for different similarity thresholds.

From the above experiment we can conclude that RCN results are competitive regarding J48 decision tree, Random Forest (RF), NBTree (NBT) and Random Tree (RT). However, our model outperforms other approaches such as Multilayer Perceptron (MLP), Naive Bayes (NB) and Support Vector Machine (SVM).

Next we study other statistics such as those extracted from the confusion matrices. True Negatives (TN) as well as True Positives (TP) correspond to correctly classified instances, that is, events that are rightly labeled as normal and attacks, respectively. Alternatively, False Positives (FP) refer to normal events being labeled as attacks while False Negatives (FN) are attack events incorrectly predicted as normal events. Table 1 shows such statistics for all classifiers used for comparison across the selected KDDTest+ dataset.

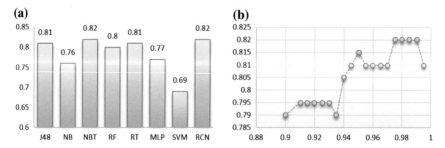

Fig. 2 Experiments using datasets KDDTrain+ and KDDTest+. **a** Accuracy of selected classifiers and **b** RCN accuracy as a function of the threshold values in Eq. (1)

Table 1 Confusion matrix associated with each classifier for the KDDTest+ dataset

	TN	FP	FN	TP	Detection rate	False alarm rate
J48	9436	275	3996	8837	0.68	0.02
NB	9010	701	4582	8251	0.64	0.07
NBT	8869	842	3257	9576	0.74	0.08
RF	9452	259	4523	8310	0.64	0.02
RT	8898	813	3011	9822	0.76	0.08
MLP	8971	740	4796	8037	0.62	0.07
SVM	8984	727	4893	7940	0.61	0.07
RCN	8891	820	3150	9683	0.75	0.08

The reader may notice that RCN ranks as the second-best algorithm regarding the number of FN patterns. In our study we are especially interested in this value since it denotes the number of abnormal patterns that the IDS was unable to detect, although most authors prefer systems with high detection rate (i.e., $TP/(TP + FN)$) and low false alarm rate which is defined as $FP/(TN + FP)$. Nevertheless, in computer networks where high security is required, reducing the false negative rate is indispensable since only those patterns having normal features will be confidently allowed.

6.2.2 Comparison with Traditional Classifiers Over KDDTest-21

The second experiment is concerned with investigating the performance of our RCN model with respect to traditional classifiers, but now using the test set called KDDTest-21. Figure 3a portrays the classification accuracy achieved for each model while Fig. 3b displays the performance of the proposed granular network for different similarity thresholds.

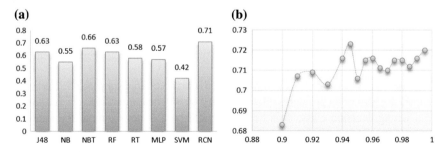

Fig. 3 Experiments using datasets KDDTrain+ and KDDTest-21. **a** Accuracy of selected classifiers and **b** RCN accuracy as a function of the threshold values in Eq. (1)

Table 2 Confusion matrix associated with each classifier for the KDDTest-21 dataset

	TN	FP	FN	TP	Detection rate	False alarm rate
J48	1879	273	3996	5702	0.58	0.12
NB	1460	692	4549	5149	0.53	0.32
NBT	1354	798	3257	6441	0.66	0.37
RF	1895	257	4523	5175	0.53	0.11
RT	1388	764	3008	6690	0.68	0.35
MLP	1426	726	4796	4902	0.50	0.33
SVM	1440	712	4893	4805	0.49	0.33
RCN	1572	580	2824	6874	0.70	0.26

It should be specified that the KDDTest-21 dataset is more complex since it involves patterns that cannot be correctly classified by all learners. Despite this fact, our model was able to compute the best accuracy (71 %), notably outperforming the remaining approaches. However, in a previous experiment the model only achieved an accuracy of 66 % due to the uncertainty present in the features during the inference stage (i.e., the overall evidence suggests accepting both decisions). To overcome this situation, we used the similarity classes pertaining to the K-nearest neighbors ($K = 3$) of the test instance O_i. In short, we adopted the similarity classes of its neighbors instead of only using the set $R(O_i)$ related to the target pattern for activating each input neuron in the network.

Table 2 shows the confusion matrix achieved by each classifier across the KDDTest-21 test set. In this case, our model computed the highest detection rate ($TP/(TP + FN) = 0.7$) and lowest false negative rate ($FN/(TP + FN) = 0.29$) which is the desired behavior. It means that the RCN will detect abnormal traffic with high accuracy, thus reducing the risk of classifying abnormal patterns as normal. In a nutshell, such statistics confirm the reliability of our granular classifier (RCN) for intrusion detection in complex computer networks. For instance, the reader may observe that if the false alarm rate is high, then the system will classify normal patterns as abnormal, but this behavior is preferable in order to avoid potential attacks.

6.2.3   Discussion

Although the above experiments show that RCNs are a suitable approach for addressing intrusion detection problems, there are cases where the inference suggests accepting a wrong decision class. This behavior could be a direct result of the strategy adopted for activating the input concepts, so other ways for estimating the activation vector could be explored. For example, in Bayesian inference one usually translates $Pr(C|[x])$ into $Pr(([x]C)Pr(C))/Pr([x])$ by the Bayes theorem, which allows a practical estimation of initial conditions required to trigger the FCM inference process.

Another aspect to be considered is related to the network weights, since rules R_1–R_5 formalize the direction (negative or positive) of each causal connection rather than its intensity. This means that the granular neural network discussed in this chapter calculates the decision class based on the initial state A^0 and the sign of causal relations, without exploiting the causal intensity. To achieve further performance gains, we are currently focused on computing this indicator via a supervised learning approach.

7 Conclusions

An important aspect in computer networks is how to detect intrusion since traditional approaches such as access control lists or firewalls are incapable of entirely protecting networks. In order to deal with such problem, several intrusion detection systems have been proposed; however, increasing the overall performance (e.g., the detection accuracy) is still an open problem for researchers. More explicitly, an essential component of intrusion detection systems is the inference algorithm used to classify network traffic patterns as either normal or abnormal. This problem could be thought of as a challenging binary classification task since modern intrusion techniques are sophisticated, so it is difficult to design models being able to distinguish between normal and abnormal patterns. As an example, frequently hackers attempt simulating trusted users in computer networks in order to gain access to remote resources. Such behavior will produce inconsistency in the collected traffic data; that is, objects that are very similar yet have been labeled as pertaining to different decision classes.

In this chapter we introduced a novel IDS based on Rough Cognitive Networks, a recently proposed granular neural network for pattern classification. Without loss of generality, we can define RCN as a Sigmoid Fuzzy Cognitive Map where input neurons represent information granules whereas output concepts denote decision classes. It should be remarked that the granulation of information is achieved by using Rough Sets, since it allows handling uncertainty arising from inconsistency. Furthermore, with the goal of increasing the reliability of the RCN-based inference process, we discussed a supervised learning methodology for automatically computing accurate similarity relations by estimating the proper parameter vector.

In order to measure the performance of our model, we adopted an improved version of the NSL-KDD dataset. From numerical simulations it is possible to conclude that our granular neural network is a suitable approach for detecting abnormal traffic patterns in computer networks. More precisely, we observed that RCNs are competitive regarding traditional classifiers such as J48 decision tree and Random Forest, across the simpler dataset (KDDTrain+). However, for the dataset KDDTest-21 the model significantly outperformed the other learners by computing the highest detection rate ($DR = 0.7$) and lowest false negative rate ($FNR = 0.29$). This confirms the reliability of the learning methodology put forth in this chapter to boost the model's performance. Future work along this front will concentrate on validating our approach on real computer networks.

References

1. Abraham, A., Falcon, R., Bello, R.: Rough Set Theory: A True Landmark in Data Analysis. Springer, Heidelberg (2009)
2. Adetunmbi, A.O., Falaki, S.O., Adewale, O.S., Alese, B.K.: Network intrusion detection based on rough set and k-nearest neighbour. I. J. Comput. ICT Res. **2**(1), 60–66 (2008)
3. Aldous, D.: The continuum random tree. I. Ann. Prob. 1–28 (1991)
4. Atzori, L., Iera, A., Morabito, G.: The internet of things: a survey. Comput. Netw. **54**(15), 2787–2805 (2010)
5. Balajinath, B., Raghavan, S.: Intrusion detection through learning behavior model. Comput. Commun. **24**(12), 1202–1212 (2001)
6. Bello, R., Falcon, R., Pedrycz, W., Kacprzyk, J.: Granular Computing: At The Junction of Rough Sets and Fuzzy Sets. Springer, Heidelberg (2008)
7. Bello, R., Verdegay, J.L.: Rough sets in the soft computing environment. Inf. Sci. **212**, 1–14 (2012)
8. Bhuyan, M.H., Bhattacharyya, D., Kalita, J.K.: Network anomaly detection: methods, systems and tools. IEEE Commun. Surv. Tutorials **16**(1), 303–336 (2014)
9. Breiman, L.: Random forests. Mach. Learn. **45**(1), 5–32 (2001)
10. Bueno, S., Salmeron, J.L.: Benchmarking main activation functions in fuzzy cognitive maps. Expert Syst. Appl. **36**(3), 5221–5229 (2009)
11. Cannady, J.: Artificial neural networks for misuse detection. In: National Information Systems Security Conference, pp. 368–81 (1998)
12. Chang, C.C., Lin, C.J.: Libsvm: a library for support vector machines. ACM Trans. Intell. Syst. Technol. (TIST) **2**(3), 27 (2011)
13. Chen, R.C., Cheng, K.F., Chen, Y.H., Hsieh, C.F.: Using rough set and support vector machine for network intrusion detection system. In: First Asian Conference on Intelligent Information and Database Systems, 2009. ACIIDS 2009, pp. 465–470. IEEE (2009)
14. Chimphlee, W., Abdullah, A.H., Noor Md Sap, M., Srinoy, S., Chimphlee, S.: Anomaly-based intrusion detection using fuzzy rough clustering. In: International Conference on Hybrid Information Technology, 2006. ICHIT'06, vol. 1, pp. 329–334. IEEE (2006)
15. Costa, K.A., Pereira, L.A., Nakamura, R.Y., Pereira, C.R., Papa, J.P., Falcão, A.X.: A nature-inspired approach to speed up optimum-path forest clustering and its application to intrusion detection in computer networks. Inf. Sci. **294**, 95–108 (2015)
16. Dickerson, J.E., Dickerson, J.A.: Fuzzy network profiling for intrusion detection. In: 19th International Conference of the North American Fuzzy Information Processing Society, 2000. NAFIPS, pp. 301–306. IEEE (2000)
17. Elkan, C.: Results of the KDD'99 classifier learning. ACM SIGKDD Explor. Newsl. **1**(2), 63–64 (2000)
18. Faraoun, K., Boukelif, A.: Genetic programming approach for multi-category pattern classification applied to network intrusions detection. Int. J. Comput. Intell. Appl. **6**(01), 77–99 (2006)
19. Feng, W., Zhang, Q., Hu, G., Huang, J.X.: Mining network data for intrusion detection through combining svms with ant colony networks. Future Gener. Comput. Syst. **37**, 127–140 (2014)
20. Gao, H.H., Yang, H.H., Wang, X.Y.: Ant colony optimization based network intrusion feature selection and detection. In: Proceedings of 2005 International Conference on Machine Learning and Cybernetics, 2005, vol. 6, pp. 3871–3875. IEEE (2005)
21. Geramiraz, F., Memaripour, A.S., Abbaspour, M.: Adaptive anomaly-based intrusion detection system using fuzzy controller. Int. J. Netw. Secur. **14**(6), 352–361 (2012)
22. Govindarajan, M.: Hybrid intrusion detection using ensemble of classification methods. Int. J. Comput. Netw. Inf. Secur. **2**, 45–53 (2014)
23. Guo, C., Zhou, Y., Ping, Y., Zhang, Z., Liu, G., Yang, Y.: A distance sum-based hybrid method for intrusion detection. Appl. Intell. **40**(1), 178–188 (2014)
24. Hofmann, A., Schmitz, C., Sick, B.: Rule extraction from neural networks for intrusion detection in computer networks. In: IEEE International Conference on Systems, Man and Cybernetics, 2003, vol. 2, pp. 1259–1265. IEEE (2003)

25. Hong, J., Baker, M.: Wearable computing. IEEE Pervasive Comput. **13**(2), 7–9 (2014)
26. Jankowski, A., Skowron, A.: Toward perception based computing: A rough-granular perspective. In: Zhong, N., Liu, J., Yao, Y., Wu, J., Lu, S., Li, K. (eds.) Web Intelligence Meets Brain Informatics. Lecture Notes in Computer Science, vol. 4845, pp. 122–142. Springer, Heidelberg (2007)
27. Jazzar, M., Bin Jantan, A.: Using fuzzy cognitive maps to reduce false alerts in SOM-based intrusion detection sensors. In: Second Asia International Conference on Modeling Simulation, 2008. AICMS 08, pp. 1054–1060 (2008)
28. John, G.H., Langley, P.: Estimating continuous distributions in Bayesian classifiers. In: Proceedings of the Eleventh conference on Uncertainty in artificial intelligence, pp. 338–345. Morgan Kaufmann Publishers Inc. (1995)
29. Karami, A., Guerrero-Zapata, M.: A fuzzy anomaly detection system based on hybrid pso-kmeans algorithm in content-centric networks. Neurocomputing **149**, 1253–1269 (2015)
30. KDD Cup 1999: KDD'99 dataset (2007). http://kdd.ics.uci.edu/databases/kddcup99/kddcup99.html
31. Khan, M.S.A.: Rule based network intrusion detection using genetic algorithm. Int. J. Comput. Appl. **18**(8), 26–29 (2011)
32. Kirkpatrick, K.: Software-defined networking. Commun. ACM **56**(9), 16–19 (2013)
33. Kohavi, R.: Scaling up the accuracy of Naive-Bayes classifiers: a decision-tree hybrid. In: KDD, pp. 202–207 (1996)
34. Kosko, B.: Fuzzy cognitive maps. Int. J. Man Mach. Stud. **24**(1), 65–75 (1986)
35. Kosko, B.: Hidden patterns in combined and adaptive knowledge networks. Int. J. Approximate Reasoning **2**(4), 377–393 (1988)
36. Kosko, B.: Fuzzy Engineering (1996)
37. Krichene, J., Boudriga, N.: Incident response probabilistic cognitive maps. In: International Symposium on Parallel and Distributed Processing with Applications, 2008. ISPA '08, pp. 689–694 (2008)
38. Kuang, F., Xu, W., Zhang, S.: A novel hybrid KPCA and SVM with GA model for intrusion detection. Appl. Soft Comput. **18**, 178–184 (2014)
39. Kuehn, A.: Extending Cybersecurity, Securing Private Internet Infrastructure: the US Einstein Program and its Implications for Internet Governance. Springer (2014)
40. Labib, K., Vemuri, V.R.: NSOM: A tool to detect denial of service attacks using self-organizing maps. Department of Applied Science University of California, Davis, California, USA, Technical Report (2002)
41. Li, L., Zhao, K.: A new intrusion detection system based on rough set theory and fuzzy support vector machine. In: 2011 3rd International Workshop on Intelligent Systems and Applications (ISA), pp. 1–5 (2011)
42. Liang, D., Pedrycz, W., Liu, D., Hu, P.: Three-way decisions based on decision-theoretic rough sets under linguistic assessment with the aid of group decision making. Appl. Soft Comput. **29**, 256–269 (2015)
43. Liu, G.G.: Intrusion detection systems. In: Applied Mechanics and Materials, vol. 596, pp. 852–855. Trans Tech Publications (2014)
44. Loganathan, G.: A new heuristic optimization algorithm: harmony search. Simulation **76**(2), 60–68 (2001)
45. Manikopoulos, C., Papavassiliou, S.: Network intrusion and fault detection: a statistical anomaly approach. IEE Commun. Mag. **40**(10), 76–82 (2002)
46. McHugh, J.: Testing intrusion detection systems: a critique of the 1998 and 1999 DARPA intrusion detection system evaluations as performed by Lincoln laboratory. ACM Trans. Inf. Syst. Secur. **3**(4), 262–294 (2000)
47. Mell, P., Grance, T.: The NIST definition of cloud computing (2011)
48. Nápoles, G., Grau, I., Vanhoof, K., Bello, R.: Hybrid model based on rough sets theory and fuzzy cognitive maps for decision-making. In: Kryszkiewicz, M., Cornelis, C., Ciucci, D., Medina-Moreno, J., Motoda, H., Ras, Z. (eds.) RSEISP 2014 (2014)
49. Pawlak, Z.: Rough sets. Int. J. Comput. Inf. Sci. **11**(5), 341–356 (1982)

50. Pedrycz, W., Al-Hmouz, R., Morfeq, A., Balamash, A.S.: Building granular fuzzy decision support systems. Knowl.-Based Syst. **58**, 3–10 (2014)
51. Pedrycz, W., Al-Hmouz, R., Morfeq, A., Balamash, A.S.: Distributed proximity-based granular clustering: towards a development of global structural relationships in data. Soft Comput. 1–17 (2014)
52. Pedrycz, W., Skowron, A., Kreinovich, V.: Handbook of Granular Computing. Wiley (2008)
53. Poongothai, T., Duraiswamy, K.: Effective cross layer intrusion detection in mobile ad hoc networks using rough set theory and support vector machines. Asian J. Inf. Technol. **12**(8), 242–249 (2013)
54. Quinlan, J.R.: C4.5: Programs for Machine Learning (2014)
55. Roh, S.B., Pedrycz, W., Ahn, T.C.: A design of granular fuzzy classifier. Expert Syst. Appl. **41**(15), 6786–6795 (2014)
56. Ruck, D.W., Rogers, S.K., Kabrisky, M., Oxley, M.E., Suter, B.W.: The multilayer perceptron as an approximation to a Bayes optimal discriminant function. IEEE Trans. Neural Netw. **1**(4), 296–298 (1990)
57. Shafi, K., Abbass, H.A.: Biologically-inspired complex adaptive systems approaches to network intrusion detection. Inf. Secur. Tech. Rep. **12**(4), 209–217 (2007)
58. Shafi, K., Abbass, H.A.: An adaptive genetic-based signature learning system for intrusion detection. Expert Syst. Appl. **36**(10), 12036–12043 (2009)
59. Shafi, K., Kovacs, T., Abbass, H.A., Zhu, W.: Intrusion detection with evolutionary learning classifier systems. Nat. Comput. **8**(1), 3–27 (2009)
60. Shrivastava, S.K., Jain, P.: Effective anomaly based intrusion detection using rough set theory and support vector machine. Int. J. Comput. Appl. **18**(3), 35–41 (2011)
61. Simmross-Wattenberg, F., Asensio-Pérez, J.I., Casaseca-de-la H.P., Martin-Fernandez, M., Dimitriadis, I.A., Alberola-Lopez, C.: Anomaly detection in network traffic based on statistical inference and alpha-stable modeling. IEEE Trans. Dependable Secure Comput. **8**(4), 494–509 (2011)
62. Siraj, A., Vaughn, R.: Multi-level alert clustering for intrusion detection sensor data. In: Annual Meeting of the North American Fuzzy Information Processing Society, 2005. NAFIPS 2005, pp. 748–753 (2005)
63. Siraj, A., Bridges, S.M., Vaughn, R.B.: Fuzzy cognitive maps for decision support in an intelligent intrusion detection system. In: Joint 9th IFSA World Congress and 20th NAFIPS International Conference, 2001, vol. 4, pp. 2165–2170. IEEE (2001)
64. Siraj, A., Vaughn, R.B., Bridges, S.M.: Intrusion sensor data fusion in an intelligent intrusion detection system architecture. In: Proceedings of the 37th Annual Hawaii International Conference on System Sciences, 2004, pp. 1–10. IEEE (2004)
65. Sivaranjanadevi, P., Geetanjali, M., Balaganesh, S., Poongothai, T.: An effective intrusion system for mobile ad hoc networks using rough set theory and support vector machine. IJCA Proc. E Governance Cloud Comput. Serv. **2**, 1–7 (2012)
66. Song, X., Wu, M., Jermaine, C., Ranka, S.: Conditional anomaly detection. IEEE Trans. Knowl. Data Eng. **19**(5), 631–645 (2007)
67. Sun, J., Yang, H., Tian, J., Wu, F.: Intrusion detection method based on wavelet neural network. In: Second International Workshop on Knowledge Discovery and Data Mining, 2009. WKDD 2009, pp. 851–854. IEEE (2009)
68. Tajbakhsh, A., Rahmati, M., Mirzaei, A.: Intrusion detection using fuzzy association rules. Appl. Soft Comput. **9**(2), 462–469 (2009)
69. Tavallaee, M., Bagheri, E., Lu, W., Ghorbani, A.A.: A detailed analysis of the KDD CUP 99 data set. In: Proceedings of the Second IEEE Symposium on Computational Intelligence for Security and Defence Applications 2009 (2009)
70. Visconti, A., Tahayori, H.: Artificial immune system based on interval type-2 fuzzy set paradigm. Appl. Soft Comput. **11**(6), 4055–4063 (2011)
71. Wang, C.M., Huang, Y.F.: Self-adaptive harmony search algorithm for optimization. Expert Syst. Appl. **37**(4), 2826–2837 (2010)

72. Wang, W., Pedrycz, W., Liu, X.: Time series long-term forecasting model based on information granules and fuzzy clustering. Eng. Appl. Artif. Intell. **41**, 17–24 (2015)
73. Wu, S.X., Banzhaf, W.: The use of computational intelligence in intrusion detection systems: a review. Appl. Soft Comput. **10**(1), 1–35 (2010)
74. Xin, J., Dickerson, J., Dickerson, J.A.: Fuzzy feature extraction and visualization for intrusion detection. In: The 12th IEEE International Conference on Fuzzy Systems, 2003. FUZZ'03, vol. 2, pp. 1249–1254. IEEE (2003)
75. Yang, H., Li, T., Hu, X., Wang, F., Zou, Y.: A survey of artificial immune system based intrusion detection. Sci. World J. **2014** (2014)
76. Yao, Y.: Three-way decision: An interpretation of rules in rough set theory. Lecture Notes in Computer Science (including subseries Lecture Notes in Artificial Intelligence and Lecture Notes in Bioinformatics) 5589 LNAI, 642–649 (2009)
77. Yao, Y.: Three-way decisions with probabilistic rough sets. Inf. Sci. **180**(3), 341–353 (2010)
78. Yong, H., Feng, Z.X.: Expert system based intrusion detection system. In: 2010 International Conference on Information Management, Innovation Management and Industrial Engineering (ICIII), vol. 4, pp. 404–407. IEEE (2010)
79. Yu, M.: A nonparametric adaptive cusum method and its application in network anomaly detection. Int. J. Advancements Comput. Technol. **4**(1), 280–288 (2012)
80. Zaghdoud, M., Al-Kahtani, M.S.: Contextual fuzzy cognitive map for intrusion response system. Int. J. Comput. Inf. Technol. **2**(3), 471–478 (2013)
81. Zhang, C., Jiang, J., Kamel, M.: Comparison of BPL and RBF network in intrusion detection system. In: Rough Sets, Fuzzy Sets, Data Mining, and Granular Computing, pp. 466–470. Springer (2003)
82. Zhang, L., Bai, Z., Luo, S., Cui, G., Li, X.: A dynamic artificial immune-based intrusion detection method using rough and fuzzy set. In: 2013 International Conference on Information and Network Security (ICINS 2013), pp. 1–7 (2013)
83. Zhong, C., Yang, F., Zhang, L., Li, Z.: An efficient distributed coordinated intrusion detection algorithm. In: 2005 International Conference on Machine Learning and Cybernetics, pp. 2679–2685 (2006)

NNCS: Randomization and Informed Search for Novel Naval Cyber Strategies

Stuart H. Rubin and Thouraya Bouabana-Tebibel

Abstract Software security is increasingly a concern as cyber-attacks become more frequent and sophisticated. This chapter presents an approach to counter this trend and make software more resistant through redundancy and diversity. The approach, termed Novel Naval Cyber Strategies (NNCS), addresses how to immunize component-based software. The software engineer programs defining component rule bases using a schema-based Very High Level Language (VHLL). Chance and ordered transformation are dynamically balanced in the definition of diverse components. The system of systems is shown to be relatively immune to cyber-attacks; and, as a byproduct, yield this capability for effective component generalization. This methodology offers exponential increases in cyber security; whereas, conventional approaches can do no better than linear. A sample battle management application—including rule randomization—is provided.

Keywords Battle management · Cybersecurity · Heuristics · Inferential reasoning · Information dominance · Military strategic planning · Transfer learning

1 Introduction

Deductive number of computational devices using embedded software is rapidly increasing and the embedded software's functional capabilities are becoming increasingly complex each year. These are predictable trends for industries such as aerospace and defense, which depend upon highly complex products that require

S.H. Rubin (✉)
Space and Naval Warfare Systems Center Pacific, San Diego, CA 92152-5001, USA
e-mail: stuart.rubin@navy.mil

T. Bouabana-Tebibel
LCSI Laboratory, École nationale Supérieure d'Informatique, Algiers, Algeria
e-mail: t_tebibel@esi.dz

© Springer International Publishing Switzerland 2016   193
R. Abielmona et al. (eds.), *Recent Advances in Computational Intelligence in Defense and Security*, Studies in Computational Intelligence 621,
DOI 10.1007/978-3-319-26450-9_8

systems engineering techniques to create. We also see consumer products as increasingly relying upon embedded software—such as automobiles, cell phones, PDAs, HDTVs, etc.

Embedded software often substitutes for functions previously realized in hardware such as custom ICs or the more economical, but slower gate arrays; for example, digital fly-by-wire flight control systems have superseded mechanical control systems in aircraft. Software also increasingly enables new functions, such as intelligent cruise control, driver assistance, and collision avoidance systems in high-end automobiles. Indeed, the average car now contains roughly seventy computer chips and 500,000 lines of code—more software than it took to get Apollo 11 to the Moon and back. In the upper-end cars, in which embedded software delivers many innovative and unique features, there can be far more code.

However, the great number of source lines of code (SLOC) itself is not a fundamental problem. The main difficulty stems from the ever-more complex interactions across software components and subsystems. All too often, coding errors only emerge after use. Worse still, even good code is increasingly the target of cyber-attacks. The software testing process must be integrated within the software creation process—including the creation of systems of systems in a spiral development. This follows because in theory, whenever software becomes complex enough to be capable of self-reference it can no longer be formally proven valid [1].

Cyber threats are growing in number and sophistication [2]. In theory, it is not possible, in the general case, to produce fault-free software [1, 3]. Attackers have shown the ability to find and exploit residual faults and use them to formulate cyber-attacks. Most software systems in use today run substantially similar software [2]. As a result, successful cyber-attacks can bring down a large number of installations running similar software. As we share more and more software (e.g., through the cloud), the situation can only get worse.

1.1 Background

Complex software underpins the GDP to the extent of about 15 % per year. Clearly, we need to devote more attention to the processes by which efficient and cyber-safe software may be created to improve the national economy.

Redundancy and Diversity in Cyber Defense. Redundancy is effective against hardware faults because such faults are random [2]. However, software faults are typically due to errors of design and/or implementation. This cannot be addressed through redundancy.

Software faults are even more serious because they represent opportunities for exploitation by cyber-attacks. Most seriously, system security software itself can thus be breached.

However, if the system software is built out of a set of diverse, but functionally equivalent components, then a single attack will be insufficient to breach the system. Again, given the same input to the diverse components, whose behavior on this input is known, one would expect the same output. If this is not the case, then a cyber-attack may be in progress.

Worms, viruses, and other infectious attacks can be countered by various types of cyber management techniques. The problem stems from the fact that software, which computes the same function does not need to have the same syntax as is currently the case. The existence of the same flaw on many computers is routinely exploited by attackers via Internet worms such as Code Red, which infected over 350,000 systems in just 13 h using a single vulnerability [4]. Hence, the goal is to introduce more diversity into computer systems. Diversity can be introduced in the software ecosystem by applying automatic program transformation, which maintains the functional behavior and the programming language semantics [2]. In essence, distinct components can compute the same function—insuring computational immunity.

Among the technologies that have the potential of mitigating the cyber-attack risks, "software redundancy" that includes "component diversity" appears to be one of the rare technologies promising an order-of-magnitude increase in system security [2]. The essential idea is to have software functionality redundantly implemented—preventing an attack against any version from being successful against the remaining versions. This also enables the detection of anomalous behaviors—including the resolution of novel solutions (i.e., by comparing multiple runs), which are not attack-based. Forrest et al. [5] argue for security enhancement through the introduction of diversity. According to Ammann et al. [6], there is a lack of quantitative information on the cost associated with diversity-based solutions and a lack of knowledge about the extent of protection provided by diversity. The security enhancement, focused on by this chapter, pertains to the synthesis and assembly of software components using delimited chance and program transformation.

Transformation-Based Diversity. Automatic program transformations can preserve functional behavior and programming language semantics [2]. There are three techniques, in practice, used to randomize code:

1. Instruction Set Randomization (ISR)—changes the instruction set of the processor so that unauthorized code will not run successfully. Cyber-attacks can't inject code if they don't know the true instruction set.
2. Address Space Randomization (ASR)—is used to increase software resistance to memory corruption attacks. ASR randomizes different regions of the process address space (e.g., stacks, arrays, strings, etc.). It has been incorporated into the Windows Vista operating system.
3. Data Space Randomization (DSR)—defends against memory error attacks by masking and unmasking data so that cyber-corrupted data will not be properly restored—implying unpredictable results, which are detectable. DSR can randomize the relative distance between two data object, unlike the ASR technique.

Combining Redundancy and Diversity. Novel and efficient intrusion detection capabilities, not achievable using standard intrusion detection techniques based on signatures or malware modeling involves the monitoring of a redundant system by comparing the behavior of diverse replicas [2]. Any difference in the output responses of the replicas implies a system failure. Most interestingly, this architecture enables the development of adaptive controllers. Our approach is adaptive too, but more sophisticated, as will be seen below.

1. N-Variant Approaches. If the same input is supplied to a set of diversified variants of the same code, then the cyber-attack will succeed on at most one variant—making the attack detectable. The problem with this approach, however, is that the type of attack must be properly anticipated so that it will succeed on at most one variant. That is, variants must vary in the way in which they will respond to a particular cyber-attack. This is increasingly unlikely in today's world.
2. Multi-Variant Code. This technique prevents cyber-attacks by using diversity. It executes variants of the same program and compares the behavior of the variants at synchronization points. Divergence in behavior suggests an anomaly and triggers an alarm. Unlike the case using the n-variant approaches, the synchronization points serve to mitigate the need for a priori knowledge of the type of attack because they provide common entry and exit points under which differences in performance can be measured.
3. Behavioral Distance. One way to beat traditional anomaly-based intrusion detection systems is to emulate the original system behavior (i.e., mimicry attacks). Behavioral distance defends against this by using a comparison between the behaviors of two diverse processes running the same input. A flag is raised if the two processes behave differently.

1.2 Related Work on Transfer Learning

The redundancy and diversity-based approaches, proposed in this chapter, pertain to transfer learning theory. They focus on the need to utilize previously-acquired knowledge to solve problems with greater rapidity and security. They differ from traditional machine learning methods in that they allow for source and target domains to be different [7]. Several survey papers on transfer learning have been published in the last few years; but few apply transfer learning based on computational intelligence (CI) [8, 9]. Transfer learning, with the support of CI formalisms such as neural networks, Bayesian networks, fuzzy systems, and genetic algorithms have been applied in real-world applications. These applications may be subdivided into the following five categories [9]: (1) Nature language processing [10–12]; (2) Computer vision [13–15]; (3) Biology [16–18]; (4) Finance [19–21]; and, (5) Business management [22–24].

Deep learning is a fundamental technique for abstract learning using neural networks [10, 14, 25]. It extracts high-level features, which offer great flexibility in transfer learning. It rests upon multiple hidden layers, where the output of one layer is the input to the next layer. Unsupervised learning is used to pre-train each layer.

Multiple task learning (MTL) is proposed for improving the learning of the target task(s). It includes a number of hidden layers, which are fewer than in deep neural networks. In MTL, information contained in other related tasks is used to promote the performance of the target task [26]. All tasks are trained in parallel using the shared input and hidden neurons. Separate output neurons, corresponding to each task, are provided [27]. This leads to redundant outputs and overlapping information. To remedy this, Silver and Poirier [28] proposed context-sensitive multiple task learning (csMTL), where only one neuron is included in the output layer; and, the input layer also contains a set of contextual inputs, which associates each training example with a particular task.

Many Bayesian-based transfer learning techniques have been developed in recent years to address the problems raised by the classifier trained on source data, which may not be predictive for the target data. To deal with this, [29] proposed a novel naïve Bayes transfer learning classification algorithm. The experimental results show that the performance of this method increases when the distribution between the source data and the target data is significantly different. Roy and Kaelbling [30], developed an approach, where the dataset is first partitioned into a number of clusters, such that the data for each cluster for all tasks has the same distribution. Next, one classifier is trained for each partition; all classifiers are then combined using a Dirichlet process. The Bayesian network is suitable for representing correlations between features in a decision region. Recently, Oyen and Lane [18] stated that it is more appropriate to estimate a posterior distribution over multiple learned Bayesian networks, rather than a single posteriori. They proposed to extend network discovery in individual Bayesian network learning, for transfer network learning, by incorporating structural bias into order-conditioned network discovery techniques.

Fuzzy logic constitutes a major component for Fuzzy Transfer Learning techniques. In [19, 20], a fuzzy-based transductive transfer learning is developed based on a distribution of data in the source domain, which differs from that in the target domain. Next, the fuzzy refinement domain adaptation method [21] is improved by developing a novel fuzzy measure to simultaneously take account of the similarity and dissimilarity in the refinement process. The emphasis is put on the advantage of fuzzy logic in knowledge transfer, where the target domain lacks critical information and involves uncertainty and vagueness. More recently, the authors of [31] proposed a framework for fuzzy transfer learning for predictive modeling in intelligent environments. Genetic algorithms and transfer learning are introduced in [32]. The approach consists in extending the transfer learning method of producing a translation function. This process allows for differing value functions, which have learned to map from source to target tasks. The transfer of inter-task mappings can reduce the time required for learning a second more complex task.

1.3 Contribution

This chapter presents an approach to counter this trend and make software more resistant through redundancy and diversity. The approach that we propose, termed Novel Naval Cyber Strategies (NNCS), addresses how to immunize component-based (functional) software. The software engineer programs defining component rule bases using a schema-based Very High Level Language (VHLL). Chance and ordered transformation are dynamically balanced in the definition of diverse components. Deviation from previously defined non deterministic I/O constraint maps indicates a likely cyber-attack. Redundancy enables simultaneous recovery in most instances; whereas, diversity prevents against the effectiveness of attacks. Moreover, the system of systems counts the relative number of diverse components yielding the same output vector and the relative number of distinct paths used in the synthesis of the mapping component. The system of systems can then be applied to previously unseen input vectors to predict output vectors along with their relative validities. The use of multiple analogies for generalization enables components to better approximate their defining semantics under a finite number of constraints. The system of systems will be shown to be relatively immune to cyber-attacks; and, as a byproduct, yield this capability for effective component generalization.

In the remainder on the paper, Sect. 2 presents the randomization technique behind the proposed approach. Section 3 shows the different aspects of the raised problem. In Sect. 4, a methodology of resolution is developed. Section 5 discusses the approach and provides some concluding remarks.

2   Randomization

Consider the following problem, where the assigned task is the lossless randomization of a sequence of integers [33]. Note that a slightly more complex (real-world) task would be to randomize a similar sequence of integers, where the error-metric (tolerance) need not be zero, but is always bounded. Such sequences arise in the need for all manner of prediction, e.g., from the path of an incoming missile to the movement of storm tracks, et al. This abstraction underpins the novel aspects of the Novel Naval Cyber Strategies (NNCS) systems (see below).

$$\text{Randomize} \quad \begin{array}{llllllll} \text{n:} & 0 & 0 & 1 & 4 & 11 & 26 & 57 \\ \text{i:} & 0 & 1 & 2 & 3 & 4 & 5 & 6 \end{array} \tag{1}$$

A randomization of (1) is given here by $n_{i+1} \leftarrow 2n_i + i$. We say that this randomization is lossless because the associated error-metric (e.g., the 2-norm) is zero. Randomizations may or may not exist given the operator, operand set, and the set error-metric bounds. Furthermore, even in cases where randomizations exist, they may not be discoverable in the allocated search time on a particular processor (s) [34]. In view of this, the general problem of randomization is inherently heuristic.

Clearly, there is no logic that can solve the inductive inference problem [34]. Rather, one needs to define a search space such that the search operators are adequately informed. The more informed the search operators, the less search that is required (i.e., successful or not). Here is one possible schema to delimit the search space in this problem:

$$n_{i+1} \leftarrow M \; \{*, /, +, -, **\} \; n_i \{*, /, +, -\} \{i, n_{i-1}\} \qquad (2)$$

Partially assigning mnemonics, this schema can be described as follows.

$$n_{i+1} \leftarrow \text{int extended} - \text{ops } n_i \text{ops} \{i, n_{i-1}\} \qquad (3)$$

But, even here, it is apparently ambiguous as to how such a schema might be found. To answer this question, consider the randomization of the even sequence, 2n, and the odd sequence, 2n + 1. The randomization of these two sequence definitions is given by $2n + j, j \in \{0, 1\}$. Next, note that "+" $\subset$ ops $\subset$ extended-ops. Each replacement, at the right, represents a level of generalization. Generalizations are not made—except to randomize two or more instances. For example, if the odd sequence were defined by 2n − 1, then a first-level randomization (i.e., based on the given mnemonics) of 2n + 1 and 2n − 1 is given by 2n ops 1. Clearly, having multiple mnemonics can greatly enlarge the search space and result in intractable solutions. An evolutionary approach to reducing the implied search time is to perform a gradient search outward from known valid points. Here, search reduction is obtained by commensurately reducing search diversity. It is claimed that this process is what enables most of us to solve inferential randomization problems such as this one, most of the time. The dual constraints of available search time on a given processor(s) versus the generality of the candidate solution space serves to dynamically contract or expand the search space.

Notice that the process of randomization not only captures existing instances in a more compact form, but in so doing embodies similar instances, which may or may not be valid. The point is that by limiting the degree of generalization, one tightens the degree of analogy and in so doing, increases the chance of a valid inference. The inferences found to be valid are fed back to the randomization process. This results in a more delimited search space and provides for multiple analogies—increasing the subsequent chance for valid inferences. Moreover, according to Solomonoff [35–37], the inference of grammars more general than regular grammars is inherently heuristic. The context-free grammar (CFG) is the lowest-level such grammar. All non-deterministic grammars may be statistically augmented—resulting in stochastic grammars [38]. Furthermore, where heuristics serve in the generation of new knowledge and that knowledge serves in the generation of new heuristics, the amplification of knowledge occurs by way of self-reference [1]! Allowing for the (self-referential) application of knowledge bases, any practical methodology, serving in the discovery of these heuristics, must be domain-general to be cost effective. The transformative search for randomization is the most general such methodology because it extracts self-referential knowledge from conditional as well as procedural knowledge in context [33, 34, 39].

3 Problem Description

The problem is to detect a cyber-attack when it happens and recover from a cyber-attack while it happens. Software needs to be subdivided into components, which map a set of input vectors to a non-deterministic set of stochastic output vectors. Components are defined in terms of other components, which are defined by rules (Fig. 1).

The behavior of a set of Boolean components or a sequence of procedural components is not unique. Thus, it is possible to synthesize a diverse set of components, which provides the desired security for an arbitrary I/O characterization.

3.1   Justification for I/O Characterization of Software Components

It is acknowledged that there is software, which cannot be sufficiently characterized by a non-deterministic stochastic I/O mapping. For example, a component might draw a picture. Here, a knowledge-based system may be applied to rank the quality of the component. In a sense, mapping input to desired output(s) is universal—it's just that intermediate evaluation code is sometimes needed. Thus, while we will not address such complexities in this paper, it is to be understood that the methodology advanced herein is completely compatible with them. In fact, it may be used to define the intermediate knowledge-based evaluation systems.

Another point of contention pertains to the use of empirical testing instead of, or in combination with, denotational or axiomatic semantics for program validation. The recursive Unsolvability of the Equivalence Problem [3] proves that in the

Fig. 1 Recursive rule-based definition of software components

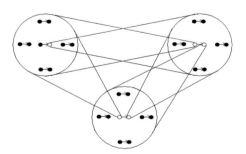

general case it is impossible to prove that two arbitrary programs compute the same function. Moreover, approaches to program validation based on computational semantics have proven to be unacceptably difficult to apply in practice. There can be no theoretical method for insuring absolute validity once a program grows to a level of complexity to be capable of self-reference [1, 3, 34].

It follows that program validation is properly based on empirical testing, the goal of which is to cover a maximal number of execution paths using a minimal number of test cases. This is none other than randomization [33, 34]. Of course, there is no need to achieve the absolute minimum here—a minimum relative to the search time required to find the test cases will suffice. In a large enough system of systems, the methodology advanced herein may be applied to the generation of relatively random test cases. Randomization serves to maximize reuse. Reuse is perhaps the best real-world technique for exposing and thus minimizing the occurrence of program bugs.

3.2 Random-Basis Testing

Each component saved in the database is associated with one or more I/O test vector pairings that serve to map a random input vector to correct non deterministic output vectors. The underpinning principle is that test vectors, which have been sufficiently randomized, are relatively incompressible. For example, consider the synthesis of a sort function using LISP (Fig. 2). There are some extraneous details such as knowing when a particular sequence will lead to a stack overflow, but these are easily resolved using an allowed execution time parameter. Impressive programs have been so synthesized—supporting the component-based concept. Notice that components can be written at any scale—from primitive statements to complex functions. Given only so much allocated search time, the system will either discover a solution or report back with failure. This is in keeping with the recursive Unsolvability of the Halting Problem [3, 34].

Consider such I/O constraints as (((3 2 1) (1 2 3)) ((3 1 2) (1 2 3))). That is, when (3 2 1) is input to the sort function, it is required to output (1 2 3). Similarly, when (3 1 2) is input to it, it is required to output the same. Clearly, there is little value in using a test set such as (((1) (1)) ((2 1) (1 2)) ((3 2 1) (1 2 3)) ((4 3 2 1) (1 2 3 4)) ...).

The problem here is that this test set is relatively symmetric or compressible into a compact generating function. A fixed-point or random test set is required instead and the use of such relatively random test sets is called, random-basis testing [40]. While the need for functional decomposition remains, under random-basis testing, the complexity for the designer is shifted from writing code to writing search schema and relatively random tests. For example, such a test set here is (((1) (1)) ((2 1) (1 2)) ((3 1 2) (1 2 3)) ((1 2 3) (1 2 3))). Many similar ones exist. One may also

```
 ((DEFUN MYSORT (S)

    (COND((NULL S) NIL)

      (T(CONS(MYMIN S(CAR S))

      (MYSORT(REMOVE(MYMIN S(CAR S)) S)))))))
? io

((((1 3 2))(1 2 3))(((3 2 1))(1 2 3))(((1 2 3))(1 2 3)))
? (pprint(setq frepos '((CRISPY'

          (DEFUN MYSORT (S)

            (COND(FUZZY((NULL S) NIL)

              ((ATOM(FUZZY S((FUZZY CAR CDR) S))) NIL))

            (T(CONS(MYMIN S(CAR S))

            (MYSORT(REMOVE(MYMIN S(CAR S)) S))))))))))
((CRISPY '(DEFUN MYSORT(S)

    (COND(FUZZY((NULL S) NIL)

      ((ATOM(FUZZY S((FUZZY CAR CDR) S))) NIL))

      (T(CONS(MYMIN S(CAR S))

      (MYSORT(REMOVE(MYMIN S(CAR S)) S))))))))
; Note that (ATOM S) was automatically programmed using the large
fuzzy function space.
? (pprint(auto frepos io))

            ((DEFUN MYSORT(S)

              (COND((ATOM S) NIL)

              (T(CONS (MYMIN S (CAR S))

              (MYSORT(REMOVE(MYMIN S(CAR S)) S)))))))
; Note that each run may create syntactically different, but semanti-
cally equivalent  functions:
? (pprint(auto frepos io))

            ((DEFUN MYSORT (S)

              (COND((NULL S) NIL)

              (T(CONS(MYMIN S(CAR S))

              (MYSORT(REMOVE(MYMIN S(CAR S)) S)))))))
```

Fig. 2 Function synthesis using random-basis testing

want to constrain the complexity of any synthesized component (e.g., Insertion
Sort, Quicksort, et al.). This can be accomplished through the inclusion of temporal
constraints on the I/O behavior (i.e., relative to the executing hardware and competing software components).

3.3 Component Definition

There are two categories of components—Boolean components, which return True or False and procedural components, which compute all other functions and can post and/or retract information to/from a blackboard. There are two blackboards—a local blackboard, which is only accessible to local component functions and procedures as well as those invoked by them and a global blackboard, which is accessible to all component functions and procedures. The blackboards dynamically augment the input vectors to provide further context.

All components are composed of rules, each of which consists of one or a conjunction of two or more Boolean components, which imply one or a sequence of two or more, procedural components—including global and local RETRACT and POST. Given an input vector and corresponding output vector(s), the rule base comprising the component must map the former to that latter at least tolerance percent of the time. The default tolerance is 100 %. Transformation may also favor the fastest component on the same I/O characterization. Notice that greater diversification comes at an allowance for less optimization.

3.4 Component Synthesis

A library of universal primitive and macro components is supplied and evolved. There are three ways that these are retrieved. First, is by name. Second is by mapping an input vector closer, by some definition (e.g., the 2-norm et al.), to a desired non deterministic output vector (i.e., hill climbing—non contracting transformations reducing the distance to a goal state with each substitution). Third is just by mapping the input vector using contracting and non contracting transformations (i.e., Type 0 transformation). Hill climbing and Type 0 transformation may be combined and occur simultaneously until interrupted. The former accelerates reaching a desired output state, while the latter gets the system off of non-global hills.

Macro components are evolved by chance. They comprise a Very High Level Language (VHLL). For example, a macro component for predicting what crops to sow will no doubt invoke a macro component for predicting the weather. Similarly, a macro component for planning a vacation will likewise invoke the same macro component for predicting the weather (i.e., reuse) [41].

Test vectors are stored with each indexed component to facilitate the programmer in their creation and diversification as well as with the overall understanding of the components function. While increasing the number of software tests is generally important, a domain-specific goal is to generate mutually random ordered pairs [40]. Components in satisfaction of their I/O test vectors are valid by definition. Non deterministic outputs are not stochastically defined for testing as it would be difficult to know these numbers as well as inefficient to run such quantitative tests.

As software gets more complex, one might logically expect the number of components to grow with it. Actually, the exact opposite is true. Engineers are required to obtain tighter integration among components in an effort to address cost, reliability, and packaging considerations, so they are constantly working to decrease the number of software components but deliver an ever-expanding range of capabilities. Thus, macro components have great utility. Such randomizations have an attendant advantage in that their use—including that of their constituent components—implies their increased testing by virtue of their falling on a greater number of execution paths [33, 34, 42]. The goal here is to cover the maximum number of execution paths using the relatively fewest I/O tests (i.e., random-basis testing [40]).

The maximum number of components in a rule, as well as the maximum number of rules in a component, is determined based on the speed, number of parallel processors for any fixed hardware capability, and the complexity of processing the I/O vectors. It is assumed that macro components will make use of parallel/distributed processors to avoid a significant slowdown. Components that are not hierarchical are quite amenable to parallel synthesis and testing.

Components may not recursively (e.g., in a daisy chain) invoke themselves. This is checked at definition time through the use of an acyclic stack of generated calls. Searches for component maps are ordered from primitive components to a maximal depth of composition, which is defined in the I/O library. This is performed to maximize speed of discovery. The components satisfying the supplied mapping characterization are recursively enumerable.

Software engineers can supply external knowledge, which is captured for the specification of components. Components are defined using a generalized language based on disjunction. This is because it is easier to specify alternatives (i.e., schemas) in satisfaction of I/O constraints than to specify single instances (e.g., A| B → C than A → C | B → C; or, A → B|C than A → B | A → C). Moreover, such an approach facilitates the automatic re-programming of component definitions in response to the use of similar I/O constraints. The idea is to let the CPU assume more of the selection task by running a specified number of rule alternates against the specified I/O constraints. This off-loads the mundane work to the machine and frees the software engineer in proportion to the processing speed of the machine. Here, the software engineer is freed to work at the conceptual level; while, the machine is enabled to work at the detailed level. Each is liberated to do what it does best. The number of (macro) Boolean components, (macro) procedural components, and alternate candidate rules is determined by the ply of each and the processing speed of the machine. Notice that the task of programming component rules is thus proportionately relaxed. Programming is not necessarily eliminated; rather, it is moved to ever-higher levels. This is randomization [33]. Furthermore, component-type rule-based languages have the advantage of being self-documenting (e.g., IF "Root-Problem" THEN "Newton-Iterative-Method").

Novel and efficient development environments can be designed to support the pragmatics of such programming.

Each run may synthesize semantically equivalent (i.e., within the limits defined by the I/O test vectors), but syntactically distinct functions (e.g., see the alternative definitions for MYSORT at the bottom of Fig. 2). Similar diversified components are captured in transformation rules. Thus, initially diversified components are synthesized entirely by chance, which of course can be very slow. Chance synthesis is a continual on-going process, which is necessary to maintain genetic diversity. But, once transformation rules are synthesized, they are applied to constituent component rules to create diversified components with great rapidity. The 3-2-1 skew may be applied to favor the use of recently acquired or fired transformation rules. It uses a logical move-to-the-head ordered search based upon temporal locality [43]. The acquisition of new components leads to the acquisition of new transforms. Note that if the system sits idle for long, it enters dream mode via the 3-2-1 skew. That is, it progressively incorporates less recently acquired/fired transforms in the search for diversified components.

Transformation rules can be set to minimize space and/or maximize speed and in so doing generalize/optimize. Such optimizations are also in keeping with Occam's Razor, which states that in selecting among competing explanations of apparent equal validity, the simplest is to be preferred. If, after each such transformation, the progressively outer components do not properly map their I/O characterization vectors, then it can only be because the pair of components comprising the transformation rule is not semantically equivalent. In this case, the transformation is undone and the transformation rule and its substituted component are expunged (i.e., since it has an unknown deleterious I/O behavior). This allows for a proper version to be subsequently re-synthesized. Components having more-specific redundant rules have those rules expunged.

Convergence upon correct components and thus correct transforms is assured. This is superior to just using multiple analogies as it provides practical (i.e., to the limits of the supplied test vectors) absolute verification at potentially multiple component levels. Such validation is not in contradiction with the Incompleteness Theorem as the test vectors are always finite as is the allowed runtime [1].

3.5 Non Monotonic Rules

Non monotonic rules are secondary rules, which condition the firing of primary rules. They have the advantage of being highly reusable—facilitating the specification of complex components. Reuse is a tenet of randomization theory [34]. Both local and global blackboards utilize posting and retraction protocols. The scope of a local blackboard is limited to the originating component and all components invoked by it. For example,

{Laces: Pull untied laces, Tie: Make bow} → GRETRACT:

(Foot − ware: Shoes are untied); GPOST: (Foot − ware: Shoes are tied)   (4)

The order of the predefined, global and local RETRACT and POST procedures is, akin to all procedural sequences, immutable.

3.6   Component Redundancy and Diversification

The pattern-matching search known as backtracking can iteratively expand the leftmost node, or the rightmost node on Open [44]. Results here are not identical, but are statistically equivalent. If one component is provided with one expansion search parameter, the other component must be provided with the same search parameter, or the resultant dual-component search will have some breadth-first, rather than strictly depth-first characteristics. This will change the semantics resulting from the use of large search spaces. Clearly, components need to be transformed with due regard for subtle context to preserve their aggregate semantics. These semantic differences become apparent on input vectors, which are outside of those used for I/O definition. Their use can result in erroneous communications via the local and/or global blackboards. The system of systems, described in the technical approach below, evolves such context-sensitive components and their transformations.

NNCSs can potentially provide exponentially more security than can a multi-compiler by finding multiple paths from start to goal states [44, 45]. Under syntactic differentiation, achieving the same results implies computing the same component semantics. Under transformational equivalence, one need not compute the same exact component semantics—only ones that achieve the same results in the context of other components. Given sufficiently large problem spaces and sufficient computational power, exponential increases in cyber security can thus be had. Syntactic differentiation can at best provide only linear increases in cyber security. Thus, our methodology offers far greater security against cyber-attacks than can conventional approaches [2].

The transformational process converges on the synthesis of syntactically distinct components, which are, to the limits of testing, semantically equivalent. Such components can be verified to be free from attack if their I/O synthesis behavior is within the specified tolerance. Even so, multiple "semantically equivalent" components may compute different output vectors on the same, previously untested input vectors. Here, diversity enables the use of *multiple functional analogies* by counting the number of diverse components yielding the same output vector. It also allows for a count of the approximate number of recursively enumerable distinct paths leading to the synthesis of each component. This *multiple analogies of derivation*, when combined with multiple functional analogies, provide a *relative validity metric* for voting the novel output vectors. These solution vectors are very important because they evidence the system capability for learning to properly generalize by way of exploiting redundancy (i.e., in both function and derivation). Furthermore, having multiple derivations provides stochastic non deterministic probabilities. This lies at the root of human imagination and knowledge.

4 Technical Approach

The more constrained the search for knowledge, not only the faster that knowledge may be discovered, but the faster heuristics aiding in the discovery of that knowledge may be obtained as well.

4.1 The System of Systems Randomization Methodology

To this point, redundancy and diversification have been discussed in the context of detecting and recovering from a cyber-attack as well as in the inductive prediction of outputs for inputs not previously supplied. The methodology that follows is depicted in Fig. 3.

Component Types. There are two categories of components—Boolean components, which return True or False and procedural components, which compute all other functions. There are two blackboards—a local blackboard and a global blackboard, which is accessible to all component functions and procedures. The blackboards dynamically augment the input vectors to provide further context. Two special predefined components are the non monotonic global and local RETRACT and POST procedures. Each carries a single well-defined argument found in the I/O library. The scope of a local blackboard is limited to the originating component and all components invoked by it. The global blackboard is visible to all components. Postings and retractions should be made by the most primitive level component as is practical (i.e., having the lowest maximal depth of composition) to facilitate efficiency and validity (i.e., minimizing the potential for deleterious side effects).

Component Structure. All components are composed of rules, each of which consists of one or a conjunction of two or more Boolean components, which imply one or a sequence of two or more, procedural components—including global and local RETRACT and POST. Given an input vector and corresponding output vector (s), the rule base comprising the component must map the former to that latter at least tolerance percent of the time. The default tolerance, t, is 100 %. A Boolean speed of computation compiler directive, s, when set means that the direction of transformation favors the component performing at least as fast on the same I/O characterization vectors. A Boolean space of computation compiler directive, a, is similar.

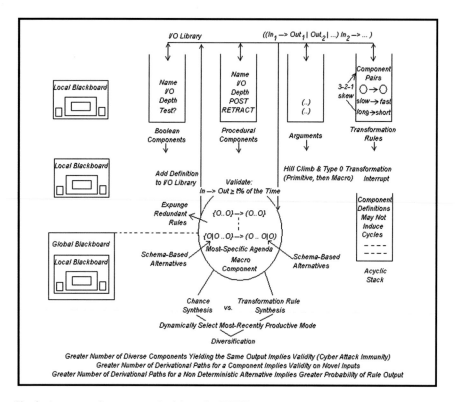

Fig. 3 A system of systems methodology for NNCS

Macro Component Library. A library consisting of at least universal primitive and macro components is supplied. I/O test vectors and the maximal depth of composition are stored with each indexed component. Components may be retrieved by name, by mapping an input vector closer, by some definition (e.g., the 2-norm et al.), to a desired non deterministic output vector (i.e., hill climbing—non contracting transformations reducing the distance to a goal state with each substitution), and/or by mapping the input vector (i.e., Type 0 transformation—contracting and non contracting transformations). Hill climbing and Type 0 transformations are interleaved, since each can benefit the other. Search is terminated upon interrupt.

Component Synthesis. Macro components are evolved by chance. Basically, Boolean and procedural components are selected from the library at chance and combined into defining rules based on software engineer defined schemas (see below). Set the maximum number of components in a rule and the maximum number of rules in a component—at the primitive level. The maximum number of such components and such rules is determined by the software engineer in consideration of the capabilities of the executing hardware, the complexity of processing the I/O vectors, and any supplied external knowledge (see below). These maximums will need to respect macro components if a sufficient number of parallel processors cannot be had. This may be accomplished by dividing this number by the maximal depth of composition found in the I/O library. The process iterates until the supplied I/O vectors are properly mapped within the specified tolerance, or an interrupt signals failure to do so (whereupon the software engineer may modify the search specification, or abandon it). Components may not recursively (e.g., in a daisy chain) invoke themselves. This is checked at definition time through the use of an acyclic stack of generated calls. Once evolved, macro components are added to the I/O library. All else being equal, search primitive components before macro components, as recursively defined (referring to the maximal depth of composition in the I/O library), for effective I/O maps.

Component Definition. Components are defined using a generalized language based on disjunction. This is because it is easier to specify alternatives (i.e., schemas) in satisfaction of I/O constraints than to specify single instances (e.g., A| B → C than A → C | B → C; or, A → B | C than A → B | A → C). The number of (macro) Boolean components, (macro) procedural components, and alternate candidate rules is determined by the ply of each and the processing speed of the machine. Furthermore, component-type languages have the advantage of being self-documenting (e.g., IF "Root-Problem" THEN "Newton-Iterative-Method"). Novel and efficient development environments can be designed to support the pragmatics of such programming.

Component Schedule. Synthesize components in satisfaction of the I/O test vectors and s, a, and t by chance. Such synthesis may lead to diverse components computing the same function. Pairings of such components form transformation rules, which are saved in a separate base and dynamically ordered using the 3-2-1 skew. Rules are logically moved to the head of their list upon acquisition or firing. Convergence upon correct components, and thus correct transforms, and so on follows with scale. A most-specific first agenda mechanism controls the firing of component rules. Redundant rules, having a more-specific (i.e., superset) of Boolean components, are expunged. The direction of transformation is determined by compiler directives, based on s, a, and t. Use of the s and/or a optimization directives minimizes the potential for diversification. Conversely, decreasing the t generalization directive maximizes the potential for diversification.

Component Computation. Diverse components are constructed by transformation, which in turn depends on random component synthesis as a source of transformation rules. The relative time spent (processors allocated) for each is dynamically given as follows.

Let, r(t) give the number of novel transformation rules yielded by components synthesized by chance, over some timeframe, t.

Let, x(t) give the number of novel component rules yielded by transformation rules, over some timeframe, t.

Note that it could potentially reduce the diversity space; and, it is otherwise redundant to self-apply transformation rules.

Then, the percent of time/resources to be spent in transformation rule synthesis is given by $\frac{r(t)+1}{r(t)+x(t)+2}$; while, the percent of time/resources to be spent in component rule synthesis is given by $\frac{x(t)+1}{r(t)+x(t)+2}$, where if $r(t) + x(t) = 0$, t is doubled else t is halved. t is initialized to 1.0. Thus, time/resources are proportionately spent where they were most-recently productive.

Component Diversification. The 3-2-1 skew favors the use of recently acquired or fired transformation rules. Transformation rules are applied to the (symmetric) rules comprising a component to yield diversified components. Diversified components are realized using at least one diversified rule, which in turn, consists of at least one diversified component. Duplicate transformation rules are logically moved to their list head. Every component substitution is verified using the local and progressively higher I/O characterization vectors and invoking components. In case of failure, the involved component rule, the involved transformation rule, and the substituted component are expunged.

Component Validation. Components are verified to be free from attack if their I/O synthesis behavior is within the specified tolerance. Even so, multiple "semantically equivalent" components may compute different output vectors on the same, previously untested input vectors. Here, diversity enables the use of Multiple Functional Analogies (MFA) by counting the number of diverse components yielding the same output vector. It also allows for a count of the approximate number of recursively enumerable distinct paths leading to the synthesis of each component. This is approximated by the number of times that it is derived—including random and transformational synthesis. This Multiple Analogies of Derivation (MAD), when combined with the MFA, provide a Relative Validity Metric (RVM) for voting the novel output vectors. Using the 3-2-1 skew, components synthesized from more recently acquired/fired transformation rules are given a higher relative validity, since they are more likely to be repeatedly derived. This makes sense because these solutions are immediately needed (i.e., just in time synthesis) and not stored for possible future use. The MAD for the ith combination of Boolean components in a rule is given by:

$$MAD(i) = \frac{card\{component_i \; synthesis\}}{(\sum_{j=1}^{|components|} card\{component_j \; synthesis\})/|components|} \tag{5}$$

The greater the MAD, the more likely the novel output vector is to be valid. The MFA for the ith combination of Boolean components in a rule is given by:

$$MFA(i) = \frac{\sum_{k=1}^{NDO} card\{component_{i,k} \; output\}}{\sum_{k=1}^{NDO}(\sum_{j=1}^{|components|} card\{component_{j,k} \; outputs\}/|components|)/NDO} \tag{6}$$

where NDO is the number of non-deterministic outputs per component. Hence, the joint RVM for the ith combination of Boolean components in a rule is given by combining (5) and (6):

$$RVM(i) = MAD(i) \cdot MFA(i) \tag{7}$$

The greater the RVM, the more likely the output is to be valid. Validity is associated with an RVM > 1. Absolute component validity is predicated on testing as absolute validity is not necessarily provable [1].

Non Deterministic Outputs. Non deterministic procedural alternatives are defined to be a member of the specified output vectors. The probability of each distinct alternative is directly proportional to the number of paths for its synthesis. This, in turn, is approximated by the number of times that it is derived—including random and transformational synthesis. Thus, the dynamic stochastic probability for the jth non deterministic selection for the ith combination of Boolean components in a rule is given by:

$$non \det prob(i,j) = \frac{card\{procedural_j \; synthesis\}}{\sum\limits_{k=1}^{NDO} card\{procedural_k \; synthesis\}} \tag{8}$$

where NDO is the number of non deterministic outputs.

4.2 Proof of Concept

This methodology will be proven to be immune to cyber-attack by two routes. First, it will be demonstrated that a significant percentage of components can be corrupted and the system of systems will autonomously discover and report this occurrence and still return correct outputs. Second, it will be demonstrated that the system of systems can generalize I/O maps in the form of diverse components that can usually properly map previously unseen inputs to correct outputs. This is accomplished by supplying NNCSs I/O vectors, letting it learn diverse component maps, supplying novel input vectors for similar problems—correct output vectors for which have never been supplied, and seeing if the system converges on finding correct non deterministic output vectors. Such a result not only solves the long-standing generalization problem in CBR [46], but the context-based knowledge extension problem, previously described in [46–49].

Contemporary components for NNCS will be taken from select newLISP functions used for the realization of the methodology (e.g., bootstrapping). The system of systems will automatically generate diverse components from them. The performance of the system will be rated as a function of scale. It will be shown that the inferential error rate is inversely proportional to scale. That is, the larger the domain-specific component base and the more processing power/time allocated, the lower the inferential error rate. A cost-benefit analysis of the protection provided by component diversification can be provided based on an empirical study and projection of its scalability.

5   CI for Tactical Battle Management

In this section, our prototype NNCS program will be demonstrated as applied to learning tactical battle management. Here, the focus will be on learning more than on actual battle management techniques. Indeed, the entire program is being reworked and moved from a rule-based approach to an algorithmic one to better capture the nuances of natural language understanding as well as those pertaining to naval battles. The problem is that although rules are universal [44] and cases are easy to manage [46–49], neither generic approach can efficiently capture the structure and knowledge applied in complex decision making.

5.1   Overview of Battle Management

The goal of battle management is to provide decision superiority to one side, which then translates into victory over an opponent(s). Successful battle management requires assessing the context—that is, what is and is not relevant to the outcome of a battle(s). It also entails logistics, or the allocation of scarce resources where and when they are needed to provide the necessary advantage. In what follows, this desired behavior is approximated through a rule-based system having a computationally intelligent capability to learn. Note that successful learning, in theory at least, is merely a matter of allocating enough time and space to the algorithm. However, in practice, a far more programmed structure becomes necessary with scale. The following example is abridged as necessary.

5.2   NNCS Interactive Learning

Here, a sample run of the first NNCS LISP program will be presented to highlight a few capabilities. Randomization is performed in "dream mode" (i.e., when the system would otherwise be idle), where the system applies learned equivalence rules to itself to randomize its knowledge bases. Not only does this facilitate contextual matching, because the context and situations are similarly normalized; but, it also economizes produced actions through randomization. As an example of the latter, it might substitute automobile for the description of a car. However, a simple randomization example requires about ten pages, which exceeds the allocated space herein. The program is cold-started and is devoid of any domain or natural language processing knowledge.

Dream Demo. Here is an early battle management executable, which demonstrates its capability for randomization in dream mode. Note that only, "shorthand keys", not the equivalent "narrative phrases", are shown to conserve space. Generally, with sizable knowledge bases, dream mode will be measured in hours, but here we use the minimum of 1 min for demo. Randomization serves to enable contextual matches that are syntactically distinct, but semantically equivalent. Since multiple transformations are used, this can be an arbitrarily complex equivalence. The randomization of actions, for example, allows one to state, "seek shelter," instead of (1) find abode with four walls and roof, (2) make path to abode, (3) ... Not only does this make for convenient communication to human or machine, but similarly enables practical transformation (e.g., shelter → aircraft hanger versus the unruly expanded equivalent).

This function uses the 3-2-1 skew to randomly transform a random rule LHS and RHS. Rules are randomly selected from [0, 0], [0, 1], [0, 2], [0, 3], ... , [0, n], and holding. Rules LHS and RHS are processed in sequence before selecting the next rule. The goal is to iteratively minimize the length of each side. This minimizes the number of questions asked and often provides the best answers. It also conserves space and thus speeds up the algorithm as well. The following three lists—left-hand side transformation rules, right-hand side transformation rules, and the rule base, respectively, were used for brief demonstration of dream mode. A fourth list, or dictionary, used for table lookups for translation is not shown. In what follows, the user repeats many queries. This is done to show how the responses evolve with learning. Also, the CI system has capabilities to reuse answers for non-monotonic reasoning, which have not been adequately demonstrated for the sake of brevity. Brief comments have been added for purposes of explanation. As in LISP, each comment is prefixed with a semicolon.

(setq LHSeqrules '((("~") ("~")) (("y" "y" "p") ("q" "t")) (("x" "p") ("y")) (("p" "x" "p") ("p" "p")) (("x" "x") ("z")))); situational transformation rules of the form, ((LHS) (RHS))

(setq RHSeqrules '((("~") ("~")) (("z" "p") ("y")) (("x" "x") ("z"))))
; action transformation rules, where the matched LHS is transformed into the RHS

(setq rulebase '((("~") ("~")) (("p" "x" "x") ("p" "x" "x")) (("x" "x" "p" "x" "x" "p" "x" "x" "p") ("x" "x" "p" "x" "x" "p" "x" "x" "p")) (("x" "x" "y" "y" "x" "y") ("u" "x" "p" "x" "x" "x" "p" "x" "x")))); main rule base, where the rules of the form, ((situation) (action)) are acquired from user interaction to correct perceived errors

> (Main)
Sun Jun 28 13:35:43 2015

Please specify the desired filename, or use "Enter" to prevent saving results
? battle management
The file, "BATTLE MANAGEMENT" is unknown. Do you wish to cold start and
create it (c), or not (n)
? c

The four knowledge bases and keylength have been initialized.
At Sea. Here is the battle management executable, illustrated for the context of
learning decision superiority for a hypothetical naval battle:
...
Can we continue (yes or no)
? y

Tell me the starting conditions and end with a <CR>
? An enemy submarine has been detected.
The matching sentence is:
An enemy submarine may launch a torpedo against our ship
Is that right
? y

You said that I was correct.
The unique key is "P" ; an automatically created unique identifier
Give me an antecedent and end with a <CR> ; antecedent is a situation
? An enemy submarine has been detected off the port bow.
The matching sentence is: ; the use of similarity metrics found a disjoint semantics
Depth charges can reach enemy submarine at distance
Am I correct
? n

You said no or that you don't know. ; entering a "dk" would produce the same
result here
    What is an equivalent consequent for:
(Press "Enter" to copy and use the shown situation for the action:)
? An enemy submarine has been detected off the port bow
? ; the action and the situation are set to the same here for simplicities sake

The associated right-hand side predicate is:
an enemy submarine has been detected off the port bow
The unique key is "i" ; we set a system parameter was set to show this unique id
Tell me the starting conditions and end with a <CR>
? ; an absent input serves to signal an end to a conjunction of antecedents

The initial (last) context was:
An enemy submarine may launch a torpedo against our ship.
- and -
An enemy submarine has been detected off the port bow.
In my judgement, the following action is likely.

This is a new action.
The fired consequent in shorthand keys is: ("f" "s")
The fired consequent in narrative phrases is:
Fire anti-submarine weapon. ; this corresponds to the "f" key
- and -
Turn ship to defend against torpedoes.; this corresponds to the "s" key
The probability is 50 percent; the system derives a probability from the match
Is that last actions meaning correct
? y

Can you give me an exactly equivalent correct context (yes, no, or dk)
? y

Tell me the starting conditions and end with a <CR>
? An enemy submarine is off the port bow.
The matching sentence is:
An enemy submarine has been detected off the port bow
Is that right
? y

You said yes. ; the echo will sometimes appear—just as in natural conversation
The unique key is "i"
Enter a contextual sentence and end with a <CR>
? The danger is that the enemy submarine may torpedo our ship.
The matching sentence is:
An enemy submarine may launch a torpedo against our ship
Is that right
? y

You said that I was correct.
The unique key is "P"
Give me an antecedent and end with a <CR>
? ; an absent input serves to signal an end to a conjunction of antecedents

Can you give me another situation having the same exact meaning (yes, no, or dk)
? n ; this is so it learns semantically equivalent situations for use in matching
Can you restate the action without changing its exact meaning (yes, no, or dk)
? y ; this is so it learns semantically equivalent actions for use in matching
Tell me a proper action, or unknown, and end with a <CR>
? Turn the ship away from the enemy submarine.
The matching sentence is:
An enemy submarine has been detected off the port bow
; this does not mean the same thing
Am I correct

? n

You said that I was incorrect or that you don't know.
What is an equivalent antecedent for:
(Press "Enter" to copy and use the shown action for the situation:)
? Turn the ship away from the enemy submarine
? ; the action and the situation are ascribed the same syntax
The associated left-hand side predicate is:; a carriage return here "duplicates"
turn the ship away from the enemy submarine
The unique key is "n"
Tell me a proper action, or unknown, and end with a <CR>

? Launch an anti-submarine weapon.
The matching sentence is:
Fire anti-submarine weapon
Am I right
? y

You said that I was correct.
The unique key is "f"
Tell me a proper action, or unknown, and end with a <CR>

??

Can you restate the action without changing its exact meaning (yes, no, or dk)
? n

There were no cycles in the action sequence.; cycles need be broken for answers
Keep going (yes, no, copy, or append (last action))
? y

Give me an antecedent and end with a <CR>
? An enemy submarine periscope was detected off the port bow.

The found matching sentence is:
An enemy submarine has been detected off the port bow
Is that correct
? y

You said yes.
The unique key is "i"
Enter a contextual sentence and end with a <CR>
?

The initial (last) context was:
An enemy submarine has been detected off the port bow.
Here is what I think is likely.
This is a new action.
The transformed and fired consequent in shorthand keys is: ("f" "s")
The transformed and fired consequent in narrative phrases is:

Fire anti-submarine weapon.
- and -
Turn ship to defend against torpedoes.
The probability is 50 percent.
Is that last action semantically correct
? y

Can you give me another context having the same exact meaning (yes, no, or dk)
? n

Can you give me an exactly equivalent correct action (yes, no, or dk)
? n

The action sequence was acyclic.
Continue (yes, no, copy, or append (last action))
? y

Enter a contextual sentence and end with a <CR>
?

The starting (last) context was:
Enemy submarine detected off starboard bow.
This is a new context for me.
The initial (last) action was:
Shoot back.
The same transformed (last) action is:
Fire back rounds.
Do these consequents have the same meaning (yes, no, or dk)
? y

Could you give me a proper action (yes, no, or dk)
(Or, initially enter "f", or "final", for a terminating cycle)
? y

Give me a consequent, or unknown, and end with a <CR>
? Sound general quarters
I recall that exact sentence.
The unique key is "h"
Enter an action sentence, or unknown, and end with a <CR>
? Launch anti-submarine weapon.
The matching sentence is:
Fire anti-submarine weapon
Is that right
? y

You agree with me.
The unique key is "f"
Tell me the starting conditions and end with a <CR>
?

The initial (last) context was:
Fire anti-submarine weapon.
Turn ship to defend against torpedoes.
Fire anti-aircraft guns.
- and -
We are under attack by an enemy submarine enemy tanks and enemy aircraft.
Is this the correct situation
? y

Here is what I think is very probable.
This is novel.
The fired consequent in shorthand keys is:
("f" "x" "N" "T")
The fired consequent in narrative phrases is:
Fire anti-submarine weapon.
Turn ship away from enemy submarine.
Fire anti-aircraft guns.
- and -
Fire agm-114r missiles at enemy tanks ashore.
The probability is 75 percent.
Is my last action correct
? y

Can you give me another proper context having the same exact meaning (yes, no, or dk)
? n

Can you give me another consequent having the same exact meaning (yes, no, or dk)
? n

There were no cycles in the action sequence.
Keep going (yes, no, copy, or append (last action))
? y

Tell me the starting conditions and end with a <CR>
? The ship is under attack by enemy aircraft, an enemy submarine, and enemy tanks.

The found matching sentence is:
We are under attack by an enemy submarine enemy tanks and enemy aircraft
Am I right
? y

You said yes.
The unique key is "R"
Tell me the starting conditions and end with a <CR>
?

The beginning (last) antecedent was:
We are under attack by an enemy submarine enemy tanks and enemy aircraft.
In my opinion, the following action is improbable.
This is novel.
The fired consequent in shorthand keys is: ("f" "x" "N" "T")
The fired consequent in narrative phrases is:
Fire anti-submarine weapon.
Turn ship away from enemy submarine.
Fire anti-aircraft guns.
- and -
Fire agm-114r missiles at enemy tanks ashore.
The probability is 25 percent.
Is my suggested action right
? y

Can you give me an exactly equivalent correct context (yes, no, or dk)
? n

Can you restate the action without changing its exact meaning (yes, no, or dk)
? n

There were no cycles in the action sequence.
Do you want me to keep going (yes, no, copy, or append (last action))
? n

May I take a dream hour now (yes or no)
? n

Do you wish to save the file, "BATTLE MANAGEMENT"
? y

Proceeding to save the file, "BATTLE MANAGEMENT".
The file, "BATTLE MANAGEMENT" has been saved in the Knowledgebases folder.
The file, "BATTLE MANAGEMENT" has been saved in the Backups folder.
Semper Fi!

5.3 Summary

Rules can be learned for a computationally intelligent system as was demonstrated above. However, the non-monotonic reasoning required with scale becomes unmanageable. Details need to be explicitly programmed—not only in the interests of efficiency, but because without a sufficient framework, such learning becomes NP-hard. This then is the pragmatic side—again, despite the universality of rule-based systems.

Case-based systems were reviewed for the intelligent capability to generalize the cases. The case-based version is a better design because it can combine statistical and formal generalizations and can insure cyber security using multiple processors vice multiple non-deterministic codes. It is also more scalable. However, even here, with increasing scale such generalization needs algorithmic knowledge to be valid, tractable, and in the needed direction. Even a transformative calculus needs heuristics to squelch combinatoric explosions. The problem is like attempting to scale the predicate calculus (Prologue) [44]. Eventually, heuristics are needed, or the back-cut and resolution mechanisms grind to a halt under the combinatorics of their own expansion [50].

6 Discussion and Concluding Remarks

The Mission critical systems are increasingly subject to operation in hostile environments, where cyber-attack is just a click away. The cost of combining redundancy and component diversity is justified by the cost of security failures in mission critical systems.

The greater the multiplicity of components derived through chance and transformation, the greater their individual reliabilities will be through the use of multiple analogies. Chance and ordered transformation are dynamically balanced in the definition of diverse components. Communication occurs, using non monotonic components, through both a global and local blackboards. Although the methodology is self-referential, it is not subject to the limitations imposed by the Incompleteness Theorem [1]. This is because it is inherently heuristic.

In theory, the only competing way to realize the results promised in this chapter is to apply knowledge to the inference of knowledge. A few approaches here have met with limited success [50]. The problem is that the knowledge, which is self-referential, needs proper context for applicability. This context cannot be generated by the formal system itself due to limitations imposed by the Incompleteness Theorem [1]. Rather, it must be externally supplied, which by definition necessarily makes it an incomplete set, or it must be heuristic in nature (e.g., multiple analogies) to avoid applicability of the Incompleteness Theorem.

A divergent multiple-analogies approach to component synthesis underpins this chapter. A theoretical consequence of this heuristic approach is that all non-trivial learning systems must embody an allowance for inherent error in that which may be learned. Thus, despite the seeming appeal of valid deduction systems (e.g., the predicate calculus and Japan's Fifth Generation project [51]), they are inherently not scalable. The Navy requires scalable software systems, which are relatively immune to cyber-attack.

The novel technology has been realized in NNCSs for the generation of symmetric software for countering cyber-attacks. The problem here pertains to the acquisition of components along with a methodology for mapping supplied input vectors to one or more desired stochastic output vectors. These maps need to be diverse to thwart cyber-attacks as well as to allow for the use of multiple analogies to better predict the (non-deterministic) mapping of previously unknown inputs. This methodology has been realized in newLISP (in view of its superior list processing capabilities) as a system of systems. It enables a relative immunity against cyber-attacks. It can also succeed against a sequence of progressively more complex problems for which no solution has been pre-programmed; although, the learning mechanism here is not very efficient. Finally, the performance of the system (i.e., the inferential error rate) is tied to the size of the transformational base as well as the processing power/time allocated in conjunction with the schema-definition language. This is unbounded, by any non-trivial metric, because the Kolmogorov complexity of computational intelligence is unbounded [52].

The value of a successful experiment is that as a result, component software systems will be able to protect themselves against cyber-attacks. Our methodology offers exponential increases in cyber security; whereas, conventional approaches can do no better than linear [45]. Moreover, intelligent software systems are able to learn outside the bounds specified by supplied I/O constraints—by inductive inference.

References

1. Uspenskii, V.A.: Gödel's Incompleteness Theorem, Translated from Russian. Ves Mir Publishers, Moscow (1987)
2. Gherbi, A., Charpentier, R., Couture, M.: Software diversity for future systems security. CrossTalk **24**, 10–13 (2011)
3. Kfoury, A.J., Moll, R.N., Arbib, M.A.: A Programming Approach to Computability. Springer, New York (1982)
4. Song, D., Reiter, M., Forrest, S.: Taking Cues from Mother Nature to Foil Cyber Attacks. NSF PR 03-130, https://www.nsf.gov/od/lpa/news/03/pr031 (2003)
5. Forrest, S., Somayaji, A., Ackley, D.H.: Building diverse computer systems. In:Workshop on Hot Topics in Operating Systems, pp. 57–72 (1997)
6. Ammann, P., Barnes, B.H., Jajodia, S., Sibley E.H. (eds): Computer Security, Dependability, and Assurance: from Needs to Solutions. IEEE Computer Society Press, Williamsburg (1998)
7. Fung, G.P.C., Yu, J.X., Lu, H.J., Yu, P.S.: Text classification without negative examples revisit. IEEE Trans. Knowl. Data Eng. **18**(1), 6–20 (2006)
8. Pan, S.J., Yang, Q.: A survey on transfer learning. IEEE Trans. Knowl. Data Eng. **22**(10), 1345–1359 (2010)
9. Lu, J., Behbood, V., Hao, P., Zuo, H., Xue, S., Zhang, G.: Transfer learning using computational intelligence: a survey. Knowl. Based Syst. **80**, 14–23 (2015)
10. Huang, J.-T., Li, J., Yu, D., Deng, L., Gong, Y.: Cross-language knowledge transfer using multilingual deep neural network with shared hidden layers. In: IEEE International Conference on Acoustics, Speech and Signal Processing (ICASSP), Vancouver, Canada (2013)
11. Swietojanski, P., Ghoshal, A., Renals, S.: Unsupervised cross-lingual knowledge transfer in DNN-based LVCSR. In: IEEE Workshop on Spoken Language Technology, Miami, USA (2012)
12. Behbood, V., Lu, J., Zhang, G.: Text categorization by fuzzy domain adaptation. In: IEEE International Conference on Fuzzy Systems, Hyderabad, India (2013)

13. Ciresan, D.C., Meier, U., Schmidhuber, J.: Transfer learning for Latin and Chinese characters with deep neural networks. In: International Joint Conference on Neural Networks, Australia (2012)
14. Kandaswamy, C., Silva, L.M., Alexandre, L.A., Santos, J.M., De Sa, J.M.: Improving deep neural network performance by reusing features trained with transductive transference. In: 24th International Conference on Artificial Neural Networks, Hamburg, Germany (2014)
15. Shell, J., Coupland, S.: Towards fuzzy transfer learning for intelligent environments. Ambient Intell. **7683**, 145–160 (2012)
16. Celiberto Jr., L.A., Matsuura, J.P., De Mantaras, R.L., Bianchi, R.A.C.: Using cases as heuristics in reinforcement learning: a transfer learning application. In: International Joint Conference on Artificial Intelligence, Barcelona, Spain (2011)
17. Niculescu-Mizil, A., Caruana, R.: Inductive transfer for Bayesian network structure learning. In: 11th International Conference on Artificial Intelligence and Statistics, Puerto Rico (2007)
18. Oyen, D., Lane, T.: Bayesian discovery of multiple Bayesian networks via transfer learning. In: 13th IEEE International Conference on Data Mining (ICDM), Dalla, USA (2013)
19. Behbood, V., Lu, J., Zhang, G.: Long term bank failure prediction using fuzzy refinement-based transductive transfer learning. In: IEEE International Conference on Fuzzy Systems, Taiwan (2011)
20. Behbood, V., Lu, J., Zhang, G.: Fuzzy bridged refinement domain adaptation: long-term bank failure prediction, Int. J. Comput. Intell. Appl. **12**(01) (2013)
21. Behbood, V., Lu, J., Zhang, G.: Fuzzy refinement domain adaptation for long term prediction in banking ecosystem. IEEE Trans. Ind. Inform. **10**(2), 1637–1646 (2014)
22. Ma, Y., Luo, G., Zeng, X., Chen, A.: Transfer learning for cross-company software defect prediction. Inf. Softw. Technol. **54**(3), 248–256 (2012)
23. Luis, R., Sucar, L.E., Morales, E.F.: Inductive transfer for learning Bayesian networks. Mach. Learn. **79**(1–2), 227–255 (2010)
24. Shell, J.: Fuzzy transfer learning. Ph.D. thesis, De Montfort University (2013)
25. Chopra, S., Balakrishnan, S., Gopalan, R.: DLID: deep learning for domain adaptation by interpolating between domains. In: ICML Workshop on Challenges in Representation Learning, Atlanta, USA (2013)
26. Caruana, R.: Multitask learning: a knowledge-based source of inductive bias. In: Tenth International Conference of Machine Learning, MA, USA (1993)
27. Caruana, R.: Multitask learning. Mach. Learn. **28**, 41–75 (1997)
28. Silver, D.L., Poirier, R.: Context-sensitive MTL networks for machine lifelong learning. In: 20th Florida Artificial Intelligence Research Society Conference, Key West, USA (2007)
29. Dai, W., Xue, G., Yang, Q., Yu, Y.: Transferring naive Bayes classifiers for text classification. In: 22nd National Conference on Artificial Intelligence, Vancouver, Canada (2007)
30. Roy, D.M., Kaelbling, L.P.: Efficient Bayesian task-level transfer learning. In: International Joint Conference on Artificial Intelligence, Hyderabad, India (2007)
31. Shell, J., Coupland, S.: Fuzzy transfer learning: methodology and application. Inf. Sci. **293**, 59–79 (2015)
32. Koçer, B., Arslan, A.: Genetic transfer learning. Expert Syst. Appl. **37**(10), 6997–7002 (2010)
33. Chaitin, G.J.: Randomness and mathematical proof. Sci. Amer. **232**(5), 47–52 (1975)
34. Rubin, S.H.: On randomization and discovery. Inform. Sciences **177**(1), 170–191 (2007)
35. Solomonoff, R.: A new method for discovering the grammars of phrase structure languages. In: International Conference Information Processing, pp. 285–290. UNESCO, Paris, France (1959)
36. Solomonoff, R.: A Formal Theory of Inductive Inference. Inf. Control **7**, 71–22 and 224–254 (1964)
37. Honavar, V., Slutzki, G. (eds.): Grammatical Inference. Lecture Notes in Artificial Intelligence, vol. 1433. Springer, Berlin (1998)
38. Fu, K.S.: Syntactic Pattern Recognition and Applications. Prentice-Hall Advances in Computing Science and Technology Series. Prentice Hall, Englewood Cliffs (1982)

39. Rubin, S.H.: Computing with words. IEEE Trans. Syst. Man Cybern. Part B **29**(4), 518–524 (1999)
40. Liang, Q., Rubin, S.H.: Randomization for testing systems of systems. In: Proceedings of the 10th IEEE International Conference Information Reuse and Integration, pp. 110–114. Las Vegas, NV, 10–12 Aug 2009
41. Bouabana-Tebibel, T., Rubin, S.H. (eds.): Integration of Reusable Systems. Springer, Switzerland (2014)
42. Kolmogorov, A.N., Uspenskii, V.A.: On the definition of an algorithm. Amer. Math. Soc. Transl. **29**(2), 217–245 (1963) (in Russian, English translation)
43. Deitel, H.M.: An Introduction to Operating Systems. Prentice Hall, Inc., Upper Saddle River (1984)
44. Nilsson, N.J.: Principles of Artificial Intelligence. Tioga Publishing Company, Palo Alto (1980)
45. Bilinski, M.: Compiler Techniques as a Defense Against Cyber Attacks. Machine Learning Series. SSC-PAC, San Diego (2014)
46. Rubin, S.H.: Case-Based Generalization (CBG) for Increasing the Applicability and Ease of Access to Case-Based Knowledge for Predicting COAs. NC No. 101366 (2011)
47. Rubin, S.H., Lee, G., Chen, S.C.: A case-based reasoning decision support system for fleet maintenance. Naval Engineers J., NEJ-2009-05-STP-0239.R1, 1–10 (2009)
48. Rubin, S.H.: multi-level segmented case-based learning systems for multi-sensor fusion. NC No. 101451 (2011)
49. Rubin, S.H.: Multilevel constraint-based randomization adapting case-based learning to fuse sensor data for autonomous predictive analysis. NC 101614 (2012)
50. Minton, S.: Learning Search Control Knowledge: An Explanation Based Approach, vol. 61. Kluwer, New York (1988)
51. Feigenbaum, E.A., McCorduck, P.: The Fifth Generation. Addison-Wesley Publishing Company, Reading (1983)
52. Rubin, S.H.: Is the Kolmogorov complexity of computational intelligence bounded above? In: 12th IEEE International Conference Information Reuse and Integration, pp. 455–461. Las Vegas (2011)

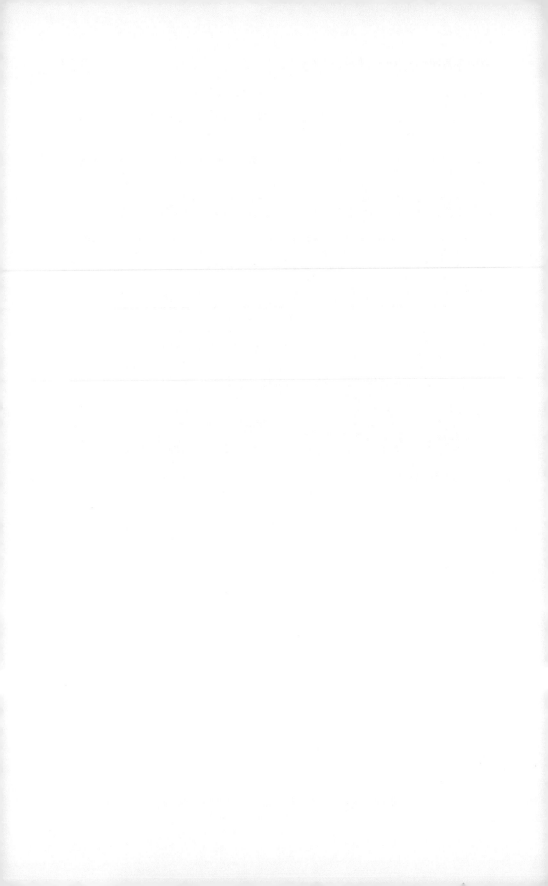

Semi-Supervised Classification System for the Detection of Advanced Persistent Threats

Fàtima Barceló-Rico, Anna I. Esparcia-Alcázar
and Antonio Villalón-Huerta

Abstract Advanced Persistent Threats (APTs) are a highly sophisticated type of cyber attack usually aimed at large and powerful organisations. Human expert knowledge, coded as rules, can be used to detect these attacks when they attempt to extract information of their victim hidden within normal http traffic. Often, experts base their decisions on anomaly detection techniques, working under the hypothesis that APTs generate traffic that differs from normal traffic. In this work we aim at developing classifiers that can help human experts to find APTs. We first define an anomaly score metric to select the most anomalous subset of traffic data; then the human expert labels the instances within this set; finally we train a classifier using both labelled and unlabelled data. Three computational intelligence methods were employed to train classifiers, namely genetic programming, decision trees and support vector machines. The results show their potential in the fight against APTs.

Keywords Advanced Persistent Threat · Anomaly Detection · Semi-supervised classification · Genetic programming · Decision trees · Support vector machines

---

FBR was partially funded with a Torres Quevedo grant from the Ministry of Economy and Competitiveness.

F. Barceló-Rico · A.I. Esparcia-Alcázar (✉) · A. Villalón-Huerta
S2 Grupo, Valencia, Spain
e-mail: aesparcia@ieee.org

F. Barceló-Rico
e-mail: fbarcelo@at3w.com

A. Villalón-Huerta
e-mail: avillalon@s2grupo.es

A.I. Esparcia-Alcázar
Present address: Universitat Politècnica de València, Valencia, Spain

© Springer International Publishing Switzerland 2016
R. Abielmona et al. (eds.), *Recent Advances in Computational Intelligence in Defense and Security*, Studies in Computational Intelligence 621,
DOI 10.1007/978-3-319-26450-9_9

225

1 Introduction

Advanced Persistent Threats (APTs) are a relatively new concept that has been defined in many ways. Most of these definitions home in the fact that they are attacks carried out with a **highly sophisticated malware**, the development of which requires **dedicated, skilled individuals** with expertise in **multiple technological fields**, as well as **significant financial resources** [1]. As befits these characteristics, APTs are generally addressed to either governments or large companies [2]. Hence, their detection and prevention are highly relevant, both economically and socially.

In addition to the above, this kind of cyberattacks is becoming both more frequent and more complex, a fact that has exposed the limitations of traditional security mechanisms, whose success in detecting such sophisticated threats has been poor. Examples of recent APTs which went undetected by current security solutions are Stuxnet [3], Duqu [4], Flame [5], Red October [6] and Miniduke [1, 7].

Although many standard web attack detection tools and apps have been adapted for the purpose, the very design of these attacks makes them extremely difficult to detect [8, 9]. Most solutions are based on expert knowledge; the techniques underlying these tools are rule-based systems, statistical and correlation methods, manual approaches and automatic blocking (black lists) [8]. One big shortcoming of these approaches is that they lack the capability to detect previously-unseen attacks [8]. Thus, building tools and algorithms to assist in the detection of *novel* APTs has become of the utmost importance for the security of both companies and states [2].

The rest of the chapter is organized as follows. In Sect. 2 we introduce related work on APT detection and the generalities of the anomaly detection (AD) methods are described. Our approach is presented in Sect. 3. A description of the characteristics of the proxy data is presented in Sect. 4. Section 4.2 includes a description of the proposed metric to assign the anomaly score to the instances. Next, in Sect. 5 the different classification methods considered are described. The specific results of these methods are presented in Sect. 8. Finally, Sects. 9 and 10 include a discussion of the results and the conclusions.

2 Related Work

2.1 Anatomy of an APT

In [8, 10] the typical APT strategy is described as follows:

- Attacker gains foothold on victim system via social engineering and malware.
- Attacker then opens a shell prompt on victim system to discover if system is mapped to a network drive.
- Victim system is connected to the network drive prompting attacker to initiate a port scan from victim system.

- Attacker will thereby identify available ports, running services on other systems, and identify network segments.
- Network map now in hand, attacker moves to targeting VIP victims with high value assets at their disposal.

In many cases the aim of the attacker once the steps described above have been accomplished, is to extract information from the source [11]. This is known as **exfiltration** and is deemed to be one of the phases where the APT can be detected [8], although, as proven by the examples given earlier, this is not always the case [1, 9].

2.2 Detecting APTs

A relevant body of work in the detection of APTs relies on monitoring an organization's network traffic -where HTTP data can be stored using a proxy- and identifying certain behaviours. This is based on the assumption that many APTs will use the HTTP protocol for the exfiltration step, given that it is supported by most organizations. For instance, [8] describes how a human expert can detect patterns that an APT might follow in order to develop countermeasures.

Frequently, detection methods work under the premise that if an APT infects a given system the behavior of the HTTP requests carried out within it will follow a different pattern than that existing in the absence of this attack, i.e. HTTP traffic will follow an *anomalous* behaviour.

Although it cannot be assumed that this will be always the case, as a cleverly designed APT will aim at disguising itself to appear as normal as possible, we can nevertheless expect that APT-induced behaviour will be closer to the anomalous rather than to the normal behaviour. The work presented here is also based on this premise.

2.3 Anomaly Detection Methods

Anomaly detection is a relevant problem that has been tackled in diverse research areas and application domains. Its importance lies with the fact that anomalies in data translate to significant, and often critical, actionable information in a wide variety of application domains. For a review of AD techniques the reader is referred to [12], which covers both those specifically developed for an application and other more generic ones. Examples of application domains of AD techniques include:

- Cybersecurity: network traffic analysis [13] to detect intrusions.
- Medicine: finding tumours in magnetic resonances [14].
- Banking: credit card fraud detection [15].
- Space: Sensor behaviour analysis to prevent spaceship failure [16].

Different technical approaches can also be found:

- Classification-based techniques [17]:

  - Artificial Neural networks (ANN) [18].
  - Bayesian Networks (BN) [19].
  - Support vector machines (SVM) [20].
  - Rules-based systems [12].

- Detection of n-neighbors [21]:

  - Distance to the closest n-neighbors [22].
  - Density of the neighborhood [23].

- Clustering-based techniques [24].
- Statistical methods [12]:

  - Parametric techniques [25].
  - Non parametric techniques [26].

- Other methods:

  - Information theory [27].
  - Spectral techniques [28].

When selecting an AD method special attention must be paid to the specific characteristics of the problem, such as type of data, type of anomalies sought, computational capacity of the device and so on. In the cases where many attributes that describe the behavior of the dataset are available, the detection method should ideally include information on all the attributes. Furthermore, anomalies can be detected by looking at attributes individually or considering them as an ensemble.

Most AD methods are based on the assignation of an anomaly score (AS) to each instance of the dataset [12]. This score represents how anomalous a given instance is. After the assignation of the AS the instances to focus on can be chosen mainly using two methods : (1) Setting a threshold (th_a) and selecting all instances whose AS is above it, and (2) selecting the N instances with larger AS.

The solution proposed in this work was developed based on the characteristics of the available data, which are explained in Sect. 4.

3 Proposed Method

As explained earlier, the method presented here utilises the premise that APT-infected HTTP traffic will tend to be anomalous to help create a classifier for suspicious/non-suspicious behaviour. This classifier will be trained using data labelled by human experts and then tested with a new set of data in order to evaluate its performance. In this way, we aim to model how experts work. The methods used to train the classifier are Genetic Programming (GP), two Decision Tree Classifiers (DTC), namely CART and Random Forests, and Support Vector Machines (SVM).

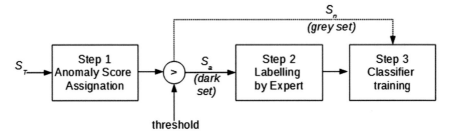

Fig. 1 General diagram of the APT detection process proposed. In the first phase, the most anomalous instances (S_a or *dark set*) are detected with the anomaly score (AS) metric defined. In the second phase the expert labels these instances as suspicious or non-suspicious. In the third phase the classifier is trained using the labelled instances from the dark set and, possibly, also instances from the (unlabelled) *grey set*

Because suspicious instances within HTTP data are rare and the amount of data available is too large to allow full labelling by the expert, we propose to look only at the anomalous instances within the available set, where we expect most suspicious instances will also be concentrated.

Our method, thus, follows these steps:

1. Select the most anomalous instances within the data available (HTTP requests registered by a proxy)—the *dark set*
2. Get the human expert to label these instances as *suspicious/non-suspicious*
3. Train a classifier using part of the labelled data in the dark set
4. Test the classifier obtained in the previous step using the remaining data in the dark set

The data that has not been labelled by the human expert (which we will refer to as the *grey set*) may also be used for testing purposes.

Step one is based on the definition of a new metric for the assignation of an anomaly score to the instances of the set. The proposed metric takes into account more information available in the data than other metrics found in the literature [29, 30].

Once the expert had studied the dark set and labelled its instances (step 2) a classification method was applied to label the instances in the grey set (step 3), based on the human knowledge included in the former. This scheme corresponds to a semi-supervised classification method, since only knowledge of a part of the data is known.

The three phases of the proposed method can be seen in Fig. 1.

4 Step One: Processing the Data

4.1 Traffic Data

The available data come from an access log of the Squid proxy application from a real organization. The records correspond to a URL session lasting several hours and

Table 1 Attributes, types and ranges

Attribute	Id	Type	Range or number of attributes
http reply code	c1	c	28
http method	c2	c	8
Main content type	c3	c	153
Bytes	c4	n	[0, 921701242]
Duration (ms)	c5	n	[0, 6674481]
Server address	–	c	7778
Date (days)	–	n	4
Squid hierarchy	–	c	2
Client address	–	c	110
URL (FQDN)	c6–c11	s	8872

The id column indicates those attributes that were selected for the proposed AS metric. *FQDN* (Fully Qualified Domain Name)

containing 1.1 million instances. Each instance, in turn, has ten associated variables which can be of three types: numerical, nominal/categorical and string, as follows:

- Numerical (n): *duration* (ms), *date* and *bytes*.
- Categorical/nominal (c): *http reply code, http method, content type, server address, squid hierarchy*, and *client address*.
- String (s): *URL*.

Since data come from a proxy register, many instances are "duplicated" in all values but the *squid hierarchy* (sh) attribute, which indicates if the query goes from the client to the proxy ($sh=DIRECT$) or from the proxy to the server ($sh=Default$ $parent$). To avoid these duplicated instances all of those with $sh=default\ parent$ value were removed, leaving the final set with 637,887 instances. The attributes contained in the final set are given in Table 1.

4.2 A New Metric for Anomaly Score Assignation

In order to define the proposed metric the most relevant attributes for detecting anomalies in the proxy data were selected by a human expert; then it was defined how to detect abnormal behaviour using these attributes. In this way we obtain a final anomaly score value for every instance.

Individual Elements

Because the aim is not to detect anomalies in the behaviour of specific individuals but rather in the group of users as a whole, we will only consider those attributes that give information on the latter, rather than on individual behaviour. The final attributes are those that have a value in the *id* column in Table 1.

The FQDN is the attribute that contributes the most information. For this reason, several characteristics contained in it were also considered relevant by an expert for the identification of anomalies in the behavior of the set:

- Top Level Domain, TLD (considered as categorical attribute) ($c6$)
- Length of the core domain (cd) ($c7$)
- Vowels/consonants ratio of the full domain ($c8$)
- Numbers/letters ratio of the full domain ($c9$)
- Sequences of elements of the *FQDN* ($c10$)
- Popular core domains that appear as subdomains, *sd* ($c11$)

For $c10$ the elements of the domains are divided into consonants, vowels, numbers and symbols.

For $c11$ the *popular* domains are extracted from Alexas lists.[1] A list consisting of 42 domains was constructed taking the 20 most popular (visited) domains of the country where data was collected (Spain), 11 more from the continent (Europe) and other 11 from the global list.

Combinations of Elements

As explained in Sect. 2, anomalies can be detected by considering each attribute independently from the others but also considering several attributes as a set. In this case anomalies were detected by identifying rare values in the combinations of attributes, independently of the individual values.

For this particular set of data the relevant attributes for detecting anomalies in their combinations are the four categorical attributes ($c12$): *http reply code, http method, main content type*, and *TLD*.

The main contribution of the proposed metric as compared to others found in the literature is the consideration of attributes and characteristics of the logs both individually and in combination.

Once the relevant attributes for detecting anomalies in these URL connections were determined, the next step was to define how the anomaly score of each instance would be assigned.

The total AS of each instance (as_i) is composed by the anomaly score of each attribute or characteristic included (as_j) weighed by its relative importance (w_j):

$$as_i = \sum_{j=c1}^{cN} w_j \cdot as_{ij} \tag{1}$$

The value of the anomaly score of each attribute (as_j) is assigned to 0 if the value adopted is considered normal and to 1 if it is considered anomalous. The question which arose now was: When are these characteristics considered to have an anomalous value? It depends on the nature of the characteristic:

---

[1] http://www.alexa.com/topsites.

- **Categorical**: when the frequency of appearance of the value adopted is very low (infrequent appearance). For instance, if any of these variables has a value that appears less than 0.01 % in the whole set, it is considered anomalous. Characteristics within this case are: $c1$, $c2$, $c3$, $c6$, $c10$, $c11$ and $c12$. The infrequent threshold for each case is different.
- **Numerical**: to find anomalous values in numerical data the parameters of the distribution are used. A value is considered to be anomalous when it lies outside the range given by the mean (m) plus (or minus) three times the standard deviation (σ)s ($x \notin [m - 3\sigma, m + 3\sigma]$). In other words, the sample has a normal distribution. Characteristics within this case are: $c4$, $c5$, $c7$, $c8$ and $c9$.

Two variations of the metric were considered:

 (i) Only the individual attributes were considered. In this case, since the expert did not point out any of the individual attributes to be more relevant than the rest, all of them contribute with the same importance to the AS metric.
(ii) In addition to the individual attributes, the anomalous combination of attributes was included. In this case, since the expert indicated that this attribute gives more information than the individual ones the weight of this characteristic is larger than those of the individual attributes in the AS metric.

5 Step Two: Data Labelling by Human Expert

After the assignation of the AS index, those instances with larger AS that could represent an APT (the *dark set*) were filtered out and an expert determined which ones were worth of further study (i.e. *suspicious* instances). Because the dark set is small compared to the whole original data set, the information given by the expert is necessarily partial.

Semi-supervised learning (SSL) methods are applicable to this type of scenario: the known information is used to reach conclusions on the unknown information. With SSL methods at least some known information about the data is considered. The hypothesis is that this information can be used for the unknown part of the set if all data follow the same structure.

6 Step Three: Building Classifiers

We used three different methods to carry out the classification of the instances, as described below.

6.1 Genetic Programming

Genetic programming (GP) [31] is a flexible and powerful evolutionary technique with some features that can be very valuable and suitable for the evolution of classifiers [32]. GP is a subclass of genetic algorithms, which uses mutation and replication to evolve structures, following Darwinian survival-of-the-fittest principles [33]. These programs are composed of nodes (mathematical functions) and terminals (inputs and constants). GP treats individual computer programs as genetic individuals potentially capable of recombining or changing to form new individuals [34].

In this work we employed the GPlab toolbox developed by Silva [35].

6.2 Tree Classification Methods

Decision Tree Classifiers (DTC) are used for classification problems in many areas. Perhaps, the most important feature of DTC's is their capability to break down a complex decision-making process into a collection of simpler decisions, thus providing a solution which is often easier to interpret [36]. Decision trees are a classification tool used for many years.

Two different methods that use DTC are considered here: **CART** approach, which employs a single tree and **random forests** which employ a set of trees and a voting mechanism.

CART

Classification and Regression Trees (CART) were proposed by Breiman [37]. With this method the tree is built by growing branches and pruning them iteratively. CART allows only either a single feature or a linear combination of features at each internal node [36]. This method is computationally very expensive as it requires the generation of multiple auxiliary trees, yet it can be a good approach since it is nonparametric and easy to apply [38].

Random Forest

Random forests are a combination of tree predictors such that each tree depends on the values of a random vector sampled independently and with the same distribution for all trees in the forest [39]. Significant improvements in classification accuracy have resulted from growing an ensemble of trees and letting them vote for the most popular class.

The common element in all of these procedures is that for the *kth* tree, a random vector Θ_k is generated, independent of the past random vectors $\Theta_1, \ldots, \Theta_{k-1}$ but with the same distribution. After a large number of trees is generated, they vote for the most popular class. These procedures are called random forests [39].

6.3 *Support Vector Machines*

Support Vector Machines (SVM) are algorithms used for classification based on the class hyperplanes [40]. It can be shown that the optimal hyperplane, defined as the one with the maximal margin of separation between the two classes has the lowest capacity. It can be uniquely constructed by solving a constrained quadratic optimization problem [41].

It is worth emphasizing one important property of this algorithm: both the quadratic programming problem and the final decision function depend only on dot products between patterns. This is precisely what lets this method to be generalized to the nonlinear case.

7 Experimental Setup

Two different frameworks for the experiments were set:

(A) Once instances in the dark set are labelled by the expert as *suspicious/non-suspicious*, the different classification methods were trained using these. The results of the training phase were then tested with the remaining instances (the gray set) and results were analysed. See Fig. 2 for a graphical representation of this approach.

(B) The training set was built using the instances marked as suspicious in the dark set plus a larger number of randomly selected instances from the grey set. The hypothesis is that instances in the grey set are *non-suspicious*. The classifiers thus obtained were tested using one third of the suspicious instances from the dark set plus a number of (unlabelled) instances from the grey set, which are assumed to be non suspicious. The latter are replaced on every iteration, so that the trained classifiers do not overfit to them, but rather learn the labelled instances. See Fig. 3 for a graphical representation of this approach.

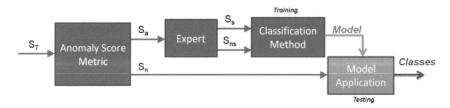

Fig. 2 Representation of **Case A** for training only with labelled data and testing with unlabelled data. S_T refers to the total set of data. S_a/S_n refer to the set of anomalous data and normal data respectively. S_s/S_{ns} refer to the suspicious and non-suspicious sets

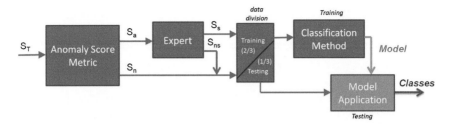

Fig. 3 Representation of **Case B** for training and testing considering that the instances not marked as suspicious by the expert are non suspicious. Original set is divided into 2/3 for training and 1/3 for testing. S_T refers to the total set of data. S_a/S_n refer to the set of anomalous data and normal data respectively. S_s/S_{ns} refer to the suspicious and non-suspicious sets

8 Results

8.1 Anomaly Score Assignation

The levels to consider each of the categorical attributes or characteristics as anomalous were:

- *Categorical attributes* ($c1$, $c2$, $c3$ and $c6$): when they appeared less than 10 times in the whole set (0.0015 %) their as_j was set to 1.
- *Sequences* of elements in the domain ($c10$): subsequences of six elements were considered to find the patters:

  - Domains with length less than 6 were directly considered anomalous.
  - Domains with a subsequence of elements that only **appear once** in the whole set were considered anomalous.

- When *popular domains* ($c11$) of the list appeared as subdomains this characteristic anomaly score was directly set to 1.
- For the *combination* of elements ($c12$):

  - Its as_j was set to 1 when a combination only **appear once** in the whole set.
  - The weight of this parameter was set to $w_{12} = 2$ to express its higher indication of an anomaly with regard to the rest of characteristics.

In the first case, where only the individual attributes were considered to form the anomaly metric and with all attributes having the same weight, the AS adopted by the instances was in the [0, 3] interval. Of the total set of 638,887 instances, 29,092 of them did have an as_i different from 0. The distribution is shown in Table 2.

In the second case, where the anomalous combination of attributes was also included in the anomaly metric in addition to all the individual attributes, the instances adopted an AS value in the interval [0,5]. Of the total set of 638887 instances, 28858 of them did have an as_i different from 0. The distribution is shown in Table 3.

Table 2 Case (i) Number of instances for each AS different than 0	As	Num. of instances
	1	28,901
	2	179
	3	12

Table 3 Case (ii) Number of instances for each AS different than 0	As	Num. of instances
	1	28,532
	2	265
	3	54
	4	6
	5	1

In both cases, to consider the instances that had at least two anomalous fields, the instances studied were those that had $as > 1$ in case (i); and $as > 2$ in case (ii), since the anomalous combinations had a value of $as = 2$. Thus, in case (i) there are 191 instances to study (S_a); and in case (ii) there are 61 instances to study (S_a).

8.2 Assignation of Labels by an Expert

Case (i): From all the instances with $as > 1$ the expert marked as instances to be further studied:

- 66 instances with $as_i = 2$ (36 %)
- 4 instances with $as_i = 3$ (33 %)

Case (ii): From all the instances with $as > 2$ the expert marked as instances to be further studied:

- 32 instances with $as_i = 3$ (59 %)
- 4 instances with $as_i = 4$ (67 %)
- 1 instance with $as_i = 5$ (100 %)

These results indicate that:

(1) Case (i) has percentages lower than 50 % while case (ii) has percentages larger than 50 %.
(2) Case (ii) has larger percentages of suspicious instances among the detected anomalous instances.
(3) In case (ii), the percentages of suspicious instances increases for larger AS values.

This indicates that metric defined in case (ii) is more trustworthy than case (i); for this reason it was chosen as final metric to set the anomaly score of the instances.

Table 4 Results GP classification only with labelled data

		GP	
		#	(%)
Training	TP	85	85.86
	FN	14	14.14
	TN	102	75.56
	FP	33	24.44

Table 5 Application of resulting GP model to rest of data (unlabelled)

	GP
	(%)
Suspicious	95.27
Non-suspicious	4.73

8.3 Application of Classification Methods to Semi-supervised Data

In order to apply the classification methods described in Sect. 6.1 in the two approaches described in Sect. 7, instances labelled by the expert were considered as known classification information for the training phase. Combined information from metrics (i) and (ii) results in 99 instances marked as *suspicious* (S_s) and 135 instances marked as *non-suspicious* (S_{ns}), since some of the instances where present in both cases.

For **Case A** the whole dark set (234 labelled instances) was used for training and the classification model obtained was then tested using the grey set (remaining unlabelled data, 638,653 instances).

For **Case B** the data considered for training was two thirds of the total set (425,925 instances), including two thirds of the labelled suspicious data (66 instances). The unlabelled data included for training and testing have been randomly selected.

Genetic Programming
First of all, GP was applied to only the dark set, or labelled data (**Case A**). Results from the best run can be seen in Table 4. Results of the application of the resulting model to the rest of data (grey set) can be seen in Table 5.

The results of the training phase with the labelled data are not very good, since many suspicious and non-suspicious instances are not well classified. This happened even though many configurations were adopted, all them with similar results. Longer run times also reached similar results. Further, the large number of potentially suspicious instances "identified" in the test phase, indicates that the results obtained are not applicable to the rest of the set, since not such a large number of suspicious instances can be present in a normal web log session like the one considered.

Considering hypothesis of **Case B**, the classes are quite unbalanced. This is an important property of the APT detection problem, since attacks only represent a small set of the total data. For this reason, options like data undersampling or oversampling would distort this property and were not appropriate for this application

Table 6 Results from GP classification in Case B

		GP	
		8k	
Training	TP	54	81.82 %
	FN	12	18.18 %
	TN	7791	97.39 %
	FP	209	2.61 %
Testing	TP	28	84.85 %
	FN	5	15.15 %
	TN	–	97.36 %
	FP	16683	2.64 %

[42]. Instead, the fitness function was modified to include weights and take into account the larger importance of classifying well the minority class, following [42].

The experiments carried out applying GP were done using a reduced dataset, due to the computational expense of using the whole training dataset. The reduced set consisted of two thirds of the *dark set* (labelled) and 8000 (8k) instances from the *grey set* (unlabelled data). Several experiments were carried out, considering different weights for the minority class. The best results can be seen in Table 6.

As has been pointed out by, e.g. [43], GP has scalability issues. Simulation with a dataset of this size was computationally expensive and experiments with a larger dataset were considered unfeasible.

CART

CART algorithm is non-parametric, which eased up its application. For **Case A** the resulting tree had 33 nodes and a depth of 8. Classification from this tree is shown in Table 7. Its application to the rest of data is shown in Table 8.

Table 7 Results of CART classification with labelled data only

		CART	
		#	(%)
Training	TP	93	93.94
	FN	6	6.06
	TN	129	95.56
	FP	6	4.44

Table 8 Application of resulting CART classifier to rest of data (unlabelled)

CART	
#	(%)
Suspicious	23.85
Non-suspicious	76.15

Table 9 Results from CART classification in Case B

	CART				
		All		12k	
Training	TP	59	89.39 %	60	90.91 %
	FN	7	10.61 %	6	9.09 %
	TN	–	99.998 %	–	99.98 %
	FP	10	0.002 %	2	0.02 %
Testing	TP	19	57.58 %	27	81.82 %
	FN	14	42.42 %	6	18.18 %
	TN	–	99.996 %	–	99.87 %
	FP	9	0.004 %	844	0.13 %

Table 10 Results of random forests classification only with labelled data

		RF	
		#	(%)
Training	TP	97	97.98
	FN	2	2.02
	TN	130	96.3
	FP	5	3.7

The high number of *suspicious* instances detected in the set of unlabelled data is too large to indicate real suspicious instances, since APTs are infrequent. In addition, the size of the set makes it unfeasible for an expert to evaluate all its instances. This indicates that this reduced set does not contain enough information on its own in order to represent the model via a single tree using the CART method.

To train the method with more information we considered the hypothesis of **Case B**. When CART algorithm is trained with two thirds of the data the training results are very good, yet the number of False Negatives (FN) of the test phase is too large, see Table 9. Several experiments modifying the size of the training set were performed, see Table 9 for best results. When the number of training instances is reduced, the obtained tree is more robust. The resulting tree had 77 nodes and depth 12.

The percentage of FN is only 0.13 % i.e. 844 instances. This small set size makes it feasible for an expert to evaluate all instances and check the quality of the results.

Random Forests

For the application of Random Forests (RF) to the **Case A**, several parameters had to be tuned. After running a few experiments, the parameters with best results were adopted. Results can be seen in Table 10. Results of the application of this classification model to the rest of the set (unlabelled) can be seen in Table 11.

These results show that the identification of the labelled data is quite good, yet the extrapolation of the classification model to the unlabelled data results in a very large amount of suspicious instances. This indicates that the model obtained with the small set of known data does not offer useful results with the remaining set.

Table 11 Application of the
resulting RF model to the
remaining data (unlabelled)

RF	
#	(%)
Suspicious	87.19
Non-suspicious	12.81

For application of **Case B** it was not possible to do it with the full identification
set (2/3 of the total set) since there were memory issues with the operations. For
this reason, several experiments have been run with different sizes of the training
set. First of all an experiment with same "best size" of the CART experiments were
performed (12k instances). This experiment did not gave good results, as can be seen
in Table 12. Different sizes were checked. Best results are also shown in Table 12.

Support Vector Machines
For this method, given that the resulting dividing hyperplane is only computed in the
projection space, the inputs were normalized in the interval [0,1] to give all them a
priori the same importance.

Results of application of SVM under approach of **Case A** is shown in Table 13,
while the results of applying the resulting classification model to the rest of the set
is shown in Table 14.

Table 12 Results from RF classification in Case B

		RF			
		12k		100k	
Training	TP	58	87.87 %	58	87.87 %
	FN	8	12.12 %	8	12.12 %
	TN	–	98.22 %	–	98.95 %
	FP	214	1.78 %	1052	1.05 %
Testing	TP	18	54.55 %	12	36.36 %
	FN	15	45.45 %	21	63.63 %
	TN	–	97.65 %	–	98.92 %
	FP	14714	2.35 %	5816	1.08 %

Table 13 Results SVM
classification only with
labelled data

		SVM	
		#	(%)
Training	TP	88	88.89
	FN	11	11.11
	TN	133	98.52
	FP	2	1.48

Table 14 Application of
resulting SVM model to rest
of data (unlabelled)

SVM	
#	(%)
Suspicious	43.65
Non-suspicious	56.35

These results show that with this method a still large number of potentially *suspicious* instances is found. Like for CART and RF, this amount is too large to represent real APTs and also for an expert to analyse all them.

For the training phase of SVM in **Case B** there were memory issues that do not allow the training with the full training set (2/3 of the total set). A reduced set was used for training then. The maximum size allowed for operations is 8k unlabelled instances plus two thirds of the suspicious ones (total of 8066 instances). Results of the training phase and the testing with the remaining data can be seen in Table 15.

Tables 16 and 17 show the best results (percentages) in cases A and B respectively for all the classification methods applied.

Comparison

In order to compare all methods tested, a number of runs were carried out so as to minimize the effects of the randomly selected variables, given that some of these methods are stochastic. For CART, RF and SVM methods, the configurations with best results were run 30 times. GP had a a different treatment, since running time

Table 15 Results SVM application

		SVM	
Training	TP	59	89.39 %
	FN	7	10.61 %
	TN	8000	100 %
	FP	0	0 %
Testing	TP	25	75.76 %
	FN	8	24.24 %
	TN	–	99.86 %
	FP	902	0.14 %

Table 16 Case A: classification results only with labelled data for different methods

		GP (%)	CART (%)	RF (%)	SVM (%)
Training	TP	85.86	93.94	97.98	88.89
	FN	14	6.06	2.02	11.11
	TN	75.56	95.56	96.3	98.52
	FP	24.44	4.44	3.7	1.48
Testing	S	95.27	23.85	87.19	43.65
	NS	4.73	76.15	12.81	56.35

Table 17 Case B: classification results for unlabelled hypothesis for different methods

		GP (%)	CART (%)	RF (%)	SVM (%)
Training	TP	81.82	90.91	87.87	89.39
	FN	18.18	9.09	12.12	10.61
	TN	97.39	99.98	98.22	100
	FP	2.61	0.02	1.78	0
Testing	TP	84.85	81.82	54.55	75.76
	FN	15.15	18.18	45.45	24.24
	TN	97.36	99.87	97.65	99.86
	FP	2.64	0.13	2.35	0.14

Table 18 Mean running time of the different configurations

	GP		CART		RF		SVM	
	2k	8k	All	12k	100k	12k	4k	8k
Training	$35 \cdot 10^3$ s	$183 \cdot 10^3$ s	185 s	186 s	1180 s	765 s	370 s	810 s
Testing	101 s	93 s	0.22 s	3.1 s	42 s	48 s	223 s	324 s

were much larger (see Table 18). To make the total set of runs feasible run time was set to one hour. The training set consisted of 1k random instances and 66 suspicions instances, a population of 50 individuals and 300 iterations.

Two errors were considered for statistical comparison of the four methods: e_1 the percentage of false negative instances and e_2 the percentage of false positive instances, both obtained during the testing phase. Since we can not assume that errors have normal distribution, we chose the *Kruskal-Wallis statistical test*, a non parametric statistic method.

Both for e_1 and e_2 the probability that the results from the different methods applied follow the same distribution is equal to 0 ($P = 0$), meaning that the applied methods are **statistically different**.

Thus, since the methods do not follow the same distribution, two parameters were used to compare them: (1) run time and (2) distribution of errors. **Run time** is important to check the feasibility of each method to be applied online. On the other hand, **errors distribution** is important to check the repeatability of each method and configuration.

Mean run time of every configuration can be seen in Table 18. It can be seen that CART is the fastest method while GP is the slowest one. However, CART, RF and SVM all have a fast enough run time that make it feasible to run many experiments and these methods could be a good option for an online application.

The error distribution of the 30 runs of each case is shown in Figs. 4 and 5 for the percentages of False Negative and False Positive instances, both for the testing phase.

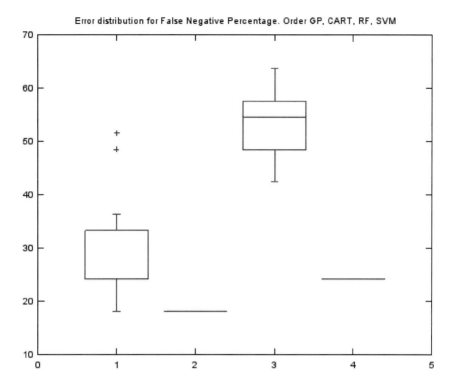

Fig. 4 Representation of error distribution for the percentage of False Negative instances **FN** in the testing phase. The order is (1) GP; (2) CART; (3) RF and (4) SVM

In the case of e_1 (FN), it is seen that CART and SVM have the lowest errors, which was already seen in the previous results. In addition, in this plot it can also be seen that these are the methods with larger repeatability (smaller dispersion of the error). In this case GP has smaller error than RF and both have a similar dispersion of the error.

In the case of e_2 (FP) results are similar to the case of e_1: CART and SVM are the methods with smaller error and smaller dispersion. For this error GP has a large error with an even larger dispersion among the runs. RF has a much smaller error than GP and a small dispersion, but its results are worse than those of CART and SVM.

9 Discussion

In this work we proposed a new Anomaly Score metric for http log instances, where the main characteristics that can indicate that an instance is anomalous are considered together. The expert indicated the interesting features to be considered in this metric.

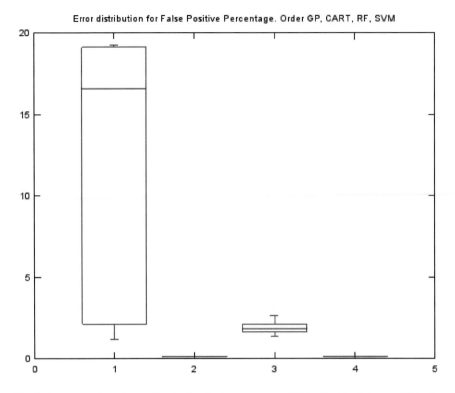

Fig. 5 Representation of error distribution for the percentage of False Positive instances **FP** in the testing phase. The order is (1) GP; (2) CART; (3) RF and (4) SVM

With this metric, the set of data containing a session of http requests was fil-tered and the most anomalous instances were selected. This subset (very reduced from original one) was studied by the expert who indicated the instances considered *suspicious*, which are the real aim of the detection.

With this subset of labelled instances two approaches for the design of classifi-cation methods were designed: considering only the labelled data; and considering only suspicious data and adopting the hypothesis that the rest are *non-suspicious*.

To these two scenarios classification is performed with several techniques: *genetic programming* (GP), *single tree classification* (CART), *random forests* (RF) and *sup-port vector machines* (SVM).

The application of **GP** showed unpromising results for case A and the extrapola-tion of the model found to the rest of the data, with 95.27 % of *suspicious* instances found. In case B, GP performs much better than in case A, having a smaller number of FP (2.64 %) and a larger percentage of TP in the test phase. This percentage could represent real suspicious instances yet the total number of resulting instances is too large to be analysed by an expert.

The **CART** method gives good results in the training phase, but when the tree found was applied to the rest of unlabelled data the number of "potentially" suspicious instances found was too large to represent real suspicious instances. In case B, several configurations are checked, considering different sizes of the training data. In case all data of the training set was considered (2/3 of total data), the results of the training phase are quite good. Yet, in the testing phase there are a 42 % of FN and a 0.004 % of FP. Thus, this configuration is good at avoiding FP, yet it is not good at detecting the TP.

The other configuration was to consider a subset for training (only 12k instances). With this configuration, the method reduces the percentage of FN to 18 % but increases the percentage of FP to 0.13 %. It can be seen that the different configurations have different performance and that getting better in one direction implies getting worse in another one. It remains for the expert to determine which part is more important, but we would suggest that detecting the known suspicious instances even at the price of detecting more FP instances is better in these cases. It is important to consider that even though in the second case the percentage of FP is larger, it still represents a small subset that can be analysed by the expert.

The application of **RF** shows that the extrapolation of the classification model trained with only labelled data gives quite bad results, since the majority of instances (87 %) are classified as *suspicious*. This indicates that this model does not represent the essence of the small set of instances labelled as suspicious.

When RF was applied in case B, something similar to the CART results happens: the larger the training set the lower the percentage of FN detected in testing phase. In this case, results both for small set for training (12k) and for a set lager (100k) give very high percentages of FN: 45 % and 63 % respectively. In addition to this fact, percentages of FP are larger than in the case of CART, reaching 2.35 % which represents a very large number of instances to be analysed by the expert.

RF have proved better than single tree like CART in many applications [39]. Nonetheless, this system is quite complex and this makes that even when both algorithms have very similar results in the training phase, the simpler one is more robust for the testing phase.

The final classification method checked was **SVM**. This method gives very similar results to CART. The extrapolation of the model of case A does not lead to good results in the unlabelled data. Meanwhile, the results of the scenario of case B gives good results both from training and testing, reaching a 76 % of TP detected and only a 0.14 % of FP in the testing set.

Since results from CART and SVM are similar and both result in subsets of potentially-suspicious instances small enough to be analysed by the expert, we have proceed in this line. An expert analysed both sets of new suspicious instances found and marked the ones considered in this way. In the subset resulting from CART method, the expert detected 66.35 % of real-suspicious instances, while in the subset of SMV method the percentage of real-suspicious was 66.19 %. This analysis indicates that still both methods are comparable in terms of performance once the results are analysed.

Considering the case A, it was seen that none of the methods applied had translatable results to the rest of the set. Training results were quite good in all cases (minimum of 80 % of well classified data), yet the small number of instances used for training phase (only a 0.037 % of the set) made the models obtained not valid for the rest of the set with a lot of information not included in the training set.

10 Conclusion and Future Work

APTs are advanced attacks against governments or large companies. For this reason they are well designed and they are difficult to detect. At the same time, it is interesting in cybersecurity to detect them or at least have a prevention method to minimize their risks.

Several classification algorithm have been applied to the detection of suspicious instances for identifying APT attacks. Two main scenarios have been checked: considering only a small set of labelled data (**case A**) and considering only a few known suspicious instances with a larger set of supposedly non-suspicious instances (**case B**).

In **case A** all methods had similar results. However, in **case B** it was shown that SVM and CART method outperformed the other two. Considering the complexity of the problem these two method obtained quite good numbers. The negative side of this experiments is that we could not make RF and GP methods perform better. The worse results from RF are probably due to the tuning of the algorithm parameters while GP issues might fall on the scalability issues of this method.

Results from both cases considered (**A** and **B**) show that hypothesis of **case A** that the small labelled set is representative of the total set is not valid, since the results point out many instances as suspicious and that large number is not realistic that happens in a real set.

On the other hand, hypothesis taken in **case B** could be more realistic, since most of the unlabelled instances are non-suspicious. Yet, with this hypothesis, some suspicious instances of the unlabelled set can be considered as non-suspicious for the training phase, biasing this the results obtained.

This makes it obvious that a better treatment of the unlabelled data could improve the results. An option is to use a recursive labelling method, considering the results obtained from a method and increasing the labelled set until the identified instances remain the same.

Another approach in this line could be to use rules defined by the expert to get the initial set of **suspicious** instances, and then use this set for the training phase and check the results.

In a different line, a future work could be to improve the used methods to adjust them to the problem considered. GP could be used if the evaluation of the fitness function is sped-up. For RF a finer tuning of the many parameters of the method could be done to refine the results. Yet, both of these improvements are time requiring and a hard work. For this reason, they remain for the future.

Experiments with other configurations or other methods are left for future work. An option could be a method considering a combination of the proposed methods. It is important to remark that the large number of instances makes it unfeasible to label all of them by a human expert. Therefore, the framework for the studies is a semi-supervised scenario. In this type of scenarios one option is to train several methods and obtain the conclusions from the results of all the methods together.

An option for overcoming the large amounts of data with which the methods have to deal. A method, already explored by some authors in other lines, is to work first with clusters of data or sets of logs, find anomalies within sets, which would mean that those sets contain anomalous instances, and then proceed to a further study of only anomalous sets. These method, as general idea, has many advantages, yet, further work is to be done to apply it to the currently considered scenario.

References

1. Virvilis, N., Gritzalis, D., Apostolopoulos, T.: Trusted computing versus advanced persistent threats: can a defender win this game?. In: IEEE 10th International Conference on Ubiquitous Intelligence and Computing and 10th International Conference on Autonomic and Trusted Computing (UIC/ATC) (2013)
2. Sullivan, D.: Beyond the hype: advanced persistent threats. Technical Report, TrendMICRO, 2011
3. Lemos, R.: Stuxnet attack more effective than bombs (2011). http://goo.gl/cnthbC
4. Symantec, W32.duqu—the precursor to the next stuxnet. (2011). http://www.symantec.com/connect/w32_duqu_precursor_next_stuxnet
5. Bencsath, B., Pek, G., Buttyan, L., Felegyhazi, M.: The cousins of stuxnet: Duqu, flame, and gauss. Future Internet **4**(4), 971–1003 (2012)
6. Labs, K.: "Red october" diplomatic cyber attacks investigation. (2013). http://goo.gl/JbLuOa
7. Tivadar, M., Balazs, B., Istrate, C.: A closer look at miniduke. (2013). http://goo.gl/YKoupm
8. Binde, B., McRee, R., OConnor, T.: Assessing outbound traffic to uncover advanced persistent threads, Technical Report, SANS Technology Institute, 2011
9. Lee, M., Lewis, D.: Clustering disparate attacks: Mapping the activities of the advanced persistent threat. In: Virus Bulletin Conference (2011)
10. Cutler, T.: The anatomy of an advanced persistent threat (2010). http://www.securityweek.com/anatomy-advanced-persistent-threat
11. Molok, N., Chang, S., Ahmad, A.: Information leakage through online social networking: opening the doorway for advanced persistence threats. In: Australian Information Security Management Conference (2010)
12. Chandola, V., Banerjee, A., Kumar, V.: Anomaly detection: a survey. ACM Comput. Surv. (CSUR) **41**(3), 15 (2009)
13. Kumar, V.: Parallel and distributed computing for cybersecurity. IEEE Distrib. Syst. **6**(10), 1–9 (2005)
14. Spence, C., Parra, L., Sajda, P.: Detection, synthesis and compression in mammographic image analysis with a hierarchical image probability model. In: IEEE Workshop on Mathematical Methods in Biomedical Image Analysis (2001)
15. Aleskerov, E., Freisleben, B., Rao, B.: Cardwatch: A neural network based database mining system for credit card fraud detection. In: IEEE Conference on Computational Intelligence for Financial Engineering (1997)
16. Fujimaki, R. Yairi, T., Machida, K.: An approach to spacecraft anomaly detection problem using kernel feature space. In: 11th ACM SIGKDD International Conference on Knowledge Discovery in Data Mining (2005)

17. Duda, R.O., Hart, P., Stork, D.: Pattern Classification, Wiley-Interscience (2001)
18. Stefano, C.D., Sansone, C., Vento, M.: To reject or not to reject: that is the question: an answer in the case of neural classifiers. IEEE Trans. Syst. Man Cybern. **30**(1), 84–94 (2000)
19. Barbara, D., Wu, N., Jajodia, S.: Detecting novel network intrusions using bayes estimators. In: 1st SIAM International Conference on Data Mining (2001)
20. Eskin, E., Arnold, A., Prerau, M., Portnoy, L., Stolfo, S.: Ageometric framework for unsupervised anomaly detection. In: Conference on Applications of Data Mining in Computer Security, Kluwer Academics (2002)
21. Tan, P., Steinbach, M.K.: Introduction to Data Mining, Addison-Wesley (2005)
22. Ramaswamy, S., Rastogi, R., Shim, K.: Efficient algorithms for mining outliers from large data sets, In: CMSIGMOD International Conference on Management of Data (2000)
23. Breunig, M., Kriegel, H. Ng, R. Sander, J.: Lof: Identifying density-based local outliers. In: ACM SIGMOD International Conference on Management of Data (2000)
24. Guha, S., Rastogi, R., Shim, K.: Rock: A robust clustering algorithm for categorical attributes. In: IEEE 15th International Conference on Data Engineering. vol. 25 no. 5 (1999)
25. Eskin, E.: Anomaly detection over noisy data using learned probability distributions, In: 17th International Conference on Machine Learning (2000)
26. Desforges, M., Jacob, P., Cooper, J.: Applications of probability density estimation to the detection of abnormal conditions in engineering, institution of Mechanical Engineers. Part C: J. Mech. Eng. Sci. **212**(8), 687–703 (1998)
27. Keogh, E., Lonardi, S., Ratanamahatana, C.: Towards parameter-free data mining. In: 10th ACMSIG-KDD International Conference on Knowledge Discovery and Data Mining (2004)
28. Agovic, A., Banerjee, A., Ganguly, A.: Ch6 Anomaly detection in transportation corridors using manifold embedding. Knowledge Discovery from Sensor Data (2007)
29. Ingham, K., Inoue, H.: Comparing anomaly detection techniques for http. Recent Advances in Intrusion Detection. Springer, Berlin (2007)
30. Kruegel, C., Vigna, G.: Anomaly detection of web-based attacks. In: 10th ACM Conference on Computer and Communications Security (2003)
31. Koza, J.R.: Genetic Programming: On the Programming of Computers by Means of Natural Selection. MIT Press, Cambridge (1992)
32. Espejo, P., Ventura, S., Herrera, F.: A survey on the application of genetic programming to classification. IEEE Trans. Syst. Man Cybern. Part C: Appl. Rev. **40**(2), 121–144 (2010)
33. Lotz, M.: Modelling of process systems with genetic programming. Master's thesis, University of Stellenbosch (2006)
34. Banzhaf, W., Nordin, P., Keller, R., Francone, F.: Genetic Programming: An Introduction, vol. 1. Morgan Kaufmann, San Francisco (1998)
35. Silva, S.: GPLAB A Genetic Programming Toolbox for MATLAB, ECOS - Evolutionary and Complex Systems Group University of Coimbra Portugal, version 3 edn
36. Safavian, S., Landgrebe, D.: A survey of decision tree classifier methodology. IEEE Trans. Syst. Man Cybern. **21**(3), 660–674 (1991)
37. Breiman, L., Friedman, J., Stone, C., Olshen, R.: Classification and Regression Trees. CRC press, Boca Raton (1984)
38. Timofeev, R.: Classification and regression trees (cart) theory and applications. Master's thesis, Humboldt University, Berlin (2004)
39. Breiman, L.: Random forests. Mach. Learn. **45**(1), 5–32 (2001)
40. Hearst, M., Dumais, S., Osman, E., Platt, J., Scholkopf, B.: Support vector machines. Intell. Syst. Appl. IEEE **13**(4), 18–28 (1998)
41. Burges, C.: A tutorial on support vector machines for pattern recognition. Data Min. Knowl. Discov. **2**(2), 121–167 (1998)
42. Alfaro-Cid, E., Sharman, K., Esparcia-Alcazar, A.: A genetic programming approach for bankruptcy prediction using a highly unbalanced database. Applications of Evolutionary Computing, pp. 169–178. Springer, Berlin (2007)
43. Thierens, D.: Scalability problems of simple genetic algorithms. Evol. Comput. **7**(4), 331–352 (1999)

A Benchmarking Study on Stream Network Traffic Analysis Using Active Learning

Jillian Morgan, A. Nur Zincir-Heywood and John T. Jacobs

Abstract Analyzing network activity as it occurs is an important task since it allows for the prevention of malicious activity on the host system and the network. In this work, we investigate the performance of different budgeting strategies, as well as an adaptive Artificial Neural Network to analyze the activities on streaming network traffic. Our results show that all of our budgeting strategies (with the exception of the fixed uncertainty strategy) are suitable candidates for classification of streaming network traffic where some of the state-of-the-art classifiers achieved accuracies in the range of 90 % or higher.

Keywords Streaming data · Active learning · Computational intelligence · Network traffic analysis

1 Introduction

Malicious network activity, such as viruses, denial-of-service attacks, and botnets, is a growing concern for businesses and the general public alike. Detection of malicious network activity as it occurs is important since it can assist the network management teams in preventing further damage on their systems and networks. Therefore, it is of interest to classify network activity as it is being streamed. With the use of computational intelligence techniques gaining popularity, many researchers propose the use of different learning techniques as well as quantifiers in

J. Morgan · A.N. Zincir-Heywood (✉)
Faculty of Computer Science, Dalhousie University, Halifax, Canada
e-mail: zincir@cs.dal.ca

J. Morgan
e-mail: Jillian.morgan@dal.ca

J.T. Jacobs
Raytheon Space and Airborne Systems, Waltham, USA
e-mail: John_T_Jacobs@raytheon.com

© Springer International Publishing Switzerland 2016
R. Abielmona et al. (eds.), *Recent Advances in Computational Intelligence in Defense and Security*, Studies in Computational Intelligence 621,
DOI 10.1007/978-3-319-26450-9_10

order to accurately analyze and classify the network traffic in streaming environments [1, 2] in order to detect malicious activity [3–5].

Even when using a machine learning algorithm, classification within a streaming environment poses many challenges. One of the main challenges is that one cannot have access to all available data in a streaming scenario at once, as in the case of non-streaming environments such as offline streaming scenarios. Thus, classification of streaming data can be quite costly in terms of resources as datasets can grow to be quite large. Furthermore, because a complete set of data cannot be viewed at once it is difficult to determine if given data instance attributes (features) are representative of others in the same class. Additionally, network traffic patterns can slowly change over time or even instantaneously. Indeed, such problems are also exhibited in other streamed datasets. To this end, active learning has often been implemented to alleviate these issues [6, 7]. Active learning is the task of selecting a data instance on which to query the true classification label and retrain the learning algorithm. Selection of labels is not a simple task as one must consider how many and which labels will represent the entire dataset and allow for the most accurate prediction of future data instances.

Many researchers in the literature [4, 6–8] determine the success of classification by measuring the overall accuracy (the number of successful classifications over the total amount of classifications) of the chosen classification algorithm and active learning strategy. This is not necessarily the best solution in determining classification success as it does not account for class distribution. If a dataset has an unbalanced distribution of classes then the resulting prediction accuracy may not be representative of the actual performance of the given classification strategy. Thus, a performance metric that factors class distribution into account is necessary.

In this work we aim to benchmark the performance of previously existing active learning and query budgeting strategies as well as an adaptive Artificial Neural Network approach when performed on network traffic flows, specifically in order to detect malicious network activity, such as botnets. In evaluating the performance of these strategies, we include two performance measures; (i) prequential accuracy and (ii) prequential detection rate.

The remainder of our chapter is organized by discussing related work in Sect. 2. Detailing the methodology employed in this research in Sect. 3. Presenting and discussing the evaluations and results in Sect. 4. Finally, we draw our conclusions and discuss the future work in Sect. 5.

2 Related Work

There are many works on the detection of malicious activity within network traffic, classification of streaming data, and active learning, as separate topics. However, to the best of our knowledge there are no works that combine all of these to evaluate and analyze their performances on the detection of malicious behaviors on network traffic. It should be noted that, some techniques have been proposed that utilize a subset of these ideas. The works most relevant to our study are described in detail below. In this case, the related works are discussed under three categories; active learning, streaming data classification, and detection of malicious activity among network traffic.

2.1 Active Learning

Wang et al. [1] proposes an active learning strategy that utilizes computational intelligence in the form of fuzzy rough sets combined with a support vector machine (SVM). This strategy is performed on both binary and multiclass benchmarking datasets where the performance is evaluated by quantifying accuracy, time costs for labeling new examples from the data, and paired Wilcoxon rank-sum tests. The results of this strategy are compared to other existing strategies such as Random Sampling, SVM Active, and QBC. The researchers conclude that this new strategy is generally successful when compared to other tested strategies in terms of accuracy and the paired Wilcoxin rank-sum tests. However, this comes at a higher time cost as labeling instances takes longer when using this strategy.

Zliobaite et al. [6], also endeavor to produce new active learning strategies. However, the active learning strategies they designed were created with the intention of being used on drifting streaming data. These methods focus on retraining a learning algorithm when confidence of successful prediction of an entity falls below a fixed threshold and randomly selecting entities. The selection of entities to train on was limited by a set budget for querying new labels on which to train the model. The strategies that were developed were tested on a series of publicly available big data sets, which were categorized as either being a prediction or textual dataset. Prediction datasets required a prediction from the classifier and textual datasets required a recommendation from the classifier. Performance of the active learning strategies was evaluated by applying these strategies to the Naive Bayes algorithm and the Hoeffding Tree algorithm. Accuracy of these techniques using different datasets [6] and labeling budgets (10 and 100 %) was measured [6]. The researchers concluded that the strategies are effective for reducing computation costs while maintaining performance.

Like the previous researchers, Zhu et al. [7] proposed another active learning strategy for the implementation of streamed datasets. The strategy they designed features a weighted-classifier ensemble framework with an emphasis on reducing variance. The researchers reported that by decreasing the classifier ensemble variance, the error rate of the classifier ensemble would decrease as well. Thus, a minimum-variance principle was introduced whereas labels were queried for instances that produced a high ensemble variance. The combination of a weighted-classifier ensemble and a minimum-variance principle were employed over three publically available prediction datasets. Performance of this strategy was evaluated by determining the accuracy and runtime when using this strategy on various data chunks and data chunk sizes. These results were then compared to the results of simpler solutions. The researchers concluded that their strategy was effective at dealing with multiclass problems in a streaming environment.

2.2 Streaming Data Classification

Dalal et al. [9] demonstrated various data mining prediction techniques used to predict user-perceived streamed media quality. The researchers proposed the use of a nearest neighbor algorithm, in which unlabelled instances in a stream are given the label (i.e., the normalized user-perceived quality rating) of the instance in the training set of the instance within the closest distance. The researchers tested this algorithm using two different types of distance metrics; summary statistics and dynamic time warping [9]. The algorithm was employed over three different data streams; a commercial, a movie trailer, and a news segment. The authors measured performance using hit rates, that is, the percentage of predictions within 0.8 standard deviations of the normalized user quality rating for the given stream. The researchers concluded that the chosen techniques performed effectively. They expanded upon their work further [10] by performing similar tests in real time using Transmission Control Protocol (TCP) based streams rather than offline using User Datagram Protocol (UDP) based streams. They conclude that the performance accuracy was hopeful (falling between 75 to 87 % accuracy) but could be improved further.

Moreover, Cunha et al. [11] evaluated the performance of Naive Bayes and C4.5 Decision Trees algorithms in classifying different failure states when streaming video data. Specifically, the researchers wanted to be able to predict whether a server failure was a performance anomaly or was caused by overloading produced by clients. Performance was measured by; True Positive Rate, False Positive Rate, Precision, Recall, F-measure, ROC Area, and Root Mean Squared Error. The researchers concluded that both algorithms were good but C4.5 performed slightly better than Naive Bayes.

Vahdat et al. [2] designed and developed a framework for employing genetic programming in order to perform classification on streamed data while maintaining a labeling budget. They employed artificially generated, as well as publically available, datasets. They measured the performance not only by the aforementioned performance metrics but also by prequential accuracy. They conclude that genetic programming with labeling budgets is an effective method for making classifications on streaming data.

2.3 Detection of Malicious Behavior Among Network Traffic

The use of flow-based network traffic in detecting malicious activity among network traffic appears to be quite popular within existing literature. In the work of

Stevanovic et al.'s [4] study, network traffic was converted into network flows in which to be classified. To evaluate the validity of using such a technique to detect malicious activity, a number of classifiers were tested; Naive Bayes, Bayesian Network, Logistic Regression, Artificial Neural Networks, Support Vector Machines with a linear kernel, C4.5 decision tree, Random Tree, and Random Forest. The proposed technique was implemented on a combination of datasets featuring traffic from Storm and Waledec botnets and normal traffic. In order to measure performance, they employed precision, recall, F-measure, and a correlation coefficient. Additionally, they measured the training and classification time when using each classifier. The researchers concluded that C4.5 Decision Tree, Random Tree and Random Forest were the most successful algorithms for their task.

Similarly, Nogueira et al. [5] proposed the use of a flow based system in order to detect botnet activity among network traffic. They employed a Neural Network model in conjunction with a flow-based system. However, the employed system also features a user interface to visualize illicit activity that was detected for further action by an administrator. In identifying botnet activity a feed-forward propagation neural network with three layers was implemented. Performance was evaluated by testing the framework on traffic generated by known safe applications such as Skype. Malicious activity was artificially generated. The authors concluded that the detection of the botnet activity using their methodology was quite successful.

Hsiao et al. [8] also proposed the use of flow-based network traffic for the purpose of detecting malicious behavior amongst said traffic. Flows were generated from network flows collected by the researchers. What differentiates this study from others is that the authors varied the number of flow attributes and which flow attributes were presented between experiments. Thus, they created four sets of attributes to be tested; NetFlow variables, Temporal Variables, Spatial Variables, and a combination of Temporal and Spatial variables. In these experiments, the classification algorithms chosen to employ on the flows were as follows: Naïve Bayes, Decision Tree and SVM algorithms. The results showed that using a combination of temporal and spatial attributes provided the best prediction accuracy.

On the other hand, Saad et al. [3] implement a slightly different approach to detecting malicious botnet behavior than the aforementioned studies that share the same goal. They used not only flow based attributes but also used host-based attributes (i.e., attributes that are exhibited in communications between hosts). They employed the following classification algorithms: Nearest Neighbor, Linear Support Vector Machine, Artificial Neural Network, Gaussian-Based Classifier, and Naïve Bayes. With these methods, the researchers aimed to satisfy three botnet-detection requirements; adaptability, novelty detection, and early detection. The authors used a combination of three datasets for their experiments. The first two datasets were generated by a botnet effected machine where all packets incoming and outgoing were captured. Each machine was affected with a different botnet; Storm or Walodec. The third dataset was made up of normal traffic. These datasets were then combined into one larger dataset in order to simulate a real world scenario. To determine the performance of the selected methodology the researchers adopted four performance metrics: training time, classification time, accuracy, and classification error. The researchers concluded that the selected methodology did not sufficiently satisfy the three stated requirements for effectively detecting botnets.

In our previous work [12], we proposed a new framework to detect HTTP-based botnet activity based on botnet behavior analysis. To achieve this, we employed machine learning algorithms on flow-based network traffic utilizing NetFlow (via Softflowd). The proposed botnet analysis system was implemented by employing two diffent learning algorithms, namely C4.5 and Naïve Bayes. Our results showed that the C4.5 learning algorithm-based classifier obtained very promising performance on detecting HTTP-based botnet activity. However, that work did not employ any streaming or budgeting strategies.

On the other hand, in this paper, we aim to apply and benchmark existing active learning strategies on network streamed traffic in order to make classification predictions for malicious network behavior. This approach differs from the aforementioned related work as the previous work has not combined active learning strategies with the streaming data classification on network traffic. In this paper, we employ such an approach, specifically to detect botnet activities under a streaming scenario. We also aim to compare the performance of these strategies with an adaptive Artificial Neural Network approach and determine which is more effective in performing the desired task.

Last but not least, we also introduce the use of performance metrics; prequential accuracy and prequential detection rate. Prequential detection rate has not been used to measure performance under Massive Online Analysis (MOA) scenarios. Prequential detection rate is a useful metric when unbalanced distributions of classes are present in a given dataset, because unlike accuracy, prequential detection rate can reflect the difference of correctly classifying data instances of the smaller classes in the data. To give an example, if a data set has 99 % of class-normal and 1 % class-malicious, by classifying everything as class-normal, a classifier can reach 100 % accuracy! Even though false positive rates may show the picture a bit clearer, in streaming environments this kind of metric can be ineffective and cannot measure the performance correctly. This is important to consider as looking at accuracy alone can skew how we perceive the performance of our algorithms on an unbalanced dataset. We also analyze the prequential values for both accuracy and detection rate as this allows for us to see how these values change over time.

3   Methodology

In this research, our goal is to utilize the algorithms as discussed in [6], and determine the success of these algorithms when classifying streamed network traffic data for detecting malicious botnet behavior. In order to achieve this, we enacted three major steps: data collection, implementation of learning algorithms in conjunction with various active learning budgeting strategies, and performance analysis.

3.1 Datasets and Features Employed

In this research, four datasets, Table 1, are employed in our evaluations. These are:

(1) KDD Cup 1999: The KDD Cup 1999 dataset is a simple to classify dataset that contains malicious and normal network traffic flows, where each instance to be classified is a connection record [13]. The malicious traffic is broken into four types; denial of service, unauthorized remote access, unauthorized access to root commands, and probing. For our purposes, we only aimed to detect whether or not a connection is malicious, thus, we combined the four attack types into one class. Even though, it is an old dataset, it is chosen to provide us with a baseline and reference point for our other results. This dataset is suitable for this purpose as it is one of the first datasets that was made publically available for benchmarking computational intelligence techniques for network security purposes.

(2) NIMS1: The NIMS1 dataset can be retrieved from the Network Information Management and Security (NIMS) [14]. The dataset is a collection of network traffic flows. Unlike the other datasets, where we aim to detect malicious behavior among network traffic, with this dataset, we aim to classify application type. Thus, this dataset is chosen to provide us a comparison of results when aiming to classify different applications on streaming network traffic as opposed to making classifications between normal and malicious network behaviors.

(3) ISOT: The ISOT (Information Security and Object Technology) dataset is a collection of publically available malicious (different botnets) and normal datasets [15]. These traffic datasets were generated by using a series of machines with different MAC addresses and IP addresses. The traffic generated was captured by the open source packet capturing tool, Wireshark,[1] in order to combine the smaller datasets into the larger ISOT dataset. Thus, we employed this dataset to predict whether or not a connection was malicious.

(4) Zeus versus Alexa: This dataset is generated by the NIMS Lab to be used for botnet detection purposes. To this end, we generated a traffic dataset that exhibited an approximate balance of malicious and normal network traffic. In order to generate this dataset, lists of valid malicious and non-malicious

Table 1 Summary of the datasets

	Instances	Attributes	Classes
KDD 99	494 021	42	2
ISOT	2 084 216	16	2
NIMS	713 851	23	11
Zeus versus Alexa	11 468	16	2

---

[1]Wireshark; https://www.wireshark.org/.

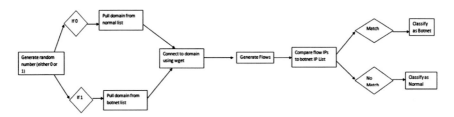

Fig. 1 Script functionality for generating web traffic

domain names were obtained [12]. We obtained the list of non-malicious (normal) domain names from Alexa, a website that ranks the top 500 websites on the Internet according to page views [16]. Because the domains listed are some of the most popular domains on the Internet, it is fair to assume that the traffic generated by accessing these domains is representative of normal, everyday, network traffic. For the malicious domains, we obtained a list of domain names that are known to belong to the Zeus botnet [17, 18]. To simulate web traffic to these domains, a script was written to randomly connect to either a normal or malicious domain using the *wget* command in Linux. These steps are detailed in Fig. 1.

Once the above datasets were obtained, we replayed the traffic on our test bed network to emulate a real-life scenario of streaming traffic. The streaming traffic is then converted to flows as the traffic runs and then, the network flows are input to the streaming classifiers. A network traffic flow is a sequence of network traffic packets with 5-tuple information over a specific period defined by the Internet Engineering Task Force [19]. This 5-tuple information includes; the source/destination IP addresses, source/destination port numbers and the protocol. Usually, in real life, the router (such as a Cisco router with NetFlow) will do this on the flow. To emulate such a scenario, we employed the following open source tools to convert the packets into flows:

- **Softflowd**: Softflowd[2] is an open source tool that accepts network packets and exports them into NetFlow[3] flows.
- **Nfcapd**: Nfcapd[4] captures the exported flows and stores them for further processing. The flow data that Nfcapd records are not in a human readable format, thus, further processing is required.
- **Nfdump**: NfDump[5] takes the recorded flow data and converts it into a human readable format (Table 1).

---

[2]Softflowd: http://www.mindrot.org/projects/softflowd/.

[3]Netflow: http://www.cisco.com/c/en/us/products/ios-nx-os-software/ios-netflow/index.html.

[4]Nfcapd: http://nfdump.sourceforge.net/.

[5]NfDump: http://nfdump.sourceforge.net/.

3.2 Learning Algorithms and Budgeting Strategies

As mentioned previously, the goal of our study is to benchmark the performance of existing active learning strategies on streamed network traffic flows. Thus, in this section we present the data stream mining tools, the budgeting strategies, and performance metrics we employed in our evaluations.

3.2.1 Massive Online Analysis

Massive Online Analysis (MOA[6]) is an open source tool for data stream mining [20]. It has proved very useful for our study as it is able to simulate a data stream with a provided input. Furthermore, MOA provides users the ability to implement the use of various machine learning algorithms and active learning strategies on the data as it is being streamed. Additionally, MOA includes an Application Programming Interface (API) suite that allows for users to create and modify the functionality of existing code to suit their own evaluation needs.

3.2.2 Labels, Budgeting, and Active Learning

In a real-world streaming network traffic environment, it is assumed that the amount of incoming data is infinite and dynamic. This means that the data attributes and how they relate to one another can change over time either slowly (concept drift) or suddenly (concept shift). Thus it can be assumed that it is of more use to train a classifier on incoming data than to use a pre-existing model. In this scenario, a classifier predicts the class of an instance based on the attributes of instances received prior. Once the prediction has been made, the classifier will query the actual class from a human provided label. The classifier will then train on the current instance with the intention of increasing prediction accuracy for future oncoming instances. In a network streaming environment where one aims to classify between normal or malicious behavior this would mean that for every flow that arrived at the network the classifier would have to be provided with its true classification label (i.e., whether the flow was normal or malicious network traffic behavior). As mentioned previously, attempting to perform classification tasks on such a large dataset can be quite costly in terms of human effort (providing true classification labels) time efficiency and hardware required to handle such large datasets. Thus, the concept of budgeting is introduced. Budgeting involves limiting the amount of queries that can be made to retrieve the true classification label of an instance in a data stream [6]. Active learning incorporates this idea of budgeting but adds a learning aspect in which the system makes an educated guess on which classifications labels are most useful to query.

---

[6]MOA: http://moa.cms.waikato.ac.nz/.

3.2.3   Budgeting and Active Learning Strategies

The budgeting strategies that were chosen for our benchmarking study were chosen based on the study performed in [6]. We chose to implement the same strategies for a multitude of reasons. Firstly, work in [6] focuses on developing strategies for streams with drifts, which is relevant for our study as most network streamed data will exhibit drifts. Secondly, using the same active learning strategies gives us an opportunity to compare how the strategies perform on streamed network traffic datasets as compared to the general prediction and textual datasets used within Zliobaite et al.'s study [6]. Because we will be comparing our results with the results of [6], we will also be using budgets of 10 and 100 %. The active learning strategies that were used in our study are described below.

- **Random**: This strategy randomly chooses data instances to query for the true label [6]. No active learning occurs with this simple budgeting strategy, so it provides an effective baseline for our evaluations.
- **Fixed Uncertainty**: Queries the true labels of the data instances with a confidence below a given threshold [6].
- **Variable Uncertainty**: Queries the true labels of the data instances with the lowest confidence within a variable time interval [6].
- **Random Variable Uncertainty**: This is a combination of the Random and Fixed Uncertainty budgeting strategies [6].
- **Select Sampling**: Queries the true labels randomly with a changing probability bias [21].

It is important to note that if a query for a true label is necessary then the training model will be trained on the instance that was queried.

3.2.4   Learning Algorithms

For our study, we selected three different algorithms to accompany our chosen active learning strategies for streaming classification. The algorithms chosen are: (i) Naïve Bayes, (ii) Hoeffding Tree, and (iii) Adaptive Artificial Neural Networks.

(i) Naïve Bayes
Naive Bayes is a simple probabilistic classifier that is known to perform quite well considering its simplicity [22]. The classifier makes predictions by assuming that all attributes of a given instance do not correlate to each other in the probability of a label being of a given class. Predicting a class using this algorithm is performed by determining which class ($C_1, C_2...C_k$ where k is the total number of classes) has the highest posterior probability based on the input x:

$$P(C_i|x) > P(C_j|x) \, for \, 1 \leq j \leq k, \, j \neq i \tag{1}$$

where:

$$P(C_i|x) = \frac{P(C_i|x)P(C_i)}{P(x)} \tag{2}$$

where $P(C_i|x)$ is the conditional probability and $P(C_i)$ is the prior probability of class C_i.

(ii) Hoeffding Tree

The Hoeffding Tree algorithm, also known as a Very Fast Decision Tree, is a more complicated algorithm that incorporates the use of decision trees. It was designed to be used on large data streams where only a subset of the data that passes through is used to find the best split for the tree. The number of samples included in this subset to achieve the desired confidence threshold is determined by a dynamic threshold called a Hoeffding Bound. Hoeglinger et al. [23] describe the Hoeffding Bound as a principle that says "with a probability of $1 - \delta$ the true mean of a variable is at least $\bar{r} - \varepsilon$" where ε is the desired error and is described as follows:

$$\varepsilon = \sqrt{\frac{R^2 \ln(1/\delta)}{2n}} \tag{3}$$

where l is the current leaf in the decision tree, R is the range of random variables, r, n is the number of independent observations made so far, and $1 - \delta$ is the error probability. The described Hoeffding Bound is then used within a decision tree to determine on which attribute to split. This is done by determining the largest gain between two attributes. If the largest calculated gain is greater than the ε then the Hoeffding Tree algorithm states that this attribute is the best attribute to split on with a probability of $1 - \delta$ [23].

(iii) Neural Networks

We also employ a well-known bio-inspired computational intelligence technique in our evaluations in order to systematically benchmark different learning techniques. To this end, we specifically use adaptive Artificial Neural Networks. Artificial Neural Networks (ANNs) are learning algorithms that are designed to imitate real-world biological neural networks. In our work, we use the Pattern Recognition network with a Multi-layer Perceptron within Matlab's[7] Neural Network Toolbox. In order for our network to work properly with streaming data we implement the use of the *adapt* function within Matlab. The *adapt* function, as we used it in our experiments, allowed for the neural network to adapt as data was being streamed. In other words, instead of training our network on a training set, the neural network would be trained on each data instance as the data (traffic) arrives. This means that a labeling budget of 100 % is used to train on each instance where the true label is queried.

---

[7]Matlab: http://www.mathworks.com/products/matlab/.

4 Evaluation and Results

For the purpose of evaluating the performance of our chosen machine learning algorithms and budgeting strategies on network datasets, two performance metrics are employed [24]: prequential accuracy and prequential detection rate.

Accuracy of a classifier is described as the total number of correct classifications over all the classification predictions made (n), that is:

$$\text{Accuracy} = \frac{tp}{tp + fn} \tag{4}$$

where, t_p denotes the true positives, and f_n denotes the false negatives.

Similarly, prequential accuracy is the total number of correct classifications over the total number of classifications made at a given point in time, that is;

$$preqACC_t = \frac{(t-1) \times preqAcc_{t-1} + C_t}{t} \tag{5}$$

where t indicates a given time instant, $t-1$ indicates the previous time instant, and C indicates whether or not the classification at the given time point was successful ($C = 1$ if the classification was correct, or $C = 0$ if the classification was incorrect).

Although accuracy is used to measure the performance in some works [1, 3, 6–10], its use could be problematic. With the use of unbalanced datasets, where the number of instances belonging to each class is significantly different, using accuracy as a measure of classification performance can be misleading. For example, if we have a dataset that consists of 98 % normal activity and 2 % malicious activity and the classification model predicts that all activity is normal then we achieve 98 % prediction accuracy. However, this result does not indicate successful classification, as no malicious activity was detected. Therefore, we want to use a performance metric that accounts for class imbalance in addition to false positive rates. Thus, the use of prequential detection rate is introduced as a performance metric for our experiments.

In this research, the detection rate at time t is calculated using Eqs. 6 and 7:

$$DR(t) = \frac{1}{Q} \sum_{q=1}^{q=Q} DR_q(t) \tag{6}$$

where:

$$DR_q(t) = \frac{tp_q(t)}{tp_q(t) + fn_q(t)} \tag{7}$$

where Q is the number of classes, q denotes a particular class, tp indicates true positives, fn indicates false negatives, and t denotes the given instant in time.

Similarly, to find the detection rate at any given time, Prequential detection rate is calculated below Eq. 8, where t denotes the given point in time.

$$preqDR_t = \frac{(t-1) \times preqDR_{t-1} + DR_t}{t} \tag{8}$$

where:

$$DR(t) = \frac{1}{Q} \sum_{q=1}^{q=Q} DR_q(t) \tag{9}$$

where:

$$DR_q(t) = \frac{tp_q(t)}{tp_q(t) + fn_q(t)} \tag{10}$$

4.1 Results of Using Learning Algorithms Together with Budgeting Strategies

Tables 2, 3, 4, and 5 show the overall results for each labeling when using different budgets with different classification algorithms (Naïve Bayes and Hoeffding Tree) over the four chosen datasets employed in this research. Classifications predictions made on the KDD 1999 Cup dataset (Table 2) generally appear to perform the same regardless of the budget or the learning algorithm chosen, with the exception of the Hoeffding Tree Algorithm using the Fixed Uncertainty Strategy whereas fewer correct classifications are made. Furthermore, we see the detection rate make a dramatic drop from the other detection rates and accuracies presented here.

Table 2 Overall prediction accuracy (ACC) and detection rate (DR) of different budgeting strategies using KDD 1999 Cup dataset

	Performance metric	Random	Fixed uncertainty	Variable uncertainty	Random variable uncertainty	Select sampling
NB	ACC	99.87	94.21	99.73	99.84	99.81
100 %	DR	99.70	92.64	99.62	99.79	99.75
NB	ACC	99.90	97.05	99.54	99.47	99.47
10 %	DR	99.14	97.12	99.37	99.29	99.38
HT	ACC	99.90	80.46	98.37	99.77	99.86
100 %	DR	99.87	50.38	96.06	99.59	99.81
HT	ACC	99.46	81.88	99.36	99.56	99.54
10 %	DR	99.15	53.99	98.84	99.38	99.38

NB indicates Naive Bias and HT indicates Hoeffding Tree

Table 3 Overal prediction accuracy (ACC) and detetion rate (DR) and of different budgeting strategies using NIMS dataset

	Performance metric	Random	Fixed uncertainty	Variable uncertainty	Random variable uncertainty	Select sampling
NB	ACC	88.73	87.20	90.41	90.72	90.47
100 %	DR	96.42	55.00	96.41	96.20	95.63
NB	ACC	82.16	89.73	89.17	91.11	89.31
10 %	DR	92.24	49.17	92.79	91.70	92.85
HT	ACC	96.38	0.33	95.33	95.30	95.54
100 %	DR	88.05	9.08	79.74	80.70	81.96
HT	ACC	93.69	0.33	94.25	94.56	94.50
10 %	DR	71	8.96	74.87	75.24	75.92

NB indicates Naive Bias and HT indicates Hoeffding Tree

We believe that this is because the differences between the malicious and non-malicious behaviors are easily separable as discussed in [25].

On the NIMS1 dataset (Table 3), all strategies that use Naïve Bayes as the classification algorithm perform similarly with each other. The same observation can be made for the strategies that use the Hoeffding Tree algorithm in terms of accuracy. We see a significant jump in performance in accuracy when using this algorithm on this dataset when compared to Naïve Bayes, except when using the Fixed Uncertainty strategy where we see a large drop in performance in terms of the detection rate.

On the ISOT dataset (Table 4), we see that Random, Fixed Uncertainty, Random Variable Uncertainty and Select Sampling perform roughly the same among all budgets and classification algorithms (excluding the fixed uncertainty strategy used in conjuction with the Hoeffding Tree algorithm). However, we see some interesting results when using the Variable Uncertainty strategy on this dataset.

Table 4 Overal prediction accuracy of different budgeting strategies using ISOT dataset

	Performance metric	Random	Fixed uncertainty	Variable uncertainty	Random variable uncertainty	Select sampling
NB	ACC	99.99	90.27	19.48	99.99	99.99
100 %	DR	99.94	94.96	58.45	99.96	99.93
NB	ACC	99.98	92.71	89.71	99.99	99.98
10 %	DR	99.84	96.18	99.77	99.88	99.88
HT	ACC	99.99	3.03	12.94	99.99	99.99
100 %	DR	99.99	49.98	55.10	99.99	99.99
HT	ACC	99.99	3.03	99.72	99.99	99.99
10 %	DR	99.94	49.99	99.79	99.96	99.96

NB indicates Naive Bias and HT indicates Hoeffding Tree

Table 5 Overal prediction accuracy of different budgeting strategies using Alexa versus Zeus dataset

	Performance metric	Random	Fixed uncertainty	Variable uncertainty	Random variable uncertainty	Select sampling
NB	ACC	98.44	97.62	97.52	98.16	97.50
100 %	DR	99.96	96.21	97.18	97.73	97.14
NB	ACC	95.80	97.50	96.15	95.40	93.85
10 %	DR	95.23	96.01	95.60	95.61	94.47
HT	ACC	98.32	31.29	93.69	97.96	97.95
100 %	DR	97.94	49.97	94.38	97.46	97.66
HT	ACC	96.47	31.21	94.84	95.22	94.34
10 %	DR	95.26	49.84	94.89	95.30	94.51

NB indicates Naive Bias and HT indicates Hoeffding Tree

Although we see very poor performance (results <60 %) when using budgets of 100 % with variable uncertainty, when we change the budget to 10 %, we actually get an increase in performance, bringing the results on par with the other strategies. This is important to note, as it shows that even though we are training on less information, we can actually train more effectively in some cases.

Lastly, we see our results for our experiments on the Alexa versus Zeus dataset in Table 5. Again, we see similar performance among all strategies, budgets, and classification algorithms except for when using the Fixed Uncertainty Strategy with the Hoeffding Tree machine learning algorithm.

In the cases where the fixed uncertainty strategy produces low rates of correct classifications, we can assume that this may be because the confidence of each instance when streaming is never low enough to invoke training. Thus, no learning is performed. This happens with the fixed uncertainty strategy in particular as the confidence threshold is fixed and never changes to adjust to incoming data.

In Figs. 2, 3, 4, 5, 6, 7, 8, 9, 10, 11, 12 13, 14, 15, 16 and 17, we can view the prequential accuracy and prequential detection rates for each instance (time point) when using machine learning algorithms (Naïve Bayes and Hoeffding Tree) in conjunction with the selected budgeting strategies.

When our budgeting experiments were performed on the KDD 1999 (Figs. 2, 3, 4, and 5) Cup dataset, we see that accuracy and detection rate increases rapidly at the beginning and then maintains a consistent performance throughout the streaming process. This statement is true for all budgeting strategies used here except for the fixed uncertainty strategy, which exhibits a different pattern. In terms of prequential accuracy, we see a slower rise when using the fixed uncertainty strategy in all cases on this dataset, except when using Naives Bayes with a budget of 10 %. In this case, the prequential accuracy initially increases rapidly but drops soon after in a concave formation. The prequential detection rate when employing the fixed uncertainty strategy remains around 50 % in all cases except when using Naives Bayes with a budget of 10 %. In that case, the prequential detection rate initially increases rapidly, but then drops soon after in a concave formation.

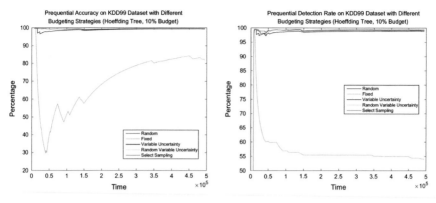

Fig. 2 Prequential accuracy (*left*) and prequential detection rate (*right*) on KDD 1999 Cup Dataset using the Hoeffding Tree Algorithm with a 10 % budget

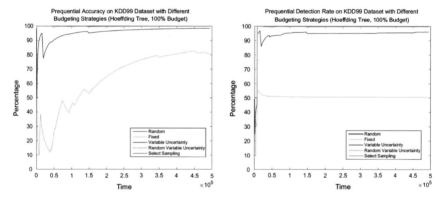

Fig. 3 Prequential accuracy (*left*) and prequential detection rate (*right*) on KDD 1999 Cup Dataset using the Hoeffding Tree Algorithm with a 100 % budget

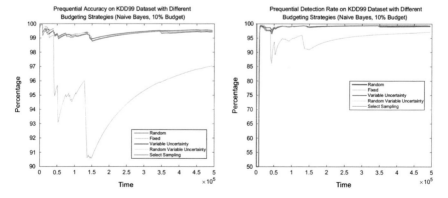

Fig. 4 Prequential accuracy (*left*) and prequential detection rate (*right*) on KDD 1999 Cup Dataset using the Naïve Bayes Algorithm with a 10 % budget

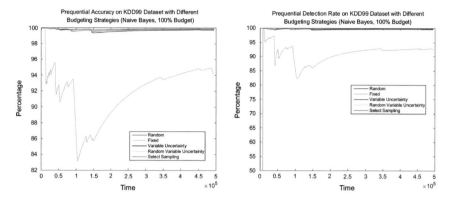

Fig. 5 Prequential accuracy (*left*) and prequential detection rate (*right*) on KDD 1999 Cup Dataset using the Naïve Bayes Algorithm with a 100 % budget

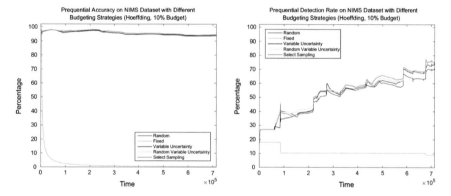

Fig. 6 Prequential accuracy (*left*) and prequential detection rate (*right*) on NIMS Dataset using the Hoeffding Tree Algorithm with a 10 % budget

When evaluating the employed machine learning algorithms in conjunction with the budgeting strategies on the NIMS dataset (Figs. 6, 7, 8, and 9), prequential accuracy tends to rise quickly and then maintains its performance. However, the detection rate appears to increase in performance more slowly, almost in a step pattern. These are consistent in all cases when using this dataset except when the fixed uncertainty strategy is utilized. In that case, the prequential accuracy rises quickly and then drops back down quickly to a very low accuracy for the rest of the stream. When measuring the prequential detection rate, under the fixed uncertainty strategy, we observe that it either remains low (when using the Hoeffding Tree Algorithm, Figs. 6 and 7) or follows the trend of the rest of the strategies but at a lower accuracy (when using the Naïve Bayes Algorithm, Figs. 8 and 9) throughout the stream.

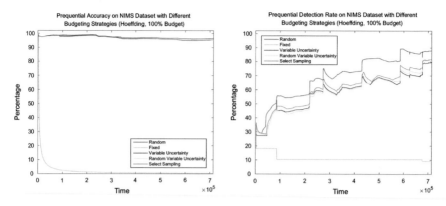

Fig. 7 Prequential accuracy (*left*) and prequential detection rate (*right*) on NIMS Dataset using the Hoeffding Tree Algorithm with a 100 % budget

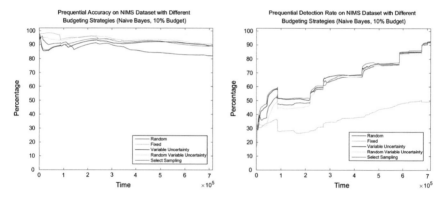

Fig. 8 Prequential accuracy (*left*) and prequential detection rate (*right*) on NIMS Dataset using the Naïve Bayes Algorithm with a 10 % budget

On the other hand, when these strategies are employed on the ISOT dataset (Figs. 10, 12, 13, and 14) we see that for the Random, Random variable Uncertainty, and Select Sampling strategies, the prequential detection rate and accuracy remain at approximately 100 % throughout the streaming process. However, with the fixed uncertainty strategy and the variable uncertainty strategy, we observe some more interesting behavior. When using a budget of 10 % with the variable uncertainty strategy, the trends exhibited follows the patterns of the Random, Random Variable Uncertainty, and Select Sampling on this dataset. However when using a budget of 100 % with either machine learning algorithm, we observe a significant drop in the performance in terms of prequential detection rate and even greater loss in terms of prequential accuracy.

When using the fixed uncertainty strategy with the Hoeffding Tree Algorithm, we observe a downwards slope in the prequential accuracy while the prequential

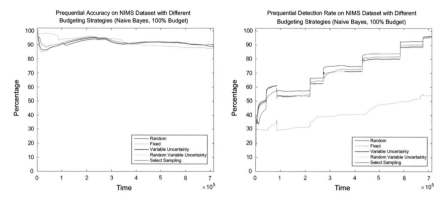

Fig. 9 Prequential accuracy (*left*) and prequential detection rate (*right*) on NIMS Dataset using the Naïve Bayes Algorithm with a 100 % budget

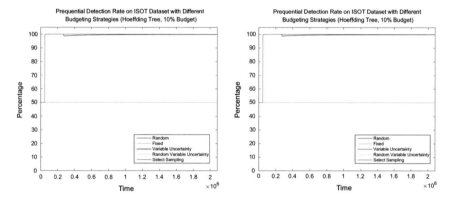

Fig. 10 Prequential accuracy (*left*) and prequential detection rate (*right*) on ISOT Dataset using the Hoeffding Tree Algorithm with a 10 % budget

detection rate maintains a steady performance around 50 %. When this strategy is used in conjunction with the Naives Bayes Algorithm an interesting pattern is exhibited where the prequential accuracy and the prequential detection rate show a sharp rise and then a bit of a fall and then a steady climb up again.

Finally, we look at the trends when using our selected strategies and chosen machine learning algorithms on the Alexa versus Zeus dataset (Figs. 14, 15, 16, and 17). Random, Random Variable Uncertainty, and Select Sampling strategies in conjunction with any budget and machine learning algorithm maintain high prequential accuracy and prequential detection rates throughout this streaming dataset. The Variable Uncertainty strategy performs similarly to the previous cases except when using the Hoeffding Tree Algorithm at a 10 % budget. In this case, the prequential accuracy and the prequential detection rate drop in performance.

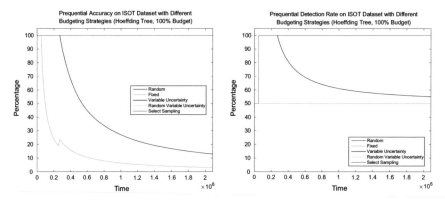

Fig. 11 Prequential accuracy (*left*) and prequential detection rate (*right*) on ISOT Dataset using the Hoeffding Tree Algorithm with a 100 % budget

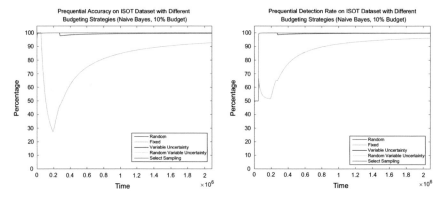

Fig. 12 Prequential accuracy (*left*) and prequential detection rate (*right*) on ISOT Dataset using the Naïve Bayes Algorithm with a 10 % budget

As with our previous experiments, the uncertainty strategy performs differently than the others. When using the Hoeffding Tree algorithm, the prequential detection rate and prequential accuracy performances remain pretty low whereas when the Naïve Bayes Algorithm is used, the performance begins to go upwards again.

Based on the similarity in performances between using 10 and 100 % budgets in all experiments, our results indicate that it is well worth to use a low budget as there appears to be little to no effect on the overall performance. Furthermore, our overall accuracy and detection rates are usually quite high when paired with the Random, Variable Uncertainty, Random Variable Uncertainty, or Select Sampling strategies with performances averaging in the 90 %'s.

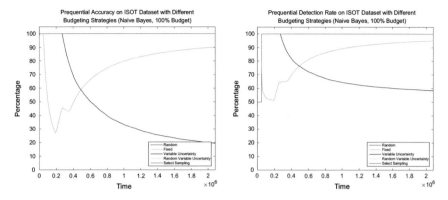

Fig. 13 Prequential accuracy (*left*) and prequential detection rate (*right*) on ISOT Dataset using the Naïve Bayes Algorithm with a 100 % budget

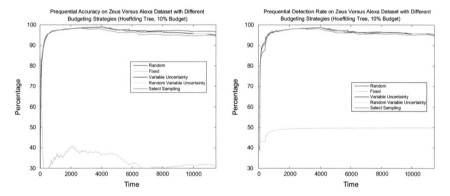

Fig. 14 Prequential accuracy (*left*) and prequential detection rate (*right*) on Alexa versus Zeus Dataset using the Hoeffding Tree Algorithm with a 10 % budget

4.2 Adaptive Artificial Neural Network Results

The overall accuracy and detection rate when applying the Adaptive Neural Network approach on the same datasets used above are presented in Table 6. Here, we observe an overall detection rate of approximately 50 % on all datasets with two classes. We also observe some very low prediction accuracies. Furthermore, in this case, the highest achieved overall accuracy and overall detection rate are ~68 % and ~50 %, respectively. These results are significantly lower than most of the results presented in the previous section.

The trend shown when looking at the prequential detection rate using this learning technique on our datasets seems to remain at approximately 50 % throughout the streaming process (Figs. 18 and 19). This consistent for tests on all datasets except for the NIMS1 dataset where detection rate is at 16.73. This is to be

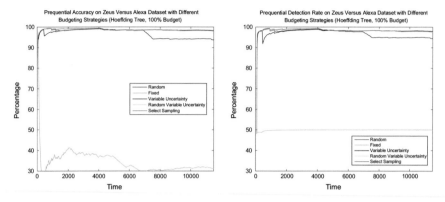

Fig. 15 Prequential accuracy (*left*) and prequential detection rate (*right*) on Alexa versus Zeus Dataset using the Hoeffding Tree Algorithm with a 100 % budget

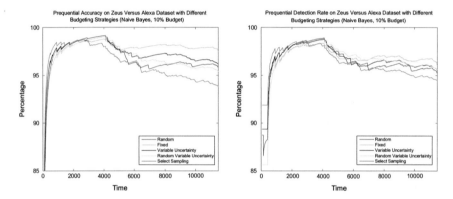

Fig. 16 Prequential accuracy (*left*) and prequential detection rate (*right*) on Alexa versus Zeus Dataset using the Naïve Bayes Algorithm with a 10 % budget

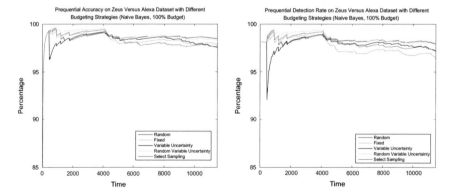

Fig. 17 Prequential accuracy (*left*) and prequential detection rate (*right*) on Alexa versus Zeus Dataset using the Naïve Bayes Algorithm with a 100 % budget

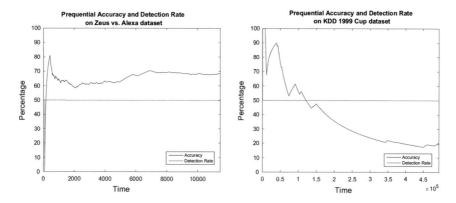

Fig. 18 Prequential accuracy versus prequential detection rate on the Zeus and Alexa dataset (*left*) and the KDD 1999 Cup dataset (*right*)

Table 6 Overall accuracy and detection rate when using Adaptive Artficial Neural Networks on various datasets

	KDD 1999 Cup	NIMS	ISOT	Zeus versus Alexa
Accuracy	19.69	1.71	6.84	68.32
Detection rate	49.97	16.73	48.48	49.74

expected however, as there are more classes. On the other hand, we observe a rapid rise in prequential accuracy at the beginning, and then either a slow drop or a steady state for the rest of the stream. This seems to indicate that the learning algorithm is not able to detect drifts in the behaviors to ask for retraining.

5 Conclusion and Future Work

In this research, we study how to classify (analyze) streaming network traffic using different machine learning algorithms under different training (budgeting) strategies. To achieve this, we analyzed the traffic using flow type features with Adaptive Artificial Neural Network, Naive Bayes and Hoeffding Tree stream classifiers under 10 and 100 % training scenarios with five different budgeting strategies to train. Furthermore, we evaluated the performance of the different combinations of these algorithms and strategies using both the standard accuracy and detection rate as well as the prequential accuracy and detection rate.

Our evaluations show that all the tested budgeting strategies perform relatively similarly (with the exclusion of the fixed uncertainty strategy) on the network datasets employed regardless of the number of different classes in the datasets (NIMS-application versus NIMS1, ISOT, and Alexa vs. Zeus datasets). The results are generally quite high, averaging in the 90 %'s especially when the Hoeffding

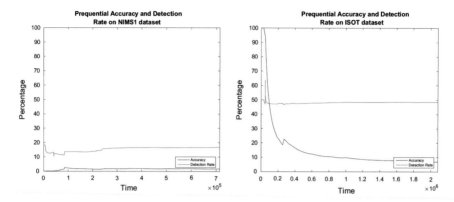

Fig. 19 Prequential accuracy versus prequential detection rate on the NIMS1 dataset (*left*) and the ISOT dataset (*right*)

Tree classifier is employed. This indicates that any of these strategies could be used successfully in classifying network traffic and detecting malicious activity.

Furthermore, we see that changing the budget to 10 % does not affect the performance of our strategies negatively, and can actually increase the performance of a given strategy.

When comparing these results to the adaptive Artificial Neural Network, we observe that this method is not effective at classifying malicious activity among streamed network traffic. Thus it is recommended that an active learning approach is used instead.

In the future, we would like to benchmark these strategies on more datasets as this will give us a better idea how these methods would perform under other real world scenarios. Additionally, it would be interesting to see which budgets would give us the highest performances as we only attempted two different budgets in this work. Finally, we would also like to improve the adaptive Artificial Neural Network used in this study.

Acknowledgement This research is supported by Raytheon SAS. The research is conducted as part of the Dalhousie NIMS Lab at https://projects.cs.dal.ca/projectx/.

References

1. Wang, R., Kwong, S., Chen, D., He, Q.: Fuzzy rough sets based uncertainty measuring for stream based active learning. In: 2012 International Conference on Machine Learning and Cybernetics (ICMLC), vol. 1, pp. 282–288 (2012)
2. Vahdat, A., Atwater, A., McIntyre, A., Heywood, M.: On the application of GP to streaming data classification tasks with label budgets. In: Proceedings in the 2014 Conference Companion on Genetic ad Evolutionary Computation Companion, GECCO Comp'14, pp. 1287–1294 (2014)

3. Saad, S., Traore, I., Ghorbani, A., Sayed, B., Zhao, D., Lu, W., Felix, J., Hakimian, P.: Detecting P2P botnets through network behavior analysis and machine learning. In: Proceedings of 9th Annual Conference on Privacy, Security and Trust (PST2011) (2011)
4. Stevanovic, M., Pedersen, J.M.: An efficient flow-based botnet detection using supervised machine learning. In: 2014 International Conference on Computing, Networking and Communications (ICNC), pp. 797–801 (2014)
5. Nogueira, A., Salvador, P., Blessa, F.: A botnet detection system based on neural networks. In: 2010 Fifth International Conference on Digital Telecommunications, pp. 57–62
6. Zliobaite, I., Bifet, A., Pfahringer, B., Holmes, G.: Active learning with drifting streaming data. IEEE Trans. Neural Netw. Learn. Syst. **25**(1), 2754 (2014)
7. Zhu, X., Zhang, P., Shi, Y.: Active learning from stream data using optimal weight classifier ensemble. IEEE Trans. Syst. Man Cybern. Part B Cybern. **40**(6), 1607–1621 (2010)
8. Hsiao, H., Chen, D., Ju Wu, T.: Detecting hiding malicious website network traffic mining approach. In: 2010 2nd International Conference on Education Technology Computer (ICETC), vol. 5, pp. 276–280 (2010)
9. Dalal, A., Musicant, D., Olson, J., McMenamy, B., Benzaid, S., Kazez, B., Bolan, E.: Predicting user-perceived quality ratings from streaming media data. In: 2007 IEEE International Conference on Communications, pp. 65–72 (2007)
10. Dalal, A., Bouchard, A., Cantor, S., Guo, Y., Johnson, A.: Assessing QoE of on-demand TCP video streams in real time, In: 2012 IEEE International Conference on Communications (ICC), pp. 1165–1170 (2012)
11. Cunha, C., Silva, L.: Separating performance anomalies from workload-explained failures in streaming servers. In: Cloud and Grid Computing (CCGrid), 2012 12th IEEE/ACM International Symposium on Cluster, pp. 292–299 (2012)
12. Haddadi, F., Morgan, J., Filho, E.G., Zincir-Heywood, A.N.: Botnet behaviour analysis using IP flows with HTTP filters using classifiers. In: 2014 28th International Conference on Advanced Information Networking and Applications Workshops, pp. 7–12 (2014)
13. Kdd Cup 1999 Data: http://kdd.ics.uci.edu/databases/kddcup99/kddcup99.html
14. NIMS1 Dataset: https://projects.cs.dal.ca/projectx/Download.html
15. ISOT Botnet Data: http://www.uvic.ca/engineering/ece/isot/datasets/
16. Alexa: http://alexa.com/topsites
17. Abuse: Zeus Tracker. https://zeustracker.abuse.ch/
18. DNS-BH-Malware Domain Blocklist: http://www.malwaredomains.com
19. Claise, B.: Specification of the IP flow information export (IPFIX) protocol for the exchange of IP traffic flow information. In: RFC 5101. http://www.rfc-editor.org/info/rfc5101 (2008)
20. MOA (Massive Online Analysis): http://moa.cms.waikato.ac.nz/
21. Cesa-Bianchi, N., Gentile, C., Zaniboni, L.: Worst-case analysis of selective sampling for linear classification. J. Mach. Learn. Res. **7**, 1205–1230 (2006)
22. Smola, A., Vishwanathan, S.V.N.: Introduction to Machine Learning. Cambridge University Press, Cambridge (2008)
23. Hoelinger, S., Pears, R.: Use of hoeffding trees in concept based data stream mining. In: Third International Conference on Information and Automation for Sustainability, ICIAFS 2007, pp. 57–62 (2007)
24. Heywood, M.I., Evolutionary model building under streaming data for classification tasks: opportunities and challenges. Genetic Programming and Evolvable Machines. **16**(3), 283–326. Springer (2015). doi:10.1007/s10710-014-9236-y
25. Kayacik, G.H., Zincir-Heywood, A.N., Heywood, M.I.: On the capability of an SOM based intrusion detection sytem. In: IEEE International Joint Conference on Neural Networks, pp. 1808–1813 (2003)

Part III
Biometric Security and Authentication Systems

Visualization of Handwritten Signatures Based on Haptic Information

Julio J. Valdés, Fawaz A. Alsulaiman and Abdulmotaleb El Saddik

Abstract The problem of user authentication is a crucial component of many solutions related to defense and security. The identification and verification of users allows the implementation of technologies and services oriented to the intended user and to prevent misuse by illegitimate users. It has become an essential part of many systems and it is used in several applications, particularly in the military. The handwritten signature is an element intrinsically endowed with specificity related to an individual and it has been used extensively as a key element in identification/authentication. Haptic technologies allow the use of additional information like kinesthetic and tactile feedback from the user, thus providing new sources of biometric information that can be incorporated within the process in addition to the traditional image-based sources. While work had been done on using haptic information for the analysis of handwritten signatures, most efforts have been oriented to the direct use of machine learning techniques for identification/verification. Comparatively fewer targeted information visualization and understanding the internal structure of the data. Here a variety of techniques are used for obtaining representations of the data in low dimensional spaces amenable to visual inspection (two and three dimensions). The approach is unsupervised, although for illustration and comparison purposes, class information is used as qualitative reference. Estimations of the intrinsic dimension for the haptic data are obtained which shows that low dimensional subspaces contains most of the data structure. Implicit and explicit mappings techniques transforming the original high dimensional data to low dimensional spaces are

J.J. Valdés (✉)
Information and Communication Technologies Portfolio,
National Research Council Canada, Ottawa, ON, Canada
e-mail: julio.Valdes@nrc-cnrc.gc.ca

F.A. Alsulaiman
College of Computer and Information Sciences, King Saud University,
Riyadh, Saudi Arabia
e-mail: falsulaiman@ksu.edu.sa

A. El Saddik
School of Electrical Engineering and Computer Science (EECS),
University of Ottawa, Ottawa, ON, Canada
e-mail: elsaddik@uottawa.ca

© Springer International Publishing Switzerland 2016
R. Abielmona et al. (eds.), *Recent Advances in Computational Intelligence
in Defense and Security*, Studies in Computational Intelligence 621,
DOI 10.1007/978-3-319-26450-9_11

considered. They include linear and nonlinear, classical and computational intelligence based methods: Principal Components, Sammon mapping, Isomap, Locally Linear Embedding, Spectral Embedding, t-Distributed Stochastic Neighbour Embedding, Generative Topographic Mapping, Neuroscale and Genetic Programming. They provided insight about common and specific characteristics found in haptic signatures, their within/among subjects variability and the important role of certain types of haptic variables. The results obtained suggest ways how to design new representations for identification and verification procedures using tactile devices.

Keywords Dimensionality reduction · Intrinsic dimension · Manifolds · Nonlinear transformations · Implicit/explicit mappings · Haptics · User authentication

1 Introduction

The problem of user authentication is a crucial component of many solutions related to defense and security. The identification and verification of users allows the implementation of technologies and services oriented to the intended user and to prevent misuse by illegitimate users. It has become an essential part of many systems and it is used in several applications, particularly in the military.

The handwritten signature is considered as one element intrinsically endowed with specificity related to an individual and therefore, it has been used extensively as a key element in individual identification/authentication. Traditional approaches have been based on the analysis of the signature as a graphic or visual element, spawning a lot of research in pattern recognition and image processing over many years. Recently introduced Haptic technologies allow the use of additional information like kinesthetic and tactile feedback from the user, thus providing new sources of biometric information that can be incorporated within the description and analysis process in addition to the aforementioned image-based sources.

One possible approach is to capture the human handwritten signature using haptic devices (Sect. 2), describe it using appropriate features and build models (computational intelligence-based for example), as a mean of user authentication. When described in relational form, the haptic information yields high dimensional datasets that require a comprehensive analysis in order to gain insight into the properties of the data and ultimately in achieving an understanding of the information contained. Additional issues are the determination of relevant features, the derivation of new, better descriptive ones, etc. In this sense, the role of visualization techniques in the knowledge discovery process is well known, particularly in the first stages of the data analysis process, although visualization is in no way restricted to that phase. A crucial advantage is the inclusion of the human element with his top pattern recognition capabilities and his problem domain knowledge. A closely related element fulfilling several roles are dimensionality reduction techniques, which enables visualization and also provide ways to understand the relation between features (original and new) and between data objects.

While work had been done on using haptic information for the analysis of hand-written signatures, most efforts have been oriented to the use of several machine learning techniques for identification/verification. Comparatively fewer targeted information visualization and data structure understanding, which is the focus of this chapter.

The chapter is organized as follows. Section 2 describes the haptic data used in the study. In Sect. 3 the topic of intrinsic dimensionality is discussed and several among the many proposed approaches for its estimation are presented. Section 4 introduces some common techniques for producing low dimensional spaces for visualization. Section 5 presents the results obtained when estimating the intrinsic dimensionality of the data and the low dimensional mappings produced by the different dimensionality reduction techniques. Conclusive remarks are outlined in Sect. 6.

2 Haptic-Signature Data and the Virtual Check Application

The experiments were performed using the Reachin Display [41] where a high quality 3D experience is combined with a haptic device. The Reachin visuo-haptic interface allows users to feel and see virtual objects at the same location in space. The sense of touch is felt through the use of the SensAble PHANTOM Desktop force-feedback device by an encoder stylus that provides 6-degree-of-freedom single contact point interaction and positional sensing.

The visual stimuli is depicted in Fig. 1, represented by a virtual pen and a virtual check on which users can record their handwritten signature. The haptic stimuli consist of force and frictional feedback that try to mimic the tactile sensations felt during signing a traditional check. More specifically, the check is built on an elastic membrane surface with particular texture features, providing the users with a user-friendly and realistic feel of the virtual object. In addition, the virtual check application records a wide array of attributes as the user sign the check. In a previous study [44], a group of 13 participants contributed with 10 instances of their signatures (handwritten on the virtual check) a total of 130 haptic-based vectors. In the present study, a more comprehensive dataset was used, composed of 20 participants, each of them contributing with 50 instance signatures for a much larger collection of 1000 haptic vectors.

The collected haptic data contains information about three-dimensional position (P), force (pressure exerted on the virtual check) (F), torque (T), and angular orientation (O). In each case, their three components along the x, y, z axis are recorded as the signature is produced. In addition, an extra feature is included (*length*) that represents the time taken to generate the signature. Each original instance consist of the aforementioned features collected over the duration of the signature production process. In practice, even for the same subject there are variations in the time taken to generate a signature. Therefore, up-sampling/down-sampling processes were per-

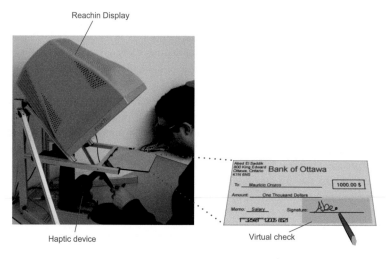

Fig. 1 Haptic-enabled virtual check application. The user writes his signature on the virtual check, visible only through the 3D-glasses and the Reachin display. He experiences the feeling of producing his signature on a solid surface as it would be the case under normal conditions

formed to make all signatures comparable in time to a common number of descriptors/signature, which resulted in versions of the original signatures described in terms of vectors of 9600 features, in addition to the *length* feature. Finally, a normalization process was performed on the dataset in order to convert each original attribute into its z-score (a variable with zero mean and unit variance). In the end the dataset consisted of 1000 signatures described by 9601 features.

3   Intrinsic Dimensionality

The analysis of high dimensional data is an increasingly common and also complex problem. The accelerated rate of development of sensor, communication and computer technologies allows the acquisition of massive amounts of data that are large not only from the point of view of the number of samples/observations, but also from the point of view of the dimensionality of the individual observations themselves. Accordingly, they are described in terms of a large collection of variables/attributes (hundreds, thousands, tens of thousands), which in the case of real world data like those from many engineering and bio-medical domains, are characterized by several kinds of mutual dependencies, redundancies and noise. While apparently the effect of the technological revolution leads to the acquisition of more information from the problems under study, in reality due to the nature of the observed variables and their complex relationships, the amount of useful information is not proportional to their cardinality, which represents the dimension of the observation space. From a data analytics perspective, the performance and efficiency of statistical and machine learning procedures degrades rapidly as the original dimension of the data increases, in what is coined as the curse of dimensionality.

It is known that in high-dimensional spaces the volumes of neighborhoods of a fixed size become large, thus requiring a number of points that grows exponentially for reliable estimation of probability density distributions. Another problem is that the high dimensional spaces associated to the representation of real world data are not filled uniformly. What is found is that very often the data concentrate in low dimension nonlinear manifolds which are embedded within the high dimensional space in which the data is represented. The dimension of those manifolds is what is considered as the intrinsic dimension. Often the dimension of the nonlinear subspace is much smaller than that of the original data representation space, which means that in fact the data is not really high dimensional. The existence of those subspaces is considered an explanation as to why data analytic methods work at all when applied to high dimensional data.

Learning these data manifolds is important and useful for understanding the internal structure of the data, as well as for improving the performance of data analytic methods like clustering, classification and regression. However, depending of the complexity of the data structure dimensionality reduction is usually a difficult task and different approaches have been proposed for finding the intrinsic dimensionality (this section) and for learning the subspace (Sect. 4).

A classical approach to estimate intrinsic dimensionality has been based on the eigenvalues obtained when performing Principal Components Analysis (PCA) [18]. In this case the dimension is determined by the number of important eigenvalues. While the simplicity of this technique makes it appealing, it has the disadvantage of the subjectivity introduced by the choice of the threshold for considering an eigenvalue important. A typical one is to retain those normalized eigenvalues larger than 0.025.

A maximum likelihood estimator (MLE) of the dimension is presented in [34]. The idea is to analyze small hyperspheres around each data point and consider the occurrence of points within as a Poisson process, under the assumption that the probability distribution of the points is constant within the sphere. A log-likelihood function is derived for the process from which an estimate for the dimension around a point $\mathbf{x}$ is obtained. This estimate is expressed in terms of the number of neighbours k as

$$\hat{m}_k(\mathbf{x}) = \left[\frac{1}{k-2} \sum_{j=1}^{k-1} \log \frac{T_k(\mathbf{x})}{T_j(\mathbf{x})} \right]^{-1}$$

where k is the number of neighbours considered and $T_k(\mathbf{x})$ is the Euclidean distance from the point $\mathbf{x}$ to its kth neighbour. The estimation for the whole dataset is obtained by averaging $\hat{m}_k(\mathbf{x})$ for all points $\mathbf{x} \in X = \{\mathbf{x}_1, \ldots, \mathbf{x}_n\}$. Since this is valid only for a certain k, several estimates are produced for a range of values of $k \in [k_1, k_2]$ and then an overall average over that range is considered as the final estimate (usual choices are $k_1 = 6$, $k_2 = 12$):

$$\hat{m}_k = \frac{1}{n} \sum_{i=1}^{n} \hat{m}_k(\mathbf{x}_i), \qquad v = \frac{1}{(k_2-k_1+1)} \sum_{k=k_1}^{k_2} \hat{m}_k \qquad (1)$$

Another approach is based on estimating the dimension of the attractor of a chaotic dynamical system, using the correlation integral [21]. The assumption is that the volume of a m-dimensional dataset scales up with its size in a power-law fashion determined by the dimension and that a similar behaviour can be observed for the number of neighbours that are less than the given size. The correlation integral is used [21] for estimating the intrinsic dimension (v) as

$$v = \lim_{s \to 0} \lim_{n \to \infty} \frac{\log(C_n(s))}{\log(s)} \qquad (2)$$

where

$$C(s) = \frac{2}{n(n-1)} \sum_{\substack{i<j}}^{n} \mathbb{1}_{\|\mathbf{x}_i-\mathbf{x}_j\| \le s}$$

n is the number of samples, composed of vectors $X = \{\mathbf{x}_1, \dots, \mathbf{x}_n\}$, s is the set dimension and $\mathbb{1}_Z = 1$ when $Z = true$ and 0 otherwise.

An algorithm for obtaining asymptotically consistent estimates of the intrinsic dimension and the Rényi α-entropy [42] using a geodesic-minimal-spanning-tree (GMST) is given by [12]. In this approach a sequence of minimal spanning trees is constructed, from which (geodesic) distances along the edges of the graphs as well as their overall lengths are used in order to simultaneously derive the dimension and entropy estimates. The procedure involves the construction of a complete graph between all pairs of data vectors (as in the Isomap procedure, Sect. 4.2), which is converted into a minimal spanning graph (the GMST) by a series of edge deletions so that the geodesic length is minimized, while still keeping all points connected. In a minimal spanning graph the overall length is given by

$$L_\gamma^{\mathbb{R}^d}(\mathcal{X}_n) = \min_{T \in \mathcal{T}} \sum_{e \in T} e^\gamma$$

where $\mathcal{X}_n = \{\mathbf{x}_1, \dots, \mathbf{x}_n\}$ is the set of objects in $\mathbb{R}^d$ (the original space with dimension d), $\mathcal{T}$ the set of spanning trees over $\mathcal{X}_n$, e an edge of T connecting two distinct objects $\mathbf{x}_i, \mathbf{x}_j$, $i \ne j$ and $\gamma \in (0, d)$ the so called edge exponent (power-weighting constant). When the Isomap algorithm is applied, an estimation of a manifold $\mathcal{M}$ embedded in $\mathbb{R}^d$ is obtained. If e_{ij} is an edge joining points i, j on $\mathcal{M}$ and $\hat{d}(e_{ij})$ the estimated length of the edge, the GMST is the minimal graph whose length is $\hat{L}_\gamma^{\mathcal{M}}(\mathcal{X}_n) = \min_{T \in \mathcal{T}} \sum_{e \in T} \hat{d}^\gamma(e)$ In [12] it is proven that the following approximation holds

$$\log(\hat{L}_\gamma^{\mathcal{M}}(\mathcal{X}_n)) = a \log(n) + b + \epsilon_n \qquad (3)$$

where $a = \frac{(m-\gamma)}{m}$, m is the estimated dimension of the manifold, b is a function related to m and the Rényi α-entropy and ϵ_n is a residual error that goes to 0 as $n \to \infty$. In practice, a collection of M bootstrap data sets are sampled from the data and for of each of them an estimated dimension $\hat{m}$ is obtained by solving Eq. 3 via least squares. The overall estimator of the intrinsic dimension $v = \sum_{i=1}^{M} \hat{m}_i/M$ is obtained by averaging over the boostraped samples.

A U-statistic approach to estimate v has been proposed in [24]. The basic idea is to use a modified version of the correlation integral where $\mathbb{1}$ is replaced by a general kernel function and the estimation is based on the convergence rate of the modified correlation integral. The U-statistic used is defined as

$$U_{n,h}(\mathcal{K}) = \frac{2}{n(n-1)} \sum_{1 \leq i < j \leq n}^{n} \mathcal{K}_h(\|\mathbf{x}_i - \mathbf{x}_j\|^2) \tag{4}$$

where $\mathcal{K}_h(\|\mathbf{x}_i - \mathbf{x}_j\|^2) = \frac{1}{h^m}\mathcal{K}(\|\mathbf{x}_i - \mathbf{x}_j\|^2/h^2)$, $\mathcal{K}$ is a measurable, non-negative, bounded kernel ($\mathcal{K} : \mathbb{R}_+ \to \mathbb{R}$), h is a kernel parameter, m is the dimension of the submanifold contained in the high dimensional space and n the number of vectors in the sample. In particular a simple kernel $\mathcal{K}(x) = (1-x)_+$ is used and five samples of sizes $\{N/5, N/4, N/3, N/2, N\}$ are considered. For them empirical estimates of the U-statistic are produced for a collection of tentative dimensions $l \in [1, l_{max}]$. Individual estimates of intrinsic dimension result from applying weighted least squares to linear fits for the obtained U-values and the tentative dimensions. The slope with the smallest absolute value is considered to represent v (see [24] for details).

Another approach from a dynamic systems perspective is the Takens estimator [48], given by

$$v = -\frac{1}{\mathcal{M}_{h_{Takens}}(\log(\|\mathbf{x}_i - \mathbf{x}_j\|/h_{Takens}))} \tag{5}$$

where $\|$ is the Euclidean norm, $h_{Takens} = \bar{d} + \sigma$, with $\bar{d}$ and σ being the mean and the standard deviation of the nearest neighbor distances. $\mathcal{M}_{h_{Takens}}$ is the mean over all distances smaller than h_{Takens}, which is a kind of maximal scale when consider neighbourhoods.

4 Dimensionality Reduction and Visualization

4.1 Sammon Mapping

The representation of distance matrices in low dimensional spaces, particularly for visualization purposes, has been the objective of multidimensional scaling methods (MDS) [10, 29, 30]. It is based on the idea of distance preservation between the original and the target spaces, which has a strong intuitive appeal. In particular, objects which are close/far in the original data space should be represented close/far from each other in the low-dimensional space and more generally, dissimilarities in the original space are considered versus distances in the target space. This is captured by minimizing objective functions like

$$\sum_{1 \leq i,j \leq N} w_{ij} \left(\mathcal{F}(\delta_{ij}^p) - d_{ij}^p \right)^2$$

where N is the number of objects, w_{ij} is a weight associated to every pair of objects i,j in either space, $\mathcal{F}$ is a monotonically increasing function, δ_{ij} is a dissimilarity measure between objects i,j in the original data space, d_{ij} their distance in the target space and p an exponent.

Different methods are derived from this general formulation, among them Sammon's nonlinear mapping [45]. It considers the transformation of vectors of two spaces of different dimension ($D > m$) by means of a transformation like $\varphi : \mathbb{R}^D \to \mathbb{R}^m$ which maps vectors $\boldsymbol{x} \in \mathbb{R}^D$ to vectors $\boldsymbol{y} \in \mathbb{R}^m$, $\boldsymbol{y} = \varphi(\boldsymbol{x})$.

$$\text{Sammon error} = \frac{1}{\sum_{i<j} \delta_{ij}} \sum_{i<j} \frac{(\delta_{ij} - d(\boldsymbol{y}_i, \boldsymbol{y}_j))^2}{\delta_{ij}}, \tag{6}$$

where typically d is an Euclidean distance in $\mathbb{R}^m$. The weight term δ_{ij}^{-1} gives more importance to the preservation of smaller distances rather than larger ones and is determined by the dissimilarity distribution in the data space. Moreover, they are fixed, which is referred to as lack of plasticity.

4.2 Isomap

Classical methods like principal components or MDS have difficulties when facing data with highly nonlinear structures. Isomap [7, 47, 49] is a flexible approach to learn a broad class of nonlinear manifolds. A key idea is to distinguish distances as measured by classical Euclidean distances (like in MDS, Sammon and related techniques), from those measured along the underlying manifold (geodesic distances). For points that are far from each other on the manifold, it may happen that they are seen as close when Euclidean distances are considered, which implies a failure from the point of view of recognizing intrinsic dimensionality. In such cases, the geodesic distances are the ones that reflect the true possibly low-dimensional geometry of the manifold. Isomap builds on MDS but aims at preserving the intrinsic geometry of the data.

The procedure consists of three steps: *(i)* Construction of the so called neighborhood graph G connecting all points i,j according to their pairwise distances $d_X(i,j)$ (usually in Euclidean metric). Only those points closer than a given distance threshold ϵ are connected with an edge having a length given by $d_X(i,j)$. This approach is called $\epsilon - Isomap$. Alternatively, a $K - Isomap$ approach can be considered, based on the number of neighbours associated to each point. In this case a number of neighbours K is given and two points i,j are connected with an edge in G if i is one of the K-neighbours of j. *(ii)* Estimation of the geodesic distances $d_M(i,j)$ between all pairs of points by computing their shortest path distances $d_G(i,j)$ in G. At the beginning $d_G(i,j)$ is $d_X(i,j)$ if there is an edge between i and j and set as ∞ otherwise. With N being the number of data points, for each $k \in [1, N]$ replace each $d_G(i,j)$ by $min\{d_G(i,j), d_G(i,k) + d_G(k,j)\}$ to obtain the shortest path distances. *(iii)* Construction of the d-dimensional embedding: This is obtained by applying classical multidimensional scaling to the D_G geodesic distance matrix, targeting a Euclidean distance matrix D_Y with d-dimensional vectors in the new space.

4.3 Locally Linear Embedding

Locally Linear Embedding (LLE) performs non linear dimensionality reduction. The mapping of high dimensional data is not approached as in MDS, nor uses shortest paths as in Isomap, but rather rely on neighbours or local points to maintain a global structure. The assumption is that data points and their neighbours are located close to a locally linear path of the manifold. This is based on a linear coefficient that is associated to neighboring points used to reconstruct the low dimensional embeddings [43]. The algorithm tries to minimize the least square error of the cost function

$\epsilon(W) = \sum_i \left| X_i - \sum_j W_{ij} X_j \right|^2$

where W_{ij} represent the contribution of data point $i_i h$ to reconstruction of $j_i h$. Assignment of W_{ij} is restricted by two constraints. First, only neighbour datapoints can be considered, otherwise W_{ij} becomes 0. Second, the sum of the contributions in a single row equals 1, specifically $\sum_j W_{ij} = 1$. This constraint, lead to weights that are steady to rotation, rescaling, and translation in reference to neighboring data points. The weight is a reflection of intrinsic geometry features of transforming high dimensional data points capturing non-linear manifold in the lower dimension space by linear mapping. The final step tries to minimize an embedding cost function in the lower dimension space while using fixed weights W_{ij} and trying to reach an optimal set of data points X_i that minimizes $\phi(W) = \sum_i \left| Y_i - \sum_j W_{ij} Y_j \right|^2$ LLE has the number of neighbors as the only algorithm parameter. An improved approach (MLLE) is introduced by considering a multiple weight vectors for each local point neighbours [56].

4.4 Spectral Embedding

Spectral embedding or Laplacian Eigenmaps [6, 38, 46, 55] is a non-linear dimensionality reduction technique that uses the laplacian notion to preserve intrinsic geometrical features between a high dimension space and a lower dimensional target space. The algorithm computes the eigenvectors of the graph Laplacian and uses Laplace Beltrami operator to construct an embedding of the manifold. The algorithms searches in the higher dimensions space for n nearest neighbors and build a heat kernel $t \in R$ which is approximated to be the Guassian to assign weights to edges $W_{ij} = e^{-\frac{\|x_i - x_j\|^2}{t}}$ where x_i, and x_j are two points of the high dimension space. Afterward, the algorithm finds the eigenvalues and eigenvectors for $Lf = \lambda Df$ where D is the constructed diagonal weight matrix that sums the weights of higher dimensional space and $L = D - W$. The algorithms finds k solutions of eigenvectors ($f(i)$ denotes the eigenvector associated with point i) and the eigenvalues λ_i that are used for dimensionality reduction, such that $x_i \rightarrow (f_1(i), \ldots, f_m(i))$. The Laplacian eigenmaps preserve local features in the lower dimensional space in which alleviate the effects of noise in the data [6].

4.5 t-Distributed Stochastic Neighbour Embedding

t-Distributed Stochastic Neighbour Embedding (t-SNE) is a non-linear dimensionality reduction technique that is an improvement to SNE [25]. SNE consider the Euclidean distance of datapoints of higher dimensional space and convert it to conditional probabilities, in which it represents similarities between datapoints. A conditional probability $p_j|i$ is the probability of datapoint x_i to choose x_j as a neighbor based on Gaussian distribution where

$$p_j|i = \frac{exp(-\|x_i - x_j\|^2 2\sigma_i^2)}{\sum_{k \neq i} exp(-\|x_i - x_k\|^2 2\sigma_i^2)}$$

where $sigma_i$ represent the Gaussian variance of datapoint x_i and k is perplexity or selected local neighbors. Similarly SNE builds conditional probabilities $q_j|i$ of datapoints x_i based on Gaussian distribution for the low dimensional space. The target is to match the probability distribution of lower dimensional datapoints to its higher counterpart. To perform this process, a cost function that minimizes the sum of Kullback-Leibler divergences is as follow:

$$C = \sum KL(P_i\|Q_i)_i = \sum_i \sum_j p_{j|i} log \frac{p_{j|i}}{q_{j|i}}$$

One drawback of SNE, is the low cost when representing widely separated points by two closely mapping points. A gradient descent method is utilized to minimize the cost function. t-SNE uses a symmetric SNE cost function by considering $p_ij = p_ji$ and $q_ij = q_ji$ and matching a joint probability distribution P of higher dimension and Q a joint probability distribution of low dimension space as follows

$$C = \sum KL(P\|Q) = \sum_i \sum_j p_{ij} log \frac{p_{ij}}{q_{ij}}$$

A simpler gradient is produced which reduce the computational overhead. To alleviate the crowding problem of SNE, the area utilized to represent map points of lower dimensional space cannot represent moderate distance points properly in comparison to close-distant datapoints. Therefore, t–SNE represent the high dimensional space with joint probabilities using Gaussian distribution while lower dimensional space is represented by joint probabilities using Student t-distribution which has a heavier tail [52].

4.6   Generative Topographic Mapping

In the Generative Topographic Mapping (GTM) method, latent variable non-linear models in a low dimensional space are used to represent the probability density $p(\mathbf{t})$ of the data in the original high dimensional space $\mathbf{t}$ of dimension D [9]. If L is the number of latent variables $x = (x_1, \ldots, x_L)$ (whose cardinality is the dimension of the latent space), a mapping from the latent space into the data space is given by the function $\mathbf{y}(\mathbf{x}; \mathbf{W})$. It maps points $\mathbf{x}$ in the latent space into corresponding points $\mathbf{y}(\mathbf{x}; \mathbf{W})$ in the data space. The mapping is controlled by a set of parameters $\mathbf{W}$. A probability distribution $p(\mathbf{x})$ on the latent-variable space, induces a distribution $p(\mathbf{y}|\mathbf{W})$ in the data space, represented as Gaussian functions with variance β^{-1} around each $\mathbf{y}(\mathbf{x}; \mathbf{W})$ given by

$$p(\mathbf{t}|\mathbf{x},\mathbf{W},\beta) = \left(\frac{\beta}{2\pi} \right)^{D/2} \exp \left\{ -\frac{\beta}{2}||\mathbf{y}(\mathbf{x}; \mathbf{W}) - \mathbf{t}||^2 \right\} \tag{7}$$

If a collection of K points in the latent space are considered, $\mathbf{x}_i, i = \{1, \ldots, K\}$, the distribution in the data space will be given by $p(\mathbf{t}|\mathbf{W},\beta) = \frac{1}{K} \sum_{i=1}^{K} p(\mathbf{t}|\mathbf{x}_i, \mathbf{W}, \beta)$ which corresponds to a Gaussian mixture model. For a dataset $\mathcal{D} = \{\mathbf{t}_1, \ldots, \mathbf{t}_N\}$ of N vectors, the log likelihood function is expressed as

$$\mathcal{L}(\mathbf{W},\beta) = \sum_{n=1}^{N} \ln \left\{ \frac{1}{K} \sum_{i=1}^{K} p(\mathbf{t}_n|\mathbf{x}_i, \mathbf{W}, \beta) \right\} \tag{8}$$

In GTM the mapping from the latent to the data space is typically chosen as a generalized linear regression model $\mathbf{y}(\mathbf{x};\mathbf{W}) = \mathbf{W}\phi(x)$, where $\phi(x)$ is a collection of basis functions (Gaussians) and $\mathbf{W}$ is a matrix of weights. The EM algorithm is used for finding $\mathbf{W}$ and β using (8) to asses convergence. Once the model parameters are determined, the latent space posterior distribution $p(\mathbf{y}|x, \mathbf{W}, \beta)$ can be obtained using Bayes theorem and from them the posterior means, which are used for the visualization. One interesting feature of GTM is its ability to estimate the so-called magnification factor [37], which measures the change between volumes in the latent and the data spaces. By looking at the distribution of this factor it is possible to get an idea about the stretching of the manifold due to the nonlinear mapping of the latent to the data space. Different variants and extensions of GTM have been proposed, notably those described in [13, 19, 20, 39, 57].

4.7 Neuroscale

The Neuroscale procedure [35, 36] is a dimension-reducing transformation for the purposes of visualization and analysis. What is sought is that the geometric structure of the data can be optimally preserved by the transformation, keeping the inter-point distances in the feature space corresponding as closely as possible the distances in the data space. In this sense, the goals are closely related to metric MDS. However, several important elements are introduced: *(i)* In contradistinction with MDS where the solution is found iteratively via an implicit mapping, Neuroscale uses a RBF neural network that is trained using the available data by optimising the network parameters in order to minimize a suitable error measure. The goal is to produce a representation of the mapping function $\varphi : \mathbb{R}^D \to \mathbb{R}^m$ from the D dimensional data space to the target space ($m < D$) in explicit form. *(ii)* It uses a flexible formulation of the error measure to optimize (stress) in terms of a convex combination of two types of objective and subjective dissimilarities in the data space, compared with the chosen metric in the target space (usually Euclidean).

If d_{ij}^* are distances between objects i, j in the original data space of dimension p and s_{ij} are (user defined) subjective dissimilarities, a general dissimilarity can be defined as $\delta_{ij} = (1 - \alpha)d_{ij}^* + \alpha s_{ij}$, where the parameter α ($\alpha \in [0, 1]$) controls the proportion of subjective information incorporated. The role of the RBF network is to provide an explicit nonlinear mapping between vectors $\mathbf{x_i}$ in the data space and vectors $\mathbf{y_i} = \mathbf{f}(\mathbf{x_i}; \mathbf{W})$ in the target space of dimension q ($q < p$), where $\mathbf{f}$ is the nonlinear transformation represented by the RBF with weights $\mathbf{W}$. In the target (feature) space the distances are given by $d_{ij} = ||\mathbf{f}(\mathbf{y_i}) - \mathbf{f}(\mathbf{y_j})||$. If $\phi_k()$ are the basis functions of the RBF (with μ_k as their centres), and w_{lk} the network weights, that distance can be expressed as

$$d_{ij}^2 = \sum_{i=1}^{q} \left(\sum_k w_{lk}[\phi_k(||x_i - \mu_k||) - \phi_k(||x_j - \mu_k||)] \right)^2$$

The RBF network is trained to minimize the error (stress) term given by

$$E = \sum_{i<j}^{N} \left(\delta_{ij} - d_{ij} \right)^2$$

4.8   Genetic Programming

Genetic Programming (GP) is an evolutionary computation technique introduced in [27, 28]. It combines the expressive high level symbolic representations of computer programs with the near-optimal search efficiency of the genetic algorithm. For a given problem, this process often results in a computer program which solves it either exactly or with acceptable approximation. Those programs which represent functions are of particular interest and can be modeled as $y = F(x_1, \ldots, x_n)$, where $\{x_1, \ldots, x_n\}$ is the set of predictor variables, and y the dependent variable, so that $\{x_1, \ldots, x_n\}, y \in \mathbb{R}$, where $\mathbb{R}$ are the reals. The function F is built by assembling functional subtrees using a set of predefined primitive functions (the function set), defined beforehand. The model describing the program is given by $y = F(\bar{x})$, where $y \in \mathbb{R}$ and $\bar{x} \in \mathbb{R}^n$. Most implementations of genetic programming for modeling fall within this paradigm.

One of the variants of GP is Gene Expression Programming (GEP) [15] which uses a simple string representation for the expression tree. In GEP the chromosomes are encoded as strings of fixed length with a head and a tail. Each chromosome can be composed of one or more genes which represent mathematical subexpressions that are linked together to form a larger one. This technique uses a variety of genetic operators such as inversion, mutation, one point recombination, two point recombination, gene recombination, root insertion sequence transposition, insertion sequence transposition, gene transposition, and random numerical constants (RNC) mutation. Advantages of GEP are the simplicity of its representation, as well as the property of its genetic operators of always producing valid expression trees.

The GEP approach was extended in [50, 51] to evolve programs that represent vector functions. $\bar{y} = F(\bar{x})$, ($y \in \mathbb{R}^m, \bar{x} \in \mathbb{R}^n$) with objective function depending of all vector components of $\bar{y}$. In this case the chromosomes are independent but evolve together (linked by a single objective function), as a population of forests such as the one needed for learning vector functions. The extension also allows the study of unsupervised problems (e.g. using a fitness function based on Sammon mapping). With the mapping function(s) in explicit form, new data can be easily transformed without having to re-generate the low dimensional space with the enlarged dataset, as is required in the vast majority of dimensionality reduction methods in which the mappings found are implicit.

A genetic programming approach to the dimensionality reduction problem has several advantages: *(i)* it produces an explicit mapping, *(ii)* it is given as a white box model in the form of closed algebraic expressions in which the relative importance of the data attributes can be easily inspected, *(iii)* the evolutionary process performs an implicit feature selection/generation as part of the search in the space of algebraic transformations. Natural disadvantages are the usually huge size of the search spaces, as well as the large number of parameters controlling the evolutionary process, which impacts the computational effort required.

4.9 Quality Measures for Embeddings

It is natural to look for ways of assessing the quality of a mapping using numeric measures. Importantly, such measures allow more objective comparisons between the many existing methods, as well as a deeper understanding of the nature of the transformation linking the high and low dimensional spaces. Several indices have been proposed and the most favored ones are those based on the analysis of neigh-bourhoods of varying sizes between the high and low dimensional spaces [3, 11, 16, 31, 32, 53, 54]. If $\mathbf{x}_i$ and $\mathbf{y}_i$ are vectors in the high and low dimensional spaces respectively (where $\mathbf{y}_i$ is the image of $\mathbf{x}_i$ in the low dimensional space), and if $\eta(\mathbf{x}_i, K)$ and $\hat{\eta}(\mathbf{y}_i, K)$ are their corresponding K-neighbourhoods, the so-called average agreement rate is defined as $Q_{NX}(K) = \frac{1}{KN} \sum_{i=1}^{N} |\eta(\mathbf{x}_i, K) \cap \hat{\eta}(\mathbf{y}_i, K)|$, where N is the number of vectors and $K \in [1, N-2]$ the neighbourhood size. A derived index based on $Q_{NX}(K)$ which measures the improvement over a random mapping is given by

$$R_{NX}(K) = \frac{(N-1)Q_{NX}(K) - K}{N - 1 - K} \tag{9}$$

which has 1 as maximum, indicating that all neighbourhoods of size K have the same composition in both spaces. This index will be the one used for comparing the haptic data mappings in Sect. 5.

5 Results

5.1 Intrinsic Dimensionality

The set of Intrinsic Dimensionality estimations for the haptic data obtained with the techniques described in Sect. 3 are shown in Table 1. Interestingly, the estima-tions fall into two well defined groups. While the values obtained with the Correla-tion Dimension, the Takens and the U-statistic methods are tightly packed around 4, those produced by the MLE, GMST and the Eigenvalues are larger, in the [7, 11.4] range. A closer look at this sort of dual behaviour of the estimated values may reflect the difference between methods that rely more on the analysis of properties within local neighbourhoods like MLE and GMST and techniques based on the study of the attractor, like Takens, the Correlation Dimension and the U-statistic (which is related technique). However, despite of these differences, when the estimated intrin-sic dimensions are compared with the dimensionality of the original dataset (9601), the differences become negligible, cleary indicating that there are indeed low dimen-sional manifolds containing most of the structure of the data.

From a PCA perspective, the distribution of the cumulative variance for differ-ent number of components is shown in Fig. 2 (left). An elbow occurs at around 50 components and while 71 of them are required to explain 90 % of the total variance, 99.5 % is explained by 461.

Table 1 Intrinsic dimensionality estimations

Method	Intrinsic dimension estimation
Eigenvalues	7.000000
Maximum likelihood estimator (MLE)	9.227160
Correlation dimension	3.987503
Takens	3.6673
Geodesic minimal spanning tree (GMST)	11.441282
U-statistic	4

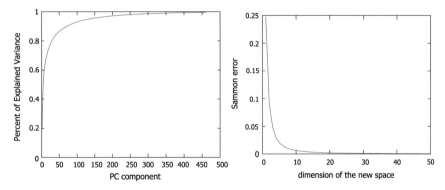

Fig. 2 *Left* Distribution of the cumulative variance in Principal Components Analysis with different number of components retained. *Right* Distribution of the nonlinear Sammon mapping error with the dimension of the target space (up to the first 50 nonlinear dimensions)

The PCA results provide an idea about the variance distribution, but not about distance preservation. Moreover, PCA embedding has the disadvantage of arbitrarily distorting pairwise distances between data objects [1]. Sometimes PCA may map two different objects to a single point in the target low-dimensional space, making them indistinguishable. Other linear embedding techniques exhibit this pairwise distances distortion behaviour. However, the problem can be circumvented by using random projections [5, 8, 14, 22, 23]. The Johnson-Lindenstrauss Lemma [26], states that a finite set of N objects in $\mathbb{R}^D$ can be linearly mapped to a subspace of dimension $M = \mathcal{O}(\log N)$ with very small pairwise distance distortion and such linear mapping can be constructed using a random matrix whose elements are chosen from a particular distribution [1, 2, 26].

The application of the Johnson-Lindenstrauss lemma to the haptic data provides different estimations of the minimum number of dimensions required for distance preservation for different values of the maximum distance distortion rate (ϵ) (Table 2). Considering a distortion rate of $\epsilon = 0.1$ the required number of dimensions (5920) represents 62 % of the attributes, which is an important reduction. Accord-

Table 2 Minimum dimension estimates for an ϵ-distance preservation mapping based on the Johnson-Lindenstrauss lemma

ϵ	0.99	0.5	0.4	0.3	0.2	0.1	0.09	0.08	0.075	0.07
Minimum nbr. of dimensions	165	331	470	767	1594	5920	7257	9121	10341	11830

ingly, a random projection operator Φ linearly mapping the data to such space on the one hand would preserve the distance structure and on the other would speed up considerably the performance of machine learning procedures applied to the data.

The nonlinear character of the haptic data subspace can be recognized in Fig. 2 (right) which shows the distribution of the Sammon error (Eq. 6) with the dimension of the target space. An elbow occurs at around 5 nonlinear new dimensions and after dimension 11 the mapping error falls to very low levels, which is in agreement with the intrinsic dimensionality estimation results.

5.2 Dimensionality Reduction Methods

From the point of visualizing the structure of the data the representation has to be made in spaces of a dimension compatible with the human perception capabilities ($\{2, 3, 4\}$), which does not necessarily coincide with the intrinsic dimensionality of the information as such. The results of the previous section indicate that the required dimension for the target spaces would exceeds what is possible, moreover using hard media. Accepting the limitations imposed by these factors, two dimensional spaces will be used for representing the results of the dimensionality reduction methods considered. In a few cases 3D snapshots are presented as well to improve the understanding of the data distributions. Section 5.3 presents a comparison of the different methods from a numeric quality assessment perspective.

The dataset contains signature contributions from 20 subjects 50 haptic signatures/subject and a user identification scenario would be a classification problem with 20 classes. Here the focus is on understanding data structure as determined by the descriptor variables and therefore, the nature of the problem is unsupervised. Even though the class distribution is known a-priori, it was not used when computing the visual spaces presented. However, for clarity purposes and for the benefit of the discussion, all objects of the same class (signatures from the same subject) are represented with the same color.

5.2.1 Sammon Mapping

Nonlinear mappings of the haptic data targeting 2 and 3 dimensional spaces were obtained using Euclidean distances in both the original and target spaces and solving Eq. 6 with the Fletcher-Reeves method [40]. In order to alleviate the problem of local minima entrapment, 10 solutions were found using different machine-generated

Fig. 3 Sammon nonlinear
mapping to a 2D target
space. Each object represents
the image of a signature and
those produced by the same
subject have the same color

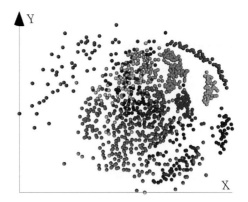

random seeds. The best solution with Sammon error = 0.0906 was used for visualization and is shown in Fig. 3.

It can be observed that in general, signatures from the same subject are close to
each other and tend to occupy definite areas of the space. Some classes are compact
while others are more spread, particularly in the periphery of the point distribution.
An important difficulty with the Sammon mapping, as well as with other methods
considered here like Isomap, LLE, Spectral Embedding and t_SNE is that they are
implicit. That is, they do not provide a mapping function by means of which new
objects can be mapped into the target space. Approaches oriented to circumvent this
limitation are explored in Sects. 5.2.7 and 5.2.8.

5.2.2 Isomap

Isomap spaces of dimension 2 and 3 were obtained using $k = \{10, 20, 100, 999\}$
neighbours and Fig. 4 shows the one corresponding to $k = 20$. An immediately distinguishing feature is that the space is much more expanded than the one produced
by the Sammon mapping. Regions with higher densities are associated with classes,
and they exhibit different degrees of homogeneity. Some classes are well differentiated regardless of their dispersion and the central area contains a large subset of
intertwined classes, similarly to Sammon mapping. However, relatively large, less
dense extruding areas are occupied by a rather well differentiated classes.

Solving identification/verification problems in those nonlinear spaces should be
beneficial. However, a difficulty with Isomap is that it is implicit. In the case of
Isomap an additional drawback is that it is not always possible to map all of the
original data space objects, as the method works only with those objects needed for
the construction of the neighborhood graph. That is, sometimes objects from the
original set are not mapped at all.

Fig. 4 Isomap 2D
representation. Each object
represents the image of a
signature and those produced
by the same subject have the
same color

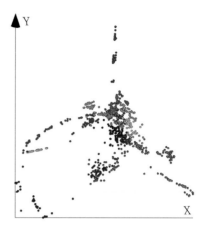

5.2.3 Locally Linear Embedding (LLE)

LLE and MLLE spaces of dimension 2 and 3 were obtained using $k = \{10, 20\}$ neighbours and the 2D one corresponding to LLE with $k = 20$ is shown in Fig. 5. The objects appear distributed along well defined directions in the space, also showing areas with high concentration (at low values of the X coordinate) and low density regions (at medium-high X values). While some classes appear well clustered and compact, others are very spread. However, they occupy low density regions and do not intersect with other classes. As opposed to the Sammon and the Isomap spaces, it is more difficult to distinguish the individual classes and there are many outlying objects. These elements suggest that the LLE's assumption that each data point and its neighbors lie on or close to a locally linear patch of the manifold does not hold for the haptic data analyzed and that even at the local level, nonlinearity persists.

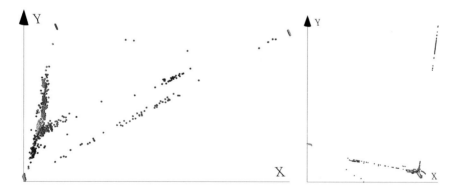

Fig. 5 *Left* Locally Linear Embedding, *Right* Modified Locally Linear Embedding (MLLE), 2D representation ($k = 20$ neighbours). Each object represents the image of a signature and those produced by the same subject have the same color

Fig. 6 Spectral Embedding
2D representation ($k = 20$
neighbours). Each object
represents the image of a
signature and those produced
by the same subject have the
same color

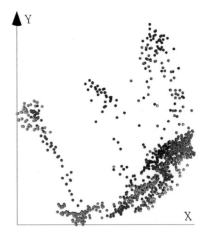

The 2D space corresponding to MLLE (the modified version of LLE) is shown in Fig. 5, where the common pattern observed with previous methods is found. Namely, a central dense region with compact, more homogeneous classes and extruding low density areas where classes are elongated along clearly preferred linear directions of the space. The latter seems suspicious, but similar results were obtained with different neighbourhood sizes. Although difficult to appreciate in Fig. 5 (Right), the classes appearing at the central region and those located at the peripheral branches coincide with those found with other methods.

5.2.4 Spectral Embedding

Spectral Embedding spaces of dimension 2 were obtained using $k = \{10, 20, 100, 999\}$ neighbours and the one for $k = 20$ is shown in Fig. 6. As with other methods, a high density area is found where classes concentrate while keeping small neighborhoods being composed mostly of elements of the same class. Also, most of the space is of lower density with less homogenous classes, oriented along preferred directions. However, class separability is low and the occurrence of outlying elements is very high, as with LLE and MLLE. From the point of view of the relative distribution of the classes, the space exhibits a similar organization as the one obtained with previous methods.

5.2.5 t-Distributed Stochastic Neighbour Embedding

t-Distributed Stochastic Neighbour Embedding spaces were computed using several perplexity parameters $\{10, 20, 30, 50\}$ and number of iterations $\{1000, 10000\}$. No substantial differences were found among them and the one corresponding to perplexity 50 is shown in Fig. 7. As with previous methods, the space follows the gen-

Fig. 7 t-Distributed
Stochastic Neighbour
Embedding 2D space. Each
object represents the image
of a signature and those
produced by the same subject
(numbered) have the same
color. A light background
covers the populated areas

eral pattern of having a central high density region where most classes concentrate
and a larger, lower density periphery occupied by fewer classes. Also, the classes
located at the central core and the periphery are the same as those found by the
previous techniques. On Fig. 7 numbers identifying the individual classes/subjects
have been added in order to clarify the association between clusters and classes. A
light background highlights populated areas and serves as an approximate pattern
for the overall distribution and can be used as reference when comparing with other
methods.

A very remarkable feature of the t-SNE result is the well defined structure of the
center of the distribution as a collection of well separated, compact clusters and their
correspondence with the known classes. Moreover, the clusters are densely packed
and are mostly unimodal (17 and 19 are clearly bivariate). There are outlying elements, but they are much fewer than those observed in the LLE, MLLE and Spectral Embedding spaces. What is distinctive for the t-SNE outliers is that very often
they appear rather far from the clusters corresponding to their core classes, distorting neighbourhood structures for some classes. In the overall, separability between
clusters/classes is remarkable.

Some peculiarities emerge from the analysis of the t-SNE as well as the spaces
produced by previous methods: *(i)* there is a sort of "common" way in which humans
tend to produce their signature (from the point of view of the physical variables associated to the use of a tactile device like the haptic pen), suggested by the existence of
a large central region where most classes appear. However, signatures from the same
subject tend to appear clustered even within that region. *(ii)* there are subjects with
clearly distinct haptic signatures. *(iii)* The bimodal nature of some individual distributions suggests that some subjects exhibit what could be considered as a kind of dual
style when producing the signature. Such styles appear somewhat consistent, considering that in all cases when such behavior is observed ({17, 19}), the two modes
are well populated. Since signing with a haptic device captures the subject's tendency to press and orient the pen in specific ways in addition to the signature itself,
the fact that the same individual has more than one way of doing it is interesting.
Whether this is something related to more permanent physio-neurological factors or
to psychological states affecting the subjects during the signature acquisition exper-

iments would be worth investigating. *(iv)* The individual class distributions exhibit great variability. For some subjects their signature instances define homogeneous, compact clusters, while for others the clusters appear spread and elongated in the nonlinear spaces. This suggest that the way in which the haptic variables interrelate for different individuals also exhibits large variation and such peculiarities could be used when designing verification/identification procedures.

5.2.6   GTM

The very high dimension of the haptic data space ($D = 9601$) poses a problem for this technique, given the much lower number of objects available (1000). Since in GTM the mapping goes from the low dimensional latent space (2 in this case) to the original, the transformation is $\psi : \mathbb{R}^2 \rightarrow \mathbb{R}^D$. Accordingly, the solution of Eq. 8 would require the estimation of a very large number of parameters, many times exceeding the number of objects.

In order to produce a reasonable approximation while keeping the problem tractable, instead of building a mapping from a latent space to the data space, the mapping targets an approximate representation of it. A good approximation is given by the first 71 principal components, which accounts for 90 % of the variance (Sect. 5.1). For the GTM a collection of points must be placed in the latent space and a Gaussian mixture model is built (Eqs. 8 and 7) by means of a Radial Basis Functions (RBF) neural network, targeting a data space of $D = 71$ which contains 90 % of the variance of the original $D = 9601$ space. A total of 225 points in the 2D latent space were distributed in a regular grid of 15×15. The number of components of the Gaussian mixture model was set to 4 in order to keep the number of parameters in the **W** matrix a few times smaller than the number of data objects. The maximum number of iterations for the EM algorithm was set to 30, but convergence was found with fewer. The latent space is shown in Fig. 8.

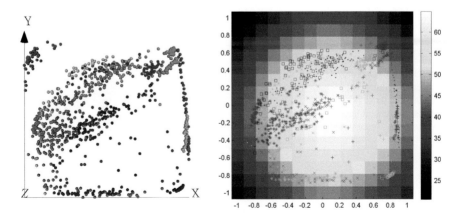

Fig. 8   Generative Topographic Mapping. Latent 2D space

The left hand side shows the posterior means for each object. The pattern defined by that distribution is characterized by a high density central region (at and above a diagonal), surrounded by lower density peripheral areas. This main feature of the GTM latent space coincides with the one defining the general structure of the spaces constructed by the previous methods. This is interesting, considering that *i)* these methods were mapping between the data space and the 2D visualization space, *ii)* GTM is using the subspace defined by the first 71 principal components instead of the whole data space and *iii)* the mapping performed by GTM goes in an opposite direction (approximation of a given data space from one consisting on latent variables). The right hand side of Fig. 8 shows the distribution of the magnification factor (as a heat map). Space stretching increases with brightness indicating that classes along the bottom and right side of the latent space are in the 71-D PC space farther from the center, like the extruding lower density structures identified by other methods (Isomap, LLE, MLLE, Spectral Embedding and t-SNE).

5.2.7 Neuroscale

This is the first of the two instances of explicit mapping presented. As was done in the case of GTM, the original data space ($D = 9601$) was replaced by the one defined by the first 71 PCs in order to keep the number of free parameters a few times smaller than the number of objects. Four centers and 100 iterations were used for the RBF network, approximating the mapping $\varphi : \mathbb{R}^{71} \to \mathbb{R}^2$.

The corresponding space is shown in Fig. 9. It has in common the distinctive features identified by previous methods: *i)* a high density region with more class agglomeration and *ii)* two lower density areas at opposite sides of the central region containing fewer classes (some appear multimodal). However, in the Neuroscale case, the more populated lower density extruding region is not as extended as in the spaces found by other techniques (e.g. Figure 7). The relative location of the classes within the space, which is an important structural feature, keeps consistency with what was found by the previous methods. This behavior of the Neuroscale approach

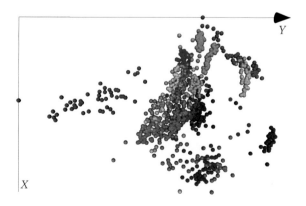

Fig. 9 Neuroscale 2D representation. Each object represents the image of a signature and those produced by the same subject have the same color

is promising, considering that it uses an intermediate PCA space, which is known to distort distance structures, precisely those defining the objective function of the procedure.

The RBF network can be used for mapping new haptic data using a composition $(\psi \cdot \varphi) : \mathbb{R}^D \to \mathbb{R}^2$ where $\psi : \mathbb{R}^D \to \mathbb{R}^{71}$ is a linear mapping using the truncated eigenvector matrix and φ is the nonlinear transformation represented by the RBF. Clearly, the first step implies an initial information loss, as the intermediate space contains only 90 of the variance with respect the original data, on top of which is the loss due to the nonlinear mapping. For practical purposes, an explicit, direct transformation is preferable as it avoids the compositional mapping, but it is not always possible.

5.2.8 Genetic Programming

In order to produce a direct mapping $\varphi : \mathbb{R}^{D=9601} \to \mathbb{R}^{m=2}$ between the data and the target spaces, the Gene Expression Programming vector functions extension was used [50, 51]. With it the m components of the mapping function $\varphi = \{\varphi_1, \varphi_2\}$ can be obtained simultaneously in either supervised or unsupervised mode. In this case the later was used (ignoring the class information associated to the haptic signatures), aiming at minimizing Sammon error (Eq. 6).

The algorithm is controlled by a large number of parameters which are associated to: *i)* the evolutionary process in general (number of generations, population size), *ii)* the structure of the populations (number of chromosomes, number of genes/chromosome), *iii)* chromosome composition (gene head size), *iv)* the type and rates for evolutionary operators (inversion, mutation, is-transposition, ris-transposition, one-point recombination, two-point recombination, rnc-mutation rate, dc-mutation rate), *v)* the use of numeric constants within the evolved equations (number of constants/gene, bounded range for the constants). Some of them are specific to Gene Expression Programming and are described in [15]. In addition, genetic programming parameters related to the algebraic structure of the evolved equations are required (function set, linking function). The function set defines the collection of elementary functions that will be made available to the evolutionary process when building candidate equations, combining data attributes, constants and other functional blocks when forming the expression tree. The linking function is the one chosen as the root of the expression tree. Weights associated to the elements of the function set allows the introduction of bias (expert knowledge) when choosing candidate functions during the evolutionary process. The set of genetic programming parameters used in the experiments is shown in Table 3.

Because of the lack of previous experiences in learning unsupervised mappings for haptic data using genetic programming, a parsimonious approach was taken. The goal was to obtain some initial mappings under relatively simple conditions in order to get some idea about their nature and behaviour. The population size chosen

Table 3 Experimental settings of the GEP algorithm

GEP parameter	Experimental values
No. generations	3000
Population size	200
No. chromosomes/individual	3
No. genes/chromosome	$\{4, 6, 8, 10, 12\}$
Gene head size	$\{4, 6, 8, 10, 12, 14\}$
Linking function	addition
No. constants/gene	2
Bounded range of Constants	$[0,10]$
Inversion rate	0.1
Mutation rate	0.044
is-transposition rate	0.1
ris-transposition rate	0.1
One-point recombination rate	0.3
Two-point recombination rate	0.3
Gene recombination rate	0.1
Gene transposition rate	0.1
rnc-mutation rate	0.01
dc-mutation rate	0.044
dc-inversion rate	0.1
Function set (e.g., $\{function_1(weight))$ $function_2(weight), \dots\})$	$\{+(1), -(1), *(1), x^2(1)\}$
No. random seeds	5

ensured a good initial genetic diversity and a modest amount of search was made (3000 generations), given the huge size of the search space. A small function set was chosen (just four functions with equal weights), at the same time composed of very simple functions (basic arithmetics and the second power as the only unary nonlinearity). The set of parameters controlling the genetic operators were chosen according to default values that proved to be effective in other genetic programming experiments, although for different problems and data. Five different initial populations were generated from machine generated random seeds and evolved with the same set of evolutionary parameters, aiming at minimizing Sammon error (Eq. 6). The best result is shown in Fig. 10.

The Sammon error obtained was 0.1356 which is higher than the one obtained with the implicit deterministic minimization of Eq. 6 in Sect. 5.2.1 (Fig. 3). The GEP space has a similar class distribution with respect to the implicit solution (translations, rotations and symmetries aside due to distance properties), but with less efficient class separability due to the higher mapping error. The overall distribution with a central higher density region flanked by two lower density areas is captured,

Fig. 10 Gene Expression
Programming (GEP) explicit
mapping (2D space). Each
object represents the image
of a signature and those
produced by the same
subject have the same color

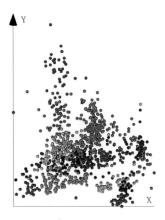

although one of them is not as extruding as with other methods. In comparison with
Fig. 7, the separated, low density area where class 16 is located appears much closer
and interlaced with classes $\{3, 9\}$. Also the lower region corresponding to classes
$\{8, 17\}$ and $\{1, 10, 19\}$ appears interlaced with the core, although with a clearly differentiated lower density. However, the relative location of the classes is mostly preserved.

Considering the inherent difficulties in learning the $\varphi : \mathbb{R}^{D=9601} \to \mathbb{R}^{m=2}$ direct
mapping, the extremely simple function set used and the modest search effort, the
result is promising.

The explicit form of $\varphi = \{\varphi_x, \varphi_y\}$ is given by Eqs. 10 and 11, which shows the
specific variables involved in the space transformation as well as their roles. Interestingly, $i)$ the number of variables involved in the mapping is very small in comparison
with the total (arities for φ_x, φ_y are 36 and 45 respectively). Since they have no intersection, the arity of φ is 81 which represents only a very small fraction of the total
number of variables (0.84 %).

$$
\begin{aligned}
\varphi_x = & (((((((((((((Oy_{9491}^2) * (((Tz_{7689} - Oy_{2747}) + (Px_{9481} + Tz_{5589})) - (Py_{8030} \\
& - Tz_{1593}))) + (Pz_{39} * ((Pz_{7215} - (Py_{4250} + Pz_{3783})) - ((Ty_{4196} + Fy_{7541}) \\
& + Fy_{8501})))) + (Oy_{9299}^2)) + Px_{8341}) + (Px_{8089}^2)) + (Oz_{7968} - ((Fx_{8344}^2)^2))) \\
& + Px_{6277}) + Tz_{573}) + ((((Py_{1886}^2) + Fz_{1998}) + (Ox_{8278} - Py_{902})) + ((k_1^2) \\
& * (Oy_{4979} + Ox_{9286})))) + ((Ty_{4460} + (((Ox_{3922} * Pz_{8859})^2) - (Oz_{8892}^2))) \\
& + Ox_{766})) + Px_{9073}) + ((Fz_{5058}^2) + ((Pz_{5979}^2) - Oy_{1055}))) \qquad (10) \\
k_1 = & \ 7.039750799831218
\end{aligned}
$$

$$\varphi_y = (((((((((((Px_{253} + ((((Px_{4921} + Oz_{2172}) - Py_{2294}) + (Tx^2_{4411})) - (Py_{2690}$$
$$+ (Ox_{3310} + Py_{1082})))) + (Fx_{9268} + (Oy_{779} + Oz_{2412}))) + (((Px_{8437}$$
$$+ (Fz_{2418} - (k_2 + Oy_{731}))) - Pz_{1995})^2)) + Fx_{3100}) + ((Oy^2_{1823})$$
$$* (((Fx^2_{2800}) - (Py_{4982} + Fy_{9221})) - Px_{37}))) + ((Tx^2_{2119})$$
$$+ (((Px_{3469} * Ox_{7606}) - (Fx_{4204} + Py_{8402})) - (Fy_{7577} + Fy_{1181}))))$$
$$+ (Oy_{8711} - Fy_{4817})) + (Oz_{7440} + Ty_{3524})) + (((((Fz_{7110} * Px_{9217})$$
$$- Px_{8221}) + (Oy_{4931} - Px_{241})) + Fz_{606})^2)) + (Oz_{2532} - Fz_{5802}))$$
$$+ ((((Oz_{5064} - Fy_{9293}) - Ox_{418}) * (Oz^2_{7260})) + Ty_{5012})) \quad (11)$$
$$k_2 = 9.767127644377611$$

With Eqs. 10 and 11 the transformation of new haptic data can be obtained. If they can be considered as coming from the same joint distribution as the data from which the mapping equations were derived, their location in the low dimensional space can be expected to share their properties, shown in Fig. 10.

5.3 Quality Assessment

Section 5.2.1 presented the individual results obtained with several mapping approaches and briefly presented their main properties (mainly from a visual point of view). However, further understanding can be obtained by using numerical descriptors, as the one presented in Sect. 4.9 using the $R_{NX}(K)$ function (Eq. 9) which compares neighbourhood differences between the data and target spaces for different neighbourhood sizes.

The higher the value of $R_{NX}(K)$ for a given method and neighbourhood size, the higher the preservation of the original neighbourhoods will be in the target space, with preservation understood as communality of neighbourhood composition. A graphic comparison of the relative qualities of the mapping techniques presented in Sect. 5.2.1 is shown in Fig. 11.

There is a subset of methods composed of t-SNE, Sammon, Isomap and GMT that exhibits better distribution of $R_{NX}(K)$ values. At the lower end are LLE (which presented a rather poor performance), MLLE and Spectral Embedding. Genetic Programming does not perform well at smaller neighbourhoods, but improves as the size increases, approaching GMT. Practically all methods have a relative maximum at low neighbourhood sizes (in the 40–60 range), most noticeably t-SNE, which up to $K = 122$ outperforms all other techniques and falls only to Sammon and Isomap for larger sizes. Those two appear as the ones with better behavior from a broader neighbourhood size perspective, despite of suffering from sensitivity to distance concentration [17] and lack of plasticity [33]. The former refers to the tendency of Euclidean distances in high dimensional spaces to concentrate; thus making all distances between pairs of objects very similar. The later refers to the ability of a method to

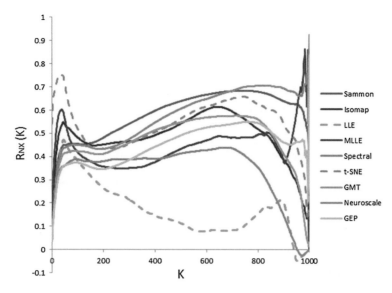

Fig. 11 $R_{NX}(K)$ quality function for different mapping procedures. (LLE): Locally Linear Embedding, (MLLE): Modified LLE, (Spectral): Spectral Embedding, (t-SNE): t-Distributed Stochastic Neighbour Embedding, (GMT): Generative Topographic Mapping, (GEP): Genetic Programming. K is the neighbourhood size

Table 4 Area under the $R_{NX}(K)$ function (in descending order)

Method	Area
Sammon	582.896
Neuroscale	573.126
t-Distributed stochastic neighbour embedding (t-SNE)	546.365
Isomap	484.893
Generative topographic mapping (GTM)	469.679
Genetic programming (GEP)	445.426
Modified locally linear embedding (MLLE)	443.231
Spectral embedding	343.365
Locally linear embedding (LLE)	174.377

break some proximities in order to improve the preservation of others. This picture is completed by looking at the overall behavior for all neighbourhoods, which is typically characterized by the area under the $R_{NX}(K)$ function, shown in Table 4.

From this point of view, Sammon, t-SNE and Isomap appear as the best performance techniques (Neuroscale does not start from the original space, but from the one defined by the subset of eigenvectors with 90 % cumulated variance). The fact that for haptic signature data non-plastic methods affected by norm concentration like Sammon mapping outperforms t-SNE which is a more state of the art technique with respect to those properties deserves further attention.

6 Conclusions

The use of haptic signatures as a biometric procedure presents many opportunities and also challenges. While it brings more information in comparison to traditional image-based signatures, its exploitation for identification/verification purposes requires finding appropriate ways for representing the haptic data and understanding of the structure of the associated information. This work represents a step in that direction, from an unsupervised point of view using class information as a qualitative aid in supporting the discussion and interpretation of the results obtained with a small set of dimensionality reduction and mapping procedures.

Within the context of the experimental data used in this study (20 subjects with 50 instance signatures/subject), it has been found that the high dimensionality of the data seems to mask an underlying embedded nonlinear space of very low intrinsic dimension (in the [4, 11] range). Even though it exceeds the dimension of spaces suitable for visual inspection, dimensionality reduction procedures consistently produce spaces sharing similar properties from the point of view of structure and class distribution. Besides the expected apriori variability due to the human and even cultural nature of a signature, it has been found that there seems to be a sort of "common" way in which a subject tend to produce the signature from the point of view of the physical variables associated to the use of a tactile device (a haptic pen), like force, torque, orientation and position. Subjects were found with highly differentiated haptic signatures.

The individual signature distributions exhibit great variability. For some subjects they appear quite homogeneous, while for others they are spread and elongated in the nonlinear spaces. This suggest that the way in which the haptic variables interrelate for different individuals also exhibits large variation. These peculiarities could be used when designing verification/identification procedures. If a "style" is conceived as a particular way in which force, torque, orientation, etc. are combined in order to produce a haptic signature, subjects were found that exhibited more than one style. It cannot be said at this stage whether this is an intrinsic property of certain individuals or it is something related to the psychological state of the subject when the signature gathering experiments were made. The haptic signature is obtained in a virtual reality space and the subject signs "on the air" and it is understandable that some subjects take more time to adapt to that environment than others.

References

1. Achlioptas, D.: Database-friendly random projections. In: Proceedings of Symposium Principles of Database Systems (PODS), Santa Barbara, CA (2001)
2. Achlioptas, D.: Database-friendly random projections: Johnson-Lindenstrauss with binary coins. J. Comput. Syst. Sci. **66**, 671–687 (2003)
3. Akkucuk, U., Carroll, J.: PARAMAP vs. Isomap: a comparison of two nonlinear mapping algorithms. J. Classif. **23**(2), 221–254 (2006)

4. Alsulaiman, F.A., Sakr, N., Valdes, J.J., El Saddik, A.: Identity verifcation based on hand-written signature with haptic information using genetic programming. ACM Trans. Multimed. Comput. Commun. Appl. (ACM TOMCCAP) **9**(2), May 2013

5. Baraniuk, R., Wakin, M.: Random Projections of Smooth Manifolds Foundations of Computational Mathematics, pp. 941–944 (2006)

6. Belkin, M., Niyogi, P.: Laplacian eigenmaps for dimensionality reduction and data representation. neural Comput. **15**(6), 1373–1396 (2003)

7. Bernstein, M., Silva, V., Langford, J.C., Tenenbaum, J.B.: Graph approximations to geodesics on embedded manifolds. Technical report, Stanford University (2000)

8. Bingham, E., Mannila, H.: Random projection in dimensionality reduction: applications to image and text data. In: Proceedings of the Seventh ACM SIGKDD International Conference on Knowledge discovery and Data Mining (KDD 01), pp. 245–250. ACM, New York (2001)

9. Bishop, C.M., Svensén, M., Williams, K.I.G.T.M.: The generative topographic mapping. Neural Comput. **10**, 215–234 (1998)

10. Borg, I.: Modern Multidimensional Scaling—Theory and Applications. Springer Series in Statistics, New York (1997)

11. Chen, L., Buja, A.: Local multidimensional scaling for nonlinear dimension reduction, graph drawing, and proximity analysis. J. Am. Stat. Assoc. **104**(485), 209–219 (2009)

12. Costa, J.A., Hero, A.O.: Geodesic entropic graphs for dimension and entropy estimation in manifold learning. IEEE Trans. Sig. Process. **52**(8), 2210–2221 (2004)

13. Cruz-Barbosa, C.R., Vellido A.A.: Geodesic generative topographic mapping. In: Geffner, H., et al. (Ed.) IBERAMIA 2008, LNAI 5290, pp. 113–122 (2008)

14. Dasgupta, S.: Experiments with random projection. In: Boutilier, C., Goldszmidt, M. (Eds.) Proceedings of the Sixteenth Conference on Uncertainty in Artificial Intelligence (UAI00), pp. 143-151. Morgan Kaufmann Publishers Inc., San Francisco (2000)

15. Ferreira, C.: Gene Expression Programming: Mathematical Modeling by an Artificial Intelligence. Springer, New York (2006)

16. France, S., Carroll, J.: Development of an agreement metric based upon the RAND index for the evaluation of dimensionality reduction techniques, with applications to mapping customer data. In: Proceedings of MLDM 2007, pp. 499-517. Springer (2007)

17. François, D., Wertz, V., Verleysen, M.: The concentration of fractional distances. IEEE Trans. Knowl. Data Eng. **19**(7), 873–886 (2007)

18. Fukunaga, K., Olsen, D.R.: An algorithm for finding intrinsic dimensionality of data. IEEE Trans. Comput. **20**, 176–183 (1971)

19. Gisbrecht, A., Mokbel, B., Hasenfuss, A., Hammer, B.: Visualizing dissimilarity data using generative topographic mapping. In: Dillmann, R., et al. (Ed.) Proceedings KI 2010. LNAI 6359, pp. 227-237. Springer (2010)

20. Gisbrecht, A., Mokbel, B., Hammer, B.: Relational generative topographic map. In: Verleysen, M. (Ed.) Proceedings of ESANN10, D-side, pp.277–282 (2010)

21. Grassberger, P., Procaccia, I.: Measuring the strangeness of strange attractors. Physica D **9**, 189–208 (1983)

22. Hegde, C., Wakin, M., Baraniuk, R.: Random Proj. Manifold Learn. Adv. Neural Inf. Process. Syst. **20**, 641–648 (2008)

23. Hegde, C., Sankaranarayanan, A.C., Yin, W., Baraniuk, R.G.: NuMax: a convex approach for learning near-isometric linear embeddings. J. Mach. Learn. Res. Preprint (2013)

24. Hein, M., Audibert, J.Y.: Intrinsic dimensionality estimation of submanifolds in Euclidean space. In: de Raedt, L., Wrobel, S. (Eds.) Proceedings of the 22nd International Conference on Machine Learning (ICML), pp. 289–296 (2005)

25. Hinton, G.E., Roweis, S.T.: Stochastic neighbor embedding. In: Advances in Neural Information Processing Systems, vol. 15, pp. 833–840. The MIT Press, Cambridge (2002)

26. Johnson, W.B., Lindenstrauss, J.: Extensions of Lipschitz mappings into a Hilbert space conference in modern analysis and probability, New Haven, CI, 1982. In: Contemporary Mathematics 26, Providence, RI: American Mathematical Society, pp. 189–206 (1984)

27. Koza, J.R.: Genetic Programming: On the Programming of Computers by Means of natural Selection. MIT Press, Cambridge (1992)
28. Koza, J.R.: Hierarchical genetic algorithms operating on populations of computer programs. In: Proceedings of the 11-th International Joint Conference on Artificial Intelligence, vol. 1, pp. 768–774 (1989)
29. Kruskal, J.: Nonmetric multidimensional scaling: a numerical method. Psychometrika **29** (1964)
30. Kruskal, J.: Multidimensional scaling by optimizing goodness of fit to a nonmetric hypothesis. Psychometrika **29** (1964)
31. Lee, J., Verleysen, M.: Quality assessment of dimensionality reduction: rank-based criteria. Neurocomputing **72**(79), 1431–1443 (2009)
32. Lee, J., Renard, E., Bernard, G., Dupont, P., Verleysen, M.: Type 1 and 2 mixtures of Kullback-Leibler divergencies as cost functions in dimensionality reduction based on similarity preservation. Neurocomputing **112**, 92–108 (2013)
33. Lee, J., Verleysen, M.: Two key properties of dimensionality reduction methods. In: Proceedings of 2014 IEEE Symposium Series on Computational Intelligence (IEEE SSCI 2014). Caribe Royale All-Suite Hotel & Convention Center. Orlando, Florida, 9–12 Dec 2014
34. Levina, E., Bickel, P.J.: Maximum likelihood estimation of intrinsic dimension. In: (Saul, L.K., Weiss, Y., Bottou, L. (Eds.) Advances in NIPS 17. Advances in Neural Information Processing Systems, vol. 17, pp 777–784. The MIT Press, Cambridge (2005)
35. Lowe, D.: Novel topographic nonlinear feature extraction using radial basis functions for concentration coding in the artificial nose. In: 3rd IEE International Conference on Artificial Neural Networks. London: IEE (1993)
36. Lowe, D., Tipping, M.E.: Feed-forward neural networks and topographic mappings for exploratory data analysis. Neural Comput. Appl. **4**, 83–95 (1996)
37. Nabney, I.: NETLAB: Algorithms for Pattern Recognition. Springer, New York (2004)
38. Ng, A.Y., Jordan, M.I., Weiss, Y.: On spectral clustering: analysis and an algorithm. In: Advances in Neural Information Processing Systems, pp. 849–856. MIT Press, Cambridge (2001)
39. Olier, I., Vellido, A.: Variational bayesian generative topographic mapping. J. Math. Model. Algoritm. **7**, 371–387 (2008)
40. Press, W., Flannery, B., Teukolsky, S., Vetterling, W.: Numeric Recipes in C. Cambridge University Press, Cambridge (1992)
41. Reachin Display AB. http://www.reachin.se/products/
42. Rényi, A.: On measures of entropy and information. In: Proceedings of the Fourth Berkeley Symposium on Mathematics, Statistics and Probability, 1960, vol 1, pp. 547–561. University of California Press, Berkeley (1961)
43. Roweis, S., Saul, L.: Nonlinear dimensionality reduction by locally linear embedding. Science **290**, 2323–2326 (2000)
44. Sakr, N., Alsulaiman, F., Valdes, J.J., El Saddik, A., Georganas, N.D.: Exploring the underlying structure of haptic-based handwritten signatures using visual data mining techniques. In: Proceedings of the IEEE 2010 Symposium on Haptic Interfaces for Virtual Environments and Teleoperator Systems, pp. 467–474. Waltham, 25–26 March 2010
45. Sammon, J.W.: A nonlinear mapping for data structure analysis. IEEE Trans. Comput. **C–18**(5), 401–409 (1969)
46. Shi, J., Malik, J.: Normalized cuts and image segmentation. IEEE Trans. Pattern Anal. Mach. Intell. **22**(8), 888–905 (2000)
47. Silva, V., Tenenbaum, J.B.: Global versus local methods in nonlinear dimensionality reduction. Advances in Neural Information Processing Systems, pp. 721–728. MIT Press, Cambridge (2003)
48. Takens, F.: On the numerical determination of the dimension of an attractor. Dyn. Syst. Bifurcat. 99–106 (1985)
49. Tenenbaum, J.B., De Silva, V., Langford, J.C.: A global geometric framework for nonlinear dimensionality reduction. Science **290**, 5500 (2000)

50. Valdes, J.J., Orchard. R., Barton, A.J.: Exploring medical data using visual spaces with genetic programming and implicit functional mappings. In: Proceedings of GECCO 2007 Conference. GECCO Workshop on Medical Applications of Genetic and Evolutionary Computation. London, UK. 7–11 July 2007
51. Valdes, J.J., Barton, A.J., Orchard, R.: Virtual reality high dimensional objective spaces for multi-objective optimization: an improved representation. In: IEEE Congress on Evolutionary Computation. pp 4191–4198. Singapore, 25–28 Sept 2007
52. van der Maaten, L.J.P., Hinton, G.: Visualizing high-dimensional data using t-SNE. J. Mach. Learn. Res. (2008)
53. Venna, J., Kaski, S.: Local multidimensional scaling. Neural Netw. **19**, 889–899 (2006)
54. Venna, J., Peltonen, J., Nybo, K., Aidos, H., Kaski, S.: Information retrieval perspective to nonlinear dimensionality reduction for data visualization. J. Mach. Learn. Res. **11**, 451–490 (2010)
55. von Luxburg, U.A.: Tutorial on Spectral Clustering. Technical report. No. TR-149. Max Planck Institute for Biological Cybernetics (2007)
56. Zhang, Z., Wang, MLLE. J.: Modified locally linear embedding using multiple weights. In: Advances in Neural Information Processing Systems 19, Proceedings of the Twentieth Annual Conference on Neural Information Processing Systems. Vancouver, British Columbia, 4–7 Dec 2006
57. Zhu, X., Gisbrecht, A., Schleif, F.M., Hammer, B.: Approximation techniques for clustering dissimilarity data. Neurocomputing **90**, 72–84 (2012)

Extended Metacognitive Neuro-Fuzzy Inference System for Biometric Identification

Bindu Madhavi Padmanabhuni, Kartick Subramanian and Suresh Sundaram

Abstract Biometrics are increasingly being used as security measures in online as well as offline systems, giving rise to more reliable and unique authentication techniques. In these systems, false positive minimization is one of the crucial requirements, which is especially critical in security sensitive applications. In this chapter, we present an Extended Metacognitive Neuro-Fuzzy Inference System (eMcFIS) based biometric identification system. eMcFIS consists of a cognitive component and a metacognitive component. The cognitive component, which is a neuro-fuzzy inference system, learns the input-output relationship efficiently. The metacognitive component is a self-regulatory learning mechanism, which actively regulates the learning in the cognitive component such that the network avoids over-fitting the training samples. Further, the learning strategies are chosen such that the network minimizes false-positive prediction. The proposed eMcFIS is first benchmarked on a set of medical datasets from machine learning databases. eMcFIS is then employed in detection of two real-world biometric security applications, signature verification and fingerprint recognition. The performance comparison with other state-of-the-art authentication systems clearly highlights the advantages of the proposed approach.

Keywords Biometrics · False positive · Metacognition · Neuro-fuzzy · McFIS

B.M. Padmanabhuni (✉)
School of Electrical and Electronic Engineering, Nanyang Technological University,
Singapore, Singapore
e-mail: padm0010@e.ntu.edu.sg

K. Subramanian · S. Sundaram
School of Computer Engineering, Nanyang Technological University, Singapore, Singapore
e-mail: KARTICK1@e.ntu.edu.sg

S. Sundaram
e-mail: ssundaram@ntu.edu.sg

© Springer International Publishing Switzerland 2016
R. Abielmona et al. (eds.), *Recent Advances in Computational Intelligence in Defense and Security*, Studies in Computational Intelligence 621,
DOI 10.1007/978-3-319-26450-9_12

309

1   Introduction

In the current world of online and offline fraud, the need for efficient security and surveillance fraud detection systems cannot be overstated. One of the main reasons for the failure in these systems is high false positive prediction. A false positive in binary classification is an error resulting from incorrectly indicating the presence of a condition while in truth it is absent. False positive reduction is important in security oriented applications like malware and intrusion detection systems as they incur huge damages. False positive minimization finds resonance in other fields too. For instance, wrongly predicting a non-malignant tumor to be malignant in computer aided disease diagnosis causes unnecessary trauma, treatment and expenditure.

Due to ever-increasing identity thefts in knowledge and token based systems (e.g., password, ID card) biometric identification systems are being progressively used in commercial and governmental organizations to establish a person's identity. In a defense establishment with biometric identification systems, if an imposter is identified as a client and given access to the utilities it causes a security breach. Similarly, in a banking system if a forged signature is identified as original and the transaction is approved, it might lead to financial loss. In all these applications, it is imperative to have a low false positive rate since the end-users are highly sensitive to misclassification on a specific class.

A classifier with high coverage may not necessarily be precise in predicting the class the end-user is sensitive to. Various works are being actively conducted to address this ever growing problem of false positive reduction. Some of the earliest works on such class-sensitive classifiers are based on Naïve Bayes algorithm. Due to their previously known robustness in the text classification domain, Naïve Bayes classifiers were used for spam mail filtering. In [1, 2] the Bayesian probability model parameters were modified to associate positive predictions of the sensitive class with high confidence for filtering spam mail. Although this modification partially handled false positive minimization it was suggested by Sahami et al. in [1] that application of SVMs could help control parameter variance during learning. Furthermore, Schneider in [3] reasoned that using cost-sensitive measures in conjunction with Naïve Bayes classifier is problematic as the probabilities computed by the classifier are unreliable and therefore proposed two statistical event models for spam mail filtering, out of which, multinomial model using feature ranking functions and taking into account word frequency information was found to result in better accuracy and less biased towards the sensitive class. Though effective, the solution is not suitable for non-text classification problems.

Support Vector Machine (SVM) based algorithms typically use parameter tuning or threshold mechanisms for false positive reduction. Parameter tuning in SVM generally involves tuning the SVM tradeoff parameter 'C' to balance between low false positives and acceptable overall accuracy. But such parameter selection mechanism [4] incurs exhaustive search for the optimum combination, which could be very time-consuming and does not really address the issue of false positives.

Parameter tuning approach by Kołcz and Alspector [5] applied different misclassification costs for classes while training the SVM classifier using prior knowledge of mail categories in spam mail filtering to reduce false positives. While the proposed method showed a clear improvement over standard SVM it needs domain-specific knowledge about the target data misclassification costs and the probability distribution of data for estimating the optimum cost factor to apply. Sculley and Wachman proposed threshold mechanism [6] to detect spam mail, such that, if the predicted score is greater than the chosen threshold, then the mails are classified as spam. The choice of good threshold is crucial in such algorithms because if the threshold is chosen to be high there are less false positives while at the same time true positive count also drops resulting in an undesirable tradeoff between false and true positives.

Boosting algorithms also were proposed for low false positive learning. An AdaBoost algorithm was proposed by Carreras et al. in [7] which used evaluated prediction confidence of a decision tree classifier for false positive reduction in spam mail filtering. Methods for automatic tuning of the classifier parameters were proposed and it was shown that deeper weak rules are apt for high precision classifier requirement as they result in a filter which only classifies messages that it is highly confident and delivers the rest to the user. Although the filter outperformed decision trees, Naïve Bayes and k-NN methods it was highlighted that further studies on effectiveness of tuning the parameters was needed as the classifier confidence depends on parameter settings. Wu et al. [8] used cascaded classifiers for face detection problem, where feature selection was carried out using asymmetric AdaBoost classifier. Each of the cascaded classifiers carried out a stage-wise rejection procedure for non-facial inputs and only those that passed through all the rejections-stages of cascade were classified as faces. Instead of a single complex classifier with low false positive rates, the cascaded classifiers with high detection and moderate false positive rate at each stage yielded a final lower false positive rate.

Yih et al. [9] used a two-stage methodology to reduce false positives. In the first stage, data from the low-false positive region is identified for training the second stage classifier so that the selected training data is characteristic of the data that is more important to the application setup. At test time, if the first stage of classifier predicted the test instance to be in the region interest of the second-stage, it was classified using the second-stage classifier and the verdict of second-stage classifier was used to predict the test instance. This two-stage filtering caused the learned classifier to be optimized for such particular data and hence enabled in lowering false-positive rate. Lynam et al. [10] proposed an ensemble approach for filtering spam mail, wherein outcomes from a set of independently developed spam filters are combined resulting in substantially better filtering than any of the individual filters. Such cascaded and ensemble classifier set-up incurs high-costs.

A compression algorithm for spam mail filtering was proposed by Bratko et al. in [11] for decreasing false-positives. Here two compression models were first built for each of the classes in the binary classification problem from the training data. Since spam mails usually contain homogenous terms, they result in high text compression rate. Given that spam mail is continuously evolving, an adaptive data compression was proposed based on online user feedback resulting in a compression model that is incrementally updateable. This proposal is only suitable for such text-based spam mail filtering and not for other applications.

Akusok et al. [12] proposed a two-stage methodology to minimize false positives in malware/anomaly detection. The malware dataset attributes were nominal in nature. Therefore a classifier with distance based machine learning technique was used in the first stage with Jaccard distance measure approximation adapted to the problem setup. The first stage of the decision process made use of a 1-Nearest Neighbor classifier. For the second stage, a classifier using modified extreme learning machine (ELM) [13], was proposed. The modified ELM classifier used information gathered from searching the nearest neighbor in the first stage for classification. While the number of false positives was almost ideal on the malware dataset, around 56 % of the test samples were classified as "Unknown" which could inhibit its usage in practical applications. It can therefore be seen from the above discussion that there is a need to develop an efficient algorithm which could reduce false positives while maintaining high accuracy. Such an algorithm should have intrinsic learning characteristics to handle false positives while maintaining high generalization ability.

In the past, fuzzy inference systems (FIS) have been extensively used in machine learning problems due to their efficiency in accurately approximating the target non-linear functions. FIS is a collection of fuzzy rules specifying mapping between input and output. The important steps in the design of a FIS are: identification of the fuzzy rule base and tuning the rule parameters, the latter of which was not well-defined and needed domain knowledge and data evidence [14]. Neural networks, on the other hand, were effective in learning from training samples. Hence, of late, FIS and neural networks are combined to make use of approximate reasoning and interpretability of FIS and learning abilities of neural networks.

Adaptive network based neuro fuzzy inference system (ANFIS) [15] is one of such first neuro fuzzy inference systems proposed which uses a gradient descent based learning algorithm for updating network parameters. ANFIS could either be used to refine fuzzy rules provided by the domain experts or could learn and generate them itself by partitioning the input space. However, it suffers when uncertainties are present in the data. In many practical applications data specifying the complete input-output mapping may not be available beforehand. This introduces additional challenges in accommodating new features or classes. To circumvent this, adaptive neuro fuzzy inference systems have been proposed which start with zero rules and build up the rule base approximating decision surface as per required accuracy. Kasabov [16] proposed one of the first adaptive neuro-fuzzy inference systems which used an incremental hybrid supervised/unsupervised online learning for evolving rules and network parameters based on the principles of resource allocation network.

Dynamic evolving neuro-fuzzy inference system (DENFIS) [17] is another such adaptive neuro-fuzzy inference systems which used evolving clustering method to partition the input space. DENFIS used offline clustering to select 'm' most activated fuzzy rules, for calculating the output which limits it from being a truly online solution. Similar to Kasabov's proposal in [16], DENFIS also uses local element tuning and hence needs more training data than ANFIS which uses global generalization. Proposals based on fuzzy support vector machines [18, 19] and fuzzy extreme learning machines [20] too exist in literature. Fuzzy SVM solutions construct large rule base with some rules being under-utilized whereas fuzzy ELM proposal needs the number of rules to be fixed a priori [21]. While all the afore mentioned neuro fuzzy inference systems possess human cognitive information processing capabilities like learning, recall, perception and problem solving, they lack the metacognitive ability of judging the knowledge existing in their cognitive component with that contained in the training samples. Therefore, they assume that knowledge is distributed uniformly in all the learning samples and learn all samples sequentially whereas human learning studies suggest otherwise [22]. Studies show that humans regulate their knowledge acquisition by judging their acquired knowledge with that of the environment. This is known as metacognition.

Recently a Metacognitive sequential learning algorithm for Neuro-Fuzzy Inference System (McFIS) was proposed in [21]. McFIS leverages on its metacognitive component in choosing a suitable learning strategy for the neuro fuzzy inference system (i.e., the cognitive component). It has been shown in [21, 23–26] that the metacognitive learning helps the fuzzy inference system to achieve better generalization performance than other existing classifiers in literature like SVM [4], eClass [27], ELM [13], Online Sequential Fuzzy Extreme Learning Machine (OS-FUZZY-ELM) [20] and Sequential Adaptive Fuzzy Inference System (SAFIS) [28]. McFIS, however, lacks the ability to handle false positives during prediction efficiently. Our proposed learning algorithm is built up on McFIS to leverage on its metacognitive learning strategies and high generalization abilities.

In this chapter we propose an extended Metacognitive Neuro-Fuzzy Inference System (eMcFIS) for minimizing false positive predictions. The metacognitive component in eMcFIS monitors the knowledge in the cognitive component and regulates the learning in it by deciding on what-to-learn, when-to-learn, and how-to-learn the presented data. In the how-to-learn strategy, eMcFIS considers the class-sensitive false positive criterion to either add a new rule to the network or update the parameters of the network such that the false positive prediction is minimized. It uses posterior probability as a measure of classifier's confidence, along with self-adaptive learning thresholds and class-specific criterion which helps in minimizing false positives. The performance of eMcFIS is first benchmarked on a set of medical datasets from the UCI machine learning repository and Mammogram database. The results clearly show that eMcFIS has lower false positive predictions along with better coverage. Statistical Friedman test followed by Benforroni-Dunn test on the classifier results over the tested data sets proves that eMcFIS performs better than other well-known classifiers. Subsequently, the generalization and false positive reduction abilities of the proposed system are validated on two real-world security sensitive biometric identification applications: signature verification and fingerprint recognition problems. The performance comparison of eMcFIS on biometric datasets with other state-of-the-art approaches effectively signifies the advantages of the proposed method.

We make the following contributions in this chapter:

- Metacognitive sequential learning algorithm for low false positive prediction
- Usage of posterior probability as a measure of classifier confidence
- False positive threshold for choosing the learning strategy to apply
- Class-specific criteria for low false positive learning

The rest of the chapter is organized as follows. Section 2 describes McFIS algorithm in brief. Section 3 presents the eMcFIS learning algorithm followed by an initial evaluation of eMcFIS on a set of medical datasets from the machine learning repositories. Section 4 presents two real-world security sensitive biometric verification problems and the performance evaluation of eMcFIS on the biometric datasets. Section 5 summarizes the chief conclusions from this study.

2 A Brief Review on Metacognitive Neuro-Fuzzy Inference System (McFIS)

In this section, we describe the McFIS algorithm in brief along with a problem summary of false positive optimization.

McFIS [21] is a metacognitive sequential learning algorithm for Neuro-Fuzzy Inference Systems. McFIS realizes metacognitive learning strategies by a two component system as depicted in Fig. 1. The primary component is a cognitive component realized by a neuro-fuzzy inference system and the second component is a metacognitive component. The cognitive component of McFIS is a four-layer neuro-fuzzy network with radial basis activation function realizing the behavior of zeroth-order TSK-type fuzzy inference system as shown in Fig. 2. Metacognitive component of McFIS monitors the knowledge in the cognitive component and decides on the suitable learning strategy to apply when presented with the input data. McFIS uses theoretically proven hinge-loss error function [29, 30] to measure the sample classification error and class-specific spherical potential derived from [31] as novelty measure of the data.

McFIS controls the learning process in the cognitive component by resolving on what-to-learn, how-to-learn and when-to-learn. These actions are realized by sample deletion, sample learning, and sample reserve strategies respectively as follows:

1. Sample deletion strategy

If the knowledge in the sample is similar to that already contained in the network then the sample is deleted without learning.

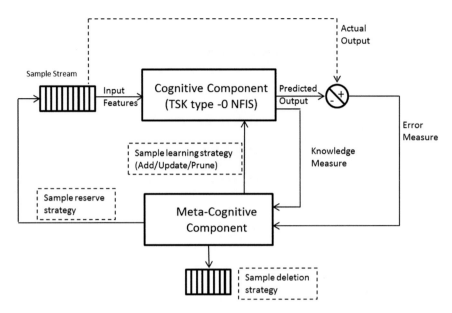

Fig. 1 Schematic diagram of McFIS and eMcFIS

2. Sample learning strategy

There are three kinds of learning strategies in McFIS namely: rule growing, rule parameter updating and rule pruning.

- Rule Growing Strategy

If the training sample contains significant information not already present in the cognitive component or if the estimated class label is incorrect, a new rule is added to capture this novel knowledge.

- Rule Parameter Update Strategy

If the predicted class label is the same as the actual class label, but the maximum hinge loss error is greater than the self-adaptive parameter update threshold, then the parameters of the nearest rule in the same class are updated using an EKF algorithm.

- Rule Pruning Strategy

McFIS removes from network rules whose contribution in the class is lesser than pruning threshold for specified consecutive samples in the class.

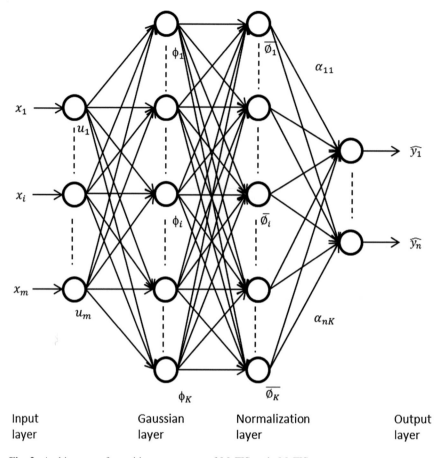

Fig. 2 Architecture of cognitive component of McFIS and eMcFIS

3. Sample reserve strategy

This represents the "what-to-learn" concept of McFIS. If the sample does not meet either the rule growth or the parameter update conditions, it is reserved for possible usage in learning at a later stage when the McFIS self-adaptive thresholds may find it conducive for the learning process.

McFIS algorithm was employed to solve various medical, energy analytics, video analytics and time-series related problems and was shown to achieve higher generalization performance over other state-of-the-art approaches [4, 13, 20, 27, 28]. McFIS however lacks the ability to handle false positives.

3 Extended Metacognitive Neuro-Fuzzy Inference System (eMcFIS)

It can be inferred from the above discussion that McFIS is a generic classification algorithm and does not specifically address the issue of false positive optimization. In this section, we describe the proposed eMcFIS learning algorithm for handling false positive reduction in binary classification problems. We first outline the cognitive component followed by the metacognitive component. The goal of the classifier is to closely approximate the decision function while minimizing the false positives incurred, rendering it to achieve high generalization performance along with low false positives.

3.1 Cognitive Component of eMcFIS

The cognitive component of eMcFIS, similar to McFIS, is a four-layer neuro-fuzzy network with radial basis activation function as depicted in Fig. 2. Consider a training data instance: (x^t, c^t), where $x^t = \left[x_1^t, \ldots, x_m^t\right]^T \in \mathfrak{R}^m$ is a m dimensional input vector of tth sample and $c^t \in (1, 2, \ldots, n)$ is its class label among n distinct classes. The class label (c^t) is converted into coded class label $\left(y^t = \left[y_1^t, \ldots, y_j^y, \ldots y_n^t\right] \in \mathfrak{R}^n\right)$ such that

$$y_j^t = \begin{cases} 1, & \text{if } j = c^t \\ -1, & \text{otherwise} \end{cases} \quad j = 1, 2, \ldots, n \tag{1}$$

The predicted output $(\hat{y}^t)$ of eMcFIS classifier is given by

$$\hat{y}^t = f(x^t, w) \tag{2}$$

where decision function $f: x \rightarrow y$ is a relationship between x^t and y^t and vector w represents classifier parameters. The first layer of neuro fuzzy network consists of m input nodes with each node representing an input feature.

The outputs from input layer (u^t) are directly transmitted to the Gaussian layer. Output of ith input node (u_i^t) is given by

$$u_i^t = x_i^t, \quad i = 1, 2, \ldots, m \tag{3}$$

Assuming that eMcFIS built K rules from previous $t - 1$ training samples, the Gaussian layer contains the rule antecedents of each of the K rules and performs rule inference to calculate the overall contribution of the rule to the input features. The membership of the kth rule is given as

$$\emptyset_k(u^t) = exp\left(\frac{-\left\|u^t - \mu_k^l\right\|^2}{2\left(\sigma_k^l\right)^2}\right), \quad k = 1, 2, \ldots, K \tag{4}$$

where μ_k^l is the center of kth Gaussian node and σ_k^l is its width (superscript l indicates the rule's class label). Normalization layer has the same number of nodes as that of Gaussian layer. The output of ith normalized node is given as follows:

$$\overline{\emptyset}_k = \frac{\emptyset_k}{\sum_{j=1}^{K} \emptyset_j}, \quad k = 1, 2, \ldots, K \tag{5}$$

The output layer has the same number of nodes as that of distinct classes. Predicted output $(\hat{y}^t)$ of the output layer is calculated as the weighted sum of the normalized output given by

$$\hat{y}_j^t = \sum_{k=1}^{K} \alpha_{jk}\overline{\emptyset}_k, \quad j = 1, 2, \ldots, n \tag{6}$$

where α_{jk} is the weight connecting kth normalized node to jth output node. The predicted class label is given by

$$\hat{c}^t = \arg\max_{j=1,\ldots,n}\left(\hat{y}_j^t\right) \tag{7}$$

3.2   Metacognitive Component of eMcFIS

The metacognitive component of eMcFIS is a self-regulatory learning mechanism. It uses the monitory signals from the cognitive component to compare the knowledge existing in the network with that in the presented sample to choose the learning strategy to apply. Similar to McFIS, eMcFIS too uses hinge-loss error function as a measure of sample error and data novelty measure derived from [31] as the knowledge-based measure. Hinge-loss error $\left(e^t = \left[e_1^t, \ldots, e_n^t\right]^T \in \mathfrak{R}^n\right)$ is defined as:

$$e_j^t = \begin{cases} 0 & if \ \hat{y}_j^t y_j^t > 1 \\ \hat{y}_j^t - y_j^t & otherwise \end{cases} \quad j = 1, 2, \ldots, n \tag{8}$$

Hinge-loss error lets the network output to go beyond ± 1. Therefore truncated hinge-loss error output is an accurate estimate of posterior probability and is given as:

$$\hat{p}(c|x^t) = \frac{min\left(max\left(\hat{y}_j^t, -1\right), 1\right) + 1}{2}, j = 1, 2, \ldots n; c = 1, 2, \ldots, n \tag{9}$$

and the maximum absolute hinge-loss error E^t is defined as

$$E^t = \max_{j=1,...,n} \left| e_j^t \right| \tag{10}$$

The classifier algorithm determines sample novelty by projecting the input feature (x^t) onto spherical feature space, S, i.e., $x^t \to \emptyset$, with center μ^l and width σ^l of the Gaussian rules describing S. Assuming K^c rules associated with class c, class-specific spherical potential is given by:

$$\varphi_c = \frac{1}{K^c} \sum_{j=1}^{K^c} \emptyset \left(x^t, \mu_j^c \right) \tag{11}$$

Smaller spherical potential suggests that the sample contains new information whereas higher value indicates that the information provided by the sample already exists in the cognitive component. Metacognitive component of eMcFIS controls the learning process in the cognitive component through sample deletion, sample learning, and sample reserve strategies using hinge-loss error function, class posterior probability and class-specific spherical potential as monitory signals.

Sample deletion strategy
If the knowledge presented by the sample is analogous to that already existing in the network, it is deleted without learning. This avoids overtraining of the network and also reduces the computational resources that would have been otherwise consumed. The sample deletion criterion is given as

$$c^t = \hat{c}^t \text{ AND } E^t < E_d \tag{12}$$

where E^t is the maximum absolute hinge error and E_d is the sample deletion threshold.

eMcFIS Sample learning strategy
eMcFIS differs from McFIS in the sample learning strategy due to its false positive optimized how-to-learn strategy.

False positive aware classifier sample learning strategy
The first step in false positive minimization is to make the classifier aware of false positive predictions during the learning process. Throughout the following discussion we represent c_1 as the positive class and c_2 as the negative class. A sample (x^t, c^t) is considered as False Positive (FP) prediction by the classifier if:

$$FP: c^t = c_2 \text{ AND } \hat{c}^t = c_1 \tag{13}$$

where c^t is the actual class label and $\hat{c}^t$ is the predicted class label.

Classifier False Positive Prediction Confidence Measure
It is important to know the confidence of the classifier while making a FP prediction to decide on a suitable learning strategy for correcting it and minimize such

subsequent false positives. We propose using posterior probability estimate of the sample belonging to positive class c_1 as confidence measure. This measure will be used as a monitory signal for handling false positives. It has been proven theoretically that truncated classifier output is an accurate estimate of posterior probability [30]. The truncated classifier output is given as:

$$T\left(\hat{y}_j^t\right) = min\left(max\left(\hat{y}_j^t, -1\right), 1\right), \ j=1,2 \tag{14}$$

where the range for $T\left(y_j^{\wedge t}\right)$ is $[-1, 1]$. The posterior probability of the input x^t belonging to positive class c_1 is given as

$$\hat{p}(c_1|x^t) = \frac{T\left(\hat{y}_j^t\right) + 1}{2} \tag{15}$$

Rule Growing Strategy

If the training sample contains significant information not already present in the cognitive component or if the estimated class label is incorrect with a high error, McFIS adds a new rule to the network. In addition to this rule growth criterion, to minimize false positive predictions, eMcFIS adds new rule when the sample results in a false positive with a positive class posterior probability estimate greater than the false positive minimization threshold.

In other words, eMcFIS combines posterior probability of predicted sample belonging to positive class c_1, $\hat{p}(c_1|x^t)$, along with hinge error and class-specific spherical potential measures proposed by McFIS to develop following rule growth criteria:

$$\left(E^t > E_a \ AND \ c^t \neq \hat{c}^t \ AND \ \varphi_c < E_S\right) \ OR \ \left(FP \ AND \ \hat{p}(c_1|x^t) > E_{FP}\right) \tag{16}$$

where E_a is self-adaptive rule addition threshold, φ_c is the class-specific spherical potential, E_S is class-specific novelty threshold, FP denotes the instance being a false positive prediction by the classifier as given in Eq. 13 and E_{FP} is the fixed false positive minimization rule addition threshold.

The term $FP \ AND \ \hat{p}(c_1|x^t) > E_{FP}$ measures the classifier's confidence of the false positive sample belonging to positive class c_1 against E_{FP}. If the confidence measured by posterior probability is greater than the false positive minimization rule addition threshold, it means that the misclassification occurs with a higher degree of confidence for positive class while the sample actually belongs to negative class. Hence this sample presents novel knowledge to be learnt for the false positive optimization. Therefore, we add it as a new rule.

Upon adding a new rule the self-adaptive rule addition threshold, E_a is adapted as follows:

$$E_a := \delta(E^t) + (1 - \delta)E_a \tag{17}$$

where δ is the slope that controls the rate of adaptation. The center (μ_{K+1}), width (σ_{K+1}) and output weight (α_{K+1}) of new rule are initialized as follows:

$$\mu_{K+1} = x^t \tag{18}$$

$$\sigma_{K+1}^l = \begin{cases} k \times d_S, & if \ \frac{d_S}{d_I} < 1.0 \\ \eta \times d_I, & if \ \frac{d_S}{d_I} > 1.0 \end{cases} \tag{19}$$

where nrS is the nearest rule in the same class *and* nrI is the nearest rule in the interclass with $d_s = ||x^t - \mu_{nrS}||$, $d_I = ||x^t - \mu_{nrI}||$ and k and η determining the overlap between the new rule and nearest rule in the same and opposite class respectively.

$$\alpha_{j,K+1} = \begin{cases} 0, & y_j^t * \frac{\sum_{i=1}^{K} \alpha_{ji}\varnothing_i}{1 + \sum_{i=1}^{K} \varnothing_i} > 1 \\ y_j^t - \frac{\sum_{i=1}^{K} \alpha_{ji}\varnothing_i}{1 + \sum_{i=1}^{K} \varnothing_i}, & otherwise \end{cases} \quad where \ j = 1, 2, \ldots, n \tag{20}$$

Rule Update Strategy
When a training sample is predicted correctly but the maximum hinge error is greater than the self-adaptive rule parameter update threshold, E_l, then McFIS updates the parameters of the nearest rule in the same class using an EKF algorithm. In addition to this criterion, if the sample results in a false positive prediction then the rule parameters of the nearest rule in opposite class, i.e., the positive class (c_1), are updated to minimize such false positive prediction errors.

Hence the rule update criterion for eMcFIS is given by:

$$(E^t > E_l \ AND \ c^t = \hat{c}^t) \ OR \ (FP) \tag{21}$$

where E_l is the self-adaptive parameter update threshold, updated as per equation below:

$$E_l := \delta E^t + (1 - \delta)E_l \tag{22}$$

If condition $(E^t > E_l \ AND \ c^t = \hat{c}^t)$ is evaluated to be true it shows that the sample was predicted correctly but maximum hinge error is greater than rule parameter update threshold. Hence the following McFIS rule update strategy equations apply wherein the nearest same class rule parameters are updated as follows:

$$w_{nrS} = w_{nrs} + G_{nrS}e^t, \ w_{nrS} = [\alpha_{nrS}, \mu_{nrS}^l, \sigma_{nrS}^l] \in \mathfrak{R}^{(m+n+1)} \tag{23}$$

where e^t is the error defined in (8), and G_{nrS} is the Kalman gain matrix given by

$$G_{nrS} = P_{nrS} a_{nrS}^t \left[R + a_{nrS}^{tT} P_{nrS} a_{nrS}^t \right]^{-1} \tag{24}$$

where $a_{nrS} \in \mathfrak{R}^{(m+n+1) \times n}$ is the output gradient with respect to the parameters (w_{nrS}), $R = r_0 I_{nxn}$ is the variance of measurement noise and $P_{nrS} \in \mathfrak{R}^{(m+n+1) \times (m+n+1)}$ is the nearest rule parameters error covariance matrix. Gradient (a_{nrS}) is given by

$$a_{nrS}^t = \begin{bmatrix} \frac{\partial \hat{y}_i^t}{\partial \alpha_{i,\,nrS}} = \overline{\varnothing}_{nrS}, i = 1, 2, \ldots, n \\ \frac{\partial \hat{y}_i^t}{\partial \mu_{nrS,j}} = 2 \varnothing_{nrS} \frac{x_j^t - \mu_{nrS,j}}{\sigma_{nrS}^2} \frac{\alpha_{i,\,nrS} - \hat{y}_i^t}{\sum_{l=1}^K \varnothing_l} \\ \frac{\partial \hat{y}_i^t}{\partial \alpha_{nrS}} = 2 \varnothing_{nrS} \frac{||x^t - \mu_{nrS}||^2}{\sigma_{nrS}^3} \frac{\alpha_{i,\,nrS} - \hat{y}_i^t}{\sum_{l=1}^K \varnothing_l} \end{bmatrix} \tag{25}$$

where $j = 1, 2, \ldots, m$ and P_{nrS} is updated as

$$P_{nrS} = \left[I_{z \times z} - G_{nrS} a_{nrS}^{tT} \right] P_{nrS} + q_0 I_{z \times z} \tag{26}$$

If (FP) condition holds true, then the parameters of the nearest rule in positive class (c_1), nrP, defined by

$$nrP = \arg \min_{l \neq 2} ||x^t - \mu^l|| \tag{27}$$

are updated as given below:

$$w_{nrP} = w_{nrP} + G_{nrP} e^t, \quad w_{nrP} = \left[\alpha_{nrP}, \mu_{nrP}^l, \sigma_{nrP}^l \right] \in \mathfrak{R}^{(m+2+1)} \tag{28}$$

where e^t is the error defined in (8) and G_{nrP} is the Kalman gain matrix given by

$$G_{nrP} = P_{nrP} a_{nrP}^t \left[R + a_{nrP}^{tT} P_{nrP} a_{nrP}^t \right]^{-1} \tag{29}$$

where $a_{nrP} \in \mathfrak{R}^{(m+2+1) \times 2}$ is the output gradient with respect to the parameters (w_{nrP}), $R = r_0 I_{2x2}$ is the measurement noise variance and $P_{nrP} \in \mathfrak{R}^{(m+2+1) \times (m+2+1)}$ is the nearest rule parameters error covariance matrix. a_{nrP} is given by

$$a_{nrP}^t = \begin{bmatrix} \frac{\partial \hat{y}_i^t}{\partial \alpha_{i,\,nrP}} = \overline{\varnothing}_{nrP}, i = 1, 2 \\ \frac{\partial \hat{y}_i^t}{\partial \mu_{nrP,j}} = 2 \varnothing_{nrP} \frac{x_j^t - \mu_{nrP,j}}{\sigma_{nrP}^2} \frac{\alpha_{i,\,nrP} - \hat{y}_i^t}{\sum_{l=1}^K \varnothing_l} \\ \frac{\partial \hat{y}_i^t}{\partial \alpha_{nrP}} = 2 \varnothing_{nrP} \frac{||x^t - \mu_{nrP}||^2}{\sigma_{nrP}^3} \frac{\alpha_{i,\,nrP} - \hat{y}_i^t}{\sum_{l=1}^K \varnothing_l} \end{bmatrix} \tag{30}$$

where $j = 1, 2, \ldots, m$ and P_{nrP} is updated as

$$P_{nrP} = \left[I_{z \times z} - G_{nrP} a_{nrP}^{tT} \right] P_{nrP} + q_0 I_{z \times z} \tag{31}$$

where $z = m + 2 + 1$ and q_0 determines the allowed step in gradient vector direction.

Rule Pruning Strategy

eMcFIS prunes the network rules whose contribution β_k in the class is lesser than the pruning threshold E_p for N_w consecutive samples in class c. Rule contribution β_k is given by

$$\beta_k = \frac{E}{\varphi_c} \emptyset_k \max |\alpha_k| \tag{32}$$

where E is the maximum hinge error, φ_c is the spherical potential of actual class c and ϕ_k and α_k are the firing strength and output weight of kth rule respectively. Whenever a new rule is added, the error covariance matrix, P, is updated as following:

$$\begin{bmatrix} P & 0 \\ 0 & p_0 I_{z \times z} \end{bmatrix} \tag{33}$$

where p_0 is the initial estimated uncertainty. Conversely, when a rule is pruned from the network the dimensionality of the matrix is reduced by removing the respective rows and columns in P.

Sample reserve strategy

If the training sample does not fulfill the rule growing or parameter update requirements, it is reserved for probable usage afterwards when eMcFIS may find it favorable for learning. For ease of understanding, the summarized pseudo-code for eMcFIS is given in Fig. 3.

3.3 Influence of Thresholds in Decision-Making of eMcFIS

In this proposal we used fixed as well as self-regulatory thresholds in the decision-making process. We hereby explain their effect and provide a guideline for their initialization. There are five important thresholds used in this study. They are: delete threshold (E_d), novelty threshold (E_S), rule addition threshold (E_a), parameter update threshold (E_l) and false positive minimization threshold (E_{FP}). For detailed information on other parameters influencing decision making capabilities of the classifier, such as class overlap factors, a reading of [21] is suggested. The proposed algorithm uses hinge-loss error function for measuring classification error. It can be seen from Eq. (10) that range of E^t is [0, 2]. There are two possibilities when a sample is classified namely, classified correctly or misclassified. The classification outcome along with the confidence of prediction and

while learning the samples **do**

1. Calculate the absolute maximum error E^t, significance φ, predicted posterior probability $\hat{p}(i|x^t)$ and predicted posterior probability of sample for positive class $\hat{p}(c_1|x^t)$ for the current input (x^t, c^t).

2. **if** $c^t = \hat{c}^t$ *AND* $E^t < E_d$**then**
    a. delete the sample without learning.
    b. Increment t by 1. Goto Step 1.

3. **else if** $(E^t > E_a$ *AND* $c^t \neq \hat{c}^t$ *AND* $\varphi_c < E_S)$ *OR* $(FP$ *AND* $\hat{p}(c_1|x^t) > E_{FP})$**then**
    a. Add new rule.
    b. Update E_a and EKF as per eqns. (17-20) and error covariance matrix as per eqn. 33
    c. **if** $(FP$ *AND* $\hat{p}(c_1|x^t) > E_{FP})$ **then**
        i. Update nearest positive class rule parameters of network according to eqns. (28-31) and E_l as per eqn. 22
    d. **end**

4. **else if** $(E^t > E_l$ *AND* $c^t = \hat{c}^t)$ *OR* $(FP$ *AND* $\hat{p}(c_1|x^t) \leq E_{FP})$**then**
    a. **if** $(E^t > E_l$ *AND* $c^t = \hat{c}^t)$ **then**
        i. Update nearest same class rule parameters of network according to eqns. (23-26) and E_l as per eqn. 22
    b. **else**
        i. Update nearest positive class rule parameters of network according to eqns. (28-31) and E_l as per eqn. 22
    c. **end**

5. **else**
    a. Reserve the sample to learn later. Increment t by 1. Goto Step 1.

6. **end**

7. **if** rule contribution β_k in the class is lesser than pruning threshold E_p for N_w consecutive samples
    a. Prune the rule from network and update error covariance matrix

8. **end**

9. Increment t by 1. Goto Step 1.

end

Fig. 3 Pseudo-code for eMcFIS classifier

class-specific criteria information is used in determining the self-regulatory thresholds.

Delete threshold E_d

When a training sample is predicted correctly with maximum absolute hinge loss error less than the delete threshold E_d it is deleted without learning, as it represents the knowledge present in the network. If E^t is close to 0, it signifies that the sample is predicted correctly with a high level of confidence as the predicted and actual output are similar. Misclassification occurs when E^t is greater than 1. If E_d is chosen close to 1 then most of the samples will be deleted whereas if it is selected close to 0 most of the samples will be used for training leading to overtraining. Therefore E_d should be selected in the range [0.9, 0.97].

Class-specific Novelty threshold E_s

This threshold measures the data sample novelty. Spherical potential is in the range [0, 1] with a value closer to 0 indicating novel knowledge and that nearer to 1 representing existing knowledge. If E_S is selected close to 0, then the algorithm does not allow for rule addition while if it is selected to be near 1, then most of the samples will be represented as novel. Therefore E_S should be in the [0.5, 0.7] range.

Rule addition threshold E_a

This is a self-regulatory threshold used to identify and learn training instances with significant knowledge first followed by fine tuning of the classifier parameters with ones that are less significant. Given that misclassification occurs when E^t is greater than 1, if E_a is chosen close to 1 then all the misclassified samples will be used for rule addition whereas if it is chosen close to 2 then fewer samples will be selected for rule addition. We therefore suggest E_a being in the range [1.2, 1.7].

Rule parameter update threshold E_l

When the sample is predicted correctly but the maximum absolute hinge loss error is greater than the self-regulatory parameter update threshold, E_l, the parameters of nearest rule in the same class are updated. If E_l is selected close to 0 then all the samples will qualify this criterion leading to over fitting. On the contrary, if E_l is selected close to 1, then most of the samples will not be selected. To let the samples with higher error rate be first used for tuning we suggest choosing E_l to be in the [0.3, 0.7] range.

False Positive Minimization threshold E_{FP}

E_{FP} is used to determine if a new rule is to be added to minimize false positives or not. A training instance is classified wrongly if it is closer to the nearest rule in opposite class than the actual class. False positive prediction occurs when the sample belongs to the negative class but is predicted as belonging to the positive class. We measure the posterior probability of a predicted FP sample belonging to the positive class c_1. If this is greater than E_{FP}, it suggests that misclassification is happening with a higher degree of confidence for the positive class. Therefore, the sample is selected for rule addition to learn the missing information about the actual class. On the contrary, if the classifier has less confidence in predicting the sample, then it suggests that there exists some knowledge in the network of the sample belonging to the actual negative class c_2. We therefore update the parameters of nearest positive class rule, nrP taking into account its distance from the false positive training sample.

All false positive instances are used for the update of nrP rule parameters. The range of posterior probability is [0, 1]. Misclassification occurs if the posterior probability for the predicted class is greater than or equal to that of the actual class. Assuming an ideal situation, the range for E_{FP} will be [0.4, 1.0]. If E_{FP} close to 0.5 is selected, then all of false positive samples will be used for rule addition resulting in over fitting the false positive instances. Similarly if E_{FP} is chosen close to 1.0, then not many samples may be available for false positive minimization rule addition. In practical scenarios posterior probability may not have such high values. Hence, the suggested E_{FP} range is [0.5, 0.7].

3.4 Performance Analysis of eMcFIS on Benchmark Datasets

In this section we evaluate the performance of the proposed eMcFIS classifier and compare it with other existing algorithms: McFIS [21], batch learning ELM [13] and standard Support Vector Machine [4] in false positive minimization using real-world binary classification benchmark data sets from University of California, Irvine (UCI) machine learning repository [32] and mammogram dataset from mammographic image analysis society digital mammogram database [33].

Benchmark Datasets
To substantially verify the performance of the proposed algorithm we have utilized data sets with small and high number of samples as well as varying dimensional features in binary classification problems. Table 1 shows the detailed specifications of the five datasets used in the study. Heart disease (HEART) is an unbalanced data set while the rest four of the benchmark datasets, PIMA Indian diabetes (PIMA), breast cancer (BC), Mammogram (MAMM) and Ionosphere (ION), are balanced. For performance comparison, we used one class as positive (P) and another as negative (N) to train the eMcFIS classifier.

Simulation Environment
We performed a ten-fold cross validation on all the benchmark data sets in Matlab R2012a environment on a Windows 7 system with 8 GB RAM by maintaining the sample sizes on each of the ten randomly generated cross validation partitions of the data set. The tunable parameters of McFIS, eMcFIS, SVM and ELM are selected using a ten-fold cross validation on training samples. The cost and kernel parameters for SVM were optimized using a grid search. We report the best parameter combination for which the algorithm produces higher validation accuracy results.

Table 1 Specification of benchmark datasets

Data set	Features	Train samples	Train samples (P)	Train samples (N)	No. of testing Samples	Test samples (P)	Test samples (N)
PIMA	8	310	155	155	368	255	113
BC	9	222	111	111	383	255	128
MAMM	9	80	40	40	11	5	6
HEART	13	70	30	40	200	110	90
ION	34	72	36	36	251	161	90

Performance Measures

Performance measures of each algorithm on ten-fold partitions of the data set are measured using class-level and global performance measures as well as incurred false positives. Elements of Confusion matrix Q are used to obtain the performance measures. Class-level performance measured by percentage classification (η_j) is defined as

$$\eta_j = \frac{q_{jj}}{N_j} \times 100\,\% \tag{34}$$

where q_{ij} is the number of correctly classified instances in class j. Global performance measures used in the evaluation are the overall classification accuracy (η_o), average per-class classification accuracy (η_a) and average false positive count (f_{mean}) defined as

$$\eta_a = \frac{1}{n}\sum_{j=1}^{n} j, \eta_o = \frac{\sum_{j=1}^{n} q_{jj}}{N} \times 100\,\%, f_{mean} = \frac{\sum_{f=1}^{10} fp}{10} \tag{35}$$

where fp denotes the number of false positives predictions in the fold f.

Performance Comparison

Table 2 gives the classifier training time, number of rules, mean and standard deviation of testing overall (η_o) and average per-class classification accuracy (η_a) along with false positive count (fp) for the tested algorithms on the benchmark data sets. It should be noted that SVM is implemented in C language and other algorithms are implemented in Matlab environment. From the performance measures reported in Table 2, it can be seen that eMcFIS achieves significant reduction in false positives along with increase in average accuracy as compared to other algorithms in all data sets. On MAMM dataset the algorithm results in zero false positives, with a performance analogous to McFIS. Even on a simple dataset like BC where all classifiers achieve similar performance with overall accuracy greater than 97 % the false positive optimization effect of the proposed eMcFIS can be seen. In addition to this, due to the metacognitive component in eMcFIS, it requires only around 21 % of the total samples in BC to develop the classifier and employs less number of rules to approximate the decision surface as compared to SVM and ELM algorithms.

In case of HEART and ION data sets, eMcFIS reports an overall accuracy comparable to the best-performing McFIS along with around 1 and 5 % improvement over SVM and ELM respectively. Except for PIMA, eMcFIS reports higher overall accuracy over SVM and ELM on all other data sets. For HEART dataset, eMcFIS reported highest overall accuracy of all the classifiers tested. It also had best average per-class classification accuracy on all the datasets. These results show that eMcFIS minimizes false positives while maintaining high generalization accuracy.

328	B.M. Padmanabhuni et al.

Table 2 Performance evaluation on benchmark datasets

Data set	Algorithm	Time (s)	Rules mean	η_o mean	η_o S.D	η_a mean	η_a S.D	f_{mean}	fp S.D
PIMA	SVM	0.006	193.2	76.17	1.64	76.00	1.41	27.6	3.24
	ELM	0.02	34.2	76.77	1.60	76.53	1.92	27.2	5.47
	McFIS	0.64	103.8	**76.79**	1.33	74.98	1.89	33.6	6.95
	eMcFIS	0.96	104.5	75.79	1.91	**77.28**	2.02	**21.3**	5.44
BC	SVM	0.003	64.3	97.21	0.56	97.30	0.74	3.1	1.79
	ELM	0.014	25.5	97.08	0.64	97.10	0.89	3.6	2.32
	McFIS	0.22	12	**97.76**	0.53	97.94	0.65	1.9	1.66
	eMcFIS	0.36	13.6	97.42	0.31	**98.00**	0.24	**0.3**	0.48
MAMM	SVM	0.003	73.6	94.55	4.69	94.33	4.92	0.2	0.42
	ELM	0.005	37.9	94.55	6.36	94.5	6.43	0.3	0.48
	McFIS	0.24	28	**96.36**	4.69	**96.00**	5.16	**0.0**	0.00
	eMcFIS	0.34	27	**96.36**	4.69	**96.00**	5.16	**0.0**	0.00
ION	SVM	0.003	50.2	91.24	2.28	89.05	2.98	16.8	6.55
	ELM	0.004	61.7	86.69	1.60	83.43	1.93	25.3	3.53
	McFIS	0.28	21.7	**92.75**	1.44	91.73	1.98	10.7	4.85
	eMcFIS	0.36	22.8	92.39	1.67	**92.65**	1.97	**5.8**	4.26
Heart	SVM	0.003	55.7	78.95	3.58	79.49	3.09	13.6	5.38
	ELM	0.003	25.7	74.7	4.08	75.09	3.96	18.9	4.91
	McFIS	0.27	26.4	78.8	3.85	78.96	3.64	17.5	4.65
	eMcFIS	0.39	29.1	**79.75**	0.92	**80.44**	0.72	**11.4**	3.41

PIMA and BC data sets were also used for evaluation in [12] which uses two-stage methodology consisting of 1-NN and ELM for false positive minimization. On PIMA dataset, the proposal reported a 3.9 % overall false positive rate (*i.e., false positive count/testing samples*) and 54.69 % overall accuracy along with classifying 25 % of testing samples as unknown. As compared to this, using the information in Tables 1 and 2 it can be calculated that eMcFIS has 5.7 % overall false positive rate and 75.79 % overall accuracy. On BC data set the algorithm reported a 0.5 % overall false positive rate and 93.16 % overall accuracy and classifies 4.2 % testing samples as unknown whereas eMcFIS has 0.08 % overall false positive rate and 97.42 % overall accuracy. It can therefore be inferred that eMcFIS is practical in reducing false positives while balancing coverage without the need for further auditing those classified as unknown.

Statistical Comparison

To statistically compare the false positive optimization performance of the proposed eMcFIS classifier with that of other classifiers on the benchmark datasets, we employed a nonparametric Friedman test followed by the Benforroni-Dunn test as described in [34]. Friedman test compares if the individual experimental mean condition differs significantly from the aggregate mean across all conditions. If the F-score measure is greater than the F-statistic at 95 % confidence level, then the equality of mean hypothesis is rejected, i.e., the hypothesis that the classifiers used in the means comparison perform alike on the benchmarked data sets is rejected. If equality of means hypothesis is rejected by nonparametric Friedman test, then pairwise post hoc test should be performed to identify which classifier algorithm mean is different from that of the other classifiers.

In this chapter we used average false positive count of four classifiers on five different datasets from Table 2 to carry out the described statistical comparison test. The F-score obtained using nonparametric Friedman test is 7.95, which is greater than the F-statistic at 95 % confidence level ($F_{3,12,0.05}$), 3.49. Since 7.95 > 3.49, mean equality hypothesis can be rejected at a confidence level of 95 %. To emphasize the performance significance of eMcFIS classifier we conduct a pairwise comparison of average rank of eMcFIS classifier with that of each of other classifiers. The critical difference is calculated as 1.14 at 95 % confidence level and the proposed eMcFIS classifier is used as control. The average rank of all the four classifiers in false positive optimization can be calculated from Table 2 and are found as eMcFIS: 1.1, McFIS: 2.5, SVM: 2.8 and ELM: 3.6. The difference in average rank between the proposed eMcFIS classifier and the other three classifiers are eMcFIS-SVM: 1.7, eMcFIS-ELM: 2.5, eMcFIS-McFIS: 1.4. The difference between the average rank of eMcFIS and each of the rest of the classifiers is greater than the CD at 95 % confidence level of 1.14. Hence it can be deduced that eMcFIS is significantly better than the rest of the classifiers in reducing false positives with a confidence level of 95 %.

4 Biometric Identification Using eMcFIS

In this section we present practical applications of the proposed eMcFIS for the security sensitive problems of biometric identification namely, fingerprint recognition [35] and forged signature identification [36].

Biometrics are automatically measurable physiological (e.g., fingerprints, facial image) or behavioral characteristics (e.g., voice, signature) that can be used to identify individuals. Biometric measurements obtained from an individual are used to create a reference template. A biometric system uses the created reference templates to automatically identify or verify a person's identity. Fingerprint biometrics use distinctive characteristics of human fingerprint whereas facial image recognition uses facial features such as location and contours of eyes, nose, cheekbones and mouth for identification. Voice based systems use speech recognition techniques while signature based biometric systems use dynamics of person's hand written signature like letter strokes, pressure, and speed of signing for identification. Biometric systems are being used as stand-alone or complimentary means of access control to conventional personal identification systems. Since such biometric identifiers are unique to an individual they are more reliable in verifying a person's identity instead of conventional identifiers like access cards for physical systems or passwords for knowledge-based authentication systems. Due to increasing online and offline fraud, several organizations from financial and banking services to government organizations are looking to biometric systems for enhanced security. We evaluate the performance of eMcFIS using two biometric datasets namely, Signatures dataset [36] and the third fingerprint verification dataset from FVC2004 [35]. A biometric system can be expressed as binary classification problem and solved using machine learning approaches.

4.1 Signature Verification Problem

Signature verification systems analyze the way a user signs his or her name. Signature verification has long been used for verifying the identification of an individual although automatic signature verification is still a new topic. Signature verification systems can be classified as online or offline systems. In offline methods, the signature is digitized using a scanner or camera and is verified by examining the overall or detailed features whereas online methods use digitizing tablet to acquire the signature dynamics in real-time during signing. In the preprocessing phase, techniques based on signal processing are generally employed to improve the acquired input data. These may include removal of noise caused by scanners using median filters, signature size normalization and thinning [37]. Segmentation of signature is another curial step in pre-processing since signatures produced by a signer may differ due to stretching or compression owing to difference in physiological or practical conditions. Segmentation techniques like those using structural analysis, connected components, dynamic time warping (DTW) were proposed in literature. Depending on offline or online signature verification system being utilized, two types of feature extraction mechanisms can be used namely: those using parameters like displacement, contour, slant, shape and ones that express the signature by characterizing it in terms of time function such as position, velocity, pressure, force, direction respectively.

   Signature verification systems can be classified into template matching, statistical or structural techniques [37]. Template matching techniques match a provided sample against known authentic or forged signature templates. When time function based features are used, DTW is most commonly used for template matching. Distance based measures, neural networks and hidden markov models (HMM) are used when employing statistical approaches. Neural network models have been widely used for signature verification due to their learning and generalization capabilities as demonstrated by the proposals using Bayesian [38], multi-layer perceptron (MLP) [39, 40], neuro fuzzy inference networks [41, 42]. In recent times, HMM based verification models [43, 44] are commonly used for signature verification as they are capable of absorbing the variability between signature patterns. SVM, due to its ability to map input vector into a higher dimensional cluster separating hyper plane, has been used in offline and online signature verification [45, 46]. Structural approaches [47, 48] are usually used in combination with other techniques to support signature structure or description match. Apart from these, decision combination schemes like majority voting and weighted averaging solutions [49] were also proposed in literature in addition to multi expert approaches combining complimentary signature verification systems like those using static as well as dynamic features [50] and global and local features [51]. While all these solutions try to improve coverage they do not address the issue of false positives.

Signatures Dataset (SD)

The MYCT SUB-CORPUS-100 dataset used in this study was generated using 100 signers [36]. Each signer produced 25 signatures. After registering the signature, forgers are supplied with the signature images of the clients to be forged. After training the forgers with these genuine signature images, forgers are asked to mimic the natural dynamics of client's signature and 25 forged signatures are collected. Together, there are 2500 (25 × 100) genuine signatures and 2500 (25 × 100) forged signatures. For each signature, 100 features are extracted by fusing the local and global information. The database is then divided into training and testing sets. The training set is composed of first five genuine signatures of each of the 100 signers and skilled forgeries for first 20 signers. The test set comprises of remaining samples of rest 80 signers (i.e., 80 × 20 genuine and 80 × 25 imposter similarity test scores) [52].

For this study we used these training and testing sets comprising of 1000 and 3600 instances respectively. In this study we took forged signature predicted to be original as being a false positive. This is critical, for example, in banking systems where forged signature transaction detected as original might lead to financial loss.

Performance Evaluation on SD

Table 3 reports the performance evaluation of McFIS (the next best classifier in minimizing false positives from statistical comparison results on medical datasets in the previous section) and eMcFIS on Signatures dataset. It can be seen from the results that eMcFIS reports overall accuracy analogous to McFIS, while its average per class accuracy is also comparable to McFIS. The false positives reported by eMcFIS were around 3.4 % (=124/3600) of total samples in contrast to 10.2 % samples reported by McFIS.

4.2 Fingerprint Verification Problem

Given that fingerprints are claimed to be unique to a person and usually remain invariant over time, they are used as biometrics for distinguishing between

Table 3 Performance evaluation on signatures dataset

Algorithm	Time (s)	Rules	η_o	η_a	*fp*	Sample usage (%)
McFIS	3.39	249	75.53	74.77	368	32.50
eMcFIS	2.52	242	75.47	73.18	**124**	30.40

individuals. There are two steps in fingerprint verification system. In the enrollment phase, fingerprints of users are acquired by a scanner to produce a raw digital representation [53] which is further processed by the feature extractor to generate user template that is stored in the biometric database. At the time of verification, the input fingerprint is compared by the feature matcher with stored template to authenticate the identity. Fingerprint is a pattern of interleaved ridges and valleys that often run in parallel and at times bifurcate and terminate. Singularities are regions in fingerprint where ridge lines have a distinct shape. These are termed as global features in fingerprint. Local features are termed as minutiae and refer to ridge discontinuities. Extracted fingerprint features typically have physical features like singularities and minutiae and could also use non-physical features like local orientation image, filter responses etc.

Fingerprint classification approaches can be categorized into rule-based, syntactic, structural, neural network and multiple classifier techniques [54]. Rule-based approaches classify fingerprints based on position and number of singularities, analogous to techniques employed by human experts for manual classification [55]. Syntactic approaches describe the fingerprint patterns using terminal symbols and production rules. Structural approaches use hierarchical organization information (e.g., trees and graphs) for classification [56]. Statistical approaches using Bayes decision rule, SVM [57], k-nearest neighbor algorithms [58] were proposed in the past. Classifiers harnessing the generalization and learning abilities of neural networks were also proposed in literature [59, 60]. Observations that some classifiers misclassify certain patterns but are effective for other patterns triggered multiple classifier based methods making use of complementary information processing capabilities of different classifiers [61].

Fingerprint Verification Competition (FVC) Dataset

A public domain fingerprint database was created for the purpose of evaluation of fingerprint recognition algorithms. Two different competitions namely, open and light were held in FVC2004 competition. We used the open category database in our experiment. In this category the data came from four different databases created using three commercially available scanners and the other using synthetic generator SFinGe. The data collection details can be found in [35]. The database comprises of 100 different fingers with eight impressions per finger. Eight impressions of each finger were matched against each other to generate 2800 genuine matching scores. First impression of each user was also matched against first impression of every other user to generate 4950 imposter matching scores. Together there are 7750 scores for each scanner resulting in total of 31,000 matching scores for all four scanners. The features of the dataset are the scores of 41 competitors of FVC2004. A random two-fold cross validation was performed to divide the FVC data set into training and testing datasets comprising of 15,500 instances each respectively to enable performance comparison of eMcFIS and McFIS classifiers with the results published in [52] and using a two-fold cross validation on the dataset.

In this study we treated imposter predicted as genuine to be false positive instance. Such assumption is relevant for instance, in a defense establishment with

Table 4 Performance evaluation on FVC dataset

Algorithm	Time (s)	Rules	η_o	η_a	f_{mean}	Sample usage (%)
McFIS	23.56	211.5	99.34	99.19	26.0	11.80
eMcFIS	09.88	83.5	99.39	99.23	**19.5**	08.14

biometric identification systems, where an imposter identified as client and given access might lead to security breaches.

Performance Evaluation on FVC Dataset

Table 4 reports the two-fold cross validation performance evaluation of McFIS and eMcFIS on FVC dataset. On this data set both McFIS and eMcFIS achieve an overall and average accuracy over 99 % with eMcFIS having a slightly better accuracy than McFIS. While the number of false positives reported by McFIS is around 0.17 % (=26/15,500) of total test samples, the effect of false positive minimization of eMcFIS can be perceived with its respective 0.13 % performance.

4.3 Performance Comparison

In this section performance of eMcFIS is compared with McFIS and the stand-alone classifier results published in [52]. Equal Error Rate (EER) was used for performance evaluation in FVC2004 [35]. EER measures the error rate when the false acceptance rate (FAR) and the false rejection rate (FRR) assume the same value. It is computed based on classifier prediction scores. Lower EER values are desirable in a biometric system. Along with proposed performance measures described in Sect. 3.4, we used EER in the performance evaluation of biometric data sets. It should however be noted that a lower EER value does not necessarily imply better coverage or lower false positives and nature of application determines the choice of performance measures.

The classifier set up in [52] generated K new training sets from the training data with K taking on values 10 and 25 for SD and FVC data sets respectively. The stand-alone classifier models were built using these modified training sets and then combined using SUM rule for final prediction decision. The performance measures of the classifiers on SD and FVC datasets are presented in Tables 5 and 6 respectively.

Table 5 EER Performance evaluation on signatures dataset

Algorithm	EER (%)
LM	36
LV	37
RV	18
McFIS	25.11
eMcFIS	22.94

Table 6 EER performance evaluation on FVC Dataset

Algorithm	EER (%)		
	FVC$_3$	FVC$_5$	FVC$_7$
LM	1.27	1.75	0.95
LV	0.94	0.73	0.76
RV	0.93	0.74	0.75
McFIS	0.715		
eMcFIS	0.789		

Random Subspace (RS) [62] ensemble was reported to outperform other ensemble methods tested in [52] namely: bagging [63], arcing [64], class switching [65] and decorate [66]. In RS individual classifiers use only a subset of all features for training and testing. RS ensemble in the paper used 50 % of features. The ensemble classifier methods were tested with Levenberg-Marquardt neural network with five hidden units (LM) [67], linear support vector machine (LV) and radial basis support vector machine (RV) as classifier and "Max rule" (MAX) and "Sum rule" (SUM) [68] as decision rules for determining final class from an ensemble of classes.

It can be seen from Table 5 that EER of eMcFIS is better than LM and LV classifiers and is next best to RV with a difference less than 5 %. When compared with ensemble classifier results reported in the paper also, eMcFIS performs significantly better than non-RV based ensemble classifier set-ups. Although the overall and average accuracy measures of McFIS are slightly better than eMcFIS, the EER of McFIS is higher than that of eMcFIS. This is because EER is calculated using classifier scores rather than enumerating correctly classified instances.

We now present performance comparison for FVC dataset. While reporting the FVC results, the authors of [52] combined the scores obtained by i best competitors of 41 used in our study, where $i = 3, 5, 7$. Table 6 shows the EER results reported for stand-alone classifiers in the paper with that of McFIS and eMcFIS.

We can see from the table that McFIS outperforms all the stand-alone classifiers while eMcFIS falls short of McFIS with a difference of 0.074 %. The paper reported that RS_LM was best ensemble classifier with EER of 0.64. The difference between it and eMcFIS is 0.15 % which is comparatively negligible given the resources and costs needed for an ensemble classifier set up. It can be inferred from the results reported in Tables 4 and 6 that although McFIS has slightly better EER than eMcFIS for the FVC dataset, the mean false positive count of eMcFIS is lower than that of McFIS which is highly desirable for the problem.

To summarize, having low false positives is highly critical in signature and fingerprint verification biometric systems due to their security requirements. While there are many biometric classifier algorithm proposals in current literature they do not explicitly address the issue of false positive reduction. eMcFIS with its intrinsic false positive handling and metacognitive self-regulatory learning is appropriate for such security sensitive biometric identification systems as it couples high accuracy with low false positive predictions.

5 Conclusion

Biometric solutions are being progressively deployed in online and offline systems to ensure reliability. False positive optimization is critical in security-sensitive biometric systems as they can result in huge damages. In this chapter, we presented an extended MetaCognitive Fuzzy Inference System (eMcFIS) with low false positive learning for biometric identification. eMcFIS employs a neuro-fuzzy inference system in the cognitive component along with a sequential evolving learning algorithm, where the data arrives sequentially and the network evolves rules to approximate the decision surface. The metacognitive component of eMcFIS decides on what-to-learn, when-to-learn, and how-to-learn the given data. It uses posterior probability as a measure of classifier's confidence, along with self-adaptive learning thresholds and class-specific criterion for low false positive learning. The performance of proposed eMcFIS classifier was evaluated initially using benchmarked medical datasets from UCI machine learning repository and mammogram database. The statistical Friedman test followed by the Benforroni-Dunn test on the benchmarked data sets highlights that the proposed eMcFIS algorithm achieves better performance than other well-known classifiers. The performance of eMcFIS in biometric identification was evaluated using two real-world security sensitive classification problems of signature and print identification. The performance comparison on the biometric datasets with other approaches clearly highlights the advantages of the proposed method in biometric identification with its intrinsic low false positive learning and efficient generalization ability.

Acknowledgement The authors would like to thank L. Nanni for sharing the biometric finger-print and signature verification datasets and the code for equal error rate computation used in performance comparison.

References

1. Sahami, M., Dumais, S., Heckerman, D., Horvitz, E.: A bayesian approach to filtering junk e-mail. AAAI Technical Report (1998)
2. Androutsopoulos, I., Koutsias, J., Chandrinos, K.V., Spyropoulos, C.D.: An experimental comparison of naive Bayesian and keyword-based anti-spam filtering with personal e-mail messages. In: Proceedings of the 23rd Annual International ACM SIGIR Conference on Research and Development in Information Retrieval, pp. 160–167. Athens (2000)
3. Schneider, K.-M.: A comparison of event models for Naive Bayes anti-spam e-mail filtering. In: Proceedings of the 10th Conference on European Chapter of the Association for Computational Linguistics, vol. 1, pp. 307–314. Budapest (2003)
4. Chang, C.-C., Lin, C.-J.: LIBSVM: a library for support vector machines. ACM Trans. Intell. Syst. Technol. 2(3), 1–27 (2011)
5. Kołcz, A., Alspector, J.: SVM-based filtering of e-mail spam with content-specific misclassification costs. In: Proceedings of the Workshop on Text Mining, pp. 1–14. San Jose (2001)

6. Sculley, D., Wachman, G.M.: Relaxed online SVMs for spam filtering. In: Proceedings of the 30th Annual International ACM SIGIR Conference on Research and Development in Information Retrieval, pp. 415–422. Amsterdam (2007)
7. Carreras, X., Marquez, L., Salgado, J.G.: Boosting trees for anti-spam email filtering. In: Proceedings of 4th International Conference on Recent Advances in Natural Language Processing, pp. 58–64. Tzigov Chark (2001)
8. Wu, J., Mullin, M.D., Rehg, J.M.: Linear asymmetric classifier for cascade detectors. In: Proceedings of the 22nd International Conference on Machine Learning, pp. 988–995. Bonn (2005)
9. Yih, W.-T., Goodman, J., Hulten, G.: Learning at low false positive rates. In: Proceedings of the 3rd Conference on Email and Anti-Spam, pp. 1–8. Mountain View (2006)
10. Lynam, T.R., Cormack, G.V., Cheriton, D.R.: On-line spam filter fusion. In: Proceedings of the 29th Annual International ACM SIGIR Conference on Research and Development in Information Retrieval, pp. 123–130. Seattle (2006)
11. Bratko, A., Cormack, G.V., Filipic, B., Lynam, T.R., Zupan, B.: Spam filtering using statistical data compression models. J. Mach. Learn. Res. **7**, 2673–2698 (2006)
12. Akusok, A., Miche, Y., Hegedus, J., Nian, R., Lendasse, A.: A two-stage methodology using k-NN and false-positive minimizing ELM for nominal data classification. Cognit. Comput. **6** (3), 432–445 (2014)
13. Huang, G.-B., Zhou, H., Ding, X., Zhang, R.: Extreme learning machine for regression and multiclass classification. IEEE Trans. Syst. Man Cybern. Part B Cybern. **42**(2), 513–529 (2012)
14. Cherkassky, V.: Fuzzy inference systems: a critical review. In: Kaynak, O., Zadeh, L.A., Türkşen, B., Rudas, I.J. (eds.) Computational Intelligence: Soft Computing and Fuzzy-Neuro Integration with Applications. NATO ASI Series, vol. 162, pp. 177–197. Springer, Heidelberg (1998)
15. Jang, J.-S.R.: ANFIS: adaptive-network-based fuzzy inference system. IEEE Trans. Syst. Man Cybern. **23**(3), 665–685 (1993)
16. Kasabov, N.: Evolving fuzzy neural networks for supervised/unsupervised online knowledge-based learning. IEEE Trans. Syst. Man Cybern. Part B Cybern. **31**(6), 902–918 (2001)
17. Kasabov, N.K., Song, Q.: DENFIS: dynamic evolving neural-fuzzy inference system and its application for time-series prediction. IEEE Trans. Fuzzy Syst. **10**(2), 144–154 (2002)
18. Lin, C.-T., Yeh, C.-M., Liang, S.-F., Chung, J.-F., Kumar, N.: Support-vector-based fuzzy neural network for pattern classification. IEEE Trans. Fuzzy Syst. **14**(1), 31–41 (2006)
19. Juang, C.-F., Chiu, S.-H., Chang, S.-W.: A self-organizing TS-type fuzzy network with support vector learning and its application to classification problems. IEEE Trans. Fuzzy Syst. **15**(5), 998–1008 (2007)
20. Rong, H.-J., Huang, G.-B., Sundararajan, N., Saratchandran, P.: Online sequential fuzzy extreme learning machine for function approximation and classification problems. IEEE Trans. Syst. Man Cybern. Part B Cybern. **39**(4), 1067–1072 (2009)
21. Subramanian, K., Suresh, S., Sundararajan, N.: A Metacognitive neuro-fuzzy inference system (McFIS) for sequential classification problems. IEEE Trans. Fuzzy Syst. **21**(6), 1080–1095 (2013)
22. Isaacson, Y.M., Fujita, F.: Metacognitive knowledge monitoring and self-regulated learning: Academic success and reflections on learning. J. Scholarsh. Teach. Learn. **6**(1), 39–55 (2006)
23. Subramanian, K., Savitha, R., Suresh, S.: A metacognitive complex-valued interval type-2 fuzzy inference system. IEEE Transa. Neural Netw. Learn. Syst. **25**(9), 1659–1672 (2014)
24. Suresh, S., Dong, K., Kim, H.J.: A sequential learning algorithm for self-adaptive resource allocation network classifier. Neurocomputing **73**(16), 3012–3019 (2010)
25. Sateesh Babu, G., Suresh, S.: Meta-cognitive Neural Network for classification problems in a sequential learning framework. Neurocomputing **81**, 86–96 (2012)
26. Subramanian, K., Suresh, S.: Human action recognition using meta-cognitive neuro-fuzzy inference system. Int. J. Neural Syst. **22**(6), 1–15 (2012)

27. Angelov, P.P., Zhou, X.: Evolving fuzzy-rule-based classifiers from data streams. IEEE Trans. Fuzzy Syst. **16**(6), 1462–1475 (2008)
28. Rong, H.-J., Sundararajan, N., Huang, G.-B., Saratchandran, P.: Sequential adaptive fuzzy inference system (SAFIS) for nonlinear system identification and prediction. Fuzzy Sets Syst. **157**(9), 1260–1275 (2006)
29. Zhang, T.: Statistical behavior and consistency of classification methods based on convex risk minimization. Ann. Stat. **32**(1), 56–85 (2004)
30. Suresh, S., Sundararajan, N., Saratchandran, P.: Risk-sensitive loss functions for sparse multi-category classification problems. Inf. Sci. **178**(12), 2621–2638 (2008)
31. Hoffmann, H.: Kernel PCA for novelty detection. Pattern Recognit. **40**(3), 863–874 (2007)
32. Blake, C., Merz, C.: UCI repository of machine learning databases, Department of Information and Computing Sciences, University of California, Irvine, CA, USA. http://archive.ics.uci.edu/ml/ (1998)
33. Suckling, J., Parker, J., Dance, D.R., Astley, S., Hutt, I., Boggis, C.R.M., Ricketts, I., Stamatakis, E., Cerneaz, N., Kok, S.-L., Taylor, P., Betal, D., Savage, J.: The mammographic image analysis society digital mammogram database. In: Exerpta Medica International Congress Series, vol. 1069, pp. 375–378 (1994)
34. Demsar, J.: Statistical comparisons of classifiers over multiple data sets. J. Mach. Learn. Res. **7**, 1–30 (2006)
35. Maio, D., Maltoni, D., Cappelli, R., Wayman, J.L., Jain, A.K.: FVC2004: third fingerprint verification competition. In: Proceedings of the First International Conference on Biometric Authentication, pp. 1–7. Hong Kong (2004)
36. Ortega-Garcia, J., Fierrez-Aguilar, J., Simon, D., Gonzalez, J., Faundez-Zanuy, M., Espinosa, V., Satue, A., Hernaez, I., Igarza, J.-J., Vivaracho, C., Escudero, D., Moro, Q.-I.: MCYT baseline corpus: a bimodal biometric database. IEE Proc. Vis. Image Signal Process. **150**(6), 395–401 (2003)
37. Impedovo, D., Pirlo, G.: Automatic signature verification: the state of the art. IEEE Trans. Syst. Man Cybern. Part C Appl. Rev. **38**(5), 609–635 (2008)
38. Xiao, X., Leedham, G.: Signature verification using a modified Bayesian network. Pattern Recognit. **35**(5), 983–995 (2002)
39. Xiao, X.-H., Leedham, G.: Signature Verification by neural networks with selective attention. Appl. Intell. **11**(2), 213–223 (1999)
40. Al-Shoshan, A.I.: Handwritten signature verification using image invariants and dynamic features. In: International Conference on Computer Graphics, Imaging and Visualisation, pp. 173–176. Sydney (2006)
41. Franke, K., Zhang, Y.-N., Köppen, M.: Static signature verification employing a Kosko-Neuro-fuzzy approach. In: Pal, N.R., Sugeno, M. (eds.) Advances in Soft Computing. Lecture Notes in Computer Science, vol. 2275, pp. 185–190. Springer, Heidelberg (2002)
42. Quek, C., Zhou, R.W.: Antiforgery: a novel pseudo-outer product based fuzzy neural network driven signature verification system. Pattern Recognit. Lett. **23**(14), 1795–1816 (2002)
43. Van, B.L., Garcia-Salicetti, S., Dorizzi, B.: On using the Viterbi path along with HMM likelihood information for online signature verification. IEEE Trans. Syst. Man Cybern. B Cybern. **37**(5), 1237–1247 (2007)
44. Yang, L., Widjaja, B.K., Prasad, R.: Application of hidden Markov models for signature verification. Pattern Recognit. **28**(2), 161–170 (1995)
45. Kholmatov, A., Yanikoglu, B.: Identity authentication using improved online signature verification method. Pattern Recognit. Lett. **26**(15), 2400–2408 (2005)
46. Ferrer, M.A., Alonso, J.B., Travieso, C.M.: Offline geometric parameters for automatic signature verification using fixed-point arithmetic. IEEE Trans. Pattern Anal. Mach. Intell. **27**(6), 993–997 (2005)
47. Han, K., Sethi, I.K.: Handwritten signature retrieval and identification. Pattern Recognit. Lett. **17**(1), 83–90 (1996)
48. Huang, K., Yan, H.: Stability and style-variation modeling for on-line signature verification. Pattern Recognit. **36**(10), 2253–2270 (2003)

49. Bovino, L., Impedovo, S., Pirlo, G., Sarcinella, L.: Multi-expert verification of hand-written signatures. In: Proceedings of Seventh International Conference on Document Analysis and Recognition, pp. 932–936. Edinburgh (2003)
50. Di Lecce, V., Dimauro, G., Guerriero, A., Impedovo, S., Pirlo, G., Salzo, A.: A multi-expert system for dynamic signature verification. In: Kittler, J., Roli, F. (eds.) Multiple Classifier Systems. Lecture Notes in Computer Science, vol. 1857, pp. 320–329. Springer, Heidelberg (2000)
51. Fierrez-Aguilar, J., Nanni, L., Lopez-Peñalba, J., Ortega-Garcia, J., Maltoni, D.: An on-line signature verification system based on fusion of local and global information. In: Kanade, T., Jain, A., Ratha, N.K. (eds.) Audio- and Video-Based Biometric Person Authentication. Lecture Notes in Computer Science, vol. 3546, pp. 523–532. Springer, Heidelberg (2005)
52. Nanni, L., Lumini, A.: An experimental comparison of ensemble of classifiers for biometric data. Neurocomputing 69(13–15), 1670–1673 (2006)
53. Maltoni, D.: A tutorial on fingerprint recognition. In: Tistarelli, M., Bigun, J., Grosso, E. (eds.) Advanced Studies in Biometrics. Lecture Notes in Computer Science, vol. 3161, pp. 43–68. Springer, Heidelberg (2005)
54. Maltoni, D., Maio, D., Jain, A.K., Prabhakar, S.: Fingerprint classification and indexing. Handbook of Fingerprint Recognition, pp. 235–269. Springer, London (2009)
55. Wang, L., Dai, M.: Application of a new type of singular points in fingerprint classification. Pattern Recognit. Lett. 28(13), 1640–1650 (2007)
56. Neuhaus, M., Bunke, H.: A graph matching based approach to fingerprint classification using directional variance. In: Kanade, T., Jain, A., Ratha, N.K. (eds.) Audio- and Video-Based Biometric Person Authentication. Lecture Notes in Computer Science, vol. 3546, pp. 191–200. Springer, Heidelberg (2005)
57. Hong, J.-H., Min, J.-K., Cho, U.-K., Cho, S.-B.: Fingerprint classification using one-vs-all support vector machines dynamically ordered with naive Bayes classifiers. Pattern Recognit. 41(2), 662–671 (2008)
58. Majumdar, A., Ward, R.K.: Fingerprint recognition with curvelet features and fuzzy KNN classifier. In: Proceedings of IASTED Conference on Signal and Image Processing, pp. 243–248. Kailua-Kona (2008)
59. Altun, A.A., Allahverdi, N.: Recognition of fingerprints enhanced by contourlet transform with artificial neural networks. In: 28th International Conference on Information Technology Interfaces, pp. 167–172. Cavtat (2006)
60. Gupta, J.K., Kumar, R.: An efficient ANN based approach for latent fingerprint matching. Int. J. Comput. Appl. 7(10), 18–21 (2010)
61. Yao, Y., Marcialis, G.L., Pontil, M., Frasconi, P., Roli, F.: Combining flat and structured representations for fingerprint classification with recursive neural networks and support vector machines. Pattern Recognit. 36(2), 397–406 (2003)
62. Ho, T.K.: The random subspace method for constructing decision forests. IEEE Trans. Pattern Anal. Mach. Intell. 20(8), 832–844 (1998)
63. Breiman, L.: Bagging predictors. Mach. Learn. 24(2), 123–140 (1996)
64. Bologna, G., Appel, R.D.: A comparison study on protein fold recognition. In: Proceedings of the 9th International Conference on Neural Information Processing, vol. 5, pp. 2492–2496. Singapore (2002)
65. Martinez-Munoz, G., Suárez, A.: Switching class labels to generate classification ensembles. Pattern Recognit. 38(10), 1483–1494 (2005)
66. Melville, P., Mooney, R.J.: Creating diversity in ensembles using artificial data. J. Inf. Fusion Spec. Issue Divers. Multi Classif. Syst. 6(1), 99–111 (2005)
67. Duda, R.O., Hart, P.E., Stork, D.G.: Pattern Classification, 2nd edn. Wiley, New York (2000)
68. Kittler, J., Hatef, M., Duin, R.P.W., Matas, J.: On combining classifiers. IEEE Trans. Pattern Anal. Mach. Intell. 20(3), 226–239 (1998)

Privacy, Security and Convenience: Biometric Encryption for Smartphone-Based Electronic Travel Documents

David Bissessar, Carlisle Adams and Alex Stoianov

Abstract We propose a new paradigm for issuing, storing and verifying travel documents that features entirely digital documents which are bound to the individual by virtue of a privacy–respecting biometrically derived key, and which make use of privacy-respecting digital credentials technology. Currently travel documentation rely either on paper documents or electronic systems requiring connectivity to core servers and databases at the time of verification. If biometrics are used in the traditional way, there are accompanying privacy implications. We present a smartphone-based approach which enables a new kind of biometric checkpoint to be placed at key points throughout the international voyage. These lightweight verification checkpoints would not require storage of biometric information, which can reduce the complexity and risk of implementing these systems from a policy and privacy perspective. Our proposed paradigm promises multiple benefits including increased security in airports, on airlines and at the border, increased traveller convenience, increased biometric privacy, and possibly, lower total cost of system ownership.

Keywords Computational intelligence · Biometrics · Border security · Defense · Privacy · Cryptography · Fuzzy extractor · Pedersen commitment · Hardware protection · Non-transferability · RSA-OAEP encryption · Digital signature

D. Bissessar (✉)
Canada Border Services Agency, Ottawa, Canada
e-mail: david.bissessar@cbsa-asfc.gc.ca

C. Adams
University of Ottawa, Ottawa, Canada
e-mail: cadams@uottawa.ca

A. Stoianov
Office of the Information and Privacy Commissioner of Ontario, Toronto, Canada
e-mail: alexstoianov@aim.com

© Springer International Publishing Switzerland 2016
R. Abielmona et al. (eds.), *Recent Advances in Computational Intelligence in Defense and Security*, Studies in Computational Intelligence 621,
DOI 10.1007/978-3-319-26450-9_13

1 Introduction

This chapter explores applications of computational intelligence and cryptography to biometric privacy in the domain of international travel and border security. Traditionally, paper-based documents have been issued to show authorization to travel. Often these have been secured with biometric images on the printed document, in a supporting database or in an embedded chip on the document itself (as in ICAO e-passports). In this chapter, we present the possibility of representing travel credentials, not as a paper document secured by an actual biometric, but rather, an electronic document secured by a privacy-respecting biometric identifier. The electronic document is stored on the traveller's smartphone and used throughout the traveller's voyage. The biometric identifier is derived in an irreversible manner from the actual biometric using computational intelligence and cryptographic techniques.

We describe the architecture and key algorithms for a notional document: the biometric-enabled electronic Travel Authority (b-TA). We use this notional b-TA as an example, to illustrate a broader class of documents: Electronic Travel Credentials. These Electronic Travel Credentials are characterized by the following:

(1) The document is issued and verified by two separate agencies
(2) The integrity of the data contained within the document must be verifiable from the time of issuance
(3) The document grants a special privilege to the individual to whom it was issued
(4) The document must not be transferable between individuals
(5) The document can be cross-referenced to an overarching traveller record (such as a passport)

Extending the ideas in this paper to a collection of credentials, stored in a secure repository on the smartphone leads to a mobile phone-based passport (an m-passport). We believe this type of credential has the possibility to significantly change the international travel, border security and mobile commerce environments.

In the scenario we put forward, the b-TA is an electronic document, using attribute-based credentials secured to the traveller through a Renewable Biometric Reference (RBR)[1] created from the traveller's fingerprint or other biometrics, and certified by the issuer's digital signature. The b-TA is cross-referenced to the traveller's e-passport. This b-TA can be seen as a privacy preserving biometric

---

[1]Given the range of terms used in the domain, we choose to follow ISO standard 24745 [41]. The concept of a RBR is described in Annex C of the 24745 standard as follows:

Renewable biometric references (RBRs) are revocable/ renewable identifiers that represent an individual or data subject within a certain domain by means of a protected binary identity (re)constructed from a captured biometric sample. A renewable biometric reference does not allow access to the original biometric measurement data, biometric template or true identity of its owner. Furthermore, the renewable biometric reference has no meaning outside the service domain.

variation of the Electronic System for Travel Authorization ESTA [18] required by The United States Customs and Border Protection (CBP) for US Visa-Exempt Countries, or Canada's Electronic Travel Authority (eTA) program under Citizenship and Immigration Canada [27]. The approach we present yields benefit in terms of security, efficiency and privacy. From a security perspective, binding the cryptographic credential to the traveller through the use of biometrics prevents many types of fraudulent activity which can occur. Operational efficiency increases since automation technologies such as kiosks and electronic gates (e-gates) become possible. From the perspective of privacy, the use of RBRs removes the need to store the biometric on a server or even on the smartphone. This can help alleviate public concern over the use of biometrics, and can simplify policy issues associated with privacy impact and personal information banks. The presented approach is scalable to accommodate b-TAs from multiple countries, and can lay foundations for m-passports. The paper presents a general b-TA scenario, elaborates a list of security and engineering requirements, puts forward a system design, and makes an assessment of the proposed system.

2 The b-TA Scenario

This section aims to present the application from a technology neutral perspective to set the context for the technical solution and engineering assessment to occur subsequently. We present the b-TA scenario and discuss its general flow. We introduce the participants and protocols, and present a set of application requirements.

At its simplest, the b-TA scenario consists of two transactions: issuance, in which the traveller applies for and is granted a b-TA, and verification, in which the traveller uses the travel authority to enter the country. We also include an optional pre-departure step in which b-TA validity is confirmed.

Internationally, it has been identified that border security and the international travel experience can benefit if certain steps in traveller processing occur prior to arrival in the country of destination [33]. As shown in Fig. 1, this can be considered as a travel continuum which extends beyond the border of the destination country.

Advanced Passenger Information (API), an approach adopted internationally, provides an example of this. In API, the airline sends an electronic manifest containing information about the passengers that have boarded a flight and will be arriving at the country of destination. This allows the ability to conduct preparation and processing in advance, to reduce border wait-times and to maintain integrity of the borders.

The model proposed in this paper can be used in a manner consistent with this philosophy of "pushing out the border". Here, a privacy-respecting check could be made of the validity and the biometric ownership of the issued document prior to boarding.

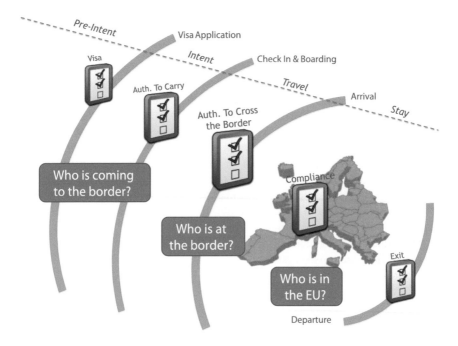

Fig. 1 FRONTEX representation of the layers of intent and information in international travel. The scenario and architecture proposed in this document actually interpose structure and information processing from the earliest pre-departure phases of the travel continuum and allows a progressive update of data throughout

3 Background

Over the last two decades there has been a rapid increase in the uptake of automated biometric systems. Biometrics is now commonly being integrated into a range of large and complex information and communication technology systems and processes. We see the use of biometric systems being implemented throughout the world in areas such as national ID systems, border security control, travel documents, crime prevention, fraud detection, forensics, war zone applications, attendance recording, access control, and financial transactions.

In particular, biometric passports [39] have become quite common. The passport uses contactless smart card technology while the biometric information is stored in a microchip embedded in the passport. While the use of biometrics greatly enhances the security of the travel documents, this also presents a new set of challenges regarding privacy and data safeguarding. Unlike passwords, biometric data are unique, permanent and irrevocable. However, the same technology that serves to threaten or erode privacy may also be enlisted to its protection, thus giving rise to "privacy-enhancing technologies" (PET). This entails the use of Privacy by Design (PbD)—embedding privacy directly into technologies and business practices, resulting in privacy, security and functionality through a "positive-sum" paradigm [51].

There have been a number of technological solutions proposed to address privacy issues in biometrics, such as Biometric Encryption (Fuzzy Extractors), Cancellable Biometrics, Cryptographic Protocols, and hardware protection (e.g., Match-on-Card). For the b-TA application, combining Fuzzy Extractors with Cryptographic Protocols within the secure architecture seems to be the most feasible approach. We also use the hardware protection to locally store the data extracted from the biometrics.

3.1 Biometric Encryption (BE) and Fuzzy Extractors

Biometric Encryption (a.k.a. biometric template protection, biometric cryptosystem, fuzzy extractor, secure sketch, etc.) was proposed as a viable approach [58] to meeting the intent of conventional biometric systems while at the same time addressing the privacy issues. BE is a group of technologies that securely bind a digital key to a biometric or generate a key from the biometric, so that no biometric image or template is stored. It must be computationally difficult to retrieve the key or the biometric from the stored BE data (often called "helper data"). The key can be recreated only if a genuine biometric sample is presented on verification. The output of the BE authentication is either a key (correct or incorrect) or a failure message. See surveys [9, 19, 21, 42, 53] for more details on BE.

Fuzzy Extractor is, in fact, a formal definition of BE introduced by Dodis et al. [31] in terms of two primitives, a fuzzy extractor and a secure sketch. The secure sketch is a helper data stored on enrollment. On verification, the exact reconstruction of the original biometric template is possible when a fresh (i.e., noisy) biometric sample is applied to the secure sketch. The fuzzy extractor is a cryptographic primitive that generates a key from the biometric and the secure sketch, for example, by hashing the reconstructed template. More details on fuzzy extractors are provided in Sect. 5.

At present, the most popular are the following BE schemes: Fuzzy Commitment, QIM, and Fuzzy Vault.

Fuzzy Commitment
Proposed in 1999 [44], this is conceptually the simplest yet the most studied BE scheme. It is applicable to ordered binary biometric templates of a fixed length, n. A k-bit key, where $k \ll n$, is bound to the template by using an Error Correcting Code (ECC). Both the template and an ECC codeword have the same length n. In total, there are 2^k possible keys and ECC codewords, such that each k-bit key directly corresponds to an n-bit codeword. The key is usually generated at random. Then the corresponding codeword is XOR-ed with the binary template to obtain an n-bit string that is stored into the helper data. The key is hashed and its hashed value is also stored either in the helper data or remotely, depending on the application. On

verification, a newly acquired biometric template will not be exactly the same as was obtained during enrollment, which will result in some bit errors. However, if the number of those errors is within the ECC bound, the key can be recovered: the fresh template is XOR-ed with the string stored in the helper data, and the ECC decoder corrects the bit errors. The k-bit key output by the ECC decoder is hashed. If the new hashed value coincides with the enrolled one, the reconstructed key is released. Otherwise, a failure is declared or, alternatively, an incorrect key is output.

Fuzzy Commitment scheme was successfully applied to most biometric modalities (see, for example, [16, 38, 62, 65]).

Quantization Index Modulation (QIM)

QIM is another scheme that is applicable to biometrics with ordered feature vectors, first proposed in [48]. Those feature vectors, b, are continuous. They are quantized using a set of quantizers. In the original version of [48], the quantizers are binary; however, other types can also be used, such as in [20]. Each continuous component of the feature vector has an offset from its quantized value. On enrollment, those offsets are stored into the helper data in the form of a correction vector, v. On verification, a newly acquired feature vector, b', is added to v. The result for each component is decoded to the nearest quantizer (e.g., in the case of binary quantizers [48], to either 1 or 0). In addition, the scheme can also employ an outer ECC to correct the remaining errors. Unlike most other BE schemes, QIM has one or more tunable parameters, the size(s) of the quantization intervals. This allows generating a Receiver Operating Characteristic (ROC) curve, like in conventional biometrics, and, therefore, tuning the security and accuracy of the scheme to the needs of a specific application.

The QIM scheme was successfully applied to face recognition [20, 61].

Fuzzy Vault

Some biometric modalities, such as minutiae-based fingerprint biometrics, do not have a template in the form of an ordered string. In this case another BE scheme, called Fuzzy Vault [43, 50, 64], can be applied. A key that is to be bound to the biometric corresponds to the coefficients of a polynomial that contains the fingerprint minutiae as its points. However, the real minutiae are buried within the set of fake minutiae called chaff points. On enrollment, the full set of points (both real and chaff) is stored into the helper data. On verification, the fresh minutiae template is matched against the full set of points. If a sufficient number of minutiae match in both sets, the correct polynomial and, thus, the key, will be reconstructed.

As already mentioned, Fuzzy Vault is used primarily for the fingerprint minutia-based biometrics.

BE Security Issues

Some BE systems may be vulnerable to low level attacks, when an attacker is familiar with the algorithm and can access the stored helper data (but not a genuine biometric sample). The attacker tries to obtain the key, or at least reduce the search space, and/or to obtain a biometric or create an approximate ("masquerade") version of it.

In recent work, several attacks against BE were identified.[2] An offline False Acceptance attack (FAR attack) is conceptually the simplest. The attacker needs to collect or generate a biometric database of a sufficient size to obtain offline a false acceptance against the BE helper data. The biometric sample (either an image or a template) that has generated the false acceptance will serve as an approximation of the genuine image/template. Since all biometric systems, including BE, have a non-zero False Acceptance Rate (FAR), the size of the offline database (that is required to crack the helper data) will always be finite. The FAR attack, and most other attacks, can be mitigated by applying a feature randomization or permutation (preferably controlled by a user's password, or even not necessarily secret [45]), by using slowdown functions, in a Match-on-Card architecture, by secure architecture with data separation, by performing verification in the encrypted domain (see the next subsection), or by other security measures common in biometrics.

There is no doubt that BE must be made resilient against attacks, and the foregoing works report substantial progress in that direction. However, BE is not solely a cryptographic algorithm and so it is inappropriate to measure its security only by a cryptographic yardstick. As with conventional biometrics, BE security should be viewed as part of the overall secure system design and its security should be assessed in the context of specific applications.

BE Products

GenKey, the Netherlands-based company, offers a broad range of BE products [34]. There are several installations of the GenKey systems for elections and digital healthcare sectors, focusing primarily on the emerging economies. GenKey products were independently tested and showed an accuracy level comparable to non-BE conventional biometrics [8, 16, 35].

The largest and, perhaps, the most notable BE deployment so far is a facial recognition with BE in a watch-list scenario for the self-exclusion program in most of the Ontario gaming sites [20].

3.2 Cryptographic Protocols for Privacy-Preserving Biometric Authentication

This is a group of emerging technologies that have several parts of a biometric system (e.g., sensor, database, and matcher) communicate via secure cryptographic protocols, so that each part learns only minimal information to ensure that the users'

---

[2]The following attacks against BE are known: Inverting the hash; False Acceptance (FAR) attack; Hill Climbing attack [3]; Nearest Impostors attack [59]; Running Error Correcting Code (ECC) in a soft decoding and/or erasure mode [59]; ECC Histogram attack [59]; Non-randomness attack against Fuzzy Vault [22]; Non-randomness attack against Mytec2 and Fuzzy Commitment schemes [59, 66]; Re-usability attack [11, 46, 54]; Blended Substitution attack [54]; and Linkage attack [21, 45, 57].

privacy is protected. Such technologies include Authentication in Encrypted Domain (see the following subsection), Secure Multiparty Computation [15], Cryptographically Secure Filtering [2, 7], and One-to-many Access Control-type System in Match-on-Card Architecture [6].

Authentication in Encrypted Domain

Authentication in the encrypted domain can be performed using homomorphic encryption [5, 14, 56, 63]. Even though a fully homomorphic encryption that supports both addition and multiplication has already been discovered, it is still impractical. Therefore, most proposed solutions deal with partially homomorphic encryption which supports either additions or multiplications but not both. Those solutions work best for binary XOR-based or distance-based biometric classifiers, such as an iris biometric with a 2048-bit binary template. For example, a component-wise Goldwasser-Micali homomorphic encryption was proposed for binary biometric data and for the Fuzzy Commitment scheme in an architecture featuring Client, Database and Matcher [5, 56]. In this scheme the Matcher never obtains biometric data, the database only obtains encrypted biometric data, and the Client never receives the private key, which is only known by the Matcher and is used in the final steps of the hamming distance calculation [56] or the key recovery [5].

Despite the progress that has been achieved in the implementation of homomorphic encryption [4, 32, 55], these schemes are still typically less practical in terms of processing speed and storage requirements than non-homomorphic approaches.

A new approach that combines Biometric Encryption with the well-known Blum-Goldwasser cryptosystem was proposed in [60]. As shown, it is possible to keep the biometric data encrypted during all the stages of storage and authentication. The solution has clear practical advantages over homomorphic encryption and is suitable for two of the most popular BE schemes, Fuzzy Commitment and QIM. A smartphone application based on the scheme of [60] is proposed in [19].

Both homomorphic encryption schemes and the Blum-Goldwasser cryptosystem are malleable; i.e., under some circumstances they are vulnerable to an adaptive chosen ciphertext attack (IND—CCA2).

3.3   Attribute-Based Credentials

We introduce some previous work in the area of credential systems and highlight the applicable approaches to non-transferability and their weaknesses. In 1985, Chaum [24] identified the privacy concerns resulting from the ability of service providers to aggregate electronic records. Chaum presents a pseudonymous system whose security is based on the discrete logarithm problem and blind signatures. Credentials in this system can be copied and transferred. In 1986, Chaum and Evertse [23] generalized the system presented in [12] to accommodate multiple credentials, from different issuers and verifiers. Following this, Damgård [29] and Chen [26] presented alternative approaches, also including a trusted-third party (TTP). In 1998–1999, Canetti et al. [17] proposed a non-transferable anonymous credential scheme which was patented by IBM in 2007 as US Patent #7222362. In Canetti's approach to non-transferability, along with the credential, the user is issued a master-key that is bound to a valuable piece of personal data which the individual is reluctant to share (such as a bank account number). To claim the privilege granted by the credential, the user must prove knowledge of the master key. However since the master key is too valuable to share, no unauthorized users are presumed to know it. As stated by the authors, this approach to non-transferability does not prevent sharing, but rather dissuades it. In 1999, Lysyanskaya et al. [47] presented an anonymous credential system which, similarly to Canetti's approach, dissuades by relying on the user's motivation to preserve a high-value secret. In 2000, Brands [13] presented the digital credentials scheme, a single-show credential system. In this scheme, the credential is provided to the verifier, and the specific attributes used are divulged during the Show protocol. Brands provided a non-transferability approach of embedding a biometric into the credential. This does not provide biometric privacy, since the biometric itself must be used and divulged. In 2001, Camenisch and Lysyanskaya [28] presented "Anonymous Credentials", a credential system using zero-knowledge proofs of knowledge (ZKPoK) to deliver multi-show credentials. These approaches all have similar weaknesses: either a trusted third party is used, or if any non-transferability is present it relies on a disincentive approach which can be circumvented if colluding users have no qualms about sharing secrets.

The protocol proposed in this paper targets the lending problem. Its integration to digital credential and anonymous credentials has been demonstrated in [10]. However, it is general enough to be added to most credential schemes.

4   Our Contribution

Our approach combines Fuzzy Extractors with Cryptographic Protocols within the secure architecture. The BE helper data (i.e., "public data") are stored locally and protected by hardware.

The idea of combining BE with cryptographic Issue and Show protocols was first proposed in [10]. It was shown that the proposed approach preserved the security of the underlying credential system, protected the privacy of the user's biometric, and could be generalized to multiple biometric modalities. In the present paper we extend the approach of [10] to smartphone-based electronic travel documents. The proposal presents the secure architecture of the entire system.

Our proposed architecture includes:

1. A biometrically secured electronic travel authority (b-TA)
2. A Traveller smartphone having an installed user-application and fingerprint sensor
3. A user-application on the cellphone which:

    3.1. Coordinates the b-TA issuance, airline check in, and customs-arrival processes
    3.2. Implements cryptographic protocols for privacy and security
    3.3. Interfaces with the programming model of the fingerprint sensor

4. Distinct touch points, progressive processing, and incremental mitigation of risk throughout the traveller continuum:

    4.1. User convenience upon b-TA application
    4.2. Verification of authority to travel pre-boarding
    4.3. Verification of identity and document integrity at arrival
    4.4. Ability to reconcile entry-exit

The protocol proposed in this paper uses fuzzy extractors to derive cryptographic keys from biometrics. To overcome the fuzzy extractor vulnerabilities (in particular, in multiple use scenarios), we enclose the derived key in a Pedersen commitment and we encrypt the public data using IND-CCA2 encryption.

Our proposed protocol includes the following features:

1. Fuzzy extractors (aka "Biometric Encryption") running on the smartphone to provide privacy protection;
2. Additional layers of encryption: Helper data are further encrypted (by RSA-OAEP [12]) and stored on smartphone; Pedersen commitment is stored by Issuer for future validation of b-TA;
3. The decryption key is not stored on the smartphone;
4. Hardware protection against tampering of the information stored on the smartphone;
5. Verification in a secure environment at a border control kiosk rather than on the smartphone.

The details of the proposed system are presented in the following Sections.

5 Supporting Algorithms

5.1 Secure Sketches and Fuzzy Extractors

A fuzzy extractor is a pair of algorithms (*gen* (...), *rep* (...)) which allow randomness to be extracted from an input string and later reproduced exactly using another input string sufficiently close to the original. Most fuzzy extractor schemes also produce helper data which is created during the initial generation step, and used to assist in reproducing the randomness.

The pair of algorithms can be represented as

$$<P, R> \; = gen \,(b) \,and \,R = rep \,(b',P),$$

where:

P is public data safe for storage, used to assist in the *rep* (...) algorithm,
R is a random string that can be used for cryptographic purposes,
B is an input string, for example a biometric template, and
b is fresh input within a certain similarity distance t from b.

5.2 Pedersen Commitments

The Pedersen commitment [30, 52] allows a sender to create a publically storable commitment on a value which irrefutably binds to the value, and also perfectly hides the value from being derived.

The Pedersen commitment scheme has two protocols $C_s = Commit \,(s, \, r) = g^s h^r$ (*mod p*) and $(s, \, r) = Open \,(C_s)$ where the secret s is a value from Z_q and random value r is uniformly drawn from Z_q. The specification of *mod p* for Pedersen commitments will be subsequently be omitted in this chapter but is implied by context.

Here, Pedersen Commitments are used to commit to the biometrically derived cryptographic key: the "hiding" property preserves privacy of the key, and the "binding" property ensures security (non-transferability).

5.3 Digital Signatures

A digital signature is a mathematical scheme for securing and demonstrating the authenticity of a digital message or document [37]. A valid digital signature gives the recipient assurances that the message was created by a known sender (authentication), and that the message has integrity (was not tampered with). It also supports the non-repudiation property (the sender cannot later deny having sent the message). Digital signatures are used in the protocols we propose to seal the attributes provided by the traveller into an electronic travel document signed by the issuing authority.

5.4   Proofs of Knowledge

A Proof of Knowledge (PoK) is an interactive protocol in which a prover P convinces a verifier V of possession of particular knowledge [36]. In general, a PoK has the properties of completeness and validity. The property of completeness states that if P holds the required knowledge, P will succeed in convincing the verifier V of that fact. The property of validity states that if verifier V accepts the proof, P really knows the required knowledge. An additional property can be added: zero-knowledge, which states that during the protocol, V learns nothing beyond the fact that P holds the required knowledge. A PoK holding completeness, validity, and zero-knowledge is called a Zero Knowledge Proof of Knowledge (ZKPoK).

5.5   Attribute-Based Credentials

Various cryptographic credential systems have been proposed in the literature; however, two of the predominant ones today are digital credentials and anonymous credentials. We discuss these at a high level and attempt to highlight the important similarities and differences.

In general, a cryptographic credential system includes three entities: the individual, the issuer, and the verifier. The individual applies for a credential from the issuer by submitting attributes. The individual receives a signed data package which is shown later to a verifier to claim a privilege.

Anonymous credentials and digital credentials are similar in some ways. The general protocol proceeds as follows:

1. An *issue* protocol, in which:

   (a) User U sends attributes X to Issuing organization I
   (b) I issues credential C to U

2. A *show* protocol, in which:

   (a) U presents C for verification of signature
   (b) U makes a claim involving attributes of X and proves it to the verifier to claim a privilege

However, the schemes differ in some important manners, including that Anonymous credentials explicitly include the concept of unlinkability across transactions, whereas digital credentials do not. In the show protocol of anonymous credentials, U does not show the values of the attributes X to verifier V, but rather uses a ZKPoK. This provides additional privacy in a multiple-show scenario.

6   Approach

We outline an approach for implementing a b-TA held on the traveller's cellphone using RBR and attribute-based credentials. We introduce the main entities involved in the scenario, describe the traveller flows, and then describe algorithms which are significant from the perspectives of security and privacy. We also describe certain special features of the algorithm including the ability to help detect passport fraud prior to boarding (optional), triangular binding of b-TA to a fingerprint RBR and an e-passport, scalability to a paperless "m-passport" application, as well as the possibility for the e-passport to be optional after the first verification. Finally, we briefly discuss how the proposed system addresses privacy and security issues.

6.1   Traveller Flows

Figure 2 captures the main system flows from a traveller's perspective. Each of these three flows, "application submission", "airport check-in", and "arrival at destination", are described in detail below.

TravellerFlow 1—Application Submission: The b-TA Issuance Process
The traveller applies online for a b-TA some time prior to departure. The traveller uses her smartphone to submit data and obtain the b-TA from the issuing organization. The entire application process is coordinated on the smartphone by a software application certified/approved by the issuer and verifier. Once issued, the b-TA is stored on the smartphone to be used during check-in and arrival processes, until it expires.

The b-TA applicant begins the application process using a smartphone equipped with specialized software (the end-user application), a biometric sensor and secure storage capability.

1   *The traveller begins the b-TA application procedure*

The end-user application coordinates user workflow including data collection, generation of an RBR, and communication with the back-end b-TA Issuing system.

The end-user application maintains secure storage which is able to store user data in a reliable manner such that this information is not released to external parties.

2   *End-user application submits b-TA application*

The end-user application provides the interface to facilitate submitting all required pieces for the online b-TA application procedure.

This information includes biographical data, biometric data and information regarding travel intent. The biometric data consists, not of a traditional biometric template or image, but of an RBR which the smartphone generates using the traveller's fingerprint, random data, and publically available group parameters.

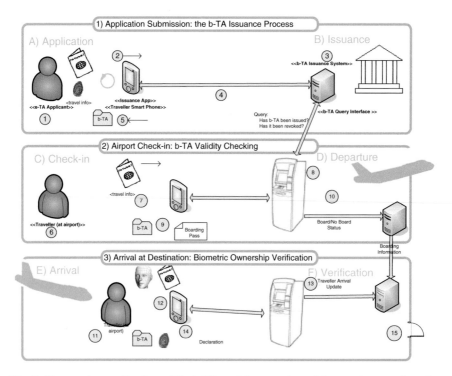

Fig. 2 Three main traveller flows. The b-TA workflow consists of three main steps from the traveller point of view: (*1*) Application Submission: The b-TA Issuance Process; (*2*) Airline Check-in: b-TA Validity Checking; (*3*) Arrival at Destination: Biometric Ownership Verification. Issuance occurs through an online application and approval process. Successful issuance provides the traveller with a digital b-TA which is stored on the smartphone. The validity of this b-TA is checked before boarding, and the biometrical linkage to the traveller is verified upon arrival in Canada. An invalid, revoked or forged b-TA can be identified at either of these steps

The end-user application uses wireless communication to transfer information to the issuance server application upon request of the traveller.

3 Issuer verifies submitted application

The issuer performs an initial check on the submitted information to see if the traveller can be granted a b-TA using the online application service.

The approval process may include queries to other core systems, record keeping, and manual processes. This attribute check occurs offline: the length of the process does not impact the protocol we present.

Upon successful completion of the attribute verification step, the issuer sends a notification to the applicant to proceed to next step: retrieval of the granted b-TA.

4 Retrieval of the b-TA

To retrieve the b-TA, the user agent and the issuance server application enter into a secure protocol in which the issuer signs the b-TA, and the traveller obtains this signed b-TA and any required helper data.

5 b-TA is stored

The signed b-TA and any associated helper data are stored on the cell phone in a secure manner, to be used at a later date (i.e., at the airport to register for boarding).

Traveller Flow 2—Airport Check-in: b-TA Validity Checking

At the departure airport, the traveller goes through the check-in procedures of the airline. These include a step in which the b-TA is verified for validity. This validity checking occurs in a wireless protocol between the end-user application on the traveller's smartphone and the airline kiosk, as well as system-to-system queries between the airline kiosk and the Issuer's b-TA query system. The kiosk verifies that travel is occurring within the allowed b-TA effective dates, that the b-TA has not been tampered with, and that travel privileges have not been revoked. No biometric check is performed at this stage. On successful completion of the check, the traveller can board the flight and proceed to the destination country.

6 The traveller begins the Check-in process

The traveller approaches the kiosk at the airport of origin to check-in and to obtain a boarding pass. The traveller has a smartphone equipped with the b-TA application and an issued b-TA.

7 The traveller submits required check-in information

The traveller enters required information at the airline check-in kiosk. This information can include name, destination, flight and b-TA number. The information can be entered automatically from the smartphone using wireless communication and using an e-passport reader. No biometric verification is performed at this stage: it occurs in the next step upon arrival in the destination country. (To note, a biometric check of fingerprint for b-TA and face for e-Passport could be added here. These would increase the cost of the check-in kiosk.)

8 The check-in kiosk processes submitted information

The airline's kiosk performs a validation of the information submitted.

This includes a verification the b-TA has indeed been correctly issued and has not been revoked. These checks can occur simply using an online query to the issuer system.

The kiosk also verifies that the expiration date has not been reached.

Important note: Passport fraud at boarding time can be detected by adding biometric verification to the airline kiosk. The traveller's face can be compared to the e-passport or a captured fingerprint can be compared against b-TA.

9 *The arrival kiosk provides next step information*

If the check is successful, the kiosk may issue a printed token or an electronic token and the traveller proceeds to "the next step" in the check-in and boarding process. This may be, for example, a security or a baggage processing step or the issuance of a boarding pass.

10 *The traveller proceeds to next step in departure*

The traveller proceeds to "the next step" as deemed by the kiosk, possibly using a token dispensed by the kiosk (such as a boarding pass, for example). The airline also maintains any departure information records as needed.

Traveller Flow 3—Arrival at Destination: Biometric Ownership Verification
At the destination country the traveller disembarks and proceeds to the verification kiosk. The verification kiosk and the smartphone engage in a protocol to verify the user's fingerprint against the fingerprint RBR in the b-TA. For first time travel, the traveller's passport and passport biometric are also verified and cross-verified with b-TA information.

11 *The traveller begins the arrival process*

The traveller arrives at the destination airport. The traveller has a smartphone on which is stored the issued b-TA and supporting data. The interaction at the airport occurs with a biometric kiosk maintained by the destination country's Border Security Agency. The kiosk will verify the authenticity of the credentials and the ownership of the b-TA through a biometric verification.

12 *The traveller provides arrival information*

Upon arrival the traveller initiates a session with the kiosk to provide passport information, b-TA information, and required biometrics to prove ownership of these.

13 *The kiosk processes the information*

The kiosk reads provided e-passport information including traveller name, passport number and the enrolment facial image captured by the issuing country. The kiosk captures another facial image and a fingerprint from the traveller.

The passport face is matched to the travellers face and the fingerprint is used to regenerate the biometric key for verification against the one sealed into the b-TA.

The kiosk may also send a copy of the face image to other systems for screening purposes depending on the requirements of the international agencies involved, and of countries of departure and arrival.

14 *Passenger arrival is recorded*

After having performed all verifications, the stored records are maintained. This can include updates to databases as well as tokens issued to the user.

15 Passenger continues through arrival process

Having cleared the verification of e-passport and b-TA, the Traveller can proceed to other arrival processes.

6.2 Key Algorithms

Algorithm: Issuance-Time RBR Capture

The traveller's smartphone is responsible for capturing the imprint of the traveller's fingerprint generating the RBR which will subsequently be submitted to the issuer to be sealed into the b-TA. This can be achieved using any RBR generation mechanism. We illustrate it in Fig. 3 using a variation of the scheme presented in [10].

The RBR is generated on the smartphone using the fuzzy extractor indistinguishability adapter (FE_{Ind}) configured in issue mode, as described in [10]. We briefly describe the mechanism here, but defer the reader to the primary literature for a full description.

First, the smartphone captures the fingerprint impression and produces biometric template b_I. This template is passed to the *gen* (...) method of the FE_{Ind} which generates the fuzzy extractor tuple $<R, P_e>$ which are obtained by encrypting the public data P obtained from the underlying fuzzy extractor with the encryption key k. The cryptographic key R obtained from the fuzzy extractor is combined with a random value r_I to produce the renewable biometric reference CR_I, a Pedersen commitment on those values.

The values of P_e, and CR_I are retained on the smartphone. The value of CR_I will be sent to the issuer as a RBR.

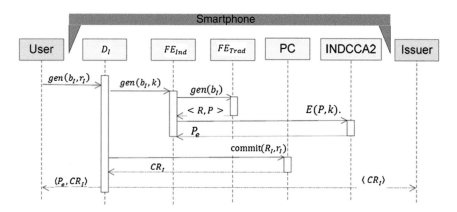

Fig. 3 Generation of the Issue-time Renewable Biometric Reference. The RBR is generated on the smartphone using the sensor and custom software embedded within it. The user supplies the biometric b_I, and accompanying random data r_I, the system provides the RBR CR_I, and accompanying public data P_e. The RBR is also provided to the issuer, who will eventually seal it into the b-TA in the issue protocol

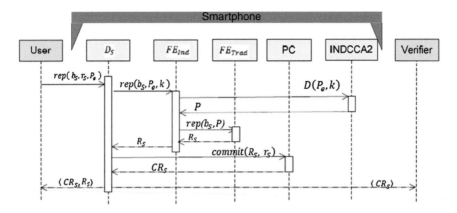

Fig. 4 Generation of the verification of RBR. Upon arrival, a fresh biometric is captured, using the kiosk. The kiosk is equipped with show time device functionality as defined in [9]

Algorithm: Generation of Verification-Time RBR

On arrival at the airport of destination, the traveller's ownership of the b-TA must be established. The first step in this process is the creation CR_S: the verification-time instance of the RBR (Fig. 4).

This algorithm creates a Pedersen Commitment on the Fuzzy Extractor R value. Here, however, rather than the biometric capture and RBR generation being performed on the smartphone, it is performed on the verification kiosk.

To create CR_S, the traveller initiates communication with the kiosk which allows P_e and r_S to be transferred. The traveller also supplies a fingerprint imprint to the kiosk. The kiosk then generates the CR_S and securely disposes of the fingerprint.

First, the smartphone and the kiosk enter into communication. The user then submits a fingerprint impression and produces a fresh biometric template b_S. This template is passed to the rep (...) method of the FE_{Ind} which regenerates a biometric key R_S. This R_S is sealed in a Pedersen Commitment CR_S which then becomes the verification time RBR [10].

CR_S and R_S are provided to the traveller to use at subsequent steps in the verification process.

Algorithm: Show Protocol

After the RBR has been regenerated by the arrival kiosk, the final step in the workflow requires verification of the b-TA, the e-passport, ownership thereof, and the claimed travel privilege. This requires the verifier to become convinced that:

1. The b-TA data package has not been tampered with,
2. The RBR sealed into the b-TA and regenerated in the previous step refer to the same biometric,
3. The e-passport number within the b-TA corresponds to that of the passport held by the traveller,

4. The traveller's face and the face image on the e-passport match, and
5. The traveller claim of privilege is valid.

The verifier checks the digital signature to verify that the digital package has not been tampered with. This signature verification protocol is defined by the underlying credential scheme. In general it is a function of the credential itself, and the issuer's public key. The public key may be installed on the kiosk itself, or can be accessed through an online connection.

Once the verification relation has been checked, the traveller proves ownership of b-TA. This is done by proving that the RBR within the b-TA and the RBR generated by the kiosk on arrival are derived from the biometric of the same individual. This is achieved using a ZKPoK *DLRep WithPC* which verified that CR_I and CR_S are commitments on the same derived key R_S. A detailed description of the protocol *DLRep WithPC* is available in [1, 9, 10].

Following proof of biometric ownership of the b-TA, the smartphone and the kiosk engage in a check that the b-TA allows entry into the country, that it is within appropriate entry dates, and it has not been revoked. This can be proven using a combination of the statement proof mechanism of the underlying credential system [13, 28] and online revocation checks with the issuing authority.

6.3 Triangular Biometric Bindings

Figure 5 illustrates how the b-TA binds to the traveller's RBR, and the traveller's e-passport. These relations enable security to be linked to an identity, and offer the potential for the b-TA to replace the e-passport after the initial verification.

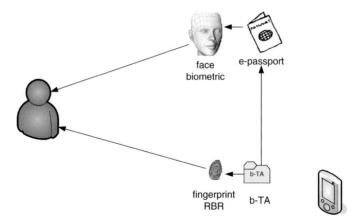

face
biometric   e-passport

fingerprint   b-TA
RBR

Fig. 5 Triangular bindings between issued documents. The b-TA is bound to the e-passport number as well as the traveller's fingerprint RBR in a tamper-proof manner. This results in two biometric linkages which provide strength of security but also enables operational efficiency and traveller convenience in terms of streamlining second passages and online application. The kiosk at arrival is responsible for both facial recognition and fingerprint RBR verification

The e-travel authority is cryptographically bound to the passport number as well as the traveller's RBR. The traveller's identity is thus bound to both the e-passport and the b-TA. The b-TA is bound to the passport number and to a RBR created using the traveller's fingerprint. The traveller's identity is bound to both the e-passport and the b-TA. The e-passport binding occurs through the issuing country's adherence to ICAO e-passport standards. While these standards are indeed quite high, and demanding, there is variability in the processes and documents issued within the international community. A domestic Government has no control over the processes, checks and balances followed by international issuers. The b-TA is bound per agreement between the issuer and the verifier.

Four important lifecycle steps in identity binding and verification are required:

0. The e-passport is issued

    Before anything occurs in the b-TA steps, the passport must be issued. The foreign country issues an e-passport to the individual, a foreign national. At the point of e-passport issuance, any number of procedures is followed including background checks and updates to the civil registry. Different countries will have different processes. The e-passport is produced in accordance with ICAO e-passport standards and includes an electronic file with the individual's passport image.

1. The b-TA is issued

    The applicant's submits informational attributes (which include at least an e-passport number and RBR) to the issuing organization. The issuer performs a background check, and issues the b-TA. The b-TA is bound to the attributes in a manner which prevents the attributes from being changed without invalidating the digital signature. At this point, the traveller's identity and ownership of biometric is assumed, but has not yet been verified.

2. First arrival:

    The first time the traveller arrives to visit the country that issued the b-TA, the biometric linkage between b-TA has not yet been verified. On first arrival, the kiosk verifies the e-Passport biometric, the b-TA biometric and the passport number contained in the b-TA. On successful verification, the verifier may record the passage in a data base, and issue a "verified entry" credential to be retained by the traveller to simplify further processing, on this trip and on subsequent trips.

3. Subsequent arrivals:

    After successful verification on the first arrival, the processing of subsequent arrivals may be streamlined. If the verifier chooses to issue "verified entry" credentials for first arrival, this can be checked on subsequent arrivals potentially making the e-passport verification optional.

6.4 m-Passports: An Electronic Wallet Approach for b-TA

What about the scenario where a traveller holds b-TAs for multiple countries? We have presented an application which manages the workflow necessary to obtain, view, modify and use a b-TA for a single destination country. This can be extended to a multi-nation scenario (Fig. 6). While beyond the scope of this study to do so, we provide a sketch to highlight the potential. We name this functionality m-passport, to mean a mobile paperless passport, stored on a smartphone.

We may consider the scenario where a home country issues an electronic identity document, and where multiple nations each issue their own b-TA to travellers and the traveller thus has a collection of b-TAs to manage. This leads to a more sophisticated user agent application which allows the traveller to store many such secured travel documents: we generically refer to this as an m-passport. This m-passport would be analogous to a passport in that it would be certified as belonging to the given traveller, issued by the country of citizenship, and a cumulative record of travel authorities and travel passages.

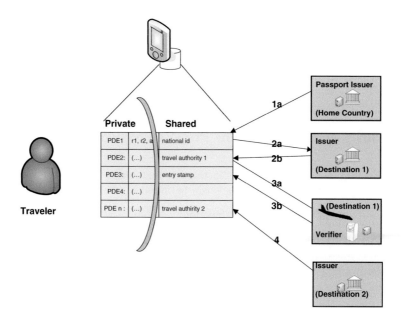

Fig. 6 m-passport schematic. The credential-based solution can be extended to multiple nations, where a national identity block is issued by the country of citizenship, travel privileges extended and verified by international destinations—and records of entry and exit are annotated. The user portion of the application would reside on the cellphone, and be akin to a mobile passport, "m-passport"

In order to make this, the following are required:

1. A more sophisticated user application is required
2. A smartphone construct allowing secure and selective read and write permissions
3. A cryptographic credential corresponding to an electronic citizenship card
4. Selective show and an ability to show across credentials issued by multiple issuers
5. Standards within and across Issuer and Verifier communities

Currently, as internet commerce reaches high levels of technical maturity, there are a number of market offerings which may lend themselves to this kind of application, including Microsoft's UProve [49], or IBM's Idemix [40], which implement the base credential schemes described in this document. Within the literature, one can refer to Chaum and Pedersen's 1992 paper [25] which sets foundations for secure wallets.

While this should be the subject of future study, there is no evident requirement for all countries to use the same attributes for the issuance and verification of b-TAs. It seems that the responsibility for the definition of these fields may be left with the issuance and verification bodies of each nation.

National certificates of identity, on the other hand, correspond to the personal identification block on a traditional e-Passport. As these will need to be read by multiple nations, it seems standardization will be required. These standards may be aligned with ICAO specifications [39].

7 Security and Privacy of the Proposed System

In this Section, we briefly outline the features that make the proposed system resilient to known attacks. More details can be found in [9, 10].

1. Resilience to data insertion and tampering:

   - RSA-OAEP encryption of the information stored on the smartphone;
   - No cryptographic keys are stored on the smartphone;
   - Hardware protection of the stored information.

2. Non-transferability:

   - Biometric enrollment and verification are performed during the Issue and Show protocols respectively;
   - The unforgeability of signatures in digital credentials;
   - The intractability of the discrete log problem.

3. No substitution or blended substitution attacks:

   - RSA-OAEP encryption of the information stored on the smartphone;
   - Hardware protection of the stored information;
   - The attacker does not know the RBR that was generated by the Fuzzy Extractor: the RBR is hidden in the Pedersen Commitment and sealed by the issuer's signature.

4. No storage or leakage of raw biometric or extracted feature set (biometric template):

   - Both raw biometric and extracted feature set are discarded at the end of the Issue process;
   - It is computationally difficult to obtain the feature set from the public data, P, that are output by the Fuzzy Extractor;
   - P is additionally encrypted by RSA-OAEP;
   - The proof of knowledge in the Show protocol keeps R a secret;
   - The attacker does not know the RBR that was generated by the Fuzzy Extractor: the RBR is hidden in the Pedersen commitment and sealed by the issuer's signature;
   - Hardware protection of the stored information;
   - During the Show process, a fresh biometric sample is obtained and verified at the kiosk, not on the smartphone.

5. Resilience to biometric spoofing and replay attacks:

   - Spoofing and replay attacks would be detected at the time of verification, which in our scenario happens in a supervised and controlled environment (in a customs-controlled area);
   - Liveness detection at the kiosk.

6. Resilience to False Acceptance attack:

   - RSA-OAEP encryption of the information stored on the smartphone;
   - Hardware protection of the stored information;
   - The attacker does not have any indication of gradual success for the attack.

7. Resilience to score-based attacks (e.g., Hill Climbing or Nearest Impostors), attacks on Error Correcting Code, and Non-randomness attacks:

   - RSA-OAEP encryption of the information stored on the smartphone;
   - Hardware protection of the stored information;
   - The attacker is not given any biometric matching score and cannot derive any such score;
   - The attacker does not have any indication of gradual success for the attack.

8. Resilience to Re-usability (aka record multiplicity) attack:

   - The RBR scheme used outputs indistinguishable data. Specifically, the encrypted public data from the fuzzy extractor are indistinguishable by the properties of RSA-OAEP. The RBR are the biometric key commitments and these are indistinguishable due to the properties of Pedersen Commitments.

9. Resilience to the database linkage attack:

   - The cryptographic key R obtained from the fuzzy extractor is different for each application, even for the same user;
   - There is no possibility for cross-database linkage based on the RBR, given computational difficulty of cracking Pedersen Commitments and their indistinguishability.

Overall, the proposed system design appears to be resilient to known attacks against biometric systems in general and fuzzy extractors in particular.

8 Conclusions

We have described a promising application of privacy-enhancing techniques as applied to biometrics in the context of a secure electronic document suitable for use as a mobile biometric-enabled electronic travel authority (b-TA) for foreign passport holders wishing to visit Canada. We have developed this scenario in detail, identifying the participating entities, transactions, and data interchanges required over the lifecycle of the document from issuance through validation, usage, and on to expiry.

As a result, all known attacks against fuzzy extractors appear to be thwarted. Given the framework, tools, and approach selected, as well as the Privacy-by-Design methodology incorporated, we expect this architecture to enhance system security, as well as the privacy and convenience of international travellers.

Acknowledgments Special thanks to the CBSA for ongoing support. Financial support from the Canadian Safety and Security Program (CSSP) of Defence Research and Development Canada (DRDC), and the Natural Sciences and Engineering Research Council of Canada (NSERC) is gratefully acknowledged.

Dedication

Daniel Patrick Bissessar

March 11, 2007–Jan 1, 2012

To perseverance and making a difference... Danny, this work that we started together is growing... You continue to inspire me every day to build stuff and make things. You taught me happiness and have enriched my life forever. Love, Papa.

References

1. Adams, C.: Achieving non-transferability in credential systems using hidden biometrics. Secur. Commun. Netw. **4**(2), 195–206 (2011)
2. Adjedj, M., Bringer, J., Chabanne, H., Kindarji, B.: Biometric identification over encrypted data made feasible. In: Prakash, A., Gupta, I.S. (eds.) Information Systems Security. LNCS, vol. 5905, pp. 86–100. Springer, Heidelberg (2009)
3. Adler, A.: Vulnerabilities in biometric encryption systems. In: Kanade, T., Jain, A., Ratha, N.K. (eds.) Audio- and Video-Based Biometric Person Authentication. 5th International Conference, AVBPA 2005, Hilton Rye Town, NY, USA, 20–22 July 2005. LNCS, vol. 3546, pp. 1100–1109. Springer, Heidelberg (2005)
4. Barni, M., Bianchi, T., Catalano, D., Raimondo, M.D., Labati, R.D., Failla, P., Fiore, D., Lazzeretti, R., Piuri, V., Scotti, F., Piva, A.: Privacy-preserving fingercode authentication. In:

Proceedings of the 12th ACM Workshop on Multimedia and Security (MMSec 2010), pp. 231–240. ACM, New York (2010)

5. Bringer, J., Chabanne, H.: An authentication protocol with encrypted biometric data. In: Vaudenay, S. (ed.) Progress in Cryptology—AFRICACRYPT 2008. LNCS, vol. 5023, pp. 109–124. Springer, Heidelberg (2008)

6. Bringer, J., Chabanne, H.: Two efficient architectures for handling biometric data while taking care of their privacy. In: Campisi, P. (ed.) Security and Privacy in Biometrics, Chapter 11, pp. 275–295. Springer, London (2013)

7. Bringer, J., Chabanne, H., Kindarji, B.: Error-tolerant searchable encryption. In: IEEE International Conference on Communications, 2009. ICC 2009, pp. 1–6 (2009)

8. Bissessar, D., Gorodnichy, D.O., Stoianov, A., Thieme, M.: Assessment of privacy enhancing technologies for biometrics. In: IEEE Symposium on Computational Intelligence for Security and Defense Applications (CISDA), pp. 1–9. Ottawa, ON, Canada, 11–13 July 2012

9. Bissessar, D.: Cryptographic credentials with privacy-preserving biometric bindings. Master's thesis, School of Electrical Engineering and Computer Science, University of Ottawa (2013)

10. Bissessar, D., Adams, C., Liu, D.: Using biometric key commitments to prevent unauthorized lending of cryptographic credentials. In: 12th International Conference on Privacy, Security and Trust (PST2014), pp. 75–83. Toronto, Canada, 23–24 July (2014)

11. Boyen, X.: Reusable cryptographic fuzzy extractors. In: Proceedings of the 11th ACM Conference on Computer and Communications Security. ACM (2004)

12. Bellare, M., Rogaway, P.: Optimal asymmetric encryption—how to encrypt with RSA. In: De Santis, A. (ed.) EUROCRYPT 1994. LNCS, vol. 950, pp. 341–358. Springer, Heidelberg (1995)

13. Brands, S.A.: Rethinking Public Key Infrastructures and Digital Certificates. MIT Press, Cambridge (2000)

14. Bringer, J., Chabanne, H., Izabachène, M., Pointcheval, D., Tang, Q., Zimmer, S.: An application of the Goldwasser-Micali cryptosystem to biometric authentication. In: Information Security and Privacy. LNCS, vol. 4586, pp. 96–106. Springer, Heidelberg (2007)

15. Bringer, J., Favre, M., Chabanne, H., Patey, A.: Faster secure computation for biometric identification using filtering. In: The 5th IAPR International Conference on Biometrics. ICB, pp. 257–264. New Delhi, India, 29 March–1 April 2012

16. Bundesamt für Sicherheit in der Informationstechnik. Study of the Privacy and Accuracy of the Fuzzy Commitment Scheme. BioKeyS III-Final Report (2011)

17. Canetti, R., Charikar, M.S., Rajagopalan, S., Ravikumar, S., Sahai, A., Tomkins, A.S.: Nontransferable anonymous credentials. U.S. Patent 7,222,362 (2007)

18. CBP ESTA Webpage. http://www.cbp.gov/travel/international-visitors/esta. Accessed 19 July 2015

19. Cavoukian, A., Chibba, M., Stoianov, A.: Advances in biometric encryption: taking privacy by design from academic research to deployment. Rev. Policy Res. 29(1), 37–61 (2012)

20. Cavoukian, A., Marinelli, T., Stoianov, A., Martin, K., Plataniotis, K.N., Chibba, M., DeSouza, L., Frederiksen, S.: Biometric encryption: creating a privacy-preserving 'Watch-List' facial recognition system. In: Campisi, P. (ed.) Security and Privacy in Biometrics, Chapter 9, pp. 215–238. Springer, London (2013)

21. Cavoukian, A., Stoianov, A.: Biometric encryption: the new breed of untraceable biometrics. In: Boulgouris, N.V., Plataniotis, K.N., Micheli-Tzanakou, E. (eds.) Biometrics: Theory, Methods, and Applications, Chapter 26, pp. 655–718. Wiley, Hoboken (2009)

22. Chang, E.C., Shen, R., Teo, F.W. Finding the original point set hidden among chaff. In: Proceedings of the 2006 ACM Symposium on Information, computer and communications security, pp. 182–188. ACM (2006)

23. Chaum, D., Evertse, J.-H.: A secure and privacy-protecting protocol for transmitting personal information between organizations. In: Odlyzko, A.M. (ed.) CRYPTO 1986. LNCS, vol. 263, pp. 118–167. Springer, Heidelberg (1987)

24. Chaum, D.: Security without identification: transaction systems to make big brother obsolete. Commun. ACM 28(10), 1030–1044 (1985)

25. Chaum, D., Pedersen, T.P.: Wallet databases with observers. In: Advances in Cryptology—CRYPTO 1992. LNCS, vol. 740, pp. 89–105. Springer, Heidelberg (1993)
26. Chen, L.: Access with pseudonyms. In: Cryptography: Policy and Algorithms, pp. 232–243. Springer, Heidelberg (1996)
27. Citizenship and Immigration Canada. eTAWebpage. http://www.cic.gc.ca/english/department/acts-regulations/forward-regulatory-plan/eta.asp. Accessed 19 July 2015
28. Camenisch, J., Lysyanskaya, A.: An efficient system for non-transferable anonymous credentials with optional anonymity revocation. In: Pfitzmann, B. (ed.) Advances in Cryptology—EUROCRYPT 2001. LNCS, vol. 2045, pp. 93–118. Springer, Heidelberg (2001)
29. Damgård, I.: Payment systems and credential mechanisms with provable security against abuse by individuals. In: Advances in Cryptology—CRYPTO88. LNCS, vol. 403, pp. 328–335. Springer, Heidelberg (1990)
30. Damgård, I.: Commitment schemes and zero-knowledge protocols. In: Damgård, I.B. (ed.) Lectures on Data Security. LNCS, vol. 1561, pp. 63–86 (1999)
31. Dodis, Y., Reyzin, L., Smith, A.: Fuzzy extractors: how to generate strong keys from biometrics and other noisy data. In: Cachin, C., Camenisch, J.L. (eds.) Advances in cryptology—Eurocrypt 2004. LNCS, vol. 3027, pp. 523–540. Springer, Heidelberg (2004)
32. Erkin, Z., Franz, M., Guajardo, J., Katzenbeisser, S., Lagendijk, I., Toft, T.: Privacy-preserving face recognition. In: PETS 2009: Proceedings of the 9th International Symposium on Privacy Enhancing Technologies, Seattle, WA, USA, 5–7 Aug 2009. LNCS, vol. 5672, pp. 235–253. Springer, Heidelberg (2009)
33. Frontex: development of capabilities for passenger analysis units. In: Operational Heads of Airports Conference 2014, Warsaw, 04–07 Feb 2014
34. www.genkey.com
35. www.genkey.com/en/news-archive/genkey-releases-biofinger-sdk
36. Goldreich, O.: Foundations of Cryptography: Basic Tools, vol. 1. Cambridge University Press, New York (2001)
37. Goldreich, O.: Foundations of Cryptography: Basic Applications, vol. 2. Cambridge University Press, New York (2004)
38. Hao, F., Anderson, R., Daugman, J.: Combining crypto with biometrics effectively. IEEE Trans. Comput. **55**(9), 1081–1088 (2006)
39. International Civil Aviation Organization: machine readable travel documents—part 1–2. Technical report. ICAO Document 9303 (2006)
40. IBM Identity Governance web page. http://www.zurich.ibm.com/security/idemix/. Accessed 19 July 2015
41. ISO/IEC IS 24745: Information Technology—Security techniques—Biometric Information Protection, June 2011
42. Jain, A.K., Nandakumar, K., Nagar, A.: Biometric template security. EURASIP J. Adv. Sig. Process. pp. 1–17 (2008). Article ID 579416
43. Juels, A., Sudan, M.: A fuzzy vault scheme. In: IEEE International Symposium on Information Theory (2002)
44. Juels, A., Wattenberg, M.: A fuzzy commitment scheme. In: Proceedings of the 6th ACM Conference on Computer and Communications Security, pp. 28–36. ACM (1999)
45. Kelkboom, E.J.C., Breebaart, J., Kevenaar, T.A.M., Buhan, I., Veldhuis, R.N.J.: Preventing the decodability attack based cross-matching in a fuzzy commitment scheme. IEEE Trans. Inf. Forensics Secur. **6**(1), 107–121 (2010)
46. Kholmatov, A., Yanikoglu, B.: Realization of correlation attack against fuzzy vault scheme. In: Proceedings of SPIE, vol. 6819, pp. 681900-1–681900-7 (2008)
47. Lysyanskaya, A., Rivest, A., Sahai, A., Wolf, S.: Pseudonym systems. In: Heys, H., Adams, C. (eds.) Selected Areas in Cryptography, pp. 184–199. Springer, Heidelberg (2000)
48. Linnartz, J.-P., Tuyls, P.: New shielding functions to enhance privacy and prevent misuse of biometric templates. In: 4th International Conference on Audio and Video Based Biometric Person Authentication, pp. 393–402. Guildford, UK (2003)

49. Microsoft research U-Prove web page. http://research.microsoft.com/en-us/projects/u-prove/. Accessed 20 July 2015
50. Nagar, A., Nandakumar, K., Jain, A.K.: Securing fingerprint template: fuzzy vault with minutiae descriptors. In: 19th International Conference on Pattern Recognition, ICPR 2008, pp. 1–4. IEEE (2008)
51. Privacy by Design Resolution of the 32nd International Conference of Data Protection and Privacy Commissioners, Jerusalem, 27–29 Oct 2010. http://www.ipc.on.ca/site_documents/pbd-resolution.pdf
52. Pedersen, T.: Non-interactive and information-theoretic secure verifiable secret sharing. In: Advances in Cryptology—CRYPTO 1991. LNCS, vol. 576, pp. 129–140. Springer, Heidelberg (1992)
53. Rathgeb, C., Uhl, A.: A survey on biometric cryptosystems and cancelable biometrics. EURASIP J. Inf. Secur. **2011**(3), 1–25 (2011)
54. Scheirer, W.J., Boult, T.E.: Cracking fuzzy vaults and biometric encryption. In: Biometric Consortium Conference, Baltimore. IEEE, Sept 2007
55. Sadeghi, A., Schneider, T., Wehrenberg, I.: Efficient privacy preserving face recognition. In: Lee, D., Hong, S. (eds.) ICISC 2009 Proceedings of the 12th Annual International Conference on Information Security and Cryptology. LNCS, vol. 5984, pp. 235–253 Springer, Heidelberg (2009)
56. Schoenmakers, B., Tuyls, P.: Computationally secure authentication with noisy data. In: Tuyls, P., Škorić, B., Kevenaar, T. (eds.) Security with Noisy Data: Private Biometrics, Secure Key Storage and Anti-Counterfeiting, pp. 141–149. Springer, London (2007)
57. Simoens, K., Tuyls, P., Preneel, B.: Privacy weaknesses in biometric sketches. In: 30th IEEE Symposium on Security and Privacy, pp. 188–203. IEEE (2009)
58. Soutar, C., Roberge, D., Stoianov, A., Gilroy, R., Vijaya Kumar, B.V.K.: Biometric encryption using image processing. In: Optical Security and Counterfeit Deterrence Techniques II, 1 Apr 1998. Proceedings of SPIE, vol. 3314, pp. 178–188 (1998)
59. Stoianov, A., Kevenaar, T., Van der Veen, M.: Security issues of biometric encryption. In: Science and Technology for Humanity (TIC-STH), 2009 IEEE Toronto International Conference. IEEE (2009)
60. Stoianov, A.: Cryptographically secure biometrics. In: SPIE Defense, Security, and Sensing. Proceedings of SPIE, vol. 7667, pp. 76670C-1–76670C-12 (2010)
61. Sutcu, Y., Li, Q., Memon, N.: Design and analysis of fuzzy extractors for faces. In: Optics and Photonics in Global Homeland Security V and Biometric Technology for Human Identification VI, 73061X, 5 May 2009. Proceedings of SPIE, vol. 7306 (2009)
62. Tuyls, P., Akkermans, A.H.M., Kevenaar, T.A.M., Schrijen, G.-J., Bazen, A.M., and Veldhuis, R.N.J.: Practical biometric authentication with template protection. In: Kanade, T., Jain, A., Ratha, N.K. (eds.) Audio- and Video-Based Biometric Person Authentication. 5th International Conference, AVBPA 2005, Hilton Rye Town, NY, USA, 20–22 July 2005. LNCS, vol. 3546, pp. 436–446. Springer, Heidelberg (2005)
63. Upmanyu, M., Namboodiri, A.M., Srinathan, K., Jawahar, C.V.: Blind authentication: a secure crypto-biometric verification protocol. IEEE Trans. Inf. Forensics Secur. **5**(2), 255–268 (2010)
64. Uludag, U., Pankanti, S., Jain, A.K.: Fuzzy vault for fingerprints. In: Kanade, T., Jain, A., Ratha, N.K. (eds.) Audio- and Video-Based Biometric Person Authentication. 5th International Conference, AVBPA 2005, Hilton Rye Town, NY, USA, 20–22 July 2005. LNCS, vol. 3546, pp. 310–319. Springer, Heidelberg (2005)
65. Van der Veen, M., Kevenaar, T., Schrijen, G.-J., Akkermans, T.H., Zuo, F.: Face biometrics with renewable templates. In: Security, Steganography, and Watermarking of Multimedia Contents VIII, 60720 J, 15 Feb 2006. Proceedings OF SPIE, vol. 6072 (2006)
66. Zhou, X., Wolthusen, S.D., Busch, C., Kuijper, A.: A security analysis of biometric template protection schemes. In: Proceedings of ICIAR 2009, pp. 429–438 (2009)

A Dual-Purpose Memory Approach for Dynamic Particle Swarm Optimization of Recurrent Problems

Eduardo Vellasques, Robert Sabourin and Eric Granger

Abstract In a dynamic optimization problem (DOP) the optima can change either in a sequential or in a recurrent manner. In sequential DOPs, the optima change gradually over time while in recurrent DOPs, previous optima reappear over time. The common strategy to tackle recurrent DOPs is to employ an archive of solutions along with information allowing to associate them with their respective problem instances. In this paper, a memory-based Dynamic Particle Swarm Optimization (DPSO) approach which relies on a dual-purpose memory for fast optimization of streams of recurrent problems is proposed. The dual-purpose memory is based on a Gaussian Mixture Model (GMM) of candidate solutions estimated in the optimization space which provides a compact representation of previously-found PSO solutions. This GMM is estimated over time during the optimization phase. Such memory operates in two modes: generative and regression. When operating in generative mode, the memory produces solutions that in many cases allow avoiding costly re-optimizations over time. When operating in regression mode, the memory replaces costly fitness evaluations with Gaussian Mixture Regression (GMR). For proof of concept simulation, the proposed hybrid GMM-DPSO technique is employed to optimize embedding parameters of a bi-tonal watermarking system on a heterogeneous database of document images. Results indicate that the computational burden of this watermarking problem is reduced by up to 90.4 % with negligible impact on accuracy. Results involving the use of the memory of GMMs in regression mode as a mean of replacing fitness evaluations (surrogate-based optimization) indicate that such learned memory also provides means of decreasing computational burden in situations where re-optimization cannot be avoided.

E. Vellasques (✉) · R. Sabourin · E. Granger
École de technologie supérieure, Montreal, Canada
e-mail: evellasques@livia.etsmtl.ca

R. Sabourin
e-mail: robert.sabourin@etsmtl.ca

E. Granger
e-mail: eric.granger@etsmtl.ca

© Springer International Publishing Switzerland 2016
R. Abielmona et al. (eds.), *Recent Advances in Computational Intelligence in Defense and Security*, Studies in Computational Intelligence 621,
DOI 10.1007/978-3-319-26450-9_14

367

Keywords Evolutionary computing · Particle swarm optimization · Dynamic optimization · Gaussian Mixture Models · Estimation of Distribution Algorithms · Surrogate-based optimization

1 Introduction

Evolutionary computing (EC) allows tackling optimization problems where the derivatives are unknown (black box optimization) by evolving populations of candidate solutions through a certain number of generations, driven by one or more fitness functions. Because of its black box nature, EC has been successfully employed in many different practical applications. The drawback is that most of the techniques available in the literature assume that the optima location does not change (static optimization) while most practical applications have a dynamic nature.

Problems where the optima change with time are known in the literature as dynamic optimization problems (DOPs). In a DOP, a change can be either of type I (optimum location changes with time), type II (optimum fitness changes with time but location remains fixed) or type III (both, the location and fitness change with time) [1] and be followed or not by a period of stasis [2]. A change in such context is subject to *temporal severity* and/or *spatial severity*. Some applications involve tracking one or more peaks moving in a sequential manner through the environment. In such scenario, changes can be of any of the three types above but the temporal severity is usually high as change occurs in short intervals of time. Spatial severity is usually low as peaks move in a smooth fashion, therefore stasis is small or inexistent.

In DOPs involving low spatial severity (e.g.: moving a robot through a 3D space), the optima usually moves within a region in the fitness landscape already surrounded by one or more candidate solutions. In such case, the most important issues to be addressed are diversity loss (in static optimization, the population tends to collapse towards a narrow region of the fitness landscape, making adaptation difficult) and outdated memory (past knowledge is very useful in EC, but it might be useless when a change occurs). Diversity loss can be mitigated by three different approaches—introducing diversity after a change occurs, maintaining diversity throughout the run or using multi-population—while outdated memory can be tackled by either erasing memory or re-evaluating memory and setting it to either previous or current value (whichever is better) [3]. Another important issue for such type of DOP is detecting when a change occurs. The most common approach is to track the fitness value of one or more sentry particles [4, 5]. An alternative is to compute a running average of the fitness function for the best individuals over a certain number of iterations [6].

In recurrent problems [7–9], the spatial severity is high as the optima re-appears abruptly in a previous location but the temporal severity is low. Such type of DOP is usually linked to applications involving optimizing system parameters for streams of data like machine learning [10], digital watermarking [11] and video surveillance [12–14]. It has been demonstrated in the literature that the use of a memory of previously seen solutions can decrease the computational cost of optimization in such recurring environments [8].

One of the main strategies to decrease the computational cost of optimization in such scenario is through the use of a memory of ready-to-use solutions, as demonstrated in [11]. The motivation for the use of ready-to-use solutions is that the optimization of streamed data (which corresponds to optimizing a stream of problem instances) can be seen as a special case of a type III change where the spatial severity is minimal (or inexistent) in the parameter space and small in the fitness space. Time severity is inexistent as the exact moment a new problem instance arrives is known beforehand. Put differently, stasis can have an indefinite length. Such case of type III change can be considered as a pseudo-type II change. Optimal solutions are interchangeable across problem instances involving pseudo-type II change. Due to its fast convergence, Particle Swarm Optimization (PSO) is preferred in such time-constrained applications.

A novel memory-based Dynamic PSO (DPSO) technique has been proposed for fast optimization of recurring dynamic problems, where a two-level memory of selected solutions and Gaussian Mixture Models (GMM) of their corresponding environments is incrementally built [15]. For each new problem instance, solutions are sampled from this memory and re-evaluated. A statistical test compares the distribution of both fitness values and the best re-evaluated solution is employed directly if both distributions are considered similar. Otherwise, L-best PSO [10] is employed in order to optimize parameters. Vellasques et al. [16] present a more detailed description of that technique, including a comprehensive experimental validation in many different scenarios involving homogeneous and heterogeneous problem streams. In the present paper, a study on the use of surrogates as a tool to decrease the computational cost of the L-best PSO is presented. The main research problem addressed in the given work is how to employ a model of the stream of optimization problems in order to at the same time (1) generate ready-to-use solutions which allow avoiding re-optimization and (2) replace costly fitness evaluations with regression when re-optimization cannot be avoided. The research hypothesis is that density estimates of historical solutions found during the optimization phase allow tackling (1) and (2) at the same time.

The application which motivates this research is the optimization of embedding parameters for digital watermarking systems in scenarios involving streams of document images. Digital watermarking allows enforcing authenticity and integrity of such type of image which is a major security concern for many different industries including financial, healthcare and legal. Since the protection provided by digital watermarks is minimally intrusive, it can be easily integrated into legacy document management systems (allowing an extra layer of security with minimum implementation costs).

The research presented in this paper is a continuation of the research presented in [15]. The main distinction, is that in the approach presented in this paper, the GMM memory of solutions serves two main purposes—(1) generating ready-to-use solutions and (2) replacing fitness evaluation with regression. Therefore, the main contribution of this paper is that here it is demonstrated that a previously learned memory of GMMs not only provides means of avoiding costly re-optimization operations but also makes possible a further decrease in computational burden in situations where re-optimization cannot be avoided. The surrogate strategy employed in these experimental simulations is based on the strategy proposed by Parno et al. [17] with the difference that in the proposed strategy, no surrogate update takes place since it is assumed that a memory of GMMs learned over a training sequence of optimization problems should provide enough knowledge about future similar problems, which would make possible replacing fitness evaluations with regression. Thus, surrogate-based optimization is formulated as an unsupervised learning problem. There are two reasons for using a surrogate in the envisioned application. The first reason is that re-optimization cannot be completely avoided and when it is made necessary, its computational cost is much higher than that of recall. Therefore, there is a considerable amount of computational cost savings to be made for such operation. The second reason is that although a surrogate can lead to such computational cost savings, it also requires costly fitness evaluations. However, in a scenario involving recurring problems, even when re-optimization cannot be avoided, previously learned GMMs can offer a good approximation of the new problem, avoiding part of the costs involved in training surrogates.

A review of DPSO is provided in Sect. 2. The proposed hybrid GMM-DPSO technique for fast dynamic optimization of long streams of problems is proposed in Sect. 3. Proof of concept simulation results and discussions are shown in Sect. 4.

2 Dynamic PSO (DPSO)

PSO [18] is an optimization heuristics based on the concept of swarm intelligence. In its canonical form, a population (swarm) of candidate solutions (particles) is evolved through a certain number of generations. As its physics equivalent, each particle i in PSO has a position $(\mathbf{x}_i)$ in a multidimensional search space and a velocity $(\mathbf{v}_i)$. The velocity of the ith particle is adjusted at each generation according to the best location visited by that particle $(\mathbf{p}_i)$ and the best location visited by all neighbors of particle i $(\mathbf{p}_g)$:

$$\mathbf{v}_i = \chi \times (\mathbf{v}_i + c_1 \times r_1 \times (\mathbf{p}_i - \mathbf{x}_i) + c_2 \times r_2 \times (\mathbf{p}_g - \mathbf{x}_i)) \qquad (1)$$

where χ is a constriction factor, chosen to ensure convergence [19], c_1 and c_2 are respectively the cognitive and social acceleration constants (they determine the magnitude of the random forces in the direction of $\mathbf{p}_i$ and $\mathbf{p}_g$ [20]), r_1 and r_2 are two different random numbers in the interval [0, 1]. PSO parameters c_1 and c_2 are set to 2.05 while χ is set to 0.7298 as it has been demonstrated theoretically that these values guarantee convergence [20]. The neighborhood of a particle can be restricted to a limited number of particles (L-Best topology) or the whole swarm (G-Best topology).

After that, the velocity is employed in order to update the position of that same particle:

$$\mathbf{x}_i = \mathbf{x}_i + \mathbf{v}_i \qquad (2)$$

Canonical PSO is tailored for the optimization of static problems. However, numerous real world problems have a dynamic nature. A DOP is either defined as a sequence of static problems linked up by some dynamic rules or as a problem that has time-dependent parameters in its mathematical formulation [21].

The three main issues that affect the performance of EC algorithms in DOPs are (1) outdated memory, (2) lack of change detection mechanism and (3) diversity loss [4, 22]. It is important to mention that for PSO, outdated memory can be easily tackled by re-evaluating the fitness function for the new problem instance [3, 5].

Change detection mechanisms either assume that changes in the environment are made known to the optimization algorithm or that they need to be detected [21]. In the proposed formulation of a DOP, each problem instance in the stream of optimization problems is static and the moment a transition between any two instances occurs is known. However, the similarity between a new and previously seen instances is unknown and the objective of change detection in this concept is to measure the similarity between new and previously seen problem instances. The most common change detection strategy is based on the use of fixed sentry particles, re-evaluated at each generation [4]. Another strategy relies on measuring algorithmic behavior with the use of a statistical test [21].

Tackling diversity loss requires more elaborate techniques, which can be categorized as [23]: increasing diversity after a change, maintaining diversity throughout the run, memory-based schemes and multi-population approach. The choice of diversity enhancement strategy is tied to properties of the dynamic optimization problem. Problems involving a single optimum drifting in the fitness landscape can be tackled by either increasing diversity after a change (e.g. re-initializing part of the swarm, a technique known as random immigrants [6]) or by maintaining diversity throughout the run. Problems involving multiple peaks drifting in the fitness landscape with the optimal position shifting between these peaks can be tackled with the use of multi-population and is for example, the assumption of the technique proposed by Parno et al. [24].

Problems involving one or more states re-appearing over time are better tackled through the use of memory-based schemes [8]. The main reason is that in such type of DOP, the transition between one or more problem instances is not smooth as assumed in [24] and the three strategies described above are of no use when the optimum moves away from the area surveyed by the swarm. In such case, as stated by Nguyen et al. [21], the optima may return to the regions near their previous locations, thus it might be useful to re-use previously found solutions to save computational time and to bias the search process. As observed by Yang and Yao [8], the main strategy to tackle such type of DOP is to preserve relevant solutions in a memory either by an implicit or an explicit memory mechanism and then, recall such solutions for similar future problems. In an implicit memory mechanism, redundant genotype representation (i.e. diploidy-based GA) is employed in order to preserve knowledge about the environment for future similar problems. In an explicit mechanism, precise representation of solutions is employed but an extra storage space is necessary to preserve these solutions for future similar problems. The three major concerns in memory-based optimization systems are: (1) what to store in the memory?; (2) how to organize and update the memory?; (3) how to retrieve solutions from the memory?

In this paper we propose a technique which is based on storing Gaussian Mixture Models (GMMs) of solutions in the optimization space along with their respective global best solutions. The main motivation for relying on GMM representation of the fitness landscape is that as stated by Nguyen et al. [21], the general assumption of a DOP is that the problem after a change is somehow related to the problem before a change, and thus an optimization algorithm needs to learn from its previous search experience as much as possible. The use of probabilistic models in EC is not new. Such approach, known as Estimation of Distribution Algorithms (EDA) [25] is an active research topic. The main advantage of EDA is that such probabilistic models are more effective in preserving historical data than a few isolated high evaluating solutions.

3   Proposed Approach

Figure 1 illustrates the proposed memory-based method. In the proposed approach, the basic memory unit is a **probe** and it contains a density estimate of solutions plus the global best solution, obtained after the optimization of a given problem instance. The first memory level is the Short Term Memory (STM) which contains

Fig. 1 Hybrid GMM/DPSO
framework

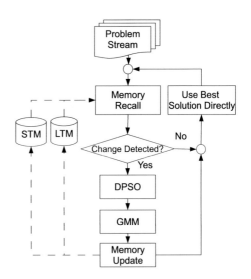

a single probe, obtained during the optimization of a single problem instance and provides fast recall for situations where a block of similar problem instances appear sequentially (e.g. a sequence of practically identical frames in a video sequence, except for noise). The second memory level is the Long Term Memory (LTM) which contains multiple probes obtained in the optimization of different problem instances and allows recalling solutions for problems reappearing in an unpredictable manner.

Given a stream of optimization problems, an attempt to recall the STM is performed first. During a recall, solutions are re-sampled from the density estimate and re-evaluated (along with the global best solution) in the new problem instance. The Kolmogorov-Smirnov statistical test is employed in order to measure the difference between the sampled and re-evaluated sets of fitness values. If they are similar, this means that the given problem instance is a recurring one and therefore, the best re-evaluated solution is employed right away, avoiding a costly re-optimization operation. If the distributions are not similar, the same process (sampling/re-evaluation/ statistical test) is repeated for each probe in the LTM until either a similarity between both distributions of fitness values is found or all probes have been tested.

In such case, the solutions sampled from the STM probe are used as a starting point for a new round of optimization. Optimization relies on a surrogated-based PSO algorithm. Such approach relies on the use of two populations—one based on exact fitness evaluations and another one which replaces exact fitness evaluations with a regression model. Both populations tackle optimization with the use of the L-best PSO (Sect. 2). The change detection module was adapted to the memory recall mechanism employed in the proposed approach. Although each problem instance is static, this PSO variant allows tackling multi-modal optimization problems and the motivation behind the proposed technique is tackling dynamic optimization of recurring problems in practical applications (which might imply in multi-modal landscapes). After that, a mixture model of the fitness landscape is estimated using the GMM approach of Figueiredo and Jain [26]. The **position** and **fitness** data of all intermediary solutions, found in all generations are employed for this purpose. The reason for using all intermediary solutions and not selected solutions (e.g. best particle positions) is that these solutions allow a more general model of the fitness landscape. That is, local best data usually results in density estimates that are over-fit to a specific problem.

This mixture model along with the global best solution will form a probe, to be stored in the STM (replacing the previous probe) and updated into the LTM. The LTM update consists of either a merge between the new probe and a probe in the memory (if they are similar) or an insert (if either the LTM is empty or no similar probe has been found). In such case, if the memory limit has been reached, an older probe is deleted (we propose deleting the probe which resulted in the smallest number of successful recalls up to that instant).

Thus, the proposed approach relies on the use of an associative memory [8]. In an associative memory approach, selected solutions are stored in a separate archive along with a density estimate that allows associating these solutions with recurring environments. However, an important distinction has to be made here. In the associative memory approaches found in the literature, the density estimates are trained using only position data while in the proposed approach, the density is trained using fitness and position data. The main motivation is that in the proposed approach, a density provides not only means of associating solutions with recurring problems, but it is also an important part of the change detection mechanism (sampled fitness values are compared with re-evaluated fitness values in order to measure the similarity between the new and previously seen problems). Therefore, instead of providing a hint of locations where good solutions are likely to be found, a memory element in the proposed approach provides a topographic map of one or more optimization problems while the change detection mechanism provides means of probing the new optimization problem in order to compare how similar both landscapes are. This GMM representation of the fitness landscape is also employed as a regression model for surrogate-based optimization.

3.1 Gaussian Mixture Modeling of Fitness Landscapes

The main reason for building and storing a model of all solutions found during the optimization of a problem instance rather than storing selected solutions is that the former allows a more compact and precise representation of the fitness landscape than individual solutions. As explained before, the GMM approach of Figueiredo and Jain [26] is employed for this purpose since it is a powerful tool for modeling multi-modal data (which makes the proposed technique robust for multi-modal optimization as well). Different than other clustering techniques such as k-means, GMM allows to model overlap among data densities, and provides more accurate (complex) modeling of data distributions. A mixture model is a linear combination of a finite number of models

$$p(\boldsymbol{x}|\Theta) = \sum_{k=1}^{K} \alpha_k p(\boldsymbol{x}|\theta_k) \tag{3}$$

where $p(\boldsymbol{x}|\Theta)$ is the probability density function (pdf) of a continuous random vector $\boldsymbol{x}$ given a mixture model Θ, K is the number of mixtures, α_j and θ_j are the mixing weights and parameters of the jth model (with $0 < \alpha_j \leq 1$ and $\sum_{j=1}^{K} \alpha_j = 1$). The mixture model parameters $\Theta = \{(\alpha_1, \theta_1), \dots, (\alpha_K, \theta_K)\}$ are estimated using the particle position $(\boldsymbol{x})$ and fitness $(f(\boldsymbol{x}))$ of all particles through all generations.

In the GMM approach employed in this framework, the mixture is initialized with a large number of components, where each component is centered at a randomly picked data point. Training is based on the use of Expectation Maximization (EM) where the E-step and M-step are applied iteratively. In the E-step, the posterior probability given the data is computed:

$$w_{i,j}^{(t)} = \frac{\alpha_j p(\boldsymbol{x}_i|\theta_j)}{\sum_{k=1}^{K} \alpha_k p(\boldsymbol{x}_i|\theta_k)} \tag{4}$$

Then, in the M-step, the model parameters are updated. Authors propose a slightly modified variant of mixing weights update, which discounts the number of parameters in each Gaussian (N):

$$N = d + \frac{d(d+1)}{2} \tag{5}$$

$$\alpha_j^{(t+1)} = \frac{max\{0, (\sum_{i=1}^{n} w_{i,j}) - N/2\}}{\sum_{k=1}^{K} max\{0, (\sum_{i=1}^{n} w_{i,k}) - N/2\}} \tag{6}$$

where d is the number of dimensions of x and n is the number of data points. The remaining parameters are updated as:

$$\mu_j^{(t+1)} = \frac{\sum_{i=1}^{n} w_{i,j}^{(t)} x_i}{w_{i,j}} \tag{7}$$

$$\Sigma_j^{(t+1)} = \frac{\sum_{i=1}^{n} w_{i,j}^{(t)} (x_i - \mu_j^{(t+1)})(x_i - \mu_j^{(t+1)})^T}{w_{i,j}} \tag{8}$$

However, during learning as the model is updated (1) components lacking enough data points to estimate their covariance matrices have their corresponding mixing weights set to zero (component annihilation) and (2) the number of components is gradually decreased until a lower boundary is achieved and then, the number that resulted in the best performance is chosen.

The use of density models in EC is not new. However, such strategy of using both phenotypic and genotypic data to estimate the models is novel. Another major difference between the proposed use of probabilistic models and those seen in the EDA literature is that in the proposed approach, all solutions found in the course of an optimization task are employed in order to estimate the model of the fitness landscape. In the EDA literature instead, selected (best fit) solutions at each generation are employed in model estimation, as a selection strategy. In the proposed approach, a density estimate is employed as a mean of matching new problems with previously seen problems.

3.2 Memory Update

The memory is due to be updated after the mixture model has been created. As mentioned before, the basic memory element in the proposed approach is a probe. Updating the STM is trivial, it requires basically deleting the current probe and inserting the new probe since the STM should provide means of recalling the last case of optimization, for situations involving a block of similar optimization problems appearing in sequence.

The LTM instead, must provide a general model of the stream of optimization problems. Since all LTM probes must be recalled before optimization is triggered, the size of the LTM must be kept to a minimum in order to avoid a situation where the cost of an unsuccessful recall is greater than the cost of full optimization. For this reason, we propose an adaptive update mechanism. In the proposed mechanism, when the LTM is due to be updated, the $C2$ distance metric [27] (which provides a good balance between computational burden and precision) is employed in order to measure the similarity between the GMM of the new probe and that of each of the probes in the LTM. The $C2$ distance between two mixtures Θ and Θ' is defined as:

$$\phi_{i,j} = (\Sigma_i^{-1} + \Sigma_j'^{-1})^{-1} \tag{9}$$

$$v_{i,j} = \mu_i^T \Sigma_i^{-1} (\mu_i - \mu_j') + \mu_j^T \Sigma_j'^{-1} (\mu_j' - \mu_i') \tag{10}$$

$$C2(\Theta, \Theta') = -\log\left[\frac{2 \sum_{i,j} \alpha_i \alpha_j' \sqrt{|\phi_{i,j}|/(e^{v_{i,j}}|\Sigma_i||\Sigma_j'|)}}{\sum_{i,j} \alpha_i \alpha_j \sqrt{|\phi_{i,j}|/(e^{v_{i,j}}|\Sigma_i||\Sigma_j|)} + \sum_{i,j} \alpha_i' \alpha_j' \sqrt{|\phi_{i,j}|/(e^{v_{i,j}}|\Sigma_i'||\Sigma_j'|)}} \right] \tag{11}$$

The new probe is merged with the most similar probe in LTM if this distance is smaller than a given threshold. Otherwise, the new probe is inserted (the probe with smallest number of successful recalls is deleted if the LTM size limit has been reached). The insert threshold is computed based on the mean minimum distance between new probes and probes on the LTM for the last T LTM updates (μ_δ^t). That is, an insert will only occur if $C2 - \mu_\delta^t$ is greater than the standard deviation for the same time-frame (σ_δ^t).

The Hennig technique, which is based on the use of Bhattacharyya distance and does not require the use of historical data, is employed in order to merge two GMMs [28]. The Bhattacharyya distance is defined as:

$$\bar{\Sigma} = \frac{1}{2}(\Sigma_1 + \Sigma_2) \tag{12}$$

$$d_B(\Theta_1, \Theta_2) = (\mu_1 - \mu_2)^T \bar{\Sigma}^{-1} (\mu_1 - \mu_2)$$

$$+ \frac{1}{2}\log\left(\frac{|\frac{1}{2}(\Sigma_1 + \Sigma_2)|}{\sqrt{|\Sigma_1||\Sigma_2|}} \right) \tag{13}$$

where μ_i is a mean vector and Σ_i is a covariance matrix.

In Hennig's approach, given a tuning constant $d^* < 1$, the two components with maximum Bhattacharyya distance are merged iteratively as long as $e^{-d_B} < d^*$ for at least one component. We propose a slight modification, which is to merge the two components with minimum distance instead, in order to get a more incremental variation in the mixture components.

If the number of mixture components after the merge operation is still greater than a limit, un-merged components from the older mixture are deleted (the old un-merged component with the highest Bhattacharyya distance from all other components is delete iteratively until the limit has been achieved).

Algorithm 1 summarizes the memory update mechanism. After the end of a round of re-optimization, a new mixture (Θ_N) is estimated based on position and fitness values of all particles from all generations during the optimization process (step 1). The estimated mixture plus the global best solution will form a probe to be added to the STM (any previous STM probe is deleted, step 2). Then, if the length of the vector containing the last n minimum $C2$ distances between new probes and probes in the LTM (δ) is smaller than T (step 3), its mean and standard deviation $(\mu_\delta^t$ and $\sigma_\delta^t)$ are initialized based on pre-defined values $(\mu_\delta^0$ and σ_δ^0, steps 4 and 5). Otherwise, μ_δ^t and σ_δ^t are computed based on δ (steps 7 and 8). After that, the minimum $C2$ distance between new probe and probes in the LTM is added to δ (steps 10 and 11). The new probe is inserted into the memory if the difference between the minimum $C2$ distance and μ_δ^t is greater than the standard deviation $(\sigma_\delta^t$, steps 12–16). It is important to notice that before the insert, the LTM probe with the smallest number of recalls is deleted if the memory limit has been reached. Otherwise the new probe is merged with the most similar probe in the LTM (steps 18 and 19). If the limit of vector δ has been reached, its first element is deleted (steps 21–23).

3.3 Memory Recall

Basically, the recall mechanism is the same for both levels of memory. The only difference is that the LTM contains more probes than the STM and for this reason, this process might be repeated for many LTM probes until either all probes have been tested of a successful recall has occurred. For a given probe, N_s solutions are sampled from its mixture model:

$$X_s = \mu_j + \Sigma_j R_s \qquad (14)$$

where X_s is a sampled solution, s is the index of a solution sampled for the component j in the mixture $(\lfloor (N_s \alpha_j) + 0.5 \rfloor$ solutions are sampled per component) and R_s is a vector with the same length as μ_j whose elements are sampled from a normal distribution $N(0, \mathbf{I})$, being $\mathbf{I}$ the identity matrix.

It is important to observe that both, position and fitness values are sampled simultaneously. Then, the sampled solutions (along with the corresponding global best) are reevaluated for the new problem instance. A Kolmogorov–Smirnov statistical test is employed in order to compare the sampled and re-evaluated fitness values. If they are below a critical value for a given confidence level, the recall is considered to be successful and the best recalled solution is employed right away, avoiding a costly re-optimization. If no probe (neither in the STM nor in the LTM) results in a successful recall, a part of STM re-sampled solutions is injected into the swarm and optimization is triggered for that problem instance.

Algorithm 1 Memory update mechanism.

Inputs:

k_{max}—maximum number of components with $\alpha_j > 0$.

$\mathfrak{M}_S$—Short Term Memory.

$\mathfrak{M} = \{\mathfrak{M}_1, \ldots, \mathfrak{M}_{|\mathfrak{M}|}\}$—Long Term Memory.

$\mathfrak{D}$—optimization history (set of all particle positions and fitness values for new problem instance).

$L_{\mathfrak{M}}$—maximum number of probes in LTM.

δ—last T minimum $C2$ distances between a new probe and probes in the LTM.

$|\delta|$—number of elements in δ.

T—maximum size of δ.

$\mu_\delta^0, \sigma_\delta^0$—initial mean and standard deviation of δ.

Output:

Updated memory.

```
 1: Estimate Θ_N using 𝔇 [26].
 2: Add Θ_N and p_g to 𝔐_S.
 3: if |δ| < T then
 4:     μ_δ^t ← μ_δ^0
 5:     σ_δ^t ← σ_δ^0
 6: else
 7:     μ_δ^t ← (1/|δ|) Σ_{i=1}^{|δ|} δ_i
 8:     σ_δ^t ← √( Σ_{i=1}^{n}(δ_i−μ_δ^t)² / |δ| )
 9: end if
10: i* ← argmin_i{C2(Θ_N, Θ_i)}, ∀_{Θ_i ∈ 𝔐}
11: δ ← δ ∪ C2(Θ_N, Θ_{i*})
12: if C2(Θ_N, Θ_{i*}) − μ_δ^t > σ_δ^t then
13:     if |𝔐| = L_𝔐 then
14:         Remove LTM probe with smallest number of successful recalls.
15:     end if
16:     Add Θ_N and p_g to 𝔐
17: else
18:     Merge(Θ_{i*}, Θ_N) (Sect. 3.2)
19:     Purge merged mixture in case number of elements exceed k_max.
20: end if
21: if |δ| > T then
22:     Remove δ_1.
23: end if
```

3.4 Surrogated-Based Particle Swarm Optimization

Since the proposed method relies on several GMMs learned with the use of a training sequence, a promising strategy to further decrease the computational cost associated in such recurring optimization problems is to employ the GMMs in regression mode whenever re-optimization is triggered. Such approach is known in the literature as surrogate-based optimization [29]. In surrogate-based optimization, the fitness landscape is sampled with the use of a sampling plan (design of experiments).

Then, during optimization a portion of the exact fitness evaluations is replaced with approximate fitness values obtained through regression and the surrogate is updated with newly sampled points as necessary.

The surrogate-based strategy described in [17] is employed in the proposed framework. This surrogate-based approach employs two populations (X_A and X_B) in parallel, one based on surrogate fitness evaluations and another one based on exact fitness evaluations. During initialization, solutions sampled from the GMM memory are injected into X_B. Optimization is first performed in X_A. Then, the best solution found in the surrogate fitness optimization ($p_{g,s2}$) is re-evaluated in the exact fitness. If it improves the neighborhood best of population X_B, that neighborhood best is replaced with $p_{g,s2}$. After that, an iteration of optimization is performed using population X_B on the exact fitness. This process is repeated until a stop criterion has been reached.

However, differently than the approach proposed in [17], no surrogate update is employed. The motivation is that a memory of previously learned GMMs already provides a valuable knowledge about new optimization problems.

In terms of regression, the proposed strategy relies on the Gaussian Mixture Regression (GMR) approach described in [30]. The advantage of Sung's approach is that it requires no modification to the proposed GMM learning. Moreover, it provides a distribution of the predicted value with $\hat{f}(x)$ as the mean and $\varepsilon^2(x)$ as the covariance matrix.

Here $\varepsilon^2(x)$ can be seen as the amount of uncertainty about the predicted value. Since it is important to allow for exploration, this quantity will be discounted from the predicted value which should direct search towards unexplored regions of the fitness landscape (higher uncertainty). More formally, costly calls to the exact fitness $f(x)$ are partially replaced by a predicted fitness $f_P(x, \Theta)$ using the strategy proposed by Torczon and Trosset [31]:

$$f_P(x, \Theta) = \hat{f}(x, \Theta) - \rho_c \varepsilon(x, \Theta) \qquad (15)$$

where $\hat{f}(x, \Theta)$ is an approximation to $f(x)$ based on model Θ, ρ_c is a constant that dictates how much emphasis will be put in exploring unknown regions of the model and $\varepsilon(x)$ is the prediction error.

4 Experimental Results

4.1 Application

The proposed fast optimization technique will be validated in the optimization of embedding parameters for a bi-tonal watermarking system [11]. This is an interesting problem because a given watermark needs to be robust against attacks but at the same time result in minimum visual interference in the host image and this trade-off can be manipulated by carefully adjusting heuristic parameters in the watermark embedder. Generally speaking, in this watermarking system, a cover bi-tonal image is partitioned into several blocks, the flippability score of each pixel is computed using a moving window, the pixels are shuffled according to shuffling seed and each bit of the given bit stream is embedded into each block of the cover image through manipulation of the quantized number of black pixels. The quantization step size (Q) determines the robustness of the watermark and since this watermarking technique allows the embedding of multiple watermarks (with different levels of robustness), we will embed two watermarks a fragile one (which can be employed to enforce image integrity) with a fixed value for its quantization step size of ($Q_F = 2$) and robust one (which can be employed to enforce image authenticity) with an adjustable quantization step size $Q_R = Q_F + \Delta_Q$ where Δ_Q is a parameter to be optimized. More details can be found in [11]. Four parameters need to be optimized: the partition block size which is an integer between 1 and the maximum attainable size for the given image as seen in [32]; the size of the window used in the flippability analysis, which can be either $3 \times 3, 5 \times 5, 7 \times 7$ or 9×9; the difference between the quantization step size for the robust and fragile watermarks (Δ_Q), which is an even number between 2 and 150; and the index of the shuffling key (we proposed using a pool of 16 possible shuffling keys, thus this index is an integer between 1 and 16).

The fitness function is a combination of the Bit Correct Ratio (BCR) between the embedded and detected watermarks (both, robust and fragile), and Distance Reciprocal Distortion Measure (DRDM) [33], which measures the quality of the watermarked image (more details can be found in [11]). The three fitness values are aggregated using Chebyshev approach [34]:

$$F(\mathbf{x}) = max_{i=1,\dots,3}\{(1 - \omega_1)(\alpha_s DRDM - r_1),$$
$$(1 - \omega_2)(1 - BCR_R - r_2),$$
$$(1 - \omega_3)(1 - BCR_F - r_3)\} \qquad (16)$$

where α_s is the scaling factor of the quality measurement $DRDM$, BCR_R is the robustness measurement of the robust watermark, BCR_F is the robustness measurement of the fragile watermark, ω_i is the weight of the ith objective with $\omega_i = \frac{1}{3}, \forall_i, r_i$ is the reference point of objective i.

Each image corresponds to an optimization problem. In situations involving streams of images with similar structure (like document images) it is very likely that some of the images in the stream will have similar embedding capacity. In such case, a stream of document images can be seen as a stream of recurrent optimization problems.

4.2 Validation Protocol

The BancTec logo (Fig. 2a), which has 26×36 pixels will be employed as robust watermark and the Université du Québec logo (Fig. 2b), which has 36×26 pixels will be employed as fragile watermark.

Fig. 2 Bi-tonal logos used as watermarks. **a** 26 × 36 BancTec logo. **b** 36 × 26 Université du Québec logo

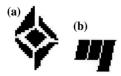

Oulu University's MediaTeam [35] (OULU-1999) database is employed as image stream. This database was scanned at 300 dpi with 24-bit color encoding. Since the baseline watermarking system is bi-tonal, the OULU-1999 database was binarized using the same protocol as in [11]. As some of its images lack the capacity necessary to embed the watermarks described above, a reject rule was applied: all images containing less than 1872 pixels with SNDM greater than zero were discarded. This rule resulted in the elimination of 15 of the original 512 images. The database was split in two subsets: a smaller one for development purposes, containing 100 images and a larger one, for validation purposes, containing 397 images. Images were assigned to these sets randomly. Table 1 shows the structure of both subsets.

Table 1 OULU-1999 database structure

Category	TRAIN	TEST
	#	#
Addresslist	0	6
Advertisement	5	19
Article	51	180
Businesscards	1	10
Check	0	3
Color segmentation	1	7
Correspondence	6	18
Dictionary	1	9
Form	9	14
Line drawing	0	10
Manual	6	29
Math	4	13
Music	0	4
Newsletter	4	37
Outline	4	13
Phonebook	4	3
Program listing	2	10
Street map	0	5
Terrainmap	2	7
Total:	100	397

(a) **(b)** **(c)** **(d)**

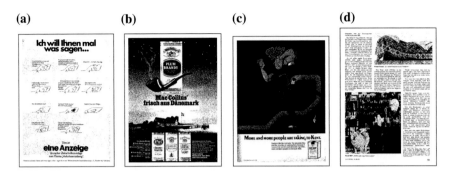

Fig. 3 Images from OULU-1999-TRAIN. **a** Image 1. **b** Image 2. **c** Image 5. **d** Image 6

As can be seen in Fig. 3, the images in this database are considerably heterogeneous.

The number of previous updates T employed to compute the adaptive threshold will be set to 10. The initial mean and standard deviation of the minimum distance were set to 361.7 and 172.3 respectively. These values were obtained by running the proposed technique in a "fill-up" mode (forcing re-optimization and LTM insert for every image in the OULU-1999-TRAIN subset). The resulting minimum $C2$ distances of these inserts were employed in order to compute these initial values.

The metrics employed in order to assess the computational performance are the average number of fitness evaluations per image ($AFPI$), the total number of fitness evaluations required to optimize the whole image stream (F_{Evals}) and the decrease in the number of fitness evaluations (DFE), computed as:

$$DFE = 1 - \frac{F_{Evals,M}}{F_{Evals,F}} \tag{17}$$

where $F_{Evals,M}$ is the cumulative number of fitness evaluations for the memory based approach and $F_{Evals,F}$ is the cumulative number of fitness evaluations for full optimization. Full optimization means applying the PSO algorithm described in Sect. 2 without resorting to neither memory recall nor to surrogates.

The reference points for the Chebyshev Weighted Aggregation were obtained through sensitivity analysis of the OULU-1999-TRAIN subset and were set to $r_1 = r_2 = r_3 = 0.01$. The scaling factor of the DRDM (α_r) was also obtained through sensitivity analysis using the training subset and was to 0.53.

During recall, 19 solutions are re-sampled from each probe, which are re-evaluated along with the global best solution, resulting in 20 sentry particles. The confidence level (α) of the KS statistic was set to the same value employed in [11], which is 0.95 and corresponds to a critical value (D_α) of 0.43. The LTM size is limited to 20 probes. The PSO parameters employed on full optimization are the same defined in [11] (20 particles, neighborhood size of 3, optimization stops if the global best has not improved for 20 generations). Cropping of 1 % of the image surface was employed during the optimization since such attack can effectively remove a non-optimized watermark. The proposed approach is compared with a previous approach (case-based), which relies on a memory of selected solutions [11] to demonstrate the main advantages of employing a memory of mixture models. In the case-based approach, the STM and LTM contain the local bests of each particle, obtained at the end of the optimization process.

In the simulations involving surrogates, re-optimization is forced for each image, as a mean of generating enough data to validate the contribution of surrogates in decreasing the computational cost of re-optimization. Recall is performed only as a mean of assigning the best model for each new problem (the one with smallest KS is chosen as the most appropriate). Therefore, the comparison to be made is between the surrogate-based approach and full optimization since the main objective here is to decrease the cost of full optimization in situations where re-optimization cannot be avoided. After each transition, 70 % of the swarm is randomized and the remaining 30 % solutions are replaced with solutions sampled from the STM. The memory is created by applying the GMM-based approach to the training stream, with full optimization activated (when re-optimization is triggered) and with the merge operator de-activated. However, since the underlying assumption of the proposed strategy is that previously learned surrogates provide valuable knowledge about new problems, it is important to define a matching strategy between previously learned surrogates and new problems. The strategy proposed here is to simply choose the GMM that results in the smallest KS value during recall.

The simulations are conducted in the heterogeneous OULU database described before. In order to understand the behavior of surrogates in more stable scenarios, simulations are also conducted using the homogeneous database of scientific documents from Computer Vision and Image Understanding (CVIU) journal. The training database (TITI-61) contains 61 pages from issues 113(1) and 113(2), split in two categories—30 pages of text and 31 pages of images—while the test database (CVIU-113-3-4) contains 342 pages of 29 complete papers from CVIU 113(3) and 113(4). More details about both databases can be found in [11].

4.3 Simulation Results

Avoidance of Re-Optimization In the first set of experiments, the surrogates are de-activated and the proposed recall mechanism works as described in Sect. 3. For each image, as soon as a case of KS value between the re-sampled and re-evaluated smaller than the critical value is found, the best recalled solution is employed directly and re-optimization is avoided. Otherwise, if no such case is found, full optimization takes place. The GMM-based approach resulted in a significant decrease in computational burden when compared to full optimization and even compared to the case-based approach (Table 2). Despite the decrease in computational burden, the watermarking performance of the GMM-based approach is practically the same of the other two approaches (Table 3).

Table 2 Computational cost of the proposed technique compared to case-based approach and full optimization

Subset	Full PSO		Case-based			GMM-based		
	AFPI	F_{Evals}	*AFPI*	F_{Evals}	*DFE* (%)	*AFPI*	F_{Evals}	*DFE* (%)
TRAIN	887(340)	88740	351(455)	35100	60.5	274(447)	27420	69.1
TEST	860(310)	341520	177(351)	70300	79.4	83(213)	32920	90.4

AFPI is the average number of fitness evaluations per image where the mean μ and standard deviation σ are presented as $\mu(\sigma)$. F_{Evals} is the cumulative number of fitness evaluations required to optimize the whole stream and *DFE* is the decrease in the number of fitness evaluations compared to full optimization

Table 3 Watermarking performance of the proposed technique compared to case-based approach and full optimization

Variant	Subset	*DRDM*	*BCR* robust	*BCR* fragile
Full PSO	TRAIN	0.03(0.03)	98.4(2.1)	99.7(0.6)
	TEST	0.03(0.04)	98.4(2.2)	99.6(0.6)
Case-based	TRAIN	0.03(0.03)	97.9(2.6)	99.6(1)
	TEST	0.03(0.03)	97.2(3.6)	99(1.6)
GMM-based	TRAIN	0.03(0.03)	97.5(2.8)	99.4(1.0)
	TEST	0.03(0.03)	96.7(4.0)	99.1(1.5)

For all values, the mean μ and standard deviation σ per image are presented in the following form: $\mu(\sigma)$. *DRDM* is presented with two decimal points and *BCR* is presented in percentage (%) with one decimal point

These results demonstrate that the memory of mixture models can cope better with the variations in the stream of optimization problems. Since the case-based probes are more tuned to the problems that generated them, they are more sensitive to small variations in a given recurring landscape caused by noise. A clear sign of this is that the smaller number of fitness evaluations was obtained even though for two of the watermarking performance metrics there was even an improvement when compared to the case-based approach. Moreover, for both cases, the watermarking performance is similar to that of full optimization, which illustrates the applicability of the proposed technique. However it is important to notice that the basic assumption is that the application can be formulated as the problem of optimizing a stream of optimization problems with some of the problems re-appearing over time, subject to noise.

Surrogate-Based Optimization Performance Table 4 shows the computational cost and watermarking performance of the simulations involving surrogates. It is possible to observe that in both cases (heterogeneous and homogeneous image streams), the use of a surrogate allowed a slight decrease in computational burden with a watermarking performance identical to that of full optimization.

The computational cost performance for the homogeneous stream is slightly better than that of the heterogeneous stream with a decrease of 12.3 % in the number of fitness evaluations when compared to full optimization. The reason is that the mem-

Table 4 Computational cost and watermarking performance of the surrogate-based strategy

Subset	AFPI	F_{Evals}	DFE(%)	DRDM	BCR robust	BCR fragile
OULU-1999-TEST	831(334)	330021	3.4	0.03(0.03)	98.4(2.2)	99.6(0.7)
CVIU-113-3-4	787(322)	269168	12.3	0.02(0.04)	98.8(2.2)	99.5(0.4)

AFPI is the average number of fitness evaluations per image where the mean μ and standard deviation σ are presented as $\mu(\sigma)$. F_{Evals} is the cumulative number of fitness evaluations required to optimize the whole stream and DFE is the decrease in the number of fitness evaluations compared to full optimization

ory learned with the use of a training sequence represents better the images found in the test database for that specific stream.

These results depict the main advantage of using a surrogate. Although the gains are not substantial as in the case where re-optimization is avoided, they are more robust in terms of watermarking performance and are a better alternative compared to performing full optimization.

4.4 Discussion

The proposed technique was assessed in a proof of concept intelligent watermarking problem. These simulation results demonstrate that the proposed technique allows decreasing computational burden of dynamic optimization with little impact on precision for applications involving the optimization of a stream of recurring problems. For example, in Fig. 4, which shows the best fitness value for each generation for both, full optimization and the proposed approach, it is possible to observe that although the proposed technique resulted in less generations, the best recalled solutions (square and triangle marks) are very close to those obtained by full optimization.

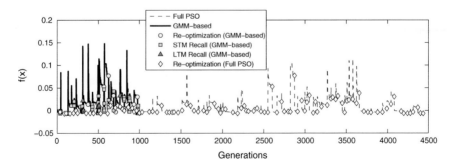

Fig. 4 Tracking of best fitness for each generation using dataset OULU-1999-TRAIN

Simulations involving the use of previously learned GMMs as surrogates demonstrate that such memory of GMMs allows not only avoiding re-optimization in situations involving recurring problems but also provides means of directing the search process for not so similar problems. Such strategy allows formulating surrogate-based optimization as a machine learning problem. One of the major concerns in surrogate-based optimization is the cost associated with probing the new environment during optimization. However, these simulation results demonstrate that it is possible to learn such a memory of surrogates in a controlled (training) environment and then, deploy this memory of surrogates to a production environment where the constraints on performance are higher. It is interesting to observe in the simulations involving the use of surrogates that the decrease in the number of fitness evaluations was higher for the homogeneous stream than for the heterogeneous stream (12.3 versus 3.4 %). Considering that in both cases re-optimization has been forcibly triggered for each image, the only possible explanation is that the train stream represents better the test stream for the homogeneous stream than for the heterogeneous stream. These results suggest that the surrogate model must be updated as optimization of new optimization problems takes place.

The trade-off between precision and computational burden is driven by (1) the number of times optimization is triggered and (2) the speed up in convergence provided by the surrogates. Therefore it is possible to improve precision by either employing a more restrictive decision threshold in the Kolmogorov-Smirnov test employed on change detection or by relying less on surrogate optimization (applying one round of surrogate optimization for every two iterations of exact optimization, for example).

5 Conclusion

A hybrid GMM/PSO dynamic optimization technique was proposed in this paper. The main objective of the proposed approach is to tackle optimization of streams of recurring optimization problems. Such formulation of dynamic optimization is applicable to many practical applications, mainly those related to optimizing heuristic parameters of systems that process streamed data such as batch processing of images, video processing, incremental machine learning.

In the proposed technique, a two-level adaptive memory containing GMMs of solutions in the optimization space and global best solutions is incrementally built. For each new problem instance, solutions are re-sampled from this memory and employed as sentries in order to (1) measure the similarity between the new problem instance and previous instances that had already resulted in optimization; (2) provide ready-to-use solutions for recurring problems, avoiding an unnecessary re-optimization.

It is worth noticing that the proposed technique resulted in a decrease of 90.4 % in computational burden (compared to full optimization) with minimum impact on accuracy in an application involving a heterogeneous stream of document images.

Although these results are still preliminary, they are comparable to results obtained in a previous version of the proposed approach already published, with the main difference that the experiments reported in this paper involved a much more challenging database which results in more varied (noisy) optimization problems.

Simulation results involving the use of the memory of GMMs as surrogates demonstrate that it is possible to use the knowledge of previously seen problems as a mean of decreasing the computational cost of re-optimization in situations where re-optimization cannot be avoided. Even though model update was not employed, the use of surrogates resulted in a decrease of 3.4 % in the number of fitness evaluations for a heterogeneous stream and 12.3 % for an homogeneous stream. To the best of our knowledge, such finding advances the state-of-the-art in the surrogate-based optimization literature by separating surrogate learning (or modeling) which can be performed in a controlled environment from prediction in a more constrained production environment. The superior performance for homogeneous database indicate that model adaptation is an important issue for heterogeneous streams of optimization problems. This means that in the same manner an adaptive memory is required to tackle recall of heterogeneous streams an adaptive surrogate is required to tackle optimization in such scenario. As a future work we propose an evaluation in a larger stream of document images and also in synthetic benchmark functions which could allow a better understanding of the mechanisms behind the proposed approach. We also propose validating the proposed technique in other applications where such stream of recurring optimization problems is applicable such as incremental learning, video processing.

Acknowledgments This work has been supported by National Sciences and Engineering Research Council of Canada and BancTec Canada Inc.

References

1. Nickabadi, A., Ebadzadeh, M.M., Safabakhsh, R.: DNPSO: A dynamic niching particle swarm optimizer for multi-modal optimization. In: IEEE World Congress on Computational Intelligence, pp. 26–32 (2008)
2. Farina, M., Deb, K., Amato, P.: Dynamic multiobjective optimization problems: test cases, approximations, and applications. IEEE Trans. Evol. Comput. **8**(5), 425–442 (2004)
3. Blackwell, T., Branke, J.: Multiswarms, exclusion, and anti-convergence in dynamic environments. IEEE Trans. Evol. Comput. **10**(4), 459–472 (2006)
4. Carlisle, A., Dozier, G.: Tracking changing extrema with adaptive particle swarm optimizer. In: Proceedings of the 5th Biannual World Automation Congress, 2002, vol. 13, pp. 265–270 (2002)
5. Hu, X., Eberhart, R.: Adaptive particle swarm optimization: detection and response to dynamic systems. In: Proceedings of the 2002 Congress on Evolutionary Computation. CEC '02, vol. 2, pp. 1666–1670 (2002)
6. Wang, H., Wang, D., Yang, S.: Triggered memory-based swarm optimization in dynamic environments, In: EvoWorkshops, pp. 637–646 (2007)

7. Barlow, G.J., Smith, S.F.: Using memory models to improve adaptive efficiency in dynamic problems, In: IEEE Symposium on Computational Intelligence in Scheduling, 2009. CI-Sched '09, pp. 7–14 (2009)
8. Yang, S., Yao, X.: Population-based incremental learning with associative memory for dynamic environments. IEEE Trans. Evol. Comput. **12**(5), 542–561 (2008)
9. Li, X., Dam, K.H.: Comparing particle swarms for tracking extrema in dynamic environments. In: Proceedings of the 2003 Congress on Evolutionary Computation. CEC '03, vol. 3, pp. 1772–1779 (2003)
10. Kapp, M.N., Sabourin, R., Maupin, P.: A dynamic model selection strategy for support vector machine classifiers. Appl. Soft Comput. **12**(8), 2550–2565 (2012)
11. Vellasques, E., Sabourin, R., Granger, E.: A high throughput system for intelligent watermarking of bi-tonal images. Appl. Soft Comput. **11**(8), 5215–5229 (2011)
12. Connolly, J.F., Granger, E., Sabourin, R.: Evolution of heterogeneous ensembles through dynamic particle swarm optimization for video-based. Pattern Recognit. **45**(7), 2460–2477 (2012)
13. Connolly, J.F., Granger, E., Sabourin, R.: An adaptive classification system for video-based face recognition. Inf. Sci. **192**, 50–70 (2012)
14. Connolly, J.F., Granger, E., Sabourin, R.: Dynamic multi-objective evolution of classifier ensembles for video-based face recognition. Appl. Soft Comput. **13**(6), 3149–3166 (2013)
15. Vellasques, E., Sabourin, R., Granger, E.: Gaussian mixture modeling for dynamic particle swarm optimization of recurrent problems. In: Proceedings of the Genetic and Evolutionary Computation Conference GECCO '12, pp. 73–80. ACM (2012)
16. Vellasques, E., Sabourin, R., Granger, E.: Fast intelligent watermarking of heterogeneous image streams through mixture modeling of PSO populations. Appl. Soft Comput. **13**(6), 3130–3148 (2013)
17. Parno, M.D., Hemker, T., Fowler, K.R.: Applicability of surrogates to improve efficiency of particle swarm optimization for simulation-based problems. Eng. Optim. **44**(5), 521–535 (2012)
18. Kennedy, J.: Some issues and practices for particle swarms. In: Swarm Intelligence Symposium, 2007, SIS 2007, pp. 162–169. IEEE (2007)
19. Blackwell, M.: Particle swarms and population diversity. Soft Comput. **9**(11), 793–802 (2005)
20. Poli, R., Kennedy, J., Blackwell, T.: Particle swarm optimisation: an overview. Swarm Intell. J. **1**(1), 33–57 (2007). June
21. Nguyen, T.T., Yang, S., Branke, J.: Evolutionary dynamic optimization: A survey of the state of the art. Swarm Evol. Comput. **6**, 1–24 (2012)
22. Blackwell, T.: Particle swarm optimization in dynamic environments. Evolutionary Computation in Dynamic Environments, pp. 29–49. Springer, Berlin (2007)
23. Yang, S.: Population-based incremental learning with memory scheme for changing environments. In: Proceedings of the 2005 conference on Genetic and evolutionary computation, GECCO '05, pp. 711–718. ACM, New York, NY, USA (2005)
24. Parrott, D.. Li, X.: A particle swarm model for tracking multiple peaks in a dynamic environment using speciation. In: Proceedings of the 2004 Congress on Evolutionary Computation. CEC '04, vol. 1, pp. 98–103 (2004)
25. Pelikan, M., Goldberg, D.E., Lobo, F.G.: A survey of optimization by building and using probabilistic models. Comput. Optim. Appl. **21**(1), 5–20 (2002)
26. Figueiredo, M.A.T., Jain, A.K.: Unsupervised learning of finite mixture models. IEEE Trans. Pattern Anal. Mach. Intell. **24**, 381–396 (2000)
27. Sfikas, G., Constantinopoulos, C., Likas, A., Galatsanos, N.P.: An analytic distance metric for gaussian mixture models with application in image retrieval. In: Proceedings of the 15th international conference on Artificial neural networks: formal models and their applications— vol. Part II, ICANN'05, pp. 835–840. Springer, Berlin (2005)
28. Hennig, C.: Methods for merging gaussian mixture components. Adv. Data Anal. Classif. **4**, 3–34 (2010)

29. Queipo, N.V., Haftka, R.T., Shyy, W., Goel, T., Vaidyanathan, R., Tucker, P.K.: Surrogate-based analysis and optimization. Prog. Aerosp. Sci. **41**(1), 1–28 (2005)
30. Sung, H.G.: Gaussian mixture regression and classification, Ph.D. thesis, Rice University (2004)
31. V. Torczon, M. W. Trosset, Using approximations to accelerate engineering design optimization, in: Proceedings of the 7th AIAA/USAF/NASA/ISSMO Multidisciplinary Analysis and Optimization Symposium, Saint Louis, USA, 1998, pp. 738–748
32. Muharemagic, E.: Adaptive two-level watermarking for binary document images, Ph.D. thesis, Florida Atlantic University (2004)
33. Lu, H., Kot, A.C., Shi, Y.Q.: Distance-reciprocal distortion measure for binary document images. IEEE Signal Process. Lett. **11**(2), 228–231 (2004)
34. Collette, Y., Siarry, P.: On the sensitivity of aggregative multiobjective optimization methods. CIT **16**(1), 1–13 (2008)
35. Sauvola, J., Kauniskangas, H.: MediaTeam Document Database II, A CD-ROM Collection of Document Images. University of Oulu, Finland (1999)

Risk Assessment in Authentication Machines

S. Eastwood and S. Yanushkevich

Abstract This work introduces an approach to building a risk profiler for use in authentication machines. Authentication machine application scenarios include the security of large public events, pandemic prevention, and border crossing automation. The proposed risk profiler provides a risk assessment at all phases of the authentication machine life-cycle. The key idea of our approach is to utilize the advantages of belief networks to solve large-scale multi-source fusion problems. We extend the abilities of belief networks by incorporating Dempster-Shafer Theory measures, and report the design techniques by using the results of the prototyping of possible attack scenarios. The software package is available for researchers.

Keywords Authentication · Security · Risks · Belief (bayesian) network · Dempster-Shafer evidence model · Border crossing automation

1 Introduction

International Air Transport Association (IATA) introduced a roadmap for 2020+ border crossing automation in [1]. The Department of Homeland Security (DHS), U.S.A. outlined [2] the breakthrough technological directions in border crossing automation. The core of both IATA's and DHS' visions is the risk assessment of the deployed technologies. Those include authentication machines (A-machines), along with their supporting infrastructure built on an intelligent platform.

An A-machine performs the human identification/verification task. It is traditionally a part of security infrastructure and management [3]. Standard [4] defines this area as *biometric identity assurance services*. The A-machine is a typical service-

S. Eastwood (✉) · S. Yanushkevich
Biometric Technology Laboratory, University of Calgary, Calgary, Canada
e-mail: sceastwo@ucalgary.ca
URL: http://www.ucalgary.ca/btlab

S. Yanushkevich
e-mail: syanshk@ucalgary.ca

© Springer International Publishing Switzerland 2016
R. Abielmona et al. (eds.), *Recent Advances in Computational Intelligence in Defense and Security*, Studies in Computational Intelligence 621,
DOI 10.1007/978-3-319-26450-9_15

Fig. 1 Groups of risk categories with respect to application scenario of A-machines

oriented architecture. The authentication is understood as the implementation of the following processes [5–7]:

(a) *Knowledge acquisition* (information about the person in various forms and from different sources including physical possessions such as keys, IDs, passports, and certificates), or/and

(b) *Biometric acquisition* (physiological and behavior appearances and characteristics of individuals that distinguish one person from another).

There are two aspects of this technology: (a) implementation of authentication techniques and supported infrastructure (A-machine), and (b) modeling and simulation that supports all phases of the A-machine life-cycle by assessing the risks of various design, testing, and deployed scenarios. Our work focuses on the design of intelligent tools for authentication risk assessment.

A general landscape of A-machine applications is introduced in Fig. 1. It addresses four key application scenarios: security of large public events, pandemic prevention, terrorist threats, and border crossing technologies [8, 9]. The lower part of each triangle corresponds to the high-risk decisions due to low information content, and the higher part corresponds to the low-risk decisions, because the accumulated information content is relatively sufficient for making reliable decisions. At the right plane of each triangle, risk categories for information accumulation are indicated in order of their priority.

This group of risk categories includes the following sources of information:

- token/ID (risks such as a counterfeit or stolen item);
- early warning biometrics (risks such as an unidentified disease);
- watchlist (risks such as non-updated data, or an attack on the database);
- advanced information (risks such as forged personal data);
- interview (risks such as an undetected lie);
- e-passport/ID (risks such as an intentional change of appearance via plastic surgery, colored eye lenses, or replacing the biometric template in an e-document); and
- surveillance (risks such as an intentional face obstruction or wrong identification).

Gathering personal information from these sources results in a decision regarding the targeted person. Initial information is provided by a token/ID, verified against a watchlist, and then additional information is continuously accumulated by means of biometric surveillance and related technologies. Risks in the A-machine application of pandemic prevention include early warning information and distance control. Other components, such as soft biometrics at distance, were reported in [12–14]. Interview (authentication) supported machines with a virtual security officer were described in [15–19]. After identifying pandemic features (geographical, as well as temperature, blood pressure, and other features), traditional technologies for human authentication can be used. Large public events are usually secured against threats using mobile identification [10, 11] (Fig. 1).

Risks in case of terrorist threats are analyzed, in particular, in [20, 21]. Traveler risk assessment in A-machines for border crossing is initiated from the moment of buying the ticket and providing personal information [22, 23].

A-machines can be potentially used to mitigate the vulnerability of domestic public transportation systems and various mass-public hubs. Passenger risk assessment in these applications is calculated using other risk categories for the following reasons. To enhance the security of mass-transit systems, specific supporting technologies for A-machines are needed [24], as passengers may not have an e-passport or e-ID. It is possible to delegate some functions of the A-machine for mass-transit public system security to the passengers' personal mobile devices. Contemporary mobile devices can authorize their holders to use various technologies [8, 9, 24]. In such an approach, the hub A-machine communicates with personal devices that constantly confirm the holder's identity, that is, each passenger is being authorized and trucked through the transit hub.

Examples of real-world A-machines for border crossing purposes are given in Fig. 2. For instance, the SmartGate is a typical A-machine which can operate only in a specific environment such as e-passport risk assessment technology based on

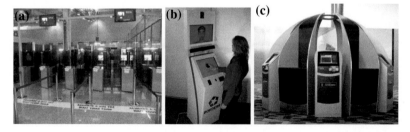

Fig. 2 A-machines for border crossing applications. **a** Faro airport, Portugal [27] http://www.fron tex.Publications/Research/Biopass; **b** Interview supported A-machine (AVATAR machine) http:// www.borders.arizona.edu/cms/sitesdefault/files/FieldTestsofanAVATARInterviewingSystemforTr ustedTraveler/; **c** New Zealand [25] http://biometrics.nist.gov/cs_links/ibpc2014/presentations/

watchlists, and pre-border risk evaluations [25, 26]. First, a traveler checks whether he/she is eligible to use the A-machine, and then undergoes the process of verifying their identity and final clearance.

The main feature of the border crossing A-machines, which distinguishes them from other authentication tools, is that they are deeply integrated into social infrastructure [28, 29]. This allows for the implementation of computational intelligence-based mechanisms for information gathering and risk assessment. For example, initial information can be provided by an ID submitted by the user, on-line biometrics, surveillance data, personal information from various databases available to the system, and interview data [20, 30]. State-of-the-art applications of A-machines for border crossing automation are reported in [31, 32].

Hence, the central procedure of authentication technology is the fusion of information provided by various sources or sensors. Data fusion is defined as *A multilevel, multifaceted process dealing with the automatic detection, association, correlation, estimation, and combination of data and information from single and multiple sources to achieve refined position and identity estimates, and complete and timely assessments of situations and threats and their significance* [33].

1.1 Problem Formulation

All phases of the life-cycle of A-machines for large-scale applications, such as border crossing automation, need knowledge of the behavior of these machines in possible operational scenarios. These scenarios address risks of the impact of complicated combinations of various factors in the performance of the A-machine that can result in a failure of the A-machine. Examples are semantic attacks (user lies in an interview) and biometric attacks (traveler uses a stolen e-passport/ID with a replaced biometric template). Risks of such threats should be taken into account at all phases of the life-cycle of the A-machine (development, prototyping, testing, deployment, and exploiting). There are two ways to solve this problem: (*a*) Risk evaluation using modeling and simulation of scenarios of interest [3, 34–36], and (*b*) Risk evaluation using a special proving ground (testing areas) [37]. A common platform for A-machine vulnerability and risk study includes simulation and modeling of the decision-making process.

Special computational intelligence-based techniques and tools for security risk assessment in various applications are called *profilers* [3]. In this work, we aim at developing a special purpose simulator, called an A-profiler, for the evaluation of risks related to A-machine applications.

1.2 Contribution

The following constitute the contributions of this work:

1. Systematic approach to developing the A-profilers for the risk assessment of
   A-machines. Instead of general terrorism risk management such as in [3], our
   work focuses on A-machine risk assessment over a library of scenarios, including cyber-attack management. However, the A-profiler can be considered as a
   part of the security portfolio management [3].
2. Two types of A-profilers are developed: (a) belief (Bayesian) network (BN) based,
   and (b) Dempster-Shafer Theory (DST) based profiler. This is a continuation and
   extension of our previous study, in particular, [34, 35].
3. We use the same computational platform, a graph model (causal network), for
   both Bayesian point estimates and DST interval estimates (pessimistic and opti-
   mistic scenarios). This approach is motivated by the large scale of the problem. In
   particular, we address this problem by using decompositions of the graph models
   and a library of risk assessment scenarios for A-machines.
4. We provide a software package "Dempster-Shafer Bayesian Network (DS-BN)"
   [38] which is a platform of the DST based profiler. This package can be used for
   various DST based multi-source fusion problems.

The main goal of our study is to increase the reliability of security risk assessment
for A-machines, using the computational intelligence-based fusion of results from
different models, metrics, and philosophies of decision-making under uncertainty.

2 Theoretical Platform of A-Profilers

Risk is defined as an event that, if it occurs, has an unwanted impact on the system's
ability to achieve its performance or outcome objectives [39] $\text{Risk} = R(p, \text{Impact})$
where p is probability of a risk. If risk and impact are identified, a risk mitigation
strategy can be developed. A risk can be expressed in the form of risk statement,
as $\text{Risk} = \text{Condition-If-Then} = Pr(A|B) - \text{Then}$ where Condition is an
event that has occurred, or is presently occurring, or will occur with certainty; Risk
is a potentially possible future event; $Pr(A|B)$ is the probability of risk event A given
event B. Risk can be evaluated via evidence accumulation and data fusion techniques.
Various data fusion paradigms can be used in an A-profiler, such as:

- *Belief (Bayesian) network* as an extension of Bayesian rule for an arbitrary directed
  acyclic graph (DAG) [40]. It is reasonable when the problem is described by DAG,
  however, the primary statistics must be available. Belief propagation is a means for
  updating the marginal distributions of variables in the DAG through knowledge
  of the values of some subset of evidentiary variables. An arbitrary A-machine can
  be represented by a DAG and modeled by a belief network.

- *Markov networks* also known as Markov random field models [40, 41]. Their unique property is that the causal description, such as in belief (Bayesian) networks, is replaced by a topology of the problem in the form of an undirected, possibly cyclic, graph.
- *Dempster-Shafer Theory* (DST) evidence model [42] and its extensions such as Dezert-Smarandache Theory (DSmT) [43]. They are useful when expert knowledge is given in imprecise form such as interval probabilities. This model may be implemented using similar belief networks but with more complicated calculations. When probabilistic intervals are small, the DST model becomes a belief network.
- *Neural networks* are a flexible heuristic technique for data fusion and statistical pattern recognition using maximum-likelihood estimation of the weight values in the model defined by the network topology [44]. Knowledge from the problem domain is incorporated into the network topology.
- *Voting logic* can be used at the low level fusion. Sensor information is used to compute detection probabilities that are combined accordingly using Boolean algebra expressions. The key idea of voting fusion is the combining of logical levels representing sensor confidence levels [51]. However, this is a noise-vulnerable fusion technique.
- *Fuzzy logic* is well suited where the boundaries between sets of values are not sharply defined or there is a partial occurrence of an event. It can be efficiency used to fuse information from multiple sources (sensors) [51].
- *Information-theory models* provide description of flows (propagation of data, uncertainty, screened customers, mismatched patterns etc.) in terms of entropy. For example, high and low entropy are associated with the A-machine input and output, respectively.

There are various combined models such as the Transferable Belief Model (TBM) [45], factor graphs which have combined properties of Bayesian and Markov networks [40, 41], fuzzy cognitive mapping [46], fuzzy neural networks [47], and fuzzy DST [48, 49], Bayesian neural networks [44], as well as the DST based neural network [50]. In this paper, the mixed model we propose is a belief (Bayesian) network with incorporated DST measures; this is similar to the TBM [45] except that our goals are different: we aim at designing a Bayesian causal network as a graphical model of an A-profiler; instead, the TBM is a fusion algorithm whose core is an updated Bayesian fusion paradigm with DST measures. Most of the aforementioned data fusion approaches are reviewed in [33, 51].

In addition, game-theory models can be used as a framework for different authentication scenarios and security personnel training, especially in training skills for attack scenarios, that is, when a customer tries to deceive personnel [52]. In these models, security personnel must anticipate what a customer will infer from the personnel's own actions or questions.

The effectiveness of each of these models depends on specific details of the problem under consideration. Moreover, they provide different interpretations of uncertainty in terms of probabilities. For example, a belief network and a Markov random field model provide different results. However, all models favour static scenarios, in which the truth values of the statements on which they rely are assumed to remain constant. Only game-theory methods explicitly model the interaction between two adversarial reasoners.

3 Risk Assessment as an Evidence Accumulation Process

Formally, authentication based on digitized data of a stochastic nature, such as biometrics, is known as *decision-making under uncertainty*. Decision-making is understood as a cognitive process leading to the selection of a course of action among several alternatives.

Any application involving A-machines incurs a risk of an incorrect authentication of a person. There is always a risk that the A-machine makes an incorrect decision. The magnitude of such risks depends on the application. For example, the error of human authentication for golf facility access, border crossings, and for nuclear plant access are associated with different risks. Risks should be evaluated using appropriate computational intelligence-based models. Various models and their metrics reflect particular aspects of decision-making under uncertainty. No methodology, model or metric exist that can be ultimately applied for risk assessment in various deployed scenarios.

The A-machine is a typical *evidence accumulation machine* because it makes decisions at various levels of the decision-making hierarchy, after accumulating a sufficient information content. The A-profiler is a modeling tool that models the

Fig. 3 **a** The causal view of the 4-state screening discipline of service over authentication resource R_i; **b** resource utilization as an evidence accumulation process (see the notation in Table 1)

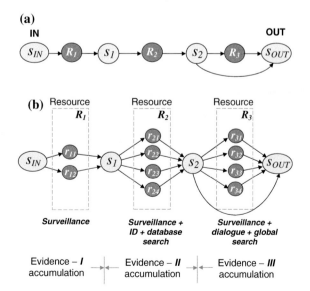

process of evidence accumulation in A-machines. The A-profiler can provide useful information about A-machine operation in various deployment scenarios, for example, when it gathers information from available sources to assess traveler risk factors [1, 4].

The platform of the risk profiler for A-machines and supporting infrastructure is the data fusion model. This model includes a *low level fusion* processes, such as data preprocessing, classification, identification, and verification, and a *high level fusion* processes, such as situation and threat assessment, as well as a fusion process refinement.

One of possible causal model of the A-machine is shown in Fig. 3a. This is a 4-state network over authentication resource R_i. The resource R_i is distributed between the three evidence accumulation phases (I, II, and III). They are modeled using the corresponding groups of variables (Table 1): phase I, state S_{IN} over authen-

Table 1 A 4-state screening discipline of service over authentication resource

State	Content
s_{IN}	Initial state, unknown (unauthorized) traveler
s_1	Traveler under visual (r_{11}) and multispectral surveillance (r_{12})
s_2	Traveler under visual (r_{21}) and multispectral surveillance (r_{22}), ID-based authentication (r_{23}), and database search (r_{24})
s_{OUT}	(*a*) Traveler at the officer desk under authentication resource R_3: visual (r_{31}) and multispectral surveillance (r_{32}), dialogue (r_{33}), global search (r_{34}), or (*b*) traveler directed from state s_2 to the authorized state s_{OUT} (resources R_3 are not activated)

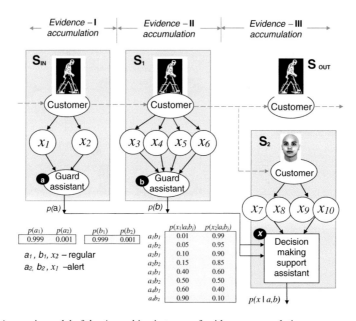

Fig. 4 A generic model of the A-machine in terms of evidence accumulation

tication resource R_1 (visual multispectral surveillance, r_{11}, r_{12}); phase II, state S_1 over authentication resource R_2 (surveillance r_{21}, r_{22}, ID identification r_{23}, and risk assessment using a watchlist database search r_{24}); and phase III, state S_2 (if necessary) over authentication resource R_3 (surveillance r_{31}, r_{32}, interview r_{33}, and additional resources for risk assessment r_{34}).

A generic model of the A-machine that collects, process, and analyzes the customer's data is illustrated in Fig. 4. As one possible description of the A-machine, this model is assumed to be a belief network with conditional probability tables that represent likelihoods based on prior information. In Fig. 4, variables x_i are associated with authentication resource as follows: $R_1 = \{r_{11}, r_{12}\} \equiv \{x_1, x_2\}$, $R_2 = \{r_{21}, r_{22}, r_{23}, r_{24}\} \equiv \{x_3, x_4, x_5, x_6\}$, and $R_3 = \{r_{31}, r_{32}, r_{33}, r_{34}\} \equiv \{x_7, x_8, x_9, x_{10}\}$. Several scenarios of evidence accumulation in border management are introduced in the ISO standard [4].

4 Graph Models for A-Profiler Design

A-profiler can be built based on various theoretical models that can describe the insufficiency of initial data or its imperfection, as well as different scenarios for modeling (causal or non-causal description). In our work, we built the A-profiler that utilizes undirected graphs, or/and directed graph models for probability distributions [40, 44].

There are scenarios in A-machine design and deployment where interactions between components have a natural directionality and can be represented by causal relations. For example, "e-passport check addresses a three-step procedure" and "failure in traveler biometric verification results in the manual control of this traveler". These scenarios should be described by directed graphs and modeled by belief (Bayesian) networks. In the cases when the interactions between components are more symmetrical, Markov random field models based on undirected graphs are preferable.

Probability distributions over a large number of variables can be denoted in a compact manner using Markov networks [40], which are undirected graphs. Let a factor F be a multi-dimensional array indexed by some subset of variables $\text{Var}(F)$. Each entry of F is non-negative, though the entries of F do not necessarily need to sum to 1. The utilized notations and formalization is given in Fig. 5. The Markov network that describes this distribution is an undirected graph over $|\mathcal{X}|$ nodes where each node denotes one of the random variables from $\mathcal{X}$. For each factor F, a clique (complete subgraph) is introduced between the nodes in $\text{Var}(F)$.

In this work, we concentrate on directed graphs such as belief networks derived from Bayesian causal models. In the next sections, we describe simple models, belief (Bayesian) networks, and their extension using DST measures.

Notations utilized in the A-profiler

- $\mathcal{X}$ will denote the set of all random variables under consideration.
- A set that contains a single variable $\{X\}$ will be written as X instead of $\{X\}$.
- For an arbitrary subset of random variables $\mathcal{Y} \subseteq \mathcal{X}$, the expression $Val(\mathcal{Y})$ will denote the set of all possible complete assignments to the variables in $\mathcal{Y}$.
- For an arbitrary assignment $V \in Val(\mathcal{Y})$ and subset of variables $\mathcal{Z} \subseteq \mathcal{X}$; $V(\mathcal{Z})$ will denote the assignment to the variables in $\mathcal{Y} \cap \mathcal{Z}$ formed by dropping from V the assignments to variables not in $\mathcal{Z}$.
- Given disjoint sets of variables $\mathcal{Y}$ and $\mathcal{Z}$; and assignments $V_\mathcal{Y} \in Val(\mathcal{Y})$ and $V_\mathcal{Z} \in Val(\mathcal{Z})$; $\langle V_\mathcal{Y}, V_\mathcal{Z} \rangle$ will denote the assignment to the variables in $\mathcal{Y} \cup \mathcal{Z}$ formed by combining the assignments $V_\mathcal{Y}$ and $V_\mathcal{Z}$.
- A Markov network "factor" F is a function over a relatively small subset of variables denoted by $Var(F)$. Given an arbitrary assignment $V \in Val(Var(F))$; $F(V)$ denotes the non-negative value returned by F.
- Given disjoint sets of variables $\mathcal{Y}$ and $\mathcal{Z}$; for any $V_\mathcal{Y} \in Val(\mathcal{Y})$ and $V_\mathcal{Z} \in Val(\mathcal{Z})$; $\Pr(V_\mathcal{Y}|V_\mathcal{Z})$ will denote the probability that the variables in $\mathcal{Y}$ will attain the values from $V_\mathcal{Y}$ given that the variables in $\mathcal{Z}$ attain the values from $V_\mathcal{Z}$. If $\mathcal{Z} = \emptyset$, then $\Pr(V_\mathcal{Y})$ simply denotes the marginal probability of the variables in $\mathcal{Y}$ attaining the values in $V_\mathcal{Y}$.

Notations utilized in the Markov network model

The product $F_1 F_2$ of two factors F_1 and F_2 is an array over the variables $Var(F_1 F_2) = Var(F_1) \cup Var(F_2)$. Each element in $F_1 F_2$ is determined by multiplying together the corresponding elements from F_1 and F_2. Letting $V \in Val(Var(F_1) \cup Var(F_2))$ be an arbitrary assignment,

$$(F_1 F_2)(V) = F_1(V(Var(F_1))) \cdot F_2(V(Var(F_2)))$$

Given a set of factors $\{F_1, F_2, \ldots, F_k\}$, the probability distribution comprised of these factors is the normalized product of all factors:

$$\forall V \in Val(\mathcal{X}) : \Pr(V) = \frac{1}{K} \prod_{i=1}^{k} F_i = \frac{1}{K} \prod_{i=1}^{k} F_i(V(Var(F_i)))$$

Where K is a normalization constant so that $\sum_{V \in Val(\mathcal{X})} \Pr(V) = 1$

Fig. 5 Notations utilized in risk assessment models

5 A-Profiler Based on Belief Networks

A sensor or a source of information can be seen as a part of the A-machine and supporting infrastructure that observes some data or evidence E, and transmits some opinion or hypothesis about the actual value of the parameter of interest B_i. The relationship between evidence E and parameter of interest B_i is represented by a probability distribution on E for each B_i, $i = 1, 2, \ldots, M$. After observing E, the sensor communicates its opinion on the value of B_i under the form of a *likelihood* vector. Let $Pr(E|B_i)$ be the likelihood of the hypothesis B_i given the evidence E. Inference on B_i is based on this likelihood and some *a priori* probabilities. The Bayesian rule enables to update our knowledge and give the *posterior* probabilities:

$$Pr(B_i|E) = \texttt{Likelihood} \times \frac{\texttt{Prior}}{\texttt{Evidence}} = Pr(E|B_i) \times \frac{Pr(B_i)}{Pr(E)} \qquad (1)$$

Bayesian updating (Eq. 1) is implemented by a belief (Bayesian) network which is a special instance of a Markov network [40, 53]. It is a directed acyclic graph where each variable/node has a corresponding factor also referred to as a conditional probability table (CPT). Let $Pa(X)$ denote the parents of node X. The factor assigned to each node X is an array over the variables $\{X\} \cup Pa(X)$ that stores the conditional probability distribution of X for each possible variable assignment to the parents of X: $Pr(X|Pa(X))$. The probability distribution denoted by the belief (Bayesian) is $\forall V \in Val(\mathcal{X}) : Pr(V) = \prod_{X \in \mathcal{X}} Pr(V(X)|V(Pa(X)))$ where the normalization constant is 1.

5.1 Simple A-Profiler

The Bayesian decision profiler is the simplest evidence accumulation technology. As an example, consider an e-passport which stores three types of biometric templates: facial, fingerprint, and iris.[1] A multi-biometric identification system can function in an accumulation mode and an update mode. Accumulation mode is defined as follows: the e-passports of various countries utilize one of three types of biometrics, B_1, B_2, or B_3 (face, fingerprints, or iris), and it is known that 30 % of e-passports utilize authentication by face, 45 % by fingerprint, and 25 % by iris (Fig. 6a).

It is known from past experience that 2, 3, and 2 % of travelers cannot be authenticated using biometric B_1, B_2, and B_3, respectively, since, defects in e-passport, aging effects (face and iris), as well as other reasons such as the corresponding templates are not available in the database. The accumulated evidence, E, that a randomly selected traveler cannot be authenticated (in probability metrics) is $Pr(E) = \sum_{i=1}^{3} Pr(B_i) \times Pr(E|B_i) = 0.006 + 0.0135 + 0.005 = 0.0245$.

Updated mode is specified as follows: assume that a randomly chosen traveler has not been authorized after the authentication procedure. The probability that it was the ith biometric, $i = 1, 2, 3$, that failed to authenticate this traveler, is called the posterior probability (Fig. 6b) and calculated using Bayesian updating (Eq. 1). So, the a priori belief about the biometrics for traveler authentication is updated as follows: probability 0.245 instead of 0.3 for B_1, 0.551 instead of 0.45 for B_2, and 0.204

---

[1]The e-passport and e-ID are defined by the ICAO standard, and are the key components of advanced border control technologies [54]. The face was recommended as the primary biometric, mandatory for global interoperability in the passport inspection systems. Fingerprint and iris were recommended as secondary biometrics.

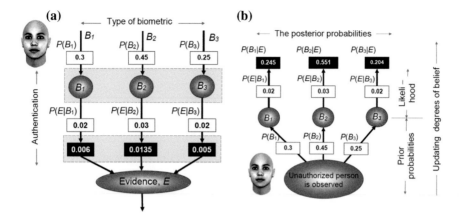

Fig. 6 The simplest A-profiler for e-passport/ID based on Bayesian inference to fusing information from multiple sources. *Left plane* **a** the evidence accumulation is provided by three types of biometrics $B_i, i = 1, 2, 3$ (face, fingerprints, and iris [54]). *Right plane* **b** Updating beliefs about the type of biometric in the A-profiler based on Bayesian rule

instead of 0.25 for B_3. This updating mechanism takes into account the likelihood of the evidence given $B_i, p(E|B_i)$, and "tunes up" the A-profiler to the evidence (the chance of seeing the evidence E if B is true).

5.2 Designing a Belief (Bayesian) Network

Causality is an efficient way to model dependencies between variables and to predict the effect of observation on the joint pdf [40, 41, 53]. A belief network locally assembles probabilistic beliefs into a coherent whole for reasoning and learning under uncertainty. While some of these beliefs could be read directly from the belief network, others require computations to be made explicit. Computing and making explicit such beliefs is known as the problem of inference in belief networks. A belief network consists of a directed acyclic graph (DAG), which represents the influences among the variables, and a set of CPTs which quantify these influences in a probability metric. Each discrete random variable (or node) has a finite number of states and is parametrized by a CPT. Inference in belief networks means to compute the conditional pdf of a hypothesis for some observed evidence.

Let us design a belief network for a causal model in which the impact on variable x_3 by variables x_1 and x_2 is described by the joint pdf $\Pr(x_1, x_2, x_3)$. Using the product rule of probability, the joint pdf is

$$\Pr(x_1, x_2, x_3) = \Pr(x_1) \Pr(x_2) \Pr(x_3|x_1, x_2) \tag{2}$$

Fig. 7 **a** The belief network
model, and formal
description by Eq. (2); **b** the
belief network model, and
formal description by Eq. (3)

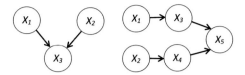

This decomposition holds for any choice of joint pdf. Equation (2) represents the
following graphical model:

1. Associate each node with the random variables x_1, x_2, and x_3, and the corresponding conditional pdf from Eq. (2).
2. Draw directed links from the nodes corresponding to the variables on which the
   pdf is conditioned.

The resulting belief network as the DAG model is given in Fig. 7a. For the con-
ditional pdf $\Pr(x_3|x_1, x_2)$, there are links from nodes x_1 and x_2 to node x_3, whereas
for the pdf $\Pr(x_1)$ and pdf $\Pr(x_2)$ there are no incoming links. If there is a link going
from a node x_1 to a node x_3, then we say that node x_1 is a *parent* of node x_3 and node
x_3 is a *child* of node x_1. A more complicated scenario is given in Fig. 7b. The joint
pdf of all 5 random variables is

$$\Pr(x_1, \ldots, x_5) = \Pr(x_1)\,\Pr(x_2)\,\Pr(x_3|x_1)\,\Pr(x_4|x_2)\,\Pr(x_5|x_3, x_4) \qquad (3)$$

Each node of the belief networks in Fig. 7 contains a CPT. The size of the CPT
is critically dependent on the DAG topology. Specifically, the number of parameters stored in the CPT is the number of joint assignments to X and $\mathrm{Pa}(X)$, that is,
$|\mathrm{Val}(\mathrm{Pa}(X))| \times |\mathrm{Val}(X)|$ [40]. This size of the CPT grows exponentially in the number of parents in the belief network. For example, the size of the CPT for 5 binary
parents of a binary variable X is $2^5 \times 2 = 64$ values; for 10 parents we need to store
$2^{10} \times 2 = 2{,}048$ values.

In general, the relationship between a given DAG and the corresponding joint pdf
over the random variables is defined as the product, over all of nodes of the DAG,
of a conditional pdf for each node conditioned on the variables corresponding to the
parents of that node in the DAG:

$$\Pr(\mathbf{x}) = \underbrace{\prod_{k=1}^{K} \Pr(x_k|x_1, x_2, \ldots, x_{k-1})}_{\text{Factored form}} = \underbrace{\prod_{k=1}^{K} \Pr(x_k|\mathrm{Pa}(x_k))}_{\text{Graphical form}} \qquad (4)$$

where $\mathrm{Pa}(x_k)$ denotes the set of parents of x_k and $\mathbf{x} = \{x_1, x_2, \ldots, x_K\}$. Equation (4)
results in the joint pdfs (2) and (3) for the $K = 3$ and $K = 5$. The corresponding
network topologies and DAG are given in Fig. 7a and Fig. 7b, respectively.

The main drawback of belief networks is that its compact graphical structure must
be supported by CPTs, which corresponds to distributed memories. The size of these

memories grows exponentially with the number of variables. To improve the computational properties of belief networks, decomposition is applied [3, 34, 55]. However, it does not always result in an acceptable number of values for the CPTs. The idea of representing CPTs by decision diagrams has been the focus of recent research [56].

6 Estimating Risk of Attacks

Figure 8 depicts a simplified belief network that models attack scenarios on an access system supporting the e-passport. This belief network will be referred to as the "scenario network". There are only three random variables: S, L, and W with the domains described below:

Variable S denotes the scenario currently taking place; $\text{Val}(S) = \{s_1, s_2, s_3, s_4, s_5\}$ where

- s_1 denotes a normal situation where the traveler holds an e-passport belonging to his/herself.
- s_2 denotes a situation where the traveler does not have an e-passport.
- s_3 denotes a situation where the traveler has lost their e-passport.
- s_4 denotes a situation where the traveler is attempting to use an e-passport that they have stolen.
- s_5 denotes a situation where the traveler is attempting to use a counterfeit e-passport, or an e-passport obtained through fraudulent means.

Variable L denotes whether or not the e-passport has been reported as lost; $\text{Val}(L) = \{l_1, l_2\}$ where l_1 indicates that the passport has been reported as lost, and l_2 indicates that the passport has not been reported as lost.

Variable W denotes whether or not the e-passport is on a watchlist for being fraudulently obtained or not; $\text{Val}(W) = \{w_1, w_2\}$ where w_1 indicates that the e-passport is on a watchlist, and w_2 indicates that the e-passport is not on a watchlist.

The conditional probabilities in Fig. 8 *do not claim to be accurate and serve only as examples.*

The probability that the e-passport was stolen when it was reported as being lost is ~0.387755. A-profiler operates with the total joint probability described by this belief network:

$$\Pr(S, L, W) = \Pr(S)\Pr(L|S)\Pr(W|S, L)$$

Risk assessments of other scenarios can be obtained by analogy. As an example of inference, imagine that it is known that the traveler's e-passport has been reported as lost, and we are trying to infer the probability that it was actually stolen. Thus, the probability that we wish to compute is: $\Pr(s_4|l_1) = \Pr(s_4, l_1)/\Pr(l_1)$. This calculation is carried out in the box in Fig. 8.

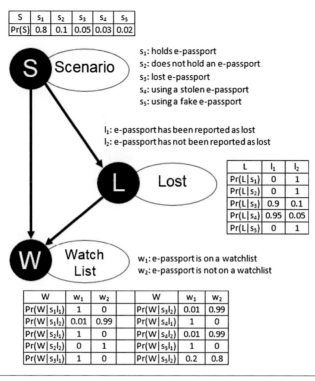

S	s_1	s_2	s_3	s_4	s_5
Pr(S)	0.8	0.1	0.05	0.03	0.02

s_1: holds e-passport
s_2: does not hold an e-passport
s_3: lost e-passport
s_4: using a stolen e-passport
s_5: using a fake e-passport

l_1: e-passport has been reported as lost
l_2: e-passport has not been reported as lost

L	l_1	l_2
$Pr(L \mid s_1)$	0	1
$Pr(L \mid s_2)$	0	1
$Pr(L \mid s_3)$	0.9	0.1
$Pr(L \mid s_4)$	0.95	0.05
$Pr(L \mid s_5)$	0	1

w_1: e-passport is on a watchlist
w_2: e-passport is not on a watchlist

W	w_1	w_2		W	w_1	w_2
$Pr(W \mid s_1 l_1)$	1	0		$Pr(W \mid s_3 l_2)$	0.01	0.99
$Pr(W \mid s_1 l_2)$	0.01	0.99		$Pr(W \mid s_4 l_1)$	1	0
$Pr(W \mid s_2 l_1)$	1	0		$Pr(W \mid s_4 l_2)$	0.01	0.99
$Pr(W \mid s_2 l_2)$	0	1		$Pr(W \mid s_5 l_1)$	1	0
$Pr(W \mid s_3 l_1)$	1	0		$Pr(W \mid s_5 l_2)$	0.2	0.8

$$\Pr(l_1) = \sum_{S,W} \Pr(S, l_1, W) = \sum_{S,W} \Pr(S) \Pr(l_1|S) \Pr(W|S, l_1)$$

$$= \sum_{S} \Pr(S) \Pr(l_1|S) \sum_{W} \Pr(W|S, l_1) = \sum_{S} \Pr(S) \Pr(l_1|S)$$

$$= \Pr(s_1) \Pr(l_1|s_1) + \Pr(s_2) \Pr(l_1|s_2) + \Pr(s_3) \Pr(l_1|s_3)$$
$$+ \Pr(s_4) \Pr(l_1|s_4) + \Pr(s_5) \Pr(l_1|s_5)$$

$$= (0.8)(0) + (0.1)(0) + (0.05)(0.9) + (0.03)(0.95) + (0.02)(0) = 0.0735$$

$$\Pr(s_4, l_1) = \sum_{W} \Pr(s_4, l_1, W) = \sum_{W} \Pr(s_4) \Pr(l_1|s_4) \Pr(W|s_4, l_1)$$

$$= \Pr(s_4) \Pr(l_1|s_4) \sum_{W} \Pr(W|s_4, l_1) = \Pr(s_4) \Pr(l_1|s_4)$$

$$= (0.03)(0.95) = 0.0285$$

$$\Pr(s_4|l_1) = \frac{\Pr(s_4, l_1)}{\Pr(l_1)} = \frac{0.0285}{0.0735} \approx 0.387755$$

Fig. 8 A belief (Bayesian) network that models attack scenarios on the e-passport system

7 Mitigating A-Profiler Complexity Using a Decomposition of Belief Networks

One significant drawback with probabilistic inference using belief networks is that inference is an NP-complete problem [40, p. 288]. To address this problem, a large belief network can be subdivided into "modeling module" [34]. A modeling module is a small belief network that may be integrated into a larger network of modeling modules. In a modeling module, some of the probability values may themselves be functions of posterior probability distributions computed in other networks. A simplified model of the A-machine for traveler risk estimation as a network of modules is shown in Fig. 9.

An example belief network that will become a non-trivial modeling module is shown in Fig. 10. This module will be named the "authentication network". This network will request a posterior distribution for the random variable S and a posterior distribution for the random variable W as input. The variable S is the "scenario"variable from the scenario network, and the variable W is the "watchlist" variable from the scenario network. The domain of each variable is described below:

Variable A denotes whether the e-passport under consideration passes authentication or not; $Val(A) = \{a_1, a_2\}$ where

- a_1 denotes the scenario where the e-passport is authenticated by the automated border control system.
- a_2 denotes the scenario where the e-passport is rejected by the automated border control system.

Variable B denotes whether the traveler who holds the e-passport passes biometric recognition; $Val(B) = \{b_1, b_2\}$ where

- b_1 denotes the scenario where the biometric data extracted from the traveler matches the data extracted from the e-passport.

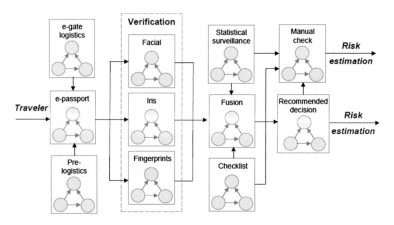

Fig. 9 A simplified model of the A-machine as a network of modules

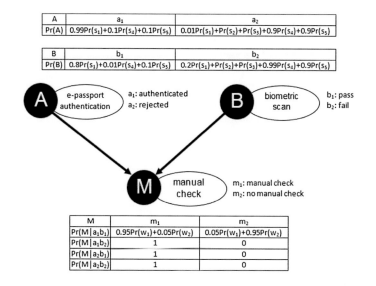

A	a_1	a_2
Pr(A)	$0.99\text{Pr}(s_1)+0.1\text{Pr}(s_4)+0.1\text{Pr}(s_5)$	$0.01\text{Pr}(s_1)+\text{Pr}(s_2)+\text{Pr}(s_3)+0.9\text{Pr}(s_4)+0.9\text{Pr}(s_5)$

B	b_1	b_2
Pr(B)	$0.8\text{Pr}(s_1)+0.01\text{Pr}(s_4)+0.1\text{Pr}(s_5)$	$0.2\text{Pr}(s_1)+\text{Pr}(s_2)+\text{Pr}(s_3)+0.99\text{Pr}(s_4)+0.9\text{Pr}(s_5)$

M	m_1	m_2
$\text{Pr}(M\,|\,a_1b_1)$	$0.95\text{Pr}(w_1)+0.05\text{Pr}(w_2)$	$0.05\text{Pr}(w_1)+0.95\text{Pr}(w_2)$
$\text{Pr}(M\,|\,a_1b_2)$	1	0
$\text{Pr}(M\,|\,a_2b_1)$	1	0
$\text{Pr}(M\,|\,a_2b_2)$	1	0

Risk of the traveler being sent to the manual check

$$\Pr(m_1) = \sum_{A,B} \Pr(A, B, m_1) = \sum_{A,B} \Pr(A)\,\Pr(B)\,\Pr(m_1|A, B)$$

$$= \sum_{A} \Pr(A) \sum_{B} \Pr(B)\,\Pr(m_1|A, B)$$

$$= \Pr(a_1)(\Pr(b_1)\,\Pr(m_1|a_1, b_1) + \Pr(b_2)\,\Pr(m_1|a_1, b_2))$$
$$+ \Pr(a_2)(\Pr(b_1)\,\Pr(m_1|a_2, b_1) + \Pr(b_2)\,\Pr(m_1|a_2, b_2))$$

$$= (0.169)((0.123)(0.5) + (0.877)(1)) + (0.831)((0.123)(1) + (0.877)(1))$$

$$\approx 0.989607$$

Fig. 10 An example of A-profiler for risk assessment of simplest scenario of e-passport authentication process in A-machine

- b_2 denotes the scenario where the biometric data extracted from the traveler does not match the data extracted from the e-passport.

Variable M denotes whether the traveler is directed to a manual check or not; $\text{Val}(M) = \{m_1, m_2\}$ where m_1 denotes the scenario where the traveler fails automated authentication and is directed to a manual check, and m_2 denotes the scenario where the traveler passes automated authentication and is not directed to a manual check.

Probabilities associated with variable A, B, and M (note the dependence on $\Pr(S)$) are given in Fig. 10. What makes modeling modules distinct from ordinary Bayesian network is the fact that the conditional probabilities may depend on posterior probability values computed in other modeling modules. The probability distributions $\Pr(S)$ and $\Pr(W)$ where S denotes the scenario type and W denotes whether of not the e-passport is on a watchlist are imported from another network.

As an example of inference, assume that the authentication network has imported the following posterior distribution for S: $\Pr(s_1) = 0.1$, $\Pr(s_2) = 0.1$, $\Pr(s_3) = 0.1$, $\Pr(s_4) = 0.3$, and $\Pr(s_5) = 0.4$; as well as the following posterior distribution for W: $\Pr(w_1) = 0.5$, $\Pr(w_2) = 0.5$. The prior probability distributions for A and B respectively become: $\Pr(a_1) = 0.169, \Pr(a_2) = 0.831, \Pr(b_1) = 0.123, \Pr(b_2) = 0.877$. The conditional probability for M given $A = a_1$ and $B = b_1$ becomes $\Pr(m_1|a_1,b_1) = 0.5$ and $\Pr(m_2|a_1,b_1) = 0.5$. The probability that the traveler will fail to cross the automated border and be sent to a manual check is therefore ~0.989607 as shown in Fig. 10.

8 DST Based A-Profiler

The main drawback of the A-profiler based on belief networks is the need of detailed initial risk statistics which are not available in practice. In most cases, these statistics can be defined as interval estimations. This means that the A-profiler should be used many times in order to calculate possible combinations of initial interval data. The result of such calculation is a point estimate which is then interpreted by an expert.

The advantage of the A-profiler based on belief networks is that it utilizes a causality paradigm based on graphs. The graphs can be decomposed to sub-graphs, and therefore can be used for modeling large-scale tasks using decomposition techniques. For such problems, we develop a technique for replacing CPTs by decision diagrams [56]. In some cases, it drastically improves the performance of the A-profiler. As our goal is to improve the robustness of the A-profiler, we need a paradigm that allows for interval estimates for the input and output data and requires less initial statistics, while retaining the advantages of a causal description of the problem. To approach this problem, we suggest the use of a DST based technique.

8.1 The Motivation and Key Idea

There are some key differences between the DST evidential techniques and belief (Bayesian) networks:

1. The DST approach can be applied when the prior probabilities and likelihood functions are unknown. That is, DST accepts an incomplete probabilistic model but Bayesian inference does not.
2. Belief (Bayesian) network can be used when the required information is available. However, when knowledge is not complete, such as ignorance exists about the prior probabilities, DST offers an alternative approach.
3. The DST well suited for fusion incomplete information from multi sources (sensors) because DST permits probabilities to be assigned directly to an uncertain event. Bayesian approach is limited in this.

The DST model and belief (Bayesian) network provide the same results when uncertainty interval is zero for all propositions and the probability mass assigned to units of propositions is zero.

There is the relationship between Markov networks (as undirected graphs) and belief (Bayesian) networks (as directed graphs) [40]. Our idea is to incorporate DST measures into belief (Bayesian) networks. This is a useful extension of both belief (Bayesian) network and DST measures because it utilizes advantages of both approaches to decision-making under uncertainty.

Our approach is different from the Transferable Belief Model (TBM) developed in [45]. Similar to our approach, the TBM model is based on the Bayesian concept of updating knowledge and the calculation of posterior probabilities: to pass from likelihoods to *posterior beliefs*. Our approaches are similar as we replace every probability function by a *DST belief function*. This novelty provides the opportunity to handle degrees of uncertainty hard to represent in a probabilistic Bayesian approach. However, the TBM is limited by the Bayesian updating rule and does not explore complex belief (Bayesian) networks. Instead our goal is to utilize causality in DST measures, resulting in a belief (Bayesian) network with DST-based CPTs. Our approach is also different from the directed graphical model introduced in [58]. Jirousek [58] does not utilize Dempster's rule of combination, nor develops the concept of conditional Dempster-Shafer tables used in this chapter. Our approach is implemented in the software package "Dempster-Shafer Bayesian Network (DSBN)" which is available for researchers [38].

8.2 DST Interval Measures

Instead of treating events directly, as it is done in belief networks, DST models are based on the concept of evidence [48, 49]. Evidence is related to data through the higher level interpretation imposed on it. A DST structure operates with sets of propositions and assigns to each of them an interval [Belief, Plausibility] in which the degree of belief must lie. Belief measures the strength of the evidence in favor of a set of propositions. It ranges from 0 (no evidence) to 1 (certainty). Plausibility is defined as Plausibility$(A) = 1 -$ Belief$(\neg A)$. It also ranges from 0 to 1 and measures the extent to which evidence exists in favor (Fig. 11).

The DST model utilizes the formalization of *imprecise* probabilities. They are captured by *belief* and *plausibility* measures, which may be interpreted as lower and upper probabilities respectively. The amount of information obtained by the action may be measured by the reduction of uncertainty that results from the action. In this sense, the amount of uncertainty and the amount of information are connected. The DST belief model provides a framework for the representation of knowledge about the value of an uncertain variable which can be used when there exists some uncertainty regarding our knowledge of the underlying measure.

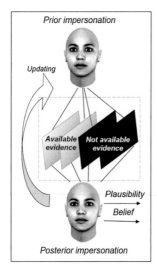

Step 1: Compute the value of ''belief'' for set C:

$$\mathrm{Bel}(C) = \sum_{B \in \mathcal{F},\, B \subseteq C} m(B)$$

$\mathrm{Bel}(C)$ is the probability that the chosen focal element forces the outcome to belong to C and so is a lower bound on the probability of the outcome belonging to C.

Step 2: Compute the value of ''plausibility'' for set C:

$$\mathrm{Pl}(C) = \sum_{B \in \mathcal{F},\, B \cap C \neq \emptyset} m(B)$$

$\mathrm{Pl}(C)$ is the probability that the chosen focal element allows the outcome to belong to C and so is an upper bound on the probability of the outcome belonging to C.

Fig. 11 Belief and plausibility computing in DST-based A-profiler

8.3 Theoretical Platform

In this and the next sections, we introduce the theoretical platform of the software package "Dempster-Shafer Bayesian Network (DSBN)" [38]. The DSBN package supports DST calculations for belief (Bayesian) networks.

Recall that $\mathcal{X}$ denotes the set of all random variables. A DST model over $\mathcal{X}$ is defined as follows [48, 49]: A set $\mathcal{F}$ of distinct non-empty subsets of $\mathrm{Val}(\mathcal{X})$, $\mathcal{F} \subseteq 2^{\mathrm{Val}(\mathcal{X})} - \{\emptyset\}$ are chosen as "focal elements". Each focal element denotes a possible range for the true outcome. A probability distribution over the set of focal elements is given. Each $B \in \mathcal{F}$ is assigned a probability value $m(B)$ which must sum to 1: $\sum_{B \in \mathcal{F}} m(B) = 1$.

A DST-based A-profiler can be interpreted as follows [48]: the set $B \in \mathcal{F}$ that contains the true outcome is chosen with probability $m(B)$. As can be seen a DST-based A-profiler does not necessarily ascribe precise probability values to each possible outcome.

Computing a lower and upper bound on the probability of the outcome belonging to a set $C \subseteq \mathrm{Val}(\mathcal{X})$ of outcomes is given in Fig. 11. If $\mathcal{F} = \{\{v\}|v \in \mathrm{Val}(\mathcal{X})\}$, then $\mathrm{Bel}(C) = \mathrm{Pl}(C)$ for all $C \subseteq \mathrm{Val}(\mathcal{X})$ and the DST model effectively denotes a probability distribution. Given DST models D_1 and D_2 over $\mathcal{X}$, we will say that $D_1 \subseteq D_2$ if $[\mathrm{Bel}_{D_1}(C), \mathrm{Pl}_{D_1}(C)] \subseteq [\mathrm{Bel}_{D_2}(C), \mathrm{Pl}_{D_2}(C)]$ for every subset $C \subseteq \mathrm{Val}(\mathcal{X})$.

Given two different DST models D_1 and D_2 over $\mathcal{X}$, the information in these models can be combined into $D_3 = D_1 \times D_2$ using Dempster's rule of combination which provides the formalism to combine probability masses from different sources of information [57]. Details are given in Fig. 12.

Combining DST models

$$m_{D_3}(B_3) = \frac{1}{M} \sum_{\substack{B_1 \in \mathcal{F}_{D_1} \\ B_2 \in \mathcal{F}_{D_2} \\ B_3 = B_1 \cap B_2}} m_{D_1}(B_1) m_{D_2}(B_2)$$

$$M = \sum_{\substack{B_1 \in \mathcal{F}_{D_1} \\ B_2 \in \mathcal{F}_{D_2} \\ B_1 \cap B_2 \neq \emptyset}} m_{D_1}(B_1) m_{D_2}(B_2)$$

DST's rule of combination. Compute the weights $m(B_3)$. The focal elements of D_3 are the non-empty elements of $\{B_1 \cap B_2 | B_1 \in \mathcal{F}_{D_1} \wedge B_2 \in \mathcal{F}_{D_2}\}$. $M \leq 1$ is a normalization constant such that $\sum_{B \in \mathcal{F}_{D_3}} m_{D_3}(B) = 1$.

Computing marginal and conditional DST models

$$m_{D_{\mathcal{Y}}}(B_{\mathcal{Y}}) = \sum_{\substack{B \in \mathcal{F}_D \\ proj(B|\mathcal{Y}) = B_{\mathcal{Y}}}} m_D(B)$$

Computing a marginalized DST model: *Let $\mathcal{Z} = \mathcal{X} - \mathcal{Y}$. For a set $S \subseteq Val(\mathcal{X})$, let $proj(S|\mathcal{Y}) = \{V_{\mathcal{Y}} \in Val(\mathcal{Y}) | \exists V_{\mathcal{Z}} \in Val(\mathcal{Z}) : \langle V_{\mathcal{Y}}, V_{\mathcal{Z}} \rangle \in S\}$. The focal elements of $D_{\mathcal{Y}}$ are $\mathcal{F}_{D_{\mathcal{Y}}} = \{proj(B|\mathcal{Y}) | B \in \mathcal{F}_D\}$.*

$$m_{D_{\mathcal{Z}|V_{\mathcal{Y}}}}(B') = \frac{1}{M} \sum_{\substack{B \in \mathcal{F}_D \\ slice(B|V_{\mathcal{Y}}) = B'}} m_D(B)$$

M is a normalization constant so that the weights sum to 1.

Computing a conditioned DST model: *Let $\mathcal{Z} = \mathcal{X} - \mathcal{Y}$. Let $V_{\mathcal{Y}} \in Val(\mathcal{Y})$ be a specific assignment that the variables in $\mathcal{Y}$ are fixed to. This evidence can be applied to model D to get the conditional DST structure $D_{\mathcal{Z}|V_{\mathcal{Y}}}$ as follows: For a set $S \subseteq Val(\mathcal{X})$, let $slice(S|V_{\mathcal{Y}}) = \{V_{\mathcal{Z}} \in Val(\mathcal{Z}) | \langle V_{\mathcal{Z}}, V_{\mathcal{Y}} \rangle \in S\}$. The focal elements of $D_{\mathcal{Z}|V_{\mathcal{Y}}}$ are the non-empty elements of $\{slice(B|V_{\mathcal{Y}}) | B \in \mathcal{F}_D\}$.*

Fig. 12 Basic operations of the DST-based A-profiler: computing the combination of DST models, marginal and conditional DST models

Given a subset of random variables $\mathcal{Y} \subset \mathcal{X}$, a DST model D over $\mathcal{X}$ can be marginalized to $\mathcal{Y}$ to get the marginal DST structure $D_{\mathcal{Y}}$ as shown in Fig. 12. In a similar manner, a conditional DST model is derived (Fig. 12).

8.4 Dempster-Shafer Analog of Markov-Networks

Like traditional Markov networks, a Dempster-Shafer Markov network (DS-MN) consists of a collection of Dempster-Shafer factors (DSFs). A DSF F is not simply an array indexed by the variables Var(F), but a DST model over the variables Var(F). Each focal element is a non-empty subset of Val(Var(F)), and the focal elements are all distinct. The restriction that the weights assigned to the focal elements sum to 1 is relaxed, similar to how the array entries in a Markov network factor do not

---

Computing the DST model from a DS-MN

For DS factors F_1 and F_2, just as with Markov networks, $Var(F_1F_2) = Var(F_1) \cup Var(F_2)$. Using the extend operation, the focal elements of both F_1 and F_2 are extended to cover the variables $Var(F_1) \cup Var(F_2)$ Letting $\mathcal{Y} = Var(F_1) \cup Var(F_2)$, the non-empty elements of $\{ext(B_1|\mathcal{Y}) \cap ext(B_2|\mathcal{Y})|B_1 \in \mathcal{F}_{F_1} \wedge B_2 \in \mathcal{F}_{F_2}\}$ are the focal elements of F_1F_2. The weight assigned to each focal element $B_3 \in \mathcal{F}_{F_1F_2}$ is:

$$m_{F_1F_2}(B_3) = \sum_{\substack{B_1 \in \mathcal{F}_{F_1}, B_2 \in \mathcal{F}_{F_2} \\ B_3 = ext(B_1|\mathcal{Y}) \cap ext(B_2|\mathcal{Y})}} m_{F_1}(B_1)m_{F_2}(B_2)$$

Given a set of factors $\{F_1, F_2, \ldots, F_k\}$, the DST model generated from the DS-MN is the product of all factors: $D = \prod_{i=1}^{k} F_i$. The weights are all divided by a normalizing constant so that they all sum to 1.

Fig. 13 Computing the weights of focal elements of DST-based A-profiler

necessarily sum to 1. DSFs are multiplied using Dempster's rule of combination. Like with Markov networks, the normalization constant is only computed after all of the factors have been multiplied together.

Multiplying DSFs over different variable sets requires the "extend" operation: let $\mathcal{Y}$ and $\mathcal{Z}$ be subsets of $\mathcal{X}$ where $\mathcal{Z} \subset \mathcal{Y}$. Given a set of assignments $S \subseteq Val(\mathcal{Z})$, S can be extended to a subset of $Val(\mathcal{Y})$ by the following: $ext(S|\mathcal{Y}) = \{\langle V_1, V_2 \rangle | V_1 \in S \wedge V_2 \in Val(\mathcal{Y} - \mathcal{Z})\}$. We describe the multiplication of DST factors F_1 and F_2 in Fig. 13.

8.5 Embedding DST Models into Belief (Bayesian) Networks

The goal of this section is to show how we can utilize the causal paradigm of belief (Bayesian) network for DST interval measures. For this, we replace the CPTs in a belief network with DST based structures to create a "Dempster-Shafer Bayesian network" (DS-BN). For each node X and for each possible assignment to the parents of X, $V \in Val(Pa(X))$, imagine that instead of a probability distribution over X, that we instead have a DST model $D_{X|V}$ over X. The DST model $D_{X|V}$ over X reflects uncertainty in the conditional probability distribution of X. For each node X, the DST models associated with each assignment to $Pa(X)$ form a DST analog of the CPT which we will refer to as a "conditional Dempster-Shafer table" (CDST).

To derive the resultant DS-MN, we now extend each DST model $D_{X|V}$ ($V \in Val(Pa(X))$) to a DST factor $F_{X|V}$ over the variables $\{X\} \cup Pa(X)$. The focal elements of $D_{X|V}$ are subsets of $Val(X)$, while the focal elements of $F_{X|V}$ are subsets of $Val(\{X\} \cup Pa(X))$. For each focal element $B \in \mathcal{F}_{D_{X|V}}$, B is extended as follows via the "Ballooning extension" [45]:

$$B \mapsto \{\langle V_X, V\rangle | V_X \in B\} \cup \left\{ V_A \in \mathrm{Val}(\{X\} \cup \mathrm{Pa}(X)) \middle| V_A(\mathrm{Pa}(X)) \neq V \right\}$$

In summary, each focal element in $D_{X|V}$ has appended onto it every assignment from $\mathrm{Val}(\{X\} \cup \mathrm{Pa}(X))$ except for those assignments that coincide with V. The weight assigned to each focal element is unchanged. The resultant DST model over $\mathcal{X}$ is the product of all DSFs $F_{X|V}$ for every X and $V \in \mathrm{Val}(\mathrm{Pa}(X))$ with the weights normalized.

8.6 Example Application of the DST-Based A-Profiler

Consider again the scenario network described in Sect. 6 and shown in Fig. 8. Instead of assigning a CPT to each node, a "conditional DST table" is assigned to each node as described in Sect. 8.5. Since variable S has 5 different values, this large domain makes the inference too complex to be followed. To simplify matters, the range of values that S can attain will be restricted to $\mathrm{Val}(S) = \{s_1, s_3, s_4\}$, omitting the case where the traveler is not enrolled in an e-passport program ($S = s_2$), and the case where the traveler is using a fraudulent e-passport ($S = s_5$).

The DST model associated with variable S is $\langle\{s_1\}, 0.8\rangle$; $\langle\{s_3\}, 0.15\rangle$; $\langle\{s_4\}, 0.05\rangle$. A DST model is expressed as a series of $\langle$ focal element, weight $\rangle$ pairs: $\langle B, m(B)\rangle$. This DST model indicates that with probability 0.8 we know that the traveler holds their own e-passport; with probability 0.15 we know that the traveler does not currently possess an e-passport; and with probability 0.05 we know that the traveler is engaged in illegal activity. The CDSTs associated with variables L and W are listed in Table 2.

Table 2 CDSTs for L and W

S		DS models for L
s_1		$\langle\{l_2\}, 1\rangle$
s_3		$\langle\{l_1\}, 0.8\rangle$; $\langle\{l_1, l_2\}, 0.2\rangle$
s_4		$\langle\{l_1\}, 0.9\rangle$; $\langle\{l_1, l_2\}, 0.1\rangle$
S	L	DS models for W
s_1	l_1	$\langle\{w_1\}, 1\rangle$
s_1	l_2	$\langle\{w_1, w_2\}, 0.02\rangle$; $\langle\{w_2\}, 0.98\rangle$
s_3	l_1	$\langle\{w_1\}, 1\rangle$
s_3	l_2	$\langle\{w_1, w_2\}, 0.02\rangle$; $\langle\{w_2\}, 0.98\rangle$
s_4	l_1	$\langle\{w_1\}, 1\rangle$
s_4	l_2	$\langle\{w_1, w_2\}, 0.02\rangle$; $\langle\{w_2\}, 0.98\rangle$

8.7 Algorithm for DST Based Belief Network

First, we derive the total DST model that is described by the above DS-BN. As described in Sect. 8.5, the total DST model is built by first deriving a DSF from each row of each CDST. There is one DSF associated with S, 3 DSFs associated with L, and 6 DSFs associated with W. The one DSF associated with S will be denoted by F_S.

Step I: The product of the 3 DST factors associated with L, denoted by F_L, is shown below. Let sl denote an arbitrary assignment from $\mathrm{Val}(\{S, L\})$, where $s \in \mathrm{Val}(S)$ and $l \in \mathrm{Val}(L)$.

Focal element	Weight	Focal element	Weight
$\{s_1 l_2, s_3 l_1, s_4 l_1\}$	0.72	$\{s_1 l_2, s_3 l_1, s_3 l_2, s_4 l_1\}$	0.18
$\{s_1 l_2, s_3 l_1, s_4 l_1, s_4 l_2\}$	0.08	$\{s_1 l_2, s_3 l_1, s_3 l_2, s_4 l_1, s_4 l_2\}$	0.02

Step II: The product of the 6 DST factors associated with W, denoted by F_W, is shown below. To simplify notation, slw denotes an arbitrary assignment from $\mathrm{Val}(\{S, L, W\})$ where $s \in \mathrm{Val}(S)$, $l \in \mathrm{Val}(L)$, and $w \in \mathrm{Val}(W)$.

Focal element	Weight	Focal element	Weight
$\{s_1 l_1 w_1, s_1 l_2 w_1, s_1 l_2 w_2, \ldots$ $\ldots, s_3 l_1 w_1, s_3 l_2 w_1, s_3 l_2 w_2, \ldots$ $\ldots s_4 l_1 w_1, s_4 l_2 w_1, s_4 l_2 w_2\}$	0.000008	$\{s_1 l_1 w_1, s_1 l_2 w_2, \ldots$ $\ldots, s_3 l_1 w_1, s_3 l_2 w_1, s_3 l_2 w_2, \ldots$ $\ldots s_4 l_1 w_1, s_4 l_2 w_1, s_4 l_2 w_2\}$	0.000392
$\{s_1 l_1 w_1, s_1 l_2 w_1, s_1 l_2 w_2, \ldots$ $\ldots s_3 l_1 w_1, s_3 l_2 w_1, s_3 l_2 w_2, \ldots$ $\ldots s_4 l_1 w_1, s_4 l_2 w_2\}$	0.000392	$\{s_1 l_1 w_1, s_1 l_2 w_2, \ldots$ $\ldots s_3 l_1 w_1, s_3 l_2 w_1, s_3 l_2 w_2, \ldots$ $\ldots s_4 l_1 w_1, s_4 l_2 w_2\}$	0.019208
$\{s_1 l_1 w_1, s_1 l_2 w_1, s_1 l_2 w_2, \ldots$ $\ldots s_3 l_1 w_1, s_3 l_2 w_2, \ldots$ $\ldots s_4 l_1 w_1, s_4 l_2 w_1, s_4 l_2 w_2\}$	0.000392	$\{s_1 l_1 w_1, s_1 l_2 w_2, \ldots$ $\ldots s_3 l_1 w_1, s_3 l_2 w_2, \ldots$ $\ldots s_4 l_1 w_1, s_4 l_2 w_1, s_4 l_2 w_2\}$	0.019208
$\{s_1 l_1 w_1, s_1 l_2 w_1, s_1 l_2 w_2, \ldots$ $\ldots s_3 l_1 w_1, s_3 l_2 w_2, \ldots$ $\ldots s_4 l_1 w_1, s_4 l_2 w_2\}$	0.019208	$\{s_1 l_1 w_1, s_1 l_2 w_2, \ldots$ $\ldots s_3 l_1 w_1, s_3 l_2 w_2, \ldots$ $\ldots s_4 l_1 w_1, s_4 l_2 w_2\}$	0.941192

Step III: Let us compute the product: $F_{S,L} = F_S \times F_L$. The result is shown below:

Focal element	Weight
$\{s_1 l_2\}$	$0.8 \cdot (0.72 + 0.08 + 0.18 + 0.02) = 0.8$
$\{s_3 l_1\}$	$0.15 \cdot (0.72 + 0.08) = 0.12$
$\{s_3 l_1, s_3 l_2\}$	$0.15 \cdot (0.18 + 0.02) = 0.03$
$\{s_4 l_1\}$	$0.05 \cdot (0.72 + 0.18) = 0.045$
$\{s_4 l_1, s_4 l_2\}$	$0.05 \cdot (0.08 + 0.02) = 0.005$

Step IV: Next, we compute the product: $F_{S,L,W} = F_{S,L} \times F_W$, as shown below:

Focal element	Weight
$\{s_1 l_2 w_1, s_1 l_2 w_2\}$	$0.8 \cdot (0.000008 + 0.000392 + 0.000392 + 0.019208) = 0.016$
$\{s_1 l_2 w_2\}$	$0.8 \cdot (0.000392 + 0.019208 + 0.019208 + 0.941192) = 0.784$
$\{s_3 l_1 w_1\}$	$0.12 \cdot (0.000008 + 0.000392 + 0.000392 + 0.019208 + \ldots$
	$\ldots 0.000392 + 0.019208 + 0.019208 + 0.941192) = 0.12$
$\{s_3 l_1 w_1, s_3 l_2 w_1, s_3 l_2 w_2\}$	$0.03 \cdot (0.000008 + 0.000392 + 0.000392 + 0.019208) = 0.0006$
$\{s_3 l_1 w_1, s_3 l_2 w_2\}$	$0.03 \cdot (0.000392 + 0.019208 + 0.019208 + 0.941192) = 0.0294$
$\{s_4 l_1 w_1\}$	$0.045 \cdot (0.000008 + 0.000392 + 0.000392 + 0.019208 + \ldots$
	$\ldots 0.000392 + 0.019208 + 0.019208 + 0.941192) = 0.045$
$\{s_4 l_1 w_1, s_4 l_2 w_1, s_4 l_2 w_2\}$	$0.005 \cdot (0.000008 + 0.000392 + 0.000392 + 0.019208) = 0.0001$
$\{s_4 l_1 w_1, s_4 l_2 w_2\}$	$0.005 \cdot (0.000392 + 0.019208 + 0.019208 + 0.941192) = 0.0049$

The above table simplifies to:

Focal element	Weight	Focal element	Weight
$\{s_1 l_2 w_1, s_1 l_2 w_2\}$	0.016	$\{s_3 l_1 w_1, s_3 l_2 w_2\}$	0.0294
$\{s_1 l_2 w_2\}$	0.784	$\{s_4 l_1 w_1\}$	0.045
$\{s_3 l_1 w_1\}$	0.12	$\{s_4 l_1 w_1, s_4 l_2 w_1, s_4 l_2 w_2\}$	0.0001
$\{s_3 l_1 w_1, s_3 l_2 w_1, s_3 l_2 w_2\}$	0.0006	$\{s_4 l_1 w_1, s_4 l_2 w_2\}$	0.0049

which is the total DST model for the DS-BN under consideration.

8.8 Example of an Inference Problem: E-Passport Lost

Let us consider the inference problem where the e-passport has been reported as being lost $L = l_1$. We wish to compute the belief and plausibility that the e-passport is simply lost and not stolen $S = s_3$. To start, we derive the conditional DST model by applying the evidence $L = l_1$ to the total DST model $F_{S,L,W}$ computed previously. This gives us the model:

Focal element	Weight
$\{s_3 w_1\}$	$\frac{1}{M}(0.12 + 0.0006 + 0.0294) = \frac{1}{0.15+0.05}0.15 = 0.75$
$\{s_4 w_1\}$	$\frac{1}{M}(0.045 + 0.0001 + 0.0049) = \frac{1}{0.15+0.05}0.05 = 0.25$

Marginalizing out variable W gives $\{s_3\} = 0.75$ and $\{s_4\} = 0.25$ We finally see that $\mathrm{Bel}(s_3|l_1) = \mathrm{Pl}(s_3|l_1) = 0.75$.

The Bayesian network with incorporated DST measures (or the DST Bayesian network) is shown in Fig. 14 (compare with Bayesian network in Fig. 8).

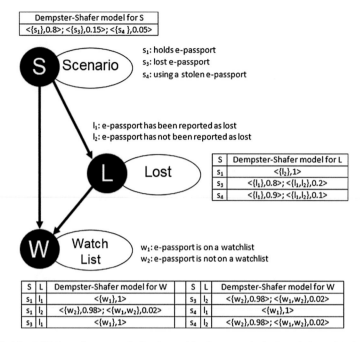

Dempster-Shafer model for S
<{s_1},0.8>; <{s_3},0.15>; <{s_4},0.05>

S Scenario

s_1: holds e-passport
s_3: lost e-passport
s_4: using a stolen e-passport

l_1: e-passport has been reported as lost
l_2: e-passport has not been reported as lost

L Lost

S	Dempster-Shafer model for L
s_1	<{l_2},1>
s_3	<{l_1},0.8>; <{l_1,l_2},0.2>
s_4	<{l_1},0.9>; <{l_1,l_2},0.1>

W Watch List

w_1: e-passport is on a watchlist
w_2: e-passport is not on a watchlist

S	L	Dempster-Shafer model for W
s_1	l_1	<{w_1},1>
s_1	l_2	<{w_2},0.98>; <{w_1,w_2},0.02>
s_3	l_1	<{w_1},1>

S	L	Dempster-Shafer model for W
s_3	l_2	<{w_2},0.98>; <{w_1,w_2},0.02>
s_4	l_1	<{w_1},1>
s_4	l_2	<{w_2},0.98>; <{w_1,w_2},0.02>

Fig. 14 The DST Bayesian network that is used in the example in Sect. 8.6. Each DS model is denoted using a list of focal element, weight pairs: $\langle B, m(B) \rangle$

8.9 Disadvantages of the DST Based Measures

In this section we will note an irregularity associated with combining DST models. DST model is well suited for representing uncertainty and combining information, especially in the case of low conflicts between sources with high beliefs. However, when there is great uncertainty in the probability values, Dempster's rule yields unexpected results.

Let D_1, D_1', D_2, D_2' be DST models over $\mathcal{X}$ such that $D_1 \subseteq D_1'$ and $D_2 \subseteq D_2'$. It is natural to expect that $(D_1 \times D_2) \subseteq (D_1' \times D_2')$, but this is not necessarily always the case as will be shown in the following counter-example. Let $\mathcal{X} = \{A\}$ where $\text{Val}(A) = \{a_1, a_2\}$; D_1 has the focal elements $\mathcal{F}_{D_1} = \{\{a_1\}, \{a_2\}\}$ with weights $m_{D_1}(\{a_1\}) = 0.1$ and $m_{D_1}(\{a_2\}) = 0.9$; and $D_2 = D_1$; D_1' has the focal elements $\mathcal{F}_{D_1'} = \{\{a_1\}, \{a_1, a_2\}, \{a_2\}\}$ with weights $m_{D_1'}(\{a_1\}) = 0.1$, $m_{D_1'}(\{a_1, a_2\}) = 0.8$, and $m_{D_1'}(\{a_2\}) = 0.1$; and $D_2' = D_1'$. It can easily be checked that $D_1 \subseteq D_1'$ and $D_2 \subseteq D_2'$.

Computing $D_1 \times D_2$ gives $m_{D_1 \times D_2}(\{a_1\}) = \frac{0.1^2}{0.1^2 + 0.9^2} = \frac{0.01}{0.82} \approx 0.012195$ and $m_{D_1 \times D_2}(\{a_2\}) = \frac{0.9^2}{0.1^2 + 0.9^2} = \frac{0.81}{0.82} \approx 0.987805$. Computing $D_1' \times D_2'$ gives $m_{D_1' \times D_2'}(\{a_1\}) = \frac{0.1^2 + 0.1 \cdot 0.8 + 0.8 \cdot 0.1}{1 - 0.1^2 - 0.1^2} = \frac{0.17}{0.98} \approx 0.173469$, $m_{D_1' \times D_2'}(\{a_1, a_2\}) = \frac{0.8^2}{1 - 0.1^2 - 0.1^2} = \frac{0.64}{0.98} \approx$

0.653061, and $m_{D_1' \times D_2'}(\{a_2\}) = \frac{0.1^2 + 0.1 \cdot 0.8 + 0.8 \cdot 0.1}{1 - 0.1^2 - 0.1^2} = \frac{0.17}{0.98} \approx 0.173469$. Hence, $(D_1 \times D_2)$ $\nsubseteq (D_1' \times D_2')$. Several alternative methods have been developed to make the DST fusion more intuitively appealing. In particular, the DSmT model proposes a new set of combination rules for information fusion [43]. The DSmT model deals with uncertain, imprecise and highly conflicting information for static and dynamic fusion. Our software package DSBN [38] can be extended for the DSmT measures.

9 Conclusion and Future Work

We introduced the concept and prototyping results of the A-profiler, a tool for risk profiling via modeling of the A-machines and supporting technologies. The key conclusions of our study are as follows:

1. An A-profiler based on belief networks can be used for large-scale tasks if historical statistics are available. In order to address the model complexity problem, we suggest that it can be decomposed and represented by a library of modeling modules. In some scenarios, the performance of the A-profiler can be significantly increased. In particular, conditional probability tables can be replaced by the more compact decision diagrams as suggested in [56].
2. The robustness of the A-profiler can be improved by using interval measures suggested by the DST model of uncertainty. We showed that the DST model can be embedded into a belief (Bayesian) network. We implemented this approach in the DS-BN-01 software package which is available for researches.

The next step in the evolution of A-profilers should be an embedding of the DSmT model [43] into a belief network.

Acknowledgments We acknowledge collaboration with *Dr. V. Shmerko* (University of Calgary, Canada). This work was partially supported by the Natural Sciences and Engineering Research Council of Canada (NSERC) through the Discovery grant "Biometric intelligent interfaces", and the Government of the Province of Alberta (Queen Elizabeth II Scholarship).

References

1. International Air Transport Association (IATA): Checkpoint of the future: Executive summary. 4th Proof. (2014). http://www.bing.com/search?q=iata
2. Department of Homeland Security (DHS): Future Attribute Screening Technology (FAST) Project, Science and Technology Directorate (2008). http://www.dhs.gov/xlibrary/assets/privacy/privacy_pia_st_fast
3. Daniels, D., Hudson, L.D., Laskey, K.B., et al.: Terrorism risk management. In: Pourret, O., Naim, P., Markot, B. (eds.) Bayesian Networks; A Practical Guide to Applications, pp. 239–262. Willey (2008)
4. ISO/IEC FDIS 30108-1:2015(E), Information technology—Biometric Identity Assurance Services—Part 1: BIAS services, International Organization for Standardization (2015)

5. Bolle, R., Connell, J., Pankanti, S., Ratha, N., Senior, A.: Guide to Biometrics. Springer, New York (2004)
6. Jain, A., Bolle, R., Pankanti, S. (eds.): Biometrics: Personal Identification in a Networked Society. Kluwer (1999)
7. Miller, B.: Vital signs of identity. IEEE Spect. **31**(2), 22–30 (1994)
8. Back, J.: Posture monitoring system for context awareness in mobile computing. IEEE Trans. Instrum. Meas. **59**(6), 1589–1599 (2010)
9. Creese, S., Gibson-Robinson, T., Goldsmith, M., et al.: Tools for understanding identity. In: Proceedings IEEE Conference Technologies for Homeland (2013)
10. Mobile Biometric Identification: white paper, Motorola, (2008). http://www.motorolasolutions.com/web/Business/Products/
11. NIST: Mobile ID Device Best Practice Recommendation Version 1.0, NISTSP 500-280 (2009)
12. Pavlidis, I., Levine, J.: Thermal image analysis for polygraph testing. IEEE Trans. Eng. Med. Biol. Mag. **6**, 56–64 (2002)
13. Poursaberi, A., Vana, J., Mracek, S., Dvora, R., Yanushkevich, S., Drahansky, M., Shmerko, V., Gavrilova, M.: Facial biometrics for situational awareness systems. IET Biom. **2**(2), 35–47 (2013)
14. Yanushkevich, S., Shmerko, V., Boulanov, O., Stoica, A.: Decision-making support in biometric-based physical access control systems: Design concept, architecture, and applications. In: Boulgouris, N.V., Plataniotis, K.N., Micheli-Tzanakou, E. (eds.) Biometrics: Theory, Methods, and Applications, pp. 599–631. IEEE Press, Wiley (2010)
15. AVATAR: Border patrol kiosk detects liars trying to enter U.S. Homeland Security News Wire (2012). http://www.homelandsecuritynewswire.com/
16. DHS (Department of Homeland Security): Future Attribute Screening Technology (FAST) Project, Science and Technology Directorate, Department of Homeland Security (2008). www.dhs.gov/xlibrary/assets/privacy/privacy_pia_st_fast
17. McBreen, H.M., Jack, M.A.: Evaluating humanoid synthetic agents in e-retail applications. IEEE Trans. Syst. Man Cybern. Part A: Syst. Hum. **31**(5), 394–405 (2001)
18. Eastwood, S.C., Yanushkevich, S.N., Drahansky, M.: Biometric intelligence in authentication machines: from talking faces to talking robots. In: Proceedings IIAI 3rd International Conference Advanced Applied Informatics, Japan, pp. 763–768 (2014)
19. Nunamaker Jr, J.F., Derrick, D.C., Elkins, A.C., Burgoon, J.K., Patton, M.W.: Embodied conversational agent-based kiosk for automated interviewing. J Manag. Inf. Syst. **28**(1), 17–48 (2011)
20. McLay, L.A., Lee, A.J., Jacobson, S.H.: Risk-based policies for airport security checkpoint screening. J. Transp. Sci. **44**(3), 333–349 (2010)
21. Nie, X., Batta, R., Drury, C.G., Lin, L.: Passenger grouping with risk levels in an airport security system. Eur. J. Oper. Res. **194**(2), 574–584 (2009)
22. Nuppeney, M.: Automated border control based on (ICAO compliant) eMRTDs, Federal Office for Information Security (2012). http://www.bsi.de
23. SITA: END-to-end border management: An integrated approach to passenger data collection, identity verification and risk management. SITA positioning paper (2012)
24. Fiondella, L., Gokhale, S.S., Lownes, N., Accorsi, M.: Security and performance analysis of a passenger screening checkpoint for mass-transit systems. In: Proceedings IEEE Conference Technologies for Homeland Security (HST), pp. 312–318 (2012)
25. Campbell, J.W.M.: New Zealand SmartGate: Using quantitative performance information to improve convenience and security. In: Proceedings International Biometric Performance Testing Conference, NIST (2014). http://www.nist.gov/itl/iad/ig/ibpc2014
26. Frontex: Best practice technical guidelines for automated border control (ABC) systems, Research and Development Unit, Frontex, Warsaw (2012). http://www.frontex.europa.eu
27. Frontex: BIOPASS II Automated biometric border crossing systems based on electronic passports and facial recognition: RAPID and SmartGate. Research and Development Unit, Frontex, Warsaw (2010). http://www.frontex.europa.eu

28. Bigo, D.S. et al.: Justice and Home Affairs Databases and a Smart Borders System at EU External Borders An Evaluation of Current and Forthcoming Proposals for European Policy Studies (CEPS), No. 52/Dec. (2012)
29. Florence, J., Friedman, R.: Profiles in terror: A legal framework for the behavioral profiling paradigm. George Mason Law Rev. 17(2), 423–481 (2010). http://www.georgemasonlawreview.org/doc/172Florenceand
30. IATA (International Air Transport Association): Checkpoint of thefuture: Executive summary. 4th Proof (2014). http://www.bing.com/search?q=iata%3A+Checkpoint+of+the+future.+Executive+summary
31. Eastwood, S.C., Shmerko, V.P., Yanushkevich, S.N., Drahansky, M., Gorodnichy, D.O.: Biometric-enabled authentication machines: A survey of open-set real-world applications. IEEE Trans. Hum. Mach. Syst. early access, May 2015
32. Yanushkevich, S.N., Eastwood, S.C., Drahansky, M., Shmerko, V.P.: Taxonomy of Impersonation Phenomenon in Authentication Machines for e-Borders. Proc. Int. Conf. Emerg. Secur. Technol. (2015)
33. Waltz, E., Llinas, J.: Multisensor Data Fusion. Artech House, MA (1990)
34. Eastwood, S.C., Yanushkevich, S.N.: Risk profiler in automated human authentication. In: Proceedings IEEE Workshop on Computational Intelligence in Biometrics and Identity Management—CIBIM, Orlando, Florida (2014)
35. Yanushkevich, S.N., Stoica, A., Shmerko, V.P.: Experience of design and prototyping of a multi-biometric early warning physical access control security system (PASS) and a training system (T-PASS). In: Proceedings of the 32nd Annual IEEE Industrial Electronics Society Conference, pp. 2347–2352. Paris, France (2006)
36. Sacanamboy, M., Cukic, B.: Combined performance and risk analysis for border management applications. In: Proceedings IEEE/IFIP Conference Dependable Systems and Networks (DSN), pp. 403–412 (2010)
37. DHS (Department of Homeland Security): DHS S&T and CBP Announce the Opening of the Maryland Test Facility (2014). http://www.dhs.gov/blog/2014/07/03/, http://www.dhs.gov/st-snapshot-new-dhs-facility-tests-biometric-technology-improves-air-entryexit-operations
38. Software package Dempster-Shafer Bayesian Network (DSBN-01). Biometric Laboratory, University of Calgary, Canada (2015). http://www.ucalgary.ca/btlab/Software
39. Aven, T.: Foundations of Risk Analysis, 2nd edn. Wiley (2012)
40. Koller, D., Friedman, N.: Probabilistic Graphical Models: Principles and Techniques, MIT Press (2009)
41. Frey, B.J., Jojic, N.: A comparison of algorithms for inference and learning in probabilistic graphical models. IEEE Trans. Pattern Anal. Mach. Intell. 27(9), 1392–1416 (2005)
42. Shafer, G.: A Mathematical Theory of Evidence. Princeton University Press, Princeton (1976)
43. Smarandache, F., Dezert, J., Tacnet, J.-M.: Fusion of sources of evidence with different importances and reliabilities In: Proceedings Workshop on the Theory of Belief Functions (BELIEF 2010), Edinburgh, UK (2010). https://hal.archives-ouvertes.fr/file/index/docid/559233/filename/GR2010-PUB00030651.pdf
44. Bishop, C.M.: Pattern Recognition and Machine Learning. Springer, New York (2006)
45. Delmotte, F., Smets, P.: Target identification based on the transferable belief model interpretation of Dempster-Shafer model. IEEE Trans. Syst. Man Cybern. Part A: Syst. Hum. 34(4), 457–471 (2004)
46. Papakostas, G.A., et al.: Fuzzy cognitive maps for pattern recognition applications. Int. J. Pattern Recognit. Artif. Intell. 22, 1461–1486 (2008)
47. Jang, J.-S.R.: ANFIS: adaptive-network-based fuzzy inference systems. IEEE Trans. Syst. Man Cybern. 23, 665–685 (1993)
48. Yager, R.R., Filev, D.P.: Including probabilistic uncertainty in fuzzy logic controller modeling using Dempster-Shafer theory. IEEE Trans. Syst. Man Cybern. 25(8), 1221–1230 (1995)
49. Yager, R.R.: Human behavioral modeling using fuzzy and Dempster Shafer theory, Social Computing, Behavioral Modeling, and Prediction, pp. 89–99. Springer, US (2008)

50. Denoeux, T.: A neural network classifier based on Dempster-Shafer theory. IEEE Trans. Syst. Man Cybern. Part A: Syst. Hum. **30**(2), 131–150 (2000)
51. Klein, L.A.: Sensor and Data Fusion: A Tool for Information Assessment and decision Making. SPIE, Bellingham (2007)
52. Bier, V.M., Azaiez, M.N. (eds.): Game Theoretic Risk Analysis of Security Threats. Springer, US (2009)
53. Barber, D.: Bayesian Reasoning and Machine Learning. Cambridge University Press, Cambridge (2012)
54. ICAO: Document 9303, part 1, vol. 2, e-passports. Retrieved 8 Sept. (2010). http://hasbrouck. org/documents/ICAO9303-pt1-vol2.pdf
55. Hwang, K., Cho, S.: Landmark detection from mobile life log using a modular Bayesian network model. Expert Syst. Appl. **36**(10), 12065–12076 (2009)
56. Eastwood, S.C., Yanushkevich, S.N., Shmerko, V.P.: Belief network support via decision diagrams. In: Proceedings of the 45th IEEE International Symposium on Multiple-Valued Logic (2015)
57. Voorbraak, F.: On the justification of Dempster's rule of combination. Artif. Intell. **48**(2), 171–197 (1991)
58. Jirousek, R.: Local computations in Dempster-Shafer theory of evidence. Int. J. Approx. Reason. **53**(8), 1155–1167 (2012)

Part IV
Situational Awareness and Threat Assessment

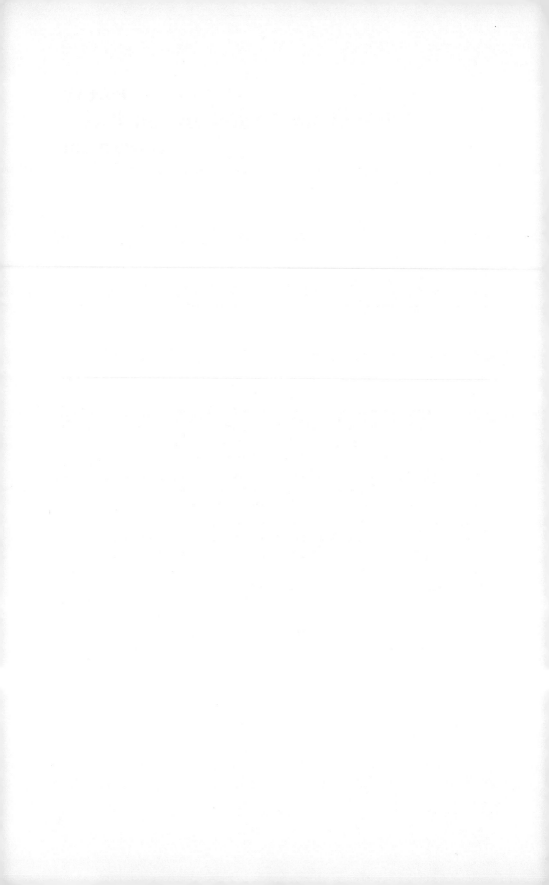

Game Theoretical Approach for Dynamic Active Patrolling in a Counter-Piracy Framework

Francesca De Simio, Marco Tesei and Roberto Setola

Keywords Game Theory · Maritime piracy · DSS · Bayesian-Dtackelberg · Security game

1 Introduction

Maritime piracy has become an important security focus area due to the influence that this phenomenon has on the global economy [1]. Commanders and policy-makers need decision-support tools for optimal resource allocation in order to plan and execute counter-piracy activities more efficiently and effectively. In this work we propose a Decision Support System (DSS) which aims to improve the allocation of mobile resources in a maritime scenarios. Specifically, a set of defender's resources have to patrol a large scenarios where several mobile targets move in the presence of mobile attackers with different and partially unknown characteristics and goals. To efficiently manage the problem, the DSS has been designed using the Game Theoretical framework in order to handle the attractiveness of targets so as modeling strategies of attackers and defenders. Indeed, Game Theory moves from the problem to find optimal solution in the presence of multiple player with different, or eventually conflicting, goals. Game theoretic models have been used for modelling scenarios in which a group of agents (generally labelled as defenders or patrollers) need to protect an area [2], a perimeter [3] or prevent an attack against one or more targets [4]. Moreover game theoretic models have been recently applied in security scenarios as airport terminals, commercial flights and ports [5–8]. These studies emphasize how Game Theory can be used to identify the best strategy for the defenders given the information and capabilities of opponents.

F. De Simio (✉) · M. Tesei · R. Setola
Complex Systems and Security Lab, Università Campus Bio-Medico di Roma, Rome, Italy
e-mail: f.desimio@unicampus.it

© Springer International Publishing Switzerland 2016
R. Abielmona et al. (eds.), *Recent Advances in Computational Intelligence in Defense and Security*, Studies in Computational Intelligence 621,
DOI 10.1007/978-3-319-26450-9_16

423

This chapter describes the architecture of the developed system, after providing an introduction about maritime piracy and a state of the art about game-theoretic patrolling.

In spite of the existing approaches which assume stationary targets and resources, we explicitly take into account the movement of targets, defenders and attackers. This introduces a spatial-temporal layer that allows to model constraints on displacements, and dynamic changes in available information. In this framework the optimal strategy is identified as the equilibrium of time-varying Bayesian-Stackelberg game.

This chapter is organized as described below. The first section provides an introduction about maritime piracy. The second section surveys the existing game-theoretic approaches for active patrolling. The third section describes the problem statement. The last three sections describe the problem formulation, the simulations carried out and the related results.

2 The Modern Piracy

Piracy is defined in the 1982 United Nations Convention on the Law of the Sea (UNCLOS) as

> any illegal acts of violence or detention, or any act of depredation, committed for private ends by the crew or the passengers of a private ship or a private aircraft, and directed against another ship or aircraft, or against persons or property on board such ship or aircraft [9].

In recent years, the problem of piracy has emerged as a major threat in the field of sea transportation expecially in some parts of the world [10]. The International Maritime Organization (IMO), a specialized agency of the United Nations (UN), has recorded 5,667 piracy attacks against international shipping since 1984 [11]. IMO reports 245 attacks only in the 2014. Similarly, the International Maritime Bureau (IMB), a branch of the International Chamber of Commerce and the industry organization, reports that worldwide there were nearly 3,000 attempted or successful maritime piracy attacks during the period 2000–2009. IMB publishes monthly, quarterly and annual piracy reports with details about names of ships attacked, position and time of attack, consequences to the crew, ship or cargo, and actions taken by the crew and coastal authorities.

The hot spots of piracy today (Fig. 1) are the Indian Ocean, East Africa and the Far East including the South China Sea, South America, and the Caribbean. In recent times, pirates have been found to be very active in the waters between the Red Sea (particularly in Gulf of Aden) and Indian Ocean, off the Somali coast, and in the Strait of Malacca. There are also reports of pirate attacks on the Serbian and Romanian stretches of the international Danube River since 2011. The Strait of Malacca remains another hot spot for piracy today, but in recent years the area has seen a dramatic downturn in piracy due to coordinated patrolling by Indonesia, Malaysia, and Singapore navy forces, and increased level of on-board security on ships. Other major piracy prone areas are the Caribbean and the Bay of Bengal in

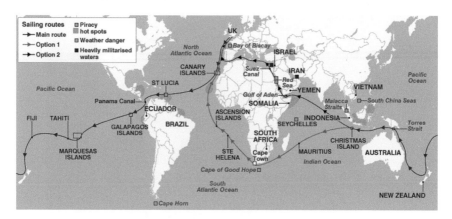

Fig. 1 Hot spots of modern piracy [13]

the Indian Ocean. According to reports, piracy in the Indian Ocean is getting more lucrative and more violent, despite an anti-piracy EU (European Union) naval force patrolling the area. The EU established Naval Force (NAVFOR) Operation ATA-LANTA on November 10, 2008, to protect World Food Program (WFP) ships delivering humanitarian supplies into Somalia. Operation ATALANTA also was authorized to protect merchant vessels in the western Indian Ocean. The EU force operates throughout an area of operations that extends south of the Red Sea, and includes the Golf of Aden, the Somali Basin, and part of the western Indian Ocean and the water surrounding the Seychelles [12].

The vast expanse of water in the Gulf of Aden (Fig. 2) combined with the large number of fishing villages on the Somali east coast, prevent effective patrolling by naval forces. Moreover the large transit distances prevent escort operations so as the co-location of pirate's bases with fishing villages inhibits military strikes on pirate bases. Finally, the instability of the Somali government and the fractured tribal structure of the fishing villages further complicate the problem and prevent diplomatic or economic solutions. U.S. Agency for International Development, through their famine early warning network, notes Somalia's increased reliance on foreign foods arriving in Somali ports and the associated decrease in regional stability. The threat of piracy further increases commodity prices, decreases income in commercial trade, and delays shipments throughout the region. The result is a cycle that increases the incentive for Somali's to turn to piracy and decreases legitimate commercial incentives.

Like legendary pirates, modern pirates are still involved in looting and hijacking ships for ransom, but their ways of operations has dramatically changed over time. Modern pirates now wear night-vision goggles, carry AK-47 s, heavy machine guns, and rocket launchers, navigate with GPS devices, and use sophisticated speedboats mounted with heavy mortars to target ships. They use radar and sonar to track their quarries, and exploit high-tech navigation equipment and communications apparatus. The small speedboats can easily overtake large ships, and can get out of sight

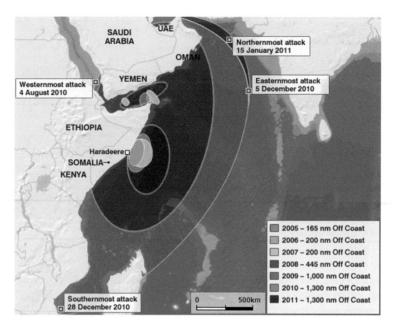

Fig. 2 Influence of piracy in Somali Basin [14]

of support ships much more easily. Modern pirates also use to organize into large groups and coordinate their attack. They attack in full daylight or at night, without regard and they often use grappling hooks to board the victim ship. They then hold the entire lot ships, cargo, and occupants for ransom. This has lead to the development of terrorism by the pirates even if their main object is robbing and kidnapping in order to make money to be invested in other criminal initiatives. This evolution in terms of technologies and tactics makes the problem of piracy an actual and unsolved threat. To combat this problem, researchers have explored various measures for getting piracy back under control and for mitigating the risks it entails.

3   State of the Art of Game-Theoretic Patrolling

A patrolling task can be defined as the act of travelling an area in order to secure it against different threats. To face this problem, the use of the Game Theory is justified by the need of a mathematical framework in order to deploy limited security resources to maximize their effectiveness. Game theory has been applied to a wide number of problem and scenarios ranging from economics (auctions, voting, bargaining, oligopolies, social network formation etc.) through political science (public choice, fair division, war bargaining etc.) and biology (mainly evolutionary Game Theory) to military operations (operations research, military planning, negotiation etc.) [15].

In the area of security where Game Theory has always been used, there now seems to be an exponential increase in interest underlined by the large and growing literature on game-theoretic models and their applications. This increase is in part due to the actual socio-technological threats that our societies face, from terrorism to drugs and crime [16]. These threats require to pay close attention to the problem of how to efficiently allocate limited resources to be effective against these everywhere dangers.

In the last few years Game Theory has been applied to patrolling problems in infrastructure security domains, in which security agencies deploy patrols and checkpoints to protect targets from terrorists and criminals. For such domains, due to limited defence resources it is not possible to cover all targets at the same time. Moreover, despite of alternative Computational Intelligence (CI) approaches, game theoretic approach provides a method to allocate security resources taking into account different weights of different targets and adversary's response to any particular protection strategy. In recent years, several studies have been conducted to fill this gap as shown in [17] in which evolutionary Game Theory is combined with a swarm intelligence method to solve a resource allocation problem.

In the specific context of counter piracy operations several allocation models have been developed [18]. Grasso et al. [19–21] proposes an operational planning system able to consider real environmental parameters such as MeTeorological and OCeanographical (METOC) information and satellite Automatic Identification System (AIS) performace surfaces. In this case a game theoretical approach would be suitable to model not only the optimal asset network planning but also the adaptive response of pirate activities.

The study of strategic interaction situations in particular, commonly named noncooperative games, has been receiving more and more attention in the security field. An interesting open strategic interaction problem is the strategic patrolling [22, 23]. This problem is characterized by a guard that decides what houses to patrol and how often and by a robber that decides what house to strike [24]. Obviously the guard will not know in advance exactly where the robber will strike. Moreover the guard does not know with certainty what adversary it is facing. A common approach for choosing a strategy for agents in such a scenario is to model the problem as a Bayesian game. A Bayesian game is a game in which agents may belong to one or more types; the type of an agent determines its payoffs. The probability distribution over agents types is common knowledge. The appropriate solution for these games is the Bayes-Nash equilibrium [15]. In [22] the authors propose a model for the strategic patrolling and an algorithm to solve it. The guard's actions are all the possible routes of houses, while the robbers action is the choice of a single house to rob. The robber can be of several types with a given probability distribution. Moreover, the rob can observe the actions undertaken by the guard and choose its optimal action on the basis of this observation. The presented model seems to be satisfactory when the guard and the robber act simultaneously. However, in real-world applications it is unreasonable to assume that robber always acts at the turn where the guard starts to patrol. This is essentially due to two reasons. Firstly, the guard cannot synchronize the beginning of its patrolling route with robbers action, since the guard cannot observe the robber.

Secondly, the robber could wait for one or more turns before choosing the house to rob in order to observe guards strategy and take advantage from this observation.

Due to the fact that the attacker can observe the defenders daily schedules, any deterministic schedule by the defender can be exploited by the attacker. One game-theoretic model that has been effective to solve these problems is a Stackelberg game between a leader (the defender) and a follower (the adversary): the leader implement a mixed strategy, which is a randomized schedule specified by a probability distribution over deterministic schedules; the follower then observes the distribution and plays a best response.

Different approaches have been used to model this kind of situations.

Agmon et al. [25] analyzes the problem of patrolling a perimeter. In their approach the environment is modelled as a circular graph, where each of the nodes is a potential target. The Patroller strategy is sought as a simple Markovian policy and as a policy with an additional state representing the facing of the agent in one of two directions. The crucial assumption here is made about the Attacker knowing the strategy used by the Patroller. Basically, the Attacker can wait unlimitedly long and observe the Patroller and thus infers his strategy. If we limit the Attackers knowledge, we get a game model analyzed in [26]. The perimeter patrol strategies cannot be directly applied on more general environment topologies. Arbitrary graphs are studied by Basilico et al. [2], where they provide a general model (termed BGA model) for finding the optimal strategy for the Patroller, which is defined as a higher-order Markovian policy. Further work extending this approach is the analysis of the impact of the Attacker's knowledge about the Patroller's policy on a general graph [27] and an extension of the model for multiple Patrollers [28].

Vanek et al. [29–31] explores how multi-agent systems, a branch of artificial intelligence, can be used to improve maritime security, with particular focus on fighting maritime piracy. Their ultimate objective is to develop an integrated set of algorithmic techniques for maximizing transit security given the limited protection resources available. They achieve this by improving the coordination of the movement of merchant vessels and naval patrols, taking into account the behaviour of pirates. In order to evaluate the proposed techniques and to gain better insight into the structure and dynamics of maritime piracy, they also employ agent-based simulation and machine learning techniques to build dynamic models of maritime transit and to model and assess piracy risk. All methods are implemented within a modular software testbed featuring a scalable simulation engine, connectors to real-world data sources and visualization front-end based on Google Earth.

Decision Support Systems based on the Bayesian-Stackelberg model have been successfully deployed in several domains [5–7, 32].

In [5] a software assistant agent called ARMOR (Assistant for Randomized Monitoring over Routes) is described. ARMOR casts the patrolling/monitoring problem as a Bayesian Stackelberg game, allowing the agent to appropriately weigh the different actions in randomization, as well as uncertainty over adversary types. It uses a solver for Bayesian Stackelberg games called DOBSS, where the dominant mixed strategies enable randomization. ARMOR has been deployed at the Los Angeles International Airport (LAX) to randomize checkpoints on the roadways entering the airport and canine patrol routes within the airport terminals. Based on pre-specified rewards they provide the custom rewards for the LAX police and the adversaries to generate a game matrix (a payoff matrix) for each adversary type. After the final game matrices are constructed for each adversary type, they are sent to the DOBSS implementation, which chooses the optimal mixed strategy over the current action space.

Whereas ARMOR handles 10 terminals at the LAX, the FAMS (Federal Air Marshals) considered in IRIS [6] must protect tens of thousands of commercial flights per day. As shown in Kiekintveld et al. [33], the DOBSS algorithm at the heart of ARMOR cannot handle problems of this magnitude. Also, in ARMOR, domain experts have to enter four payoff values for each of the 10 targets in the domain. IRIS models the problem as a Stackelberg game, with FAMS as leaders that commit to a flight coverage schedule and terrorists as followers that attempt to attack a flight. It uses ERASER-C as a solver for the Stackelberg game. In particularly, IRIS combines three key elements: it uses the ERASER-C solver for this class of security games, that exploits symmetries in the payoff structure; it models the problem with definition of actions for Defenders and Attackers that allow us to efficiently handle the scheduling constraints inherent in the domain; it includes an attribute-based preference elicitation system for calculating risk values for targets to alleviate the need for users to enter risk values for each target individually. IRIS makes also use of algorithmic advances in multi-agent systems research to solve the class of massive security games with complex constraints that were not previously solvable in realistic time–frames.

In [7] a software system, labelled Game-theoretic Unpredictable and Randomly Deployed Security (GUARDS), is described. It utilizes a Stackelberg framework to aid in protecting the airport transportation network. From an application perspective, the fundamental novelty in GUARDS is the potential national scale deployment at over 400 airports.

A game-theoretic security application to aid the United States Coast Guard (USCG), called Port Resilience Operational/Tactical Enforcement to Combat Terrorism (PROTECT) is presented in [32]. It uses an attacker-defender Stackelberg game framework, with USCG as the defender against terrorist adversaries that patrols before potentially launching an attack. The aim of PROTECT solution is to provide a mixed strategy, i.e. randomized patrol patterns taking into account the importance of different targets, the surveillance of the adversaries and the anticipated reaction to USCG patrols. GUARDS and PROTECT introduce many new features and challenges beyond the previous applications at LAX and FAMS, mainly due to the potential large scale deployment: the strategy space of the agents may exponentially increase with the number of security activities, attacks, and resources.

These methods provide an optimal mixed strategy for the single security activity such as assigning checkpoints or air marshals; the randomized solutions produced avoid deterministic strategies that are easily exploitable and remove the human element in randomization since humans are well known to be poor randomizers. Unfortunately previous methods are specific to the stand-alone location they consider and thus cannot directly be applied in an anti-piracy framework.

4 Problem Statement

Focusing on piracy framework, the aim of our research is to optimally (and dynamically) allocate the position of l Patrollers in order to protect t mobile targets with respect to an unknown number of enemies. Targets are supposed to move on known routes inside a generally quite huge maritime area. Moreover, despite the classical formulation, in this framework we need to explicitly take into account the time dimension to consider the travelling time.

To make the problem computationally manageable, we approach it as a two-stage repeated game (as illustrated in Fig. 3). Specifically at each turn the patroller, solving a Bayesian-Stackelberg game, identifies the optimal target to cover on the basis of the position, type and route of the targets so as any other relevant information about the enemies (e.g. suspected presence as illustrated in the next section). Similarly each enemy identifies the target to attack solving a deterministic Stackelberg game on the basis of position, type and routes of the targets taking also into account information about strategies of the patroller. These data represent the input for a path-planning module which identifies the best route for the players considering their characteristics (e.g. speed) so as the maritime scenario (e.g. presence of obstacles, forbidden areas, weather, etc.). Then the players start to move on planned paths and, after a defined time, the algorithm repeats the optimization procedure starting from the updated information. Subsequently the payoff matrix that describes the possible outcome of the game is dynamically constructed during the evolution of the game through an iterated procedure as illustrated in Fig. 4. Starting from the characterization of the scenarios in terms of continuous network of an arbitrary topology, the problem is presented as a two-players Stackelberg Security Game (SSG).

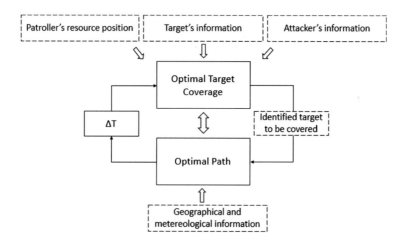

Fig. 3 Illustration of the two-stage repeated game

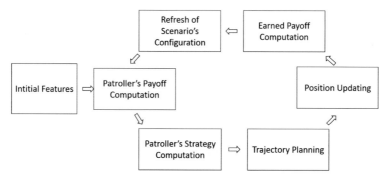

Fig. 4 Illustration of the system's seven-step workflow

At each step the patroller determines, on the basis of his actual location and information, the optimal Bayesian-Stackelberg solution expressed as the best target to protect. After updating the position of the players and computing the actual earned payoffs, the configuration of the scenario is updated taking into account new possible information available from external sources (e.g. radio communication systems, sightings, etc.) and the new positions of the players. The procedure is then repeated starting from the computation of the payoff of the patroller on the basis of updated information.

More in details the patroller and the attacker are characterized by the following features.

Features of the patroller:

- *Position*: actual position of the patroller within the scenario (expressed in latitude and longitude).
- *Velocity*: how fast a patroller can move within the scenario (expressed in knots).
- *Range of coverage*: the dimension of the circular area, centred on actual position of the patroller, inside which it is able to effectively contrast any attacks (expressed in nautical miles).

Features of the attacker:

- *Position*: actual position of attacker (expressed in latitude and longitude).
- *Velocity*: how fast an attacker can move within the scenario (expressed in knots).
- *Attack Capabilities*: a parameter that express the effectiveness of the offensive instruments (weapons, electronic instruments, etc.) present in the boat of the attacker.

In our framework, targets are not players of the game, but they are defined as time varying parameters in the scenario. All the data related to targets and to the scenario are assumed to be known by all the players.

More in detail, each target is characterized by the following parameters:

- *Position*: actual position of the target (expressed in latitude and longitude).
- *Value*: expressed in terms of intrinsic value of the target. This feature represent the payoff earned by the players in case of capture (from attacker) or protection (from patroller).
- *Velocity*: how fast a target can move within the scenario (expressed in knots).
- *Route*: trajectory, assumed fixed and known, covered by a target while it is travelling within the scenario.
- *Defence Capabilities*: a parameter that express the effectiveness of defensive instruments (water cannons, electronic instruments, etc.) in order to cope with an attack.

On the other side the maritime scenario is described in terms of:

- *Geographical constraint*: expresses the characteristics of the area in terms of obstacles, forbidden areas, etc. These data impose constraints to the path planning module, but do not effect the coverage area.
- *Weather condition*: information about the actual marine weather which can influence the behaviour of the players (e.g. due to severe weather conditions a vessel transits within an area with reduced velocity).
- *Critical Area*: expressed in terms of geographical locations where historically piracy attacks are more frequent due to peculiarities (e.g. presence of hideaways, etc.). Consequently a target travelling through a critical area has a higher payoff than its intrinsic value both for the patroller (the target is in a dangerous area) and the attacker (the target is in a valuable area).
- *Suspected Area*: expressed in terms of geographical location considered vulnerable by the patroller. While critical areas are statically defined by statistical information, suspected areas are defined at run-time on the basis of available information (e.g. a satellite image indicating the presence of an unidentified vessel). Notice that only patroller has information about suspected areas hence this element influences only the patroller's payoff.

The aim of the attacker is to conquer the targets with the highest value (or highest payoff) avoiding to be intercepted by the patroller. To defeat a target, an attacker has to be sufficiently close to the target (e.g. the target has to be in the nearness of the attacker) for a sufficiently long period of time. This period depends on the difference between attack and defence capabilities: higher is the attack capability less time is needed to defeat the target.

On the other side, patroller wants to avoid any piracy attack preferring, at the same time, the protection of those targets with the highest payoff. To be protected a target should be inside the coverage area of a Patroller.

As mentioned before we assume that target and scenario (with the only exception to the suspected areas) are known both to the attacker and the patroller. Patroller may have no information on the positions of the attacker, except for those indirectly deducted from the suspected area, at the same time attackers know the strategy of the Patrollers in the current round. Consequently the solution has been evaluated in the Stackelberg-Bayesian framework through an algorithm that equalizes the interest associated to the most relevant targets, providing for each single resource a most convenient possible mixed strategy.

5 Problem Formulation

The proposed solution follows multiple steps, as will be further explained below. Specifically, it will be illustrated how Patrollers and Attackers determine, at each iteration, the corresponding payoffs and strategies and how the different contextual variables are updated. As mentioned before, Patrollers (and Attackers) evaluate at each iteration the best strategy estimating, on the basis of the actual information, the payoff associated with the different targets presents within the scenario. These computations are then repeated after a sampled period of time τ. In this context we will consider, without loss of generality, $\tau = 1\,h$. In the follow we will identify with the subscript L (as Leader) the Patroller, with the F (as Follower) the Attackers and with T all the target contributions. Without loss of generality, we will consider to have l Leaders (or resources of the team of the Patrollers), f Followers (or, also in this case, resources of the team of Attackers) and t targets. Concerning the mathematical exposition, an $\underline{\underline{X}}$ will be considered as a matrix while $\underline{X}$ will be a column vector. For the sake of simplicity, in the follow we assume to explicitly indicate the iteration index $k\tau$ with $k \in N^+$ except when this may creates confusion.

5.1 Payoff Computation for Patrollers

The evaluation of the payoff of the Patrollers is performed taking into account characteristics of the scenario, value of targets, position of other Patrollers and information about Attackers.

Defined an identity row vector $\underline{1} = [1, 1, \ldots, 1]$ with l elements, the matrix of payoffs is defined by:

$$\underline{\underline{U}}_L = \gamma_1 \underline{\underline{U}}_{TL}\underline{1} + \gamma_2 \underline{\underline{U}}_{SL}\underline{1} + \gamma_3 \underline{\underline{U}}_{DL} + \gamma_4 \underline{\underline{U}}_{FL} \qquad \underline{\underline{U}}_L \in \mathfrak{R}^{t \times l} \qquad (1)$$

where γ_i $(i = 1, \ldots, 4)$ are weights, in order to model the relevance of each element in the computation of the payoffs. It will be assumed as boundary condition that:

$$\sum_{i=1}^{4} \gamma_i = 1 \qquad (2)$$

In Eq. (1), the first contribution rules the importance of each target in terms of its actual position and intrinsic relevance (probably a cargo ship with fruit has less

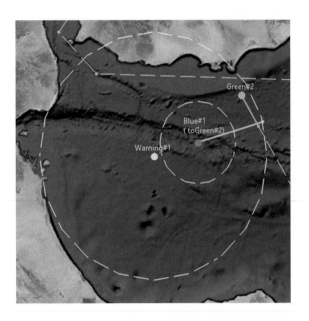

Fig. 5 Example of a risk area

importance than a cargo ship with electronic components). The contribution can be mathematically modeled as:

$$\underline{U}_{TL} \in \mathfrak{R}^t \quad : \quad u_{TL}(i) = u_T(i)\left(1 + \alpha_T^C(i)\right) \qquad \forall i \in [1, \dots, t] \qquad (3)$$

with:

$$\alpha_T^C(i) = \begin{cases} \alpha^C \text{ if } P_T(i) \in A^C \\ 0 \text{ otherwise} \end{cases} \qquad (4)$$

where $P_T(i)$ is the actual position of the i target. Equation (5) illustrates how the intrinsic value of the target is, eventually, amplified by a constant coefficient $\alpha^C \in (0, 1]$ in the case the target is inside a critical area A^C. As mentioned before, a critical area is a zone where there is an high probability of piracy attacks. This assumption is based on historical and statistical data (Fig. 5).

The second contribution is similar to the previous one, but it increase the payoff associated to the i target in the case the target is inside a suspected area. As illustrated in the previous section, a suspected area is defined on the basis of information at disposal of the Patrollers about the presence of suspected boats. Known the suspected area A^S, the contribution can be modeled as:

$$\underline{U}_{SL} \in \mathfrak{R}^t \quad : \quad u_{SL}(i) = u_T(i)\,\alpha_T^S(i, \tau) \qquad \forall i \in [1, \dots, t] \qquad (5)$$

where:

$$\alpha_T^S(i, \tau) = \begin{cases} \alpha^S(\tau) \text{ if } P_T(i) \in A^S(\tau) \\ 0 \text{ otherwise} \end{cases} \qquad (6)$$

The parameter $\alpha^S(\tau)$, is time dependent so as the boundary of the suspected area $A^S(\tau)$: this because they change on the basis of actual information (e.g. satellite images, etc.). Notice that $A^S(\tau)$ and $\alpha^S(\tau) \in [0,1]$ are calculated by a dedicated DSS module whose description is beyond the scope of this chapter.

The third contribution models the attractivity of the targets for the Patrollers, with respect to their mutual distance: e.g. if a target is too far, it will be more convenient to cover another nearer ship. More specifically, to take into account the effective velocities of each Patroller, this contribution considers the time needed by the Patroller to cover the actual target position. This quantity is evaluated by the path-planner module taking into account the actual position P_T of the i target and the P_L position of the j Patroller, the velocity of the Patroller v_L and considering both geographical constraints and weather conditions.

$$\underline{\underline{U}}_{DL} \in \mathfrak{R}^{t \times l} \quad : \quad u_{DL}(i,j) = \begin{cases} \frac{\delta_D - \Delta(i,j)}{\delta_D} & \text{if } \Delta(i,j) \leq \delta_D \\ 0 & \text{otherwise} \end{cases} \tag{7}$$

$\forall i \in [1,\dots,t], \forall j \in [1,\dots,l]$. $\Delta(i,j)$ is the reaching time (for the j Leader and the i target) calculated by the path-planning algorithm and δ_D is a threshold coefficient that allows to set the sensitivity of the factor in relation to time. Specifically, if the time needed to reach the target is greater than δ_D, the contribution is zero (i.e. no contribution to the actual payoff). Otherwise, it ranges from 0 to 1 with 1 that means that the Patroller reaches the target instantaneously (i.e. $\Delta(i,j) = \delta_D$).

The fourth and last contribution rules the effect of the eventual presence of recognized Attackers in the scenario. Specifically, it takes into account the possibility that an enemy attacks the i target. In order to evaluate this quantity, it is important to remember that the Patrollers are considered as "Leaders" of the Stackelberg Game, so they can use only the data (more specifically, the mixed strategies of the Attackers) derived in the previous turns. To manage the impossibility to estimate the effective parameters of the Attackers (i.e. its velocity and offensive capabilities) we consider this as a Bayesian contribution with various pre-setted models of Attackers obtained from statistical data. Known the f^* number of models F^* of prototypes of Attacker and f recognized Attackers on the field, the Attacker contribution on the payoff of the Patroller is based on the modeled mixed strategies of the Attacker, assuming the worst possible condition. This aspect will be explained more in detail in the following sections. Considering the contribution of the Attacker as a vector, this will be expressed as:

$$\underline{\underline{U}}_{FL} \in \mathfrak{R}^t \quad : \quad u_{FL}(i) = u_T(i) \left(\sum_{j}^{f} \omega_F(i,j) \right) \qquad \forall i \in [1,\dots,t] \tag{8}$$

where each element $\omega_F(i,j)$ represents the j mixed strategy of the Attacker for the i target, obtained from the previous round.

5.2 Payoff Computation for Attackers

As mentioned before, when there aren't recognized Attackers on the field the DSS system works to prevent attacks, modifying the hypothetical payoff of Attackers in order to minimize risks for most valuable targets and to optimize the strategies of the Patrollers. When an Attacker is recognized in the field the DSS system models the payoffs and the possible strategies of the Attacker too. Similarly to the Patrollers, also the utilities of the Attackers are expressed as a sum of different contributions, assuming that Attackers have a perfect knowledge of the targets behavior, in terms of actual position P_T, route, intrinsic value u_T and defense capabilities D_T. However there are some differences, specifically:

- We assume to have, instead of a single type of Attacker, f^* different classes of Attackers each one characterized by their own peculiarities in terms of offender capabilities C_F and velocity v_F;
- Coherently with the leader-follower paradigm, we assume that Attackers have knowledge about the choice done in the same round by the Patrollers: more specifically the Attackers know which target each Patroller wants to cover so as the time requested to the Patroller to reach the target;
- For Attackers there is not the term related to the suspected area.

To manage the presence of multiple classes of Attackers, they are modeled as a Bayesian entity which express the mixed strategy associated to each class of Attackers. Consequently the Attacker payoff will be a three dimensional matrix $\underline{\underline{U}}_F \in \Re^{t \times f \times f^*}$ where the entry $u_F(i, j, k)$ represent the payoff associated to the i target for the j Attacker in the case this latter belong to the k class of Attackers. The payoff for each single class of Attackers is expressed as:

$$\underline{\underline{U}}_F(k) = \gamma_1 \underline{\underline{U}}_{TF}(k) + \gamma_2 \underline{\underline{U}}_{DF}(k) + \gamma_3 \underline{U}_{LF}^T \underline{1}(k) \qquad \forall k \in F^* \qquad (9)$$

where γ_i^F ($i = 1, 2, 3$), are weight coefficients that model the different relevance of each element in the total payoff computation, with the constraint that:

$$\sum_i \gamma_i = 1 \qquad (10)$$

Similarly to Patrollers, the first contribution rules the importance of each target in terms of its actual position and intrinsic relevance:

$$\underline{U}_{TF} \in \Re^t : \quad u_{TF}(i) = u_T(i)\left(1 + \alpha_F^C(i)\right) \qquad \forall i \in T \qquad (11)$$

taking into account the Attacker point of view, with:

$$\alpha_F^C(i) = \begin{cases} \alpha^C & \text{if } P_T(i) \in A^C \\ 0 & \text{otherwise} \end{cases} \qquad (12)$$

In this case the intrinsic value of the target is, eventually, amplified by a constant coefficient $\alpha_F^C \in (0, 1]$ in the case the target is inside a critical area A^C. Notice that the boundary of the critical area A^C is the same of those used for the Patrollers, but the amplification coefficient α_F^C may be different expressing an higher (or lesser) interest for the Attackers to perform an attack preferentially in well know area. Notice that this contribution does not depend on the specif class of the Attackers, i.e. it is the same for all the f^* classes.

The second contribution models the attractivity of a target as function of its distance from the Attackers. As for the Patrollers case, this contribution weights the intrinsic value of targets in relation to the time needed by each class of Attackers to complete the attack considering: the actual target, the Attackers position P_F and the relative defender and offender capabilities. Also for the Attacker, this quantity is evaluated by the path-planning module. In this case the calculus is performed f^* times considering, for each calculation, the characteristic of each class of Attacker in terms of velocity and offender capabilities. Specifically:

$$\underline{\underline{U}}_{DF} \in \mathfrak{R}^{t \times f \times f^*} \ : \ u_{DT}(i,j,k) = \begin{cases} \frac{\delta_D - \Delta(i,j)}{\delta_D} & \text{if } \Delta(i,j) \leq \left(\delta_D - \frac{C_D(i)}{C_F(j,k)}\right) \\ 0 & \text{otherwise} \end{cases} \tag{13}$$

$\forall i \in [1, \dots, t], \forall j \in [1, \dots, f], \forall k \in [1, \dots, f^*].$

where $\Delta(i,j)$ is the reaching time (for the j Attacker and the i target) calculated by the path-planning algorithm and δ_D is the threshold coefficient: if the time to reach the target overcome this threshold the target is considered no attractive. In spite of the analogous contribute of the Patroller, here the threshold is modulated taking into account the offensive and defensive capability of Attackers and targets. Specifically in the case the Attacker has an huge offender capability with respect to the target defender capability, i.e. $C_F(j,k) \gg C_D(i)$ then the threshold takes into account exclusively the time needed to reach the target. However, if $C_F(p,f) \ll C_D(i)$, i.e. the defender capabilities of the target overcome the offender capabilities of the Attacker, the latter has reduced interest, hence, to be attractive, the target should be really very close to the Attacker.

The third and last contribution is necessary in a SSG context: the Attacker knows the strategies of the Patroller. So the third contribution $\underline{\underline{U}}_{LF}$ considers the actions of the Patrollers, made in the round, in order to influence the payoff of the Attacker. Essentially the third contribution models the discouraging effect related to the Patrollers, reducing the attractiveness of those targets that has been selected to be covered. Known the coverage vector of the team of Patrollers, the contribution vector will be:

$$\underline{U}_{LF} \in \mathfrak{R}^t \ : \ u_{LF}(i) = \begin{cases} 0 & \text{if } i \text{ is covered} \\ u_T(i) & \text{otherwise} \end{cases} \quad \forall i \in T \tag{14}$$

5.3   Strategy Computation for Patrollers

Each Patroller, in each round, has to identify its best target to cover. To this end all Patrollers have to compare their own strategy in order to identify those that globally optimize the allocation, taking also into account the strategies of the Attackers.

Concerning the Patrollers, the possible strategies can be grouped in a strategy matrix as:

$$\underline{\underline{S}}_{L}(i,j) = \begin{pmatrix} \omega_L(1,1) & \omega_L(1,2) & \cdots & \omega_L(1,l) \\ \omega_L(2,1) & \omega_L(2,2) & \cdots & \omega_L(2,l) \\ \vdots & \vdots & \ddots & \vdots \\ \omega_L(t,1) & \omega_L(t,2) & \cdots & \omega_L(t,l) \end{pmatrix} \quad \forall i \in T \ \forall j \in L \qquad (15)$$

where $\omega_L(i,j)$ represents the option that j Patroller covers the i target, e.g. $\omega_{1,2}$ represents the strategy "the Patroller 2 goes to cover the target 1". Each strategy is associated with a payoff, and it is determined as mentioned in the previous paragraphs.

To identify the best strategies that each one of the Patrollers has to play at the current round, i.e. to select which target the Patroller has to cover, we develop a two steps solution. Firstly, we find the solution of the Stackelberg problem using an ORIGAMI-based algorithm [34]. The ORIGAMI (Optimizing Resources in Game Using Maximal Indifference) solver acts in order to reduce the attractiveness of the most valuable targets for the Attackers, trying to "equalize" the payoffs, i.e. Attackers have not a preferred target to attack. The solution of the ORIGAMI algorithm is a mixed strategy that converts each single resource in a set of most convenient possible strategies. The second algorithm called MORESCO (MOst RElevant Strategy in COoperative teams) evaluates all the mixed strategies generated by ORIGAMI, and assigns a single pure strategy to each patroller, in order to simulate both the cooperation between elements and the (automatic) evolution of the scenario. The MORESCO takes into account the variance between each mixed strategy for each Patroller, and models the "determination" of each element to cover each target.

5.4   Strategy Computation for Attackers

Also in this case we use the ORIGAMI algorithm to calculate the mixed strategy of each Attacker. As introduced before, each Attacker can have multiple possible payoffs, each one related to a single Attacker class. So the real mixed strategy for each Attacker is a Bayesian correlation between all possible mixed strategies, assuming that it can be represented with all possible Attacker models simultaneously. The strategy matrix, considering all possible classes, is:

$$\underline{\underline{S}}_F (i,j,k) \in \mathfrak{R}^{t \times f \times f^*} : \underline{\underline{S}}^*_F (i,j) = \begin{pmatrix} \omega_F (1,1) & \omega_F (1,2) & \cdots & \omega_F (1,f) \\ \omega_F (2,1) & \omega_F (2,2) & \cdots & \omega_F (2,f) \\ \vdots & \vdots & \ddots & \vdots \\ \omega_F (t,1) & \omega_F (t,2) & \cdots & \omega_F (t,f) \end{pmatrix} \tag{16}$$

$\forall k \in \left[1, \ldots, f^*\right].$

Due to the third dimension of the payoff matrix of Attackers, it is evident that for each target there are f^* possible strategies associated to f recognized Attackers in the field. Defining any possible set of mixed strategies for a target, for each Attacker, as $\omega_T (i,j,k)$, $\forall i \in [1, \ldots, t]$ $\forall j \in \left[1, \ldots, f\right]$ $\forall k \in \left[1, \ldots f^*\right]$, in order to simulate the worst possible situation, the most relevant possible mixed strategy for each Attacker will be considered. So it is possible to derive the final matrix of mixed strategies as:

$$\underline{\underline{S}}_F (i,j) = \begin{pmatrix} \omega_F (1,1) & \omega_F (1,2) & \cdots & \omega_F (1,f) \\ \omega_F (2,1) & \omega_F (2,2) & \cdots & \omega_F (2,f) \\ \vdots & \vdots & \ddots & \vdots \\ \omega_F (t,1) & \omega_F (t,2) & \cdots & \omega_F (t,f) \end{pmatrix} \tag{17}$$

$\forall i \in [1, \ldots, t]$ $\forall j \in \left[1, \ldots, f\right]$

where:

$$\omega_F (i,j) = max_k \left(\omega_T (i,j)\right) \quad \forall i \in [1, \ldots, t] \ \forall j \in \left[1, \ldots, f\right] \tag{18}$$

i.e. for each i-j combination, the j attacker will be represented by the more efficient class of attackers k that maximize $\omega_T (i,j)$.

6 Simulations

In order to validate the proposed approach a set of simulations has been performed. To evaluate the effectiveness of the strategies of Patrollers we consider a fixed set of targets (30 targets) with assigned routes (selected from a set of 10 pre-defined routes) and randomly introduced in the scenario within a time horizon of 30 h. In each simulation, 5 Patrollers ($l = 5$) and 5 Attackers ($f = 5$) are present in the scenario. The starting point of each Patroller is selected randomly from a set of 10 pre-defined realistic starting points (i.e. Port of Aden, Port of Djibouti, etc.). The starting point and the instant of appearance of the Attackers are randomly defined within the first 5 h of the simulation. Concerning the intrinsic parameters of each entity, the targets will be defined with a random combination of:

- Intrinsic payoff—$u_t = 0.5 + 0.5\eta_1$;
- Velocity—$V_T = \left(16 + 4\eta_2\right)$ knots;
- Target defense capability—$C_D = 2 + 1\eta_3$.

The Patrollers are modeled with a random combination of:

- Operational range—$R_L = (11 + 3\eta_4)\ NM$;
- Velocity—$V_L = (24 + 4\eta_5)$ knots;

Where η_i are uniform random numbers in $[-1, 1]$.

Concerning the Attackers, as mentioned before they are expressed in terms of Bayesian combinations, so the effective attacker velocity v_F and the offensive capability C_F are defined choosing (randomly) a model from a representative set of 3 classes $(f^* = 3)$ of Attackers with the following characteristics:

$$class\ F^* (1) = \begin{cases} v_F = 20\,\text{knots} \\ \quad\quad C_F = 4 \end{cases} \tag{19}$$

$$class\ F^* (2) = \begin{cases} v_F = 22\,\text{knots} \\ \quad\quad C_F = 3 \end{cases} \tag{20}$$

$$class\ F^* (3) = \begin{cases} v_F = 18\,\text{knots} \\ \quad\quad C_F = 5 \end{cases} \tag{21}$$

The parameters contained in the model, more specifically the weight parameters γ_i and the threshold limits δ, will be imposed as $\gamma_i = 0.25$ (in the computation for Leaders), $\gamma_i = 0.33$ (in the computation for Followers) and $\delta = 10$. Referring to (2) and (10) this particular setting of the parameters will equalize the weight of all the contribution.

The criteria used for the model validation are:

- The average number of covered targets N_C simultaneously covered during the simulation;
- The average number of attacked targets N_A simultaneously attacked during the simulation;
- The efficiency of the action of the Patrollers expressed as the sum of intrinsic payoffs of the covered targets with respect to the actual total payoff of all the targets in the scenario:

$$Eff = \frac{\sum_{i=1}^{t_{covered}} u_T (i)}{\sum_{j=1}^{t} u_T (j)} \tag{22}$$

- The average payoff values P_C in relation to the group of covered targets;

$$P_C = \frac{\sum_{i=1}^{t_{covered}} u_T (i)}{N_C} \tag{23}$$

- The average payoff values P_A in relation to the group of attacked targets.

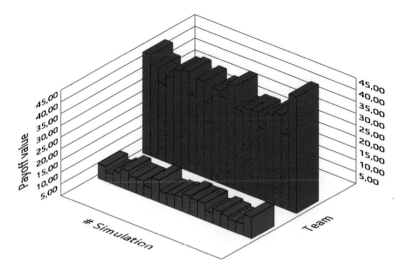

Fig. 6 Evolution of team payoffs P_L (*in blue*) and P_F (*in red*)

$$P_A = \frac{\sum_{i=1}^{t_{attacked}} u_T(i)}{N_A} \quad (24)$$

- The total payoff earned by each team (P_L for the leader, P_F for the follower) during a simulation, expressed as mean of the total payoff earned in each round.

$$P_L = \frac{\sum_{i=1}^{n_{rounds}} P_C(i)}{n_{rounds}} \quad (25)$$

$$P_F = \frac{\sum_{i=1}^{n_{rounds}} P_A(i)}{n_{rounds}} \quad (26)$$

20 simulations were carried out obtaining the following values for the indicators:

$$Eff : 56\% \quad N_T = 10.37 \quad N_C = 5.02 \quad N_A = 0.63 \quad P_C = 6.37 \quad P_A = 3.31 \quad (27)$$

Figure 6 reports the average payoffs P_L and P_F of both teams during each simulation.

It is interesting to note that with this model, the Leaders cover the most relevant targets during the time ($P_C \geq P_A$), allowing Followers to attack only second choices. Moreover, a single Patroller covers at least one target in the whole simulation ($N_C \geq l$), while an Attacker can hit at most only one target in each simulation ($P_C \leq 1$). The average efficiency means that the Patrollers are able to control directly is the 56% of the total goods present on the scenario.

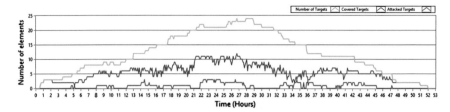

Fig. 7 Number of targets in the scenario (*green*), covered targets (*blue*) and attacked targets (*red*) in a single run

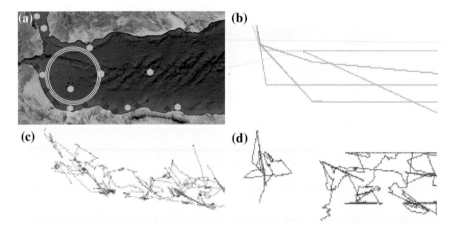

Fig. 8 **a** Scenario Map: Gulf of Aden. In *red* the Attackers starting points; in *cyan* the Patrollers starting points; The *yellow circle* is the critical area. **b** Routes of the Targets: heat map. **c** Routes of the Patrollers: heat map. **d** Routes of the Attackers: heat map

For a more detailed analysis Fig. 7 reports the time history obtained by a single simulation.

Heat plots (Fig. 8) show the behavior of targets and players on the scenario during a single run: the targets follow pre-defined and imposed routes (Fig. 8b). Concurrently, the teams act their strategies in order to cover/attack the best target as possible relating to the evolution of the scenario. In the Fig. 8c it is possible to note how the dislocation of Patrollers limits the movement of Attackers (especially in the middle of the map, where the Patrollers are more present). In the Fig. 8d it is possible to note that the attackers prefer to attack ships on the critical area only if it is possible. This is due to the fact that in this area the targets' route are more scattered hence patrollers have more difficulties to cover all the possible routes. To this end, it is also important to note that patroller are mostly present in the south-eastern region of the scenario where they can cover a larger number of routes and move from a target routes to an other to protect different targets. On the other side the attackers, especially in the critical area of the scenario, seem to follow more clearly the target's routes.

7 Conclusions

This chapter describes an optimization model that uses a game theoretical approach to solve the problem of patrolling of mobile targets. The problem is solved considering an iterative two stages approach where, in the first stage, the leader (the patroller) and the follower (the attacker) select the best target to cover and attack respectively solving a Bayesian-Stackelberg game. In the second stage a dedicated path-planning module identifies the route that patrollers and attackers have to follow to perform their selected strategies. The strict cooperation between the path-planning module and the game theoretical model allows to consider typical variables present in piracy context such as routes of targets, distances, dynamic parameters (i.e. velocities), geographical boundary conditions, weather conditions, etc. Future researches will be dedicated to generalise the framework introducing further constraints related to the behaviour of the targets.

References

1. Bowden, A., et al.: The economic costs of maritime piracy. Oceans Beyond Piracy, One Earth Future Foundation, Technical Report (2011)
2. Basilico, N., Gatti, N., Amigoni, F. : Leader-follower strategies for robotic patrolling in environments with arbitrary topologies, In: AAMAS (2009)
3. Agmon, N., Kraus, S., Kaminka, G.A.: Multi-robot perimeter patrol in adversarial settings. In: ICRA, pp. 2339-2345 (2008)
4. Vanek, O., Bosansky, B., Jakob, M., Pechoucek, M.: Transiting areas patrolled by a mobile adversary. Comput. Intell. Games (CIG) (2010)
5. Pita, J., et al.: ARMOR Software: A Game Theoretic Approach for Airport Security (2008)
6. Tsai, J., Rathi, S., Kiekintveld, C., Ordez, F., Tambe, M.: IRIS–A Tool for Strategic Security Allocation in Transportation Networks (2009)
7. An, B., Pita, J., Shieh, E., Tambe, M.: GUARDS and PROTECT: Next Generation Applications of Security Games (2011)
8. Tambe, M.: Security and Game Theory Algorithms, Deployed Systems, Lesson. Springer, New York (2012)
9. Article 101 of the United Nations Convention on the Law of the Sea (UNCLOS) (1982)
10. Kraska, J.: Contemporary Maritime Piracy International Law, Strategy, and Diplomacy at Sea. Praeger, New York (2011)
11. IMO Doc. MSC.4/Circ.164, Reports on Acts of Piracy and Armed Robbery against Ships, at 2, 3 Dec 2010
12. European Union Naval Force. http://eunavfor.eu/ (2015. Accessed 12 Feb 2015
13. BBC News Magazine. http://news.bbc.co.uk/2/hi/uk_news/magazine/8388222.stm (2015). Accessed 12 Feb 2015
14. BBC News Magazine. http://www.bbc.co.uk/news/10401413 (2015). Accessed 12 Feb 2015
15. Fudenberg, D., Tirole, J.: Game Theory. MIT Press, Cambridge (1993)
16. Bologna, S., Setola, R.: The need to improve local self-awareness in CIP/CIIP. In: First IEEE International Workshop on Critical Infrastructure Protection. IEEE (2005)
17. Leboucher, C., Chelouah, R., Siarry, P., Le Menec, S.: A swarm intelligence method combined to evolutionary game theory applied to resources allocation problem. In: International Conference on Swarm Intelligence (2011)

18. Woosun, An., Ayala, D.F.M., Sidoti, D., Mishra, M., Xu, Han., Pattipati, K.R., Regnier, E.D., Kleinman, D.L., Hansen, J.A.: Dynamic asset allocation approaches for counter-piracy operations. In: Proceedings of the 15th International Conference on Information Fusion (FUSION), pp.1284-1291, 9-12 July 2012
19. Grasso, R., Braca, P., Osler, J., Hansen, J.: Asset network planning: integration of environmental data and sensor performance for counter piracy. EUSIPCO (2013)
20. Grasso, R., Braca, P., Osler, J., Hansen, J., Willet, P.: Optimal asset network planning for counter piracy operation support, part 1: under the hood. IEEE A& E Syst. Mag. (2013)
21. Grasso, R., Braca, P., Osler, J., Hansen, J., Willet, P.: Optimal asset network planning for counter piracy operation support, part 2: results. IEEE A& E Syst. Mag. (2013)
22. Paruchuri, P., Pearce, J.P., Tambe, M., Ordonez, F., Kraus, S.: Ancefficient heuristic approach for security against multiple adversaries. In: Proceedings of AAMAS, pp. 311-318. Honolulu, USA (2007)
23. Paruchuri, P., Tambe, M., Ordonez, F., Kraus, S.: Security in multiagentvsystems by policy randomization. In: Proceedings of AAMAS, pp. 273-280. Hakodate, Japan (2006)
24. Gatti, N.: Game theoretical insights in strategic patrolling: model and algorithm in normal-form. In: Proceedings of the ECAI (2008)
25. Agmon, N., Kraus, S., Kaminka, G.: Multi-robot perimeter patrol in adversarial settings. In: IEEE International Conference on Robotics and Automation, ICRA 2008, pp. 2339-2345 (2008)
26. Agmon, N., Sadov, V., Kaminka, G., Kraus S.: The impact of adversarial knowledge on adversarial planning in perimeter patrol. In: Proceedings of the 7th International Joint Conference on Autonomous Agents and Multiagent Systems, International Foundation for Autonomous Agents and Multiagent Systems, vol. 1, pp. 55-62 (2008)
27. Basilico, N., Gatti, N., Rossi, T., Ceppi, S., Amigoni, F.: Extending algorithms for mobile robot patrolling in the presence of adversaries to more realistic settings. In: Proceedings of the 2009 IEEE/WIC/ACM International Joint Conference on Web Intelligence and Intelligent Agent Technology, vol. 02 (2009)
28. Basilico, N., Gatti, N., Villa, F.: Asynchronous Multi-Robot Patrolling Against Intrusions in Arbitrary Topologies (2010)
29. Vanek, O., Hrstka, O., Pechoucek, M.: Improving group transit schemes to minimize negative effects of maritime piracy. IEEE Intell. Trans. Syst. (2014)
30. Vanek, O., Pechoucek, M.: Dynamic group transit scheme for corridor transit. In: Proceedings of the 5th International Conference on Modeling, Simulation and Applied Optimization (2013)
31. Vanek, O., Jakob, M., Hrstka, O., Pechoucek, M.: Agent-based Model of Mariime Traffic in Piracy-affected Waters. Trans. Res. Part C: Emerg. Technol. (2013)
32. Shieh, E., An, B., Yang, R., Tambe, M., Baldwin, C., Di Renzo, J., Maule, B., Meyer, G.: PROTECT: an application of computational game theory for the security of the ports of the united states.In: Proceedings of the Conference on Artificial Intelligence (AAAI) Spotlight Track (2012)
33. Kiekintveld, C., Jain, M., Tsai, J., Pita, J., Tambe, M., Ordonez, F.: Computing Optimal Randomized Resource Allocations for Massive Security Games (2009)
34. Kiekintveld, C., et al.: Computing optimal randomized resource allocations for massive security games. In: Proceedings of the 8th International Conference on Autonomous Agents and Multiagent Systems, vol. 1, pp. 689-696 (2009)

mspMEA: The Microcones Separation Parallel Multiobjective Evolutionary Algorithm and Its Application to Fuzzy Rule-Based Ship Classification

Marco Cococcioni

Abstract This chapter presents a new parallel multiobjective evolutionary algorithm, based on the island model, where the objective space is exploited to distribute the individuals among the processors. The algorithm, which generalizes the well-known *cone separation* method, mitigates most of its drawbacks. The new algorithm has been employed to speed-up the optimization of fuzzy rule-based classifiers. The fuzzy classifiers are used to build an emulator of the Ship Classification Unit (SCU) contained in modern influence mines. Having an accurate emulator of a mine's SCU is helpful when needing: (i) to accurately evaluate the risk of traversal of a mined region by vessels/AUVs, (ii) to assess the improvements of ship signature balancing processes, and (iii) to support in-vehicle decision making in autonomous unmanned mine disposal.

Keywords Parallel multi-objective evolutionary algorithms · Fuzzy rule-based classifiers · Influence mine modeling · Ship classification

1 Introduction

The use of Multiobjective Evolutionary Algorithms (MEAs) for optimizing fuzzy rule-based systems has attracted wide interest over the last decade [1, 3, 6, 8–10, 17]. An extensive list of contributions to this research area, usually referred to as *multiobjective genetic fuzzy systems*, can be found in [11]. This family of tools is powerful in regression and classification problem for many reasons: (i) the fuzzy rules are automatically extracted from data, (ii) they are human interpretable; (iii) the optimization in performed globally (and thus the risk of getting stuck into a

M. Cococcioni (✉)
Department of Information Engineering, University of Pisa, Largo Lucio Lazzarino, 1, 56122 Pisa, Italy
e-mail: m.cococcioni@iet.unipi.it

© Springer International Publishing Switzerland 2016
R. Abielmona et al. (eds.), *Recent Advances in Computational Intelligence in Defense and Security*, Studies in Computational Intelligence 621,
DOI 10.1007/978-3-319-26450-9_17

445

local optimum is low), and; (iv) the possibility to include multiple objectives allows designing more accurate systems.

A hot topic in current literature about multiobjective optimization of fuzzy system is the one related to the handling of *high-dimensional* and/or *large* datasets. There are many different ways to speedup multiobjective optimizations of fuzzy rule-based systems. For instance, concerning the Takagi-Sugeno fuzzy model, fitness approximation techniques can be used [6, 8–10]. When the dataset is large, only a subset of the available samples for the evolutionary optimization of the model can be utilized, where this subset is evolved using another evolutionary algorithm in a co-evolutionary scheme, as done in [1] and in [17].

Of course, MEAs parallelization on multiple processors is a viable option. While there are many parallel MEAs, most of them use parallelizing techniques borrowed from the parallelization of single objective EAs [16, 20, 26]. Such an approach does not exploit the multiobjective nature of the problem and thus can be less effective in approximating the Pareto optimal front. One of the few exceptions is the Cone Separation method introduced in [2] (hereafter referred to as cone separation parallel MEA—cspMEA—), where the authors have shown the superiority of using a divide-and-conquer approach based on the objective space over random assignment (i.e., without exploiting the objective space).

In this work we focus on two-objective problems and, after highlighting the main limitations of cspMEA, we introduce an extension of it, based on the concept of *microcones*, which mitigates such limits.

We have then employed this new algorithm, named mspMEA (microcones separation parallel MEA), to learn fuzzy rule-based classifiers in a ship classification problem. The aim of building such a classifier is to build a stochastic emulator of the ship classification unit (SCU) found on modern influence mines. This unit is demanded to decide if the mine has to detonate, by discriminating the presence of target ships from non-target ones. Of course, to make this decision the influence mine is equipped with a number of sensors (acoustic, magnetic, pressure, etc.) along with a complex logic. Being able to emulate this logic has a number of defense applications, such as: (i) to a accurately compute the risk associated with the traversal of a mined region by vessels/AUVs [12, 18, 23], (ii) to assess the benefits of a ship signature balancing process [14, 21], and (iii) to support in-vehicle decision making in unmanned mine disposal operations [22, 24]. The latter is an enabling technology in autonomous mine countermeasures approached using AUVs, a hot topic in current mine countermeasures strategies.

This chapter is organized as follows. In Sect. 2 the cspMEA is described, along with its main limitations. In Sect. 3 we introduce the concept of microcones and then we outline the whole mspMEA algorithm. Section 4 describes the problem at hand, namely the building of a stochastic emulator for SCUs by means of fuzzy rule-based classifiers. The result of the optimization of the fuzzy classifiers by means of cspMEA and mspMEA are provided in Sect. 5, while Sect. 6 draws conclusions.

2 The Cone Separation Parallel Multiobjective Evolutionary Algorithm (cspMEA)

The aim of multiobjective optimization (when approached according to the Pareto optimality criterion) is to find an approximation to the Pareto front. The Pareto front is the set of non-dominated solutions that are non-dominated by any other feasible solution. A solution is said to dominate another one if and only if it is equal or better on all the objectives and is better than the others at least of one objective. The numerical solution of a multiobjective problem having a continuous Pareto front is aimed at finding a finite number of equally distributed solutions along the Pareto optimal front. MEAs are especially suited to solve these kinds of problems, since they are population based: with only one run, they are able to find a set of non-dominated solution close to the Pareto front.

Among many others (PAES [19], (2 + 2)M-PAES [9], SPEA2 [28], AMGA2 [25], CHEA [5], MOEA/D [27], etc.) NSGA-II [15] has had a wide influence on many multiobjective evolutionary algorithms developed since its introduction (besides being one of the algorithms to compare to when proposing new ones).

While the parallelization of Multiobjective EAs could be made by resorting to general purpose technique exploited to parallelize single objective EA [16, 26], specific methods can be devised to better parallelize them exploiting their multi-objective nature.

The cone separation method described in [2] (that here we have referred to as cspMEA to distinguish it from the mspMEA) is one of such attempts, and is built upon the concepts introduced in NSGA-II.

2.1 The Idea of Dividing the Objective Space into Regular Cones

The cspMEA introduced in [2] is a parallel version of NSGA-II [15], which uses the island model with migration (the number of islands corresponds to the number of processors).

To describe the working principle of cspMEA let us focus on a bi-objective (let x be the first objective and y the second), where both the conflicting objectives have to be minimized. Given the current population, the non-dominated solutions are first computed. Let also consider the solutions achieving the best value (the minimum in our case) on the first objective $^1B \equiv (^1B_x, \, ^1B_y)$, and the one obtaining the minimum on the second objective $^2B \equiv (^2B_x, \, ^2B_y)$. We are now able to compute the position of the *nadir point* (the diamond in Fig. 1b, c, d), that is the virtual point having coordinates: $(^2B_x, \, ^1B_y)$. Let us also define another virtual point, called the *ideal point* (the triangle in Fig. 1b, c, d), as the point having coordinates $(^1B_x, \, ^2B_y)$. We can now normalize the objective space, in such a way the rectangle in Fig. 1b

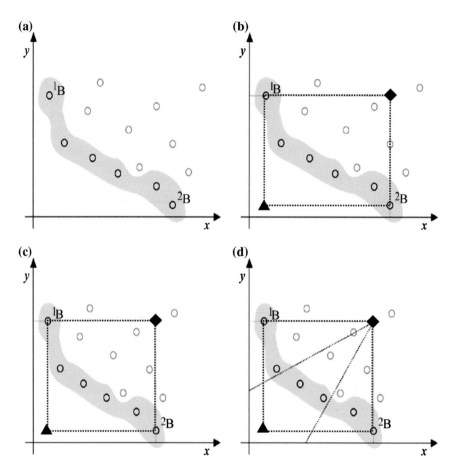

Fig. 1 How the cspMEA computes the *unit square* and divides it into *p* cones. The considered problem is a minimization problem with two objectives, *x* and *y*. *Gray circles* dominated solutions; *black circles* non-dominated solutions; *diamond* the *nadir point*; *triangle* the *ideal point*. The rectangle in (**b**) becomes a *unit square* (dotted square) in (**c**), after normalizing the objective space. Finally in (**d**) the unit square is divided into 3 regions (the *cones*)

between the nadir point and the ideal point becomes a square (Fig. 1c): this square is called the *unit square* [2].

The basic idea in cspMEA is to divide the unit square of Fig. 1c into *n* cones (Fig. 1d), *n* being the number of processors, and to assign the candidate solutions falling within the *p*th cone to the *p*th processor. Therefore the strategy is to have each island focused on a specified cone, where the borders of the cones are treated as hard constraints, i.e., handled by resorting to the *constrained domination principle*. This principle assumes that all solutions outside the designated region are dominated by all solutions within it. When the offsprings generated on a processor fall outside the associated cone, they are *always* migrated to the appropriate one.

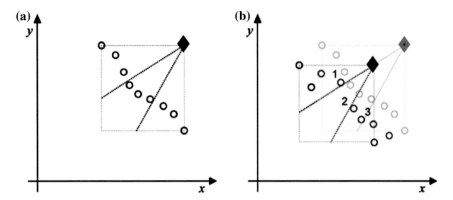

Fig. 2 Unit square update from previous (**a**) to next generation (**b**) in cspMEA

Figure 2 shows the re-computation of the unit square when the fitness of the next generation is available. An important aspect to notice is that to compute the unit square the best fitness must be shared among the processors (more precisely, exchanging only the extremes found by each processor is enough [2]).

An appealing property of cspMEA is that the final Pareto front (the *global* one) is approximated by the union of the Pareto front approximations computed by each processor (*local* Pareto fronts). Unfortunately, there is no guarantee that the union of the local Pareto front contains only non-dominated solutions. For this reason, as a final step, the dominated solutions are discarded from the front, thus obtaining the true global Pareto front approximation.

Since cspMEA is based on NSGA-II, it ranks the population using the non-dominated sorting. Then individuals having the same rank are ordered by crowding distance (see next subsections).

2.2 Non-dominated Sorting

On each cone the cspMEA uses the non-dominated sorting operator (the same one adopted by NSGA-II). Non-dominated sorting consists in computing rank 1 solutions, i.e., all the non-dominated solutions. Then, it discards rank 1 solutions, and on the remaining ones, it computes again those that are non-dominated (and assigns them the rank 2). It then continues until no further solutions are present.

2.3 The Crowding Distance

The crowding distance is the operator allowing to selecting a subset of solutions among the ones in the same rank, say k, when more than needed are present. It performs this selection trying to widening the Pareto front approximation as much as possible, and distributing solutions as uniformly as possible along the front. Its definition can be found in [15].

2.4 The cspMEA Algorithm

In summary, the cspMEA algorithm is outlined in Algorithm 1.

Algorithm 1: Outline of cspMEA [2]

Initialize the different sub-populations
Normalize fitness values
Determine region constraints
Non-dominated sorting
REPEAT
    Generate Offspring
    IF (migration)
            Normalize fitness values
            Determine region constraints
    Migrate individuals violating constraints
    Non-dominated sorting
    Prune population to original size
UNTIL stop-condition

The flowchart of cspMEA is provided in Fig. 3 (the operations that cannot be parallelized are marked with a circled "C").

2.5 Limits of cspMEA

In [2] it has been shown that cspMEA is superior to a naïve parallelization approach based on random assignment to the processors (i.e., without exploiting the objective space). However, the algorithm has three weaknesses: (i) local Pareto fronts might not belong to the global Pareto front, and (ii) the solutions of the global Pareto front can be unevenly distributed, and (iii) the migration rate can be too high, being it unbounded. Let us see these limits more in detail in next subsections.

2.5.1 Local Pareto Fronts Might Not Belong to the Global Pareto Front

In case of discontinuous global Pareto fronts, it may happen that solutions of one local Pareto front are dominated by some contained in other cones, as depicted in Fig. 4. This is a serious problem, because it could lead to a significant waste of

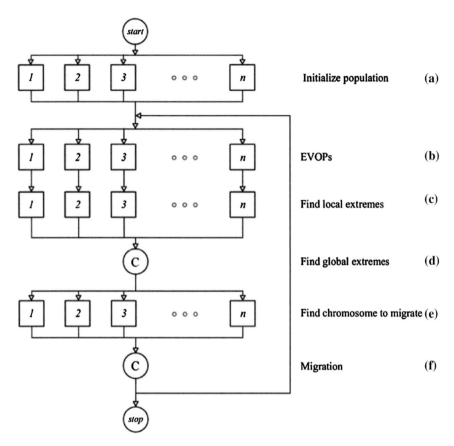

Fig. 3 A flowchart of the cspMEA introduced in [2] (EVOPs stands for evolutionary operations: fitness evaluation, ranking, selection, crossover and mutation). The operation that cannot be parallelized (finding the global extremes and migration) are marked with a circled "C"

computing resources. For example, Fig. 4 shows how the work done by processor 4 is completely useless, since its computed local Pareto front will be discarded at the end, when the global Pareto front is computed (in the considered case, 20 % of the computing power is therefore totally wasted).

2.5.2 The Solutions of the Global Pareto Front can be Unevenly Distributed

To illustrate this weakness, we provide two situations. The first is one is when a local Pareto front is discontinuous (see Fig. 5). In this case, while the crowding distance of NSGA-II would distribute solution evenly along the front in a single processor (Fig. 5b), cspMEA generates solutions unevenly distributed along the front (Fig. 5a). In Sect. 3 we will discuss how to mitigate this problem.

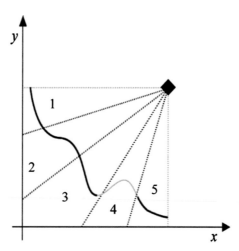

Fig. 4 The local Pareto front computer by the 4th processor (*cone 4*) is completely dominated by solutions of the local Pareto front computed by the 3rd processor (*cone 3*). In particular, the local Pareto front on cones 1, 2, 3 and 5 belongs to the global Pareto front, while the local Pareto front of the 4th cone is completely absent from the global Pareto front

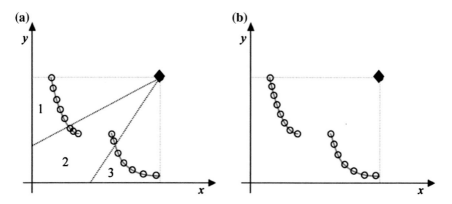

Fig. 5 Suppose we have the global Pareto front depicted in (**a**) and to have 3 processors (*3 cones*), each with a population size equal to 5. Since the local Pareto front of the second cone is discontinuous, the density of solutions on local Pareto front 2 is higher than that of solutions of local Pareto fronts 1 and 3 (assuming each processor has a population size equal to 5). In (**b**) we can see the distribution of the solutions of the $(3 \times 5 = 15)$ solutions we would have obtained by running NSGA-II on a single processor: here the use of the crowding distance applied to the *whole* Pareto front would have given a uniform of solutions. This means that dividing the Pareto front into cones (as done by cspMEA) can lead to an uneven distribution of the candidate solutions

The second situation when a non-uniform distribution of solutions can occur is when the global Pareto front is far from the unit circle (either being it convex or concave), as discussed in Fig. 6.

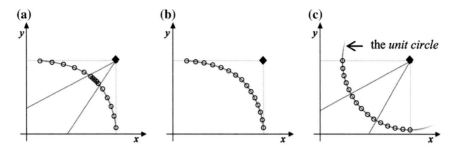

Fig. 6 In **c** we have a global Pareto front which is very close to the *unit circle* centered at the nadir point: in this case the solutions are uniformly distributed by cspMEA along the front. However, this is not the case in **a**, i.e., when the global Pareto front is far from the *unit circle*: in this case the density of solutions on cone 2 is higher than those on cones 1 and 3 (the three local crowding distances independently computed by cspMEA fail to generate a globally uniform distribution). This phenomenon is due to cspMEA, as NSGA-II is not affected by it: in **b** the distribution made by NSGA-II (using one crowding distance on the whole front) is shown. It can be seen easily how in this case the solutions are uniformly distributed along the front even when the Pareto front is far from the *unit circle*

2.5.3 The Migration Rate can be too High, Being it Unbounded

When a new offspring is computed, it can happen that it does not fall within the cone associated to the parents involved on its generation. In this case the offspring is moved to the right cone (i.e., to its associated processor). In the worst case, all the offsprings might have to be migrated to other processors. This means that in cspMEA we have no control on the number of migrations. As we will see in the following, this stems from the fact that belonging to a cone is handled as a *hard constraint*.

In next section we will provide a new pMEA able to mitigate the three limits of cspMEA discussed above.

3 The Microcones Separation Parallel Multiobjective Evolutionary Algorithm (mspMEA)

3.1 The Idea of Microcones

We propose here a new parallel MEA, still based on the idea of cone separation. In this case, however, we not only divide the unit square into n cones, but we also uniformly divide each cone into N_{group} microcones (numbered from 1 to N_{group}, from top left to bottom right). Then the algorithm assigns all the microcones with number 1 to processor 1, all the microcones with number 2 to processor 2, etc. (see Fig. 7). Obviously, when $N_{group} = 1$, mspMEA and cspMEA coincide.

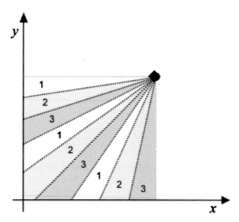

Fig. 7 Working principle of microcones separation parallel MEA (mspMEA) in the case of 3 cones divided into 3 microcones each: the 3 microcones numbered 1 will be assigned to processor 1, the 3 with number 2 to processor 2, and the 3 with number 3 to processor 3

The mspMEA has an important property that helps mitigating the problem discussed in Fig. 4. To show it, we provide another example, still made of 3 cones and 3 microcones (see Fig. 8). In this case we have also displayed the global Pareto

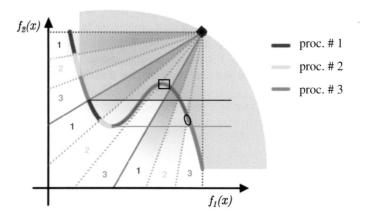

Fig. 8 Another example of the working principle of mspMEA: glocal Pareto front of proc. #1 (*red*), proc. #2 (*yellow*) and proc. #3 (*blue*). The clear advantage of mspMEA over cspMEA is that the second microcone of proc.#3 knows that its front is not Pareto optimal (the *black rectangle*) because of the presence of his solutions (the one above the *purple horizontal line*) falling within its first microcone. This does not mean that the union of the 3 glocal Pareto front approximation gives the global Pareto front, because some solutions on the glocal fronts can be still dominated by others of other glocal fronts (e.g., the blue part of the glocal front portion found on the third microcone by the 3rd processor falling within the black ellipsoid is dominated by solutions found by the second microcone of the second processor, i.e., by the ones above the *green horizontal line*). This means that not all the problems discussed in Fig. 4 are solved, but at least they have been heavily mitigated

front. The interesting point to note is that each processor is partially aware of the whole global Pareto front, having part of it. For this reason, we have named it *glocal*, (a hybridization of *local* and *global*), for remarking the fact that the Pareto front approximation locally computed by each processor contains information about the global Pareto front (although only partially).

However, this clear advantage comes with a drawback: the offspring obtained by recombining two parents belonging to two (in general distinct) microcones is more likely to fall outside all the microcones assigned to the current processor. Thus, the chance of having to migrate the offsprings is slightly higher for mspMEA than for cspMEA. To overcome this problem, we have considered the constraint of belonging to processor p as a *soft constraint* instead of a *hard constraint* (as done by cspMEA). This means that we tolerate the presence of individuals falling on microcones different from those associated to the pth processor. However we will give priority to those individuals that *do* belong to the right microcones (the ones associated to the current processor). At each generation we migrate a fixed number of individuals to their right processors, randomly chosen among the ones violating the soft constraint. The important aspect here is that with the mechanism of random selection we are also able to control (and bound) the migration rate. To further privilege the recombination of solutions that belongs to the right set of microcones we have added the binary tournament selection for selecting the pair of individuals to mate.

Before providing the outline of the whole mspMEA algorithm we have still to add an improvement to the selection criterion, since the one adopted by NSGA-II and cspMEA (the crowding distance) can be inappropriate if utilized alone when working with microcones. For this reason we have introduced the *microcone assignment* technique, as explained in the next.

3.2 The Microcone Assignment

The microcone assignment plays the same role of the crowding distance in NSGA-II and cspMEA. It is used to select which offsprings to copy in the next generation, for each processor. Since a processor covers multiple, disjoint regions of the Pareto front (the microcones), we need a way to determine which are the best candidate solutions for each microcone. Then, the best solutions on each microcone can be iteratively picked up, microcone after microcone, until the size of the next population is equal to the size of the previous population. Clearly, within each microcone, the best solutions are those falling within that microcone, *if there is/are any*. Furthermore, the best ones within the ones that fall in a microcone can be chosen according to the crowding distance. Doing so we have a criterion to give priority to solutions falling within the microcone. However, as Fig. 9 shows, it can happen that the number of required solutions from a given microcone is larger than the number of available solutions falling within that microcone (or, in the worst case, there are none at all). In this case, we decide to allow selecting solutions

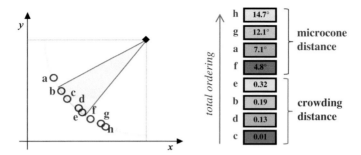

Fig. 9 An example of the working principle of the combined use of crowding distance and microcone distance (microcone assignment), when considering a single microcone. For solution falling within the microcone (*b*, *c*, *d*, and *e*) the crowding distance sorting is employed. For solutions outside the microcone (*a*, *f*, *g* and *h*) the microcone distance sorting is utilized. Solutions within the microcone as selected first

falling outside the considered microcone, *but close to it*. Of course, we need to privilege solutions as closer as possible to the microcone at hand. To do this, we introduce the *microcone distance*, which is the angular distance between a solution and the considered microcone. Figure 9 shows an example of the combined use of crowding distance and microcone distance, for a microcone.

In Fig. 10 we show an example of the benefits of the combined use of crowding distance and microcone distance when considering *all* the microcones associated to a processor. In particular, in the example we have 2 processors and 2 microcones ($N_{group} = 2$) per processor. Suppose that the microcones of processor 1 are those in gray. Microcone 1 of processor 1 (the gray one on top) has no solutions, while the second of the same processor has four. If the population size is four, using the crowding distance only (and considering the hard constraint approach) the four solutions to copy in the next generation of processor 1 are the four ones of the

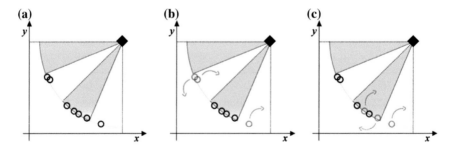

Fig. 10 An example of the working principle of the combined use of crowding distance and microcone distance (microcone assignment), when considering all the microcones of a processor. The four selected solutions by microcone assignment are the black ones in (**c**), while the four that would have been selected using only the crowding distance (with hard constraints) are the black ones in (**b**). The *gray arrows* indicate solutions violating the soft constraint (and thus the ones candidate to migrate)

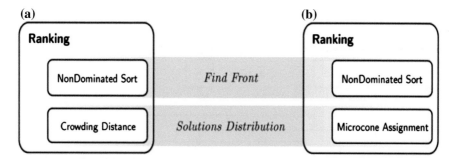

Fig. 11 Ranking adopted by NSGA-II and cspMEA (**a**). Ranking adopted by mspMEA (**b**). Both use the non-dominated sorting operator. They only differ on which rank 1 solutions are selected to be copied in next generation

second microcone, highlighted in black in Fig. 10b. This means that one microcone would have no solutions at all. As a consequence, the glocal Pareto front generated by processor 1 would be quite narrow. On the contrary, when considering the microcone assignment, the algorithm starts by selecting a solution from microcone 1. Since it has no solutions falling within it (and thus the crowding distance set of solutions is empty), the solution with the lowest microcone distance is chosen, i.e., the one on the top. Then the algorithm seeks one solution from the second microcone. In this case, the one with the lowest crowding distance is chosen. Then, since we have to pick four solutions in total, one more is added from the first microcone (the one with the second smallest microcone distance, i.e., the second from the top). Then another one is selected for the second microcone of processor 1 (the one with the second smallest crowding distance). In summary, the four selected solutions for processor one are those in black in Fig. 10c (the others are those candidate for migration). As it can be seen, the glocal Pareto front approximation is wider adopting microcone assignment (Fig. 10c) than adopting crowding distance (Fig. 10b), giving a better distribution and coverage of the global Pareto front.

Summarizing, in our mspMEA we have still adopted the non-dominated sorting operator as in NSGA-II and in cspMEA. The only difference is the fact that we have combined it with the microcone assignment (instead of the pure crowding distance as in cspMEA), to select the subset of individuals to copy in next generation (see Fig. 11).

3.3 The mspMEA Algorithm

The final mspMEA algorithm is given in Fig. 12. In next section we introduce the problem we have solved with the help of mspMEA.

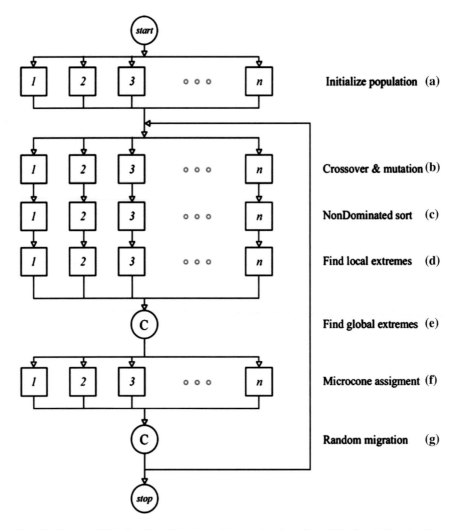

Fig. 12 The mspMEA algorithm. The deterministic migration of cspMEA is substituted with a random migration (g) of a prefixed number of individuals chosen among the ones violating the soft constraint (the ones falling within wrong microcones)

4 The Application: Modeling the Ship Classification Unit of an Influence Mine

4.1 Modern Influence Mines

While (old) contact mines detonate in case of contact (both with target and non-target ships) modern influence mines, equipped with a series of sensors, are

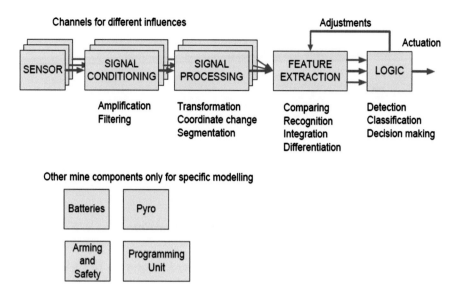

Fig. 13 General representation of a modern influence mine. Each box is a subsystem of the mine. The classification is performed by the SCU

able to classify ships and to detonate only when they are targets. Furthermore, they are able to compute the optimal detonation time. Figure 13 illustrates a general representation of a modern influence mine. The classification step shown in the figure is performed by a specific unit: the ship classification unit (SCU).

4.2 Modeling Influence Mines

Building a model for an influence mine is useful in many cases, as discussed in the introduction. The mine logic can be modeled either deterministically or stochastically [13]. A deterministic model is an accurate model of every subsystem of an influence mine. When building such a model, the modeler has to choose the usage criteria, in order to decide which subcomponents needs to be modeled and which ones could be disregarded (see Fig. 14).

Deterministic modeling requires a lot of efforts and time. Furthermore, once the model has been built for a specific mine, it cannot be reused for different mines (meaning that the process has to restart from scratch [13]). In some situations an accurate, deterministic model is not strictly necessary for the user's goals. In such situations, building a stochastic one (able to describe the behaviour of the mine *on average*) could be enough. In particular, the construction of a stochastic emulator of the SCU is adequate in many applications. In next section we describe how we have built such an emulator using fuzzy rule-based classifiers optimized by means of the mspMEA algorithm.

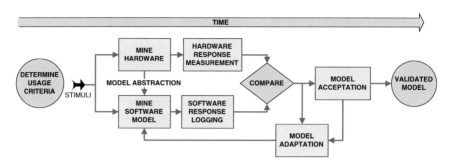

Fig. 14 Deterministic modeling process of mines

4.3 The Ship Classification Unit Fuzzy Emulator (SCUFE)

We have developed a Ship Classification Unit Fuzzy Emulator (SCUFE), which is a fuzzy rule-based classifier able to classify ships as the mine would do, once trained using MEAs in the appropriate way. In particular, the input of the fuzzy classifier is a set of features extracted from one or more signatures. Modern influence mines use one or more sensors for the different influences contained in the signature of a ship. These influences can be acoustic, magnetic, pressure, electric, seismic or even flow. The mines not only try to classify the ship but also try to determine the optimal detonation time. The mine logic actually translates the specific signature features in characteristic properties that describe the ship and its location. The developed SCUFE mine model only models the classification and the mine counter measures (MCM) rejection capabilities of influence mines. The functionality of the SCUFE is indicated by the dashed box in Fig. 15, and his diagram is detailed in Fig. 16.

4.4 Fuzzy Rule-Based Classifiers

The role of the fuzzy rule-based classifier within the SCUFE is to classify the features extracted from signatures in order to stochastically mimic the ship classi-fication unit of the influence mine to model. The fuzzy rules have to be learned from a labeled dataset, containing samples of both target and non-target ships.

The inputs used for the fuzzy classifiers are the same ones specified in [13], i.e., the following 18 features extracted from the static magnetic (SM) signature:

- Global minima and maxima of the x, y and z components (6 features)
- Rate of rise of the x, y and z components (3 features)
- Global maximum of the total field (1 feature)
- Pseudo period (1 feature)
- Integral values of the x, y and z components and the total field (4 features)
- Number of zero crossings of the x, y and z components (3 features)

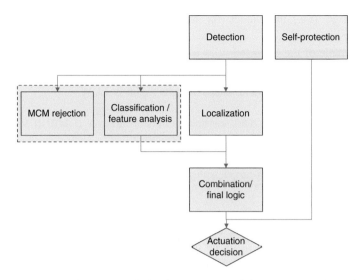

Fig. 15 Capabilities and modules typically found in modern influence mines. The *dashed box* is unit emulated by SCUFE. The detection, self-protection, localization and the combinational/final logic modules are not covered by the SCUFE

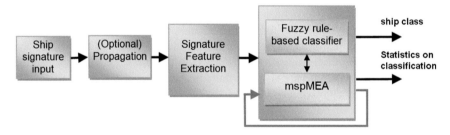

Fig. 16 Detailed diagram of the SCUFE model

The employed fuzzy rules are of the form:

$$R_i: \textbf{if } X_1 \textit{ is } A^1_{\delta_{i,1}} \quad \textbf{and } \dots X_F \textit{ is } A^F_{\delta_{i,F}}$$
$$\textbf{then } Y \textit{ is } C_1 \dots C_M \quad \textit{with} \quad \gamma^1_i \dots \gamma^M_i \tag{1}$$

where: X_f is the fth input feature ($f = 1 \dots F, F = 18$); A^f_t is the tth fuzzy set defined over the fth input variable (assuming to have uniformly partitioned the domain of the fth variable into T Gaussian fuzzy sets); $C_1 \dots C_M$ are the M classes (in this work, $M = 2$: C_1 is the target class and C_2 the non-target class), and; γ^m_i are the probabilities of belonging to the mth class associated with the ith rule [4]. In particular, the $\delta_{i,f}$ used in (1) specifies the fuzzy set involved in the ith rule for the fth variable. The mspMEA algorithm has to optimize the number of rules, the fuzzy sets involved (i.e., the values for the $\delta_{i,f}$) and the associated probabilities $(\gamma_i^1$ and $\gamma_i^2)$.

An example of the extracted rules is the following:

> **if**
>> the TOTAL INTEGRAL MAGNETIC FIELD VALUE is **LOW**
>
> **and**
>> the NUMBER OF ZERO CROSSINGS IN THE Z COMPONENT is **MEDIUM**
>
> **then**
>> the ship is a **target**,   with probability **0.87** and
>>
>> the ship is a **non-target**, with probability **0.13**

The crossover and mutation operators are the ones described in [10]. The training and test sets are the same utilized in [13], consisting of 56 samples of target ships and 432 of non-target ships, for a total of 488 samples in the 18-dimensional feature space.

5   Results

Two objective functions have been used in MEA optimization: the true positive rate and the false positive rate (the former to be maximized and the latter to be minimized). Since the performance of a continuous classifier can be charted on the Receiver Operating Characteristic (ROC) space, in this work we have adopted the Area Under the ROC curve (AUC) as the sole indicator of the goodness of the whole Pareto front approximation. Doing so, we have been able to easily compare the performances of different MEAs: the higher the AUC the better. The AUC (which ranges between 0 % and 100 %), is particularly interesting because it has a "physical" meaning: it is equal to the probability of correctly classifying a pair of samples, when one belongs to the target class and the other to the non-target one.

As regards the settings of the FRBCs, we have employed 5 fuzzy sets (VERY LOW, LOW, MEDIUM, HIGH, and VERY HIGH) on each input feature and a maximum number of rules equal to 40.

Concerning the parallel MEAs, we have run cspMEA with a number of cones equal to the number of available processors (8 in our case). The mspMEA has been run with 4 cones and 4 microcones per cone (this turned out to be the best tradeoff on the problem at hand), and the number of solutions to randomly migrate per epoch was set to 20. We have compared the accuracy and the speedup of these two algorithms with a sequential NSGA-II running on a single processor. The population size and the number of epochs for the sequential NSGA-II have been set to 200 and 5000, respectively. On the contrary, the cspMEA and mspMEA have been run with a population size of 25 for each of the 8 processors, and the same number

Table 1 Results in terms of accuracy (AUC) and speedup

	AUC	Speedup
NSGA-II (1 processors)	**96 % ± 0.28 %**	-
cspMEA (8 processors)	96 % ± 0.52 %	6.39 ± 0.12
mspMEA (8 processors)	96 % ± 0.45 %	**7.45 ± 0.24**

of epochs. In this manner, the number of fitness evaluations has been the same for the three algorithms.

Table 1 shows the AUC on test set and the speedup obtained by the parallel versions against the sequential one (NSGA-II), after averaging the results of ten runs. As confirmed by a t-test with 95 % of statistical significance, the three AUC values reported in Table 1 are not statistically different. This means that the three algorithms are statistically equivalent from the accuracy point of view. However, there is a statistically significant difference in the speedup achieved by mspMEA (7.45) with respect to that achieved by cspMEA (6.39). In particular, mspMEA has a significant higher speedup (16.58 % of improvement), reaching a value rather close to 8, the theoretical maximum value. The difference in the speed is also due to the well-known cold start problem of cspMEA, when a high, uncontrolled number of migrations occur. In addition, mspMEA has other degrees of freedom that help tuning the parallel algorithm for the problem at hand (the number of microcones and the migration rate).

6   Conclusions

We have presented the mspMEA algorithm, an extension of the well-known cspMEA algorithm. While the latter is built upon the concept of *cones*, the former is built upon the concepts of both cones and *microcones*. The use of microcones gives to each processor a wider view of the global Pareto front: this prevents wasting time working on portions of the fronts that are dominated by solutions contained in other microcones of the same processor. It also allows a better distribution of the solutions along the global Pareto front in case of discontinuous fronts. In addition the new algorithm has a greater flexibility over cspMEA: the user can tune the number of microcones and the migration rate for the problem at hand.

We have shown how the mspMEA outperforms cspMEA in the considered application, namely, the fuzzy rule-based emulation of the ship classification unit within modern influence mines. This has led to a better use of the 8 processors with a speedup improvement of 16.58 % over cspMEA. As a future work, we will investigate whether the concept of microcones is effective in parallelizing other types of MEAs, and how to further speedup the learning phase of the ship classification emulator by moving the fitness evaluation on GPGPUs (following the approach discussed in [7]).

References

1. Antonelli, M., Ducange, P., Marcelloni, F.: A new approach to handle high dimensional and large datasets in multi-objective evolutionary fuzzy systems. In: Proceedings of the 2011 IEEE International Conference on Fuzzy Systems, pp. 1286–1293, 27–30 June 2011
2. Branke, J., Schmeck, H., Deb, K., Reddy, S.M.: Parallelizing multi-objective evolutionary algorithms: cone separation. In: Proceedings of the 2004 Congress on Evolutionary Computation, vol. 2, pp. 19–23 (2004)
3. Cococcioni, M., Corsini, G., Lazzerini, B., Marcelloni, F.: Solving the ocean color inverse problem by using evolutionary multi-objective optimization of neuro-fuzzy systems. Int. J. Knowl. Based Intell. Eng. Syst. (KES) 12(5–6), 339–355 (2008)
4. Cococcioni, M., D'Andrea, E., Lazzerini, B.: Providing PRTools with fuzzy rule-based classifiers. In: Proceedings of the 2010 IEEE International Conference on Fuzzy Systems, pp. 1–8 (2010)
5. Cococcioni, M., Ducange, P., Lazzerini, B., Marcelloni, F.: A new multi-objective evolutionary algorithm based on convex hull for binary classifier optimization. In: Proceedings 2007 IEEE Congress on Evolutionary Computation (IEEE-CEC 2007), pp. 3150–3156. Singapore, 25–28 Sept 2007
6. Cococcioni, M., Grasso, R., Rixen, M.: A hybrid continuity preserving inference strategy to speed up Takagi-Sugeno multiobjective genetic fuzzy systems. In: Proceedings of the 5th IEEE International Workshop on Genetic and Evolutionary Fuzzy Systems (GEFS 2011), pp. 66–72. Paris, 11–15 Apr 2011
7. Cococcioni, M., Grasso, R., Rixen, M.: Rapid prototyping of high performance fuzzy computing applications using high level GPU programming for maritime operations support. In: Proceedings of the 2011 IEEE Symposium on Computational Intelligence for Security and Defense Applications (CISDA 2011), pp. 17–23. Paris, 11–15 Apr 2011
8. Cococcioni, M., Lazzerini, B., Marcelloni, F.: Fast multiobjective genetic rule learning using an efficient method for Takagi-Sugeno fuzzy systems identification. In: Proceedings of 8th International Conference on Hybrid Intelligent Systems, pp. 272–277. Barcelona, Spain (2008)
9. Cococcioni, M., Lazzerini, B., Marcelloni, F.: Towards efficient Takagi-Sugeno multi-objective genetic fuzzy systems for high dimensional problems. In: Tenne, Y., Goh, C.-K. (eds.) Computational Intelligence in Expensive Optimization Problems. Studies in Evolutionary Learning and Optimization, pp. 397–422 (2009)
10. Cococcioni, M., Lazzerini, B., Marcelloni, F.: On reducing computational overhead in multi-objective genetic takagi-sugeno fuzzy systems. Appl. Soft Comput. 11(1), 675–688 (2011)
11. Cococcioni, M.: The evolutionary multiobjective optimization of fuzzy rule-based systems bibliography page. http://www.iet.unipi.it/m.cococcioni/emofrbss.html
12. Connors, W.A., Fox, W.L.J.: Performance-based planning and evaluation for modern naval mine countermeasures systems. NATO STO. Technical report (CMRE-FR-2014-025)
13. de Jong, A., Cococcioni, M.: NATO fuzzy logic generic mine model. In: Proceedings of Undersea Defence Technology Europe (UDT-EUROPE 2007), pp. 1–15. Naples, Italy, 5–7 June 2007
14. De Jong, C.A., Quesson, B.A., Ainslie, M.A., Vermeulen, R.C. Measuring ship acoustic signatures against mine threat. In: Proceedings of Meetings on Acoustics, vol. 17, p. 070059 (2012)
15. Deb, K., Pratap, A., Agarwal, S., Meyarivan, T.: A fast and elitist multiobjective genetic algorithm: NSGA-II. IEEE Trans. Evol. Comput. 6(2), 182–197 (2002)
16. Dorronsoro, B., Danoy, G., Nebro, A.J., Bouvry, P.: Achieving super-linear performance in parallel multi-objective evolutionary algorithms by means of cooperative coevolution. Comput. Oper. Res. 40(6), 1552–1563 (2013)

17. Fazzolari, M., Giglio, B., Alcalá, R., Marcelloni, R., Herrera, H.: A study on the application of instance selection techniques in genetic fuzzy rule-based classification systems: accuracy-complexity trade-off. Knowl. Based Syst. **54**, 32–41 (2013)
18. Grasso, R., Cecchi, D., Cococcioni, M., Trees, C., Rixen, M., Alvarez, A., Strode, C.: Model based decision support for underwater glider operation monitoring. In: Proceedings of the 2010 Oceans MTS/IEEE Seattle Conference (OCEANS/Seattle 2010), pp. 1–8. Seattle, Washington, USA, 20–23 Sept 2010
19. Knowles, J., Corne, D.W.: Approximating the nondominated front using the Pareto archived evolution strategy. Evol. Comput. **8**(2), 149–172 (2000)
20. Lee, D.S., Morillo, C., Bugeda, G., Oller, S., Onate, E.: Multilayered composite structure design optimisation using distributed/parallel multi-objective evolutionary algorithms. Compos. Struct. **94**(3), 1087–1096 (2012)
21. McIntosh, D.J.: Real variability in ship systems' noise and vibration. Design and through-life management implications for underwater noise and habitability. In: Proceedings of the INTER-NOISE and NOISE-CON Congress and Conference. Institute of Noise Control Engineering, vol. 249, issue 1, pp. 6298–6307 (2014)
22. Nielsen, P.L., Fox, W.L.J.: Seabed characterization for mine hunting sonar performance and mine burial predictions. NATO STO Technical report (CMRE-FR-2013-002) (2013)
23. Percival, A.M., Couillard, M., Midtgaard, Ø., Fox, W.L.J.: Unmanned systems, autonomy, and side-looking sonar: a framework for integrating contemporary systems into the operational MCM architecture. NATO STO CMRE Technical report (CMRE-FR-2013-013) (2013)
24. Song, K., Chu, P.C.: Conceptual design of future undersea unmanned vehicle (UUV) system for mine disposal. IEEE Syst. J. **8**(1), 43–51 (2014)
25. Tiwari, S., Fadel, G., Deb, K.: AMGA2: Improving the performance of the archive-based micro-genetic algorithm for multi-objective optimization. Eng. Optim. **43**(4), 377–401 (2011)
26. Van Veldhuizen, D.A., Zydallis, J.B., Lamont, G.B.: Considerations in engineering parallel multiobjective evolutionary algorithms. IEEE Trans. Evol. Comput. **7**(2), 144–173 (2003)
27. Zhang, Q., Hui, L.: MOEA/D: a multiobjective evolutionary algorithm based on decomposition. IEEE Trans. Evol. Comput. **11**(6), 712–731 (2007)
28. Zitzler, E., Laumanns, M., Thiele, L.: SPEA2: improving the strength Pareto evolutionary algorithm. Technical report 103, Computer Engineering and Networks Laboratory (TIK), ETH Zurich, Zurich, Switzerland (2001)

Synthetic Aperture Radar (SAR) Automatic Target Recognition (ATR) Using Fuzzy Co-occurrence Matrix Texture Features

Sansanee Auephanwiriyakul, Yutthana Munklang and Nipon Theera-Umpon

Abstract Synthetic aperture radar (SAR) image classification is one of the challenging problems because of the difficult characteristics of SAR images. In this chapter, we implement SAR image classification on three military vehicles types, i.e., T72 tank, BMP2 armored personnel carriers (APCs), and BTR70 APCs. The texture features generated from the fuzzy co-occurrence matrix (FCOM) are utilized with the multi-class support vector machine (MSVM) and the radial basis function (RBF) network. Finally, the ensemble average is implemented as a fusion tool as well. The best detection result is at 97.94 % correct detection from the fusion of twenty best FCOM with RBF network models (ten best RBF network models at $d = 5$ and other ten best RBF network models at $d = 10$). Whereas the best fusion result of FCOM with MSVM is at 95.37 % correct classification. This comes from the fusion of ten best MSVM models at $d = 5$ and other ten best MSVM models at $d = 10$. As a comparison we also generate features from the gray level co-occurrence matrix (GLCM). This feature set is implemented on the same classifiers. The results from FCOM are better than those from GLCM in all cases.

Keywords MSTAR data set · Synthetic aperture radar (SAR) image · Fuzzy co-occurrence matrix · Support vector machine · Radial basis function · Automatic target recognition · Texture feature extraction

S. Auephanwiriyakul (✉) · Y. Munklang
Department of Computer Engineering, Faculty of Engineering, Chiang Mai University, Chiang Mai, Thailand
e-mail: sansanee@ieee.org

Y. Munklang
e-mail: yutthanamunklang@gmail.com

N. Theera-Umpon
Department of Electrical Engineering, Faculty of Engineering, Chiang Mai University, Chiang Mai, Thailand
e-mail: nipon@ieee.org

© Springer International Publishing Switzerland 2016
R. Abielmona et al. (eds.), *Recent Advances in Computational Intelligence in Defense and Security*, Studies in Computational Intelligence 621,
DOI 10.1007/978-3-319-26450-9_18

1 Introduction

One of the most challenging problems in image classification is to classify synthetic aperture radar (SAR) since SAR images do not look similar to optical images at all. Example of optical images and SAR images of three military vehicles (T72 tank, BMP2 armored personnel carriers (APCs), and BTR70 APCs) are shown in Fig. 1. Even for human eyes sometimes it is difficult to classify objects in SAR images. There are several studies involving SAR image classification and target detection in SAR images [1–13]. Some of these methods provide a good to very good detection results. However, some of them require pre-processing steps of SAR images, template creation process, or segmentation process. Recently, there have been a few attempts in increasing the detection performance by using data fusion or feature fusion [14–16]. Although, they provide better detection results than that without fusion methods, their results are approximately 95 %. In [17], the detection result is around 98 %, but they need to pre-process the SAR images before performing any detection.

Fig. 1 Example of optical images (*left*) and SAR images (*right*) of **a** T72, **b** BMP2, and **c** BTR70

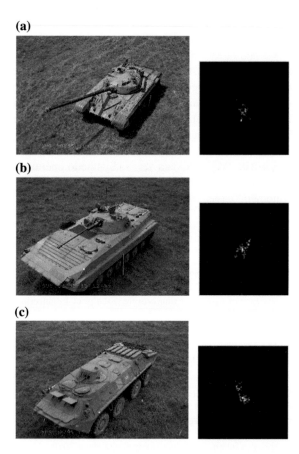

In this chapter, we will use texture features generated from the fuzzy co-occurrence matrix [18, 19] without any pre-processing method to detect three type of military vehicles, i.e., T72 tank, BMP2 APCs, and BTR70 APCs—with different orientations in SAR images from the MSTAR public release data set collected by the DARPA and Wright Laboratory. We previously utilized these features in texture classification and abnormality detection in mammograms [19]. Very good results were achieved in both classification and detection problems. Since different types of military vehicles in SAR images have different textures, it might be reasonable to try to detect these vehicles using texture features. After we generate features from the fuzzy co-occurrence matrix, we utilize support vector machine (SVM) with one against all scheme [20, 21] and radial basis function network (RBF) network [22]. We implement an ensemble average on the outputs [23] from 10 best models from SVM and other 10 best models from RBF networks as well. We also compare the result with the texture features extracted from the gray level co-occurrence matrix (GLCM) [24].

2 Methodology

We compare our result with those from GLCM (a second-order statistics of an image) [24]. Here we will briefly describe GLCM. The joint probability of occurrence of two gray level values with a particular distance (d) and orientation (θ) (shown in Fig. 2) is counted as $P(i, j, d, \theta)$. Suppose the size of an image is $N_x \times N_y$. Let $L_x = \{1, 2, …, N_x\}$ and $L_y = \{1, 2, …, N_y\}$ be the horizontal and vertical spatial domains, respectively, and $G = \{1, 2, …, N_g\}$ be the set of N_g quantized gray tones. In the experiment, we varies N_g from 2, 3, 4, 5, 6, 7, 8, 16, and 32. The image I can be represented as a function which assigns each gray tone in G to each resolution cell or a pair of coordinates in $L_y \times L_x$.

Fig. 2 GLCM orientation assignment

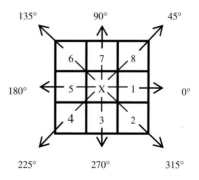

The joint probability of occurrence of two gray level values in each distance and direction is calculated as:

$$
\begin{aligned}
P(i,j,d,\,0) &= \#\{((k,l),\,(m,n)) \in (L_y \times L_x) \times (L_y \times L_x) \\
&\quad |k-m|=0, |l-n|=d, I(k,l)=i, I(m,n)=j\} \\
P(i,j,d,\,45) &= \#\{((k,l),(m,n)) \in (L_y \times L_x) \times (L_y \times L_x) \\
&\quad |k-m|=d, |l-n|=-d, I(k,l)=i, I(m,n)=j\} \\
P(i,j,d,\,90) &= \#\{((k,l),(m,n)) \in (L_y \times L_x) \times (L_y \times L_x) \\
&\quad |k-m|=d, |l-n|=0, I(k,l)=i, I(m,n)=j\} \\
P(i,j,d,\,135) &= \#\{((k,l),(m,n)) \in (L_y \times L_x) \times (L_y \times L_x) \\
&\quad |k-m|=d, |l-n|=d, I(k,l)=i, I(m,n)=j\} \\
P(i,j,d,\,180) &= \#\{((k,l),(m,n)) \in (L_y \times L_x) \times (L_y \times L_x) \\
&\quad |k-m|=0, |l-n|=-d, I(k,l)=i, I(m,n)=j\} \\
P(i,j,d,\,225) &= \#\{((k,l),(m,n)) \in (L_y \times L_x) \times (L_y \times L_x) \\
&\quad |k-m|=d, |l-n|=-d, I(k,l)=i, I(m,n)=j\} \\
P(i,j,d,\,270) &= \#\{((k,l),(m,n)) \in (L_y \times L_x) \times (L_y \times L_x) \\
&\quad |k-m|=d, |l-n|=0, I(k,l)=i, I(m,n)=j\} \\
P(i,j,d,\,315) &= \#\{((k,l),(m,n)) \in (L_y \times L_x) \times (L_y \times L_x) \\
&\quad |k-m|=d, |l-n|=d, I(k,l)=i, I(m,n)=j\}
\end{aligned}
\tag{1}
$$

where # denotes the number of elements in the set. In the experiment, d is varied between 1, 2, 3, 5, and 10 with $\theta = 0$ or $\theta = 0, 45, 90$, and 135 degrees. We compute 14 GLCM features [24], i.e., contrast (f_1), correlation (f_2), energy (f_3), homogeneity (f_4), variance (f_5), sum average (f_6), sum variance (f_7), sum entropy (f_8), entropy (f_9), difference variance (f_{10}), difference entropy (f_{11}), information measure of correlation1 (f_{12}), information measure of correlation2 (f_{13}), and maximum probability (f_{14}). We also compute an average and standard deviation of each feature from all directions to produce $\mu_1, \mu_2, \ldots, \mu_{14}$ and $\sigma_1, \sigma_2, \ldots, \sigma_{14}$, respectively. Then we have 6 sets of features, called GLCM1, GLCM2, ..., GLCM6 with different combinations of features and different θ. Table 1 shows the number of dimensions of each feature vector generated from each combination set. Although, N_g and d are varied, the number of dimensions is still similar to the one shown in Table 1.

To reduce the domination of feature, each feature is normalized using

$$
x' = \frac{x - \mu_x}{\sigma_x}.
\tag{2}
$$

Since our fuzzy co-occurrence matrix (FCOM) is built by incorporating the fuzzy C-means (FCM) clustering [25, 26] with the GLCM. We will briefly describe the FCM first. Let $X = \{\mathbf{x}_1, \mathbf{x}_2, \ldots, \mathbf{x}_N\}$ be a set of vectors, where each vector is a p-dimensional vector. The update equation for FCM is as follows [26]

Table 1 GLCM feature sets

Name		θ	Feature	Name		θ	Feature
GLCM1	Combination	0°	f_1–f_4	GLCM4	Combination	0°	f_1–f_{14}
	No. of dimensions	4			No. of dimensions	14	
GLCM2	Combination	0°, 45°, 90°, and 135°	f_1–f_4	GLCM5	Combination	0°, 45°, 90°, and 135°	f_1–f_{14}
	No. of dimensions	16			No. of dimensions	56	
GLCM3	Combination	0°, 45°, 90°, and 135°	μ_1–μ_4, σ_1–σ_4	GLCM6	Combination	0°, 45°, 90°, and 135°	μ_1–μ_{14}, σ_1–σ_{14}
	No. of dimensions	8			No. of dimensions	26	

$$u_{ij} = \frac{1}{\sum_{k=1}^{Cl} \left[\frac{\|x_i - c_j\|}{\|x_i - c_k\|} \right]^{\frac{2}{m-1}}}. \tag{3}$$

$$c_j = \frac{\sum_{i=1}^{N} u_{ij}^m x_i}{\sum_{i=1}^{N} u_{ij}^m} \tag{4}$$

where, u_{ij} is the membership value of vector x_j belonging to cluster j, c_j is the center of cluster j, and m is the fuzzifier. The algorithm of the FCM is:

```
Fix the number of clusters (Cl)
Initiate prototypes
Do {
        Update membership values using (3)
        Update prototypes using (4)
} Until prototypes stabilize
```

We implement the FCM on an original gray scale image I with $m = 2$. The number of clusters (Cl) is varied between 2, 3, 4, 5, 6, 7, 8, 16 and 32. Each pixel will be assigned to the cluster with the maximum membership value. Finally, FCOM plane will be created in each direction ($\theta = 0, 45, 90$, and 135 degrees). The number of FCOM planes in each direction is equal to the number of clusters. In the experiment θ is set to 0 degree or 4 directions mentioned above and d is varied between 1, 2, 3, 5, and 10. The fuzzy co-occurrence matrix algorithm is as follows:

```
For each direction (θ)
  For each pixel (p)
    Find a pixel (q) that is d apart from p
    Find the assigned cluster of pixel q
    Suppose assigned cluster of p is k and assigned
      cluster of q is l
    Set all eight FCOMs to zero
    For each cluster (i)
      FCOM(i,k,l) = FCOM(i,k,l) + u_pi + u_qi
      (where u_pi and u_qi are the membership values of
      pixels p and q in cluster i)
    End For i
  End For p
End For θ
```

From the algorithm, the number of clusters indicates the size of FCOM. For example, if $Cl = 8$, each FCOM plane will have the size of 8×8. After the FCOM plane is produced, we compute 14 features similar to those from GLCM as following:

for $1 \leq k \leq Cl$

$$fc_1: contrast_k = \sum_{i,j} |i-j|^2 FCOM(k,i,j), \tag{5}$$

$$fc_2: correlation_k = \sum_{i,j} \frac{(i-\mu_i^k)(j-\mu_j^k)FCOM(k,i,j)}{\sigma_i^k \sigma_j^k}, \tag{6}$$

$$fc_3: energy_k = \sum_{i,j} FCOM(k,i,j)^2, \tag{7}$$

$$fc_4: homogeneity_k = \sum_{i,j} \frac{FCOM(k,i,j)}{1+|i-j|}, \tag{8}$$

$$fc_5: variance_k = \sum_{i,j} (i-\mu_i^k)(j-\mu_j^k)FCOM(k,i,j), \tag{9}$$

$$fc_6: sumAve_k = \sum_{i,j} (ij)FCOM(k,i,j), \tag{10}$$

$$fc_7: sumVar_k = \sum_{i,j} (ij - sumAve_k)^2 FCOM(k,i,j), \tag{11}$$

$$fc_8: sumEnt_k = - \sum_{i,j} FCOM_{x+y}(k,i,j) \log\left(FCOM_{x+y}(k,i,j)\right), \tag{12}$$

$$fc_9: entropy_k = -\sum_{i,j} FCOM(k,i,j) \log(FCOM(k,i,j)), \tag{13}$$

$$fc_{10}: dif\_var_k = (varFCOM_{x-y}(k)) \tag{14}$$

$$fc_{11}: difEnt_k = -\sum_{i,j} FCOM_{x-y}(k,i,j) \log(FCOM_{x-y}(k,i,j)), \tag{15}$$

$$fc_{12}: InfoCor1_k = \frac{entropy_k - HXY1}{\max(HX, HY)} \tag{16}$$

$$fc_{13}: InfoCor2_k = (1 - \exp[-2.0(HXY2 - entropy_k)])^{1/2} \tag{17}$$

$$fc_{14}: MaxCorCoef_k = (\text{second largest eigenvalue of } Q)^{1/2} \tag{18}$$

where, for $1 \leq k \leq Cl$

$$FC_x^k(i) = \sum_j FCOM(k,i,j),$$

$$FC_y^k(j) = \sum_i FCOM(k,i,j),$$

$$\mu_i^k = \sum_i iFC_x^k(i),$$

$$\mu_j^k = \sum_j jFC_y^k(j),$$

$$\sigma_i^k = \sum_i (i - \mu_i^k)^2 FC_x^k(i),$$

$$\sigma_j^k = \sum_j (j - \mu_j^k)^2 FC_y^k(j),$$

$$FCOM_{x+y}(k) = \sum_{\substack{i,j \\ i+j=l}} FCOM(k,i,j), \text{ for } l = 2,3,\ldots,2C \tag{19}$$

$$FCOM_{x-y}(k) = \sum_{\substack{i,j \\ |i-j|=l}} FCOM(k,i,j), \text{ for } l = 0,1,\ldots,C-1$$

$$HX = \text{entropy of } FC_x^k$$

$$HY = \text{entropy of } FC_y^k$$

$$HXY1 = -\sum_{i,j} FCOM(k,i,j) \log(FC_x^k(i)FC_y^k(j)),$$

$$HXY2 = -\sum_{i,j} FC_x^k(i)FC_y^k(j) \log(FC_x^k(i)FC_y^k(j)),$$

$$\text{and } Q(i,j) = \sum_l \frac{FCOM(k,i,l)FCOM(k,j,l)}{FC_x^k(i)FC_y^k(l)}.$$

Again, similar to the GLCM, we also compute the average and the standard deviation of each feature from all directions, i.e., $\mu_{fc1}, \mu_{fc2}, \ldots, \mu_{fc14}$, and $\sigma_{fc1}, \sigma_{fc2},$

Table 2 FCOM feature sets

No. of clusters	No. of dimensions					
	FCOM1	FCOM2	FCOM3	FCOM4	FCOM5	FCOM6
	$\theta = 0°$ using fc_1-fc_4	$\theta = 0°$, $45°$, $90°$, and $135°$ using fc_1-fc_4	$\theta = 0°$, $45°$, $90°$, and $135°$ using $\mu_{fc1}-\mu_{fc4}$, $\sigma_{fc1}-\sigma_{fc4}$	$\theta = 0°$ using fc_1-fc_{14}	$\theta = 0°$, $45°$, $90°$, and $135°$ using fc_1-fc_{14}	$\theta = 0°$, $45°$, $90°$, and $135°$ using $\mu_{fc1}-\mu_{fc14}$, $\sigma_{fc1}-\sigma_{fc14}$
2	8	32	16	28	112	56
3	12	48	24	42	168	84
4	16	64	32	56	224	112
5	20	80	40	70	280	140
6	24	96	48	84	336	168
7	28	112	56	98	392	196
8	32	128	64	112	448	224
16	64	256	128	224	896	448
32	128	512	256	448	1792	896

..., σ_{fc14}, respectively. We create 6 sets of features similar to the features generated from GLCM, called FCOM1, FCOM2, ..., FCOM6 with different combinations of features, different numbers of clusters and different θ. Table 2 shows the number of dimensions of each feature vector generated from each combination set. Since the number of clusters will indicate the number of FCOM planes, the number of feature dimensions will be different. Again, even though d is varied, the number of feature dimensions is similar to the one shown in Table 2. We also normalized each feature using Eq. (2) similar to the one from GLCM.

Now, we will briefly describe the multi-class SVM (MSVM) [20] which is a method that assigns a class label to a vector which belongs to one of several classes. In this chapter, we utilize one-versus-all strategy. Suppose we have an optimum discriminant function ($D_i(\mathbf{x})$ for $i = 1, ..., Class$). From the support vector machine (SVM) [21], we have an optimum hyperplane at $D_i(\mathbf{x}) = 0$ that will separate class i from all the others. Hence, each SVM classifier gives $D_i(\mathbf{x}) > 0$ for vectors in class i, and $D_i(\mathbf{x}) < 0$ for those in all other classes. Then the classification rule is

$$\mathbf{x} \text{ is assigned to class } i \text{ if } \quad i = \arg \max_{j=1,..,n} D_j(x) \tag{20}$$

The SVM used in the experiment is the one with soft margin optimization. To ease our training process, we set $C = 1$ and $\varepsilon = 1 \times 10^{-3}$ in the experiment. The radial basis function used in each SVM is

$$K(\mathbf{x}_i, \mathbf{x}_j) = \exp\left(\frac{-\|\mathbf{x}_i - \mathbf{x}_j\|^2}{2\sigma^2}\right) \tag{21}$$

Another classifier used in this chapter is the RBF network [22]. The output of node i in the hidden layer is not calculated from the inner product between input vector $\mathbf{x}$ with weights, but instead it is calculated from

$$y_i = \exp\left(\frac{-\|\mathbf{x} - \mathbf{t}_i\|^2}{(\sigma 0.8326)^2}\right) \tag{22}$$

where $\mathbf{t}_i$ is the center of node i and σ is the spread of node i. The output of node j in the output layer will be the inner product between inputs coming to node j with their weights. For the weights training, the linear least square method [22] is implemented.

One might want to improve the result by fusing the output of the best models. We implement the simple ensemble average by averaging the output from all the output [23]. To fuse the output from multi-class SVM, we only use $D_i(\mathbf{x})$ for $i = 1, 2,$ and 3 and then compute the average of those values over the M best multi-class SVM models by

$$F_i(\mathbf{x}) = \frac{1}{M}\sum_{j=1}^{M} D_i^j(\mathbf{x}) \tag{23}$$

where $D_i^j(\mathbf{x})$ is the discriminant function of class i from model j for input vector $\mathbf{x}$, and $F_i(\mathbf{x})$ is the fused output of class i for input vector $\mathbf{x}$. For the fusion of the M best RBF network models, we also implement a similar scheme.

3   Experiment Results

The MSTAR public data set was collected by the DARPA/WRIGHT laboratory Moving and Stationary Target Acquisition and Recognition (MSTAR) program. It contains a high resolution SAR data. The data set utilized in this chapter contains two types of military vehicles, i.e., tanks and armored personnel carriers (APCs). However, there are two types of APCs, i.e., BMP2 and BTR70, three different T72 tanks, and three different BMP2 APCs indicated by their serial numbers. However, in this chapter, we only run the experiment on the T72 tank with serial number S7, the BMP2 APC with serial number C21, and the BTR70 APC with serial number C71. Hence, there are 233 training BMP2 images, 196 blind test BMP2 images, and 233 training BTR70 images, 196 blind test BTR70 images, and 228 training T72 images, 191 blind test T72 images. These images are with the size of 128×128. Since, the objects is approximately at the center of each SAR image, we use a square window of size $M \times M$ centered at (64, 64) to extract the features. In the experiment, we vary M to 21, 31, and 35. Example of object in each window size is shown in Fig. 3.

(a)

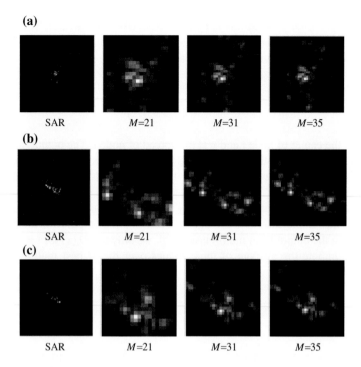

(b)

(c)

Fig. 3 Example of objects in different window sizes for **a** BMP2, **b** BTR70, and **c** T72. Please be noted that the images are not of the same scale, we zoom in some images for the display purpose in this figure

In the experiment with FCOM and GLCM feature generation, we set the same parameters in MSVM and RBF network. That is $\sigma = 0.01$, 0.25, 0.50, 0.75, and 1 to 50 with the step size of 0.5 for both MSVM and RBF network experiments. For the RBF network, we let the number of hidden nodes equal to the number of training feature vectors and use those feature vectors to be the centers of hidden nodes. To create generalized network, we implement the 10-fold cross validation on the training images set. For each d, we compute ensemble average of the 10 best validation MSVM models, and ensemble average of 10 best validation RBF network model. Table 3 shows the 10 best validation set for FCOM with $d = 1, 2, 3, 5$, and 10 from MSVM. These 10 best models come from different FCOM feature sets and different M and σ. Table 4 shows the blind test result from these 10 best models. Table 5 shows the 10 best validation set for FCOM with $d = 1, 2, 3, 5$, and 10 from RBF network.

Again, these 10 models are with different FCOM feature sets, M and σ. Table 6 shows the blind test result from these 10 best models. Table 7 shows the ensemble average results on blind test set from 10 best MSVM models and 10 best RBF network models with different d. We can see that best validation detection from FCOM with MSVM is 97.14 % whereas the blind test result for this model is

Table 3 Ten best MSVM classification rates on validation sets using FCOM for $d = 1, 2, 3, 5,$ and 10

Model no.	d = 1		d = 2		d = 3		d = 5		d = 10	
	Cl	Validation result	Cl	Validation result	Cl	Validation result	Cl	Validation result	Cl	Validation result
1	32	**85.71**	16	**91.30**	7	**94.20**	8	**97.14**	7	**97.14**
2	7	82.61	8	**91.30**	16	**94.20**	6	97.10	7	**97.14**
3	32	81.43	7	89.86	8	**94.20**	16	97.10	16	**97.14**
4	7	81.16	6	89.86	16	92.86	32	95.71	16	**97.14**
5	6	80.00	7	89.86	7	92.75	6	95.71	32	**97.14**
6	7	79.71	7	89.86	8	92.75	8	95.71	8	97.10
7	8	79.71	7	89.86	8	92.75	16	95.65	8	97.10
8	16	79.71	8	88.41	16	92.75	7	95.65	8	97.10
9	32	78.57	6	88.41	7	91.30	7	95.65	5	95.71
10	3	78.57	16	88.41	8	91.30	7	94.29	7	95.71
Average		80.72		89.71		92.91		95.97		96.84

Table 4 The classification rates on blind test sets from 10 best MSVM models in Table 3 using FCOM for $d = 1, 2, 3, 5,$ and 10

Model no.	$d = 1$	$d = 2$	$d = 3$	$d = 5$	$d = 10$
1	69.81	79.25	82.68	85.59	85.25
2	67.07	79.25	80.96	85.08	80.62
3	62.26	77.02	81.99	83.88	**89.02**
4	67.58	75.81	78.90	83.53	87.48
5	64.15	77.02	82.68	82.33	88.16
6	62.95	81.65	79.42	86.11	84.91
7	62.44	74.61	84.91	81.82	84.56
8	68.27	80.96	83.53	80.62	87.65
9	60.55	75.81	78.22	81.99	83.70
10	59.35	79.42	83.53	79.59	81.82
Average	64.44	78.08	81.68	83.05	85.32

Table 5 Ten best RBF network classification rates on validation sets using FCOM for $d = 1, 2, 3,$ 5, and 10

Model no.	$d = 1$		$d = 2$		$d = 3$		$d = 5$		$d = 10$	
	Cl	Validation result	Cl	Validation result	Cl	Validation result	Cl	Validation result	Cl	Validation result
1	7	**82.61**	6	**90.00**	7	**92.86**	16	**98.55**	6	97.14
2	8	81.43	16	89.86	8	**92.86**	5	97.14	8	**97.14**
3	32	78.57	8	88.57	8	92.75	6	97.10	16	**97.14**
4	8	78.26	7	88.57	7	92.75	8	97.10	7	97.10
5	8	77.14	8	88.57	8	91.43	7	95.71	7	95.71
6	6	77.14	7	88.41	32	91.43	5	95.65	7	95.71
7	32	77.14	8	88.41	8	91.30	5	95.65	5	95.71
8	7	76.81	16	88.41	7	91.30	5	94.29	7	95.71
9	16	75.71	7	88.41	32	90.00	7	94.29	8	95.71
10	6	75.71	8	87.14	6	90.00	8	94.29	7	95.65
Average	78.05		88.63		91.67		95.98		96.28	

Table 6 The classification rates on blind test sets from 10 best RBF network models in Table 5 using FCOM for $d = 1, 2, 3, 5,$ and 10

Model no.	$d = 1$	$d = 2$	$d = 3$	$d = 5$	$d = 10$
1	65.52	73.76	74.61	85.93	89.02
2	65.69	73.24	81.48	81.30	88.16
3	57.12	73.93	78.90	84.05	87.82
4	57.98	77.02	75.99	83.53	**89.88**
5	60.21	74.10	80.10	84.05	83.53
6	63.12	74.61	82.33	84.39	87.99
7	66.04	77.02	79.07	83.36	84.05
8	59.52	81.82	84.22	81.30	86.11
9	66.55	75.81	76.33	76.50	83.70
10	57.29	69.98	73.93	79.25	89.37
Average	61.90	75.13	78.70	82.37	86.96

Table 7 Ensemble averge classification rates using FCOM from 10 best MSVM models and 10 best RBF network models for different *d* on blind test sets

D	MSVM	RBF network
1	70.33	72.56
2	83.53	85.08
3	87.31	88.16
5	89.19	92.80
10	**93.14**	**96.23**

85.25 %. The best validation detection and blind test from FCOM with RBF network are 97.14 % and 89.02 %, respectively. We can see that when *d* increases, the detection performance also increases. This might be because the histogram of object in SAR image is dense around the bright area. Hence when we compute the values of FCOM plane with low distance (*d*), they will not be much different.

The example of FCOM planes with $Cl = 8$, $M = 35$, $\theta = 0°$, and with different *d* is shown in Fig. 4. From the figure, we can see that there is more information when *d* increases. For example, in the 4th cluster there are more peaks and the values of the peaks are higher for larger *d*. The values of FCOM planes have similar characteristics in the other clusters as well.

The best ensemble results on blind test sets from MSVM and RBF network from Table 7 are 93.14 and 96.23 %, respectively. These detection results are better than the normal average of all 10 best models in each classifier. However, one might wonder how the ensemble average will perform if the ensemble average is implemented over the outputs from different *d*. We implement the ensemble average on 50 MSVM best models with 10 best model from each *d*, and that on 40 MSVM best models with 10 best models from $d = 2, 3, 5$, and 10, and that on 30 MSVM best models with 10 best models from $d = 3, 5$, and 10, and finally that on 20 MSVM best models with 10 best models from $d = 5$ and 10. We also implement the same scheme with the RBF network ensemble average. Table 8 shows the blind test results from the ensemble average of MSVM best models from different *d* and that of RBF network best models from different *d*. The best ensemble average detection results in this case from MSVM and RBF network are 95.37 % and 97.94 %, respectively. This type of ensemble average produces better detection results than the previous ensemble average. This might be again because

Table 8 Blind test classification rates using FCOM from ensemble average of SVM best models with different *d* and that of RBF network best models with different *d* and with 10 best models from each *d*

Classifier	$d = 1, 2, 3, 5$, and 10	$d = 2, 3, 5$, and 10	$d = 3, 5$, and 10	$d = 5$ and 10
SVM	92.11	93.65	95.03	95.37
RBF network	94.85	96.74	97.43	**97.94**

Table 9 Confusion matrix of the best ensemble average in Table 8 from MSVM

			Algorithm output		
			BMP2	BTR70	T72
Desired output		BMP2	178	5	13
		BTR70	3	192	1
		T72	5	0	186

Table 10 Confusion matrix of the best ensemble average in Table 8 from RBF network

			Algorithm output		
			BMP2	BTR70	T72
Desired output		BMP2	189	2	5
		BTR70	0	195	1
		T72	4	0	187

of larger d gives more information than smaller d. Hence fusing the outputs from inputs with more information should give better results.

The confusion matrices from MSVM and RBF network in this case are shown in Tables 9 and 10, respectively. The reason of misclassification in any experiments mentioned previously might be because the generated FCOM plane of each object is similar to the other objects. For example, BMP2 and BTR70 that are misclassified as T72 has similar FCOM plane in each cluster to those from T72 with $\theta = 0°$ and $Cl = 8$ as shown in Fig. 5.

The GLCM detection results are shown in Tables 11, 12, 13, 14 and 15. Tables 11 and 12 show the 10 best MSVM models on validation and blind test sets, respectively. Tables 13 and 14 show the 10 best RBF network models on validation and blind test sets, respectively. Finally Table 15 shows the ensemble average results from both MSVMs and RBF networks. Again, the 10 best models are from different GLCM feature sets, different M, and different σ. The best validation result and blind test result from MSVM are 92.75 and 79.59 %, respectively. Whereas the best validation and blind test detection results from RBF network are 92.86 and 82.50 %, respectively. Again, if distance (d) increases, the detection result also increases with the same reason as in FCOM. The GLCM plane of each d with $\theta = 0°$ and $N_g = 8$ of BMP2 in Fig. 4a is shown in Fig. 6. Again, there is more information (in terms of the number of pair-pixels) provided with larger d.

The best ensemble average results on blind test sets from 10 best MSVM models and 10 best RBF network models are 85.25 % and 92.80 %, respectively. These best results are from $d = 10$. We also implement the ensemble average of the outputs of MSVM models and RBF network models from different d similarly to the one for FCOM. Tables 16, 17 and 18 show the ensemble results in this case. The best ensemble average results in this case for MSVM and RBF network fusion are 87.31 % and 94.68 %, respectively. Again output fusion from $d = 5$ and 10 provides the best detection results in both MSVM and RBF network. The reason is the same as mentioned earlier in the case of FCOM. The misclassification is caused by the generated GLCM plane which is similar to that of other objects.

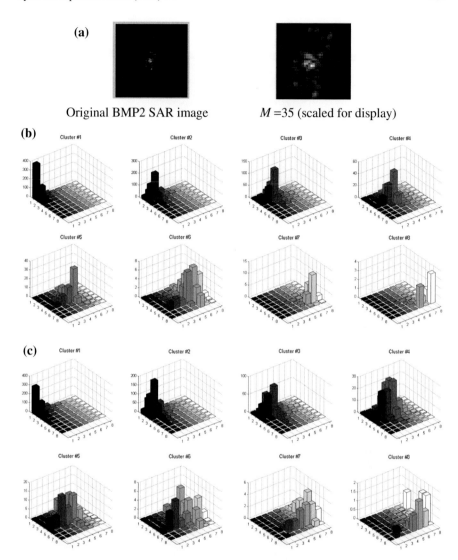

Fig. 4 **a** original BMP2 SAR image and its corresponding subimage ($M = 35$), the value of each FCOM plane for **b** $d = 1$, **c** $d = 2$, **d** $d = 3$, **e** $d = 5$, and **f** $d = 10$

From all the results, we can see that FCOM performs better than GLCM in all the cases in the experiment as we have shown in different applications [18, 19]. This might be because, with the same setting, i.e., d, θ, Cl (N_g), the FCOM plane has more information than the GLCM plane. In addition, the number of FCOM planes is equal to the number of clusters whereas there is only one GLCM plane. This will provide more information than the one given by the GLCM plane. An example of GLCM planes (and FCOM planes) with $d = 10$, $\theta = 0°$ with N_g (and Cl) = 2, 4, and 8 are

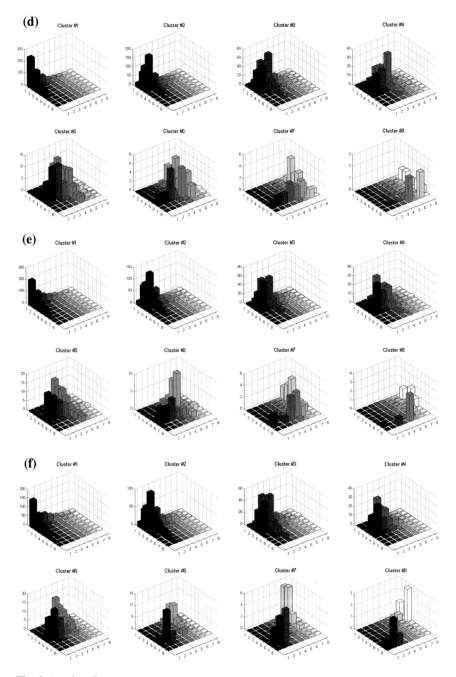

Fig. 4 (continued)

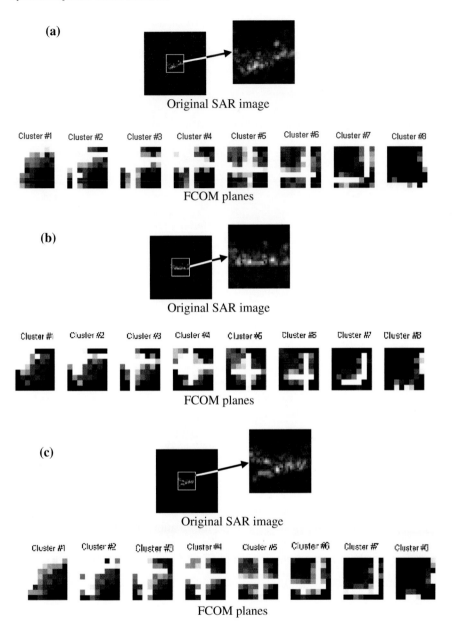

Fig. 5 Example of similar FCOM planes between different objects **a** BMP2, **b** BTR70 and **c** T72

484 S. Auephanwiriyakul et al.

Table 11 Ten best SVM classification rates on validation sets using GLCM for $d = 1, 2, 3, 5$, and 10

Model no.	$d = 1$		$d = 2$		$d = 3$		$d = 5$		$d = 10$	
	N_g	Validation result	N_g	Validation result	N_g	Validation result	N_g	Validation result	N_g	Validation result
1	32	**82.86**	32	**85.71**	32	**84.26**	32	**85.71**	32	**92.75**
2	32	**82.86**	32	**85.71**	16	82.86	32	85.51	16	89.86
3	32	81.43	16	84.06	32	82.86	32	84.29	32	87.14
4	16	75.36	32	82.61	32	81.43	32	84.08	32	87.14
5	32	74.29	32	81.43	32	80.00	32	82.86	16	85.51
6	32	72.86	32	80.00	32	79.71	32	82.86	32	85.51
7	6	72.86	32	78.57	16	78.57	32	82.61	32	85.51
8	7	72.86	32	78.57	32	78.26	16	82.61	32	84.29
9	32	72.46	32	78.26	32	78.26	8	81.43	5	84.06
10	32	71.43	32	77.14	16	77.14	16	81.43	32	82.86
Average	75.93		81.21		80.34		83.34		86.46	

Table 12 The classification rates on blind test sets from 10 best SVM models in Table 3 using GLCM for $d = 1, 2, 3, 5$, and 10

Model no.	$d = 1$	$d = 2$	$d = 3$	$d = 5$	$d = 10$
1	67.24	62.95	75.30	78.04	**79.59**
2	68.27	67.07	72.04	72.21	76.84
3	59.52	64.49	75.81	71.18	71.18
4	57.29	69.13	70.50	72.21	76.67
5	56.09	64.49	68.61	70.15	74.96
6	57.98	58.66	68.27	73.93	79.76
7	53.17	56.43	70.67	63.98	75.64
8	52.32	61.41	65.69	68.61	74.61
9	58.15	62.26	69.30	64.15	64.49
10	57.46	63.29	66.72	66.04	73.76
Average	58.75	63.02	70.29	70.05	74.75

shown in Fig. 7. Also from this figure, we can see that with larger N_g (and Cl), the GLCM plane (and FCOM planes) has more information than the case with smaller N_g (and Cl). This might be the reason why the detection performance is better with increasing number of gray scales in case of GLCM or number of clusters in case of FCOM.

Table 13 Ten best RBF network classification rates on validation sets using GLCM for $d = 1, 2,$ 3, 5, and 10

Model no.	$d = 1$		$d = 2$		$d = 3$		$d = 5$		$d = 10$	
	N_g	Validation result	N_g	Validation result	N_g	Validation result	N_g	Validation result	N_g	Validation result
1	32	79.71	32	87.14	32	88.41	32	89.86	32	92.86
2	32	75.36	32	79.71	32	88.41	32	87.14	32	90.00
3	32	72.86	32	78.57	32	84.29	32	86.96	16	90.00
4	32	72.86	16	78.26	32	81.16	32	85.71	32	89.86
5	32	71.43	32	76.81	8	80.00	32	85.71	16	88.41
6	32	71.01	16	76.81	32	79.71	16	85.51	8	87.14
7	32	71.01	32	74.29	32	78.26	8	84.29	32	87.14
8	32	70.00	16	73.91	16	78.26	16	84.29	16	85.71
9	6	70.00	32	73.91	32	78.26	16	82.86	6	85.71
10	32	67.14	32	72.46	16	77.14	32	82.86	32	85.71
Average	72.14		77.19		81.39		85.52		88.25	

Table 14 The classification rates on blind test sets from 10 best RBF network models in Table 5 using GLCM for $d = 1, 2, 3, 5,$ and 10

Model No.	$d = 1$	$d = 2$	$d = 3$	$d = 5$	$d = 10$
1	63.12	65.69	69.13	78.73	**82.50**
2	64.15	64.84	74.27	74.27	77.87
3	59.35	64.67	71.53	73.58	80.27
4	56.26	61.23	71.87	67.07	79.59
5	60.03	57.29	62.26	76.16	76.84
6	56.60	51.63	64.67	66.04	70.67
7	59.18	63.81	51.97	68.95	77.36
8	56.95	53.00	67.58	72.04	68.95
9	48.03	58.83	68.78	65.18	69.47
10	52.83	58.49	63.46	62.95	76.50
Average	57.65	59.95	66.55	70.50	76.00

Table 15 Ensemble averge classification rates using GLCM from 10 best SVM models and 10 best RBF network models for different d on blind test sets

d	SVM	RBF network
1	68.27	75.81
2	72.21	77.70
3	80.45	83.71
5	84.39	89.02
10	**85.25**	**92.80**

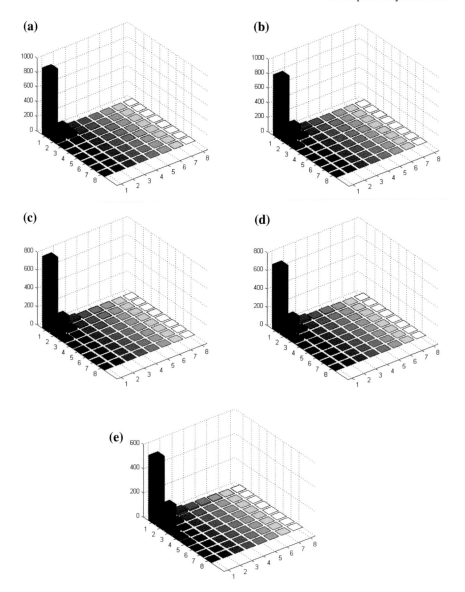

Fig. 6 GLCM plane of BMP2 in Fig. 4a with **a** $d = 1$, **b** $d = 2$, **c** $d = 3$, **d** $d = 5$, and **e** $d = 10$

Table 16 Classification rates on bilnd test sets using GLCM from ensemble average of SVM best models with different d and that of RBF network best models with different d with 10 best models from each d

Classifier	$d = 1, 2, 3, 5,$ and 10	$d = 2, 3, 5,$ and 10	$d = 3, 5,$ and 10	$d = 5$ and 10
SVM	85.08	85.59	86.79	87.31
RBF network	93.31	93.14	95.88	94.68

Table 17 Confusion matrix of the best ensemble average in Table 16 from MSVM

		Algorithm output		
		BMP2	BTR70	T72
Desired output	**BMP2**	158	18	20
	BTR70	7	183	6
	T72	13	5	173

Table 18 Confusion matrix of the best ensemble average in Table 16 from RBF network

		Algorithm output		
		BMP2	BTR70	T72
Desired output	**BMP2**	185	5	6
	BTR70	6	186	4
	T72	4	1	186

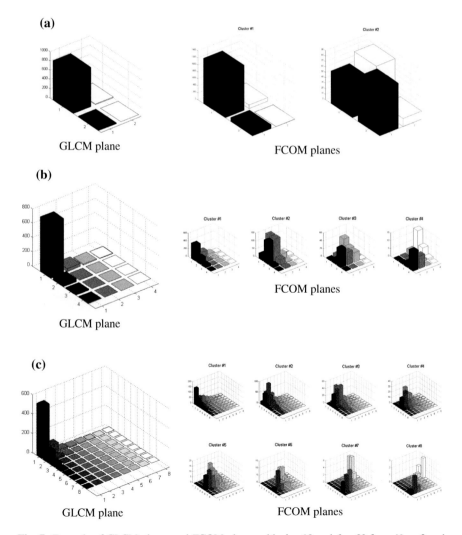

Fig. 7 Example of GLCM planes and FCOM planes with $d = 10$ and $\theta = 0°$ for **a** $N_g = 2$ and $Cl = 2$, **b** $N_g = 4$ and $Cl = 4$, and **c** $N_g = 8$ and $Cl = 8$

4 Conclusion

In this chapter, the texture features generated from the fuzzy co-occurrence matrix (FCOM) are implemented in the synthetic aperture radar (SAR) image classification. We implement these features on the multi-class support vector machine (MSVM) and the radial basis function (RBF) network. The ensemble average is utilized as an information fusion tool. There are two types of fusion, i.e., the fusion of the outputs from ten best models from each classifier in each distance d, and the fusion of the outputs from several best models from several d of each classifier. We found out that the best detection results is 97.94 % correct detection from the fusion of twenty best FCOM with RBF network models (ten best RBF network models at $d = 5$ and other ten best RBF network models at $d = 10$). Whereas the best fusion result of FCOM with MSVM is 95.37 % correct classification. This comes from the fusion of ten best MSVM models at $d = 5$ and other ten best MSVM models at $d = 10$. The detection results from FCOM are far better than that from gray level co-occurrence matrix (GLCM) in any cases.

Acknowledgement The authors would like to thank the Sensor ATR Division of the U.S. Air Force Research Laboratory and Veridian Corporation, especially to Mark Axtell, for providing the MSTAR data set.

References

1. Zhai, Y., Li, J., Gan, J., Xu, Y.: A novel SAR image recognition algorithm with rejection mode via biomimetic pattern recognition. J. Inf. Comput. Sci. **10**(11), 3363–3374 (2013)
2. Suvorova, S., Schroeder, J.: Automated Target recognition using the Karhunen-Loéve transform with invariance. Digit. Sig. Process. **12**, 295–306 (2002)
3. Du, C., Zhou, S., Sun, J., Zhao, J.: Feature extraction for SAR target recognition based on supervised manifold learning. In: 35th International Symposium on Remote Sensing of Environment (ISRSE35), IOP Conference Series: Earth and Environmental Science, vol. 17 (2014)
4. Theera-Umpon, N.: Fractal dimension estimation using modified differential box-counting and its application to MSTAR target classification. In: Proceedings of the 2002 IEEE International Conference on System, Man and Cybernetics, pp. 537–541 (2002)
5. Theera-Umpon, N., Khabou, A.M., Gader, D.P., Keller, M.J., Shi, H., Li, H.: Detection and classification of MSTAR objects via morphological shared-weight neural networks. In: Proceedings of SPIE 3370, Algorithms for Synthetic Aperture Radar Imagery V (1998)
6. Yuan, X., Tand, T., Xiang, D., Li, Y., Su, Y.: Target recognition in SAR imagery based on local gradient ratio pattern. Int. J. Remote Sens. **35**(3), 857–870 (2014)
7. Park, S., Smith, J.T.M., Mersereau, M.R.: Target recognition based on directional filter banks and higher-order neural networks. Digit. Sig. Process. **10**(4), 297–308 (2000)
8. Anagnostopoulos, C.G.: SVM-besed target recognition from synthetic aperture radar images using target region outline descriptors. Nonlinear Anal. **71**, e2934–e2939 (2009)
9. Bhanu, B., Jones III, G.: Object recognition results using MSTAR synthetic aperture rader data. In: IEEE Workshop on Computer Vision Beyond the Visible Spectrum Methods and Applications, pp. 55–62 (2000)

10. O'Sullivan, J., Devore, M.D., Kedia, V., Miller, M.I.: SAR ATR performance using a conditionally Gzussian model. IEEE Trans. Aerosp. Electron. Syst. **37**(1), 91–108 (2001)
11. Thiangarajan, J.J., Ramamurthy, K.N., Knee, P., Spanias, A., Berisha, V.: Sparse representations for automatic target classification in SAR images. In: Proceedings of the 4th International Symposium on Communications, Control and Signal Processing (2010)
12. Ye, X., Gao, W., Wang, Y., Hu, X.: Research on SAR images recognition based on ART2 neural network. In: 2012 7th IEEE Conference on Industrial Electronics and Applications, pp. 1888–1891 (2012)
13. Ni, J.C., Xu, Y.L.: SAR automatic target recognition based on a visual cortical system. In: 2013 6th International Congress on Image and Signal Processing, pp. 778–782 (2013)
14. Cui, Z., Cao, Z., Yang, J., Feng, J.: A hierarchical propelled fusion strategy for SAR automatic target recognition. EURASIP J. Wirel. Commun. Netw. **2013**, 39 (2013)
15. Srinivas, U., Monga, V., Raj, G.R.: SAR automatic target recognition using discriminative graphical models. IEEE Trans. Aerosp. Electron. Syst. **50**(1), 591–606 (2014)
16. Liu, H., Li, S.: Decision fusion of sparse representation and support vector machine for SAR image target recognition. Neurocomputing **113**, 97–104 (2013)
17. Ruohong, H., Keji, M., Yanjing L., Jiming Y., Ming, X.: SAR target recognition with data fusion. In: 2010 WASE Internation Conference on Information Engineering, pp. 19–23 (2010)
18. Munklang, Y., Auephanwiriyakul, S., Theera-Umpon, N.: A novel fuzzy co-occurrence matrix for texture feature extraction. In: Murgante, B., Misra, S., Carlini, M., Torre, C., Nguyen, H. Q., Taniar, D., Apduhan, B., Gervasi, O. (eds.) Computational Science and Its Applications— ICCSA 2013. Lecture Notes in Computer Science (LNCS), pp. 246–257. Springer, Heidelberg (2013)
19. Munklang, Y., Auephanwiriyakul, S., Theera-Umpon, N.: Examination of mammogram image classification using fuzzy co-occurrence matrix. Int. J. Tomogr. Simul. **28**(3), 96–103, (2015)
20. Abe, S.: Support Vector Machines for Pattern Classification. Advances in Pattern Recognition. Springer, London (2005)
21. Cristianini, N., Shawe-Taylor, J.: An Introduction to Support Vector Machines and Other Kernel-Based Learning Methods. Cambridge University Press, Cambridge (2000)
22. Hagan, M.T., Demuth, H.B., Beale, M.H., Jesus, O.D.: Neural Network Design, 2nd edn. (2014)
23. Liu, Y., Yao, X.: Ensemble learning via negative correlation. Neural Netw. **12**, 1399–1404 (1999)
24. Haralick, R.M.: Statistical and structural approaches to texture. Proc. IEEE **67**(5), 786–804 (1979)
25. Dunn, J.: A fuzzy relative of the ISODATA process and its use in detecting compact, well-separated clusters. J. Cybern. **3**(3), 32–57 (1973)
26. Bezdek, J.: Pattern Recognition with Fuzzy Objective Function Algorithms. Plenum Press, New York (1981)

Text Mining in Social Media for Security Threats

Diana Inkpen

Abstract We discuss techniques for information extraction from texts, and present two applications that use these techniques. We focus in particular on social media texts (Twitter messages), which present challenges for the information extraction techniques because they are noisy and short. The first application is extracting the locations mentioned in Twitter messages, and the second one is detecting the location of the users based on all the tweets written by each user. The same techniques can be used for extracting other kinds of information from social media texts, with the purpose of monitoring the topics, events, emotions, or locations of interest to security and defence applications.

Keywords Information extraction · Natural language processing · Social media · Text mining · Automatic text classification · Conditional random fields · Deep neural networks

1 Introduction

There is a huge amount of user-generated content available over the Internet, in various social media platforms. An important part of this content is in text form. Humans can read only a small part of these texts, in order to detect possible threats to security and public safety (such as mentions of terrorist activities or extremist/radical texts). This is why text mining techniques are important for security and defence applications. Therefore, we need to use automatic methods for extracting information from texts and for detecting messages that should be flagged as possible threats and forwarded to a human for further analysis.

Information extraction from text can target various pieces of information. The task could be a simple key phrase search (with focus on key phrases that could be

D. Inkpen (✉)
School of Electrical Engineering and Computer Science, University of Ottawa,
800 King Edward, Ottawa, ON K1N 6N5, Canada
e-mail: Diana.Inkpen@uOttawa.ca

© Springer International Publishing Switzerland 2016
R. Abielmona et al. (eds.), *Recent Advances in Computational Intelligence in Defense and Security*, Studies in Computational Intelligence 621,
DOI 10.1007/978-3-319-26450-9_19

491

relevant for detecting terrorist threats) or a sophisticated topic detection task (i.e., to classify a text as being about a terrorism-related topic or not). Topic detection was studied by many researchers, while only a few focused on social media texts [32]. Emotion detection from social media texts could also be of interest to security applications, in particular anger detection. Messages that express anger at high intensity levels could be flagged as possible terrorist threats. Combined with topic detection, anger detection could lead to more accurate flagging of the potential threats. Emotion classification was tested on social media messages, for example on a blog dataset [14] and on the LiveJournal dataset [21].

Location detection from social media texts is the main focus of in chapter. There are two types of locations: location entities mentioned in the text of a message and the physical locations of the users. We present experiments that show that location mentions can be extracted from Twitter messages: in particular, what cities, states/provinces, or countries are mentioned in a tweet [20]. This is useful in order to detect events or activities located in specific places that are mentioned by people. For example, potential terrorist plots can target specific geographic areas. For the second kind of locations, we present experiments that predict the physical location of a Twitter user based on all the messages written by the user [26]. Only a few users declare their location in their Twitter account profile. We used this data (tweets annotated with user location) as training data for a classifier that can be used to prediction the location of any user. The classifiers catch subtle differences in the language (dialect) and the types of entities mentioned. User location can be of interest to defence applications in cases when many disturbing messages are posted by a user, in order to estimate the possible location of this user.

The first task discussed (called task 1 bellow) detects location mentioned and it needs to extract spans of one or often several words. Another example of task that extracts spans of text is risk detection. In particular, information about maritime situation awareness, from textual reports (risk spans, type of risk, type of vessel, location, etc.) was extracted using a similar technique [33].

The second task that we present in detail in this chapter (called task 2 bellow) is detecting the location of the Twitter users based on their messages. This is a classification task in which the classifier needs to choose one of the possible locations for which the classifier was trained. It is not a sequential classification task, this is why we experimented with standard classifiers that choose one class for each text, as well as with new models based on Deep Neural Networks.

The two tasks together would allow an intelligent system to analyze information posted on Twitter in real time. It can analyze each tweet in order to spot locations mentioned in it (task 1) and to visualize these locations on a map. If many tweets at a given time mention a specific location, it might be the case that some event (such as natural disaster or terrorist attack) just happened somewhere in the world. Or the system can monitor only a region of interest. The system could also keep track of individual users that might have a suspicious behaviour (for example possible terrorist activities, or cyber bullying). If the user does not have a declared location in his/her Twitter profile, the system can collect all the recent tweets from that user, then apply our models from task 2 in order to compute the location of the user.

The novelty of the computational intelligence methods proposed in this chapter consists in the way we address task 1, via a sequence-based classifier followed by disambiguation rules, and the way we address task 2, via Deep Neural Networks, which were not applied yet to this task.

2 Proposed Computational Intelligence Solution

2.1 Extracting Expressions Using Conditional Random Fields

Early information extraction techniques were based on identifying patterns that can extract information of interest [7]. The patterns were often manually formulated, though it is possible to automatically learn patterns. Modern methods of information extraction are based on the latter idea, and even deeper on automatic text classification [1]. In this chapter, we discuss on the latest advances in information extraction from text, based on classifiers such as Support Vector Machines (SVM) [9], Deep Neural Networks [5], and Conditional Random Fields (CRF) [22].

The first two classifiers are applied to a text as a whole and are able to predict a class from a set of pre-determined classes. SVM classifiers were shown to obtain high performance on text data, including the emotion classification tasks. The deep neural networks were very recently applied on text data with high success rate [16].

The CRF classifiers were designed specifically for sequence classification. They can be applied to detecting spans of text that are of interest, by classifying each word into one of the following classes: beginning of a span, inside a span, and outside a span. In this way, CRF learns spans of interested from the annotated training data, and can be applied to detect similar spans in new test data. We used this technique for location expressions detection, due to the sequential nature of the task (an expression contains one or more words, and often a city name, followed by a state/province, followed by a country name).

Before using these classification techniques, we applied Natural Language Processing (NLP) techniques to pre-process the texts in order to extract the features needed for the classification. Examples of features are: words, n-grams (sequences of 2, 3, or more words), part-of-speech tags (such as nouns, verbs, adjectives, adverbs), and syntactic dependency relations.

Social media text is particularly difficult because the current NLP tools are trained on carefully edited texts, such as newspaper texts. Therefore they need to be adapted before being able to run on social media texts that are more informal, ungrammatical, and full of abbreviations and jargon. There are two ways to adapt tools to social media texts. One is to normalize the texts, which is rather difficult without loosing useful information about the people who post messages on social media. The second way, that we use here, was to train all the tools and methods on social media texts, in addition to shallow forms of text normalization, which we used in order to extract better features for the classification tasks. We also added new types of features specific to social media texts, such as hashtags for Twitter messages, emoticons, etc.

A CRF is a *undirected graphical model*. In a CRF, or more specifically, a linear chain CRF, if we denote the input variables by X and the output labels Y, the conditional probability distribution $P(Y|X)$ obeys the *Markov property*:

$$P(y_i|y_1, y_2, \ldots, y_{i-1}, y_{i+1}, \ldots, y_n, x)$$
$$= P(y_i, y_{i-1}, y_{i+1}, x) \tag{1}$$

Given some specific sequence of input variables $\mathbf{x}$, the conditional probability of some sequence of output label $\mathbf{y}$ is:

$$P(\mathbf{y}|\mathbf{x}) =$$
$$\frac{1}{Z(\mathbf{x})} exp(\sum_{i,k} \lambda_k t_k(y_{i-1}, y_i, x, i) + \sum_{i,l} \mu_l s_l(y_i, x, i)) \tag{2}$$

where $Z(\mathbf{x}) = \sum_y exp(\sum_{i,k} \lambda_k t_k(y_{i-1}, y_i, x, i) + \sum_{i,l} \mu_l s_l(y_i, x, i))$ is the *normalizing constant*, t_k and s_l are *feature functions*, λ_k and μ_l are the corresponding weights.

2.2 Classifying Texts Using Deep Neural Networks

In this section, we present the artificial neural network architectures that we will employ in the task of detection user locations based on their tweets. The reason we chose this technology is that other methods, such as SVM classifiers, were already applied for this task in related work (that we will compare with).

A feedforward neural network usually has an input layer and an output layer. If the input layer is directly connected to the output layer, such a model is called a *single-layer perceptron*. A more powerful model has several layers between the input layer and the output layer; these intermediate layers are called *hidden layers*; this type of model is known as a *multi-layer perceptron* (MLP). In a perceptron, neurons are interconnected, i.e., each neuron is connected to all neurons in the subsequent layer. Neurons are also associated with activation functions, which transform the output of each neuron; the transformed outputs are the inputs of the subsequent layer. Typical choices of activation functions include the identity function, defined as $y = x$; the hyperbolic tangent, defined as $y = \frac{e^x - e^{-x}}{e^x + e^{-x}}$ and the logistic sigmoid, defined as $y = \frac{1}{1+e^{-x}}$. To train a MLP, the most commonly used technique is *back-propagation* [35]. Specifically, the errors in the output layer are back-propagated to preceding layers and are used to update the weights of each layer.

An artificial neural network (ANN) with multiple hidden layers, also called a Deep Neural Network (DNN), mimics the deep architecture in the brain and it is believed to perform better than shallow architectures such as logistic regression models and ANNs without hidden units. The effective training of DNNs is, however, not achieved until the work of [5, 18]. In both cases, a procedure called *unsupervised pre-training* is carried out before the final supervised fine-tuning. The pre-training significantly decreases error rates of Deep Neural Networks on a number of ML tasks such as object recognition and speech recognition. The details of DNNs are beyond the scope of this chapter; interested readers can refer to [5, 18, 39] and the introduction from [4].

Data representation is important for machine learning [11]. Many statistical NLP tasks use hand-crafted features to represent language units such as words and documents; these features are fed as the input to machine learning models. One such example is emotion or sentiment classification which uses external lexicons that contain words with emotion or sentiment prior polarities [2, 15, 24, 28]. Despite the usefulness of these hand-crafted features, designing them is time-consuming and requires expertise. We also used hand-crafted features for task 1, while here, for task 2, we let the DNN choose the features automatically, since this in one of the advantages of the method.

A number of researchers have implemented DNNs in the NLP domain, achieving state-of-the-art performance without having to manually design any features. The most relevant to ours is the work in [16], who developed a deep learning architecture that consists of stacked denoising auto-encoders and apply it to sentiment classification of Amazon reviews. Their stacked denoising auto-encoders can capture meaningful representations from reviews and outperform state-of-the-art methods; due to the unsupervised nature of the pre-training step, this method also performs domain adaptation well.

In the social media domain, [38] extracted representations from Microblog text data with Deep Belief Networks (DBNs) and used the learned representations for emotion classification, outperforming representations based on Principal Component Analysis and on Latent Dirichlet Allocation.

Huang and Yates [19] showed that representation learning also helps domain adaptation of part-of-speech tagging, which is challenging because POS taggers trained on one domain have a hard time dealing with unseen words in another domain. They first learned a representation for each word, then fed the learned word-level representations to the POS tagger; when applied to out-of-domain text, it can reduce the error by 29 %.

3   Datasets

3.1   Location Expressions Data

Annotated data are required in order to train our supervised learning system. Our work is a special case of the Named Entity Recognition task, with text being tweets and target Named Entities being specific kinds of locations. To our knowledge, a corresponding corpus does not yet exist.[1]

We used the Twitter API[2] to collect our own dataset. Our search queries were limited to six major cell phone brands, namely iPhone, Android, Blackberry, Windows Phone, HTC and Samsung. Twitter API allows its users to filter tweets based on their languages, geographic origins, the time they were posted, etc. We utilized such functionality to collect only tweets written in English. Their origins, however, were not constrained, i.e., we collected tweets from all over the world. We ran the crawler from June 2013 to November 2013, and eventually collected a total of over 20 million tweets.

The amount of data we collected is overwhelming for manual annotation, but having annotated training data is essential for any supervised learning task for location detection. We therefore randomly selected 1000 tweets from each subset (corresponding to each cellphone brand) of the data, and obtained 6000 tweets for the manual annotation (more data would have taken too long to annotate).

We have defined annotation guidelines to facilitate the manual annotation task. Mani et al. [27] defined spatialML: an annotation schema for marking up references to places in natural language. Our annotation model is a sub-model of spatialML. The process of manual annotation is described next.

A gazetteer is a list of proper names such as people, organizations, and locations. Since we are interested only in locations, we only require a gazetteer of locations. We obtained such a gazetteer from GeoNames,[3] which includes additional information such as populations and higher level administrative districts of each location. We also made several modifications, such as the removal of cities with populations smaller than 1000 (because otherwise the size of the gazetteer would be very large, and there are usually very few tweets in the low-populated areas) and removal of states and provinces outside the U.S. and Canada; we also allowed the matching of alternative names for locations. For instance, "ATL", which is an alternative name for Atlanta, will be matched as a city.

We then used GATE's gazetteer matching module [10] to associate each entry in our data with all potential locations it refers to, if any. Note that, in this step, the only information we need from the gazetteer is the name and the type of each location. GATE has its own gazetteer, but we replaced it with the GeoNames gazetteer which serves our purpose better. The sizes of both gazetteers are listed in Table 1.[4] In addition to a larger size, the GeoNames contains information such as population, administrative division, latitude and longitude, which will be useful later in Sect. 4.4.

---

[1][25] recently released a dataset of various kinds of social media data annotated with generic location expressions, but not with cities, states/provinces, and countries.
[2]https://dev.twitter.com.
[3]http://www.geonames.org.
[4]The number of countries is larger than 200 because alternative names are counted; the same for states/provinces and cities.

Table 1 The sizes of the gazetteers

Gazetteer	Number of countries	Number of states and provinces	Number of cities
GATE	465	1215	1989
GeoNames	756	129	163285

The first step is merely a coarse matching mechanism without any effort made to disambiguate candidate locations. For example, the word "Georgia" would be matched to both the state of Georgia and the country in Europe.

In the next phase, we arranged for two annotators, who are graduate students with adequate knowledge of geography, to go through every entry matched to at least one of locations in the gazetteer list. The annotators are required to identify, first, whether this entry is a location; and second, what type of location this entry is. In addition, they are also asked to mark all entities that are location entities, but not detected by GATE due to misspelling, all capital letters, all small letters, or other causes. Ultimately, from the 6000 tweets, we obtained 1270 countries, 772 states or provinces, and 2327 cities.

We split the dataset so that each annotator was assigned one fraction. In addition, both annotators annotated one subset of the data containing 1000 tweets, corresponding to the search query of Android phone, in order to compute an inter-annotator agreement, which turned out to be 88 %. The agreement by chance is very low, since any span of text could be marked, therefore the kappa coefficient that compensates for chance agreement is close to 0.88. The agreement between the manual annotations and those of the initial GATE gazetteer matcher in the previous step was 0.56 and 0.47, respectively for each annotator. The fully-annotated dataset (as well as our source code for task 1) can be obtained through this link.[5]

Annotation of True Locations Up to this point, we have identified locations and their types, i.e., geo/non-geo ambiguities are resolved, but geo/geo ambiguities still exist. For example, we have annotated the token "Toronto" as a city, but it is not clear whether it refers to "Toronto, Ontario, Canada" or "Toronto, Ohio, USA". Therefore we randomly choose 300 tweets from the dataset of 6000 tweets and further manually annotated the locations detected in these 300 tweets with their actual location. The actual location is denoted by a numerical ID as the value of an attribute named *trueLoc* within the XML tag. An example of annotated tweet is displayed in Table 2.

3.2 User Location Data

For the second task, we need data annotated with users' locations. We choose two publicly available datasets which have been used by several other researchers. The

---

[5]https://github.com/rex911/locdet.

Table 2 An example of annotation with the true location

> Mon Jun 24 23:52:31 +0000 2013
> <location locType='city', trueLoc='22321'>Seguin </location>
> <location locType='SP', trueLoc='12'>Tx </location>
> RT @himawari0127i: #RETWEET#TEAMFAIRYROSE #TMW #TFBJP
> #500aday #ANDROID #JP #FF #Yes #No #RT #ipadgames #TAF #NEW
> #TRU #TLA #THF 51

first one is from [13].[6] It includes about 380,000 tweets from 9,500 users from the contiguous United States (i.e., the U.S. excluding Hawaii, Alaska and all off-shore territories). The dataset also provides geographical coordinates of each user. The second one is much larger and we obtained it from [34].[7] It contains 38 million tweets from 449,694 users, all from North America. We regard each user's set of tweets as a training example (labelled with location), i.e., $(x^{(i)}, y^{(i)})$ where $x^{(i)}$ represent all the tweets from the ith user and $y^{(i)}$ is the location of the ith user. Meta-data like user's profile and time zone will not be used in our work.

4 Task 1: Detecting Location Expressions

For this subtask, we propose to use methods designed for sequential data, because the nature of the problem is sequential. The different parts of a location such as country, state/province and city in a tweet are related and often given in a sequential order, so it seems appropriate to use sequential learning methods to automatically learn the relations between these parts of locations. We decided to use CRF as our main machine learning algorithm, because it achieved good results in similar information extraction tasks.

4.1 Designing Features

Features that are good representations of the data are important to the performance of a machine learning task. The features that we design for detecting locations are listed below:

- Bag-of-Words: To start with, we defined a sparse binary feature vector to represent each training case, i.e., each token in a sequence of tokens; all values of the feature vector are equal to 0 except one value corresponding to this token is set

---

[6]http://www.ark.cs.cmu.edu/GeoTwitter.

[7]https://github.com/utcompling/textgrounder/wiki/RollerEtAl_EMNLP2012.

to 1. This feature representation is often referred to as *Bag-of-Words* or unigram features. We will use *Bag-of-Words Features* or *BOW features* to denote them, and the performance of the classifier that uses these features can be considered as the baseline in this work.

- Part-of-Speech: The intuition for incorporating Part-of-Speech tags in a location detection task is straightforward: a location can only be a noun or a proper noun. Similarly, we define a binary feature vector, where the value of each element indicates the activation of the corresponding POS tag. We later on denote these features by *POS features*.
- Left/right: Another possible indicator of whether a token is a location is its adjacent tokens and POS tags. The intuitive justification for this features is that locations in text tend to have other locations as neighbours, i.e., "Los Angeles, California, USA"; and that locations in text tend to follow prepositions, as in the phrases "live in Chicago", "University of Toronto". To make use of information like that, we defined another set of features that represent the tokens on the left and right side of the target token and their corresponding POS tags. These features are similar to Bag-of-Words and POS features, but instead of representing the token itself they represent the adjacent tokens. These features are later on denoted by *Window features* or *WIN features*.
- Gazetteer: Finally, a token that appears in the gazetteer is not necessarily a location; by comparison, a token that is truly a location must match one of the entries in the gazetteer. Thus, we define another binary feature which indicates whether a token is in the gazetteer. This feature is denoted by *Gazetteer feature* or *GAZ feature* in the next sections.

In order to obtain BOW features and POS features, we preprocessed the dataset by tokenizing and POS tagging all the tweets. This step was done using the Twitter NLP and Part-of-Speech Tagging tool [29].

For experimental purposes, we would like to find out the impact each set of features has on the performance of the model. Therefore, we test different combinations of features and compare the accuracies of resulting models.

4.2 Experiments

Evaluation Metrics We compute the precision, recall and F-measure, which are the most common evaluation measures used in most information retrieval tasks. Specifically, the prediction of the model can have four different outcomes: true positive (TP), false positive (FP), true negative (TN) and false negative (FN), as described in Table 3 with respect to our task. We will present separate results for each type of locations (cities, states/provinces, and countries).

Table 3 Definitions of true positive, false positive, true negative and false negative

Model	Ground truth	
	Location	¬ Location
predicted as location	TP	FP
predicted as ¬ location	FN	TN

Precision measures how correctly the model makes predictions; it is the proportion of all positive predictions that are actually positive, computed by:

$$precision = \frac{TP}{TP + FP} \tag{3}$$

Recall measures the model's capability of recognizing positive test example; it is the proportion of all actually positive test examples that the model successfully predicts, computed by:

$$recall = \frac{TP}{TP + FN} \tag{4}$$

Once precision and recall are computed, we can therefore calculate the F-measure by:

$$F = \frac{1}{\alpha \frac{1}{P} + (1 - \alpha)\frac{1}{R}} \tag{5}$$

where P is the precision and R is the recall; α is the weighting coefficient. In this work, we shall use a conventional value of α, which is 0.5; one can interpret it as equally weighting precision and recall.

We report precision, recall and F-measure for the extracted location expressions, at both the token and the span level, to evaluate the overall performance of the trained classifiers. A token is a unit of tokenized text, usually a word; a span is a sequence of consecutive tokens. The evaluation at the span level is stricter. In other words, if a token belongs to the span and is tagged by the classifier the same as the location label, we count it as a true positive; otherwise, we count it as false positive; the same strategy is taken for the negative class. At the span level, we evaluate our method based on the whole span; if our classifiers correctly detects the start point, the end point and the length of the span, this will be counted as a true positive; however, if even one of the three factors was not exact, we count it as a false positive. It is clear that evaluation at the span level is stricter.

In our experiments, one classifier is trained and tested for each of the location labels *city*, *SP*, and *country*. For the learning process, we need to separate training and testing sets. We report results for 10-fold cross-validation, because a conventional choice for n is 10. In addition, we report results for separate training and test data (we chose 70 % for training and 30 % for testing). Because the data collection took several months, it is likely that we have both new and old tweets in the dataset;

therefore we performed a random permutation before splitting the dataset for training and testing.

We would like to find out the contribution of each set of features in Sect. 4.1 to the performance of the model. To achieve a comprehensive comparison, we tested all possible combinations of features plus the BOW features. In addition, a baseline model which simply predicts a token or a span as a location if it matches one of the entries in the gazetteer.

We implemented the models using an NLP package named MinorThird [8] that provides a CRF module [36] easy to use; the loss function is the log-likelihood and the learning algorithm is the gradient ascent. The loss function is convex and the learning algorithm converges fast.

4.3 Results for Location Expressions

The results are listed in the following tables. Table 4 shows the results for countries, Table 5 for states/provinces and Table 6 for cities. To our knowledge, there is no previous work that extracts locations at these three levels, thus comparisons with other models are not feasible.

Discussion The results from Tables 4, 5 and 6 show that the task of identifying cities is the most difficult, since the number of countries or states/provinces is by far smaller. In our gazetteer, there are over 160,000 cities, but only 756 countries and 129 states/provinces, as detailed in Table 1. A lager number of possible classes generally indicates a larger search space, and consequently a more difficult task. We also observe that the token level F-measure and the span level F-measure are quite similar, likely due to the fact that most location names contain only one word.

Table 4 Performance of the classifiers trained on different features for countries

Features	Token			Span			Separate train-test sets	
	P	R	F	P	R	F	Token F	Span F
Baseline-Gazetteer Matching	0.26	0.64	0.37	0.26	0.63	0.37	–	–
Baseline-BOW	0.93	0.83	0.88	0.92	0.82	0.87	0.86	0.84
BOW+POS	0.93	0.84	0.88	0.91	0.83	0.87	0.84	0.85
BOW+GAZ	0.93	0.84	0.88	0.92	0.83	0.87	0.85	0.86
BOW+WIN	0.96	0.82	0.88	0.95	0.82	0.88	0.87	0.88
BOW+POS+GAZ	0.93	0.84	0.88	0.92	0.83	0.87	0.85	0.86
BOW+WIN+GAZ	0.95	0.85	0.90	0.95	0.85	0.89	0.90	0.90
BOW+POS+WIN	0.95	0.82	0.88	0.95	0.82	0.88	0.90	0.90
BOW+POS+WIN+GAZ	0.95	0.86	0.90	0.95	0.85	0.90	0.92	0.92

Column 2 to column 7 show the results from 10-fold cross validation; the last two columns show the results from random split of the dataset where 70 % are the train set and 30 % are the test set. (The same in Tables 5 and 6)

Table 5 Performance of the classifiers trained on different features for SP

Features	Token			Span			Separate train-test sets	
	P	R	F	P	R	F	Token F	Span F
Baseline-Gazetteer Matching	0.65	0.74	0.69	0.64	0.73	0.68	–	–
Baseline-BOW	0.90	0.78	0.84	0.89	0.80	0.84	0.80	0.84
BOW+POS	0.90	0.79	0.84	0.89	0.81	0.85	0.82	0.84
BOW+GAZ	0.88	0.81	0.84	0.89	0.82	0.85	0.79	0.80
BOW+WIN	0.93	0.77	0.84	0.93	0.78	0.85	0.80	0.81
BOW+POS+GAZ	0.90	0.80	0.85	0.90	0.82	0.86	0.78	0.82
BOW+WIN+GAZ	0.91	0.79	0.84	0.91	0.79	0.85	0.83	0.84
BOW+POS+WIN	0.92	0.78	0.85	0.92	0.79	0.85	0.80	0.81
BOW+POS+WIN+GAZ	0.91	0.79	0.85	0.91	0.80	0.85	0.84	0.83

Table 6 Performance of the classifiers trained on different features for cities

Features	Token			Span			Separate train-test sets	
	P	R	F	P	R	F	Token F	Span F
Baseline-Gazetteer Matching	0.14	0.71	0.23	0.13	0.68	0.22	–	–
Baseline-BOW	0.91	0.59	0.71	0.87	0.56	0.68	0.70	0.68
BOW+POS	0.87	0.60	0.71	0.84	0.55	0.66	0.71	0.68
BOW+GAZ	0.84	0.77	0.80	0.81	0.75	0.78	0.78	0.75
BOW+WIN	0.87	0.71	0.78	0.85	0.69	0.76	0.77	0.77
BOW+POS+GAZ	0.85	0.78	0.81	0.82	0.75	0.78	0.79	0.77
BOW+WIN+GAZ	0.91	0.76	0.82	0.89	0.74	0.81	0.82	0.81
BOW+POS+WIN	0.82	0.76	0.79	0.80	0.75	0.77	0.80	0.79
BOW+POS+WIN+GAZ	0.89	0.77	0.83	0.87	0.75	0.81	0.81	0.82

We also include the results when one part of the dataset (70 %) is used as training data and the rest (30 %) as test data. The results are slightly different to that of 10-fold cross validation and tend to be lower in terms of f-measures, likely because less data are used for training. However, similar trends are observed across feature sets.

The baseline model not surprisingly produces the lowest precision, recall and f-measure; it suffers specifically from a dramatically low precision, since it will predict everything contained in the gazetteer to be a location. By comparing the performance of different combinations of features, we find out that the differences are most significant for the classification of cities, and least significant for the classification of states/provinces, which is consistent with the number of classes for these two types of locations. We also observe that the simplest features, namely BOW features, always produce the worst performance at both token level and span level in all three tasks; on the other hand, the combination of all features produces the best performance in every task, except for the prediction of states/provinces at span level. These results are not surprising.

We conducted t-tests on the results of models trained on all combinations of features listed in Tables 4, 5 and 6. We found that in *SP* classification, no pair of feature combinations yields statistically significant difference. In *city* classification, using only BOW features produces significantly worse results than any other feature combinations at a 99.9 % level of confidence, except BOW+POS features, while using all features produces significantly better results than any other feature combinations at a 99 % level of confidence, except BOW+GAZ+WIN features. In *country* classification, the differences are less significant; where using all features and using BOW+GAZ+WIN features both yield significantly better results than 4 of 6 other feature combinations at a 95 % level of confidence, while the difference between them is not significant; unlike in *city* classification, the results obtained by using only BOW features is significantly worse merely than the two best feature combinations mentioned above.

We further looked at the t-tests results of *city* classification to analyze what impact each feature set has on the final results. When adding POS features to a feature combination, the results might improve, but never statistically significantly; by contrast, they always significantly improve when GAZ features or WIN features are added. These are consistent with our previous observations.

Error Analysis Some of the predictions errors were due to partial detection of some names, for example "Korea" was predicted as a country, instead of "South Korea". Another source of errors was due to misspellings and to non-standard nicknames that were not in our gazetteers. We went through the predictions made by the location entity detection model, picked some typical errors made by it, and looked into the possible causes of these errors.

Example 1:

> Mon Jul 01 14:46:09 +0000 2013
> Seoul
> yellow cell phones family in South Korea #phone #mobile #yellow #samsung
> http://t.co/lpsLgepcCW

Example 2:

> Sun Sep 08 06:28:50 +0000 2013
> minnesnowta.
> So I think Steve Jobs' ghost saw me admiring the Samsung Galaxy 4 and now is messing with my phone. Stupid Steve Jobs. #iphone

In Example 1, the model predicted "Korea" as a country, instead of "South Korea". A possible explanation is that in the training data there are several cases containing "Korea" alone, which leads the model to favour "Korea" over "South Korea". In Example 2, the token "minnesnowta" is quite clearly a reference to "Minnesota", which the model failed to predict. Despite the fact that we allow the model to recognize nicknames of locations, these nicknames come from the GeoNames gazetteer; any other nicknames will not be known to the model. On the other hand, if we treat "minnesnowta" as a misspelled "Minnesota", it shows that we can resolve the issue of unknown nicknames by handling misspellings in a better way.

4.4 Location Disambiguation

In the previous section, we have identified the locations in Twitter messages and their types; however, the information about these locations is still ambiguous. In this section, we describe the heuristics that we use to identify the unique actual location referred to by an ambiguous location name. These heuristics rely on information about the type, geographic hierarchy, latitude and longitude, and population of a certain location, which we obtained from the GeoNames Gazetteer. The disambiguation process is divided into 5 steps, as follows:

1. **Retrieving candidates**. A list of locations whose names are matched by the location name we intend to disambiguate are selected from the gazetteer. We call these locations candidates. After step 1, if no candidates are found, disambiguation is terminated; otherwise we continue to step 2.
2. **Type filtering**. The actual location's type must agree with the type that is tagged in the previous step where we apply the location detection model; therefore, we remove any candidates whose types differ from the tagged type from the list of candidates. E.g., if the location we wish to disambiguate is "Ontario" tagged as a city, then "Ontario" as a province of Canada is removed from the list of candidates, because its type *SP* differs from our target type. After step 2, if no candidates remain in the list, disambiguation is terminated; if there is only one candidate left, this location is returned as the actual location; otherwise we continue to step 3.
3. **Checking adjacent locations**. It is common for users to put related locations together in a hierarchical way, e.g., "Los Angeles, California, USA". We check adjacent tokens of the target location name; if a candidate's geographic hierarchy matches any adjacent tokens, this candidate is added to a temporary list. After step 3, if the temporary list contains only one candidate, this candidate is returned as the actual location. Otherwise we continue to step 4 with the list of candidates reset.
4. **Checking global context**. Locations mentioned in a document are geographically correlated [23]. In this step, we first look for other tokens tagged as a location in the Twitter message; if none is found, we continue to step 5; otherwise, we

disambiguate these context locations. After we obtain a list of locations from the context, we calculate the sum of their distances to a candidate location and return the candidate with minimal sum of distances.

5. **Default sense**. If none of the previous steps can decide a unique location, we return the candidate with largest population (based on the assumption that most tweets talk about large urban areas).

4.5 Experiments and Results for Actual Locations

We ran the location disambiguation algorithm described above. In order to evaluate how each step (more specifically, step 3 and 4, since other steps are mandatory) contributes to the disambiguation accuracy, we also deactivated optional steps and compared the results.

Example 3:

> Fri Jul 19 16:35:29 +0000 2013
> NYC and San Francisco
> You Have to See this LEOPARD phone HTC 1 case RT PLS http://t.co/Ml6zH3Yp2b

The results of different location disambiguation configurations are displayed in Table 7, where we evaluate the performance of the model by accuracy, which is defined as the proportion of correctly disambiguated locations. By analyzing them, we can see that when going through all steps, we get an accuracy of 95.5 %, while by simply making sure the type of the candidate is correct and choosing the default location with the largest population, we achieve a better accuracy. The best result is obtained by using the adjacent locations, which turns out to be 98.2 % accurate. Thus we conclude that adjacent locations help disambiguation, while locations in the global context do not. Therefore the assumption made by [23] that the locations in the global context help the inference of a target location does not hold for Twitter messages, mainly due to their short nature.

Error Analysis Similar to Sect. 4.3, this section presents an example of errors made by the location disambiguation model in Example 3. In this example, the disambigua-

Table 7 Location disambiguation results	Deactivated steps	Accuracy (%)
	None	95.5
	Adjacent locations	93.7
	Global context	**98.2**
	Adjacent locations + context locations	96.4

tion rules correctly predicted "NYC" as "New York City, New York, United States"; however, "San Francisco" was predicted as "San Francisco, Atlantida, Honduras", which differs from the annotated ground truth. The error is caused by step 4 of the disambiguation rules that uses contextual locations for prediction; San Francisco of Honduras is 3055 km away from the contextual location New York City, while San Francisco of California, which is the true location, is 4129 km away. This indicates the fact that a more sophisticated way of dealing with the context in tweets is required to decide how it impacts the true locations of the detected entities.

5 Task 2: Detecting User Locations

We define our work as follows: first, a classification task puts each user into one geographical region (see Sect. 5.5 for details); next, a regression task predicts the most likely location of each user in terms of geographical coordinates, i.e., a pair of real numbers for latitude and longitude. We present one model for each task.

5.1 Models

Model 1 The first model consists of three layers of denoising auto-encoders. Each code layer of denoising auto-encoders also serves as a hidden layer of a multiple-layer feedforward neural network. In addition, the top code layer works as the input layer of a logistic regression model whose output layer is a softmax layer.

Softmax Function
The softmax function is defined as:

$$softmax_i(\mathbf{z}) = \frac{e^{\mathbf{z}_i}}{\sum_{j=1}^{J} e^{\mathbf{z}_j}} \tag{6}$$

where the numerator z_i is the ith possible input to the softmax function and the denominator is the summation over all possible inputs. The softmax function produces a normalized probability distribution over all possible output labels. This property makes it suitable for multiclass classification tasks. Consequently, a softmax layer has the same number of neurons as the number of possible output labels; the value of each neuron can be interpreted as the probability the corresponding label given the input. Usually, the label with the highest probability is returned as the prediction made by the model.

In our model, mathematically, the probability of a label i given the input and the weights is:

$$P(Y = i | x^N, W^{(N+1)}, b^{(N+1)})$$

$$= softmax_i(W^{(N+1)} x^N + b^{(N+1)})$$

$$= \frac{e^{W_i^{(N+1)} x^N + b_i^{(N+1)}}}{\sum_j e^{W_j^{(N+1)} x^N + b_j^{(N+1)}}} \qquad (7)$$

where $W^{(N+1)}$ is the weight matrix of the logistic regression layer and $b^{(N+1)}$ are its biases. N is the number of hidden layers, in our case $N = 3$. x^N is the output of the code layer of the denoising auto-encoder on top. To calculate the output of ith hidden layer $(i = 1 \ldots N)$, we have:

$$x^i = s(W^{(i)} x^{i-1} + b^{(i)}) \qquad (8)$$

where s is the activation function, $W^{(i)}$ and $b^{(i)}$ correspond to the weight matrix and biases of the ith hidden layer. x^0 is the raw input generated from text,[8] as specified in Sect. 5.5. We return the label that maximizes Eq. (7) as the prediction, i.e.:

$$i_{predict} = \arg \max_i P(Y = i | x^N, W^{(N+1)}, b^{(N+1)}) \qquad (9)$$

We denote this model as SDA-1.

Model 2 In the second model, a multivariate linear regression layer replaces a logistic regression layer on top. This produces two real numbers as output, which can be interpreted as geographical coordinates. Therefore the output corresponds to locations on the surface of Earth. Specifically, the output of model 2 is:

$$y_i = W_i^{(N+1)} x^N + b_i^{(N+1)} \qquad (10)$$

where $i \in \{1, 2\}$, $W^{(N+1)}$ is the weight matrix of the linear regression layer and $b^{(N+1)}$ are its biases, x^N is the output of the code layer of the denoising auto-encoder on top. The output of ith hidden layer $(i = 1 \ldots N)$ is computed using Eq. (8), which is the same as Model 1. The tuple (y_1, y_2) is then the pair of geographical coordinates produced by the model. We denote this model as SDA-2. Figure 1 shows the architecture of both models.

5.2   Input Features

To learn better representations, a basic representation is required to start with. For text data, a reasonable starting representation is achieved with the *Bag-of-N-grams* features [4, 16].

---

[8]Explained in Sect. 5.2.

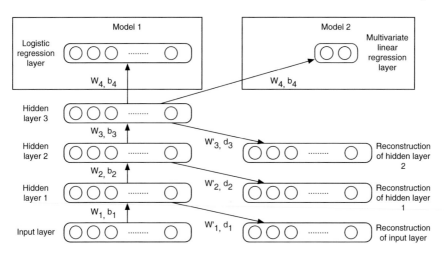

Fig. 1 Illustration of the two proposed models (with 3 hidden layers). The models differ only in the output layers. The neurons are fully interconnected. A layer and its reconstruction and the next layer together correspond to a denoising auto-encoder. For simplicity, we do not include the corrupted layers in the diagram. Note that models 1 and 2 are not trained simultaneously, nor do they share parameters

The input text of Twitter messages is preprocessed and transformed into a set of Bag-of-N-grams **frequency** feature vectors. We did not use binary feature vectors because we believe the frequency of n-grams is relevant to the task at hand. For example, a user who tweets *Senators* 10 times is more likely to be from Ottawa than another user who tweets it just once. (The latter is more likely to be someone from Montreal who tweets *Senators* simply because the Canadiens happen to be defeated by the Senators that time.) Due to computational limitations, we consider only the 5000 most frequent unigrams, bigrams and trigrams.[9] We tokenized the tweets using the *Twokenizer* tool [29].

5.3   Statistical Noises for Denoising Auto-Encoders

An essential component of a DA is its statistical noise. Following [16], the statistical noise we incorporate for the first layer of DA is the masking noise, i.e., each active element has a probability to become inactive. For the remaining layers, we apply Gaussian noise to each of them, i.e., a number independently sampled from the Gaussian distribution $\mathcal{N}(0, \sigma^2)$ is added to each element of the input vector to get the corrupted input vector. Note that the Gaussian distribution has a 0 mean. The

---

[9]Not all of these 5000 n-grams are necessarily good location indicators, we don't manually distinguish them; a machine learning model after training should be able to do so.

standard deviation of the Gaussian distribution σ decides the degree of corruption; we also use the term *corruption level* to refer to σ.

5.4 Loss Functions

Pre-training In terms of training criteria for unsupervised pre-training, we use the squared error loss function:

$$\ell(x, r) = ||x - r||^2 \tag{11}$$

where x is the original input, r is the reconstruction. The squared error loss function is a convex function, so we are guaranteed to find the global optimum once we find the local optimum.

The pre-training is done by layers, i.e., we first minimize the loss function for the first layer of denoising auto-encoder, then the second, then the third. We define the decoder weight matrix as the transposition of the encoder weight matrix.

Fine-Tuning In the fine-tuning phase, the training criteria differ for model 1 and model 2. It is a common practice to use the *negative log-likelihood* as the loss function of models that produce a probability distribution, which is the case for model 1. The equation for the negative log-likelihood function is:

$$\begin{aligned} \ell(\theta &= \{W, b\}, (x, y)) \\ &= -\log(P(Y = y | x, W, b)) \end{aligned} \tag{12}$$

where $\theta = \{W, b\}$ are the parameters of the model, x is the input and y is the ground truth label. To minimize the loss in Eq. (12), the conditional probability $P(Y = y | x, W, b)$ must be maximized, which means the model must learn to make the correct prediction with the highest confidence possible. Training a supervised classifier using the negative log-likelihood loss function can be therefore interpreted as maximizing the likelihood of the probability distribution of labels in the training set.

On the other hand, model 2 produces for every input a location $\hat{y}(\hat{lat}, \hat{lon})$, which is associated with the actual location of this user, denoted by $y(lat, lon)$. Given latitudes and longitudes of two locations, their great-circle distance can be computed by first calculating an intermediate value $\Delta\sigma$ with the Haversine formula [37]:

$$\Delta\sigma = \arctan$$

$$\left(\frac{\sqrt{\left(\cos\phi_2 \sin\Delta\lambda\right)^2 + \left(\cos\phi_1 \sin\phi_2 - \sin\phi_1 \cos\phi_2 \cos\Delta\lambda\right)^2}}{\sin\phi_1 \sin\phi_2 + \cos\phi_1 \cos\phi_2 \cos\Delta\lambda} \right) \tag{13}$$

Next, calculate the actual distance:

$$d((\phi_1, \lambda_1), (\phi_2, \lambda_2)) = r\Delta\sigma \qquad (14)$$

where ϕ_1, λ_1 and ϕ_2, λ_2 are latitudes and longitudes of two locations, $\Delta\lambda = \lambda_1 - \lambda_2$, r is the radius of the Earth. Because d is a continuously differentiable function with respect to ϕ_1 and λ_1 (if we consider (ϕ_1, λ_1) as the predicted location, then (ϕ_2, λ_2) is the actual location), and minimizing d is exactly what model 2 is designed to do, we define the loss function of model 2 as the great-circle distance between the estimated location and the actual location:

$$\begin{aligned} \ell(\theta &= \{W, b\}, (x, y)) \\ &= d(Wx + b, y) \end{aligned} \qquad (15)$$

where $\theta = \{W, b\}$ are the parameters of the model, x is the input and y is the actual location.[10] Now that we have defined the loss functions for both models, we can train them with back-propagation [35] and Stochastic Gradient Descent (SGD).

5.5   Experiments

Evaluation Metrics We train the stacked denoising auto-encoders to predict the locations of users based on the tweets they post. To evaluate SDA-1, we follow [13] and define a classification task where each user is classified as from one of the 48 contiguous U.S. states or Washington D.C. The process of retrieving a human-readable address including street, city, state and country from a pair of latitude and longitude is known as *reverse geocoding*. We use MapQuest API[11] to reverse geocode coordinates for each user. We also define a task with only four classes, the West, Midwest, Northeast and South regions, as per the U.S. Census Bureau.[12] The metric for comparison is the classification accuracy defined as the proportion of test examples that are correctly classified. We also implement two baseline models, namely a Naive Bayes classifier and an SVM classifier (with the RBF kernel); both of them take exactly the same input as the stacked denoising auto-encoders.

To evaluate SDA-2, the metric is simply the mean error distance in kilometres from the actual location to the predicted location. Note that this is the distance on the surface of the Earth, also known as the great-circle distance. See Eqs. (13)–(14) for its computation. In Sect. 5.6, we applied two additional metrics, which are the median error distance and the percentage of predictions less than 100 miles away from the true locations, to comply with previous work. Similarly, we implement a baseline model which is simply a multivariate linear regression layer on top of the

---

[10]Alternatively, we also tried the loss function defined as the average squared error of output numbers, which is equivalent to the average Euclidean distance between the estimated location and the true location; this alternative model did not perform well.

[11]http://www.mapquest.com.

[12]http://www.census.gov/geo/maps-data/maps/pdfs/reference/us_regdiv.pdf.

input layer. This baseline model is equivalent to SDA-2 without hidden layers. We denote this model as baseline-MLR. After we have obtained the performance of our models, they will be compared against several existing models from previous work.

Early Stopping We define our loss functions without regularizing the weights; to prevent overfitting, we adopt the early-stopping technique [41]; i.e., training stops when the model's performance on the validation set no longer improves [3].

To make the comparisons fair, we split the Eisenstein dataset in the same way as [13] did, i.e., 60 % for training, 20 % for validation and 20 % for testing. The Roller dataset was provided split, i.e., 429,694 users for training, 10,000 users for validation and the rest 10,000 users for testing; this is the split we adopted.

Tuning Hyper-parameters One of the drawbacks of DNNs is a large number of hyper-parameters to specify [3]. The activation function we adopt is the sigmoid function $y = \frac{1}{1+e^{-x}}$, which is a typical choice as the non-linear activation function. For the size (the number of neurons) of each hidden layer, usually a larger size indicates better performance but higher computational cost. Since we do not have access to extensive computational power, we set this hyper-parameter to 5000, which is equal to the size of the input layer. As for the corruption level, the masking noise probability for the first layer is 0.3; the Gaussian noise standard deviation for other layers is 0.25. These two values are chosen because they appear to work well in our experiments based on the validation dataset. The Mini-batch size chosen for stochastic gradient descent is 32, which is a reasonable default suggested by Bengio [3]. For the learning rates, we explore different configurations in the set {0.00001, 0.0001, 0.001, 0.01, 0.1} for both pre-learning rate and fine-tuning learning rate. Lastly, the pre-training stops after 25 epochs, which usually guarantees the convergence. Fine-tuning stops after 1000 epochs; because of the early stopping technique described in Sect. 5.5, this number is rarely reached.

Implementation Theano [6] is a scientific computing library written in Python. It is mainly designed for numerical computation. A main feature of Theano is its symbolic representation of mathematical formulas, which allows it to automatically differentiate functions. We train our model with stochastic gradient descent which requires the computation of gradients, either manually or automatically. Since Theano does automatic differentiation, we no longer have to manually differentiate complex functions like Eq. (13). We implemented SDA-1, SDA-2[13] and the baseline multivariate linear regression model with Theano. Scikit-learn [30] is a machine learning package written in Python. It includes most standard machine learning algorithms. The two baseline models compared against SDA-1 (Naive Bayes and SVM) are implemented using the Scikit-learn package.

---

[13]Our code is available at https://github.com/rex911/usrloc.

Table 8 Classification accuracy for SDA-1 and other models

	Model	Classif. Acc. (%)	
		Region (4-way)	State (49-way)
Eisenstein et al. (2010)	Geo topic model	58	24
	Mixture of unigrams	53	19
	Supervised LDA	39	4
	Text regression	41	4
	kNN	37	2
Our models	**SDA-1**	**61.1**	**34.8**
	Baseline-Naive Bayes	54.8	30.1
	Baseline-SVM	56.4	27.5

5.6   Results for User Locations

Evaluation on the Eisenstein Dataset The SDA-1 model yields an accuracy of 61.1 % and 34.8 %, for region classification and state classification, respectively. The results of all models are shown in Table 8. Among all previous works that use the same dataset, only [13] report the classification accuracy of their models; to present a comprehensive comparison, all models from their work, not just the best one, are listed. Student's t-tests suggest that the differences between SDA-1 and the baseline models are statistically significant at a 99 % level of confidence.[14]

It can be seen that our SDA-1 model performs best in both classification tasks. It is surprising to find that the shallow architectures that we implemented, namely SVM and Naive Bayes, perform reasonably well. They both outperform all models in [13] in terms of state-wise classification. A possible explanation is that the features we use (frequencies of n-grams with $n = 1, 2, 3$) are more indicative than theirs (unigram term frequencies).

Table 9 shows the mean error distance for various models trained on the same dataset. The difference between SDA-2 and the baseline model is statistically significant at a level of confidence of 99.9 %.[15] Our model has the second best results and performs better than four models from previous work. In addition, the fact that SDA-2 outperforms the baseline model by a large margin shows the advantages of a deep architecture and its ability to capture meaningful and useful abstractions from input data.

Evaluation on the Roller Dataset Table 10 compares the results from various models on the Roller dataset. The model in [17], which included extensive feature engineering, outperformed other models. In addition it achieves the best results by uti-

---

[14]We are unable to conduct t-tests on the Eisenstein models, because of the unavailability of the details of the results produced by these models.

[15]We are unable to conduct t-tests on the other models, because of the unavailability of the details of the results produced by these models.

Table 9 Mean error distance of predictions for SDA-2 and models from previous work

Model	Mean error distance (km)
Eisenstein [12]	845
SDA-2	**855.9**
Priedhorsky [31]	870
Roller [34]	897
Eisenstein [13]	900
Wing [40]	967
Baseline-MLR	1268

Table 10 Results from SDA-2 and the best models of previous work

Model	Mean error (km)	Median	Acc. error (km)
Roller [34]	860	463	34.6
Han [17]	NA	260	45
Han [17] using top 3 % features (6420)	NA	NA	10
SDA-2	733	377	24.2

NA indicates *Not Available*

lizing about 90 % of all 214,000 features; when using the top 3 % (6420) features, the Accuracy was 10 %.[16] The SDA-2 model, despite the computational limitation, achieved better results than [34] using just 5,000 features.

Error Analysis The datasets we used for this task do not have a balanced distribution. Users are densely distributed in the West Coast and most part of the East, whereas very few are located in the middle. Such label imbalance has a negative effect on statistical classifiers, and adversely affects regression models because many target values will never be sampled. This would explain some of the prediction errors made by our models.

6   Conclusion and Future Work

In this chapter, we looked at techniques that allow us to extract information form texts. This information can be useful ins security applications by allowing to monitoring od locations, topics, or emotions mentioned in texts. Of particular interest are social media messages, which are more difficult to process than regular texts.

We examined two tasks in detail. The first task was extracting location entities mentioned in tweets. We extracted different types of features for this task and did experiments to measure their usefulness. We trained CRF classifiers that were able to achieve a very good performance. We also defined disambiguation rules based on a few heuristics which turned out to work well. In addition, the data we collected and

---

[16]Only this metric was reported by the author in the top 3 % features configuration.

annotated for task 1 is made available to other researchers to test their models and to compare with ours.

We identify two main directions of future work. First, the simple rule-based disambiguation approach does not handle issues like misspellings well, and can be replaced by a machine learning approach, although this requires more annotated training data. Second, since in the current model, we consider only states and provinces in the United States and Canada, we need to extend the model to include states, provinces, or regions in other countries as well.

For the second task, user location detection, we proposed models based on DNN. Our experimental results show that our SDA-1 model outperformed other empirical models; our SDA-2 model's performance is reasonable. We demonstrate that a DNN is capable of learning representations from raw input data that helps the inference of location of users without having to design any hand-engineered features. The results also show that deep learning models have the potential of being applied to solve real business problems that require location detection, in addition to their recent success in natural language processing tasks and to their well-established success in computer vision and speech recognition.

We believe a better model can yet be built. For example, our exploration for hyper-parameters is by no means exhaustive, especially for the mini-batch size and the corruption levels, due to the very high running time required. It would be interesting to find out the optimal set of hyper-parameters. More computational capacity also allows the construction of a more powerful DNN. For example, in our SDA the hidden layers have a size of 5000, which is equal to the size of input layer; however, a hidden layer larger than the input layer learns better representations [4].

In terms of improvement in the future, we plan to collect a dataset uniformly distributed geographically, and the locations do not have to be limited to the contiguous United States. Alternatively, one may notice that the distribution of users is similar to that of the U.S. population, therefore it is possible to use the U.S. census data to offset such skewed distribution of users. In addition, the input of our system consists only of tweets, because we are mostly interested in recovering users' location from the language they produce; however, real applications require a higher accuracy. To achieve this, we could also incorporate information such as users' profiles, self-declared locations, time zones and interactions with other users. Another type of stacked denoising auto-encoder is one that only does unsupervised pre-training, then the output of the code layer is regarded as input into other classifiers such as SVM [16]. It would be interesting to compare the performance of this architecture and that of an SDA with supervised fine-tuning with respect to our task.

The most important direction of future work in regard to applications for security and defence is to use the information extracted from texts, about topics, emotions, and locations, in order to flag social media messages for possible security threats. These pieces of information can be combined via a rule-based approach. Alternatively, classification techniques, such as the ones we used for task 2, can be employed; but annotated training data (marked with security threat labels) would be needed.

Acknowledgments I would like to thank my collaborators who helped with the location detection project: Ji Rex Liu, Diman Ghazi, Atefeh Farzindar and Farzaneh Kazemi for task 1 and Ji Rex Liu for task 2.

References

1. Aggarwal, C., Zhai, C.: A survey of text classification algorithms. In: Aggarwal, C.C., Zhai, C. (eds.) Mining Text Data, pp. 163–222. Springer (2012). http://dx.doi.org/10.1007/978-1-4614-3223-4_6
2. Aman, S., Szpakowicz, S.: Identifying expressions of emotion in text. In: Text, Speech and Dialogue, pp. 196–205. Springer (2007)
3. Bengio, Y.: Practical recommendations for gradient-based training of deep architectures. Neural Networks: Tricks of the Trade 7700, pp. 437–478 (2012). http://link.springer.com/chapter/10.1007/978-3-642-35289-8_26
4. Bengio, Y., Courville, A., Vincent, P.: Representation learning: a review and new perspectives. Pattern Anal. Mach. Intell. **35**(8), 1798–1828 (2013). http://ieeexplore.ieee.org/xpls/abs_all.jsp?arnumber=6472238
5. Bengio, Y., Lamblin, P.: Greedy layer-wise training of deep networks. Adv. Neural Inf. Process. Syst. **19**(153) (2007). https://papers.nips.cc/paper/3048-greedy-layer-wise-training-of-deep-networks.pdf
6. Bergstra, J., Breuleux, O., Bastien, F., Lamblin, P., Pascanu, R., Desjardins, G., Turian, J., Warde-Farley, D., Bengio, Y.: Theano: a CPU and GPU math expression compiler. In: Proceedings of the Python for Scientific Computing Conference (SciPy). vol. 4, p. 3 (2010)
7. Chang, C.H., Kayed, M., Girgis, M.R., Shaalan, K.F.: A survey of web information extraction systems. Knowl. Data Eng. IEEE Trans. **18**(10), 1411–1428 (2006)
8. Cohen, W.W.: Minorthird: methods for identifying names and ontological relations in text using heuristics for inducing regularities from data (2004)
9. Cortes, C., Vapnik, V.: Support-vector networks. Mach. Learn. **20**(3), 273–297 (1995). http://dx.doi.org/10.1007/BF00994018
10. Cunningham, H., Maynard, D., Bontcheva, K., Tablan, V.: A framework and graphical development environment for robust nlp tools and applications. In: Proceedings of the 40th Anniversary Meeting of the Association for Computational Linguistics (ACL'02). Association for Computational Linguistics (2002)
11. Domingos, P.: A few useful things to know about machine learning. Commun. ACM **55**(10), 78–87 (2012)
12. Eisenstein, J., Ahmed, A., Xing, E.P.: Sparse additive generative models of text. In: Proceedings of the 28th International Conference on Machine Learning (ICML'11), pp. 1041–1048 (2011)
13. Eisenstein, J., O'Connor, B., Smith, N.A., Xing, E.P.: A latent variable model for geographic lexical variation. In: Proceedings of the 2010 Conference on Empirical Methods in Natural Language Processing, pp. 1277–1287. ACL (2010)
14. Ghazi, D., Inkpen, D., Szpakowicz, S.: Hierarchical versus flat classification of emotions in text. In: Proceedings of the NAACL HLT 2010 workshop on computational approaches to analysis and generation of emotion in text, pp. 140–146. Association for Computational Linguistics, Los Angeles, (June 2010). http://www.aclweb.org/anthology/W10-0217
15. Ghazi, D., Inkpen, D., Szpakowicz, S.: Prior and contextual emotion of words in sentential context. Comput. Speech Lang. **28**(1), 76–92 (2014). http://dx.doi.org/10.1016/j.csl.2013.04.009
16. Glorot, X., Bordes, A., Bengio, Y.: Domain adaptation for large-scale sentiment classification: a deep learning approach. In: Proceedings of the 28th International Conference on Machine Learning (ICML'11). pp. 513–520 (2011)

17. Han, B., Cook, P., Baldwin, T.: Text-based Twitter user geolocation prediction. Artif. Intell. Res. **49**(1), 451–500, (Jan 2014). http://dl.acm.org/citation.cfm?id=2655713.2655726

18. Hinton, G.E., Osindero, S., Teh, Y.W.: A fast learning algorithm for deep belief nets. Neural Comput. **18**(7), 54–1527, (Jul 2006). http://dl.acm.org/citation.cfm?id=1161603.1161605

19. Huang, F., Yates, A.: Exploring representation-learning approaches to domain adaptation. In: Proceedings of the 2010 Workshop on Domain Adaptation for Natural Language Processing, pp. 23–30 (2010). http://dl.acm.org/citation.cfm?id=1870530

20. Inkpen, D., Liu, J., Farzindar, A., Kazemi, F., Ghazi, D.: Detecting and disambiguating locations in Twitter messages. In: Proceedings of the 16th International Conference on Intelligent Text Processing and Computational Linguistics (CICLing 2015). Cairo, Egypt (2014)

21. Keshtkar, F., Inkpen, D.: A hierarchical approach to mood classification in blogs. Nat. Lang. Eng. **18**(1), 61–81 (2012)

22. Lafferty, J.D., McCallum, A., Pereira, F.C.N.: Conditional random fields: Probabilistic models for segmenting and labeling sequence data. In: Proceedings of the Eighteenth International Conference on Machine Learning, ICML '01, pp. 282–289. Morgan Kaufmann Publishers Inc., San Francisco (2001). http://dl.acm.org/citation.cfm?id=645530.655813

23. Li, H., Srihari, R.K., Niu, C., Li, W.: Location normalization for information extraction. In: Proceedings of the 19th international conference on Computational linguistics, vol. 1, pp. 1–7. Association for Computational Linguistics, Morristown, (Aug 2002). http://dl.acm.org/citation.cfm?id=1072228.1072355

24. Li, T., Zhang, Y., Sindhwani, V.: A non-negative matrix tri-factorization approach to sentiment classification with lexical prior knowledge. In: Proceedings of the Joint Conference of the 47th Annual Meeting of the ACL and the 4th International Joint Conference on Natural Language Processing of the AFNLP. pp. 244–252. Association for Computational Linguistics, (Aug 2009). http://dl.acm.org/citation.cfm?id=1687878.1687914

25. Liu, F., Vasardani, M., Baldwin, T.: Automatic identification of locative expressions from social media text: A comparative analysis. In: Proceedings of the 4th International Workshop on Location and the Web. LocWeb '14, pp. 9–16. ACM, New York (2014). http://doi.acm.org/10.1145/2663713.2664426

26. Liu, J., Inkpen, D.: Estimating user locations on social media: a deep learning approach. Technical Report. University of Ottawa (2014)

27. Mani, I., Hitzeman, J., Richer, J., Harris, D., Quimby, R., Wellner, B.: SpatialML: annotation scheme, corpora, and tools. In: Proceedings of the 6th international Conference on Language Resources and Evaluation (2008), p. 11 (2008). http://www.lrec-conf.org/proceedings/lrec2008/summaries/106.html

28. Melville, P., Gryc, W., Lawrence, R.D.: Sentiment analysis of blogs by combining lexical knowledge with text classification. In: Proceedings of the 15th ACM SIGKDD International Conference on Knowledge Discovery and Data Mining (KDD'09), p. 1275. ACM Press, New York (June 2009). http://dl.acm.org/citation.cfm?id=1557019.1557156

29. Owoputi, O., O'Connor, B., Dyer, C., Gimpel, K., Schneider, N., Smith, N.A.: Improved part-of-speech tagging for online conversational text with word clusters. In: Proceedings of NAACL-HLT, pp. 380–390 (2013)

30. Pedregosa, F., Varoquaux, G., Gramfort, A., Michel, V., Thirion, B., Grisel, O., Blondel, M., Prettenhofer, P., Weiss, R., Dubourg, V., Vanderplas, J., Passos, A., Cournapeau, D., Brucher, M., Perrot, M., Duchesnay, E.: Scikit-learn: machine learning in Python. J. Mach. Learn. Res. **12**, 2825–2830 (2011)

31. Priedhorsky, R., Culotta, A., Del Valle, S.Y.: Inferring the origin locations of tweets with quantitative confidence. In: Proceedings of the 17th ACM Conference on Computer Supported Cooperative Work & Social Computing (CSCW '14), pp. 1523–1536. ACM Press, New York (Feb 2014). http://dl.acm.org/citation.cfm?id=2531602.2531607

32. Razavi, A.H., Inkpen, D., Brusilovsky, D., Bogouslavski, L.: General topic annotation in social networks: A latent dirichlet allocation approach. In: Advances in Artificial Intelligence, Lecture Notes in Computer Science, vol. 7884, pp. 293–300. Springer, Berlin (2013). http://dx.doi.org/10.1007/978-3-642-38457-8_29

33. Razavi, A.H., Inkpen, D., Falcon, R., Abielmona, R.: Textual risk mining for maritime situational awareness. In: 2014 IEEE International Inter-Disciplinary Conference on Cognitive Methods in Situation Awareness and Decision Support (CogSIMA), pp. 167–173. IEEE (2014)
34. Roller, S., Speriosu, M., Rallapalli, S., Wing, B., Baldridge, J.: Supervised text-based geolocation using language models on an adaptive grid. In: Proceedings of the 2012 Joint Conference on Empirical Methods in Natural Language Processing and Computational Natural Language Learning. pp. 1500–1510. Association for Computational Linguistics (Jul 2012). http://dl.acm.org/citation.cfm?id=2390948.2391120
35. Rumelhart, D.E., Hinton, G.E., Williams, R.J.: Learning internal representations by error propagation. Technical Report. DTIC Document (1985)
36. Sarawagi, S., Cohen, W.W.: Semi-markov conditional random fields for information extraction. NIPS **17**, 1185–1192 (2004)
37. Sinnott, R.W.: Virtues of the haversine. Sky Telesc. **68**, 158 (1984)
38. Tang, D., Qin, B., Liu, T., Li, Z.: Learning sentence representation for emotion classification on microblogs. Natural Language Processing and Chinese Computing, vol. 400, pp. 212–223 (2013). http://link.springer.com/chapter/10.1007/978-3-642-41644-6_20
39. Vincent, P., Larochelle, H., Bengio, Y., Manzagol, P.A.: Extracting and composing robust features with denoising autoencoders. In: Proceedings of the 25th International Conference on Machine Learning (ICML'08), pp. 1096–1103 (2008). http://portal.acm.org/citation.cfm?doid=1390156.1390294
40. Wing, B.P., Baldridge, J.: Simple supervised document geolocation with geodesic grids. In: Proceedings of the 49th Annual Meeting of the Association for Computational Linguistics: Human Language Technologies (ACL HLT '11), pp. 955–964. Association for Computational Linguistics (June 2011). http://dl.acm.org/citation.cfm?id=2002472.2002593
41. Yao, Y., Rosasco, L., Caponnetto, A.: On early stopping in gradient descent learning. Constr. Approx. **26**(2), 289–315 (2007)

DroidAnalyst: Synergic App Framework for Static and Dynamic App Analysis

Parvez Faruki, Shweta Bhandari, Vijay Laxmi, Manoj Gaur
and Mauro Conti

Abstract Evolution of mobile devices, availability of additional resources coupled with enhanced functionality has leveraged smartphone to substitute the conventional computing devices. Mobile device users have adopted smartphones for online payments, sending emails, social networking, and stores the user sensitive information. The ever increasing mobile devices has attracted malware authors and cyber-criminals to target mobile platforms. Android, the most popular open source mobile OS is being targeted by the malware writers. In particular, less monitored third party markets are being used as infection and propagation sources. Given the threats posed by the increasing number of malicious apps, security researchers must be able to analyze the malware quickly and efficiently; this may not be feasible with the manual analysis. Hence, automated analysis techniques for app vetting and malware detection are necessary. In this chapter, we present DroidAnalyst, a novel automated app vetting and malware analysis framework that integrates the synergy of static and dynamic analysis to improve accuracy and efficiency of analysis. DroidAnalyst generates a unified analysis model that combines the strengths of the complementary approaches with multiple detection methods, to increase the app code analysis. We have evaluated our proposed solution DroidAnalyst against a reasonable dataset consisting real-world benign and malware apps.

P. Faruki (✉) · S. Bhandari · V. Laxmi · M. Gaur
Department of Computer Science & Engineering, Malaviya National
Institute of Technology, Jaipur, India
e-mail: parvezfaruki.kg@gmail.com; 2012rcp9518@mnit.ac.in

S. Bhandari
e-mail: 2014rcp9508@mnit.ac.in

V. Laxmi
e-mail: vlaxmi@mnit.ac.in

M. Gaur
e-mail: gaurms@mnit.ac.in

M. Conti
Department of Mathematics, University of Padua, Padova, Italy
e-mail: conti@math.unipd.it

© Springer International Publishing Switzerland 2016   519
R. Abielmona et al. (eds.), *Recent Advances in Computational Intelligence
in Defense and Security*, Studies in Computational Intelligence 621,
DOI 10.1007/978-3-319-26450-9_20

Keywords Android security · Malware analysis · Synergic analysis framework · Static and dynamic analysis · Statistically robust features · Machine learning

1 Introduction

The tremendous success of Android Mobile devices (smartphone, tablets) due to its open nature has resulted in vast number of third-party developer apps. The apps are distributed from Google Play, the official Android market, and other third party markets. The emergence of the third party and improper app vetting results in benign apps with software vulnerabilities or bugs that can be exploited to steal personal information and user privacy-invasive practices. The software flaws may also result due to logical errors, in time delivery pressures and lack of security issues and experience on the emerging mobile platform [17]. Furthermore, the software vulnerabilities are a part of Android security model. The same questions apply to the popular apps downloaded and used by a majority of the users based on trust and ranking at online app marketplaces. In [20], the authors, conducted a comprehensive review of different analysis and detection methods and stressed the necessity of multiple detection methods on Android.

The Android security model is a permission-based security model to prevent the resources not requested apps resources they never requested. For example, The Internet is accessible only if the app developer explicitly declares the INTERNET permission; SMS permitted if SMS_SEND is declared in the Application PacKage (APK) manifest. Furthermore, availability of Internet does not implicitly guarantee the data security. The app user decides to install the APK with all required permissions even if some app has an approval leading to undesirable consequences. Hence, a naive user may not be able to judge the appropriateness of security requirements. Cyber criminals and malware authors are targeting Android due to the enormous user base and availability of more than 1.5 million apps on Google Play with millions of downloads each day [40]. The exploding number of third-party developer apps necessitate an automated analysis system to analyze malware menace. Mobile platform usage has complicated the security issues as personal mobile devices are also being used in companies for professional work. Thus, the Bring-Your-Own-Device (BYOD) brings in additional security concerns such as security of data, privacy issues, and user data invasion. Moreover, Fernandes et al. [26] developed an FM radio frequency based attack model and reported the mobile OS vulnerability against such threats.

Malicious app detection and mitigation are important concerns for anti-malware research. The app detection approaches are either static, dynamic or hybrid. However, Android mobile platform faces the analysis challenge within limited memory-constrained processing and limited power availability. Static app analysis is performed by inspecting the complete code without executing it. Dynamic analysis generates temporal or spatial snapshots of processor execution, memory, network activity, system call logs, SMS messages sent, phone calls. Static approaches are useful for resource constrained Android devices. This approach can be defeated by employing repackaging, polymorphic or code transformed malware app. Dynamic analysis overcomes the limitation of static methods against obfuscated transforms.

Both the approaches have their strengths and weakness. In [16, 28, 57], authors have implemented static, dynamic or hybrid analysis. However, a few approaches provide a complete solution with post-processing machine learning approaches. Hence, we have proposed and implemented DroidAnalyst, an automated web-based analysis system to analyze bytecode on the Dalvik Virtual Machine (DVM), and native execution based on system calls. The proposed approach takes into account the conditions and interactions necessary to reveal the hidden behavior. DroidAnalyst is one of the few frameworks implementing various static, anti *anti-analysis* techniques to detect advanced and targeted malware.

The remaining part of this chapter is organized as follows: Sect. 2 discusses the most relevant static and dynamic analysis approaches. Section 3 is a brief account of the DroidAnalyst system and necessity of multiple analysis modules. We briefly discuss the Android system and necessity of the proposed framework. In Sect. 4, we discuss static analysis modules ApPRaISE, AndroSimilar, AbNORMAL and DePLORE and the implementation details classification system. Furthermore, we identify the necessity of multiple analysis and detection methods for improved code coverage. Section 5 details the working of the dynamic analysis module and integration of novel analysis techniques. In Sect. 6, we evaluate the DroidAnalyst framework based on relevant parameters. Finally, we conclude the chapter in Sect. 7 discussing the enhancements and future work.

2 Related Work

Androguard [13] is a recursive disassembly based static analysis tool to detect similar and classes and methods and identify cloned apps. The signature-based modules identify known malicious applications. Zheng et al. [55] proposed the static analysis stress test to experiment and evaluate the performance of anti-malware against code transformations. Authors concluded that static analysis methods can be circumvented with simple obfuscation techniques. There is an exponential rise of obfuscated Android malware eluding the static analysis methods [21]. Hence, dynamic analysis and detection techniques are gaining prominence on Android platform. Further Rocha et al. [43] proposed a hybrid and lightweight analysis model enforcing realistic policies on the mobile platform with low overhead.

Droidbox [19] is an Android dynamic analysis tool providing emulation, taint analysis, and API monitoring capabilities. It extends TaintDroid [52] with new taint sources and sinks. Droidbox records app behavioral such as file operations, telephony operations (i.e., SMS and phone calls), cryptography operations and network traffic monitoring to identify sensitive information invasive practices. DroidScope [53] is an offline and dynamic Virtual Machine Introspection (VMI) to detect malicious apps. However, the approach requires OS modification. Bläsing et al. [14] proposed AASandbox, an app profiling framework logging system calls with Loadable Kernel

Modules. The authors employ clustering to predict the app behavior. Jiang et al. [56] classified the Android malware family dataset identifying their characteristics and released the samples under Genome Project. In [48] authors analyzed more than 6,000 Android malware and clustered them among families using the VirusTotal API [10].

Andrubis [33] is a web-based interface for analyzing malicious apps using both static and dynamic analysis. It is developed using Droidbox, Taintdroid, Androguard and APK Tool. Thus, Android framework is modified to an extent for Taint analysis and API monitoring. CopperDroid [42] is a system call based analysis framework to monitor inter-process communication through VMI. Authors of CopperDroid evaluated Android Malware Genome Project malware families [9]. Mobile Sandbox [49] is a static, and dynamic analysis web service using the existing analysis methods combined with novel system call tracing for app analysis.

However, Petsas et al. [39] demonstrate advanced malware apps thwarting virtual/emulated environment to hinder dynamic analysis. Authors patched existing malware apps with *anti-analysis* features to show the weakness of majority of existing frameworks already discussed in [14, 33, 42, 54]. Furthermore, Vidas et al. [50] proposed a system to identify the emulated Android environment. The authors identified the difference in behavior, performance evaluation, presence/absence of smartphone hardware and software capabilities. Such a system highlights the importance of employing anti *anti-analysis* techniques among the sandbox environment. To overcome this limitations, Faruki et al. [23] proposed a platform-invariant *anti* anti-emulation sandbox to detect the stealth Android malware. Thus, the synergy of static and dynamic analysis can be combined to create an effective analysis.

Andrubis [2, 33] and Mobile Sandbox [49] are two known systems similar to the proposed DroidAnalyst. Andrubis employs Droidbox, a dynamic app analysis tool and TaintDroid, privacy leakage detection tool for automated analysis. However, Andrubis app analysis is limited to the Android API version 9, i.e. Android OS, Gingerbread version 2.3. Mobile Sandbox improved the Droidbox application to support the Android versions up to API level 17 (Android Jellybean OS version 4.2). Hence, APK files up to the Android Jellybean 4.2 version can be analyzed. However, both the above services employ the existing tools that modify the Android platform. However, rather than depending upon the OS modifications, DroidAnalyst can be configured to support the latest version of Android OS supported applications. Hence, DroidAnalyst can support the devices with higher API versions for which the APK are developed by the third party developers.

Furthermore, Andrubis and Mobile Sandbox cannot detect advanced malware equipped with *anti-analysis* techniques discussed in [39]. The dynamic analysis module of DroidAnalyst is developed to be a scalable anti *anti-analysis* framework to detect analysis environment-reactive malware. There are a very few analysis frameworks using the synergy of static and dynamic approaches. Furthermore, known frameworks Andrubis [34, 51], MobileSandbox [49] and Foresafe [50] can be circumvented by the advanced and targeted malware. Hence, we propose and implement DroidAnalyst, an extension of our proposed research [23].

3   Our Contribution: DroidAnalyst

The existing web-based analysis systems incorporate the static and dynamic analysis techniques [39]. Furthermore, the DroidAnalyst improves existing methods by including *anti-analysis* malware detection capability discussed in [33, 50]. In particular, our contributions are:

1. An automated framework incorporating multiple static and dynamic analysis detection methods for Android APK vetting, and malware analysis. Proposed DroidAnalyst, combines the synergy of existing state of the art with multiple analysis detection methods.
2. We implement various user interface (UI) stimulation techniques to force the APK file reveal the malicious behavior. Furthermore, the dynamic analysis module integrates a range of static, anti *anti-analysis* techniques to coax the environment aware advanced Android malware reveal hidden behavior.
3. We employ machine learning methods to generate a set of features with multiple detection methods based on the static and dynamic module analysis.
4. The proposed research is available to the researchers through a web-based interface http://www.droidanalyst.org [35].

Static analysis module has multiple detection methods: (i) Parse the Androidmanifest.xml with Android Permission Risk Model (ApPRaISe), (ii) Statistically robust signature method (AndroSimilar [22]) to detect app malware variants, (iii) vulnerable component detection (INVAsiON), and (iv) Inter-component control flow dependence analysis (AbNORMAL). Further, we (v) test the input APK files against commercial anti-malware with VirusTotal API [10], and (vi) apply dynamic analysis methods to analyze the analysis ware threats.

The static component generates APK meta-data information to get an overview of the app. In the first stage, app package name, MD5, SHA signature are produced. The meta-data information such as the presence of the cryptographic, dynamic and native code is ascertained. Simultaneously, the VirusTotal API are used to scan the APK file to identify the known malware. Furthermore, the Android manifest is parsed to determine the permissions requested, and components declared in the APK. Vulnerable components such as activities, broadcast receivers, content providers, and services are identified. The app is decompiled into Dalvik bytecode to identify any suspicious functionality. The bytecode is used to generate a Control Flow Graph (CFG), an asynchronous control flow graph to detect misuse of sensitive features such as SMS, Call, picture click, and audio recording.

The dynamic analysis module has multiple analysis components that execute the APK file on Android emulator and record the operations of the APK. The default emulator is modified to represent a real Android device. Default information such as IMEI, IMSI, device identifiers, network information, device name are rebuilt to resemble an actual Android device.

The system logs interaction of Dalvik bytecode and system call interaction at the native level. Static analysis module generates a complete report of the submitted

APK. The multiple analysis techniques classify APK either as normal or malicious. The collected features are categorized with tree-based classifiers Random Forest [15], and J48. The most relevant features are identified with minimum redundancy Maximal Relevance (mrMR) [38] to maintain feature relevance. DroidAnalyst identifies *anti-analysis* malware evading known systems Andrubis [51], Mobile Sandbox [49], and Copperdroid [37, 50]. To evaluate the analysis framework we crawled 36,788 collected apps from Google Play, and other third-party markets like Anzhi, HiApk, appchina, androidbest.ru and 13,462 malware APK from different malware repositories.

3.1 Android Background

Android is being developed under Android Open Source Project (AOSP), maintained by Google and promoted by the Open Handset Alliance (OHA). Android apps are developed in Java. Android OS is built on top of Linux kernel due to its robust driver model, efficient memory and process management, and networking support for the core services. Android user app, written in Java language is translated into Dalvik bytecode that runs under newly created runtime, the *Dalvik Virtual Machine* (DVM). Android protects the sensitive functionality such as telephony, GPS, network, power management, radio and media as system services using the mandatory permission-based model.

3.1.1 App Components

An Android app is composed of one or more components discussed below:

- *Activity*: It is the user interface component of an app. Any number of activities are defined in the manifest depending on the developer requirement and app functionality.
- *Service*: Service component performs background tasks without any UI
- *Broadcast Receiver*: This component listens to the Android system generated events. For example, BOOT_COMPLETED, SMS_RECEIVED are system events.
- *Content Provider*: Content provider also known as the *data-store* provides a consistent interface for data access within and between different apps.

Android security protects apps and data with the combination of system-level and Inter Component Communication (ICC) [25]. Android middleware mediates the ICC between application and components. Access to a component is restricted by assigning access permission label. The developer assigns permission labels through the manifest within an app. The developer defines the app security policy. However, permission(s) allocated to the components in an application specifies an access policy to protect app resources.

3.1.2 Android Permission Model

To restrict an app from accessing the sensitive functionality such as telephony, network, contacts/SMS/sdcard and GPS location, Android provides permission-based security model within the application framework. Developer must declare the required permissions to access a resource using the `<uses-permissions>` tag inside `AndroidManifest.xml`. Android controls the individual apps to mitigate the undesirable effects on the system apps or third-party developer apps using the app sandboxing. The process enforces the restrictions at the install time. Android permission protection is discussed below [25]:

1. *Normal*: These permissions have a minimum risk. The Normal permissions are granted by default during installation.
2. *Dangerous*: These permissions fall into the high-risk group as they access the private data and relevant sensors of the device. A user must explicitly permit the usage of dangerous permissions before installing the APK.
3. *Signature*: Signature permissions are available with the system apps. They are granted automatically at the time of installation.
4. *SignatureOrSystem*: These permissions are granted if the requesting app is signed with the same certificate as the Android system image or with an app that declared such permission.

Android permissions are coarse-grained. For example, the `INTERNET` permission does not have the capability to restrict access to a particular Uniform Resource Locator (URL). During the APK installation, the user is forced to grant either all permissions or deny the app installation. Hence, the dangerous permissions cannot be avoided at the install time. Moreover, the users cannot differentiate between the necessity and its imperative misuse that may expose them for exploitation [47].

4 DroidAnalyst: System Description

This section gives a broad view of the DroidAnalyst. We first discuss the Droid-Analyst architecture and various modules involved in creating the analysis system. It is followed by a brief discussion of its components in detail. This section covers the automated analysis beginning with the static analysis module, followed by the dynamic analysis to complement the static analysis. The dynamic module is designed to coax the APK to execute on a modified Android emulator(s).

4.1 *DroidAnalyst: Architecture*

Figure 1 illustrates the synergic framework utilizing the static and dynamic analysis approaches to generate analysis report. The following sections discuss the multiple analysis and detection methods integrated together in DroidAnalyst.

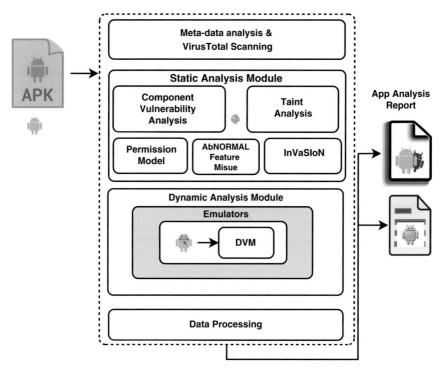

Fig. 1 DroidAnalyst overview

4.2 Static App Analysis

The static analysis module is formed of multiple analysis components (Permission risk model, Component vulnerabilities detection, taint analysis, ICC-based control flow analysis and signature-based malware variant detection model). The initial impression of the APK is generated by calculating the corresponding hash value is matched against the VirusTotal database. In this step, our system compares the MD5 and SHA1 of inspected APK with the hash signatures of VirusTotal malware database. Detection ratio of suspect malware is tested against 56 commercial anti-malware. This step is an indicator for identifying already known malware APK. Simultaneously, indicators such as the use of reflection code, the presence of the cryptographic code, dynamic code loading, and the presence of native code inside APK. The APK file is then reverse engineered to its Dalvik bytecode and gain access to the Android components. We analyze the `Androidmanifest.xml` to avail the list of permissions requested by the APK. We use APK Tool [11] for the same. Apart from permissions, we extract components, intents, services and broadcast receivers for the vulnerability analysis. SDK version of the APK is also listed to ensure the compatible APK is given to the dynamic analysis module.

We further analyze the executable code, i.e., the Dalvik bytecode in the `classes.dex`. It is reverse engineered to an intermediate smali [12] format to perform automated parsing. DroidAnalyst is capable of identifying the advertisement libraries from the Dalvik bytecode to ensure proper identification of executable code and hence the analysis report remains accurate. The smali file corresponding to each class is inspected for malicious functionality or dangerous methods.

4.3   Android Permissions RISk Modeling (ApPRaISe)

Android permission model is an important security measure to prevent unauthorized access to sensitive resources. Permissions classified dangerous in the Android framework use resource that costs money. Dangerous permissions used by the APK developer need to be explicitly accepted and installed during APK installation. Thus, users have to understand the impact of dangerous permissions. For example, it is uncommon for a game app to request `SEND_SMS` permissions. Unfortunately, the majority users ignore to check required permissions during installation. Furthermore, naive users cannot identify the potential misuse of app permission. Hence, we propose ApPRaISe, an app permission assessment model that determines the risk associated with dangerous permissions. Further, the presence of essential bytecode features such as native, dynamic, cryptographic and Reflection code is identified.

Individual dangerous permissions may not be harmful in themselves; for example, an app using `INTERNET` or `READ_SMS` permission individually may not pose a risk to the user device. However, if the app requests both the permissions together in an app, those permissions can be misused to harm the user privacy. Hence, we perform the `n-Set` permission usage analysis to identify the dangerous permissions risk.

4.3.1   n-Set Permissions Usage Analysis

Comparison of dangerous permissions between benign and malware apps considering the combination of 2 permissions (2-Set) combination of 3 permissions (3-Set) and 4-Set is performed on a reasonable size dataset. Figures 2 and 3 gives the comparative analysis for 2-Set and 3-Set permissions respectively. In fact, we analyzed 4-Set and 5-Set permissions, but the number of apps that cover those sets were not significant. In 2-Set (Fig. 2), `INTERNET/READ_PHONE_STATE` and `INTERNET/READ_SMS` are requested by more malware than benign apps. Similarly, the Figure illustrates many such combinations requested by malware compared to benign apps. From the total permissions of both data sets, top 70 permission combinations are selected as features. Additionally ApPRaISe considers following Dalvik bytecode features:

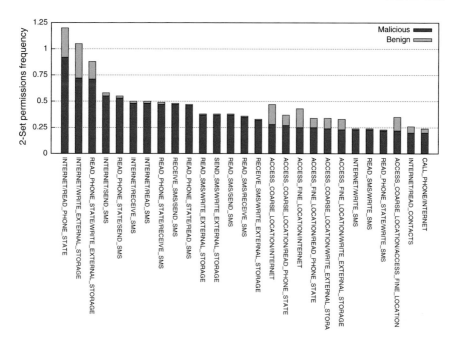

Fig. 2 Comparative analysis for 2-Set permission

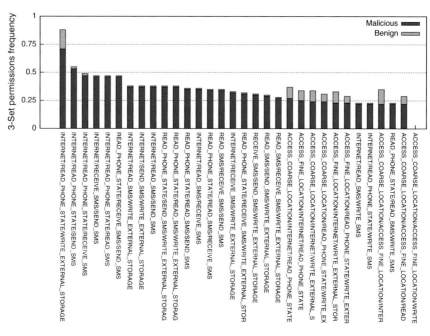

Fig. 3 Comparative analysis for 3-Set permission

1. **Cryptographic Code**: This feature is present if the strings are encrypted in an APK file.
2. **Native Code**: This feature is present if an app contains native (C/C++) code embedded as library or executable.
3. **Dynamic Code**: This feature is present if an app can load external Java classes at runtime.
4. **Reflection Code**: This feature is present if an app uses Java Reflection API in the code.

The proposed model also identifies over-privileged permissions declared in the `Androidmanifest.xml`, absent in the Dalvik bytecode. To determine the unused permissions ApPRaISe:

1. Lookup APIs, intent-filters or content providers corresponding to the permission from mapping database and find their usage in the app. If usage is not found, we conclude that permission is unused, otherwise follow next step.
2. Perform reverse path reachability analysis using synchronous and asynchronous control flow and locate the bytecode where the permission is used.
3. If the topmost method found in the path is an entry point method called by the Android framework (e.g., onClick, onCreate, onBind); we conclude that permission declared in the manifest file has been used in the app code; hence justified use declared permission. Otherwise, permission is treated as declared but unused, hence over-privileged.

4.3.2 Results

We have evaluated the ApRaiSE model on 73 features (permissions and bytecode features) extracted from 11,639 apps and trained with Random Forest [15] decision tree algorithm. Random Forest is chosen as the machine learning model to balance the performance trade-off. Permission modeling results are illustrated in Table 1, enlisting acceptable low false positives.

Random Forest machine learning classifiers [15] are nearest neighbor predictors for regression and classification to construct multiple decision trees during training and testing. Random forests form a strong learner classifier from a group of weak learners as it brings weak classifiers together [15]. Hence, the Tree-based Classifiers are considered for accurate classification. The machine learning model used k-fold cross–validation to discriminate malicious APK from the normal files.

Table 1 Performance of ApPRaISe on 73 features

Actual class	Predicted class	
	Malware (%)	Benign (%)
Malware	**81.3**	18.7
Benign	2.8	**97.2**

Actual class	Predicted class	
	Malware (%)	Benign (%)
Malware	**84.6**	15.4
Benign	7.7	**92.3**

Table 2 Performance of ApPRaISe on top 20 features

A good feature selection algorithm must have: (1) simplified model; (2) short training time; and (3) reduced variance. The ApPRaISe model further identified top 20 permission features with minimum redundancy Maximal Relevance [38] feature selection algorithm. mrMR employs mutual information, correlation, or similarity to identify and remove the redundant features. The mrMR maximizes the mutual information of the selected attributes with a combinatorial estimation [7] to extract robust probabilities and thus improves the relevant features for accurate classification.

The evaluation results discussed in the Table 2 illustrate 85 % accuracy based on permissions and bytecode features. The test phase the benign and malware apps, it was observed that the apps using ACCESS_NETWORK_STATE always paired with the INTERNET permission. More than 72 % apps requested ACCESS_COARSE_LOCATION AND ACCESS_FINE_LOCATION permissions. Genuine Wallpaper APK files also included such permission. Some applications were over-privileged or require excess permission such as reading the browsing history, bookmarks, reading phone storage, even though the app category did not necessitate this. Hence, the ApPRaISe model developed the n-set model to detect risky apps based on the user private data invasive practices.

4.4 INVAsiON—IdeNtifying Vulnerable App cOmpoNents

Android framework allow developers to divide their apps into one or more logical components such as *activities*, *services*, *broadcast receivers* and *content providers*. Besides, the framework also provides API to facilitate inter-app communication. An app can export its several components to allow other third-party apps for re-use. To illustrate a real-world scenario, we developed an app that captures pictures by communicating with device camera app. Internally, inter-app communication happens between components of two different apps using higher level abstractions such as `Intent`. Unfortunately, the exported components of an app can be vulnerable to attack by adversaries if proper care is not exercised. Android framework recommends that a developer should protect the exported app components using permissions.

Grace et al. [29] developed *Woodpecker* to identify vulnerable components in preloaded system apps within stock Android versions of OEMs. However, the authors perform reachability analysis on exported components using the control-flow to investigate the execution paths using dangerous permission. An exported component can be utilized by malicious apps to exploit an un-intentional benign app to elevate the malicious privileges. For example, a benign app component sends SMS

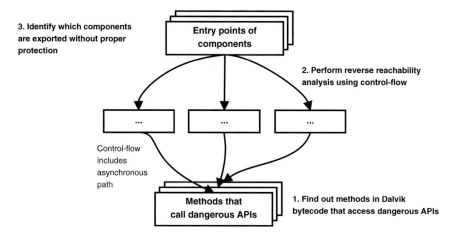

Fig. 4 Working of INVAsiON

by accepting the phone number and message text as arguments of an Intent. If this component is exported without proper protection, malicious apps can leverage it to send SMS messages without requesting SEND_SMS permission in its manifest. Thus, it is important for developers to vet their code to ensure the protection of declared components. Figure 4 illustrates the steps performed by the proposed INVAsiON to identify the exported and vulnerable components.

As illustrated with steps in Fig. 4, the first step is to search for Dalvik bytecode methods that access the dangerous permissions. In step 2, we perform the reverse reachability analysis using control flow to check if the execution path is initiated from an entry point of some exported component. In contrast to the *Woodpecker*, we identify feasible paths from entry points that can lead to usage of sensitive permission. An advantage is that the information about the permission that can be misused is also identified. Further, unlike the *Woodpecker*, INVAsiON does not perform data-flow analysis to find an exact relation between the arguments of an exported components' entry points to API calls related to dangerous permissions. The report points to the vulnerable component and potential permission misuse.

Further, we also consider the asynchronous API calls. For example, consider the java.lang.Thread class. This class is used to implement threads, which apps generally use to perform background tasks for better UI responsiveness. A developer just have to simply extend this class, implement the run() method, and then call the start() method to schedule the thread. However, if we analyze Dalvik bytecode of an app, the run() method does not appear to be reachable from start(), despite the fact that after the start() method is called, control-flow goes through the Dalvik VM to the underlying thread scheduler and eventually to the run() method (See Fig. 5). INVAsiON considers asynchronous APIs like these to keep continuity in building control-flow graph. Particularly, we currently consider Thread, Runnable, TimerTask, CountDownTimer, AsyncTask and Handler APIs.

Fig. 5 Discontinuity in
control-flow due to
`java.lang.Thread`

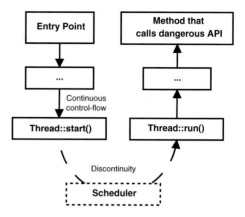

We recommend that, unless it is necessary, vulnerable exported components of
an app must be protected by either *signature* or *signatureOrSystem* level permissions to limit their exposure. The problem limit the malicious app from accessing
the vulnerable component due to the unavailability of same certificate.

4.5 DEPLorE—DEtecting Privacy LEakage

In the following, we enumerate the challenges and requirements for developing a
precise static taint analysis algorithm [27]:

1. The app components follows a *lifecycle* imposed by the Android framework.
   Component lifecycle has pre-defined or user-defined callback methods. The components are invoked at different times during app execution. Precise control-flow
   generation requires an app to model the *lifecycle* accurately.
2. Algorithm should be context sensitive in nature. A context-insensitive analysis
   joins the analysis results for a method *m* at all call sites to *m* even if the arguments
   to *m* at those call sites differ.
3. Algorithm should be object-sensitive in nature. That is algorithm should be able
   to distinguish between instances of the same object (i.e., class).
4. Similarly, the algorithm should be field sensitive in nature.
5. Algorithm must consider the semantics of several API and Dalvik bytecode constructs like *arrays*.

4.5.1   Implementation Details

DEPLorE, like FlowDroid [27], is *context–*, *object–*, *field–* and *flow–sensitive*,
aliasing aware implementation of static taint analysis algorithm to detect privacy

Table 3 Partial list of taint *source* and *sink* APIs

API	Type
`android.location.Location::getLatitude()`	LOCATION
`android.telephony.TelephonyManager::getDeviceId()`	DEVICE_INFO
`android.telephony.TelephonyManager::getSubscriber()`	DEVICE_INFO
`android.accounts.Account::name`	ACCOUNT_INFO
`java.io.FileOutputStream::write(Byte[])`	FILE
`java.net.URL::URL(String, String, int, String)`	INTERNET

`android.telephony.SmsManager::sendTextMessage`
`(String, String, String, PendingIntent,`
`PendingIntent)`
	SMS

Fig. 6 An overview of DEPLorE

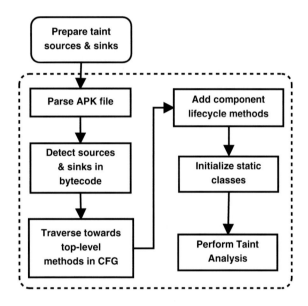

invasion in Android apps. We consider `location, device information, installed packages, Wi-Fi, and Bluetooth` details and account information as taint *sources*. Further, File, Internet and SMS are our taint *sinks*. Table 3 lists the present *source* and *sink* API calls. We have Formalized the Dalvik bytecode semantics into an abstract state machine based on the ScanDal [32] approach.

Figure 6 illustrates the proposed taint analysis DEPLorE functionality. App bytecode may contain a large number of execution paths; so instead of traversing every possible path, we prune it by searching the methods in which *source* and *sink* APIs have been accessed. Furthermore, with the help of control-flow graph (CFG), we traverse towards top-level methods, which gives us an opportunity to find common execution paths for both *sources* and *sinks*. Before starting actual taint analysis, we initialize abstract state machine environment by creating instances of static classes in

Table 4 Accuracy of DEPLorE against DroidBench test suite

Test case	# of Apps	Detected by DEPLorE
Android specific	6	5
Callbacks	13	8
Field and object sensitivity	3	2
General Java	10	8
Implicit flows	4	2
Inter-app communication	3	1
Lifecycle	10	8
Reflection	4	3

a heap and also prepend lifecycle callback methods for every component of interest in the list of top-level methods. Then, we perform inter-procedural taint analysis for all the Dalvik methods in the list by employing the modified fixed-point algorithm [18]. It must be noted that among top-level Dalvik methods that are user triggered callback methods, privacy leak may happen in a particular sequence during execution. To prune the search for privacy leak, we strictly perform taint analysis of *source* methods before the *sink* methods.

4.5.2 Results

We have evaluated DEPLorE against *DroidBench*,[1] an open-source, comprehensive test suite to assess the effectiveness and accuracy of taint analysis tools for Android [27]. DroidBench consists of some apps that cover a broad range of privacy invasive scenarios, including false leakage scenarios. Table 4 illustrates the detection rate of DEPLorE against DroidBench test suite.

4.5.3 Limitations

DEPLorE is still to handle important cases such as location-related API callbacks, reflection, inter-component communication, APIs related to byte streams, bytecode array semantics and class hierarchy analysis. Further, additional taint *sources* such as browsing history, contacts and SMS messages can be added to strengthen the taint analysis.

---

[1] https://github.com/secure-software-engineering/DroidBench.

4.6 AndroSimilar—Robust Statistical Malware Signature

We have integrated AndroSimilar [24], our proposed statistical malware variant detection method, as a part of DroidAnalyst to detect obfuscated malware variants. As the proposed approach is effective against code obfuscation, repackaged malware of known benign apps, we have integrated the approach with DroidAnalyst. AndroSimilar is a byte analysis method that finds regions of statistical similarity with known malware to detect those unknown, zero-day samples. We also show that syntactic similarity considering whole app, rather than just embedded DEX file is more efficient, contrary to known fuzzy hashing approach. A brief summary of the proposed approach is given below.

- Submit Google Play, third-party or an obfuscated malicious app as input to AndroSimilar.
- Generate entropy values for every byte sequence of fixed size in a file and normalize them between [0, 1000].
- Select statistically robust byte features as per the similarity digest hashing scheme.
- Extracted features are stored in the sequence of Bloom Filters to generate the APK signature.
- Compare the generated signature with existing malware signature database. Similarity score beyond a given threshold marks the sample as repackaged malware.

The proposed methodology needs a fewer number of signatures, the variants of a single malware family have a strong code co-relation. Further, we choose to cluster variants representing robust bytecode similarity and discard the other app signatures. Signature clustering is performed with SDHash [44–46] similarity. The distance between the signatures within a cluster is quite small (i.e., high similarity) according to an empirically chosen *inter-app similarity threshold*. From each group, we choose a single point capable of representing all the apps in the cluster. App feature that are dissimilar to others files is described as a separate cluster. However, reduced signatures may miss the low similarity unknown variant. Nevertheless, such techniques are suitable for mobile devices.

4.7 AbNORMAL—ANdrOid FeatuRe Misuse AnaLysis

SMS Trojan apps that send SMS or make calls to premium-rate numbers have an alarming share in overall malware outbreak. An increased number of SMS Trojan apps is due to significant monetary gains. An innocent user cannot judge the appropriateness of permissions requested by an app according to its functionality [30]. The first SMS Trojan app, `Trojan-SMS.AndroidOS.FakePlayer`, disguised as a music player, was reportedly distributed from Google Play in 2010 [6]. Android Malware Genome project [9] reported many more SMS Trojan families such as `HippoSMS`, `DogWars`, `GGTracker` and `YZHCSMS`. In 2012, the anti-malware industry got hold of other families such as `SMS.Boxer`, `SMSFoncy` and `SMStado` [3–5].

In particular, we identify malicious activity such as sending SMS or make calls to premium-rate numbers, recording audio/video and taking pictures without user's knowledge or consent as *feature misuse* in general. AbNORMAL prototype performs inter-component control-flow dependence analysis to detect feature abuse in Android apps. AbNORMAL generates a precise and complete Control-Flow Graph (CFG) using static program analysis of Dalvik bytecode. The technique incorporates execution path of sensitive resource access (sending SMS, taking pictures), to reach towards the *top-level methods*. Then, all the *top-level methods* are examined to identify if they are user-triggered events (e.g., `onClick`) or entry points (e.g., `onCreate`) triggered by the Android framework. We assume the hypothesis that the legitimate access to sensitive resources in apps originate from user triggered inputs. However, we observe that not all Trojan apps send SMS messages without initiated from user-triggered events. For example, the more recent variant of `FakeInst` family tricks users to purchase paid contents by clicking on UI, and instead they send SMS to premium-rate numbers. We believe such cases are very difficult to detect and consider them out of scope.

4.7.1 Inter-component Control-Flow Graph

Our analysis for *feature misuse* in apps depends upon the quality of control flow graph (CFG). CFG is a directed graph, where a node represents a code block (i.e., the sequence of bytecode instructions), and an edge represents a path dependency between two nodes. An edge can be one of the following types:

- *Conditional*: It represents a conditional branching (i.e., `if`, `while`, `case`, `for` etc.) from one node to another node within a method.
- *Synchronous*: It represents a direct flow from one node in method $m1$ to another node in method $m2$ in the graph. The method invocation in Java is an example of a synchronous flow.
- *Asynchronous*: It represents an indirect flow from one node in method $m1$ to another node in method $m2$ in the graph. For example, forking a `java.lang.Thread` or scheduling a `java.util.TimerTask`.
- *ICC*: It represents an interaction between different components of an app. For example, an *activity* component can interact with a *service* component using the `startService()` API.

Each type of the edge can be visualized in the CFG as illustrated in Fig. 7.

We have developed AbNORMAL in Python on top of state of the art static program analysis tool*Androguard* [13]. *Androguard* decompiles the Dalvik bytecode of an app and generates programmable structures including inter-procedural CFG. We augment this CFG with asynchronous control-flow information. Particularly,

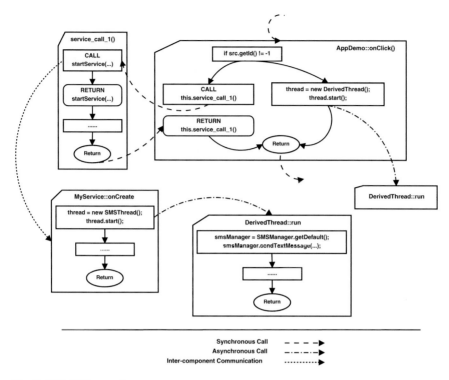

Fig. 7 ICC CFG example

we currently consider `Thread`, `Runnable`, `TimerTask`, `CountDownTimer`, `AsyncTask` and `Handler` APIs.

We further augment CFG with *ICC* information. We build CIG of an app by applying interprocedural data-flow analysis [18] on the ICC API. The node of the CIG represents a component of an app, whereas edge from one node $n1$ to another node $n2$ represents the fact that component $n1$ launches $n2$. Initially, we search for the source components in Dalvik bytecode that use *ICC* related APIs. Then we perform interprocedural constant propagation [31] to identify possible argument values of these APIs at runtime. These constant argument values are further analyzed to resolve possible target components launched from the source components. As *content providers* in an app provide only data related services, we currently exclude while building CIG. The procedure is listed below:

4.7.2 Implementation Details

1. Identify the methods in Dalvik bytecode that access sensitive features of interest (i.e., sending SMS, recording video).
2. Build a precise and complete CFG by taking into account asynchronous APIs as well as CIG.

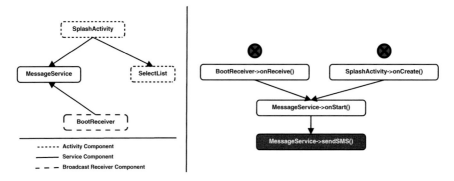

Fig. 8 CIG and feature misuse in HippoSMS malware

3. Perform reverse path reachability in CFG to ascend towards *top-level* methods.
4. Classify *top-level* methods into user triggered and entry point callbacks.

4.7.3 Results

In this section, we evaluate the effectiveness of AbNORMAL by analyzing some notable SMS Trojans and Spyware apps. We have also evaluated AbNORMAL against benign apps from Google Play store to verify legitimate usage of sensitive features.

4.7.4 HippoSMS

HippoSMS malware family was found in Chinese app stores in July 2011. It sends SMS messages to hard-coded premium-rated numbers. Moreover, it can also block/remove reply messages from service providers to prevent the user from knowing about the charges incurred. Partial CIG of an app of this malware family (left) and SMS feature misuse (right) is illustrated in Fig. 8.

We observe that the `sendSMS()` function of `MessageService` component is sending SMS messages to premium-rate numbers. The component gets invoked from the entry point of same component. We can observe from the CIG that `MessageService` component is launched by both `SplashActivity` and `BootReceiver` components. Ascending up toward the top, we can conclude that this app sends SMS at two instances: (1) Whenever `SplashActivity` is launched, and (2) The `BootReceiver` is executed due to broadcast event.

4.7.5 Evaluation

Table 5 illustrates AbNORMAL evaluation. The existing analysis techniques evaded by the advanced malware is identified with the proposed AbNORMAL. Promising

Table 5 Sample evaluation of CONFIDA

Package name	Sensitive feature(s)	# of sensitive feature paths		
		Total	Without user triggered events	Correctly detected by CONFIDA
Malware apps				
com.ku6.android.videobrowser (HippoSMS)	SMS	2	2	2 ✔
org.me.androidapplication1 (FakePlayer)	SMS	1	1	1 ✔
kagegames.apps.DWBeta (Dog Wars)	SMS	1	1	1 ✔
t4t.power.management (GGTracker)	SMS	1	1	1 ✔
com.talkweb.ycya (RogueSPPush)	SMS	1	1	1 ✔
com.mobile.app.writer.zhongguoyang (Pjapps)	SMS, Audio	3	3	3 ✔
com.software.application (SMS Boxer)	SMS	7	3	3 ✔
com.parental.control.v4 (Dendroid)	SMS, Call, Audio, Video, Photo	6	6	6 ✔
Benign apps				
cn.menue.superredial[1]	Call	2	1	2 ✗
com.wn.message	SMS	3	0	0 ✔
polis.app.callrecorder[1]	Audio	1	1	1 ✗
com.me.phonespy[1]	Photo	1	1	1 ✗
com.Rainbow.hiddencameras	Photo	1	0	0 ✔

✔ depicts correct detection. ✗ depicts True Negatives (TNs) and False Positives (FPs) in malware and benign category respectively
Google Play apps: https://play.google.com/store/apps/details?id=cn.menue.superredialhl=en, https://play.google.com/store/apps/details?id=com.appstar.callrecorderhl=en

results of the approach motivates the use of AbNORMAL as an app vetting framework. Table 5 enlists Google Play apps identified malicious with the proposed approach.

Static analysis is an effective approach that analyzes applications without executing them. If the application is encrypted, obfuscated or requires some input to execute, it has to be executed to analyze its runtime behavior. We leverage the existing Android emulator provided with the SDK for analysis. Emulator being a software can be easily reset to its clean state once the app under execution has been tested. In the following section, we discuss the necessity and integration of dynamic analysis module in the DroidAnalyst.

5 Dynamic Analysis

In this section we detail the scalable, and a dynamic analysis module to analyze, and detect advanced Android malware and resource hogger apps draining the Android devices. Privacy risk apps may leak user information such as smartphone identification number (IMEI), subscriber identification (IMSI) without the device owner knowledge and consent. The advanced malicious apps like Anserverbot, Pincer [41], roguelemon [41] have an inbuilt capability to identify the analysis environment based on the existing static properties. If the malicious app identifies the analysis environment, it behaves benign. We identify such properties as *anti-analysis* techniques.

To counter such advanced malware and reveal their malicious behavior, we execute the apps in an emulated environment enriched with static, anti *anti-analysis* capabilities to detect the environment reactive malware. The anti *anti-analysis* techniques are defined as the modification of static parameters within the Android virtual device to resemble them as real Android devices. The proposed modifications are successful in revealing the hidden behavior as they identify the proposed Sandbox as real Android devices. Proposed Sandbox monitors file operations, app downloads, suspicious payload installation, encrypted strings, premium rate SMS and voice calls. The proposed approach also monitors aggressive app behavior such as contacting URLs that exploit the network bandwidth.

The essence of the proposed dynamic analysis sandbox is its multiple analysis methods as illustrated in Fig. 9. When an app is submitted to the Sandbox, a clean, isolated environment is initialized with a refreshed Android emulator consisting clean OS snapshot. Android Virtual Device (AVD) manager [1] allows creation, execution, saving and restores and load the emulator. The emulator is customized by: (1) installing the appropriate system apps (Google Market app, default Google apps); (2) wallpaper settings are modified; (3) default user settings are changed by adding, phone numbers, SMS messages; and (4) user settings are customized. The modified device (emulator) is launched each time as the new APK file is submitted for analysis. The Sandbox starts the emulator(s) with a save-to-snapshot state to resemble it as a real device by adding wallpaper, messages, contacts and setting custom device settings. Each time an app is submitted for analysis, clean emulator snapshot is loaded.

As illustrated in Fig. 9, the Framework core controls all the components for essential feature collection, facilitates the AVD loading, and generates the analysis report. Dalvik Dynamic Instrumentation (DDI) hooking libraries hook their methods for behavior monitoring. Analysis module results are summarized to predict malicious behavior, resource hogging activity or benign behavior. Proposed sandbox employs DDI to identify resource hoggers and privacy risk apps.

The analysis environment sets up static, anti *anti-analysis* features to modify static emulator properties to resemble it as a real device. Proposed Sandbox is scalable as we employ a transparent functionality without changing the Android platform. Resource Hogger App detection is based on anomalous consumption of CPU,

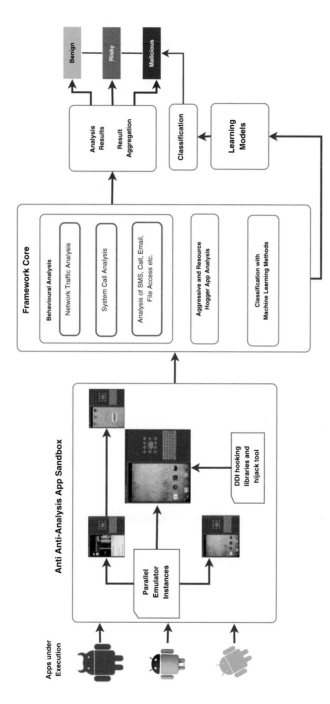

Fig. 9 Proposed dynamic analysis approach

memory or network resource consumption in comparison to benign apps. The proposed approach finds a strong link between malware and its resource usage pattern. App tagging is performed based on its behavior.

5.1 Anti Anti-analysis Capabilities

Persistent, advanced Android malware families like AnserverBot, Bgserv and FakenetFlix defend themselves against analysis environment and obtain more propagation time to avoid revealing malware behavior on the emulator. We modify the IMEI, IMSI, serial number, phone number defaults that are easy to verify an emulator identity. Changes are made to geolocation properties, system time, E-mail account configuration, wallpaper apart from adding images and audio/video files to change the default look. With all these random changes, it would be difficult even for an advanced Android malware to detect the virtual environment. Table 6 compares the static parameters used by *anti-analysis* malware to fingerprint the analysis environment. As the default emulator is used to achieve better performance, malware authors detect the listed static defaults and hide the original malicious behavior. Hence, we extracted the corresponding device parameters of Samsung, Nexus 7 Tablet, Micromax and Karbonn devices. Furthermore, we modified the default Android emulator and replaced the Android defaults with the real Android device. This technique had a desired effect on environment aware malware. Anserver, Bgserv, Dendroid malware samples that do not execute on the existing web services, displayed the malicious behavior assuming that they are running on a real device.

5.2 App Stimulation

Many apps avoid their normal functioning on the standard emulator. However, when we Design a behavioral analysis framework, we need their actual functioning to detect malicious activity. Hence, we need to lure apps by providing the required environment. So we find all the Intents needed by the app and generate them with the Android Debug Bridge (ADB); to start corresponding components (i.e., activity, service or broadcast receiver). The Intents may be developer defined or broadcast. We try to generate some of the Intents implicitly by performing the task like SMS_SENT or NEW_OUTGOING_CALL and explicit generation of other Intents. In case of the time triggered actions, we change emulator's system time to future time with some fixed interval. For example, incrementing an hour 24 times with 3 seconds interval during the app execution at some point in time. To simulate user events, we use Android Monkey [1], a command line tool that provides random user inputs to the Android package with a reasonable control over the type and the number of injected events. The framework initiates activities and services from android debug bridge (Adb) to lure a persistent threat to begin malicious service if present.

Table 6 Comparison of static parameters of android AVD and real android device

Property value	Default AVD	Real android device
IMEI	000000000000000	911315462535214
IMSI	310260000000000	925117254763458
Phone Number	15555525554	121314115554
Serial Number	98101430121181100000	54215E52C54525851254
Network	Android	Tmobile
ro.build.id	ICS MR0	IMM76I
ro.build.display.id	sdk-eng 4.0.2 ICS MR0229537 testkeys	TBW592226 8572 V000225
ro.build.version. incremental	229537	TBW592226 8572 V000225
ro.build.version.sdk	19	17
ro.build.version. release	4.0.2	4.2.2
ro.build.date	Wed Nov 23 22:46:18 UTC 2011	2013 01 25 15:53:21 CST
ro.build.date.utc	1322088378	1359100401
ro.build.type	eng	user
ro.build.user	android-build	ccadmin
ro.build.host	vpbs2.mtv.corp.google.com	BUILD14
ro.build.tags	test-keys	test-keys
ro.product.model	sdk	msm7627a
ro.product.brand	generic	qcom
ro.product.name	sdk	msm7627a
ro.product.device	generic	msm7627a
ro.product.board		7×27
ro.board.platform		msm7627a
ro.build.product	generic	msm7627a
ro.build.description	sdk-eng 4.0.2 ICS MR0 229537testkeys	msm7627a-user 4.0.4 IMM76ITBW5922268572 V000225 testkeys
ro.build.fingerprint	generic/sdk/generic:4.0.2/ICS MR0/ 229537:eng/test-keys	qcom/msm7627a/msm7627a: 4.0.4/IMM76I/ TBW592226 8572 V000225: user/test-keys
net.bt.name	Android	Airtel

5.3 Behavioral Analysis

After recording the app actions, we analyze them with behavioral analysis with `logcat` results to detect additional installations, new process spawns and SMS sent. We scan the traffic (`.pcap`) files to analyze malicious URL or sensitive information leakage. Analysis of system calls relates file and network related activities. Additionally we use Dalvik Dynamic Instrumentation (DDI) to keep track of dynamic operations. Using DDI we use string monitor to monitor string operations and track some external operations like SMS sent/received, and phone calls.

Proposed analysis reports few system calls (bind and connect) prominently visible among malware apps, hence an app with such calls is considered risky. Proposed Sandbox marks actions like sending SMS, e-mail(s) without user consent as covert misuse of existing facilities not seen among normal apps. Sending private user data such as call logs, contacts, existing SMS and e-mails, encrypting sensitive user data (contacts, SMS), GPS coordinates activities not visible in normal apps. An app is termed Potential Risk if its behavior deviates from normal:

- Sending Device data (IMEI and IMSI)
- Use of executable and shell files during execution
- Permission gain for any of its files
- Prohibits app removal once installed on the device
- Use of certain prominent System calls not used by normal apps

5.4 Network Activity

Android emulator natively supports the network traffic capturing. The proposed approach integrates TCPDump [8] to capture the network traffic to a PCAP file. The data such as some bytes sent, bytes received, URLs contacted, messages sent to hard coded numbers are monitored to identify suspicious activities. Further the host connected to, the port used for connection and data sent is identified. Malicious Domains are identified with the URL blacklists. PCAP content classifies the network activity during the post processing.

5.5 Resource Hogger or Aggressive Malware Analysis

We analyzed comparative resource usage pattern of number of benign and malicious apps to construct a threshold pattern that would be reasonably greater than the benign usage (e.g., twice the benign usage) but lesser than the peak malware usage. Resources consumption may change in the course of execution, so we fetch maximum usage by a given app category, i.e., the difference between the resource usage before, during and after completing the app execution.

5.6 Dalvik Dynamic Instrumentation (DDI)

DDI [36] hooks itself to classes and methods of Dalvik Virtual Machine (DVM). We use DDI to observe various runtime strings to detect encrypted malware. Framework Instruments hooks for SMSManager class to keep track of messages sent by an app and Intent class to monitor phone calls and e-mails.

5.7 Machine Learning Analysis Model

As a preliminary step, representative Google Play and different families of malware apps are chosen. These apps are used to train the framework by recording behavioral information. It is converted to features and are marked according to their respective category. Using these feature vectors, we train the machine learning algorithms to test the potential app under execution. Random Forest machine learning classifiers [15] are nearest neighbor predictors for regression and classification to construct multiple decision trees during training and testing. Random forests form a strong learner classifier from a group of weak learners bringing them together [15]. Hence, Tree-based Classifiers are considered for accurate classification. The machine learning model used k-fold cross–validation to discriminate malicious APK from the normal files.

5.7.1 Analysis Tools

We use some tools available on emulator that are as follows:

- **Logcat**: Tool to access Android system log. It provides a mechanism for collecting and viewing system debug output.
- **Tcpdump**: Command line tool to intercept the packets being received and transmitted over the network.
- **Monkey**: Command line tool available on emulator/device to simulate prespecified number and type of user events.
- **Strace**: Command line tool available on Android emulator/device for system call tracing.
- **Dumpsys**: It dumps the state of the various components of Android emulator. It is also available as a command-line tool. It is capable of dumping the emulator state and its snapshots.

5.8 System Call Monitoring

The existing approaches Droidbox, TaintDroid or the analysis systems using them only analyze the Dalvik bytecode. However, Android apps also allow the usage of native code getting executed on the processor. Thus, the native code tracing is not possible. To track the native code we have included the modified strace tool for Android to intercept the system calls of the app being monitored. Once the APK is launched in the emulator, strace instance is initiated and gets attached the Dalvik process of the executing APK file. Then, the native calls are recorded for further analysis. Analysis of System calls helps to understand file operations and network related app actions. The framework fetches the files written/modified during the execution and scan for any traces of user data. During the analysis on a number of Android apps (benign and malicious). Specific system calls such as `bind` and `connect` have been visible among the malware APK. Hence, the system call analysis tags an app as Risky once it identifies the usage of such system calls.

6 Evaluation

The motive of DroidAnalyst is to provide the researchers an automated app analysis framework with an APK report including the results of the static and dynamic analysis. It is further used to interpret the Android app as a normal or malicious, rather than the automatic identification by the system. We evaluate DroidAnalyst to verify the web-based system provides enough information to an analyst. We have analyzed the following aspects during evaluation: (i) app stimulation, (ii) system correctness, (iii) performance evaluation, (iv) detect environment aware apps, (v) detectability, and (vi) scalability analysis.

DroidAnalyst has been released for researchers. The static analysis approach takes average 2 min to analyze and generate the response. The dynamic module takes an average 7–12 minute execution time. The maximum time limit is 15 min within the Sandbox. The emulator takes about 50 seconds to reboot the clean state, load stimulation techniques, and anti *anti-analysis* features to handle the subsequent execution. The Android Monkey spends an average 5 min for app interaction by inserting the gestures and stimulation techniques. The post processing techniques need about 2 min to extract the features from execution logs (i.e., Network traffic, System calls, UI gestures and stimulation methods, files created, written). The extracted information is submitted to the machine-learning algorithms for post-processing analysis techniques. The DroidAnalyst is capable of running eight emulation instances in parallel; However, it can be scaled further according to the analysis requirements.

6.1 App Stimulation Effects

The core functionality of DroidAnalyst is to generate a complete APK interaction report rather than detection. Hence, the system needs to stimulate the app reveal as much as possible. Hence, to test the stimulation engine effect, we chose 172 Google Play store and 165 malicious apps from different families. Furthermore, we applied techniques to initiate UI component, i.e. the activity component for interaction. The app not responding to the system were separated.

At the first stage individual stimulation techniques were applied (invoke only; main activity, monkey tool, app stimulation, and DDI hooking). The complete picture is illustrated in the Fig. 10. The first bar illustrates the initiating the main activity. The second one is a combination of monkey gestures and main activity. The third bar displays response to differently implemented stimulation techniques. The bar labeled DroidAnalyst illustrates the results of multiple stimulation techniques put

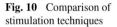

Fig. 10 Comparison of stimulation techniques

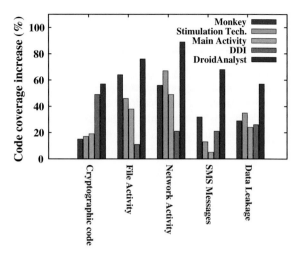

together. The results shows the effect of combined stimulus resulting in high code coverage. For example, random clicks generated by the Android Monkey triggers SMS sending activities. However, services are triggered with the service iterators. It is also interesting to note that combination of multiple techniques has high coverage as compared to one single method. Hence, the integration of the synergy of static and dynamic analysis is justified.

6.2 System Correctness

The system correctness methodology ensures that the logs of an APK is generated only when such action deemed to have been performed. We chose random 10 samples from known families performing varied malicious activities. Anserverbot malware checks for `IMEI` and `Model.build` to verify the presence of analysis environment. If the app detects emulator based on static properties, it hides the malicious behavior. If the malicious APK identifies random numeric values, it sends the IMEI to a remote server. `FAkeInstaller` malware sends premium rate `SMS` messages without user consent. `RootSmart` malware employs root exploits and executes native calls to exploit the device. `FakeInst` and `AdSMS` sends premium `SMS` messages apart from sending the `IMEI` and `IMSI` numbers to the remote server. The `TapSnake` malware family leaks user location to the remote server for targeted advertisements. Dendroid malware clicks pictures, records audio and sends call history to the remote server.

The above information is verified and reported by the leading commercial anti-malware and researchers. Hence, we evaluate the effectiveness of DroidAnalyst against the known malware. The evaluation reports provide additional information evading the commercial anti-malware. The reason for the improved effect is the combination of static and dynamic analysis methods.

6.3 Environment Reactive Malware Detection

Android malware is following the trends of detecting analysis systems; i.e. it identifies the emulated/analysis environment and behaves normally to evade the analyzer. Various static and dynamic techniques exist to detect the emulation. Variants of FakeNetFlix, Anserver, and Dendroid are equipped with static *anti-analysis* techniques to evade the analysis systems. To the best of our knowledge, DroidAnalyst is the first system to have integrated a range of static, anti *anti-analysis* techniques to coax the advanced and targeted Android malware reveal hidden behavior.

6.4 Scalability

The third parameter tests the scalability. Can DroidAnalyst handle the large-scale analysis response to thousands of hundreds of submissions. Detection of scalability is evaluated against data crawled between February 2013 and December 2014. A total 47,342 apps were crawled from Google Play, Anzhi, and other third party Asian markets; 26,469 malicious are downloaded from virus share, contagiominidump, and other third party markets. We also received 312 new malware samples based on user uploads. Out of the total, we randomly selected 6,743 benign and 2,786 from the malware dataset.

We performed the analysis for 16 days on intel core i7 processor, with 8 GB memory. Out of the total benign labeled apps, the proposed multiple methods approach tagged 217 apps as malicious. These apps were already labeled benign by the commercial anti-malware. The benign data labeled normal with anti-malware is classified malicious with DroidAnalyst. The majority of the samples belonged to FakeInstaller, Anserver, Kmin and FakePlayer families that either send premium SMS messages or steals the user information and send them to the remote servers.

6.5 Limitations

DroidAnalyst is not a panacea for APK analysis and has a share of limitations. When virtual machines are executed on the x86 architecture, fingerprinting methods to detect execution environment do exist. DroidAnalyst has successfully integrated the static, anti *anti-analysis* techniques to coax the advanced malware, but new methods to detect analysis environment become available. Further, the absence of device sensors can be used to detect the emulator. Google Bouncer has limited the instances of malware outbreak from Google Play. Hence, it is a challenge for malware authors to adopt smart techniques to evade the analysis environment.

7 Conclusions and Future Work

In this chapter, we discuss the intricacies of a user-driven, fully automated Android app assessment framework: DroidAnalyst. Proposed research framework combines the synergy of static and dynamic analysis techniques to increase the code coverage for analyzing emerging malware threats. While illustrating the tool, we have also elaborated the techniques embedded in the DroidAnalyst framework. A user scans the APK file to receive a detailed analysis report consisting meta-data information, permission modeling, bytecode analysis. Furthermore, DroidAnalyst generates app execution information, component vulnerability analysis, and signature-based malware variant detection method. DroidAnalyst explores multiple detection techniques like robust statistical signatures, identification of overprivileged apps, detection of vulnerable components, detection of covert feature misuse within the static analysis module. Furthermore, the dynamic analysis module counters the obfuscated and encrypted payloads, performs taint analysis to identify sensitive data leakage and cryptographic operations used by the advanced malware threats. DroidAnalyst can be deployed as an App Store verification and vetting tool to protect Android marketplaces. DroidAnalyst is a modular research framework and currently under active development for feature expansion.

In future, we would like to develop an interface through which an Android APK shall be able to upload the samples from target Android device. In future, we plan to simulate additional sensor features and strengthen the emulator based analysis including intelligent and efficient custom classifiers that can perform in near real time within the mobile platform resource constraints.

Acknowledgments The work of Parvez Faruki, Manoj Singh Gaur, and Vijay Laxmi was partially supported by Department of Information Technology, Government of India project grant "SAFAL-Security Analysis Framework for Android pLatform vide grant number: 12(7)/2014-ESD". Mauro Conti was supported by the Marie Curie Fellowship PCIG11-GA-2012-321980, funded by the European Commission for the PRISM-CODE project. This work has been partially supported by the TENACE PRIN Project 20103P34XC funded by the Italian MIUR, and by the Project Tackling Mobile Malware with Innovative Machine Learning Techniques funded by the University of Padua. We thank Ammar Bharmal and Vijay Kumar for their contributions in the development of "www.droidanalyst.org" as M.Tech scholars at the Computer science and engineering department, MNIT Jaipur. We appreciate the efforts made by Rohit Gupta, Jitendra Saraswat, and Lovely Sinha for their contributions in maintaining the DroidAnalyst.

References

1. Android tools: ADB, emulator, AVD manager, android, mksdcard, monkey, logcat. http://developer.android.com/tools/help
2. Andrubis: a tool for analyzing unknown android applications. http://anubis.iseclab.org/ (2014). Accessed July 2014
3. F-secure malware threat report 2012 q4. http://www.f-secure.com/static/doc/labs_global/Research/Mobile%20Threat%20Report%20Q4%202012.pdf (2014). Accessed November 2014

4. F-secure malware threat report 2013 q3. http://www.f-secure.com/static/doc/labs_global/Research/Mobile_Threat_Report_Q3_2013.pdf (2014). Accessed July 2014
5. F-secure malware threat report 2014 q1. http://www.f-secure.com/static/doc/labs_global/Research/Mobile_Threat_Report_Q1_2014_print.pdf (2014). Accessed June 2014
6. First Sms Trojan for Android. http://www.securelist.com/en/blog/2254/First_SMS_Trojan_for_Android (2014). Accessed 2013
7. Minimum redundancy feature selection-wiki. https://en.wikipedia.org/wiki/Minimum_redundancy_feature_selection (2014). Accessed August 2014
8. Tcpdump/libcap public repository. http://www.tcpdump.org/ (2014). Accessed July 2014
9. Android Malware Genome Project. http://www.malgenomeproject.org/ (2014). Accessed 11 February 2014
10. VirusTotal. https://www.virustotal.com/ (2014). Accessed 11 February 2014
11. APKTool. Reverse Engineering with ApkTool. https://code.google.com/android/apk-tool (2012. Accessed 20 March 2012
12. BakSmali. Reverse Engineering with Smali/Baksmali. https://code.google.com/smali (2014). Accessed 20 March 2014
13. BlackHat. Reverse Engineering with Androguard. https://code.google.com/androguard (2013). Accessed 29 March 2013
14. Bläsing, T., Batyuk, L., Schmidt, A.-D., Çamtepe, S.A., Albayrak, S.: An android application sandbox system for suspicious software detection. In: MALWARE, pp. 55–62 (2010)
15. Breiman, L.: Random forests. Mach. Learn. **45**(1), 5–32 (2001)
16. Burguera, I., Zurutuza, U., Nadjm-Tehrani, S.: Crowdroid: behavior-based malware detection system for android. In: Proceedings of the 1st ACM Workshop on Security and Privacy in Smartphones and Mobile Devices, SPSM'11, pp. 15–26, New York. ACM (2011)
17. Conti, M., Dragoni, N., Gottardo, S.: Mithys: mind the hand you shake - protecting mobile devices from SSL usage vulnerabilities. CoRR, abs/1306.6729 (2013)
18. Cousot, P., Cousot, R.: Abstract interpretation: a unified lattice model for static analysis of programs by construction or approximation of fixpoints. In: Proceedings of the 4th ACM SIGACT-SIGPLAN Symposium on Principles of Programming Languages, pp. 238–252. ACM (1977)
19. Desnos, A., Lantz, P.: Droidbox: an android application sandbox for dynamic analysis (2011)
20. Faruki, P., Bharmal, A., Laxmi, V., Ganmoor, V., Gaur, M., Conti, M., Rajarajan, M.: Android security: a survey of issues, malware penetration, and defenses. Commun. Surv. Tutor. IEEE **17**(2), 998–1022, Secondquarter (2015)
21. Faruki, P., Bharmal, A., Laxmi, V., Gaur, M.S., Conti, M., Rajarajan, M.: Evaluation of android anti-malware techniques against dalvik bytecode obfuscation. In: 13th IEEE International Conference on Trust, Security and Privacy in Computing and Communications, TrustCom 2014, Beijing, China, 24–26 September 2014, pp. 414–421 (2014)
22. Faruki, P., Ganmoor, V., Laxmi, V., Gaur, M.S., Bharmal, A.: Androsimilar: robust statistical feature signature for android malware detection. In: Proceedings of the 6th International Conference on Security of Information and Networks, SIN'13, pp. 152–159, New York. ACM (2013)
23. Faruki, P., Ganmoor, V., Vijay, L., Gaur, M., Conti, M.: Android platform invariant sandbox for analyzing malware and resource hogger apps. In: Proceedings of the 10th IEEE International Conference on Security and Privacy in Communication Networks (SecureComm 2014), Beijing China, 26–28 September 2014. Securecomm (2014)
24. Faruki, P., Laxmi, V., Bharmal, A., Gaur, M., Ganmoor, V.: Androsimilar: robust signature for detecting variants of android malware. J. Inf. Secur. Appl. (2014)
25. Felt, A.P., Chin, E., Hanna, S., Song, D., Wagner, D.: Android permissions demystified. In : Proceedings of the 18th ACM Conference on Computer and Communications Security, CCS'11, pp. 627–638, New York. ACM (2011)
26. Fernandes, E., Crispo, B., Conti, M.: FM 99.9, radio virus: exploiting FM radio broadcasts for malware deployment. IEEE Trans. Inf. Forensics Secur. **8**(6), 1027–1037 (2013)
27. Fritz, C., Arzt, S., Rasthofer, S., Bodden, E., Bartel, A., Klein, J., le Traon, Y., Octeau, D., McDaniel, P.: Highly precise taint analysis for android applications. Technical Report EC SPRIDE, TU Darmstadt (2013)

28. Grace, M.C., Zhou, W., Jiang, X., Sadeghi, A.-R.: Unsafe exposure analysis of mobile in-app advertisements. In: Proceedings of the Fifth ACM Conference on Security and Privacy in Wireless and Mobile Networks, WISEC'12, pp. 101–112, New York. ACM (2012)
29. Grace, M.C., Zhou, Y., Wang, Z., Jiang, X.: Systematic detection of capability leaks in stock android smartphones. In: NDSS. The Internet Society (2012)
30. Kelley, P.G., Consolvo, S., Cranor, L.F., Jung, J., Sadeh, N.M., Wetherall, D.: A conundrum of permissions: installing applications on an android smartphone. In: Blythe, J., Dietrich, S., Camp, L.J. (eds.) Financial Cryptography Workshops, Lecture Notes in Computer Science, vol. 7398, pp. 68–79. Springer (2012)
31. Kildall, G.A.: A unified approach to global program optimization. In: Proceedings of the 1st annual ACM SIGACT-SIGPLAN symposium on Principles of programming languages, pp. 194–206. ACM (1973)
32. Kim, J., Yoon, Y., Yi, K., Shin, J., Center, S.: ScanDal: static analyzer for detecting privacy leaks in android applications. In: Proceedings of the Workshop on Mobile Security Technologies (MoST), in Conjunction with the IEEE Symposium on Security and Privacy (2012)
33. Lindorfer, M.: Andrubis: a tool for analyzing unknown android applications. http://blog.iseclab.org/2012/06/04/andrubis-a-tool-for-analyzing-unknown-android-applications-2/ (2012)
34. Lindorfer, M., Neugschwandtner, M., Weichselbaum, L., Fratantonio, Y., van der Veen, V., Platzer, C.: Andrubis—1,000,000 apps later: a view on current android malware behaviors. In: Proceedings of the the 3rd International Workshop on Building Analysis Datasets and Gathering Experience Returns for Security (BADGERS) (2014)
35. MSWG. Department of computer science and engineering, malaviya national institute of technology, Jaipur. https://www.droidanalyst.org (2014). Accessed July 2014
36. Mulliner, C.: Dalvik dynamic instrumentation. http://www.mulliner.org/android/feed/mulliner_dbi_hitb_kul2013.pdf (2013). Accessed October 2013
37. Neuner, S., Van der Veen, V., Lindorfer, M., Huber, M., Merzdovnik, G., Mulazzani, M., Weippl, E.: Enter sandbox: android sandbox comparison. In: Proceedings of the IEEE Mobile Security Technologies Workshop (MoST), vol. 5. IEEE (2014)
38. Peng, H., Long, F., Ding, C.: Feature selection based on mutual information criteria of max-dependency, max-relevance, and min-redundancy. IEEE Trans. Pattern Anal. Mach. Intell. 27(8), 1226–1238 (2005)
39. Petsas, T., Voyatzis, G., Athanasopoulos, E., Polychronakis, M., Ioannidis, S.: Rage against the virtual machine: hindering dynamic analysis of android malware. In: Proceedings of the Seventh European Workshop on System Security, p. 5. ACM (2014)
40. Play, G.: Official Android Market. https://market.android.com/ (2013). Accessed 17 June 2013
41. Rasthofer, S., Arzt, S., Miltenberger, M., Bodden, E.: Harvesting runtime data in android applications for identifying malware and enhancing code analysis. Technical Report TUD-CS-2015-0031, EC SPRIDE, February (2015)
42. Reina, A., Fattori, A., Cavallaro, L.: A system call-centric analysis and stimulation technique to automatically reconstruct android malware behaviors. In: Proceedings of the 6th European Workshop on System Security (EUROSEC 2013), Prague, Czech Republic (2013)
43. Rocha, B.P.S., Conti, M., Etalle, S., Crispo, B.: Hybrid static-runtime information flow and declassification enforcement. IEEE Trans. Inf. Forensics Secur. 99(8) (2013)
44. Roussev, V.: Building a better similarity trap with statistically improbable features. In: 42nd Hawaii International Conference on System Sciences, 2009. HICSS'09, pp. 1–10. IEEE (2009)
45. Roussev, V.: An evaluation of forensic similarity hashes. Dig. Investig. 8, S34–S41 (2011). Aug
46. Roussev, V.: Data fingerprinting with similarity hashes. Adv. Dig. Forensics (2011)
47. Sanz, B., Santos, I., Laorden, C., Ugarte-Pedrero, X., Bringas, P.G., Álvarez, G.: Puma: permission usage to detect malware in android. In: International Joint Conference CISIS12-ICEUTE'12-SOCO'12 Special Sessions, pp. 289–298. Springer (2013)
48. Spreitzenbarth, M., Freiling, F.: Android Malware on the Rise. Technical report (2012)

49. Spreitzenbarth, M., Freiling, F., Echtler, F., Schreck, T., Hoffmann, J.: Mobile-sandbox: having a deeper look into android applications. In: Proceedings of the 28th Annual ACM Symposium on Applied Computing, SAC'13, pp. 1808–1815, New York. ACM (2013)
50. Vidas, T., Christin, N.: Evading android runtime analysis via sandbox detection. In: Proceedings of the 9th ACM Symposium on Information, Computer and Communications Security, ASIA CCS'14, pp. 447–458, New York. ACM (2014)
51. Weichselbaum, L., Neugschwandtner, M., Lindorfer, M., Fratantonio, Y., van der Veen, V., Platzer, C.: Andrubis: Android Malware Under The Magnifying Glass. Technical Report TR-ISECLAB-0414-001, Vienna University of Technology (2014)
52. William, E., Peter, G., Byunggon, C., Landon, C.: TaintDroid: an information flow tracking system for realtime privacy monitoring on smartphones. In: USENIX Symposium on Operating Systems Design and Implementation, USENIX (2011)
53. Yan, L.K., Yin, H.: Droidscope: seamlessly reconstructing the OS and dalvik semantic views for dynamic android malware analysis. In: Proceedings of the 21st USENIX Security Symposium (2012)
54. Zheng, C., Zhu, S., Dai, S., Gu, G., Gong, X., Han, X., Zou, W.: Smartdroid: an automatic system for revealing UI-based trigger conditions in android applications. In: Proceedings of the Second ACM Workshop on Security and Privacy in Smartphones and Mobile Devices, SPSM'12, pp. 93–104, New York. ACM (2012)
55. Zheng, M., Lee, P.P.C., Lui, J.C.S.: ADAM: an automatic and extensible platform to stress test android anti-virus systems. In: DIMVA, pp. 82–101 (2012)
56. Zhou, Y., Jiang, X.: Dissecting android malware: characterization and evolution. In: Proceedings of the 33rd IEEE Symposium on Security and Privacy, Oakland 2012. IEEE (2012)
57. Zhou, Y., Wang, Z., Zhou, W., Jiang, X.: Hey, you, get off of my market: Detecting malicious apps in official and alternative android markets. In: NDSS. The Internet Society (2012)

Part V
Strategic/Mission Planning and Resource Management

Design and Development of Intelligent Military Training Systems and Wargames

D. Vijay Rao

Abstract Today's military teams are required to operate in environments that are
increasingly complex. Such settings are characterized by the presence of
ill-structured problems, uncertain dynamics, shifting and ill-defined or competing
goals, action/feedback loops, time constraints, high stakes, multiple players and
roles, and organizational goals and norms. Warfare scenarios are real world systems
that typically exhibit such characteristics and are classified as Complex Adaptive
Systems. To remain effective in such demanding environments, defence teams must
undergo training that targets a range of knowledge, skills and abilities. Thus
oftentimes, as the complexity of the transfer domain increases, so, too, should the
complexity of the training intervention. The design and development of such
complex, large scale training simulator systems demands a formal architecture and
development of a military simulation framework that is often based upon the needs,
goals of training. In order to design and develop intelligent military training systems
of this scale and fidelity to match the real world operations, and be considered as a
worthwhile alternative for replacement of field exercises, appropriate Computational Intelligence (CI) paradigms are the only means of development. A common
strategy for tackling this goal is incorporating CI techniques into the larger training
initiatives and designing intelligent military training systems and wargames. In this
chapter, we describe an architectural approach for designing composable,
multi-service and joint wargames that can meet the requirements of several military
establishments using product-line architectures. This architecture is realized by the
design and development of *common* components that are reused across applications
and *variable* components that are customizable to different training establishments'
training simulators. Some of the important CI techniques that are used to design
these wargame components are explained swith suitable examples, followed by
their applications to two specific cases of Joint Warfare Simulation System and an
Integrated Air Defence Simulation System for air-land battles is explained.

D. Vijay Rao (✉)
Institute for Systems Studies and Analyses, Defence Research and Development
Organisation, Metcalfe House, Delhi 110054, India
e-mail: doctor.rao.cs@gmail.com

© Springer International Publishing Switzerland 2016
R. Abielmona et al. (eds.), *Recent Advances in Computational Intelligence
in Defense and Security*, Studies in Computational Intelligence 621,
DOI 10.1007/978-3-319-26450-9_21

555

Keywords Military systems analysis · Modeling and simulation · Constructive simulations · Computational intelligence techniques · Scenario-based exploratory analysis

1   Introduction

Warfare is changing, perhaps more rapidly and fundamentally today, than at any point in history. The emergence of new operational drivers such as asymmetric threats, urban operations, joint and coalition operations and the widespread use of military communications and information technology networks have highlighted the importance of providing warfighters with competencies required to act in a coordinated, adaptable manner, and to make effective decisions in environments characterized by large amounts of sometimes ambiguous information. Warfare systems are characterized by the presence of ill-structured problems, uncertain dynamics, shifting and ill-defined or competing goals, action/feedback loops, time constraints, high stakes, multiple players and roles, and organizational goals and norms. Such systems are typically classified as Complex Adaptive Systems. While the beginnings of understanding warfare as a complex adaptive system dates more than 2500 years to the writings of Sun Tzu, recently, a growing body of literature describes the broader aspects of defence systems and operations as a complex systems science. Complexity results from the inter-relationships, inter-actions and inter-connectivity of elements within a system and between a system and its environment. They are dynamic systems that are able to *adapt in* and *evolve with* a changing environment. Sir Smith's thesis [1] that the world entered a new paradigm of conflict at the end of the 20th and beginning of the 21st centuries, which he calls "war amongst the people", and that Western, industrialized armies are ill-suited to the new style of warfare, is noteworthy.

With rapid advances in technology and increasingly complex defense systems in operation, substantial effort and resources are spent on training for their effective usage. To improve the efficiency, effectiveness, usage and safety of training, organizations and user agencies are investing heavily into developing computer-based training simulators. While investment in new technologies can make available new opportunities for action, it is only through effective training that personnel can be made ready to apply their tools in the most decisive and discriminating fashion. The infeasibility of replicating the environment under which such systems are deployed and operated, coupled with resource constraints and environmental hazards are forcing military organizations worldwide to invest heavily on computer based systems for their training needs. A recent empirical study on the impact of computer based training (CBT) on maintenance costs and actions in a sonar system operations found that CBTs use has adversely influenced parts costs, actions, and labor costs associated with operating and maintenance of the AN/SQQ-89(v) sonar system and has negatively impacted sailor performance on ships. It also suggested that explicit costs are traded for an obscured cost in terms of parts, maintenance actions, labor-hours readiness [2]. Such studies along with the lessons learnt during several training sessions of military training simulators and wargames emphasize that Modeling and Simulation (M&S) is one of the critical success factors that should be mainstay in designing and developing training simulators if they are to be effective. In order to design and develop intelligent military training systems of industry strength scale with fidelity to match the real world operations, and to be considered as a worthwhile alternative for replacement of field exercises, appropriate computational intelligence (CI) paradigms are the only means of development. We describe an architectural approach for designing composable multi-services and joint wargames, and explain the applicability of the various CI techniques in every aspect of the architectural design. These are explained with suitable examples along with potential applications for military systems design are discussed throughout the chapter.

2 Modeling, Simulation and Military Systems Analysis

Operations Research (OR) and Systems Analysis (SA) are the two related methods of logically attempting to solve complex problems having a quantitative analytical component [3, 4]. Figure 1 identifies the various ways of studying a real-world system. Warfare has many facets of study and analysis and generally classified as complex, adaptive systems that are nonlinear, dynamic, and show emergence behaviors. Thus in order to obtain closed form solutions for such systems in their entirety is elusive. Modeling and Simulation (M&S) is thus becoming the main approach used by defense organizations to study warfare systems for analysis and training, model complex military operations, design training and analysis systems of existing and proposed defense systems. (*We shall use the term wargames to also mean the military training systems that are specific instances of the generic class of wargames that are implemented for training purposes while the former also encompasses analytical and research wargames*).

One of the major challenges of military systems analysis is to identify the models that are suitable to the problems at every level of the pyramid. The choice primarily depends upon the purpose, resolution, and objectives of the study and can be classified as strategic level, tactical level and operational level games. All the approaches to force-on-force analysis are underpinned by theories of combat. Combat is an exceedingly complex and simultaneous interaction of several factors that are typically classified under complex adaptive systems. Force-on-Force campaign analysis that use combat models are not intended to predict accurately who is likely to win or lose and an engagement, a battle, a campaign, or a war but to predict whether one system, tactic force structure, or course of action is likely to perform *roughly* better or worse than another. As shown in Fig. 1, there is a continuum of techniques and applications from operations research to force-on-force campaigns and net assessment that spans the complete range of defence decision-making problems. Broadly

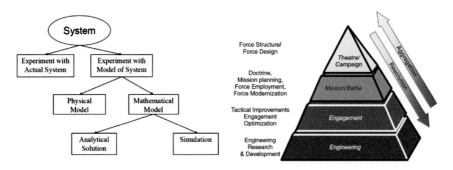

Fig. 1 Ways to study a system [4] and military systems analyses hierarchy [5]

speaking, modeling and simulation may be applied at the tactical, operational or strategic levels to meet the functional requirements of training, operational planning, force structuring to include force development, and, strategy formulation. Simulation is also the only way to test and train for some special environments, such as nuclear events, biological and chemical contamination, and operations that require large-scale mobilization and movement. Analytical simulations are used to study problems like force composition, weapons effectiveness, and logistics issues. Examples to illustrate are:

- What would be the survivability and cost-effectiveness of an unmanned combat system as compared to a manned combat mission?
- Which is the best weapon mix, force mix of aircraft-weapons types and configurations that can achieve the maximal damage to given target with a minimal acceptable loss of own resources?
- Given a set of military resources, what is the optimal deployment of these resources against a given threat scenario.
- In what specific scenarios are joint services operations synergistic? What are the core and critical factors that ensure synergy in jointness?

2.1 Modeling and Simulation Techniques for Training and Analyses

Modern methods of training are being introduced with enhanced use of modeling and simulation [6]. Modeling and simulation refers to the use of models, including emulators, prototypes, simulators, and stimulators, either statically or over time, to develop data as a basis for making managerial or technical decisions and training. The terms "modeling" and "simulation" are often used interchangeably. Simulation is the imitation of the operation of a real-world process or system over time. Simulation not only helps them in learning the given scenarios, but teaching themselves replicate real-life experiences to relive and recreate what they have seen, on their own missions. Basic applications that evolve from the core Modeling and simulation domains of *Engineering*, *Training* and *Analysis* areas into Simulators, Wargames and Performance Evaluation systems are shown in Fig. 2. Depending upon the goals of the study, analytical techniques process simulations, trace driven simulations, or discrete event simulation systems are employed in each of these three areas [7].

Simulation is the imitation of the operation of a real-world process or system over time. A simulation of a system is the operation of a model of the system. The model can be reconfigured and experimented with; usually, this is impossible, too expensive or impractical to do in the system it represents. The operation of the model can be studied, and hence, properties concerning the behavior of the actual system or its subsystem can be inferred. Simulation can be used before an existing system is altered or a new system built, to reduce the chances of failure to meet specifications, to eliminate unforeseen bottlenecks, to prevent under or over-utilization of resources, and to optimize system performance. To simulate is to mimic a real system so that we can explore it, perform experiments on it, and understand it before implementing it in the real world. This becomes extremely important, especially when the real system cannot be engaged, because it may not be accessible, or it may be dangerous or unacceptable to engage, or it is being designed but not yet built, or it may simply not exist [4].

When we simulate, we are first required to develop a *mathematical model* of the original entity (weapon, equipment or process) wherein, the model so developed represents the key characteristics or behaviors of the selected physical or abstract system or process. The model represents the system itself, whereas the simulation represents the repetitive operation of the processes of the system, over a period of time. This could be to simulate the behavior of a weapon/equipment or a group of entities (platoon/company/combat team) in a particular scenario. Military simulations are seen as a useful way to develop tactical, strategic and doctrinal solutions. The term military simulation can cover a wide spectrum of activities, ranging from full-scale field-exercises, to abstract computerized models that can proceed with little or no human involvement as shown in Fig. 2. The simulations have been universally identified to be of three types—*live*, *virtual* and *constructive* [3].

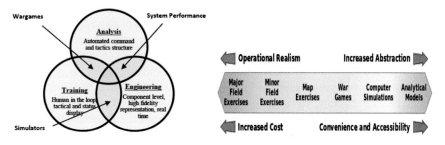

Fig. 2 Application areas of military modeling and simulation: performance evaluations, simulators, and wargames

Live simulation refers to a simulation that involves real people in real systems. For example, two pilots can be trained for dog fighting by using real aircrafts in the air. In this case, the aircrafts and pilots are real, but the interaction between the aircrafts are simulated and simulation decides how effective the pilots and aircrafts are against each other. Similarly, all the weapon systems can be equipped with emitters, and all the equipment and personnel can be equipped with sensors. If the weapons are aimed and fired correctly, the emission by the emitter can be sensed by the sensors, which indicates a hit and a kill.

Virtual simulation involve humans and/or equipment where actual players use simulation systems in a computer generated synthetic or virtual environment. The running time can be real or in discrete steps, allowing users to concentrate on the key training objective. These represent a specific category of devices that utilize simulation equipment (which exactly replicates the controls of the original equipment) to create a simulated world for the user. In this manner, the system can accept input from the user (e.g., body tracking, voice/sound recognition, physical controllers) and produce output to the user (e.g., visual display, aural display).

Constructive Simulation, also known as *wargaming*, derives its name from the fact that the pieces operating on the battlefield are not individual tanks and aircraft but a construction of many different types of equipment into a single aggregated unit like an armor company, artillery battery.

Wargames are physical or electronic simulation of military operations designed to explore the effects of warfare or testing strategies or an operational concept without actual combat. Wargame is the employment of military resources in training for military operations, either exploring the effects of warfare or testing strategies without actual combat. It is the most cost-effective methodology for training as it creates a realistic environment to generate near-real responses to various contingences as well as handling of complex weapon systems. The main advantage of using wargames is to enable the users to take another look at specific events from a stress free environment and enhance their performance for the given event.

The first two types of simulations are used to train individuals operating equipment, this equipment is in turn controlled by leaders in command posts who see the battle in a more abstract form. Constructive simulations allow these commanders to face situations and make decisions under the stress of time and limited resources just as they will during actual combat. Constructive simulations immerse these commanders in a situation where the enemy is highly trained, experienced, and just as determined to win the war. Here soldiers discover whether the tactics they have been taught really work, here they develop confidence in their ability to operate as a team and win wars. These simulations have emerged as one of the powerful tools of system analysis in military applications. They have been used extensively for training, planning, analysis and decision support purposes. A wide range of wargames has been developed at various resolution levels to support different objectives. Training wargames allows analyzing various aspects of tactics at lower levels. Higher level wargames can be used for evaluation of various employment/deployment plans, different course of action and also evaluation of weapon systems [7, 8].

The new reality of military operations is characterized by complex interactions, adaptivity and nonlinearity, with an increase of uncertainty and risk, explicitly or implicitly, in all dimensions of warfare [9–13]. Uncertainty is the inability to determine a variable value or system state (or nature) or to predict its future evolution. Uncertainty is a fact (that is certain): real-world data will be uncertain, incomplete, ambiguous, contradictory, and vague, and this uncertainty can never be reduced completely; but can only be managed. While uncertainty can only be managed, the real objective of studies is reducing risk in taking informed decisions. Computational Intelligence (CI) paradigms encompass a collection of heuristic techniques to imitate or represent aspects of cognitive and biological processes in nature, which have been successfully used to model and manage these inherent uncertainties in the design and development of wargames.

We propose a number of CI techniques to design intelligent military training simulators and wargames. In contrast to the organisation-specific, training-specific monolithic system development, we propose a product-line, layered approach to design large-scale intelligent wargames that can be easily customized to specific requirements of organisations. Such an architectural framework based approach has its basis in software reuse and component based system development. The various components common to a family of wargame solutions and the variable components that are customised to meet a specific end product is described. All these CI techniques have been integrated in a Discrete Events Simulation Specification (DEVS) framework to design specific end-products to meet the training requirements of military schools [14–18]. These predominant CI techniques that have been successfully used to develop intelligent military training simulators are described in the sections that follow.

3   An Architectural Approach to Design and Development of Wargames

The design and development of large scale simulators, software testbeds and wargames, demands a formal architecture and development of a military simulation framework that is often based upon the needs, goals of training and resolution of the wargames. The Joint Warfare Simulation System (JWSS) is a constructive simulation based software testbed that is designed to cater for *Analysis*, *Training* and preliminary studies of *Engineering* design. The JWSS is designed based on the operational foundations of the military domain; conceptual foundation required for the modeling and theoretical foundations of implementing and composing simulation system. JWSS system design is highly influenced by (i) *scope, resolution of the entities involved,* (ii) *level of wargame, trainee audience and objectives* (iii) *number and types of entities being addressed/modeled* (iv) *resolution of the battlefield entities and fidelity of the combat entity models* (v) *number and types of players and their hierarchy configurations* (vi) *area of operations* (vii) *terrain and environmental features* (viii) *time advancement and resolution.* Conventional approaches to designing and developing Wargames are based on a monolithic homogeneous design and development as any software development system. This approach imposes difficulties in developing and maintaining these systems as they must keep evolving to be useful. In JWSS an Inner-Sourcing, hybrid approach to designing common components and an agent-oriented approach to designing wargame components is proposed. Organizations leveraging open-source development practices for their in-house software development is called Inner Source [19].

An Agent Based Modeling Approach to Wargame Development

Agent-oriented system development aims to simplify the construction of complex systems by introducing a natural abstraction layer on top of the object-oriented paradigm composed of autonomous interacting actors [20–22]. It has emerged as a powerful modeling technique that is more realistic for today's dynamic warfare scenarios than the traditional models which were deterministic, stochastic or based on differential equations. These approaches provide a very simple and intuitive framework for modeling warfare and are very limited when it comes to representing the complex interactions of real-world combat because of their high degree of aggregation, multi-resolution modeling and varying attrition rate factors. The effects of random individual agent behavior and of the resulting interactions of agents are phenomenon that traditional equation-based models simply cannot capture. Figure 3a, b shows the agent based architecture of a virtual warfare training simulator [22]. In agent-based modeling (ABM), a system is modeled as a collection of autonomous decision-making entities called agents. Each agent individually assesses its situation and makes decisions on the basis of a set of rules. Agents may execute various behaviors appropriate for the system they represent—for example, producing, consuming, or selling. Repetitive competitive interactions between agents are a feature of agent-based modeling, which relies on the power of computers to explore dynamics out of the reach of pure mathematical methods. At the simplest level, an agent-based model consists of a system of agents and the relationships between them. Even a simple agent-based model can exhibit complex behavior patterns and provide valuable information about the dynamics of the real-world system that it emulates. In addition, agents may be capable of evolving, allowing unanticipated behaviors to emerge. Sophisticated ABM sometimes incorporates neural networks, evolutionary algorithms, or other learning techniques to allow realistic learning and adaptation. The benefits of ABM over other modeling techniques can be captured in three statements: (*i*) ABM captures emergent phenomena; (*ii*) ABM provides a natural description of a system; and (*iii*) ABM is flexible. It is clear, however, that the ability of ABM to deal with emergent

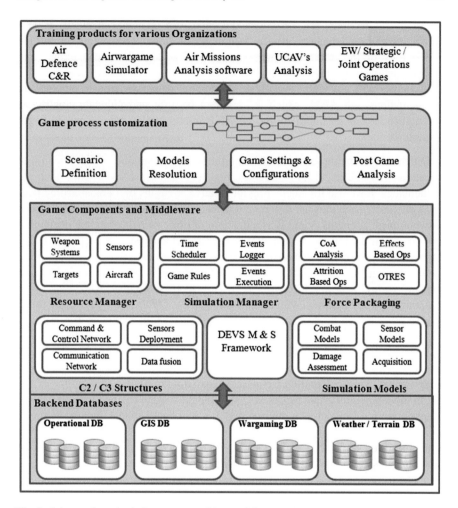

Fig. 3 Joint warfare simulation system architectural framework

phenomena is what drives the other benefits. ABM appear to represent complex, adaptive systems where, non-linearity, interactions and emergence are an inherent nature of systems, such as warfare systems. One may want to use ABM [23] when there is potential for studying emergent phenomena, i.e., when:

- Individual behavior is nonlinear and can be characterized by thresholds, if-then rules, or nonlinear coupling. Describing discontinuity in individual behavior is difficult with differential equations.
- Individual behavior exhibits memory, path-dependence, and hysteresis, non-Markovian behavior, or temporal correlations, including learning and adaptation.

- Agent interactions are heterogeneous and can generate network effects. Aggregate flow equations usually assume global homogeneous mixing, but the topology of the interaction network can lead to significant deviations from predicted aggregate behavior.
- Averages will not work. Aggregate differential equations tend to smooth out fluctuations, not ABM, which is important because under certain conditions, fluctuations can be amplified: the system is linearly stable but unstable to larger perturbations.

Differential equations are a fundamental modeling technique, which finds place in warfare modeling. Land wargames typically use Lanchester laws that are mathematical formulae for calculating the relative strengths of a predator/prey pair. The Lanchester equations are differential equations describing the time dependence of two armies' strengths A and B as a function of time, with the function depending only on A and B. During World War I, Frederick Lanchester devised a series of differential equations to demonstrate the power relationships between opposing forces. Among these are what is known as *Lanchester's Linear Law* (for ancient combat) and *Lanchester's Square Law* (for modern combat with long-range weapons such as firearms). Pursuit-Evasion games, Differential games, air to air combat models such as Adaptive Maneuvering Logic (AML) all have their basis as DE, and have been an active area of research in Warfare modeling [24, 25].

In many cases, such as wargames, ABM is most natural for describing and simulating a system composed of *behavioral* entities [26–28]. Each agent is implemented using different CI techniques depending upon its purpose. For example, in modeling air warfare tactics in JWSS, the pilot agent can be modeled using a simple behavioral model, a cognitive model, a rule-based model, control-theoretic model, or a neuro-fuzzy model [29–34]. Computer-generated forces and semi-automated forces have an important role to play in modeling counter insurgency operations, terrorist attacks, and operations other than war [35–37]. These are very efficiently modeled using ABMs. In order to design and develop training simulators for such operations, the opponents are modeled using agents governed by simple rules, and emergent phenomenon. Simulators are built for operations, tactics and strategies training using CGF and SAF. Epistemic states are often used to represent an actual or a possible cognitive state that drives the human-like behavior of an agent. Commonly used models are propositional, probabilistic and possibilistic world models, where methods of knowledge representation, reasoning and inferencing about the various mental constructs of the agent, including beliefs, desires, goals, intentions, and knowledge are used to simulate its human-like cognitive states. Validation of CGFs and SAFs is an area of concern; and drawing lessons from these simulations is difficult and caution needs to be exercised.

Main components of Joint Warfare Simulation System (JWSS) Architectural Framework are as follows:

Backend Databases

All the JWSS data is classified into static data (such as resources, weapons, their characteristics) and dynamic (run-time results that are generated by gaming the mission plans) data, stored in the data servers that form the backend for the entire game. The database is designed using MySql and SQLServers that are ADODB and ODBC compliant databases. Backend Databases are partitioned into four functional clusters to improve the performance and maintenance of the servers: *Operational Database, GIS Database, Wargaming Database and Weather/Terrain Database. Operational Database* contains data that changes frequently during the game and is dependent on the scenario being simulated. Data about missions, resources deployed, event logs generated during simulation, and results are stored in this database. *GIS Database* contains geographical data such as raster and vector maps of different themes, DTED and DEM data of maps of the Theatre and Area of Operation (AOP) during game customization and game initialization phase. An open source GIS has been customized for the military (Mil-GIS) to depict the various theatres of warfare in JWSS. *Wargaming Database* contains the data which is read only and can be changed only by the controller during the initialization phase. Performance parameters of an aircraft and their configurations, resource specifications, sensor details, target information, force structure, network and game settings all forms part of this database. *Weather/Terrain Database* contains data such as weather and terrain information in an enclosed region specified by the users. Location-based intelligent services (LBIS) are developed on these databases to generate the military intelligence, information, and data that is utilized by the players to decide the course of actions during the wargame exercises. These LBIS along with fuzzy linguistic variables to represent the uncertainty of information obtained from various intelligence sources are generated, collated and inferred to generate the Fog of War during the training exercises. For example, *Suitability of terrain conditions to troops and logistics movement along the selected route to meet the objective is low.*

Resource Manager

Resources of the game include the various types of entities involved in the game such as weapon systems, sensors, platforms such as aircraft, naval ships, air defence artillery, infantry, armour brigades, and the hierarchy of organization and their compositions, platform-weapon configurations, types of targets, expected damages to targets, weapon-target matching and their primary and secondary effects of damages. Weapon systems include all kinds of air-launched and surface-launched weapons like Bombs, Guns, Rockets and Missiles. All types of Aircraft such as *Fighters, Air-to-Air Re-fuellers, Transport, Unmanned combat systems, Unmanned aerial vehicles,* their operational performances, effective radii of action, and effectiveness against targets are stored in the resource databases. Air Defense (AD) units such as Radars, air defense Guns, Mobile Observation Posts, and Aerostats form a part of the Resource Folder. The *Target Folder* contains relevant information of all the static targets and RCS of various aircraft, EW emitter signatures obtained from ESM missions in the game. Targets include Bridges, Airfields, Refineries, Oil Depots and other such vulnerable areas and vulnerable points (VA/VP's). The information in the resource folder is updated as and when new resources are inducted into the services and may not change frequently during the wargame simulation. *Target folder* is updated whenever any new intelligence inputs about targets are made available. Creation and administration of the Resource Folder and Target Folder as per the game scenario is the primary task of the resource manager. *Searching*, *Matching* and *Retrieval* of "similar" images from the target folder to the acquired target from an unmanned system uses fuzzy image retrieval and fuzzy inferencing techniques [38].

Force Packaging

Before planning, study of the anticipated air and ground threats is essentially required to get the near real time information of the battlefield in time and space domain. Anticipated threats determine ingress/egress tactics and techniques to minimize risk, aircraft selection, weapon configuration, weapon delivery mode and other operational factors. Data from many sources such as HUMINT, COMINT and SIGINT enables the most effective use of available resources to destroy or neutralize adversary's assets.

Many critical decisions are taken in the force planning phase. How large a force package should be and which types of aircraft with what weapons configuration is best suited for the mission objective? Should there be a need for escorts in the force package? What likely threats can be encountered and how does one mitigate these risks? Which routes would optimize use of terrain masking? At what time target should be shot down? While planning planner must consider assets availability, range to target, C2/C3 connectivity, and tactics. For tactical routing, planner must focus on suppressing threats using SEAD/AD/ECM/Stand-off escorts and identifying/acquiring the target. In the target area, terrain masking and high speed are used to minimize the threat exposure to adversary's low level air defense. Optimal Deployment of resources against a perceived threat is obtained using Genetic algorithms, and Weapon Target matching used for force packaging the campaigns is done using Genetic algorithms.

C2/C3 Structures

A Command, Control and Communication (C3) structure is an information system employed within a military organization. Command is the functional exercise of authority, based upon knowledge, to attain an objective or goal. Control is the process of verifying and correcting activity such that the objective of command is accomplished. Communication is the ability and function of providing the necessary liaison to exercise effective command between tactical and strategic units of command. Thus the C3 structure can be succinctly defined as a knowledgeable exercise of authority in accomplishing military objectives and goals.

The C3 structure is implemented in JWSS with the Air Defense Direction Centre (ADDC) being responsible for providing air defense to assigned VA/VP using air defense systems against air threats. ADDC is also responsible for communicating the change in status of Control order and status of target. It passes these messages to the Base Air Defense Centre (in case of airbases) or to the Air Defense Command Post (in case of other VA/VP's). The ADDC could control one or more BADC. BADC on receipt of the message from the ADDC activates its own radar and transmits the message on the basis of its track information to engage the threat.

Simulation Manager

The Simulation Manager coordinates and controls the whole entire wargame. It starts the game by starting a simulation clock and initializing the game parameters, target folder, resource folder, war date and block. It allocates resources for the block of game, sets up weather conditions and monitors the game. The missions of various teams are processed by the simulation manager using various simulation models and the results are generated in a quantitative manner. The damage caused to the aircraft, airbase, and other VA/VP's is computed using weaponeering techniques based damage assessment models [6]. Mathematical models have been developed for computing weapon trajectories, wind effects, and atmospheric parameters. Statistical distributions are used to generate various events and for damage assessment. All these events and the effects caused are recorded by the Event Logger to be viewed and analyzed later. Extensive set of fuzzy game rules are used to simulate realistic war scenario such *as if takeoff runway is non-operational then abort the mission, otherwise allow aircraft to take-off, if aircraft in a mission is unserviceable then abort the mission, unserviceable factor of 3 % before take-off and 2 % after take-off is acceptable, if storm is present at takeoff base then abort the Mission, if landing runway is non-operational, then the landing aircraft is made available after the runway available time added to turn around service time (TRS) of the aircraft.*

Simulation Models

The core functionality of the JWSS framework lies in the various simulation models that are being used extensively by the simulation manager during the course of the game. Some of the classes of models are explained below:

Damage Assessment Models

The assessment of damage has its basis in the various mathematical models developed based on the weaponeering principles, force structure planning, weapon planning directive and the weapon-target matching documents. These models help in determining the optimal quantity of a specific type of weapon required to achieve a specific level of damage to ground targets considering weather, terrain, target vulnerability, weapon effects, munitions delivery errors, damage criteria, probability of kill, weapon detonation reliability, weapon release conditions and other operational factors. An Over the Target Requirement Estimation System (OTRES) tool has been developed based on these principles that estimate the damage caused and generates the courses of action for a planner [21]. The planner selects one course of action (mission plan) which is gamed against the threat (*perceived enemy threat* during mission assessment and *actual enemy threat* during gaming the missions) using the JWSS test-bed to assess the mission effectiveness. Target damage assessment using weaponeering principles gave realistic results when used in field training and deployment of the system. Computation of damages for ground-ground, ground-air, air-ground, and air-air engagements uses physics based, logic based, probabilistic and fuzzy logic based models [21, 24, 25, 39].

Sensor Models

Electronic warfare (EW) is the art and science of denying enemy force, the use of the electromagnetic spectrum while preserving its use for friendly forces. It involves radars, electronic sensors, jammers used in conjunction with traditional weapon systems as part of the warfare. The EW module in JWSS includes firstly modeling of the performance of radars in the presence of different weather and terrain conditions considering the Line of Sight between Radar and the target, secondly electronic countermeasures and electronic, counter- countermeasures and lastly modeling the effect of different types of jammer on the radar performance. The parameters considered in these models are radar characteristics, target characteristics, environment parameters, jammer parameters, threshold detection level.

Combat Models

Combat models are developed for mathematical analysis of attrition process of the forces in combat. Combat is an engagement or a series of engagements between two conflicting forces, which causes attrition. An engagement can be defined as a set of actions within a particular region, over a particular time period, and with a given force structure. In the context of land wargames, it is a complex system involving men, machines, materials, money, environment, terrain and their complicated interactions. Involvement of quantitative and qualitative factors gives rise to different degrees of complexities to the system. Training, battle fatigue, fear, morale, leadership are some qualitative factors which govern human behavior during the battle. Many of these attributes are intangible and we may not be able to give a specific number. Neuro-Fuzzy linguistic variables are used to model these qualitative factors that play an important role in wargames. In the case of naval platforms, trajectories of the weapons launched and the incidence of impact, the warhead tonnage of explosive, and physics based impact dynamics decide the extent of damage caused to the platforms. In air combat scenarios, a weighting factor that is derived from the static combat potential of the packages of both sides is used to assess the aircraft attrition, Adaptive Maneuvering Logic (AML) for one-one, pursuit evasion modes [24] is modeled or a more detailed model that considers the aerodynamics, aircraft weapon load, EW, RWR, MWR, frequencies of operation, on board missiles, and their ranges to compute the dynamic combat potential for gaming and computing the attritions. These entire processes can be simplified by deriving probabilistic game rules that can also be used for quick statistical analysis of the air campaign. The inherent complexity of the combat process leads to great complexity in the operational models of *combat attrition*, and *combat effects*.

Target Acquisition Models

In cases of bad weather, terrain and other environmental conditions, and height of the air attacks, the target may not be acquired by the on-board sensors, and weapons are not released. Mission planners need an estimate of the probability the target would be acquired in order to get a better measure of mission success. As for many air to ground missions, rules of engagement require that the pilot makes a direct visual or instrumental acquisition of the target before weapons are employed. This is true for conventional bombs, rockets, guns and some guided missiles. The choice of tactics and weapons and the estimation of the effectiveness of the mission should include the probability of target acquisition for the successful attack. Mathematical models have been developed to predict the probability that targets can be detected which is applicable not just to the human eye, but to wide variety of on board electro-optical sensors operating in different parts of the electromagnetic frequency spectrum [38, 40]. These models are executed at run time to determine if the target is acquired and record the decisions made by the pilot for analysis.

DEVS Framework

Discrete Event System Specification (DEVS) is a universal formalism for discrete event dynamical systems [16, 17]. DEVS offers an expressive framework for modelling, design, analysis and simulation of autonomous and hybrid systems. Because of its system theoretic basis, DEVS is a universal formalism for discrete event dynamical systems. The DEVS framework enables a large system to be specified by hierarchically decomposing the system into modules called *Atomic* or *Component Models*, each having the internal structure and the state transition [15, 16]. The specification of the coupling between the component models and the hierarchy structure of the atomic models corresponds to the *Network* or *Coupled model* of the DEVS formalism [14–18]. DEVS environments are implemented over middleware systems such as HLA, RMI, and CORBA. DEVS exhibits concepts of systems theory and modeling and supports capturing the system behavior from the physical and behavioral perspectives that are implemented using CI techniques (Fig. 4).

A mission objective (goal) set by the instructor is designed within a contextual setting and also describing the scenario and settings within which the training is imparted and the trainees are assessed. The lesson plans are designed using all the four types of learning depending upon the nature of lessons and training to be imparted [41, 42]. The lesson plans are designed based on the domain knowledge that is explicitly represented by ontology of the warfare resources, aircraft, weapons, performance characteristics, constraints, weather, and terrain information. The lesson plans are dynamically adapted by asking relevant questions on the concepts of learning from the ontology and reasoning based on the answers to change the lesson plans accordingly. The goals are decomposed as tasks, and sub-tasks in a

Fig. 4 Agent architecture for JWSS

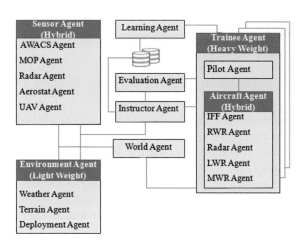

hierarchical manner, indicating the roles of the armed services, support organizations and people who would be collaborating to collectively achieve the objective. Case based Planning and Reasoning can be used to retrieve the past cases and adapted to the future mission plans [43–49]. The sequence and timing diagrams of the tasks are generated and these are associated with the resource constraints and resolution of conflicts. The assessment of the trainees is done by evaluating the plans made by the trainees to meet the goals. The Learning Management sub-system (LMS) in this simulator architecture (JWSS) is responsible for planning the lessons for the trainees, storing and updating the contents, evaluate the trainees and also learn from the behavior of the trainees for further lesson planning. The LMS consists of three prominent agents: Instructor agent, Learning agent and Evaluation agent. The Instructor agent is composed of a Lesson Planner that identifies a goal for the trainees, composes the lesson plan from the learning objects and given to all the trainees. The trainees decompose the task into a number of independent tasks that are to be achieved by each of the teams, in order to achieve the objectives of the goal. The Instructor agent updates the state of a lesson plan and creates a scenario that is based upon the information received from weather, terrain and deployment agent and provides an information service to the world agent after its own process of reasoning. This information is then used by other agents such as Manual Observation Post (MOP), Pilot, Unmanned Air Vehicle (UAV), Identification Friend/Foe (IFF), Radar Warning Receiver (RWR), Missile Warning Receiver (MWR), Laser Warning Receiver (LWR), Mission Planning, Sensor Performance, Target Acquisition and Damage Assessment and Computation (Fig. 3a, b).

4 Computational Intelligence Techniques in Designing JWSS

JWSS has been designed in the military domain for training, analysis, to generate strategic scenarios for forecasting, creating what-if scenarios and evaluating effectiveness of military operations and procedures.

4.1 Design of a Joint Services Military Ontology

Ontologies are specifications of the conceptualisation and corresponding vocabulary used to describe a domain [18, 50]. It is an explicit description of a domain and defines a common vocabulary as a shared understanding. It defines the basic concepts and their relationships in a domain as machine understandable definitions. We design a military ontology consisting of a formal and declarative representation which includes the vocabulary (or names) for referring to the terms of army, navy and airforce and the logical statements that describe what the terms are, how they are related to each other, and how they can or cannot be related to each other (Figs. 5, 6). Ontology therefore provides a vocabulary for representing and communicating knowledge about some aspect of military training and a set of relationships that hold among the terms in that vocabulary. The main purpose of ontology is, however, not to specify the vocabulary relating to an area of interest but to capture the underlying conceptualisations [51–56]. Noy and McGuinness [41] have identified five reasons for the development of ontology:

- to share common understanding of the structure of information amongst people or software agents;
- to enable reuse of domain knowledge;
- to make domain assumptions explicit;
- to separate domain knowledge from the operational knowledge;
- to analyse domain knowledge.

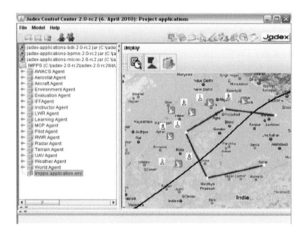

Fig. 5 Mission planning simulator in agent-oriented architectures

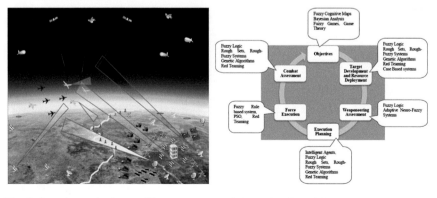

Fig. 6 Steps in conducting a military wargame and the application of CI techniques

Three main challenges in designing reusable learning objects are (i) intelligence; (ii) sharable; and (iii) dynamic. This is overcome by developing semantic metadata for providing intelligence to learning objects; developing content packaging for enhancing the sharability of learning objects and developing learning object repository with ontologies and Semantic Web technologies for making learning objects more dynamic [57–60]. To meet these challenges the following methodological steps are followed to design and develop the online environment of learning object repository.

- *Stage 1*: To develop a metadata framework which integrates the most suitable metadata as well as proposed pedagogical and military metadata elements that can be applied to a variety of learning objects.
- *Stage 2*: To apply a content packaging standard that packages learning objects together so they can be exported to and retrieved from various learning management systems.
- *Stage 3*: To identify the ontology (i.e. a common vocabulary of terms and concepts) for construction education and to develop a Semantic Web environment that will increase sharability of objects within construction domains.

In the design of the JWSS, military domain knowledge is represented and stored as ontology in Protégé (Fig. 7a, b). Protégé is a freely available, open-source platform that provides a suite of tools to construct domain models and knowledge-based applications that use ontologies [61–63]. At its core, Protégé implements a rich set of knowledge-modeling structures and actions that support the creation, visualization, and manipulation of ontologies in various representation formats (including the Web Ontology Language, OWL and Resource Description Framework (RDF)). Protégé can be customized to provide domain-friendly support for creating knowledge models and entering data. Further, Protégé can be extended by way of a plug-in architecture and a Java-based Application Programming Interface (API) for building knowledge-based tools and applications (Fig. 8).

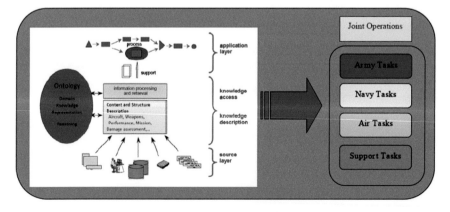

Fig. 7 Architecture of the LMS for dynamically composing joint operations lesson plans

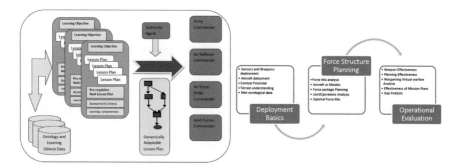

Fig. 8 Ontology based Instructor agent to dynamically plan adaptable lessons based on competency gaps

Protégé can load OWL/RDF ontologies, edit and visualise classes and properties; execute reasoners such as description logic classifiers and edit OWL individuals for SemanticWeb. Protégé is widely used for modelling of simple applications to high-tech, high-powered applications. It also offers support to ontology libraries and OWL language. Military ontology for joint operations is designed and developed for the JWSS using Protégé. While metadata of learning objects describe the artifacts of learning objects that are shared by diverse domains, an ontology represents a knowledge domain that shares the relationships of learning objects within a specific context.

Reasoning the Ontology
One of the main reasons for building an ontology-based application is to use a reasoner to derive additional truths about the concepts. A reasoner is a piece of software able to infer logical consequences from a set of asserted facts or axioms. The notion of a semantic reasoner generalises that of an inference engine, by providing a richer set of mechanisms to work with. The inference rules are commonly specified by means of an ontology language, and often a description language. Many reasoners use first-order predicate logic to perform reasoning; inference commonly proceeds by forward chaining and backward chaining. In the JWSS, reasoning helps in formulating questions for testing the understanding of related concepts. This is used to evaluate the competency of trainees, lessons planned and perform a gap analysis so that new lessons can be generated to fill the gaps [64–67].

4.2 Strategic Planning Wargames Using Fuzzy Cognitive Maps

Strategic Planning is a multi-dimensional assessment of a situation where several geo-political, economic and military dimensions are evaluated before arriving at course of actions. Existing relationships between countries can be described from a variety of perspectives, such as historical, respectful, friendly, neighboring, cultural, traditions, ideological, religious, trade, political, and economic aspects. One way to build these relationships is to strengthen the economic relationships, wherein the decision maker takes into consideration many factors and variables that influence the promotion of these relationships, prominent among them being economic relationship. This information and these factors are diversified and may involve different dimensions and the challenges in Strategic Planning lie in recognizing, finding and extracting the underlying relations and strengths of influence of these different variables. A conscientious decision maker who takes responsibility for promoting and strengthening bilateral economic relationships needs access to information that is fuzzy and qualitative. Military options are usually the method of last resort, and a brute-force approach that is often the result of a trigger that evolves over states and time domain when the geo-political options fail. This basic *concept* of the information is represented as a linguistic variable whose values are words rather than numbers across different domains including the political and investment domains. Due to nature of the problem, data from different domains that is imprecise, ill-structured, uncertain and ambiguous needs to be modeled. A fuzzy ontology that represents the geo-political, historical, respectful, friendly, neighboring, cultural, traditions, ideological, religious, trade, and political, military and economic ties is constructed using Protégé software. This Ontology is useful for acquiring and sharing knowledge, building a common consensus and constructing knowledge-based systems that can be used to build sub-schemas to represent the perspectives of the stakeholders, and reason the ontology for hidden and underlying relationships and their strengths of influence. Fuzzy Cognitive Maps (FCM) are fuzzy graph structures for representing causal reasoning with a fuzzy relation to a causal concept [68]. Fuzzy cognitive maps are especially applicable in the soft knowledge domains such as political science, military science, history, international relations, and Strategic Planning Wargames. Fuzzy logic generated from fuzzy theory and FCM is a collaboration between fuzzy logic and concept mapping. FCM is used to demonstrate knowledge of the causality of concepts to define a system in a domain starting with fuzzy weights quantified by numbers or words [69]. As a soft-system modeling and mapping approach, FCM combines aspects of qualitative methods with the advantages of quantitative (causal algebra) methods. In a FCM, the positive (+) and the negative (−) signs above each arrowed line provide a causal relationship whereby each fuzzy concept is linked with another one. In this sense, the FCM is a cognitive map of relations between the elements (e.g., concepts, events, resources) that enables the computation of the impact of these elements on each other, where the theory behind that computation is fuzzy logic. Since FCMs are signed fuzzy non-hierarchic digraphs [69], metrics can be used for further computations, and causal conceptual centrality in cognitive maps can be defined with adjacency-matrix [68] (Fig. 9).

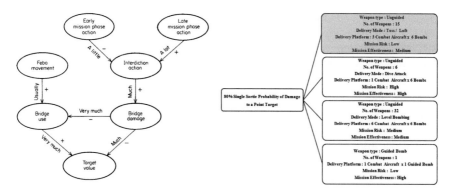

Fig. 9 Assessing the bridge target value in forward edge of battle area (FEBA) operation and course of action analysis

The various steps in constructing an FCM for representing a crisis situation are:

(1) Identification of factors and representing them as concepts
(2) Specification of relationships among concepts
(3) Defining the levels of all factors,
(4) Defining the intensities of causal effects,
(5) Identify changeable factors versus dependent factors,
(6) Simulating the fuzzy cognitive map,
(7) Modifying the fuzzy cognitive map,
(8) Simulating the modified fuzzy cognitive map, and
(9) Deriving the Conclusions from Reasoning.

FCMs for various crisis situations are constructed and used for assessing value of targets, prioritization of targets, and evaluating effects-based operations. In the JWSS, strategic planning considers all the concepts represented in fuzzy ontology, qualitative attributes of selecting the Course of Action (CoA), weaponeering principles of damage computation, reasoning the ontology and generating the course of actions.

4.3 Adaptive Lesson Plans Using Game Trees

The adaptability of the game playing depending on the background and training needs of different users is selected by two main cognitive criteria memory and learning. The competency level required by the lesson plan is compared with the competency level of the trainee. The gap is reduced by reasoning the ontology concepts and choosing the lesson plans that are represented in a *concept-map* and implemented as a *concept-graph*. The trainee is now switched to a new lesson that matches with the competency of the lessons. The lesson plans for the Army, Navy and Air Force and Joint operations which demand inter-disciplinary domain knowledge are organized as a concept graph. The starting node for the trainee is identified based on the competency level and based on the various preliminary questions, the Instructor agent reasons from the ontology *concept-graph* and composes the new lesson plan by traversing it to reduce the competency gap (Fig. 10).

Consider a training exercise for military operations in which the trainees from different branches of specializations with different skills, prior training and field operations are assigned tasks of a campaign (Table 1). These tasks are assigned to the trainees with the intent of teaching concepts, examples, and field cases which are then evaluated in the field training. The prior training is used to compute the trainee competency factor, and the lesson plan initially assigned has the training competency level. The gap which is the difference between the two values is used to decide the switched lesson plan so that the semantic distance is minimised. The military ontology is used to traverse the concept-map that is implemented as a concept graph, and is used to adapt the lesson plans for the trainee with the goal of minimising the semantic gap in the lessons chosen. The quantitative answers for the different tasks given to the trainees are calculated by wargaming the tasks and

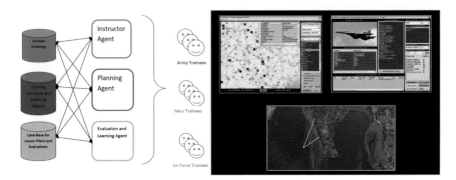

Fig. 10 Architecture of agents in JWSS and screens depicting the army, air and naval tasks for trainees

Table 1 Classification of live, virtual and constructive simulations

People	Systems	Operation	Simulation
Real	Real	Simulated	Live
Real	Simulated	Simulated	Virtual
Simulated	Simulated	Simulated	Constructive

generating the mission success factors for the two lessons: one that is conventionally computed using databases, and the other that uses ontology [41]. Consider the cases of two trainees (named Trainee 1 and Trainee 3) with different game scenarios and lesson plans from the same training system.

Trainee 1: To understand and evolve different strategies to gather Location based Intelligence necessary as pre-curser to destroy the target (Fig. 11).

Trainee 3: To understand the concepts in Mission Planning and Air Tasking operations (Fig. 12).

The mission success factor for Trainee 1 increased from 7.2 to 9.3 and from 5.3 to 9.8 on running the JWSS wargame by using the military ontology. The ontology requirements found an importance in military simulators mainly because of the Joint Warfare operations that are introduced in the course of training. These values may or may not have increased as much with the individual service wargames. This gives an intuitive indication of synergy in joint wargames that demand a much greater understanding of the warfare concepts and applying them in joint missions that surpass the boundaries of individual wargames (Table 2).

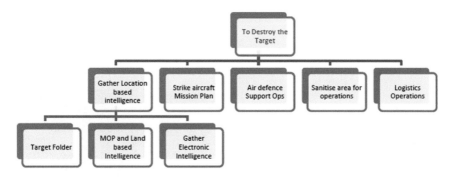

Fig. 11 T-001 Game scenario lesson plans generated for Trainee 1

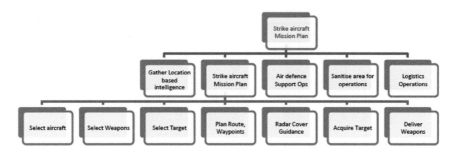

Fig. 12 T-003 Game scenario lesson plans generated for Trainee 3

Table 2 Fuzzy antecedents (factors) to generate training lesson plans

Mission_ID	Trainee	Service	Training and field operations skills	Lesson plan (Task)	Trainee competency factor	Training competency factor	Competency gap	Mission success factor (1–10) conventional	Mission success factor (1–10) ontology and adaptive lessons
#001	T-001	Intelligence	Low	ESM and location based intelligence gathering	7	8	1	7.2	9.3
#002	T-002	Navy	High	Anti-submarine warfare	2	5	3	4.6	6.4
#003	T-003	AirForce	High	Strike and air-air combat	7	8	1	5.3	9.8
#004	T-004	Logistics	Low	Mobilization of logistics and road move plan	5	8	3	6.8	7.4
#005	T-005	Army	Very high	Air defence guns deployment	6	8	2	7.2	9.3

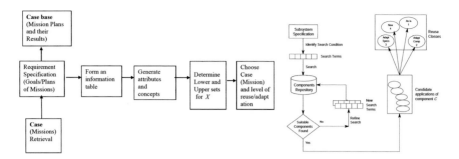

Fig. 13 Case base of missions plans and retrieval of relevant missions using rough-fuzzy techniques

4.4 Case Based Planning and Adaptation of Missions

In case-based planning (CBP), previously generated plans are stored as cases in memory and can be reused to solve similar planning problems in the future. CBP can save considerable time over planning from scratch (generative planning), thus offering a potential (heuristic) mechanism for handling intractable problems. One drawback of CBP systems has been the need for a highly structured memory that requires significant domain engineering and complex memory indexing schemes to enable efficient case retrieval. Computational intelligence techniques based on Rough-Fuzzy hybridization techniques are used to retrieve past mission plans that meet the military goals and/or effects to meet the present military objectives and software components that are stored as a case-base. These are implemented in the JWSS system for effective planning (Fig. 13).

4.5 Optimal Deployment of Resources and Weapon-Target Optimisation Using Genetic Algorithms

In deriving an optimal deployment strategy of resources against a perceived threat scenario, a intelligent strategy that mitigates the risk of the threat while minimizing the damage to own resources is developed using genetic algorithms. Another sub-system that uses GA for optimization is the platform-weapon-target matching that maximizes the estimated damage to a given target. Genetic algorithms provide an evolutionary approach towards solving the problem at hand by aiming to improve the fitness of each successive generation of possible solutions, mimicking the process of natural selection on a suitably simplistic scale. Unlike other search-based optimization procedures such as Hill Climbing or Random Search, GAs have consistently achieved good performance in terms of balancing between the two conflicting objectives of any search procedure, which are the *exploitation* of the best solution and the *exploration* of the search space. An initial population of individuals (positions of deployment, considering the constraints of performance, detection probability of sensors, terrain and weather conditions; Platform-Weapon-Target damage effectiveness based on the physical hard-points and effectiveness of the weapons against a given target) is required as a starting point for the optimization process. To this effect, we seed a uniformly distributed population within the desired bounds of the solution space. A fitness function (maximizing the effective target detection) is then defined, which assigns a score to every member of the current generation based on the evaluation of relevant characteristics, which in this case are the total volume enclosed by the network of sensors, with given detection and communication capabilities. The fittest individuals from this pool are identified for creating the next generation via the chosen implementations of selection and crossover functions, which dictate the process of reproduction and survival of individuals. A mutation factor is also specified for the genome to reduce the chances of the solution converging towards a local maximum. For every individual, a convex hull is stretched over the point cloud formed by the nodes in three dimensional space to form a polyhedron. The volume of this polyhedron not only serves as the initial score for the individual prior to constraint checking, but its visualization can also be used to identify shadow zones as well as highlight nodes surplus to requirements in achieving the given objective. For the duration of the optimization phase, we assume that the transmission and detection ranges for all nodes are omni-directional and isotropic by default. Node and medium characteristics are then used to compute the distortions to the coverage spheres of each node. As Delaunay triangulation is used to form the convex hull for each individual, every node has three neighbors, without accounting for any redundancy requirements. Once the optimization is complete and a solution is obtained, the direction vectors of antennas on each node can be specified, taking into consideration its neighbors and face coverage specifications.

Since shadow zones in such a scenario are essentially holes in the coverage shell, penalties are required on the raw scores to discourage such arrangements from participating in the evolution of the genome. In every successive generation, the score of the best-fit individual is expected to improve due to selective breeding. As the score stagnates with respect to average change in fitness, generation, or time, the algorithm terminates with an optimal solution as its output.

4.6 Red Teaming Using Intelligent Agents and Computer Generated Forces

Red teaming is the practice of viewing a problem from an adversary or competitor's perspective. The goal of most red teams is to enhance decision making, either by specifying the adversary's preferences and strategies or by simply acting as a devil's advocate. Red teaming may be more or less structured, and a wide range of approaches exists. These techniques help analysts and policy-makers stretch their thinking through structured techniques that challenge underlying assumptions and broaden the range of possible outcomes considered. Alternative analysis includes techniques to challenge analytic assumptions (e.g. 'devil's advocacy'), and those to expand the range of possible outcomes considered (e.g. 'what-if analysis,' and 'alternative scenarios'). Collective behavior is the result of evolutionary processes that shape behavior to modify and respond to environmental conditions (Gordon 2014). Investigating how these algorithms evolve can show how diverse forms of collective behavior arise from their function in diverse environments. An assessed deployment of the enemy's ground defence, air defence, with inputs from ESM missions, ground picture images from unmanned systems, and other human, and electronic intelligence, developing triggers from strategic games, a military-geo-political, economic, trades, and cultural map using fuzzy cognitive maps is developed. Strategic Course of Action (CoA) analysis is developed by considering the plausible CoA of the Red teaming analysis Missions are then planned against targets that are prioritized against the back-drop of the assessed red team. An assessment of the situation is made by using a game theoretic framework is built which is then given to the commanders as a specific scenario for analysis. The CI techniques used in modeling the Red teaming are behavioral game theory, cognitive process modeling, multi-agent systems, Markov decision process and social networks modeling. These factors are used in conjunction with the FCMs to predict the plausible next CoAs the adversary would take in order to react to the developing scenario. Bayesian Belief networks, Dempster-Shafer theory, Belief-Desire-Intention model to represent the epistemic states of red teaming agents, Influence diagrams for decision making, modal logics and deduction techniques are used in red teaming's possible world assessment.

4.7   Automatic Target Recognition by Unmanned Systems

In the JWSS, a list of targets and information obtained from various sensors, ELINT, COMINT and HUMINT is stored in a specialized database called *Target Folder*. Automatic Target Recognition (ATR) refers to the use of computer processing to detect and recognize target signatures in sensor data. The sensor data are usually an image from a forward-looking infrared camera, electro-optic sensor, synthetic-aperture radar, television camera, or laser radar, although ATR techniques can be applied to non-imaging sensors as well. ATR has become increasingly important in modern defense strategy because it permits precision strikes against certain tactical targets with reduced risk and increased efficiency, while minimizing collateral damage to other objects. If computers can be made to detect and recognize targets automatically, the workload of a pilot can be reduced and the accuracy and efficiency of the pilot's weapons can be improved. An overview of the CI techniques that are used in ATR is shown in Fig. 14. An image enhancement technique based on Blind De-convolution algorithm to improve the image quality followed by edge enhancement algorithms that adaptively enhance the edges and

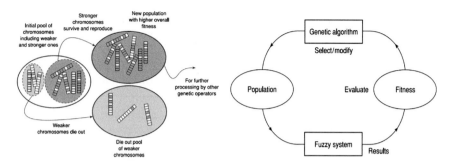

Fig. 14 Process of GA and integrating GA with Fuzzy logic system

wipe off blurriness in the image is implemented in the JWSS. The de-blurring results of the proposed algorithm and retrieval of plausible matching images using Content based image retrieval (CBIR) proved better than the conventional techniques. Using either the color, shape or texture features separately to compare and retrieve images with crisp equal weights was found to be ineffective. Instead, a fuzzy combination of the color, shape and texture features to design a better query retrieval system is implemented, where the feature weights are assigned depending upon the different conditions when the image was taken. This methodology based on Fuzzy techniques was very effective in identifying the target images obtained from UAV missions and has been implemented in the JWSS to model the effectiveness of UAV missions [38] (Fig. 15).

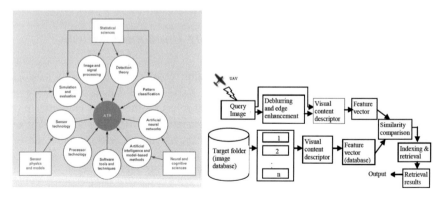

Fig. 15 Computational Intelligence techniques in ATR, and image recognition and detection in UAV/UCAV missions in JWSS

4.8 Design of Game Rules Using Fuzzy Rule Based Systems

Crisp rules use the cookie cutter function to identify the damages caused from the weapons. We design and develop a fuzzy cookie cutter damage function that generates rules that are used to game the missions and assess the damages caused to targets and own resources. Traditional approaches to wargame simulations use classical logic for damage assessment. Classical two-valued logic system, crisp set theory and crisp probability on which the damage assessment is based, are inadequate and insufficient for dealing with real-life war scenario that involves complexity and different sources of uncertainty. Damage assessment for a target done using cookie-cutter function gives the probability of damage of a target. Crisp cookie-cutter function states that a target is damaged inside a circle of specified radius r, and no damage occurs outside it

$$d(x, y) = 1, \quad x^2 + y^2 \leq r^2$$
$$0, \quad \text{otherwise}$$

where, $d(x, y)$ is the damage function of the point target by a weapon whose point of impact is (x, y) [3] The target is assumed to be completely damaged within the circle of radius r and no damage occurs outside r. The notion of probability stems from, and depends on, the idea of repeated trials. Under identical and repeatable laboratory conditions conducted on simple models, this probabilistic notion readily applies; but, in real-world systems, experiments are rarely identical and repeatable. Therefore, for the subjective assessment of complex military systems, probability has its limitations. Fuzzy Logic is the logic behind approximate reasoning instead of exact reasoning. As knowledge acquisition in wargames design and development is obtained from pilots and defence analysts, it is usually true that facts and rules are neither totally certain nor totally consistent due to the varied experience sets of the pilots. This leads to the reasoning processes used by experts in certain situations as approximate. The theory of fuzzy sets is used to help assess uncertain information derived from this approximate reasoning process. Structural damage can be considered as a linguistic variable with values such as "severely damaged," or "moderately damaged." These are meaningful classifications but not clearly defined. With the use of fuzzy sets, however, we can quantify such terminology and apply it in a meaningful way to help solve a complex problem. An evident advantage of the fuzzy set approach is the possibility of representing numeric and linguistic variables in a uniform way and of using a formalized calculus to manipulate these variables. For example, consider a large area-target of size of 550 ft to be attacked, where the fuzzy variables *target-ground contrast* 80 %, the *terrain*, rated 8, is fairly smooth, aircraft altitude is 900 ft, aircraft range is 5000 ft is flying at 100 knots speed. The *target identification factor* for this target is seen as "good" with value 7.3295. In this mission, on firing the rules for inference, the offset from the desired point of impact is 29 m, considered "less"(i.e. fairly accurate targeting); *weapon-target match* is 6 (average), "good" *target identification factor*

7.3295, the relative damage caused is 28.9187 which is a "moderate" damage to the target.

In the JWSS, AI based techniques such as a fuzzy rule-based system to design the game rules in a mission planning and evaluation system [70]. The conventional crisp cookie cutter function used to compute the probabilistic damage caused to a target is replaced by a fuzzy cookie-cutter function, which takes into account many physical parameters before assessing the possibilistic damage caused to the target. This methodology of damage assessment computation of targets using fuzzy rule bases gave realistic results, comparable with the experts' judgements, in field training.

4.9   Environment Modeling in JWSS

It receives information from weather, terrain and deployment agent and provides an information service to the world agent after its own process of reasoning. This information is then used by other agents such as Manual Observation Post (MOP), *Pilot*, Unmanned Air Vehicle (UAV), Identification Friend/Foe (IFF), Radar Warning Receiver (RWR), Missile Warning Receiver (MWR), Laser Warning Receiver (LWR), *Mission Planning*, *Sensor Performance*, *Target Acquisition* and *Damage Assessment and Computation*. The weather agent is an important agent that that has functions such as *Get_Visibility()*, *Get_Temperature()* and *Get CloudCover()*. The weather agents's reasoning has been designed using ANFIS, a neuro-fuzzy hybridization technique that is used to predict the Mission_Success_Factor(), considering the weather conditions along the mission route [50, 71].

Surface aviation weather observations include weather elements and forecasts pertaining to flying. A network of airport stations provides routine up-to-date surface weather information. Upper-air weather data is received from sounding balloons (radiosonde observations) and pilot weather reports that furnish temperature, humidity, pressure, and wind data. Aircraft in flight also report turbulence, icing and height of cloud tops. The weather radar provides detailed information about precipitation, winds, and weather systems. Doppler technology allows the radar to provide measurements of winds through a large vertical depth of the atmosphere. Terminal Doppler weather radars are used to alert and warn airport controllers of approaching wind shear, gust fronts, and heavy precipitation which could cause hazardous conditions for take-off, landing and diversion. Low-level wind shear alert systems provide pilots and controllers with information on hazardous surface wind conditions (on and near airbases) that create unsafe operational conditions. Visible, infrared and other types of images of clouds are taken from weather satellites in orbit. Weather is a continuous, multi-dimensional, spatio-temporally data intensive, dynamic and partly chaotic process. Traditionally, two main approaches for weather forecasting are followed: Numerical Weather Prediction and Analogue forecasting. For the JWSS application, it is needed to consider the past weather conditions at given places of operation and predict the weather for simulation of mission tasks in real-time. In this paper, the ANFIS neuro-fuzzy hybridization technique is used to predict the weather conditions along the mission route and study the effects of weather in the virtual warfare scenario analysis in terms of pilot decisions in mission planning, performance of sensors, and target identification and damage assessment.

A Neuro-Fuzzy Hybridization Approach to Weather Prediction

The weather agent has been designed using ANFIS to give the predicted Mission_Success_factor in weather constraints. In the following section, the neuro-fuzzy hybridization approach will be discussed. Both neural networks and fuzzy systems are dynamic, parallel processing systems that estimate input–output functions [6–8]. They estimate a function without any mathematical model and learn from experience with sample data. It has also been proven that (1) any rule-based fuzzy system may be approximated by a neural net and (2) any neural net (feed-forward, multilayered) may be approximated by a rule-based fuzzy system. Fuzzy systems can be broadly categorized into two families. The first includes linguistic models based on collections of IF–THEN rules, whose antecedents and consequents utilize fuzzy values. The Mamdani model falls in this group where the knowledge is represented as it is shown in the following expression.

$$R^i: \text{If } X_1 \text{ is } A_1^i \text{ and } X_2 \text{ is } A_2^i \ldots\ldots\ldots \text{and } X_n \text{ is } A_m^i, \text{ then } y^i \text{ is } B^i$$

The second category, which is used to model the Weather prediction problem is the Sugeno-type and it uses a rule structure that has fuzzy antecedent and functional consequent parts. This can be viewed as the expansion of piece-wise linear partition represented as shown in the rule below.

$$R^i: \text{If } X_1 \text{ is } A_1^i \text{ and } X_2 \text{ is } A_2^i \ldots\ldots\ldots \text{and } X_n \text{ is } A_m^i, \text{ then } y^i = a_0^1 + a_1^i X_1 + \ldots\ldots + a_n^i X_n$$

$$\tilde{A} \cap \tilde{B} = \left\{ \left(x, \mu_{\tilde{A} \cap \tilde{B}}(x) \right) \middle| \mu_{\tilde{A} \cap \tilde{B}}(x) = \mu_{\tilde{A}}(x) {}^{\wedge} \mu_{\tilde{B}}(x) = \min\left(\mu_{\tilde{A}}(x), \mu_{\tilde{B}}(x) \right) \right\} \quad (1)$$

The conjunction "and" Operation between fuzzy sets known as *Linguistics*, for the implementation of the Mamdani rules is done by employing special Fuzzy Operators called T-Norms [6]. The ANFIS uses by default the Minimum T-Norm which is the case here and it can be seen in the above equations. The approach approximates a nonlinear system with a combination of several linear systems, by decomposing the whole input space into several partial fuzzy spaces and representing each output space with a linear equation. Such models are capable of representing both qualitative and quantitative information and allow relatively easier application of powerful learning techniques for their identification from data. They are capable of approximating any continuous real-valued function on a compact set to any degree of accuracy. This type of knowledge representation does not allow the output variables to be described in linguistic terms and the parameter optimization is carried out iteratively using a nonlinear optimization method.

Fuzzy systems exhibit both symbolic and numeric features. Neuro-fuzzy computing is a judicious integration of the merits of neural and fuzzy approaches, enables one to build more intelligent decision-making systems. Neuro-fuzzy hybridization is done broadly in two ways: a neural network equipped with the capability of handling fuzzy information [termed fuzzy-neural network] and a fuzzy system augmented by neural networks to enhance some of its characteristics like flexibility, speed, and adaptability [termed neural-fuzzy system]. ANFIS is an adaptive network that is functionally equivalent to a fuzzy inference system and referred to in literature as "adaptive network based fuzzy inference system" or "adaptive neuro-fuzzy inference system" (Fig. 3). In the ANFIS model, crisp input series are converted to fuzzy inputs by developing triangular, trapezoidal and sigmoid membership functions for each input series. These fuzzy inputs are processed through a network of transfer functions at the nodes of different layers of the network to obtain fuzzy outputs with linear membership functions that are combined to obtain a single crisp output the predicted Mission_Success_Factor, as the ANFIS method permits only one output in the model. The following Eqs. 2–4 correspond to triangular, trapezoidal and sigmoid membership functions (Figs. 16, 17).

$$\mu_s(X) = \begin{cases} 0 \text{ if } X < a \\ (X-a)/(c-a) \text{ if } X \in [a,c) \\ (b-X)/(b-c) \text{ if } X \in [c,b] \\ 0 \text{ if } X > b \end{cases} \tag{2}$$

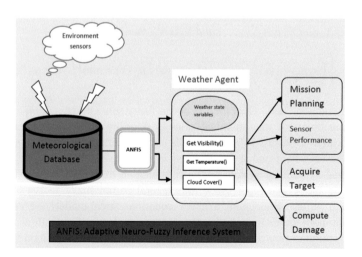

Fig. 16 Weather agent's architecture and behaviors

Fig. 17 ANFIS architecture
to design the weather agent

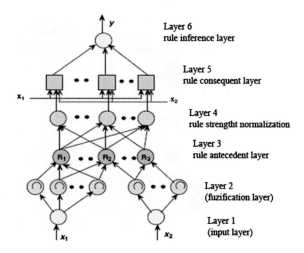

$$\mu_s(X) = \begin{cases} 0, & \text{if } X \leq a \\ (X-a)/(m-a), & \text{if } X \in (a, m) \\ 1, & \text{if } X \in [m, n] \\ (b-X)/(b-n), & \text{if } X \in (n, b) \\ 0, & \text{if } X \geq b \end{cases} \tag{3}$$

$$f(x; a, c) = \frac{1}{1 + e^{-a(x-c)}} \tag{4}$$

    Weather conditions of interest to JWSS [19, 21, 22, 71] are classified as Precipitation (Drizzle, Rain, Snow, Snow-grains, Ice crystals, Ice pellets and Hail), Obscuration (Mist, Fog, Dust, Sand, Haze, Spray, Volcanic ash, Smoke) and Others (Dust/Sand whirls, Squalls, Funnel cloud, Tornado or Water spout, Sandstorm, Dust-storm). Temperature, Clouds, Height of cloud base, Wind speed and direction, Icing, Precipitation, Visibility, Fog, Mist, Rain, Thunderstorm, Haze, dust/sand whirls and squall speeds are quantified using linguistic fuzzy variables. Target Identification factor: Rapid and certain target detection and identification are the dominant factors in the success of all air-to-ground attacks. The ability of tactical fighters to penetrate enemy defenses and to acquire and identify ground targets successfully within weather constraints is a keystone of success in a mission. It has been observed that aerial observers respond to targets in a manner indicating that detection/ identification represents a continuum rather than discrete phenomena. At one extreme the response is based on the ability to merely discriminate the existence of a military object among non-military objects (detection) [26–28]. At the other extreme the observer can describe the object in precise detail (identification). Factors considered for computing the Target Identification factor are target size, percent contrast, illumination, terrain, weather conditions, altitude and speed of the aircraft at time of target acquisition. **Target Size:** As target size increases, probability of correct target identification increases. It may vary from small to large tactical targets, including personnel, trucks, and tanks to big targets as bridges, runways and taxi-tracks. **Contrast:** Target/Ground Brightness Contrast is expressed as a percentage. **Illumination:** Detection performance increases as illumination increases. Effects of decreases in illumination occurring after sunset and before sunrise are very important and need to be considered. **Terrain:** Types of terrain have been defined in terms such as number of slope changes per unit area and average slope change. Four different terrain types have been defined-fairly smooth, moderately rough, rough, and very rough. As the roughness of terrain increases, percent terrain view decreases, and decrease in detection performance is observed. **Weather:** Temperature, humidity, and wind effects the performance of sensors (such as Radars) deployed, where as conditions such as Precipitation, icing, wind, visibility, fog, rain, date and time of operation, clouds, and storm effect the pilots' decisions in planning and executing the missions. **Altitude:** The relationship between altitude and target detection/identification is normally one in which there is assumed to be an optimal altitude; above and below this optimum altitude, detection is reduced. As altitude increases, detection performance decreases. As altitude is increased beyond an optimal point, detection probability falls off rapidly.

Data on all these factors are collected from meteorological department databases, handbooks and experimental field trials and heuristic knowledge from experts and defense analysts (in questionnaire form) are collected and recorded. They are then represented as decision matrices and decision trees which form the basis to design the membership functions and rules. The rules are then executed in the mission processing module and defuzzified to obtain the damage to target. These results are then compared to the expected output and fine-tuned before storing in the rule base. A decision to include the new rule or not is provided to the commander. Missions and results of the missions are stored as a case-base for retrieval and reuse of missions plans in new situations. The fuzzy linguistic variables used in the design of the game rules are as follows:

Mission_Success_Factor (with weather constraints): **[1–10]** {Very Low: **[0.0–3.5]**; Low with Moderate Risk **[2.5–5.5]**; Medium with Controllable Risk **[4.5–7.5]**; High with Moderate Risk **[6.5–8.0]**; Very High with Low Risk **[7.5–10.0]**} **Temperature:** [Very Low, Low, Moderate, High, Very High] **Fog-Haze:** [Shallow, Patches, Low Drifting, Blowing, Showers, Thunderstorm, Freezing, Partial] **Wind-Speed:** [Light, Moderate, Heavy] **Clouds/Cloud Base**: [Shallow, Patches, Low Drifting, Blowing, Showers, Thunderstorm, Partial]; [Height (ft)] **Visibility:** [Low, Medium, Clear] **Turbulence:** [Clear, Low, Medium, Heavy] **Storm/Squalls:** [Clear, Low, Medium, Heavy] **Sky Cover:** [Clear, Few, Scattered, Broken, Overcast, Vertical Visibility] **Terrain:** **[1–100]** {Fairly Smooth **[0–22]**; Moderately Rough **[14–49]**; Rough **[45–81]**; Very Rough **[75–100]**}.

Target Size (in feet): {Very small: [0–100]; Small: [70–190]; Medium sized: [160–300]; Large: [270–400]; Fairly Large: [360–500]; Extremely Large: [450–900]} **Damage:** Offset (in meters): {Very Less:[0–23]; Less: [16–36]; Medium: [34–57]; Large: [56–80]; Very Large [78–100]} **Weapon Target Match:** [0–10] {Poor: [0–3.6]; Average: [3.36–6.669]; Good: [6.73–14.2]}

Target Identification Factor: [0–10] {Very poorly identified: [0–1.19]; Poorly identified [0.96–2.43]; Average identification [2.34–5.61]; Good identification [5.43–7.55]; Excellent identification [7.35–10]} **Relative Damage (Damage relative to intended damage):** [0–100] {Mild: [0–18]; Moderate: [16–36]; Average: [34–57]; Severe: [56–80]; Fully Damaged: [78–96]}.

Data from meteorological database is used to train the network to apply a hybrid method whose membership functions and parameters keep changing until the weather forecast error is minimized (Fig. 5a, b). Then the resulting model is applied to the test data of the mission time and places en-route from take-off base, target and landing base.

5   Results Discussion

The fuzzy variables are used to calculate the Mission success factor based on the prevailing weather conditions generated by the ANFIS model, target identification factor and firing of the rules to compute the relative damage to the target. Offset is calculated using actual altitude, actual vertical flight path angle, actual wind speed and observed altitude, observed altitude, observed vertical flight path angle, observed wind speed by the weapon system trajectory calculation module and the aircraft speed as the input variables (Table 5). Offset is a measure of induced error, wind induced error, and vertical flight path angle induced error.

Case Mission ID # 001: Consider a large area-target of size of 550 ft to be attacked, where the fuzzy variables *target-ground contrast* 80 %, the *terrain*, rated 8, is fairly smooth, aircraft altitude is 900 ft, aircraft range is 5000 ft is flying at 100 knots speed. The *target identification factor* for this target is computed as "good" with value 7.32. (*In the tables below * denotes the Missions planned and executed when considering the Weather conditions.*)

In this mission, on firing the rules for inference, the offset from the desired point of impact is 29 m, considered "less"(i.e. fairly accurate targeting); *weapon-target match* is 6 (average), "good" *target identification factor* 7.32, the relative damage caused is 28.92 which is a "moderate" damage to the target. We consider two scenarios of weather conditions at the given place and time or the mission plan (Fig. 2). Weather conditions are identified based on the place and time of missions. The ANFIS model computes the Mission_Success_Factor as 8.4 when no weather conditions are considered, and reduces to 3.7 when weather conditions are considered in the JWSS (Table 3). These conditions also reduce the Relative Damage from 28.91 to 13.55 (Table 5) and *offset* of the weapon hitting away from the intended target increased from 29.03 to 37.54 (Table 6).

Table 3 Fuzzy rules to determine the mission success factor in weather conditions

MissionID	Temperature	Fog-Haze	Wind speed (m/s)	Clouds/base	Visibility	Turbulence	Storm/squalls	Mission success factor
#001	Moderate	Clear	Low-drifting	Clear	Clear	Low	Clear	8.4
#001*	Very low	Moderate	High	Low	Poor	High	Clear	3.7
#002	Moderate	Clear	Moderate	Scattered	Clear	Low	Clear	9.8
#002*	Very high	Haze	High	Low	Poor	Low	Squall	7.1

*denotes the Missions planned and executed when considering the Weather conditions

Table 4 Fuzzy rules to determine the target identification factor

MissionID	Target size(ft)	Target-ground contrast %	Illumination (foot candles)	Terrain	Weather_mission success factor	Aircraft altitude (feet)	Aircraft range (feet)	Aircraft speed (knots)	Target identification factor
#001	550	80	40	8	8.4	900	5000	100	7.32
#001*	550	45	20	8	3.7	900	5000	80	5.67
#002	550	80	60	7	9.8	750	4000	80	8.03
#002*	550	45	30	7	7.1	750	4000	60	6.43

*denotes the Missions planned and executed when considering the Weather conditions

Table 5 Fuzzy rules to compute the *Relative damage to target*

MissionID	Offset (meters)	Target radius (km)	Weapon-target match	Weapon delivery mode	Target identification factor	Relative damage
# 001	29.03	0.09	6	6	7.32	28.91
#001*	37.54	0.09	6	6	5.67	13.55
# 002	6.07	90.0	9	9	8.03	88.74
#002*	12.65	90.0	9	9	6.43	65.92

denotes the Missions planned and executed when considering the Weather conditions

Case Mission ID # 002: Another mission planned by the commander where a similar target is chosen with the fuzzy variables as shown in Tables 3 and 4. While the offset has reduced to 6 m, considered "very less" (i.e. very accurate targeting), choosing a different weapon system and delivery improved the weapon-target match to 9 ("good"), and mode of weapon delivery 9, the *target identification factor* also improved to 8.033 (considered "excellent"), and the relative damage caused is 88.74, which is a "substantial" damage to the target (Tables 4 and 5). Weather conditions are again identified based on the place and time of missions. The ANFIS gives the Mission_Success_factor as 9.8 when no weather conditions are considered, and reduces to 7.1 when weather conditions are considered in the JWSS (Tables 3 and 4). These conditions also reduce the Relative Damage from 28.91 to 13.55 (Table 5) and *offset* of the weapon hitting away from the intended target increased from 6.07 to 12.05 (Table 6). These attributes form the antecedents of the fuzzy rule and the consequent is shown in the last column of the tables. For all the missions that the pilots plan in the wargame exercises, these fuzzy game rules are used to infer the expected damage caused to the target. These missions form a part of a case-base which is used as part of the '*learning*' by the system for future instructional use.

5.1 Modeling Pilot Agents in Air Warfare Simulation System

Advances in combat aircraft avionics and onboard automation, information from onboard and ground sensors and satellites, pose a threat in terms of data and cognition overload to the pilot. Under these conditions, decision making becomes a difficult task.

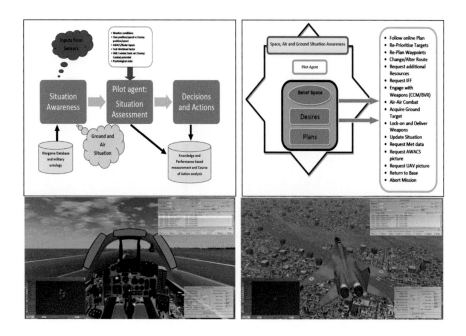

The factors identified in Table 7 are representative of the two pilots P1 and P2, who differ mainly in *Information Processing* and decision making, *Risk taking* and *Reaction to stress* which are typically identified personality traits. Data collected using clinical and psychometric tests for all the pilots are stored in the Pilot's database. These (fuzzy) attribute values from the pilot's database are fuzzified and used to determine the pilot's personality as one of the inputs to the ANFIS tool (Table 8).

5.2 Data Mining Techniques and Reasoning in Wargame Results Analytics

Having designed and developed the JWSS as an exploratory, battlefield experi-mentation, test-bed using an inner-sourcing, product-line architecture that supports multi-resolution models and a wargame process customization script to cater to various military training establishments, this test-bed serves as a platform for mission analysis, and doctrine analysis, using data mining and pattern analysis techniques. Digital Battlefield simulation and experimentation uses data mining techniques such as association, clustering, classification, learning, decision trees and rules that provide insights into the doctrines and their effectiveness [72]. Each of these methods use CI techniques, in turn, to arrive at realistic rules for doctrine assessment and evaluation (Fig. 18).

Table 6 Fuzzy attributes to determine the *offset* of the weapon from the intended target

Mission ID #	Apparent altitude (km)	Apparent angle (degrees)	Apparent wind velocity (km/hr)	Actual altitude (km)	Actual angle (degrees)	Actual wind velocity (km/hr)	Aircraft speed (km/hr)	Offset (m)
001	1.65	−26.9	−25.24	1.67	−26.9	−25.24	829.8	29.03
002*	1.65	−26.9	−23.9	1.67	−26.9	−25.24	820.1	37.54
002	1.65	−25.2	−28	1.65	−25.2	−30.4	830.2	6.07
002*	1.65	−25.2	−22	1.65	−25.2	−30.4	824.7	12.65

*denotes the Missions planned and executed when considering the Weather conditions

Table 7 Pilot's attributes considered in the ANFIS

Pilot id	Personality type	Risk taking	Info processing and risk taking	Aviat or skills/experience	Firing skills/experience	Sensor-motor abilities	Personality leadership	Motivation	Reaction to stress	Physiolog ical/medical health
P1	A	High	High	Very High	Very High	High	Excellent	High	Composed	Cat 1
P2	B	Low	Low	High	High	High	Excellent	High	Stressed	Cat 1

Table 8 Pilot attributes and decisions to determine the mission success factor in different training situations

MissionID	Combat aircraft	Mission Comdr	Sorties flown/day	Enemy air/ground defence Threat	Situation awareness	Entropy	Information overload	Combat potential ratio (MAUT) (Own: En)	Combat utility factor (ANFIS) (Own: En)	Subjective mental workload NASA-TLX	Situational decision of pilot (AWSS)	Mission success factor
#001 Sit 1	Multi-Role	P1	3	High	Very high	High	High	Static: 1:0.78 Dyn: 1:13	1:1.4	Low	Re-prioritize targets	8.7
#001 Sit 1	Multi-Role	P2	3	High	Very high	High	High	Static: 1:0.78 Dyn: 1:1.3	1:1A	High	Request additional resources	4.3
#002 Sit 2	Multi-Role	P1	2	Very low	Low	Low	Low	Static: 1:0.44 Dyn: 1:0.65	1:-0.3	Low	Lock-on and deliver weapons	7.6
#002 Sit 2	Multi-Role	P2	2	Very Low	Low	Low	Low	Static: 1:0.44 Dyn: 1:0.65	1:-0.3	High	Look for secondary targets	3.2

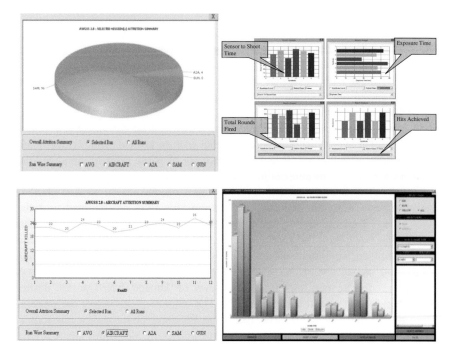

Fig. 18 Battlefield experimentation, results analysis and doctrine analysis using data mining techniques

6 Case Study: Joint Warfare Analyses and Integrated Air Defence

Joint Warfare Simulation System has been designed and developed to meet the training and operational analysis requirements of military officers. It provides a platform for deployment of resources, weapon target matching, weaponeering assessment, force planning, force execution, damage assessment, quantitative results analysis and displaying reasoning for generating outcomes. As a training platform it can be used to train military officers in various roles in formulating and evaluating strategies and decision making processes, at tactical and operational levels of warfare. For operational analysis version, this can be used to find out the effectiveness and performance of various weapon systems, weapon delivery platforms, force multipliers, and sensors in a simulated battlefield scenario between two or more opposing forces.

JWSS has been designed and developed as a test-bed to simulate wide range of military air operations such as counter air missions, counter surface force operations, air defense missions, and combat support operations [2] between two or more opposing forces. It provides a platform for deployment of resources, weapon target matching, weaponeering assessment, force planning, force execution, damage assessment, quantitative results analysis and displaying reasoning for generating outcomes which is crucial for debriefing and learning purpose. It also computes the attrition rates, statistics of various operations & in-depth history of various events generated during the simulation which helps in analysis and validation of tactics and various operational objectives. This application when configured in training mode, trains military officers in planning missions to meet the objectives such as destruction of a synthetically generated target like an airfield, vital bridge, nodal point, or army installations. It uses extensive set of game rules to simulate wide range of operations. This application can also be configured as an analytical tool for operational analysis at the tactical levels for decision making. In the first phase of conducting any wargame exercise, the mission objectives are outlined to define which facets of enemy activity are to be affected by the mission. Based on these tactical objectives, with the study of target folder, suitable target damage criteria, force and ordnance requirements are defined to achieve the desired level of damage. In the third phase, combat models are used to define the type and quantity of weapons needed to produce the required level of damage, based on the desired mean point of impact and target elements. The outputs of this phase are essential inputs to execution planning. The Execution Planning phase assign missions to specific units, perform attack and support force packaging, determine attack timing, and outline communications and coordination requirements (C2/C3 structures). Detailed mission planning is also done in this phase. In Force Execution and Combat assessment phase, the missions are executed in simulation manager using game rules, acquisition models and damage assessment models and the results are assessed to determine if the objectives have been met, or re-strike is required. Some of the salient features of this simulation system are: Multi-Team War Scenario Analysis System, Training Toolkit, Tactical Deployment Evaluation & Decision Making Tool, Operational Analysis Tool to Evaluate Performance & Effectiveness of Aircraft, sensors, Weapon Systems and Missions, Electronic Warfare & Logistics Support, Weather and Terrain effects with customized GIS, Quantitative Evaluation of Mission Objectives and Plans, "What-If" Scenario Analysis. Screen shots of the JWSS software are show in Fig. 19.

Fig. 19 Joint warfare simulation system as a common test-bed for multiple training products

7   Conclusions

In this paper, we have presented a Joint Warfare Simulation System, a digital battlefield test-bed, in which a war scenario between two or more opposing forces can be simulated. The planners plan various offensive and defensive emissions which are gamed against the perceived threat using the JWSS to assess the mission effectiveness. Some of the major components of the design are the joint forces scenario databases, military ontology, resource databases that contain the performance characteristics of weapons, sensors, missiles, aircraft, naval platforms, air and ground defence systems, target analysis, resource deployment, mission planning, target damage assessment and results analysis. Fuzzy sets and systems are used to represent the damage assessment techniques, game rules are designed using adaptive neuro-fuzzy systems, software agents are used to build on-board intelligent pilot model, where the pilot agent uses mission ontology to plan his mission automatically. An automated decision tool to generate the decisions of the pilot in various situations is also built in the system by considering the cognitive and behavioural characteristics of the selected pilot from the pilot database. Fuzzy Cognitive maps to represent the pilots' plans are implemented and NNs to model the pilots decision making ability. Mission Plans are automatically generated by specifying the mission objective, which in turn generates all the plausible courses of action, with varying mission costs and risks and utilising resource in the inventory. Military logistics planning is also done by the system to automatically generate the most efficient routes using AI techniques. This utilises the entire road/rail/air data during peace times and war times. Generation of automated mission plans, evaluating the effectiveness of mission plans, Red-teaming, Threat perception of the enemy, modeling counter-insurgency operations, are designed using computer generated forces and semi-automated forces with propositional, probabilistic and possible world approaches and evolutions using swarm optimisation techniques. Ontology to represent the common knowledge-base and representation of the adaptive lesson plans is implemented as part of the system. The conventional damage models built using cookie-cutter approximation are replaced by Fuzzy damage functions. These models are built and implemented as Joint Warfare Simulation systems that can be customised to meet the varying needs of the trainees. Weather and Environment variables are an important factor to consider in the assessment of mission plans and generate results. These are modeling using ANFIS. A Weapon-Platform-Target system that helps in matching and assigning the resources to targets is an important part of the Wargames that is matched using fuzzy inference systems. A Weaponeering tool that estimates the number of weapons to destroy a target with a given assurance level is developed. A Weapon target matching using GA and a tool to aid the planner in Optimal Deployment of sensors and resources is implemented using GA.

References

1. Smith, R.: The Utility of Force: The Art of War in the Modern World. Alfred A. Knopf, New York (2007)
2. McNab, R.B., Angelis, D.I.: Does computer based training impact maintenance costs and actions? An empirical analysis of the US Navy's AN/SQQ-89(v) sonar system. Appl. Econ. **46** (34), 4256–4266 (2014)
3. Tolk, A.: Engineering Principles of Combat Modeling and Distributed Simulation. Wiley, New Jersey (2012)
4. Law, A., Kelton, W.: Simulation Modelling & Analysis. McGraw-Hill, New York (1991)
5. Jaiswal, N.K.: Military Operations Research: Quantitative Decision Making. Kluwer Academic Publishers, USA (1997)
6. Best, C., Galanis, G., Kerry, J., Sottilate, R., Fundamental Issues in Defence Training and Simulation. Ashgate Publishing Ltd., Surrey, GU9 7PT, England (2013)
7. Rao, D.V.: The Design of Air Warfare Simulation System. Technical report, Institute for Systems Studies and Analyses (2011)
8. Ryan, A., Grisogono, A.M.: Hybrid complex adaptive engineered systems: a case study in defence. In Proceedings of the International Conference on Complex Systems (2004)
9. Ilachinski, Andy: Irreducible semi-autonomous adaptive combat (ISAAC): an artificial-life approach to land combat. Mil. Oper. Res. **5**(3), 29–46 (2000)
10. Ilachinski, Andy: Towards a Science of Experimental Complexity: An Artificial-Life Approach to Modeling Warfare. Center for Naval Analyses, undated, Alexandria, VA (1999)
11. Dockery, J.T., Woodcock, A.E.R. (eds.): The Military Landscape: Mathematical Models of Combat. Woodhead Publishing, Cambridge (1993)
12. Ilachinski, A.: Land Warfare and Complexity, Part I: Mathematical Background and Technical Sourcebook (No. CNA-CIM-461.10). Center for Naval Analyses, Alexandria VA (1996)
13. Ilachinski, A.: Land Warfare and Complexity, Part Ii: An Assessment of the Applicability of Nonlinear Dynamics and Complex Systems Theory to the Study of Land Warfare (No. CNA-CRM-96-68). Center for Naval Analyses, Alexandria, VA (1996)
14. Mittal, S., Martin, J.L.R.: Netcentric System of Systems Engineering with DEVS Unified Process. CRC Press, Boca Raton (2013)
15. Wainer, G.A.: Discrete-Event Modeling and Simulation: A Practitioner's Approach. CRC Press, Boca Raton (2009)
16. Zeigler, B.P., Praehofer, H., Kim, T.G.: Theory of Modeling and Simulation: Integrating Discrete event and Continuous Complex Dynamic Systems, 2nd edn. Academic Press, San Diego, CA, USA (2000)
17. Zeigler, B.P., Sarjoughian, H.S.: Guide to Modeling and Simulation of Systems of Systems: Simulation Foundations, Methods and Applications. Springer, London (2013)
18. Zeigler, B.P., Hammonds, P.E.: Modeling and Simulation-Based Data Engineering: Introducing Pragmatics into Ontologies for Net-Centric Information Exchange. Elsevier Academic Press, USA (2007)
19. Stol, K.J., Avgeriou, P., Babar, M., Lucas, Y., Fitzgerald, B.: Key factors for adopting inner source. ACM Trans. Softw. Eng. Method. (TOSEM), **23**(2) (2014)
20. Rafferty, L.A., Stanton, N.A., Walker, G.H.: The Human Factors of Fratricide. Ashgate Publishing Ltd., Surrey, GU9 7PT, England, (2012)
21. Rao, D.V.: Damage estimation and assessment modelling of ground targets in air-land battle simulations. J. Battlefield Technol. **18**(1) (2015)
22. Rao D.V., Saha B.: An Agent oriented Approach to Developing Intelligent Training Simulators. In: SISO Euro-SIW Conference, Ontario, Canada (2010)
23. Bonabeau, E.: Agent-based modeling: methods and techniques for simulating human systems. Proc. Natl. Acad. Sci. U.S.A. **99**(Suppl 3), 7280–7287 (2002). doi:10.1073/pnas.082080899
24. Burgin, G.H., Fogel L.J.: Air-to-Air combat tactics synthesis and analysis program based on an adaptive maneuvering logic. J. Cybern. **2**(4), (1972)

25. Fogel, L.J.: Rule based air combat simulation. DTIC Report, http://www.dtic.mil/dtic/tr/fulltext/u2/a257194.pdf
26. Weiss, G. (ed.): Multiagent Systems (Intelligent Robotics and Autonomous Agents series), 2nd edn. The MIT Press, Cambridge 02142, USA (2013)
27. Ilachinski, A.: Artificial War: Multiagent-Based Simulation of Combat. World Scientific, Singapore (2004)
28. Wooldridge, M.: An Introduction to MultiAgent Systems, 2nd edn. Wiley, UK (2009)
29. Taher, J., Zomaya, A.Y.: Artificial Neural Networks in Handbook of Nature-Inspired and Innovative Computing. In: Zomaya, A.Y. (ed.) Integrating Classical Models with Emerging Technologies, pp. 147–186. Springer, USA (2006)
30. Jang, J.S.R.: ANFIS: Adaptive Network-based fuzzy inference systems. IEEE Trans. Syst. Man Cybern. 23(3), 665–685 (1993)
31. Jang, J.S.R., Sun, C.T., Mizutani, E.: Neuro-Fuzzy and Soft Computing: A Computational approach to Learning and Machine Intelligence. Prentice Hall, New Jersey (1997)
32. Mitra, S., Yoichi, Hayashi: Neuro-Fuzzy rule generation: survey in soft computing framework. IEEE Trans. Neural Netw. 11(3), 748–768 (2000)
33. Goodrich, K.H., McManus, J.W.: Development of a tactical guidance research and evaluation system (TGRES). In: AIAA Paper 893312 (1989)
34. Kurnaz, S., Cetin, O., Kaynak, O.: Adaptive neuro-fuzzy inference system based autonomous flight control of unmanned air vehicles. Expert Syst. Appl. 37(2), 1229–1234 (2010)
35. Moon, I.C., Carley, K.M.: Modeling and simulating terrorist networks in social and geospatial dimensions. IEEE Intell. Syst. 22(5), 40–49 (2007)
36. Lee, G., Oh, N., Moon, I.C.: Modeling and simulating network-centric operations of organizations for crisis management. In Proceedings of the 2012 Symposium on Emerging Applications of M&S in Industry and Academia Symposium, p. 13. Society for Computer Simulation International (2012)
37. Moon, I.C., Oh, A.H., Carley, K.M.: Analyzing social media in escalating crisis situations. In: 2011 IEEE International Conference on Intelligence and Security Informatics (ISI), pp. 71–76. IEEE (2011)
38. Rao, D.V., Singh, S.: Recognition and identification of target images using feature based retrieval in UAV missions. In: IEEE National Conference on Computer Vision, Pattern Recognition, Image Processing and Graphics (NCVPRIPG 2013) IIT Jodhpur (2013)
39. Fowler, C.A. Bert: Asymmetric warfare: a primer. IEEE Spectrum http://spectrum.ieee.org/aerospace/aviation/asymmetric-warfare-a-primer (2006)
40. Kokar, M.M., Wang, J.: An example of using ontologies and symbolic information in automatic target recognition. In: Sensor Fusion: Architectures, Algorithms, and Applications VI, pp. 40–50. SPIE. (2002)
41. Rao, D.V., Shankar, R., Iliadis L., Sarma, V.V.S.: An Ontology based approach to designing adaptive lesson plans in military training simulators. In: Proceedings of the 13th EANN (Engineering Applications of Neural Networks), Springer LNCS AICT, vol. 363(1), pp. 81–93 (2012)
42. Woelk, D.: e-Learning technology for improving business performance and lifelong learning. In: Singh, Muninder P. (ed.) The Practical Handbook of Internet Computing. Chapman & Hall, USA (2005)
43. Hammond, K.F.: Case based planning: a frame-work for planning from experience. Cogn. Sci. 14(3), 385–443 (1990)
44. Hanks, S., Weld, D.S.: A domain-independent algorithm for plan adaptation. J. Artif. Intell. Res. 2, 319–360 (1995)
45. Kambhampati, S., Hendler, J.A.: A validation-structure-based theory of plan modification and reuse. Artif. Intell. 55(2), 193–258 (1992)
46. Munoz-Avila, H., Cox, M.: Case-based plan adaptation: an analysis and review. IEEE Intell. Syst. 23, 75–81 (2007)

47. Ram, A., Francis, A.: Multi-plan retrieval and adaptation in an experience-based agent. In: Leake, D.B. (ed.) Case-Based Reasoning: Experiences, Lessons, and Future Directions. AAAI Press (1996)
48. Spalazzi, L.: A survey on case-based planning. Artif. Intell. Rev. **16**(1), 3–36 (2001)
49. Sugandh, N., Ontanon, S., Ram, A., Online Case-based plan adaptation for real-time strategy games. In: Proceeding Twenty Third AAAI Conference on AI, pp. 702–707 (2008)
50. Rao, D.V., Iliadis L., Spartalis S.: A Neuro-fuzzy hybridization approach to model weather operations in a virtual warfare analysis system. In: Proceedings of the 12th EANN (Engineering Applications of Neural Networks), Springer LNCS AICT, vol. 363(1), pp. 111–121 (2011)
51. Uschold, M., Gruninger, M.: Ontologies: principles, methods and applications. Knowl. Eng. Rev. **11**(2) (1996)
52. Gomez-Perez, A., Corcho, O., Fernandez-Lopez, M.: Ontological Engineering: with examples from the areas of Knowledge Management. e-Commerce and the Semantic Web. Springer, Berlin (2004)
53. Neches, R., Fikes, R., Finin, T., Gruber, T., Patil, R., Senator, T., Swartout, W.: Enabling technology for knowledge sharing. AI Magazine Fall **12**(3) (1991)
54. Fisher, M., Sheth, A.: Semantic enterprise content management. In: Singh, Muninder P. (ed.) The Practical Handbook of Internet Computing. Chapman & Hall, USA (2005)
55. Meliza, L.L., Goldberg, S.L., Lampton, D.R.: After action review in simulation-based training, RTO-TR-HFM-121-Part-II, Technical report (2007)
56. Raju P., Ahmed. V.: Enabling technologies for developing next generation learning object repository for construction. Autom. Constr. vol 22 (2012)
57. Berry, J., Benlamri, R., Atif, Y.: Ontology-Based framework for context-aware mobile learning. In: IWCMC'06, Canada (2006)
58. Sabou, M., Wroeb, C., Goble, C., Stuckenschmidt, H.: Learning domain ontologies for semantic web service descriptions. J. Web Semant. vol. 3 (2005)
59. Brusilovsky, P., Wolfgang, N.: Adaptive hypermedia and adaptive web. In: Singh, Muninder P. (ed.) The Practical Handbook of Internet Computing. Chapman & Hall, USA (2005)
60. Aparicio IV, M., Singh, M.P.: Concepts and practice of personalisation. In: Singh, Muninder P. (ed.) The Practical Handbook of Internet Computing. Chapman & Hall, USA (2005)
61. http://protege.stanford.edu/doc/users.html#papers
62. http://www-ksl-svc.stanford.edu:5915/doc/frame-editor/what-is-an-ontology.html
63. http://www.cs.man.ac.uk/~sattler/reasoners.html
64. Lenat, D.B., Guha, R.V.: Building Large Knowledge-Based Systems, Reading. Addison-Wesley, MA (1990)
65. Gruber, T.: Toward principles for the design of ontologies used for knowledge sharing. In: Guarino, N. (ed.) International Workshop on Formal Ontology, Padova, Italy (1993)
66. Gruber, T.: Toward principles for the design of ontologies used for knowledge sharing. In: Guarino, N., Poli, R. (eds.) Formal Ontology in Conceptual Analysis and Knowledge Representation. Kluwer Academic, Dordrecht (1995)
67. Noy, N.F., McGuinness, D.L.: Ontology development 101: a guide to creating your first ontology. Stanford Knowledge Systems Laboratory, Technical Report KSL-01–05 and Stanford Medical Informatics Technical Report SMI-2001-0880 (2001)
68. Cox, E.: The Fuzzy Systems Handbook, 2nd edn. Academic Press, New York (1999)
69. Banks, J.: Handbook of Simulation: Principles, Methodology, Advances, Applications, and Practice. Wiley, New York (1998)
70. Rao D.V., Jasleen, K.: A Fuzzy rule-based approach to design game rules in a mission planning and evaluation system. In: 6th IFIP Conference on Artificial Intelligence Applications and Innovations. Springer, Berlin (2010)

71. Rao D.V., Iliadis L., Spartalis S., Papaleonidas A.: Modelling environmental factors and effects in virtual warfare simulators by using a multi agent approach. Int. J. Artif. Intell. **9** (A12), pp. 172–185 (2012)
72. Kim, J., Moon, I.C., Kim, T.G.: New insight into doctrine via simulation interoperation of heterogeneous levels of models in battle experimentation. Simulation **88**(6), 649–667 (2012)
73. Rao, D.V., Mehra, P.K.: A methodology to evaluate the combat potential and military force effectiveness for decision support. J. Battlefield Technol. **16**(1) (2013)
74. Mendel, J.M.: Uncertain Rule-Based Fuzzy Logic Systems-Introduction and New Directions. Prentice Hall PTR, Upper Saddle River (2001)
75. Ilachinski, A.: EINSTein. In: Artificial Life Models in Software, pp. 259–315. Springer, London (2009)
76. Ilachinski, A.: EINSTein: a multiagent-based model of combat. In: Artificial Life Models in Software, pp. 143–185. Springer, London (2005)

Improving Load Signal and Fatigue Life Estimation for Helicopter Components Using Computational Intelligence Techniques

Catherine Cheung, Julio J. Valdés and Jobin Puthuparampil

Abstract The accurate estimation of helicopter component loads is an important factor in life cycle management and life extension efforts. This chapter explores continued efforts to utilize a number of computational intelligence algorithms, statistical and machine learning techniques, such as artificial neural networks, evolutionary algorithms, fuzzy sets, residual variance analysis, and others, to estimate some of these helicopter dynamic loads. For load prediction using indirect computational methods to be practical and accepted, demonstrating slight over-prediction of these loads is preferable to ensure that the impact of the actual load cycles is captured by the prediction and to incorporate a factor of safety. Subsequent calculation of the component's fatigue life can verify the slight over-prediction of the load signal. This chapter examines a number of techniques for encouraging slight over-prediction and favoring a conservative estimate for these loads. Estimates for the main rotor normal bending on the Australian S-70-A-9 Black Hawk helicopter during a left rolling pullout at 1.5 g manoeuvre were generated from an input set consisting of thirty standard flight state and control system parameters. The results of this work show that when using a combination of these techniques, a reduction in under-prediction and increase in over-prediction can be achieved. In addition to load signal estimation, the component's fatigue life and load exceedances can be estimated from the predicted load signal. In helicopter life cycle management, these metrics are more useful performance measures (as opposed to mean squared error or correlation of the load signal), therefore this chapter describes the process followed to calculate these measures from the load signal using Rainflow counting, material specific fatigue data (S-N curves), and damage theory. An evaluation of the proposed techniques based on the fatigue life estimates and/or load exceedances is also made.

C. Cheung (✉) · J.J. Valdés · J. Puthuparampil
National Research Council Canada, 1200 Montreal Rd, Ottawa, ON K1A 0R6, Canada
e-mail: catherine.cheung@nrc-cnrc.gc.ca

© Springer International Publishing Switzerland 2016
R. Abielmona et al. (eds.), *Recent Advances in Computational Intelligence
in Defense and Security*, Studies in Computational Intelligence 621,
DOI 10.1007/978-3-319-26450-9_22

605

1 Introduction

Operational requirements are significantly expanding the role of aircraft fleets in many countries. Particularly in military helicopter fleets, this expansion has resulted in aircraft flying missions that are beyond the original design usage. Therefore, the current life usage estimation for the fatigue critical components may no longer have the required low probability of failure, or conversely, operational lives could be expanded if actual mission flight load spectra were found to be considerably less severe. Due to such change in usage, there is a need to monitor individual aircraft usage, compare it with the original design usage spectrum to more accurately determine the life of critical components.

For helicopters, which constantly operate in a complex dynamic and highly vibratory loading environment, there are a number of structural components that have a specified fatigue life, that is, the length of time that the component can be safely operated with minimal or acceptable risk of failure. These components experience low-magnitude high-cycle fatigue as well as large-magnitude low-cycle fatigue caused by the oscillating loads of the main rotor and/or tail rotor systems. In the dynamic system of helicopters, typically encompassing the rotor system, flight control linkages and the rotor masts (but excluding the engines, driveshafts, and transmission), the fatigue life of the components is known as the component retirement time. These retirement times are traditionally determined using the safe life methodology which assumes that the probability of a crack forming in a component during its service lifetime remains below an acceptable level [1].

Monitoring loads on individual aircraft (or platform) in a fleet has the potential to lead to more reliable and safer exploitation of the platform, optimized and extended usage of the component and platform operational lives, better platform usage and fleet management, and reduced maintenance and life cycle costs. Currently, load monitoring and fatigue life monitoring in helicopters is rarely performed directly. The most common method for tracking helicopter usage is simply by monitoring flight hours. The loads on the components are then assumed to be the same as those derived during aircraft certification based on full-scale fatigue tests and instrumented flight tests. However, these load assumptions can differ from reality on an individual platform basis, and may even be aggravated when platforms assume different roles than the ones for which they were initially designed and tested for thus affecting safety and economic efficient usage. Therefore, monitoring loads on individual aircraft in a fleet has the potential to lead to an increased and safer exploitation of each platform, with increased efficiency, better platform usage and fleet management, and reduced life cycle costs.

While measuring dynamic component loads directly is possible through installed sensors, the installation and operation of a sensor suite on an individual aircraft basis is challenging and expensive, and therefore seldom implemented. Traditional measurement systems for dynamic components include slip rings or telemetry systems; however, these methods have not proven reliable—they introduce a considerable amount of noise, are difficult to maintain, and pose the risk of loss of sensor signal during flight. Therefore, a robust and accurate process to indirectly estimate these loads could be a practical alternative, which can be used to complement existing methods or supplement sensor data in the future when these measurement methods are proven reliable. The implementation of computational intelligence techniques to this problem is a natural fit, given the complexity of the load signals and influence of numerous factors (e.g. aircraft altitude, aircraft speed, engine torque, etc.). Load estimation methods can utilize data obtained from existing aircraft instrumentation, such as standard flight state and control system (FSCS) parameters, to minimize the need for additional sensors and accordingly avoiding the substantial costs associated with installing, maintaining and monitoring additional instrumentation.

Although the use of computational intelligence algorithms (most commonly neural networks and regression methods) has been explored by others to indirectly (i.e. without the use of additional sensors) estimate loads or fatigue lives in aircraft components, the developed methodologies have not been reliable and accurate for an in-service application. The methodology presented in this chapter enables the prediction of the load signals on components of a helicopter, using existing flight data and avoiding the use of additional sensors. We refer to the time-varying measurement of the load by the helicopter instrumentation as the load signal or load-time signal in this work. The prediction is performed through the use of computational intelligence algorithms, statistical and machine learning techniques, such as artificial neural networks, evolutionary algorithms, fuzzy sets, residual variance analysis, and others. There is an important pre-processing stage including data synchronization, appropriate standardization, time dependencies assessment, sampling, relevance evaluation (feature selection), cleaning, and load signals information content characterization. The importance of these steps is crucial and previous modelling attempts that did not consider them failed. The predicted load signals then form the basis for estimating the fatigue life of the component.

Preliminary work exploring the use of computational intelligence techniques for estimating helicopter loads on the Australian S-70A-9 Black Hawk (Fig. 1) showed that reasonably accurate and correlated predictions could be obtained [2, 3]; however, these efforts had a tendency to under-predict the target signal. If these predictions were then used for calculating component retirement times, an under-predicted

Fig. 1 S-70A-9 Australian Army Black Hawk

load signal would indicate a less severe loading and therefore the fatigue damage associated with that load signal would be underestimated, posing a potential safety risk. While an over-predicted signal would provide a conservative estimate for the component's remaining life, to avoid excessive reductions to the component's fatigue life, an overly conservative estimate is also undesirable; in order for load prediction and fatigue life estimation using computational intelligence techniques to be useful and accepted, demonstrating slight over-prediction is preferable to ensure that the impact of the actual load cycles is captured by the prediction. In this chapter, a number of techniques are introduced and explored to improve the load prediction, directly addressing the under-prediction tendency previously encountered. The fatigue life of the component can then be estimated from the predicted load signal. In helicopter life cycle management, this metric is a more useful performance measure (as opposed to mean squared error or correlation of the load signal). This work therefore evolved to attain a complete approach to load monitoring which includes estimating the load-time signal and calculating the subsequent fatigue life, being particularly devoted to rotating components and avoiding the use of additional sensors, which is specifically challenging when dynamic components are considered. Initial results calculating the fatigue damage accumulation based on load signal predictions using a simplified methodology established a process to evaluate load signal predictions [4, 5].

This chapter describes continued efforts to improve the load-time signal predictions by using several different techniques to encourage some over-prediction and thereby favor conservative estimates. These techniques include (i) altering the sampling scheme of the training and testing sets, and (ii) implementing alternative error functions. The specific problem was to estimate the main rotor normal bending (MRNBX) on the Australian Army Black Hawk helicopter using only flight state and control system (FSCS) variables during the rolling pullout flight condition. The process followed to calculate the fatigue life from the load signal using Rainflow counting, material specific fatigue data (S-N curves), and damage theory is also described. An evaluation of the proposed techniques based on fatigue life measures is made. The objective of this work is to explore the effectiveness of these techniques to generate slightly over-predicted load signals and conservative component fatigue life estimates.

The chapter is organized as follows: Sect. 2 describes the test data, Sect. 3 explains the methodology that was followed, Sect. 4 describes the two techniques explored for favoring conservative estimates, Sect. 5 details the computational intelligence techniques used for search and modelling, Sect. 6 explains the fatigue life estimation process, Sect. 7 highlights the key results and Sect. 8 presents the conclusions.

2 Black Hawk Flight Loads Survey Data

A joint program between the United States Air Force and the Australian Defence Force, implemented in the year 2000, consisted of a flight loads survey for the S-70A-9 Black Hawk (UH-60 variant). The flight tests covered a broad range of steady state and transient flight conditions at various gross weights, altitudes and air-

Table 1 Black Hawk flight state and control system (FSCS) parameters

Mnemonic	Description
VCASBOOM	Air speed (boom)
LOADFACT	Vertical acceleration, load factor at CG
ATTACK	Angle of attack (boom)
SIDESLIP	Sideslip angle (boom)
PITCHATT	Pitch attitude
PITCHRAT	Pitch rate
PITCHACC	Pitch acceleration
ROLLATT	Roll attitude
ROLLRAT	Roll rate
ROLLACC	Roll acceleration
HEAD180	Heading
YAWRAT	Yaw rate
YAWACC	Yaw acceleration
LGSTKP	Longitudinal stick/cyclic position
LATSTKP	Lateral stick/cyclic position
PEDP	Directional pedal position
COLLSTKP	Collective stick position
STABLAIC	Stabilator position
NR	% of max main rotor speed
ERITS	Retreating tip speed
MRQ	Main rotor shaft torque
TRQ	Tail rotor drive shaft torque
NO1QPCT	No. 1 Engine torque
NO2QPCT	No. 2 Engine torque
NO1T45	No. 1 Eng power lever (temp)
NO2T45	No. 2 Eng power lever (temp)
HBOOM	Barometric altitude (boom)
FAT	Temperature (Kelvin)
HD	Altitude (height density)
ROCBOOM1	Barometric rate of climb (boom)

craft configurations. A total of 65 hours of flight test data was obtained from the survey, which included the measurements from 321 strain gauges (249 on the airframe and 72 on dynamic components). Flight State and Control System (FSCS) parameters were also simultaneously acquired, as well as data from accelerometers and other sensors installed at numerous locations on the aircraft. The data used for this work were the FSCS parameters and the strain gauge measurements on the dynamic components of the aircraft. Full details of the instrumentation and flight loads survey are provided in [6].

One of the key aspects of this research is to rely solely on data from the FSCS parameters to determine helicopter dynamic loads and component fatigue life, as these FSCS parameters are already recorded by the flight data recorder found on most helicopters. There were thirty FSCS parameters recorded on the Black Hawk helicopter during the flight loads survey. The thirty FSCS parameters on the Black Hawk that were examined are listed in Table 1. The main rotor normal bending (MRNBX) was selected as the target output. This gage measured the normal bending on the main rotor blade cuff. From over 50 flight conditions, rolling left pullout at 1.5 g was selected for this work. This manoeuvre is a relatively severe and dynamic flight condition that should present a greater modelling challenge since there is considerable variation in the parameter values through each recording. The FSCS parameters were sampled at 52 Hz, while the main rotor sensors recorded at a higher frequency of 416 Hz, so the main rotor sensor data was downsampled to match the FSCS parameter frequency.

3 Methodology for Load Signal Prediction

The overall goal of this work was to develop computational models to generate accurate predictions for helicopter loads using flight state and control system (FSCS) parameters. The use of existing flight data (FSCS parameters) means that the costly installation and maintenance of additional sensors can be avoided. A variety of computational intelligence techniques, including statistical and machine learning methods, were implemented for load signal prediction. The general process followed for the load time signal prediction is depicted in Fig. 2. From the instrumented aircraft data, the FSCS parameters and sensor measurements were extracted as the input parameters and the target parameter respectively. Computational intelligence techniques were then applied to learn models which predict the component load signal from the FSCS parameters.

The application of computational intelligence and machine learning techniques to develop these models for load signal prediction occurred in two phases: (i) data exploration: characterization of the internal structure of the data and assessment of the information content of the predictor variables (in particular their relation to the target variable); and (ii) modelling: build models relating the dependent and the predictor variables. The load signal predictions were then used to calculate the fatigue life of the component. This methodology is illustrated in Fig. 3.

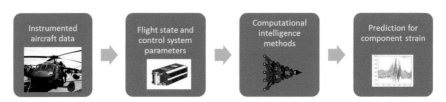

Fig. 2 Component load signal prediction from FSCS parameters

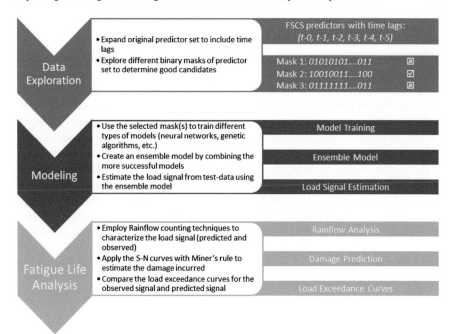

Fig. 3 Methodology for load signal prediction and fatigue life estimation

The data exploration phase is a key pre-processing stage that includes data synchronization, appropriate standardization, time dependencies assessment, sampling, relevance evaluation (feature selection), cleaning, and load signals information content characterization. These steps are important in order to extract the most relevant predictive information from the data. Typically the recorded FSCS parameters are not chosen with load signal estimation in mind and from this point of view they contain noise, spurious interactions and other unwanted effects. They distort and bury useful dependencies which also are often weak and indirect. Data exploration is essential in order to identify these dependencies and extract load signal properties for improving the results of the modelling process. Moreover, it provides knowledge which allows a better understanding of the inner fabric and particularities of the physical system.

In the data exploration phase, the time dependencies within the FSCS parameters and the target variables were explored through the use of phase space methods and residual variance analysis (the Gamma test). These techniques were applied in order to find subsets of predictor variables with relevant function relationships with respect to the target variable, as well as for determining time-lagged structures involving those subsets. They are important for assessing how past events influence future target variable values. Using the Gamma test, it was determined in [3, 7] that 5 time lags (i.e. for a given time t, lagged values at $\{t, t-1, \ldots, t-5\}$) were required to best estimate the behavior of the system from the 30 FSCS parameters. From this result, a total of 180 predictor variables (30 parameters × 5 time lags) would be necessary to train the machine learning algorithm. Five time lags at the FSCS sampling rate of 52 Hz corresponds to about 180° or half of one revolution of the main rotor's rotation. The next stage in the process was to explore the relevant information in the data using the Gamma test and multi-objective genetic algorithms (MOGA). Through the application of MOGA, binary masks were generated which identify which of the 180 predictor values have the greatest influence on the target signal. The most promising subsets (or masks) were then selected as inputs and used as a base for model search in the modelling stage.

During the modelling stage, computational techniques were implemented to build models relating the target variable to the subset of predictor variables (as identified in the data exploration stage). In particular, neural networks were used, trained with evolutionary computation techniques (differential evolution (DE)). Initial work [2, 3] explored the use of different modeling techniques for this application, including M5 model trees, local linear regression, genetic programming, and neural networks, but the neural network framework consistently provided better results and so this technique was pursued.

3.1 Gamma Test (Residual Variance Analysis)

The Gamma test, developed by [8–10], is an algorithm that assists in the creation of data-driven models of smooth systems. This technique aims to estimate the amount of noise present in a given dataset (its variance), wherein noise is recognized as any source of variation in the output (target) variable that cannot be explained by a smooth transformation (model) relating the output with the input (predictor) variables. The critical piece of information provided by the Gamma test is whether it is possible to successfully find (fit) a smooth model to the dataset. Since model search through data mining is a time consuming, computationally expensive procedure, it is beneficial to have an estimate of the amount of determinism carried by the variables involved in the process and a quantitative upper bound for the best error that could be expected from a smooth model using the selected predictors, that is, an indication as to whether it is hopeful or hopeless to try to learn a model using the given predictors. The Gamma test can also help gauge the complexity of the underlying data and if it is necessary for more explanatory variables to be incorporated, in order to provide an improved model. If the test outputs a small gamma estimate value, it suggests that a smooth deterministic dependency can be expected.

If $X = \{x_0, \ldots, x_n\}$ is a set of candidate predictor variables of a target variable y their functional relationship could be expressed as $y = f(x_0, \ldots, x_n) + r$ where f is a (deterministic) model and r a residual variable representing noise, random variation or unaccounted determinism not captured by X. Under the assumptions of continuity and smoothness (bounded partial derivatives), the Gamma test produces an estimate of the variance associated to r using only the available (training) data. Normalization of this variance into a parameter referred to as the vRatio (V_r) allows for the comparison of *r*-variances across datasets. Another important value calculated by the Gamma test is the so-called gradient, G, which is a measure of the complexity associated with the system.

The Gamma test is a versatile tool which can be employed in numerous ways during data exploration. In this work, it was used to determine the number of time lags which are relevant in the prediction of the future target sensor values, and most notably, to determine the subset of lagged FSCS variables with the largest prediction potential (therefore, the best candidates for building predictive models). The Gamma test techniques were used for an extensive data exploration for different target sensors under different flight conditions, in an effort to find the optimal subsets of the lagged FSCS variables while simultaneously minimizing V_r (large prediction power), G (low complexity), and number of predictor variables (denoted by #). This task is accomplished by means of an evolutionary computation framework using multi-objective genetic algorithms with $<V_r, G, \#>$ as three simultaneous objectives to minimize.

3.2 Simplified Methodology

While the application of these techniques in the data exploration phase enables much useful information to be extracted from the data, this approach is fairly complex and computationally intensive. Therefore, recent efforts [4] incorporating the fatigue life estimation steps simplified the whole methodology to expedite the entire estimation process. The simplified version of the methodology is shown in Fig. 4, where the original 30 FSCS parameters are used as inputs without any time lags to a scaled conjugate gradient (SCG) neural network model to predict the main rotor normal

Fig. 4 Simplified
methodology

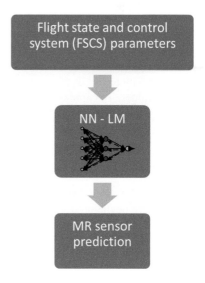

bending. Application of this simplified methodology is the most direct and straight-forward method of modelling the target load signal and estimating the component fatigue life from the 30 FSCS parameters. Results from both the full and simplified methodologies are presented in this work.

4 Techniques for Conservative Estimates

4.1 Training Set Construction

The dataset used for training a neural network should capture all of the fundamental dynamics of the system, if the resulting learnt model is to be capable of making accurate predictions with previously unseen data. A common practice in machine learning to eschew this problem is to use 50–90 % of the available data for training, with the remaining for testing. However, this greedy approach is highly impractical with tremendously large data sets, as is the case in this study. To keep computing times realistic, two alternative methods were used in constructing the training sets: the k-leaders sampling, and the biased sampling.

The k-means clustering [11] method is employed to consolidate a training set of practical size ensuring that the sample is still statistically representative of the entire dataset. A training set of 2000 clusters was formed via the k-means algorithm applied using Euclidian distance, from which the data vector closest to the centroid of each cluster was chosen as the k-leader, resulting in 2000 k-leaders. As every data vector is part of a cluster with a representative k-leader, this algorithm ensures that every multivariate vector in the initial dataset is represented in the training sample while producing samples of computationally manageable sizes.

A second sampling technique to form the training and testing sets was explored. A biased sampling technique was introduced in previous work [12, 13] and is used again here. In an unbiased sampling scheme, the statistical distribution of the training and testing sets would ideally be equivalent. For the helicopter load estimation problem though, this need not be the case, as upper and lower values in the target signal contribute significantly more damage towards the component than the inter-mediate values. To capture this difference in importance, a biased sampling scheme was devised where the training set purposely contains far more tuples associated with these classes of special interest than would be present in the natural distribution of the data. This, in theory, would (i) drive the learning process towards capturing these extremities more accurately, and (ii) lead the network to over-predict the intermedi-ate values.

The threshold defining these regions, or classes, was set as a function of the mean and standard deviation of the target signal: the data points exceeding the mean plus 1 standard deviation belonged to the 'high' class, those whose values were below the mean minus 1 standard deviation fell in the 'low' class, and the remaining data was assigned to the 'medium' class. The distribution between the high, medium, and low

Fig. 5 Distribution of training data. The x-axis plots the peak values in terms of the number of standard deviations from the mean (mean at 0)

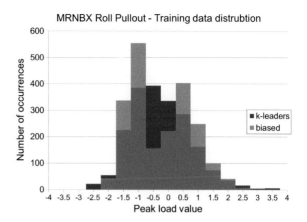

classes in the training set was set as 0.4, 0.2 and 0.4 respectively with the purpose of steering the process towards the extreme classes. For MRNBX rolling pullout, the distribution of the data points in the training set for the two sampling schemes (k-leaders and biased) is shown in Fig. 5.

For the simplified version of the methodology, a traditional 90/10 % partitioning between training and testing sets was used without any structural sampling. Models were typically trained from a 90 % subset of the data, while the remaining 10 % was used for testing.

4.2 Alternative Error Functions

The second method attempted to encourage conservative estimates for helicopter loads uses alternate error functions instead of mean squared error (MSE) for the neural network to minimize during its training.

An asymmetric fuzzy-based error function was introduced in [13] and is described in Eq. 1:

$$\mathscr{E}(x, T) = \frac{1}{S(x, T)} - 1 \tag{1}$$

where $S(x, T)$ is a membership function of predicted values x with respect to the class defined by a target value T. For $T \geq 0$, $S(x, T)$ is defined piecewise by Eq. 2:

$$S(x, T) = \begin{cases} (1 + (\alpha_u |x - T|))^{-1} & \text{if } x < T \\ 1 & \text{if } x \geq T \ \& \ |x - T| <= \varepsilon \\ (1 + (\alpha_o |x - T|))^{-1} & \text{if } x \geq (T + \varepsilon) \end{cases} \tag{2}$$

Fig. 6 Fuzzy membership function. The *top plot* shows the fuzzy membership function with two sets of parameter values ($\varepsilon = 0.15, \alpha_u = 10, \alpha_o = 1$ and $\varepsilon = 0.05, \alpha_u = 5, \alpha_o = 3$) to illustrate the effect of changing each parameter. The *bottom plot* shows the resulting error function for each set of parameters

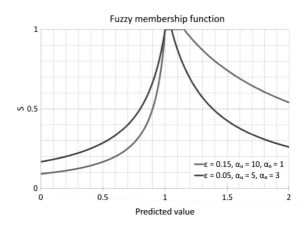

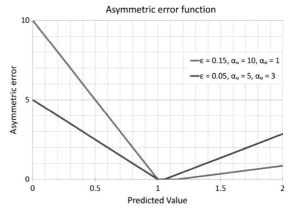

The parameters of the S function ($\{\varepsilon, \alpha_u, \alpha_o\} \in \Re^+$) can be varied in order to stimulate slight to moderate over-prediction (that is, conservative estimates, provided that the target variable represents a helicopter load). The membership function, S, with two sets of parameter values and the resulting error functions can be seen in Fig. 6. This asymmetric error function contrasts with the standard mean squared error (MSE) measure which is symmetric and therefore does not differentiate under and over-prediction (Fig. 7). Variants of the asymmetric error function were attempted with the parameter settings listed in Table 2.

Another error function that was implemented in this work is the symmetric modified absolute percentage error (SMAPE) given by Eq. 3,

$$SMAPE = n^{-1} \sum_{i=1}^{n} \frac{|y_i - f_i|}{\frac{|y_i| + |f_i|}{2}} \tag{3}$$

where y_i is the ith observed data point, and f_i is the corresponding prediction [14]. This function is not symmetric despite its name and over-prediction results in a

Fig. 7 Comparison of three error functions: mean squared error (MSE), symmetric modified absolute percentage error (SMAPE), and fuzzy-based asymmetric error function ($\varepsilon = 0.15, \alpha_u = 10, \alpha_o = 1$)

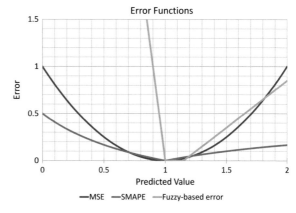

Table 2 Parameter values for asymmetric error function

Parameter	Values
ε	$0.15, 0.1, 0.05, 0$
α_u	$10, 5, 3$
α_o	$10, 5, 3, 1$

smaller error than under-prediction by the same amount (e.g. for $y = 100$ and $f = 110$ yields $SMAPE = 9.5\,\%$, while for $y = 100$ and $f = 90$ yields $SMAPE = 10.3\,\%$). The three error functions are plotted in Fig. 7.

5 Neural Network Training Methods

Training neural networks involves an optimization process, typically focused on minimizing an error measure. This operation can be done using a variety of approaches ranging from deterministic methods to stochastic, evolutionary computation (EC) and hybrid techniques. Since asymmetric and other error functions were used in this study, traditional deterministic (gradient descent) methods could not be used, as the partial derivatives of the error surface are not provided. Consequently, differential evolution (an EC technique) was the primary learning technique used in this work.

Differential evolution (DE) [15, 16] is a type of evolutionary algorithm which uses populations of individual real-valued vectors and subjects them to an evolution process. Although less frequently utilized than genetic algorithms, it has proven its effectiveness in complex optimization problems, outperforming other approaches [17, 18]. The general methodology is as follows:

step 0 Initialization: Create a population $\mathscr{P}$ of random vectors in $\mathfrak{R}^n$, and decide upon an objective function $f : \mathfrak{R}^n \to \mathfrak{R}$ and a strategy $\mathscr{S}$, involving vector differentials.

step 1 Choose a target vector from the population $x_t \in \mathscr{P}$.

step 2 Randomly choose a set of other population vectors $\mathscr{V} = \{x_1, x_2, \ldots\}$ with a cardinality determined by strategy $\mathscr{S}$.

step 3 Apply strategy $\mathscr{S}$ to the set of vectors $\mathscr{V} \cup \{x_t\}$ yielding a new vector $x_{t'}$.

step 4 Add x_t or $x_{t'}$ to the new population according to the value of the objective function f and the type of problem (minimization or maximization).

step 5 Repeat steps 1–4 to form a new population until termination conditions are satisfied.

There are several variants which can be classified using the notation $DE/x/y/z$, where x specifies the vector to be mutated, y is the number of vectors used to compute the new one and z denotes the crossover scheme. Let F be a scaling factor, $\mathscr{C}_r \in \mathfrak{R}$ be a crossover rate, D be the dimension of the vectors, $\mathscr{P}$ be the current population, $N_p = card(\mathscr{P})$ be the population size, v_i, $i \in [1, N_p]$ be the vectors of $\mathscr{P}$, $b_{\mathscr{P}} \in \mathscr{P}$ be the population's best vector w.r.t. the objective function f and $r, r_0, r_1, r_2, r_3, r_4, r_5$ be random numbers in $(0, 1)$ obtained with a uniform random generator function $rnd()$ (the vector elements are v_{ij}, where $j \in [0, D)$). Then the transformation of each vector $v_i \in \mathscr{P}$ is performed by the following steps:

step 1 Initialization: $j = (r \cdot D), L = 0$

step 2 $while(L < D)$

step 3 $if((rnd() < \mathscr{C}_r)||L == (D - 1))$
/* create a new trial vector. For example, as: */
$t_{ij} = b_{\mathscr{P}j} + F \cdot (v_{r_1j} + v_{r_2j} - v_{r_3j} - v_{r_4j})$

step 4 $j = (j + 1) \bmod D$

step 5 $L = L + 1$

step 6 goto 2

step 7 stop

Many particular strategies have been proposed that differ in the way the trial vector is constructed (step 3 above). In this chapter the $DE/rand/1/exp$ strategy was used as it has worked well for most problems (see Sect. 5.1 for experimental settings). This strategy is formulated as shown in step 3 above.

A powerful deterministic optimization technique for training neural networks is the scaled conjugate gradient (SCG) algorithm, which aims to minimize mean squared error as its objective function. The classical conjugate gradient method is based on constructing vectors that satisfy orthogonality and conjugacy conditions and do not require the Hessian matrix of second partial derivatives [19]. Line minimization is used with the purpose of estimating the step size along the chosen direction. However, the scaled conjugate gradient method introduces an adaptive parameter to modify the step size by considering the behavior of the second derivative information. The result of this adaptive parameter is to scale the step size, hence the name, scaled conjugate gradient [20].

Table 3 Experimental settings for DE

Parameter	Value
Population size	20
F	0.5
C_R	0.8
r_i	rnd([0, 1])
Range of x_i^0	[−3, 3]
Strategy	DE/rand/1/exp

5.1 Experimental Settings

Single-hidden layer neural networks with either 10 or 11 neurons in the hidden layer were trained and tested. A linear transfer function was used for the output layer and a hyperbolic tangent transfer function for the hidden layer. The networks were trained using DE with the settings listed in Table 3. The configurations and settings for the neural networks were selected on the basis of recommended values from literature, previously used settings that led to good results, or simply to cover the allowable parameter range.

Similar to the procedure conducted in past studies, the data goes through a standardization procedure as a pre-processing stage [7]. All of the input variables (FSCS parameters) are converted to z-scores with all variables having a mean of zero and a standard deviation of one. If x_i is a variable with mean, $\overline{x_i}$, and standard deviation, s_i, then the z-score transform, z_i, is given by Eq. 4:

$$z_i = \frac{x_i - \overline{x_i}}{s_i} \tag{4}$$

Standardization allows the values of all variables to be measured in units of their own standard deviation, with respect to a common mean of zero, which makes direct comparisons easy. Moreover, since the variance of all variables is the same (i.e. one), the influence of each variable in similarities, distances, etc. is the same. The target variable undergoes the same standardization procedure; however, once the model has been created and applied to the testing data, the target variable goes through the reverse process to become unstandardized so that it can be used for damage estimation.

6 Fatigue Life Estimation and Load Exceedance Curves

Based on the load-time signal predictions, the resulting fatigue life and load exceedance curves were calculated. These estimates followed the traditional algorithm of estimating fatigue damage: counting the number of load cycles at relevant amplitudes, relating the load amplitudes and number of cycles to fatigue failure from

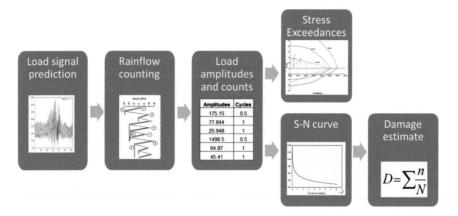

Fig. 8 Fatigue life analysis

material data (S-N curve), and finally applying fatigue damage accumulation theory (i.e. Palmgren-Miner's Rule). The process described in [4] was followed, in particular, Rainflow counting, the material specific S-N curve, and Palmgren-Miner's linear damage rule as shown in Fig. 8.

The cycle counting method that was used in this work identified all relevant load amplitudes and respective cycle counts directly from the component's load-time signal. The information obtained from the cycle counting method could then be mapped directly to the S-N curve model to determine the cycles to failure and/or used to construct a load exceedance plot.

6.1 Fatigue Damage Accumulation

The developed methodology approximates the S-N curve through the use of the Defence Science and Technology Organisation's (DSTO)/Sikorsky's S-N curve model [21] and respective required parameters. This model was based on component fatigue testing carried out at a particular mean or steady stress and constant amplitude loading. This model was adopted since it was considered to be the closest to a working model which could be easily and practically applied. The DSTO/Sikorsky model uses the parametrization in Eq. 5 to evaluate the lifetime of components [21]:

$$\frac{S}{S_E} = 1.0 + \frac{\beta}{N^\gamma} \tag{5}$$

where:

- S is the stress amplitude, or in this case using Black Hawk data, the load amplitude [in-lbs];
- N is the lifetime (number of cycles) associated with a load amplitude representing ($N * 10^6$) cycles;

- S_E is the endurance/fatigue limit described in the same units as S [in-lbs]. Load amplitudes (S) below the value of S_E are not considered to be damaging to the material. For MRNBX installed on the main rotor blade cuff, the mean endurance limit was 54,000 in-lbs [6];
- β is a positive constant associated with the material; $\beta = 0.05$ for the Beta Titanium material of the main rotor blade cuff [21]; and
- γ is another positive constant associated with the fatigue behavior of the material; $\gamma = 1.2$ for Beta Titanium [21]. The component retirement time is then the reciprocal of the total damage.

The Palmgren-Miner's rule is a commonly used theory to compute damage accumulation for high cycle fatigue loading cases, combining multiple cycles of different load amplitudes and being used in conjunction with a selected S-N curve model. The resulting damage rate reflects the effective damage accumulation as a fraction of allowable damage accumulation to failure of the component material, i.e., consumed to total life to failure of the component. This damage rate can be solved from the several groups of counted load cycles of different amplitudes and the corresponding cycles to failure from the S-N curve given by Eq. 6,

$$D_{new} = D_{old} + \sum_{S_i \in S} \frac{n(S_i)}{N(S_i)} \tag{6}$$

where D_{new} is the damage rate, D_{old} is any previous damage accumulated on the material, n is the number of cycles counted with the Rainflow cycle counting at a given load amplitude (S_i), and N is the number of cycles to failure for the load amplitude exerted on the material (S_i), determined from the S-N curve.

It should be noted that there are two major assumptions to the Palmgren-Miner's rule. The first of which is that the sequence of loading is not considered important. The second assumption is that the rate of damage accumulation is considered to be independent of stress level, even though it is well known that mean stress has an important effect on fatigue life, especially with regards to crack nucleation and propagation. Despite these two important shortcomings, the Palmgren-Miner's rule is still the most commonly used theory to determine damage accumulation for helicopter components because of its simplicity and since more complicated models do not always generate more accurate predictions.

6.2   Load Exceedance Curves

Load or stress exceedance curves plot the number of times a load range or amplitude has been exceeded in a given amount of time [22]. These curves represent the sequence of loads or stresses experienced by a component or structure during its service and are commonly analyzed in aircraft life cycle management. The information for the load amplitude and number of exceedances is derived from the Rainflow counting step in the fatigue life estimation process. These plots were used to evaluate the load-time signal predictions since the fatigue damage accumulation was minimal.

7 Results

The original set of 30 FSCS parameters was expanded to 180 predictors to include their time history data (as discussed in Sect. 3). Two different sampling schemes (k-leaders and biased) were used to form the training and testing sets. Through multi-objective genetic algorithms and the Gamma test, reduced subsets (masks) of predictor variables were identified and the most promising masks were used to build models estimating the main rotor normal bending (MRBNX) for the left rolling pull-out flight condition. The models were feed-forward neural networks that used DE to select the network weights with the exception of the models using the simplified methodology which used SCG to train the network. Ensemble models were formed for the neural networks using DE by a simple average of the top performing individual models, while the results for the simplified method represent the output of only the top performing model.

7.1 Training Set Construction Results

Previous work presented results of applying the biased sampling scheme to the load predictions for the main rotor normal bending (MRNBX) and for the main rotor pushrod (MRPR1) for the rolling pullout flight condition [12]. That work differed slightly in that it used deterministic methods, particle swarm optimization, and hybrids of these methods for determining the neural network weights, whereas in this work we have restricted neural network learning to DE because of the alternative error functions implemented.

It should be noted that the data exploration phase, generated different masks that varied with the type of training set used. For MRNBX rolling pullout, the most promising mask used for modelling contained 17 variables (of the 180 possible variables) for the k-leaders training set; in contrast, the biased training set's mask contained 68 variables (of the 180 possible variables). As seen from Table 4 listing these predictors, only 5 different FSCS parameters and their time lags are included in the k-leaders masks while 27 of the 30 FSCS parameters and their time lags are included in the biased sampling masks. Another noteworthy point is that the collective stick position (COLLSTKP) and No. 1 engine torque (NO1QPCT) are the only two FSCS parameters common to both sets of masks. Although the main rotor shaft torque (MRQ) and tail rotor shaft torque (TRQ) feature prominently in the k-leaders mask, these two parameters are noticeably absent from the biased sampling mask, which is particularly interesting given that previous studies examining the masks generated by the MOGA-Gamma test stressed the importance of these two parameters [7].

Figure 9 compares the load predictions for the baseline case of k-leader sampling and biased sampling using DE for training and MSE as the error function. The dashed vertical lines differentiate between different flight recordings collected over several days. It is evident through the significant amount of variation in the time signal,

Table 4 Comparison of masks from k-leaders and biased sampling

K-leaders sampling	Biased sampling	
17 FSCS parameters	68 FSCS parameters	
COLLSTKP (t, t-2)	COLLSTKP (t, t-1, t-3, t-4)	PEDP (t-1, t-5)
MRQ (t, t-1, t-2, t-3, t-4, t-5)	NO1QPCT (t-3, t-4, t-5)	STABLAIC (t-1, t-2)
TRQ (t, t-2, t-3, t-4, t-5)	NO2QPCT (t-1, t-3, t-4, t-5)	NR (t-1)
NO1QPCT (t-2, t-3)	PITCHATT (t-1, t-4, t-5)	ERITS (t-1, t-3)
HD (t, t-4)	PITCHRAT (t, t-2, t-3, t-4, t-5)	TRQ (t-3, t-5)
	ROLLATT (t, t-1, t-2, t-3, t-5)	NO1T45 (t, t-1, t-2)
	ROLLRAT (t, t-3, t-4)	NO2T45 (t-2)
	ROLLACC (t, t-2, t-3, t-4, t-5)	HBOOM (t-3)
	YAWRAT (t, t-1, t-4)	FAT (t-1)
	YAWACC (t-3)	LOADFACT (t-5)
	ROCBOOM1 (t-1, t-2, t-3, t-4, t-5)	ATTACK (t-2)
	SIDESLIP (t-1, t-4, t-5)	HEAD180 (t, t-4)
	LGSTKP (t-4)	VCASBOOM (t-5)
	LATSTKP (t-2, t-3, t-5)	

Refer to Table 1 in Sect. 2 for description of FSCS parameter abbreviations

that the rolling pullout manoeuvre is prone to considerable dissimilarity between each flight recording, due to factors such as aircraft configuration, pilot, and environmental conditions. Additionally, it can be seen that individual flight records are non-homogeneous; in other words, data from a flight record is not just one flight condition exclusively. This is a consequence of recording highly dynamic manoeuvres, as there is a high probability that the recording includes the helicopter's steady state condition immediately prior to and following the manoeuvre, and the transitions into and out of the manoeuvre (in addition to the manoeuvre itself).

Table 5 shows the statistics calculated for the load signal predictions in all the cases explored in this work. The normalized RMSE is simply the RMSE normalized by the absolute range (difference between maximum and minimum) of the observed values, $RMSE/(y_{max} - y_{min})$. In a cursory effort to help better evaluate the different techniques in terms of the goal of slight over-prediction for this work, some 'over-prediction statistics' were generated. The first measure, N'/N, is the ratio of number of over-predicted test samples (N') to total number of test samples (N). Ideally this measure would have a value $N'/N = 1$, meaning that every test sample was over-predicted. The second measure, $RMSE'$, is the RMSE for only the samples (N') that were over-predicted. $RMSE$ reflects the average magnitude of the over-prediction and it should ideally be a 'moderate' value, signifying a slight amount of over-prediction. Additional work should be conducted to understand what constitutes a 'moderate' $RMSE$. These two measures were determined by mean-centering the observed and predicted signals, then comparing the absolute value of the resulting predicted and observed signals.

Fig. 9 Effect of different sampling schemes. These load signal plots show the variation of the MRNBX moment with time over a series of flight records, the boundaries of which are indicated by the *dashed vertical lines*. The load values vary cyclically between 0–50,000 in-lb. The sampling frequency is 52 Hz, so this plot shows about 96 s of data. *Top plot* shows the baseline results for MRNBX rolling pullout with k-leaders sampling in the training set and neural network learning by DE. *Bottom plot* shows the results for MRNBX rolling pullout using biased sampling and NN learning by DE

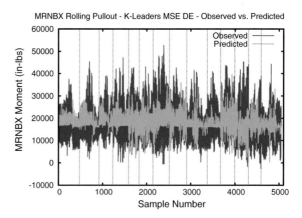

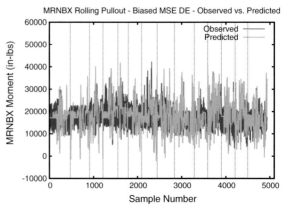

As seen in Fig. 9, the baseline case (k-leaders training with an MSE error function) yields a prediction for MRNBX rolling pullout that largely under-predicts the target signal and has poor coverage. The prediction from the biased sampling using MSE as the error function shows significant improvement at the peak values of large magnitude and with improved coverage. It is not surprising that the peak values with smaller magnitude were not predicted as well since the training set was constructed to favor the large magnitude values. Figure 10 shows magnified sections of these two cases. From these plots, it is evident that the phase of the load signal estimates using k-leaders is much more accurate than the phase prediction using biased sampling, a behavior which is also reflected in the lower correlation values of the biased sampling (Table 5). This pattern is consistently observed across the three different error functions. This behavior is not surprising since the k-leaders training set was formed to preserve the probability distribution and to closer represent statistically the properties of the original data. The biased sampling training set, on the other hand, was formed to skew the probability distribution of the original data so that the extreme values could be learned more easily and consequently the training set was

Table 5 Load signal prediction metrics

Sampling method	Error function	NN training	Correlation	Normalized RMSE (%)	Over-prediction Statistics	
					N'/N	RMSE'
Biased	MSE	DE	0.091	18.0	0.55	5709
Biased	SMAPE	DE	0.142	33.7	0.85	11449
Biased	Asymmetric	DE	0.302	11.8	0.45	2812
K-leaders	MSE	DE	0.323	13.5	0.24	3503
K-leaders	SMAPE	DE	0.303	22.4	0.64	10871
K-leaders	Asymmetric	DE	0.423	13.9	0.44	4734
90/10 split	MSE	SCG	0.277	18.1	0.06	3529

Fig. 10 Magnified plots. These load signal plots show the variation of the MRNBX moment with time over a series of flight records, the boundaries of which are indicated by the *dashed vertical lines*. The load values vary cyclically between 0–35,000 in-lb. The sampling frequency is 52 Hz, so this plot shows about 2 s of data. The *top plot* shows a magnified section of MRNBX rolling pullout with k-leaders sampling in the training set and neural network learning by DE. The *bottom plot* shows a magnified section for MRNBX rolling pullout using biased sampling and NN learning by DE

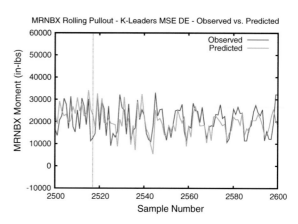

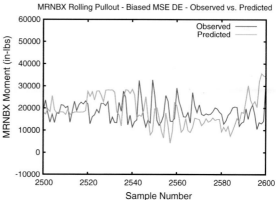

not a statistical representation of the data and was less likely to promote learning of the phase of the load signal.

Comparing the values in Table 5, it appears that there is an increase in the over-prediction ratio (N'/N) when the biased sampling training (0.55) is used instead of the k-leaders training (0.24). However the normalized RMSE decreases using biased training, suggesting that the overall prediction is less accurate than the k-leaders prediction, even though visual inspection of the plots would support the contrary. MSE is affected by both amplitude and phase differences between the target and predicted signals. Even if the amplitudes are of equivalent magnitude, a small phase shift or misalignment causes a large increase in MSE values (as well as a drop in correlation). However, if the target and predicted signals are in phase, MSE is mostly affected by amplitude differences, with smaller impact on correlation values. Since the k-leaders prediction is more correlated and in phase with the target signal, we believe that its overall RMSE is relatively low even though it is clear that many of the amplitudes are significantly under-predicted. Since the prediction using biased training is less correlated and not as closely in phase with the target signal, the RMSE errors due to this shift are more prominent leading to a higher overall RMSE. We have seen these trends in previous work [12, 13], where the better prediction did not always result in a lower RMSE, which is in part why it was important to implement more accurate and appropriate measures to evaluate the different techniques, including the fatigue life estimation.

7.2 Alternative Error Functions

The fuzzy-based asymmetric-error function described in Sect. 4.2 was introduced briefly in previous work [13] using Cartesian genetic programming. In this work, we explore in more depth the parameters of this function and also introduce the SMAPE error function as an alternative error function. The predictions for MRNBX rolling pullout using these two error functions with k-leaders training are shown in Fig. 11.

For the asymmetric error function, the predictions show significant improvement over the baseline case (Fig. 9 top plot) in terms of coverage although under-prediction of the peak values of large magnitude was still common. For the SMAPE function, the predictions show much more over-prediction but the level of over-prediction is somewhat erratic and extreme at times. Unfortunately, there are no parameters to vary in the SMAPE function to try to better control the model in terms of level of over-prediction. From Table 5, again the number of over-predicted points (N') increases using the alternative error functions. The large $RMSE'$ value for the SMAPE error function reflects the significant levels of over-prediction seen in the load signal plot.

7.3 Combinations of Techniques

Naturally, it is possible to combine the two suggested methods: biased sampling with SMAPE, and biased sampling with the fuzzy asymmetric error function. These results are shown in Fig. 12. Combining the two techniques did not necessarily pro-

Fig. 11 Effect of alternative error functions. These load signal plots show the variation of the MRNBX moment with time over a series of flight records, the boundaries of which are indicated by the *dashed vertical lines*. The load values vary cyclically between 0–50,000 in-lb. The sampling frequency is 52 Hz, so this plot shows about 96 s of data. *Top plot* shows the predictions using the asymmetric error function with k-leaders training. *Bottom plot* shows the results for MRNBX rolling pullout using SMAPE with k-leaders training

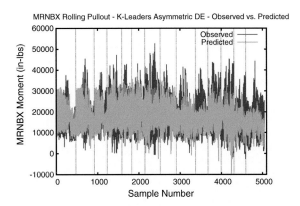
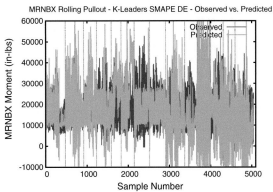

vide a more accurate prediction. The predictions with biased training and the asymmetric error function show considerable under-prediction of all peak values. In contrast, the predictions with the SMAPE function and biased training, much like the results using SMAPE and k-leaders, shows much over-prediction with considerable margins at the large amplitude peak values. From Table 5, the large $RMSE'$ value of the over-predicted points reflects the significant level of over-prediction by the models trained with the SMAPE error function and the biased training set. In both cases (biased training with SMAPE and biased training with fuzzy asymmetric error) the number of over-predicted points increased over the baseline case (k-leaders training with MSE). It is interesting that while the load signal prediction for the models using biased training and asymmetric error did not seem satisfactory upon visual inspection, the 'over-prediction statistics' seem to indicate quite a good result with almost 1/2 of the points over-predicted and a low to moderate $RMSE'$. It is very likely that the points forming the set of over-predicted points (N') did not include many of the peak values and the classification of N' points does not differentiate between points that are large magnitude peaks, small magnitude peaks, or non-peak values at all. In future, this measure could be refined further to focus solely on the peak values.

Fig. 12 Effect of combining alternative error functions with biased training. These load signal plots show the variation of the MRNBX moment with time over a series of flight records, the boundaries of which are indicated by the *dashed vertical lines*. The load values vary cyclically between 0–50,000 in-lb. The sampling frequency is 52 Hz, so this plot shows about 96 s of data. The *top plot* shows the predictions using the asymmetric error function with biased training. The *bottom plot* shows the results using SMAPE with biased training

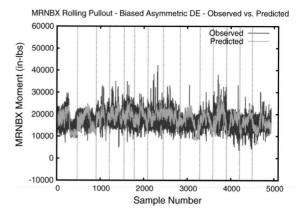

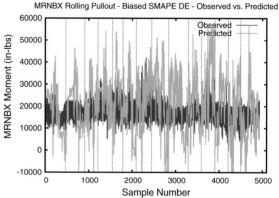

7.4 Simplified Methodology Results

For the simplified version of the methodology, a traditional 90/10 % partitioning between training and testing sets was used without any structural sampling. Models were typically trained from a 90 % subset of the data, while the remaining 10 % was used for testing. Figure 13 shows the results of the load prediction using the simplified version of the methodology with the standard 30 FSCS parameters and scaled conjugate gradient neural network training. The number of samples plotted in Fig. 13 is considerably lower than the number plotted for the other models since the testing set is only 10 % of the whole data set. The prediction is relatively weak with many of the magnitude peaks under-predicted and the coverage is poor. From Table 5, the number of over-predicted points are very few ($N'/N = 0.06$) which is less than the baseline case of k-leaders training with MSE. As mentioned earlier, the application of this simplified methodology is the most direct method of modelling the target load signal and estimating the component fatigue life from the 30 FSCS parameters. We

Fig. 13 Load prediction for simplified method and SCG learning. These load signal plots show the variation of the MRNBX moment with time over a series of flight records, the boundaries of which are indicated by the *dashed vertical lines*. The load values vary cyclically between 0–40,000 in-lb. The sampling frequency is 52 Hz, so this plot shows about 15 s of data

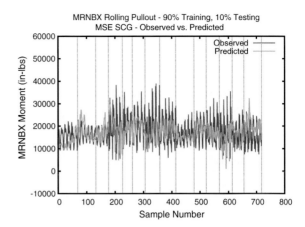

included these results to show the results that could be obtained using a relatively straight-forward method in a computational intelligence framework.

7.5   Load Exceedance Plots

Based on the load signal predictions in the different cases, the fatigue damage accumulation was calculated by following the process described in Sect. 6. However, the endurance limit of the Black Hawk main rotor blade cuff where the MRNBX gage was located is 54,000 in-lbs. As can be seen from the load signal plots, most of the load values were well below this endurance limit, and while the fatigue damage is based on the load amplitudes as opposed to peak load values, very few of the load cycles experienced during the analyzed flights were damaging. Therefore the resulting fatigue damage accumulation in all cases was minimal and consequently calculation of fatigue damage errors is not useful in this particular case and these values are not provided.

Instead of the fatigue damage accumulation, the load exceedance curve was generated for each case to illustrate the stress history. The load exceedance curves generated for the different cases allow for a better assessment of the techniques employed to improve the load signal predictions, similar to what the fatigue damage accumulation results would have demonstrated. These plots are shown in Fig. 14. Since the training and testing sets were different for each sampling method (k-leaders, biased, 90/10), there are separate plots for each sampling method.

These load exceedance plots show on a semi-log scale the number of times the MRNBX moment exceeded certain load amplitudes in the testing set. The small magnitude, frequent loads appear at the left side of the plot progressing to the large magnitude, less frequent loads at the right hand side of the plot. Normally, a load exceedance plot would represent a significant amount of aircraft flying hours (e.g. 1000 flight hours), but in this case, it is limited to approximately 100 s at best of flight

Fig. 14 Load exceedance
plots. The *top plot* shows the
load exceedance curves for
k-leaders training using
MSE, SMAPE, and
asymmetric error functions.
The *middle plot* shows the
load exceedance curves for
biased training using MSE,
SMAPE, and asymmetric
error functions. The *bottom
plot* shows the load
exceedance curve for the
simplified method using
SCG training and MSE

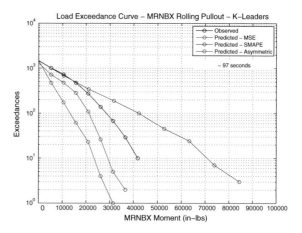

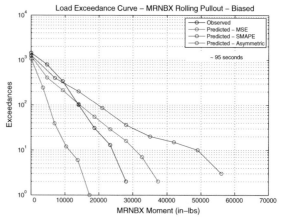

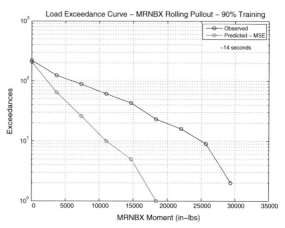

in the left rolling pullout manoeuvre, due to the fact that only one flight condition is included in the study at the moment, resulting in a reduced data set. In future, the full range of flight conditions could be analyzed using this methodology to develop a more accurate and complete picture of the component loads.

The features of importance in predicting the load exceedance curve are a close matching, slightly over-predicted curve particularly in the large magnitude load amplitude range, since these are the loads that would be damaging. Accuracy in predicting the curve in the small magnitude load amplitude range is not so critical. The results from the model generated by the simplified method using SCG training, MSE, and the 30 FSCS parameters (Fig. 14 bottom plot) yield a curve that under-predicts the load amplitude exceedances, more noticeably as the load amplitudes increase. These results are consistent with the load signal plots that showed significant under-prediction of the peak magnitudes.

The results from the ensemble models generated by k-leaders training (Fig. 14 top plot) show under-predicted curves for both MSE and fuzzy asymmetric error functions, again with obvious underestimation of the number of occurrences of large amplitude loads. The models using SMAPE were successful in over-predicting the load exceedance curve and thus providing a 'conservative' estimate, however, as the load amplitudes increased these models predicted a large number of additional load exceedances that were not present in the observed signal. Such significant overestimation of the load magnitudes would result in extremely conservative (i.e. reduced) fatigue lives for the component.

The ensemble models built using the biased training set (Fig. 14 middle plot) showed more promising results. While the models using biased training and the fuzzy asymmetric error function showed significant under-prediction of the load exceedance curve which is consistent with the load signal plots, the curves generated using MSE and SMAPE both demonstrated over-prediction of the load exceedance curve and most importantly at the larger load amplitudes. The results using MSE and biased training showed a close matching load exceedance curve with some over-prediction, while the curve from SMAPE and biased training showed much larger margins of over-prediction. Both of these sets of models under-predicted the number of exceedances at the smaller amplitude loads, but again this result is not considered important since the loads are small and mostly non-damaging. Overall, visual inspection of the load exceedance plots indicate that the most closely predicted curve with some over-prediction was generated by the models using biased training and the MSE error function.

To compare the different cases using a quantified measure, several error measures were calculated to provide a more complete understanding of the model behavior. Evaluation of each model is conducted through a pairwise comparison of the load exceedance of a prediction against its observed counterpart using 2 different metrics: normalized RMSE and an integral error. The integral error of the load exceedance curves is calculated according to Eq. 7.

Table 6 Load exceedance plot results

Sampling method	Error function	NN training	Normalized RMSE (%)	Integral error (%)
Biased	MSE	DE	11.9	−28.9
Biased	SMAPE	DE	7.1	−4.0
Biased	Asymmetric	DE	29.7	−67.9
k-leaders	MSE	DE	25.9	−56.9
k-leaders	SMAPE	DE	5.3	10.5
k-leaders	Asymmetric	DE	12.9	−34.4
90/10	MSE	SCG	24.3	−52.4

$$error = \frac{\int y_{predicted} - \int y_{observed}}{\int y_{observed}} \qquad (7)$$

In each error measure, for each pairwise comparison, only the range of moments common to the pair is used in the evaluation. In other words, the comparison range is from 0 in-lbs to the highest moment value present in both the predicted and the observed load exceedance. These results are shown in Table 6.

Looking at the error values in Table 6, it would appear that the models using SMAPE yielded the smallest errors (both normalized RMSE and integral error). However it is important to highlight that both error measures evenly weight the different load amplitude magnitudes, so that the low magnitude load amplitudes are considered equally with the large magnitude load amplitudes, even though in practice as was stated earlier, the large magnitude amplitudes are more critical since they are the loads that would be most damaging. Furthermore since the errors could only be calculated for load amplitudes common to both the observed load exceedance curve and the particular predicted curve, the largest load amplitudes included were 28,000 in-lbs for the biased training curves, 42,000 in-lbs for the k-leaders training curves, and 18,000 in-lbs for the simplified training curve. Consequently the large margin of over-prediction of the large magnitude load amplitudes by the SMAPE models was not captured in these error values.

8   Concluding Remarks

In continuing efforts to improve aircraft usage and availability, reduce the cost of maintenance, and improve safety of aircraft fleets, the ability to monitor and track loads and fatigue life on critical components accurately is essential. Indirect methods to estimate loads on these components is a practical alternative to costly and maintenance-intensive measurement systems. The application of computational intelligence algorithms, statistical and machine learning techniques, such as artificial neural networks, evolutionary algorithms, fuzzy sets, and residual variance analysis to estimate some of these helicopter dynamic loads has led to promising results. This work explored a number of different computational techniques to encourage slight over-prediction of helicopter load signals, in particular different training set construction techniques and minimizing several different error functions in the model training process. The results found that the presented techniques certainly provided some improvement over the previous results, particularly an increase in the number of over-predicted points and more accurate prediction of the peak values in the load signal. There was also considerable improvement over the results obtained by the simplified method which did not incorporate any structural sampling for forming the training set. The results achieved in this study are quite encouraging as reasonably accurate and correlated models were found for the specified flight condition in spite of the complexity of the manoeuvre. In addition, the overall quality of the predictions improved through the use of the explored techniques as under-prediction of the signal was less frequent.

It was hoped that continuing the methodology to include estimation of the fatigue life would provide a practical and quantifiable metric to evaluate the different techniques; however, the dataset analyzed in this work did not contain enough damaging load cases to result in any significant fatigue damage. Without having fatigue damage that could be calculated, it was a challenge to find suitable error measures to quantitatively capture the desired behavior of the models since the objective of this particular study was building models that slightly over-predicted the load signal. The load exceedance plots that were generated though, still provided valuable insight into the ability of the different models to generate accurate predictions of the component loads.

Future work will aim to expand the amount of data analyzed, including more flight conditions and in particular flight conditions featuring load amplitudes exceeding the component's endurance limit to result in some fatigue damage. A sensitivity study could be implemented to gauge the change in output for a given change in individual input. It would be interesting to compare the results of this study with the masks generated through the data exploration phase of the methodology. Continuing work will also be directed at further improving the predictions by exploring additional error functions and combinations of techniques to encourage conservative load and fatigue life predictions.

Acknowledgments This work was supported by Defence Research and Development Canada. Access to the Black Hawk data was granted by the Australian Defence Science and Technology Organisation.

References

1. Lombardo, D.: Helicopter structures—a review of loads, fatigue design techniques and usage monitoring. Technical Report ARL-TR-15, Defence Science and Technology Organisation, 1993

2. Valdés, J.J., Cheung, C., Wang, W.: Evolutionary computation methods for helicopter loads estimation. In: 2011 IEEE Congress on Evolutionary Computation (CEC), pp. 1589–1596. June 2011
3. Valdés, J.J.: Computational intelligence methods for helicopter loads estimation. In: The 2011 International Joint Conference on Neural Networks (IJCNN), pp. 1864–1871. July 2011
4. Cheung, C., Rocha, B., Valdés, J.J., Kotwicz-Herniczek, M., Stefani, A.: Expanded fatigue damage and load time signal estimation for dynamic helicopter components using computational intelligence techniques. In: Proceedings of the American Helicopter Society 70th Annual Forum, Montreal, Canada, May 2014
5. Cheung, C., Rocha, B., Valdés, J.J., Stefani, A., Li, M.: An approach to fatigue damage estimation of helicopter rotating components using computational intelligence techniques. In: Proceedings of American Helicopter Society 69th Annual Forum, Phoenix, Arizona, May 2013
6. Georgia Tech Research Institute: Joint USAF-ADF S-70A-9 flight test program, summary report. Technical Report A-6186, Georgia Tech Research Institute, 2001
7. Cheung, C., Valdés, J.J., Li, M.: Exploration of flight state and control system parameters for prediction of helicopter loads via gamma test and machine learning techniques. In: Abou-Nasr, M., Lessmann, S., Stahlbock, R., Weiss, G.M. (eds.) Real World Data Mining Applications. Annals of Information Systems, vol. 17, pp. 359–385. Springer, Switzerland (2015)
8. Jones, A., Evans, D., Margetts, S., Durrant, P.: The gamma test. In: Sarker, R., Abbass, A., Newton, C. (eds) Heuristic Optimization for Knowledge Discovery, pp. 142–168. Idea Group, Hershey, PA (2002)
9. Stefánsson, A., Končar, N., Jones, A.: A note on the gamma test. Neural Comput. Appl. **5**, 131–133 (1997)
10. Evans, D., Jones, A.: A proof of the gamma test. Proc. Roy. Soc. Lond. A **458**, 1–41 (2002)
11. Anderberg, M.: Cluster Analysis for Applications. Wiley, London (1973)
12. Valdés, J.J., Cheung, C., Li, M.: Towards conservative helicopter loads prediction using computational intelligence techniques. In: The 2012 International Joint Conference on Neural Networks (IJCNN), pp. 1–8. June 2012
13. Cheung, C., Valdés, J.J., Li, M.: Use of evolutionary computation techniques for exploration and prediction of helicopter loads. In: 2012 IEEE Congress on Evolutionary Computation (CEC), pp. 1–8. June 2012
14. Chen, Z., Yang, Y.: Assessing forecast accuracy measures. http://www.stat.iastate.edu/preprint/articles/2004-10.pdf (2004)
15. Storn, R., Price, K.: Differential evolution-a simple and efficient adaptive scheme for global optimization over continuous spaces. Technical Report, TR-09012, ICSI, 1995
16. Price, K.V., Storn, R.M., Lampinen, J.A.: Differential Evolution. A Practical Approach to Global Optimization. Natural Computing Series. Springer, Heidelberg (2005)
17. Kukkonen, S., Lampinen, J.: An empirical study of control parameters for generalized differential evolution. Technical Report 2005014, Kanpur Genetic Algorithms Laboratory (KanGAL), 2005
18. Gämperle, R., Müller, S., Koumoutsakos, P.: A parameter study for differential evolution. In: Grmela A., Mastorakis, N.E. (eds.) Advances in Intelligent Systems, Fuzzy Systems, Evolutionary Computation, pp. 293–298. WSEAS Press (2002)
19. Pres, W., Flannery, B., Teukolsky, S., Vetterling, W.: Numeric Recipes in C. Cambridge University Press, New York (1992)
20. Moller, M.F.: A scaled conjugate gradient algorithm for fast supervised learning. Neural Netw. **6**, 525–533 (1993)
21. King, C., Lombardo, D.: Black Hawk helicopter component fatigue lives: sensitivity to changes in usag. Technical Report DSTO-TR-0912, Defence Science and Technology Organisation, 1999
22. Broek, D.: The Practical Use of Fracture Mechanics. Kluwer, Dordrecht (1989)

Evolving Narrations of Strategic Defence and Security Scenarios for Computational Scenario Planning

Kun Wang, Eleni Petraki and Hussein Abbass

Abstract Defence and security organisations rely on the use of scenarios for a wide range of activities; from strategic and contingency planning to training and experimentation exercises. In the grand strategic space, scenarios normally take the form of linguistic stories, whereby a picture of a context is painted using storytelling principles. The manner in which these stories are narrated can paint different mental models in planners' minds and open opportunities for the realisation of different contextualisations and initialisations of these stories. In this chapter, we review some scenario design methods in the defence and security domain. We then illustrate how evolutionary computation techniques can be used to evolve different narrations of a strategic story. First, we present a simple representation of a story that allows evolution to operate on it in a simple manner. However, the simplicity of the representation comes with the cost of designing a set of linguistic constraints and transformations to guarantee that any random chromosome can get transformed into a unique coherent and causally consistent story. Second, we demonstrate that the representation being utilised in this approach can simultaneously serve as the basis to form a strategic story as well as the basis to design simulation models to evaluate these stories. This flexibility fulfils a large gap in current scenario planning methodologies, whereby the strategic scenario is represented in the form of a linguistic story, while the evaluation of that scenario is completely left for the human to subjectively decide on it.

K. Wang (✉)
School of Engineering, Ocean University of China, Qingdao, China
e-mail: wangkunouc@163.com

E. Petraki
Faculty of Arts and Design, University of Canberra, Canberra, Australia
e-mail: Eleni.Petraki@canberra.edu.au

H. Abbass
School of Engineering and IT, University of New South Wales, ADFA, Sydney, Australia
e-mail: hussein.abbass@gmail.com

© Springer International Publishing Switzerland 2016
R. Abielmona et al. (eds.), *Recent Advances in Computational Intelligence in Defense and Security*, Studies in Computational Intelligence 621,
DOI 10.1007/978-3-319-26450-9_23

1 Introduction

Scenario planning is a powerful tool for defence and security organisations and to develop robust and adaptive system and strategies. Computational scenario planning attempts to fully automate the scenario planning process where work can be grouped into tactical level and strategic level planning [1].

The tactical level scenario planning relies on building computational models of complex environments; and new scenarios are subsequently generated by sampling the parameter space. Recent years have reported notable progress in scenario planning at this level using computational intelligence (CI) techniques including multi-agent simulation [2], or incorporated with evolutionary computation (EC) to better explore the parameter space in which applications include MEBRA [3] and Alam's work [4] in risk assessment, Bui's work in military logistic management [5], Xiong's work in project scheduling [6], and Amin's work in air-traffic control [7].

At the strategic level, scenarios normally take the form of linguistic stories, whereby a picture of a context is painted using storytelling principles. The manner these stories are narrated can paint different mental models in planners' minds and open opportunities for the realisation of different contextualisations and initialisations of these stories. Therefore, creating a scenario of strategic level is similar to writing a story about some possible future [1]. DARPA established the Narrative Networks project to study how storytelling can affect human cognition and behavior thus national and military security. Existing computational storytelling techniques have been applied and attracted significant attention in training and education [8], entertainment industry [9, 10], and related strategic planning [1, 11–15]. Section 2 provides a literature review in computational storytelling techniques in which few work have been observed in automatically generating story-like scenarios in the defence and security domain [1, 11].

However, a bottleneck in this process is that strategic scenarios normally take the form of a story written in a natural language which are developed manually. To transform this story into a computational environment, or to fully automate the process of creating different narrations of that story to better explore the strategic space, is currently a holy grail of computational scenario planning.

This chapter overcomes this challenge by proposing an evolutionary computation approach to automate the narrations of a strategic story. The proposed solution offers an automated method that can create variations and different narrations of strategic stories. This will allow different contextualisation of the same stories and produce a tool to better explore the subset of the strategic space of interest.

One difficulty is to design a simple representation of a strategic story that allows evolution to operate on it in a simple manner. However, the simplicity of the representation comes with the cost of designing a set of linguistic constraints and transformations to guarantee that any random chromosome can get transformed into a unique coherent and causally consistent story.

We overcome this difficulty in two steps: (1) extracting the linguistic constraints of a strategic story in our interested domain and representing them computationally; (2) encoding a story narration into a genome and transforming it into a unique text-form story through genotype-phenotype mapping, in which the above linguistic constraints serve as reference of coherence and causal consistence.

Another difficulty is to evaluate thus evolve these strategic stories from human planners' perspective, while humans are fatigued easily. To alleviate human fatigue, methods including improving evaluation or fitness input interfaces, and reducing population and generation sizes of EC have been used. However, this can deteriorate the performance of EC [16]. So a promising solution is still to predict human evaluations using surrogate models [16, 17].

We overcome the above difficulty in another two steps: (3) conducting a human-based evaluation experiment to collect human subjective evaluations of some generated strategic stories; (4) building surrogate models based on the above-collected data.

The representation utilized in step (2) can simultaneously serve as the basis for step (3), which is to form a strategic story as well as the basis to design simulation models to evaluate these stories. This flexibility fulfil a large gap in current scenario planning methodologies, whereby the strategic scenario is represented in the form of a linguistic story, while the evaluation of that scenario is completely left for the human to subjectively decide on it.

This chapter is organised as follows:

Section 2 reviews existing work in computational storytelling and, especially, the human-guided evolutionary storytelling field.

Section 3 addresses step (1) by proposing a story parsing method that can extract the linguistic constraints of a strategic story in the form of a dependence network in our domain of interest.

Section 4 addresses step (2) by encoding a story narration as a linear permutation of events and the temporal and spatial information involved, and proposing a genotype-phenotype mapping mechanism. This mapping transforms a random permutation into a unique text-form story, in which the linguistic constraints extracted in step (1) serve as a reference of coherence and causal consistency.

Section 5 defines the objective and subjective metrics of strategic stories for the surrogate models. To obtain subjective metrics data, a human evaluation experiment in step (3) is discussed in Sect. 6.1. Different story narrations are generated based on the permutation representation proposed in step (2), and evaluated by humans. Section 6.2 addresses step (4) by building surrogate models of human story evaluation based on the data collected in the above human evaluation experiment.

Finally, step (1)–(4) are synthesized to serve an automatic and evolutionary story narration process discussed in Sect. 7, whereby EC is designed to better explore the subset of strategic space. An experimental study is carried out to test the performance of this evolutionary story narration approach in its capability to evolve good narrations of strategic scenarios.

2   Related Work

Techniques used in existing computational storytelling applications and related work to the human-guide evolutionary storytelling approach are discussed.

2.1   Techniques in Computational Storytelling

A knowledge perspective generalised from Ciarlini's work in 2010 [18] is adopted to classify existing computational storytelling applications. The template-, rule-, and formal grammar-based approaches can be categorised under the declarative way of specifying story control knowledge, and the transitional search, CBR and population-based approaches under the embedded approaches.

Template-Based Approach. A template refers to a linear sequence of world states and/or actions that must be implemented by the storytelling system, represented in ontology [19] or logic [18, 20–25]. A story produced using this approach is the instantiation of a certain template. A template describes knowledge about "features that enhance user experience" [25], "authorability and authorial intent" [24], or "landmarks to decompose story generation in order to address scalability issues" [25]. Templates may appear as: "thematic frames" [20], "frameworks of emotional states in audience" [21], "linear scripts for scenes" [22], "sets of state propositions" [24], "theme-based planning operators" [19], "state trajectory constraints" [25], "pre-defined plots" [18].

Rule-Based Approach. In a storytelling system that use AI planning, a rule serves as the search operator to inform the system as to how to achieve certain goals. This approach has become the key to realise "interactivity" in interactive storytelling (IS) or interactive drama—a highly thriving research area in computational storytelling in which the audience plays an active part in the composition of a story [26]—and enjoyed wide applications [10, 18–25, 27–33].

Formal Grammar-Based Approach. A story grammar explicitly expresses the operations and the semantic relationships of the story building blocks (such as the states of the objects, actions performed by the characters in the story) in formal grammar [34]. By deriving the story grammar, different strings of story elements can be generated. Then, by instantiating each building block in a derived string with the corresponding entities and relationships in the knowledge of the story world—the static schema [18]—a story can be composed. Applications of this type include the works of Prince [35], Colby [36], Lee [37] and, recently, of Bui [1, 12] and Wang [38], with the work of Bui [1] focusing on generating user preference scenarios for strategic planning of military logistics systems and future air traffic management.

Traditional Search Approach. This approach regards storytelling as problem solving using a classical search [39], such as searching through a tree or network [40, 41], or searching guided by heuristics [32, 33, 42]. Every solution found during the search—an action sequence or trajectory of the search—is a generated story.

Case-Based Reasoning (CBR) Approach. A story generated using this approach is a sequence of cases—a case is a structured representation of existing stories— or subordinate building blocks in a case recorded during the CBR process [9, 11, 43–45].

Population-Based Approach. This approach applies different types of EC to collect stories—the individuals—with desirable features, such as coherence, creativity and interestingness [1, 12–14, 38].

Declarative approaches have advantages in terms of "ease of understanding, conciseness of expression, modularity and ease of validation" [25] and authors can directly express their authorial intent, which is lacking in embedded approaches. However, the implicitness in story knowledge representation in the embedded approaches can relieve the author's burden to some degree. For instance, the challenge of "quantifying the qualitative"—explicitly and formally representing authentic story features—can be avoided by implicitly embedding the constituents of these features in the search, reasoning or learning strategies.

The CBR approach and the template-based approach can produce human-like stories, while the stories may be lacking in diversity [12, 38, 46]. The rule-, traditional search- and formal grammar-based approaches can produce comparatively diversified stories and stories with certain authentic features: believability [29] and interactivity [10] with the rule-based approach; creativity [28], tension [33] and suspense [32] etc. in the traditional search-based approach; and coherence using the formal grammar-based approach [1, 12, 38]. However, the quality of the stories may vary: the rule-based approach can hardly produce interesting stories by simply recording simulated events [46]; the formal grammar-based approach may produce dull stories without other mechanism to guide the grammar derivation [12, 38]; and the authentic features captured by objective measures (such as heuristic functions) in the traditional search-based approach "need to be assessed, either singly or in combination, by human readers" [47].

The population-based approach can benefit from the implicit self-feedback loop introduced by EC in which stories generated in one iteration contribute to subsequent generations of improved stories and this process can rely on human feedback to guide evolutionary dynamics. Therefore, this chapter adopts a population-based approach.

2.2 Existing Work on Human-Guided Evolutionary Storytelling

Regular Grammar-Based Evolutionary Storytelling. Bui [12] represented an existing "fabula model" [48] into a regular grammar [34]—the story grammar— which describes the causal relations between different types of events in stories. Then, simple stories can be generated and evolved by deriving the story grammar and further applying grammar evolution guided by human-in-the-loop evaluation of generated stories. The results show that this system can evolve stories with accumulated good features such as interestingness and creativity. However, only simple stories with a single character can be generated because capturing long-distance causal dependency is a challenge for regular grammar while, in a complex story which possesses multiple characters and branches, it is usual that one story line is interrupted by another before returning to the original one and continuing to unfold whereby a long-distance causal relationship emerges.

Tree Adjoining Grammar-Based Evolutionary Storytelling. Wang [38] used tree adjoining grammar (TAG) and TAG-guided genetic programming (TAG3P) [49] to generate and evolve complex stories in order to deal with the problem that regular grammar is incapable of capturing long-distance dependency. However, the results indicate that it might not be sufficient to use a story structure that holds only one level of complexity—one that only focuses on the causality between events. Also, this approach applies full human-in-the-loop evaluation and, consequently, problems such as human fatigue and user inconsistency emerge.

GL-2 Grammar-Based Evolutionary Storytelling. Bui [1] proposed an expressive GL-2 grammar for representing practical story-like scenarios to address the problem in his previous work [12] that regular grammar only captures causal relationships between events from the point of view of a single story character. He introduced a set of scenario building blocks: Events, Time, Location, Objects, Actions and Relations. The task of scenario generation then becomes a task of generating the networks of these scenario building blocks and relationships. This approach can evolve natural disaster scenarios for future air traffic management from a human's perspective.

However, as has been mentioned by the author, GL-2 is not expressive enough to represent relations between two groups of events as possible in causal relations. Also, the problem of full human-in-the-loop evaluation still remains unsolved.

Common to all the above grammar-based approach is the difficulty in confining the generate stories in a particular domain. We conjecture that a possible solution is to establish a pragmatic story parsing (aka analysing) tool that can assist the finding of the structure of stories (or the linguistic constraints)—one that is easy for EC to manipulate—in any domain of interest.

Implementation of Evolutionary Story Narrating in Children Story Domain. To overcome the above problems, Wang [13–15] proposed an evolutionary story narrating approach and demonstrated an implementation of this approach in children's stories. The research highlighted the effectiveness of this approach in terms of accumulating good story narrations with minimum human involvement during the interactive story evolution. The evolution is guided by surrogate models of human evaluation which are built from the data collected from a human-based evaluation experiment. The surrogate models can also incorporate diversified human opinions in story evaluation. While this evolutionary story narrating approach can be applied to other domains, the above implementation focused on children's narration. The chromosome may need to incorporate different information of events when representing stories in another domain. For instance, the previous chromosome may not be suitable for representing practical scenarios since "a practical scenario may not require a character at all" [1] or it would not produce coherent and causally-related events by manipulating the characters.

This chapter extends the above implementation and attempts to apply the evolutionary story narrating approach to strategic scenario generation in defence and security domain.

3 Parsing Strategic Stories into Dependence Networks

This section addresses step (1) of our EC approach to automate the narrations of a strategic story. We propose to extract the linguistic constraints of a strategic story in our domain of interest and represent them computationally in a dependence network.

The proposed method can extract a dependence network from an existing English text-form story. Based on a strong consensus among narratologists, a story is represented as a sequence of events [50]. These events—the nodes in the network—are connected by dependence relations, the links. Although we can further group events thus extract hierarchical dependence networks [13, 15], we focus on story narrating on the event level in this chapter.

An event-level dependence network is obtained in two steps: parameterised event extraction and dependent relation building.

3.1 Parameterised Event Extraction

The event definition in [13, 15] is used in this chapter: an event is a predicate that denotes an action, state, or occurrence in a story; it is bound by a position in the temporal dimension, possesses a spatial situation in the story world and has participants as parameters.

A clause can be regarded as the minimum unit of our defined event (505, 506 in [51]). Therefore, events can be extracted by tracing the verbs or verb phrases in each of the clauses in a English text-form story. We revise the TimeML event annotation guidelines [52] and provide rules for event recognition in Appendix A.

An ontology of event parameters is proposed in [15] and demonstrated as follows to extract further information related to an event, which covers "when, where, who and what are involved in an event". The event parameters also serve as identifiers for dependent relation building.

- **Time** is a temporal expression that denotes when the event happened.
- **Space** is the spatial expression denoting where this story happened in which a concrete thing is involved, such as "the forest" in "in the forest".
- **Character** is an active participant that can perform actions to change the states of the story world.

- **Description** is a description of a concrete thing in the story world that indicate an participant's identity or state.
- **Topic** is an abstract thing or concept mentioned in the story which can be represented by a sequence of events in the context, such as a new situation.
- **Object** is a concrete thing in the story world that is not a character, a description, a concrete thing in a time parameter.

"Time" and "Space" parameters are firstly extracted by tracking particular noun phrases (NPs), prepositional phrases (PPs), adverbial phrases (ADVs) and subordinating clauses (SBARs) listed in [15], Chap. 4.

"Character", "Description", "Topic" and "Object" parameters are denoted by noun phrases (NPs) (44, 54 in [51]). Extracting them requires a process of coreference resolution, or a Character parameter may not be identified merely because it does not appear in the same label in the story, for instance "soldiers" may appear as "army" or "land forces". Although the participants of events are not shuffled in the final generated narrations, these parameters are still extracted as identifiers for dependent relation building.

Due to the strategy to reduce redundancy in natural language, some extracted events may be missing the temporal and spatial background information. This may pose difficulties for the reader to make sense of the whole story when we produce new story narrations by shuffling these events. Therefore, we also add the implicit Time parameters[1] to the events to complete the missing background information. The implicit Time parameter of an event is assigned to that of the previous narrated event in the original story, or that of the next event if it is the first event in the story.

3.2 Dependent Relation Building

A dependent relation is defined upon a pair of events in which one event serves as one of the enabling conditions of the other. It can be identified from a counter-factual test [53].

To minimise the effect of human subjectivity on the dependent relation building process to some degree, we propose a set of general rules for dependent relation building. For event 2 to be determined as being dependent on event 1, three major constraints need to be met: firstly, event 1 must occur before event 2 in the story; secondly, event 2 must share at least one participant with event 1; thirdly, event 1 must be the nearest event to event 2, and any previous event—may be more than one—that is the nearest event that shares one of the participants of event 2 will be identified as an enabling event.

The above general rules may be subject to revision and may facilitate automatic implementation of building the dependent relations between events. However, this

---

[1] In this chapter, the implicit Space parameter of an event is not added because it involves a concrete thing—a participant including a Character or an Object—in the story, while the participants of events are left untouched in a generated narration.

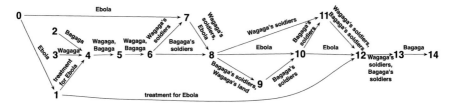

Fig. 1 An example of story dependence network extracted from a made-up strategic story: the nodes labeled by integers denote the events labeled by their occurrence order in the story; and each directed line labeled by the shared participant(s) denotes a dependent relation from one of the enabling conditions of an event to the events

can be a challenging task. For one thing, time reasoning is required to infer the temporal order of all the events in the story. The occurrence order of events cannot be determined based on their representation order in the text-form story because the events in a story may not be narrated in the order of their actual occurrence in the story world. In linguistics, this phenomenon is referred to as "two basic temporalities of narrative" [54]. For another, the shared participant in event 2 should be aware of the happening of event 1. The complexity of this problem provides challenge that requires further research. As a result, a manual dependent relation building method based on the general rules has been adopted at this stage.

3.3 Preliminary Event Grouping

An event is further combined with the events that serve as some grammatical components in order to make its meaning complete and compact. These events serve as its subject, object, complement, appositive, modifier, time-denoting adverbial, space-denoting adverbial, or direct cause or effect represented in non-finite adverbial subclauses which include -ing clauses, -ed clauses and to-infinitive clauses. Chapter 4 in [15] provides the detailed rules.

Figure 1 illustrates an example of a dependence network extracted from a made-up strategic scenario in defence and security domain discussed in this chapter.

3.4 Computational Representation of Dependence Network

The story dependence network is represented in the following data structures: "event_relation", "time string" and "space string" list.

Event Information. The event information is described in the ID, TIME, SPACE members in the "event_relation" structure, and the "event string", "time string", "space string" structures.

Fig. 2 An example instance
of "event_relation" data
structure

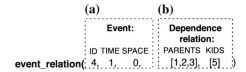

Every object of "event_relation" structure (see Fig. 2) is based on an event in the story. ID denotes the occurrence order (beginning from 0) of this event in the story; TIME is assigned to the occurrence order (beginning from 1) of the Time parameter of this event in the story; SPACE is assigned to the occurrence order (beginning from 1) of the Space parameter of this event in the story unless this event contains no explicit Space parameter in which case the SPACE is assigned to 0 (see the second "event_relation" object in Fig. 3 for an instance).

Every object of "event string", "time string" and "space string" structure provides the text-form of the major part, the Time parameter and Space parameter of this event, respectively. The integer label at the beginning of the string corresponds to the ID, TIME and SPACE member in the "event_relation" structure, respectively.

Dependent Relation Information. The "event_relation" structure also includes two members that represent the information of dependent relations in the story: PARENTS and KIDS which incorporates the IDs of the events that enabled this event and were enabled by this event, respectively.

A dependent relation is expressed and built up from the basis event of an "event_relation" instance to each of the events whose ID have been included in the KIDS member, or from each of the events whose ID have been included in the PARENTS member to the basis event. Four dependent relations are expressed in Fig. 2:

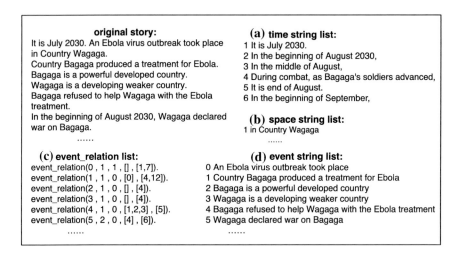

Fig. 3 An example of computational representation of story dependence network

the ones from event No. 1, 2 and 3 to 4, each described in the PARENTS member; and the one from event No.4 to 5, which is described in the KIDS member.

Figure 3 illustrates the computational representation of the story dependence network.

4 Story Narrating as Permutation Problem

This section addresses step (2) of our EC approach to automate the narrations of a strategic story. We encode a story narration with flashback[2] into a linear permutation representation of events and their temporal and spatial information.

Different narrations can be automatically produced by shuffling the event order and the temporal and spatial information in the original story. Time and space profoundly influence the way in which we build mental models of a story. Sternberg [55] suggests that we should consider the story-discourse relationship (the time of what is told and the telling) in terms of the universals of suspense, curiosity, and surprise. Therefore, automatic story narrating can open opportunities for the realisation of different contextualisations and instantiations of stories.

Also, we propose a genotype-phenotype mapping mechanism to transform a random permutation into a unique text-form story, whereby the linguistic constraints extracted in Sect. 3 (step (1) of the EC-based automatic story narration approach) serve as reference of coherence and causal consistence.

4.1 Encoding Story into Genome

We impose the following constraints to control the coherence in the generated story narrations, so that humans can provide a comparatively objective evaluation of them. A story is told by combining a forward narration and flashback of the events in the story dependence network. Firstly, randomly choose a layer in the story dependence network as a threshold layer which is the meeting point of the forward and flashback narration; secondly, all events with smaller layer values are narrated in the forward direction, which means an event will only be narrated when all the events in its PARENTS events have been narrated; thirdly, events with bigger layer values are narrated in a flashback way, which means an event will only be narrated when all the events in its KIDS events have been narrated; finally, the narration will end at the event with the threshold layer value.

---

[2]A flashback is a story played backward. It can be a full flashback in which the whole story is played backward, or a partial flashback in which a subset of events are played backward, with the rest of the events played forward.

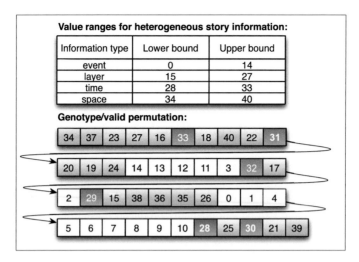

Fig. 4 An example of genotype: the *white* genes denote the events in the dependence network; the highlighted *light grey* genes denote the layers, the *dark grey* genes the temporal information in the story, and the *green* genes the spatial information

We use a linear permutation to encode a new story narration with flashback generated from the story dependence network. This representation can incorporate information about the order of events, temporal and spatial information, and layer threshold that indicates the conjunction of forward and flashback narration. This can be achieved by assigning unique value ranges to different types of information. Figure 4 provides a genotype of a generated narration example.

4.2 Obtaining Text-Form Story from Genotype-Phenotype Mapping

However, the order of events in the simply generated permutation may not conform to the above constraints. Therefore, we design a genotype-phenotype mapping that transforms a random chromosome into a unique coherent and causally consistent story.

A random permutation is firstly transformed into a valid genome that conforms to the above constraints. The dependent relation information represented in the dependence network's event_relation list—the PARENTS and KIDS members—serves as the reference for checking if the constraints are fulfilled. The pseudo code for the transformation process is illustrated in Fig. 5.

The text-form story can then be obtained by extracting the first layer gene, the list of event genes, time and space genes, then enumerating the text representation of each of the events one after another in the order of their positions in the event list.

```
INIT
      SET ``OK'' list to empty
      SET ``NG'' list to empty
      SET ``valid permutation'' list to empty
      SET ``new valid gene'' to -1
      SET ``current NG gene'' to -1
READ the first gene in the Permutation
SET ``current gene'' to the first gene in Permutation
WHILE ``current gene'' is not the first layer gene
      READ the next gene in the Permutation
      SET ``current gene'' to the next gene in the Permutation
ENDWHILE
SET the threshold layer to ``current gene''
SET ``validity constrains'' based on the threshold layer
SET ``current gene'' to the first gene in Permutation
REPEAT
      Step 1: add to ``OK'' list the genes corresponding to the chains in the dependency network
              which recently met the ``validity constrains'' based on the ``new valid gene''
      SET ``current NG gene'' to the first item in ``NG'' list
      REPEAT
          IF ``current NG gene'' is in ``OK'' list THEN
                add ``current NG gene'' to ``valid permutation'' list
                delete ``current NG gene'' from ``NG'' list
                GOTO Step 1
          ELSE
                SET ``current NG gene'' to the next item in ``NG'' list
          ENDIF
      UNTIL ``current NG gene'' is the last item in ``NG'' list
      IF ``current gene'' is a layer gene or parameter gene THEN
          add ``current gene'' to ``valid permutation''
          SET ``new valid gene'' to ``current gene''
      ELSE
          IF ``current gene'' is in ``OK'' list THEN
                delete ``current gene'' from ``OK'' list
                add ``current gene'' to ``valid permutation''
                SET ``new valid gene'' to ``current gene''
          ELSE
                add ``current gene'' to ``NG'' list
          ENDIF
      ENDIF
      READ the next gene in the Permutation
      SET ``current gene'' to the next gene
UNTIL ``current gene'' is the last gene in the Permutation
```

Fig. 5 Pseudo code of permutation transformation process

The text-form of the Time and Space parameter—the corresponding "time string" and "space string"—is added to the beginning and end of the "event string", respectively. The text-form of the shared Time parameter in adjacent events is omitted in the latter event to avoid redundancy. This process is illustrated in Figs. 6 and 7.

Fig. 6 Story information
extracted from example
genotype

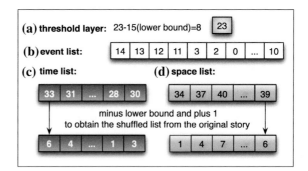

Fig. 7 Extraction of text
representation of event from
example genotype

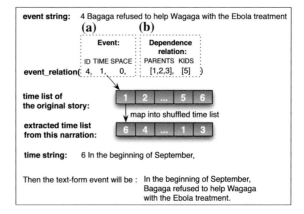

5   Story Metrics Selection

It is a challenging task to quantify the quality of a story [12], so it is practical to apply
human subjective evaluation. We conjecture that human subjective evaluation of a
story is affected by the underlying objective metrics that reflect the properties of this
story. This section defines the objective and subjective metrics for story evaluation
for the human evaluation experiment in step (3).

5.1   Subjective Story Metrics

The story quality is further classified into the following four subjective story metrics. We stop at this level of sub-categorisations considering the efficiency and data
fusion problems. **Coherence** reflects "a global representation of story meaning and
connectedness," [56] and makes a story understandable to the reader [57]. **Novelty**
reflects the unexpectedness and rule-breaking degree of a story. If we say that the
above two metrics reflect a readers's global impression of a story after understanding
is achieved, **interestingness** may indicate the dynamics of a human's appreciation
of a story before full comprehension is achieved [58].

5.2 Objective Story Metrics

Four objective story metrics representing the quantitative features of a story narration are defined.

disOfFlashback is defined as the distance of the flashback feature, DOF, of a story narration compared with the original story. Let $DNLN$ the dependence network layer number, and TL the threshold layer; DOF is calculated as

$$DOF = \frac{DNLN - TL}{DNLN} \qquad (1)$$

consistEventOrder is the consistency of the event order, CE, of a story narration with the event order of the original story. Let $SCOSEO$ the sorting cost to the original story's event order, and n is the number of events in the dependence network. The sorting cost is calculated using bubble sort.

$$CE = \frac{SCOSEO}{n \times (n - 1)/2} \qquad (2)$$

consistT refers the consistency of the arrangement of temporal information, CT, of a story narration with that of the original story. Let $TTCOS$ be the times of temporal information change from the original story, and NTS number of temporal expressions in the story.

$$CT = 1 - \frac{TTCOS}{NTS} \qquad (3)$$

consistS refers to the consistency of the arrangement of spatial information, CS, of a story narration with that of the original story. Let $TSCOS$ be the times of spatial information change from the original story, and NSS number of spatial expressions in the story.

$$CS = 1 - \frac{TSCOS}{NSS} \qquad (4)$$

These metrics serve as the mechanism that manipulate the degrees of mental picture change of a reader [54] during story narration.

6 Surrogate Models of Human Story Evaluation

This section addresses steps (3) and (4) to solve the second difficulty in the EC-based automatic story narration approach in terms of how to evaluate thus evolve strategic stories from human planners' perspective with reduced human involvement.

The permutation-based representation utilized in Sect. 4 (step (2) of the EC-based automatic story narration approach) serves as the basis for forming a strategic story narration as well as the basis for designing simulation models to evaluate these stories. In step (3), different story narrations are generated and evaluated by humans to collect subjective metrics of them. In step (4), surrogate models of human evaluation are built by mapping the objective story metrics to the subjective story metrics whose data is collected from this human evaluation experiment.

6.1 Human Evaluation Experiment

Story Narration Sample Story Narration Sample are generated based on the dependence network extracted from a made-up strategic scenario in the defence and security domain. The story has been inspired by real life events and was strategic relevance.

It is July 2030. An Ebola virus outbreak took place in Country Wagaga. Country Bagaga produced a treatment for Ebola. Bagaga is a powerful developed country. Wagaga is a developing weaker country. Bagaga refused to help Wagaga with the Ebola treatment. In the beginning of August 2030, Wagaga declared war on Bagaga. In the middle of August, land forces from Wagaga and Bagaga started to engage. A number of soldiers from Wagaga became infected with the virus. During combat, Wagaga soldiers transferred the virus to Bagaga's soldiers as Bagaga's soldiers advance in Wagaga's land. It is end of August, Bagaga's soldiers advanced deep in Wagaga's lands. The Ebola virus spread very fast in Bagaga's soldiers. Wagaga's soldiers managed to go behind Bagaga's lines and cut their logistic supply of the virus treatment. It is beginning of September, the Bagaga's army is surrounded with Wagaga's army. Bagaga need to design strategies to manage the new situation.

16 story narrations are selected: 13 from randomly generating 3000 story narrations, 2 from intentionally fixing the consistT and consistS metrics,[3] and one original story. This sample incorporates big variance in the values of their objective metrics in the following steps: firstly, the value space [0, 1] of the objective metrics is firstly divided into two ranges: "LOW" for values in [0,0.5) and "HIGH" for those in [0.5, 1]; and, then, 16 story narrations with all the possible value range permutations are selected. The permutations and values objective metrics of the 16 selected story narrations in the sample are listed in Tables 1 and 2.

Subjective Evaluation Data Collection. The subjective evaluation data is collected through the scores given by one human participant who is presented with a printed version of the story narration sample.

The participant is required to read the story narrations in the sample one after another and provide his evaluations of the particular story narration after reading it. This is completed in a continuous time slot of a day where a break less than 5 min is allowed. The experimental location is consistent in the same quiet room in order to maintain constant environmental situation; and the participant is not constrained in terms of the time he needs to finish reading one narration. The order in which the story narrations in the sample are read is shuffled to minimise any learning effects.

---

[3]Using pure random generation, slim chances ($\frac{4}{7! \times 6!}$) can be expected to obtain a narration that possesses comparatively high values of consistT and consistS at the same time.

Table 1 Permutations of objective metrics in story narration sample in human-based evaluation experiment

No.	DOF	CE	CT	CS
1	LOW	LOW	LOW	LOW
2	LOW	LOW	LOW	HIGH
3	LOW	LOW	HIGH	LOW
4	LOW	LOW	HIGH	HIGH
5	LOW	HIGH	LOW	LOW
6	LOW	HIGH	LOW	HIGH
7	LOW	HIGH	HIGH	LOW
8	LOW	HIGH	HIGH	HIGH
9	HIGH	LOW	LOW	LOW
10	HIGH	LOW	LOW	HIGH
11	HIGH	LOW	HIGH	LOW
12	HIGH	LOW	HIGH	HIGH
13	HIGH	HIGH	LOW	LOW
14	HIGH	HIGH	LOW	HIGH
15	HIGH	HIGH	HIGH	LOW
16	HIGH	HIGH	HIGH	HIGH

Table 2 Values of objective metrics in story narration sample in human-based evaluation experiment

No.	DOF	CE	CT	CS
1	0.333	0.476	0.000	0.143
2	0.417	0.419	0.333	0.571
3	0.417	0.486	1	0.143
4	0.417	0.381	0.667	0.571
5	0.000	0.952	0.167	0.286
6	0.333	0.638	0.167	0.714
7	0.167	0.743	0.667	0.143
8	0.000	1.000	1.000	1.000
9	1.000	0.019	0.000	0.286
10	0.667	0.305	0.167	0.714
11	0.917	0.143	0.667	0.000
12	0.583	0.438	0.667	0.714
13	0.667	0.562	0.000	0.000
14	0.583	0.714	0.167	0.714
15	0.750	0.505	0.667	0.286
16	0.750	0.571	0.667	0.571

The human participant is required to give a score (ranging from 0 to 10 in which 0 denoting an extremely undesirable story narration and 10 a great one from the human participant's perspective) to each of the subjective metrics of a story narration.

The first narration in the above sample with comparatively low scores—2 for coherence, 3 for novelty and 3 for interestingness—is presented as follows.

In the middle of August, Bagaga need to design strategies to manage the new situation. The Bagaga's army is surrounded with the Wagaga's army.

It is July 2030. Wagaga's soldiers cut Bagaga's soldiers' logistic supply of the virus treatment. Wagaga's soldiers managed to go in Bagaga's soldiers.

In the beginning of September, Wagaga is a developing weaker country. Bagaga is a powerful developed country. An Ebola virus outbreak took place in Country Wagaga. Country Bagaga produced a treatment for Ebola. Bagaga refused to help Wagaga with the Ebola treatment.

During combat, as Bagaga's soldiers advances, Wagaga declared war on Bagaga. It is end of August. Land forces started to engage in Wagaga's land. A number of soldiers became infected with the virus behind Bagaga's lines.

In the beginning of August 2030, Wagaga soldiers transferred the virus to Bagaga's soldiers deep in Wagaga's land.

It is July 2030. Bagaga's soldiers advanced from Wagaga. The Ebola virus spread very fast from Wagaga and Bagaga.

6.2 Building Surrogate Models

The collected subjective metrics data is firstly normalised to remove difference in value ranges. For each subjective metrics, the scores of all the 16 narrations in the story narration sample are divided by his score of the eighth story narration in the story narration sample (the original story).

We build a set of individual surrogate models for story evaluation, each capturing the mapping between a particular subjective metrics and all the objective metrics of a story narration.

The notation used in the multiple linear regression model represented in Eq. (5) is explained as follows: Y is an $n \times 1$ vector representing n cases of observed data about a subjective story metrics which is collected from the above experiment in the form of human evaluation scores for n story narrations in the sample; β is a 5×1 vector of regression coefficients each of which denotes an objective story metrics' weight in determining the value of a subjective story metrics, including the intercept; X is a matrix that gives all the observed values of the objective story metrics; e is the $n \times 1$ vector of statistical errors; and $y_i = x_i'\beta + e_i$ is the ith row of Eq. (5), where i denotes the ith case of the observed data—data of the ith generated story narration.

$$Y = X\beta + e. \tag{5}$$

Table 3 shows the results of the surrogate model for each subjective metrics.

Table 3 Linear regression models of subjective metrics as surrogate model

Subjective metrics	DOF	CE	CT	CS	Intercept
Coherence	0.483	0.985	3.465	3.240	0.988
Novelty	3.261	3.493	2.665	1.176	1.954
Interestingness	2.791	3.557	2.962	1.515	1.828

7   Multi-objective Evolutionary Story Narration

In this section, EC is designed to better explore the subset of strategic space, with the step (1)–(4) discussed above synthesized to serve an automatic and evolutionary story narration process.

Step (1): The linguistic constraints extracted in Sect. 3 from a strategic story serve as reference for coherence and causal consistence for the genotype-phenotype mapping in step (2).

Step (2): The permutation-based representation discussed in Sect. 4 serves as the genotype, with a genotype-phenotype mapping provided to guarantee that any random chromosome can get transformed into a unique coherent and causally consistent story. This representation serves as the basis to form a strategic story as well as the basis to design simulation models to evaluate stories for step (3).

Step (3): The evaluation of strategic stories is completely left for the human to subjectively decide on it. A human evaluation experiment discussed in Sect. 6.1 collected data of subjective metrics of a story sample set from humans, with the corresponding objective metrics of the story recorded.

Step (4): The surrogate models were obtained in Sect. 6.2 based on the data collected in step (3). They serve as the objective functions of EC. Each of the surrogate models can predict a human's subjective evaluation of a story narration regarding a particular metrics from the corresponding objective metrics.

The details of the EC are given as follows.

7.1   Elitism Strategy

The elitism strategy in NSGA-II [59] is applied to maintain elitist solutions in the population during the evolutionary story narrating process. A binary tournament selection based on the crowding distance is used to select parents from the population for crossover and mutation.

7.2   Genetic and Search Operators

The genetic and search operators used in the multi-objective evolutionary story narrating process in this paper are listed below.

- **Crossover operators**: the partially mapped (PMX), order (OX) and cycle (CX) crossovers in [60] which can maintain the validity of the permutation after crossover.
- **Mutation operators**: the inversion, insertion, displacement and reciprocal exchange operators explained in [60].

7.3 Experimental Study of Evolutionary Process

An experimental study is discussed to test the performance and understand the effects of the multi-objective evolutionary story narrating process elaborated above. After testing the performance of the story evolutionary process under different parameter settings, the following are selected. Comparatively big population size and diversified crossover and mutation operators are selected to incorporate diversity in the objective metrics discussed in Sect. 5.2 given the redundancy in the layer genes.

- **Population size**: 1000
- **Generation limit**: 5000
- **Crossover rate**: 0.8 where the PMX, OX and CX crossover operators share equal probabilities
- **Mutation rate**: 0.2 where the inversion, insertion, displacement and reciprocal exchange mutation operators share equal probabilities

Figure 8 presents a comparison of the distributions of objective metrics of the story narration individuals in the initial and final population annotated by "-init" and "-final", respectively. From CE, CT and CS, the surrogate model of human evaluation tends to guide the evolutionary process to converge to story narrations whose orders of events, temporal and spatial information are more consistent with those of

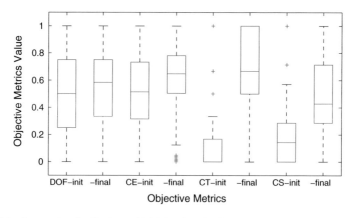

Fig. 8 Objective metrics distributions of initial and evolved story narrations

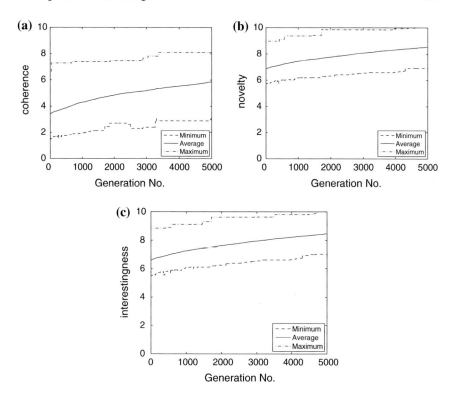

Fig. 9 Transition of values of subjective metrics during evolutionary process, **a** coherence, **b** novelty, **c** interestingness

the original story. This effect is especially obvious in CT and CS whereby random narrations in the initial population hardly achieve high values while the majority of the evolved narrations possess high values. It is also interesting to notice the diversity for the DOF metrics, which indicates a possible preference in flashback in the evolved narrations.

The transition of the coherence, novelty and interestingness subjective metrics during the evolutionary process is illustrated in Figs. 9 and 10. Figure 9 includes plots of the minimum, average and maximum values of the objective functions among the story narrations in the population of each generation as evolution proceeds; and Fig. 10 demonstrates the transition of the non-dominated fronts during the evolutionary process whereby non-dominated fronts in the subsequent generations are denoted by the cross points with bigger size.

The evolutionary process succeeded in evolving story narrations, i.e., collecting story narrations with improved quality in terms of the approximated coherence, novelty and interestingness subjective metrics, reflected in the increasing trend in the plots in the two figures.

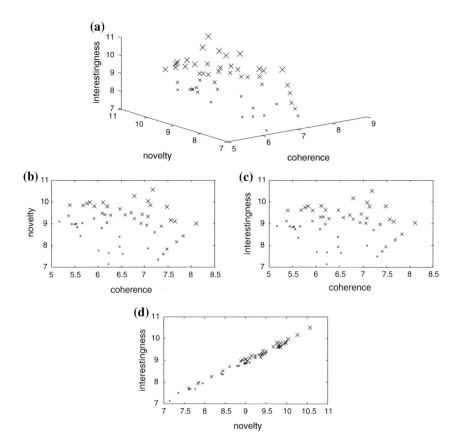

Fig. 10 Transition of non-dominated fronts in subjective metrics dimension during evolutionary process: non-dominated fronts in subsequent generations are denoted by cross points with bigger sizes, **a** coherence-nonvelty-interestingness, **b** coherence-nonvelty, **c** coherence-interestingness, **d** nonvelty-interestingness

Figure 10 reveals possible trade-offs between coherence and novelty, and between coherence and interestingness, while a possible linear relationship between novelty and interestingness in human story evaluation.

What follows below is one of best evolved narrations corresponding to a non-dominated front in the final population of evolution. The values of the subjective metrics assigned by the surrogate models are 7.477 for coherence, 9.773 for novelty and 9.803 for interestingness, with the values of the objective metrics 0.583 for DOF, 0.714 for CE, 1.0 for CT and 0.714 for CS.

Bagaga is a powerful developed country. It is July 2030. An Ebola virus outbreak took place in Country Wagaga. Country Bagaga produced a treatment for Ebola. Wagaga is a developing weaker country. Bagaga refused to help Wagaga with the Ebola treatment.

In the beginning of August 2030, Wagaga declared war on Bagaga. Land forces started to engage from Wagaga and Bagaga.

In the beginning of September, Bagaga needed to design strategies to manage the new situation. The Bagaga's army was surrounded with the Wagaga's army. It is end of August. Wagaga's soldiers cut Bagaga's soldiers' logistic supply of the virus treatment. Wagaga's soldiers managed to go behind Bagaga's lines. The Ebola virus spread very fast in Bagaga's soldiers. Bagaga's soldiers advanced in Wagaga's land.

During combat, as Bagaga's soldiers advanced, Wagaga soldiers transferred the virus to Bagaga's soldiers deep in Wagaga's land.

In the middle of August,a number of soldiers became infected with the virus from Wagaga.

Compared with the original story presented in Sect. 6.1, this narration possesses a certain degree of flashback and thus it may trigger human curiosity about possible causes and effects of an event or possible current states of the world. The change of temporal information further changes the order of events thus the context of the scenario. For instance, a scenario is depicted from "In the beginning of September, Bagaga needed to design strategies to manage the new situation." to "Bagaga's soldiers advanced in Wagaga's land." which says Bagaga's army was surrounded, his supply of medical treatment was then cut, followed by the army chased and unfortunately infected with Ebola virus. It may be inferred that when Bagaga's army were surrounded then cut supply of Ebola treatment, they had not been infected with the virus yet, which is different from the original story whereby they were surrounded after being infected with the virus and possibly because of ill soldiers who slowed down the army's progress.

8   Conclusion

This chapter presents the first automation of the narration of a strategic defence and security scenario, written in the form of a natural language story. By designing appropriate computational models to explore a strategic space defined by a story in natural language, it will be simpler to automate the computational scenario planning cycle. In other words, we will be able to automate the evaluation process of defence and security scenarios on multiple levels of resolution, starting from a grand strategic level down to a tactical level. Results show the effectiveness of this approach in collecting story narrations with improved quality in coherence, novelty or interestingness. This work holds enormous potential in anticipating and resolving further risks, outbreaks, and potential conflict, which could assist in awareness raising and training of military defence organisations.

Further work may include synthesising multiple strategic stories or scenarios in our domain of interest into one dependence network and generate thus evolve scenarios using this evolutionary-based automatic storytelling approach with more diversity accordingly.

Acknowledgments This work was supported by the Scientific Research Starting Foundation for Returned Overseas Chinese Scholars, the Shandong Postdoctoral Innovation Foundation under Grant 140880, and the Chinese Fundamental Research Funds for the Central Universities under Grant 201513001.

Appendix A: Event Recognition

The following grammatical components in a clause are identified as events.

(E1) VERB without predicative complement: all "tensed or untensed verbs" [52] and idiomatic expression (e.g., show up), prepositional verb (e.g., depend on) and phrasal-prepositional verb (e.g., catch up with) [51].

(E2) Predicative complement with NP as the head: predicative complement whose head is a noun phrase (NP) and the verb belongs to one of the following classes: copulative predicates (e.g., to be, seem), inchoative predicates (e.g., become), aspectual predicates (e.g., begin, continue, end, finish), change of state predicates (e.g., retire, appoint, elect, resign), predicates of evaluation and description (e.g., consider, describe, depict, evaluate) [52]. For instance, "became a good guy" in "He became a good guy." or "was elected president of the country" in "She was elected president of the country."

(E3) Predicative complement with ADJECTIVE as the head: predicative complement whose head is an adjective and the verb belongs to one of the following classes: the above-mentioned copulative predicates, inchoative predicates, aspectual predicates, change of state predicates and predicates of evaluation and description, as well as causative predicates (e.g., cause, make) and predicates of perception (e.g., look, hear) [52]. For instance, "looks nice" in "The dress looks nice." or "made happy" in "The cake made the boy happy."

(E4) Predicative complement with PP as the head: predicative complement whose head is a prepositional phrase (PP) and the verb belongs to the classes that belong to copulative predicates, inchoative predicates, aspectual predicates, change of state predicates and predicates of evaluation and description, causative predicates and predicates of perception as has been listed in TimeML event annotation standard [52]. For instance, "is in good mood" in "She is in good mood.", "was behind the tree" in "Cinderella was behind the tree".

References

1. Bui, V., Bender, A., Abbass, H.: An expressive gl-2 grammar for representing story-like scenarios. In: IEEE Congress on Evolutionary Computation (CEC), 2012, pp. 1–8. IEEE (2012)
2. Alam, S., Abbass, H., Barlow, M.: Atoms: air traffic operations and management simulator. IEEE Trans. Intell. Transp. Syst. **9**(2), 209–225 (2008)
3. Abbass, H.A., Alam, S., Bender, A.: MEBRA: multiobjective evolutionary-based risk assessment. IEEE Comput. Intell. Mag. **4**(3), 29–36 (2009)

4. Alam, S., Tang, J., Abbass, H.A., Lokan, C.: The effect of symmetry in representation on scenario-based risk assessment for air-traffic conflict resolution strategies. In: IEEE Congress on Evolutionary Computation, 2009, CEC'09, pp. 2180–2187. IEEE (2009)
5. Bui, V., Bui, L., Abbass, H., Bender, A., Ray, P.: On the role of information networks in logistics: An evolutionary approach with military scenarios. In: IEEE Congress on Evolutionary Computation, 2009. CEC'09, pp. 598–605. IEEE (2009)
6. Xiong, J., Liu, J., Chen, Y., Abbass, H.: An evolutionary multi-objective scenario-based approach for stochastic resource investment project scheduling. In: IEEE Congress on Evolutionary Computation (CEC) 2011. pp. 2767–2774. June 2011
7. Amin, R., Tang, J., Ellejmi, M., Kirby, S., Abbass, H.: An evolutionary goal-programming approach towards scenario design for air-traffic human-performance experiments. In: IEEE Symposium on Computational Intelligence in Vehicles and Transportation Systems (CIVTS), 2013, pp. 64–71. April 2013
8. Gelin, R., d'Alessandro, C., Le, Q.A., Deroo, O., Doukhan, D., Martin, J.C., Pelachaud, C., Rilliard, A., Rosset, S.: Towards a storytelling humanoid robot. In: AAAI Fall Symposium Series (2010)
9. Gervás, P., Díaz-Agudo, B., Peinado, F., Hervás, R.: Story plot generation based on CBR. Knowl. Based Syst. **18**(4–5), 235–242 (2005)
10. Chang, H.M., Soo, V.W.: Planning-based narrative generation in simulated game universes. IEEE Trans. Comput. Intell. AI Games **1**(3), 200–213 (2009)
11. Keppens, J., Zeleznikow, J.: A model based reasoning approach for generating plausible crime scenarios from evidence. In: Proceedings of the 9th international conference on Artificial intelligence and law, ACM, pp. 51–59 (2003)
12. Bui, V., Abbbass, H., Bender, A.: Evolving stories: Grammar evolution for automatic plot generation. In: IEEE Congress on Evolutionary Computation (CEC), 2010, pp. 1–8. IEEE (2010)
13. Wang, K., Bui, V., Petraki, E., Abbass, H.: From subjective to objective metrics for evolutionary story narration using event permutations. In: Bui, L., Ong, Y., Hoai, N., Ishibuchi, H., Suganthan, P. (eds.) Simulated Evolution and Learning. Lecture Notes in Computer Science, vol. 7673, pp. 400–409. Springer, Berlin (2012)
14. Wang, K., Bui, V., Petraki, E., Abbass, H.: Evolving story narrative using surrogate models of human judgement. In: Kim, J.H., Matson, E.T., Myung, H., Xu, P. (eds.) Robot Intelligence Technology and Applications 2012. Advances in Intelligent Systems and Computing, vol. 208, pp. 653–661. Springer, Berlin (2013)
15. Wang, K.: Human-guided evolutionary-based linguistics approach for automatic story generation. Ph.D. thesis, University of New South Wales 12 (2013)
16. Babbar-Sebens, M., Minsker, B.S.: Interactive genetic algorithm with mixed initiative interaction for multi-criteria ground water monitoring design. Appl. Soft Comput. **12**(1), 182–195 (2012)
17. Deb, K., Sinha, A., Korhonen, P.J., Wallenius, J.: An interactive evolutionary multiobjective optimization method based on progressively approximated value functions. IEEE Trans. Evol. Comput. **14**(5), 723–739 (2010)
18. Ciarlini, A.E., Casanova, M.A., Furtado, A.L., Veloso, P.A.: Modeling interactive storytelling genres as application domains. J. Intell. Inf. Syst. **35**(3), 347–381 (2010)
19. Ong, E.: A commonsense knowledge base for generating children's stories. In: Proceedings of the 2010 AAAI Fall Symposium Series on Common Sense Knowledge, pp. 82–87 (2010)
20. de Sousa, R.: Artificial intelligence and literary creativity: Inside the mind of BRUTUS, a storytelling machine. Comput. Linguist. **26**(4), 642–647 (2000)
21. Shim, Y., Kim, M.: Automatic short story generator based on autonomous agents. Intelligent Agents and Multi-Agent Systems, pp. 151–162. Springer, Berlin (2002)
22. Spierling, U., Grasbon, D., Braun, N., Iurgel, I.: Setting the scene: playing digital director in interactive storytelling and creation. Comput. Graph. **26**(1), 31–44 (2002)
23. Theune, M., Rensen, S., op den Akker, R., Heylen, D., Nijholt, A.: Emotional characters for automatic plot creation. Technologies for Interactive Digital Storytelling and Entertainment, pp. 95–100. Springer, Berlin (2004)

24. Riedl, M.O.: Incorporating authorial intent into generative narrative systems. In: AAAI Spring Symposium: Intelligent Narrative Technologies II, pp. 91–94 (2009)
25. Porteous, J., Cavazza, M., Charles, F.: Applying planning to interactive storytelling: Narrative control using state constraints. ACM Trans. Intell. Syst. Technol. (TIST) **1**(2), 111–130 (2010)
26. Bates, J.: Virtual reality, art and entertainment. Presence **1**(1), 133–138 (1992)
27. Meehan, J.: Inside Computer Understanding: Five Programs Plus Miniatures, Tale-spin, pp. 197–226 (1981)
28. Turner, S.R.: Minstrel: a computer model of creativity and storytelling. Ph.D. thesis, Los Angeles, CA, USA (1993). UMI Order no. GAX93-19933
29. Bates, J., Loyall, A.B., Reilly, W.S.: An architecture for action, emotion, and social behavior. Artificial Social Systems, pp. 55–68. Springer, Berlin (1994)
30. Magerko, B.: A proposal for an interactive drama architecture. Ann Arbor **1001**, 48109 (2000)
31. Szilas, N.: Idtension: a narrative engine for interactive drama. In: Proceedings of the Technologies for Interactive Digital Storytelling and Entertainment (TIDSE) Conference, vol 3, pp. 187–203 (2003)
32. Cheong, Y.G., Young, R.M.: A computational model of narrative generation for suspense. In: AAAI, pp. 1906–1907 (2006)
33. Mateas, M., Stern, A.: Writing façade: A case study in procedural authorship, Second Person: Role-Playing and Story in Games and Playable Media, pp. 183–208 (2007)
34. Chomsky, N.: Three models for the description of language. IRE Trans. Inf. Theory **2**(3), 113–124 (1956)
35. Prince, G.: A Grammar of Stories: An introduction, vol. 13. Walter De Gruyter Inc, Berlin (1973)
36. Colby, B.N.: A partial grammar of eskimo folktales. Am. Anthropol. **75**(3), 645–662 (1973)
37. Lee, M.G.: A model of story generation. Ph.D. thesis, University of Manchester, Manchester (1994)
38. Wang, K., Bui, V., Abbass, H.: Evolving stories: Tree adjoining grammar guided genetic programming for complex plot generation. In: Deb, K., Bhattacharya, A., Chakraborti, N., Chakroborty, P., Das, S., Dutta, J., Gupta, S., Jain, A., Aggarwal, V., Branke, J., Louis, S., Tan, K. (eds.) Simulated Evolution and Learning. Lecture Notes in Computer Science, vol. 6457, pp. 135–145. Springer, Berlin (2010)
39. Russell, S., Norvig, P.: Artificial Intelligence: A Modern Approach. Prentice Hall Series in Artificial Intelligence, Prentice Hall (2010)
40. Hoang, H., Lee-Urban, S., Muñoz-Avila, H.: Hierarchical plan representations for encoding strategic game AI. In: Artificial Intelligence and Interactive Digital Entertainment, pp. 63–68 (2005)
41. Kelly, J.P., Botea, A., Koenig, S.: Offline planning with hierarchical task networks in video games. In: Proceedings of the Fourth Artificial Intelligence and Interactive Digital Entertainment (AIIDE) Conference, pp. 60–65 (2008)
42. Bailey, P.: Searching for storiness: Story-generation from a reader's perspective. Technical Report, AAAI Technical Report 1999
43. Fairclough, C., Cunningham, P.: An interactive story engine. Artificial Intelligence and Cognitive Science, pp. 171–176. Springer, Berlin (2002)
44. Díaz-Agudo, B., Gervás, P., Peinado, F.: A case based reasoning approach to story plot generation. Advances in Case-Based Reasoning, pp. 142–156. Springer, Berlin (2004)
45. Swartjes, I.: The plot thickens : bringing structure and meaning into automated story generation. Ph.D. thesis, University of Twente (2006)
46. Theune, M., Faas, S., Nijholt, A., Heylen, D.: The virtual storyteller: story creation by intelligent agents. In: Technologies for Interactive Digital Storytelling and Entertainment (TIDSE) Conference, Darmstadt, Citeseer, pp. 204–215 (2003)
47. y Pérez, R.P., Sharples, M.: Three computer-based models of storytelling: Brutus, minstrel and mexica. Knowl. Based Syst. **17**(1), 15–29 (2004)
48. Swartjes, I., Theune, M.: A fabula model for emergent narrative. Technologies for Interactive Digital Storytelling and Entertainment, pp. 49–60. Springer, Berlin (2006)

49. Hoai, N.X.: A flexible representation for genetic programming from natural language process-
    ing. Ph.D. thesis, Australian Defence force Academy, University of New South Wales, Can-
    berra (2004)
50. Ryan, M.L.: Toward a definition of narrative. In: Herman, D. (ed.) Cambridge Companion
    To Narrative. Cambridge Companions to Literature, pp. 22–35. Cambridge University Press,
    Cambridge (2007)
51. Leech, G.N., Leech, G., Svartvik, J.: A Communicative Grammar of English. Prentice Hall,
    Hemel Hempstead (2002)
52. Saurı, R., Goldberg, L., Verhagen, M., Pustejovsky, J.: Annotating Events in English: TimeML
    Annotation Guidelines (2009)
53. Trabasso, T., Van den Broek, P., Suh, S.Y.: Logical necessity and transitivity of causal relations
    in stories. Discourse Process. $12(1)$, 1–25 (1989)
54. Bridgeman, T.: Time and space. In: Herman, D. (ed.) Cambridge Companion to Narrative.
    Cambridge Companions to Literature, pp. 52–65. Cambridge University Press, Cambridge
    (2007)
55. Sternberg, M.: Universals of narrative and their cognitivist fortunes (i). Poet. Today $24(2)$,
    297–396 (2003)
56. Karmiloff-Smith, A.: Language and cognitive processes from a developmental perspective.
    Lang. Cogn. Process. $1(1)$, 61–85 (1985)
57. Young, R.M.: Computational creativity in narrative generation: utility and novelty based on
    models of story comprehension. In: AAAI Spring Symposium. vol SS-08-03. Stanford, CA,
    pp.149–155 (2008)
58. Campion, N., Martins, D., Wilhelm, A.: Contradictions and predictions: two sources of uncer-
    tainty that raise the cognitive interest of readers. Discourse Process. $46(4)$, 341–368 (2009)
59. Deb, K., Pratap, A., Agarwal, S., Meyarivan, T.: A fast and elitist multiobjective genetic algo-
    rithm: NSGA-II. IEEE Trans. Evol. Comput. $6(2)$, 182–197 (2002)
60. Michalewicz, Z., Fogel, D.B.: How to Solve it: Modern Heuristics. Springer, New York (2004)

A Review of the Use of Computational Intelligence in the Design of Military Surveillance Networks

Mark G. Ball, Blerim Qela and Slawomir Wesolkowski

Abstract This chapter is a review of how computational intelligence methods have been used to help design various types of sensor networks. We examine wireless sensor networks, fixed sensor networks, mobile ad hoc networks and cellular networks. The goal of this review is to describe the state of the art in using computational intelligence methods for sensor network design, to identify current research challenges and suggest possible future research directions.

Keywords Sensor network · Surveillance · Wireless sensor network · MANET · Cellular network · Computational intelligence · Evolutionary optimization · Fuzzy logic · Neural networks

1 Introduction

A key challenge in military operations is the ability to carry out intelligence, surveillance and reconnaissance (ISR). ISR can be achieved from fixed assets such as long range radars or surveillance cameras, or moving assets such as aircraft, satellites or unmanned aerial vehicles (UAVs), or a combination of both. There is a large variety of sensors enabling the creation of sophisticated systems of systems (where the lower-level system is each sensor) such as sensor networks (SNs). In general, an SN is a network of nodes which allows the monitoring of the environment via each node's one or more sensors. Sensors perceive their environment via a variety of sensors from video cameras to motion sensors to various radars.

M.G. Ball (✉) · B. Qela · S. Wesolkowski
Defence Research and Development Canada, Ottawa, Canada
e-mail: mark.ball@forces.gc.ca

B. Qela
e-mail: bqela@ieee.org

S. Wesolkowski
e-mail: s.wesolkowski@ieee.org

© Springer International Publishing Switzerland 2016
R. Abielmona et al. (eds.), *Recent Advances in Computational Intelligence in Defense and Security*, Studies in Computational Intelligence 621,
DOI 10.1007/978-3-319-26450-9_24

SNs such as wireless sensors networks (WSNs) [1, 2], Mobile Ad hoc Networks (MANETs) [3, 4] and cellular networks (CNs) [5] have been extensively studied in the open literature. Fixed sensor networks (FSNs) have not been studied to a great degree, given their primary military application domain. WSNs, MANETs and FSNs are critical for military ISR. Given some similarities between CNs and FSNs, we will also examine relevant CN research.

Enabling technologies are important in devising and managing sensor networks. A recurring theme in sensor network research is how to obtain the best overall situation awareness (SA) or picture from a variety of surveillance systems working cooperatively. SA may be improved by ensuring that sensor resolution is appropriate to the intended target type by scheduling different sensors to provide complimentary coverage (notably by using data from one sensor to queue another) or by maximizing the size of the area covered by the sensor network.

For WSNs, given each sensor node's limited size, another important consideration is sensor power optimization. Computational intelligence (CI) methods are used in a variety of these sensor technologies, sensor coordinating technologies and systems of systems analyses [6]. Operations research and analysis has been used in the systems of systems analysis of sensor networks such as sensor placement, number of sensors, type of sensors, energy-aware protocols, power efficiency and optimization in sensor networks, network topology control, as well as sensor-embedded efficient clustering-based algorithms for data aggregation, and routing [7–12]. This chapter will summarize the state of the art in the use of computational intelligence to carry out operational analysis of SNs and will illustrate the importance of this work in the military and security domains. This survey will also summarize the types of problems studied and identify research gaps by suggesting new research directions.

This chapter is organized in the following manner. In Sect. 2, we will provide a general overview of the field and define most terms. In Sect. 3, we will discuss WSNs given that they distinguish themselves from other networks by their need to conserve battery power. Section 4 will summarize Large Sensor Networks (LSNs) which will group three similar groups of sensor networks: FSNs (e.g., the North Warning System in Canada and the United States), CNs (e.g., AT&T's cellular network in the United States) and MANETs (e.g., Survivable, Adaptive Networks known as SURAN initiated by the Defense Advanced Research Projects Agency— DARPA [13]). Section 5 concludes with a discussion on linkages between SNs and common research challenges.

2 General Overview of Sensor Networks

In this section, we will discuss SN categorization, as well as define the common terms used in the paper.

2.1 Types of Networks

First, we will define and discuss several different sensor network types. The sensor networks we will examine include WSNs, FSNs, MANETs, and CNs.

Wireless Sensor Networks consist of a large number of miniaturized electronic devices equipped with wireless communication capabilities and processing power. These small devices, namely sensor nodes, can sense, actuate, process information, communicate among themselves thus providing significantly a higher sensing capability compared to each individual sensor node. Individual sensor nodes are generally equipped with non-rechargeable batteries and are considered expendable i.e., sensor nodes are typically not recovered when their batteries are depleted. A WSN usually needs one or more data sink nodes which are powerful transmission nodes with high computational power and energy resources, enabling them to reach a destination node or base station. These sink nodes could be mobile depending on the specific application. Taking into account the scarce energy resources of typical sensor nodes, a major WSN challenge is the requirement to extend the network lifetime by exploiting energy-aware design principles and power optimization schemes.

Fixed sensors are the surveillance and reconnaissance assets most common to military operations which operate over large distances (from kilometers to thousands of kilometers). These include any stationary sensor, such as primary radar installations. We also include satellite based sensors in this category. Even though these sensors are in motion, their trajectory cannot be altered as part of normal sensing operations. This results in repeated coverage pattern analogous to a very large, though slowly repeating, fixed sensor. When several of these sensor nodes are used together to provide improved SA, they become a Fixed Sensor Network.

Mobile Ad hoc Networks are dynamic, self-configurable and highly adaptive multi-node networks equipped with mobile devices connected by wireless links. MANETs are rapidly deployable, autonomous networks, which do not require a fixed infrastructure. Mobile nodes are free to move independently in any direction over large areas. Thus, they can be deployed and used in remote areas (e.g., to help with disaster relief), and battlefields of various sizes. FSNs and MANETs can be considered large networks as compared to WSNs. Thus, LSNs will encompass FSNs and MANETs.

CNs can also be considered LSNs due to many similarities they share with FSNs. CNs are made up of linked cellular base stations. Cellular telephones connect wirelessly to cellular base stations, which are in turn connected to a larger telephone network (of wired and cellular telephones). Each base station has a range from one to ten kilometers depending on its location, and the network of base stations in aggregate provides coverage of an entire service area. While our focus is on surveillance rather than communications networks, the CN coverage problem is similar and, therefore, a review of the methods used to address this problem in the cellular industry will be carried out highlighting salient points relevant to LSNs.

The foremost metric by which sensors, sensor networks, or cellular networks are measured is network coverage. Three types of coverage will be studied [14]: blanket, barrier and sweep. Blanket coverage is the total surface area covered and is constant in time as long as all sensors remain functional. Ideally blanket coverage would encompass the entire area of interest (AOI). Barrier coverage is obtained by a line of sensors with some amount of overlap such that a target is not able to pass through the line undetected. The North Warning System (NWS) [15] is an example of an FSN which provides barrier coverage. Sweep coverage begins with barrier coverage but moves the barrier across an AOI over time, resulting in a total area covered that is akin to blanket coverage. An example could be a MANET helping in search and rescue; i.e., the search starts at the last known location of a missing plane and then expands in various directions in a sweeping action. Blanket coverage is the easiest form of coverage to measure as it is simply the total surface area within range of the SN.

Figure 1 illustrates a few common examples of sensor network types useful in the military and security domain. Figure 1a shows the barrier coverage provided by the NWS. The figure shows the area covered by the NWS radars based on publicly available radar locations [15] and ranges [16, 17].

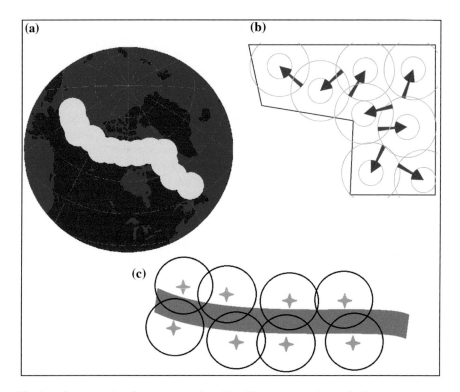

Fig. 1 A few examples of sensor networks with military and security applications

Figure 1b represents one example of a WSN: AOI covered by pan-tilt-zoom (PTZ) cameras that have a limited range. Cameras may detect activity in the AOI depending on the target size and type, given a rotatable restricted field of view for each camera [18]. In this figure, each camera has a "zoomed out" and "zoomed in" range represented by the inner and outer circles respectively. The zoomed out instantaneous field of view (FOV) is shown in blue and the zoomed in FOV in red. These sensors are attempting to provide blanket coverage, although some gaps in coverage are visible, and only a small portion of the AOI is covered at any given time. Sensors are often modelled as unchanging circular projections on a two dimensional map. This approximation does not necessarily hold (depending on the application) for sensors similar to a PTZ which have a FOV that is non-circular and moveable in three dimensions.

Figure 1c illustrates another WSN example: small wireless magnetic sensors spread along a dirt road and used for vehicle detection [19]. This WSN essentially provides blanket coverage of the road; however, assuming the target is travelling on the road from one direction or the other, this WSN can also be seen as a series of barriers. In this case fairly large coverage gaps could be allowed while still being able to detect a truck passing through. In contrast, a set of PTZ cameras intended to detect a person on city streets (e.g., in London, United Kingdom) would have to be able to cover a very large portion of a potentially large AOI.

2.2 Network Characteristics

In this section, we will discuss SN characteristics and the importance of each in analysis. Table 1 summarizes the characteristics of studied sensor networks by extending Table 1 from [2] and adopting most of their terminology. We subdivide sensor networks into WSNs, FSNs, CNs and MANETs.

Table 1 first examines the sensor and base station characteristics. Miniaturization is a key technology enabling WSNs: sensor size is on the order of centimeters or smaller [2, 20, 21]. Cellular base stations are antennae or groups of antennae positioned on top of a cell tower or a building, while large sensors vary in size from a handheld camera to an antenna array the size of a large field [22]. WSNs compensate for their small size (and accordingly limited power) by being deployed in large numbers (typically hundreds or thousands [23]) over a relatively small area (up to a few city blocks). Fixed sensor networks are typically made up of sensors that were designed to be used individually to cover a large area (up to thousands of square kilometers). Cellular networks cover entire countries and the number of base stations required to do so is consequently large, even though the range of individual stations can be relatively large (on the order of tens of square kilometers). Passive sensors, such as cameras, only receive information while active sensors, such as radars, send out a pulse and wait for a return. Heterogeneous networks are made up of multiple types of sensors ideally providing complementary information. The ad hoc nature of MANETs leads them to be heterogeneous and cellular base stations

Table 1 Network characteristics (modelled on Table 1 in [2])

		WSN	Fixed nets	Cell nets	MANETs
Sensors and base stations	Size	Small	Medium to large	Medium	Medium
	Spatial coverage	Dense	Sparse	Sparse	Sparse or dense
	Number	Large	Small	Large	Various
	Type	Passive and active	Passive and active	Active	Passive and active
	Mix	Heterogeneous or homogeneous	Heterogeneous or homogeneous	Homogeneous	Heterogeneous
	Deployment	Random, ad hoc or fixed/planned	Fixed/planned	Fixed/planned	ad hoc
	Dynamics	Stationary or mobile	Stationary or mobile	Stationary	Mobile
Entities of interest	Extent	Distributed or localized	Distributed or localized	Localized	Distributed or localized
	Nature	Cooperative or non-cooperative	Cooperative or non-cooperative	Cooperative	Cooperative or non-cooperative
	Mobility	Static or dynamic	Mostly dynamic	Dynamic	Dynamic
Operating environment	Threat level	Low to high	Low to high	Low	Low to high
	Size of area	Small	Medium to large	Large	Large
Communication	Networking	Wireless	Wired	Wired	Wireless
	Bandwidth	Low	High	High	High
Processing architecture		Distributed or hybrid	Centralized	Hybrid	Distributed or hybrid
Energy available		Constrained	Unconstrained	Unconstrained	Partially constrained

must be homogeneous to communicate with phones i.e., use the same communication protocols. The mix of sensors in WSNs and LSNs depends on the application type. Fixed sensor nodes and cellular base stations are always placed at predetermined locations, while MANET nodes may be located anywhere given that they are mobile [24, 25]. WSN node locations may be predetermined but the nodes are typically deployed in a large group and often spread out randomly [1]. Once deployed, MANET nodes remain mobile, while cellular bases stations are fixed. WSN nodes are sometimes capable of autonomous movement, limited by their power supply. Other WSN nodes may be stationary or may be transported by the medium they are embedded in. Fixed sensor nodes are typically stationary but may be moved between uses [26], or in some cases, as part of their use (such as synthetic aperture radar [27]). Satellite or air-based sensors begin to blur the line between fixed sensor networks and MANETs as they are collecting data while in motion, though their movements are planned.

Sensor networks may be used to study entities that are distributed (weather) or localized (individuals), cooperative or not. Cellular nodes are the exception as they make contact with individual phones that want to be connected. Similarly sensor networks may be used in low or high threat environments i.e., from cities in countries at peace to battlefields. Cellular networks do exist in conflict zones where they may be attacked, but they are not designed to withstand an attack. Most targets of interest for each network are most likely mobile although WSNs may be embedded in an entity to monitor changes in that entity.

Communication refers to the link between the individual sensor and its network. For cellular networks the communication is between base stations and the communication backbone, which is wired, as opposed to the wireless communications it enables. WSNs and MANETs rely on wireless communications as part of their operations [24]. FSNs are typically wired, though wireless communication (e.g., via satellites) may be part of the chain.

Large fixed sensors typically send their data to some central repository for processing. Cellular base stations do some of the processing, but rely on the network switching subsystem to make a connection. In the case of WSNs and MANETs, at least some of the processing is expected to take place at the nodes though it may be distributed.

Fixed sensors and cellular base stations have either their own power sources or use power from an electric grid ensuring continuous operation. WSN nodes are usually powered by small batteries that have a limited lifetime. MANET sensors are powered by the platform that carries them, which typically needs to be refuelled periodically; therefore, energy-awareness is important although it is not a primary concern as it is in WSNs [21, 28].

2.3   Discussion

While the focus of the remainder of this paper will be on the research carried out in each of the network categories, this section will provide the overall context by discussing linkages across the various network types. We will also provide a summary of numerous CI techniques that have been applied to each network type. The goal is to highlight similarities and differences between SN types.

Sensor networks are often treated as synonymous with WSNs. WSNs are an emerging technology that is receiving much attention in research and development. On the other hand, the concept of large-scale sensors, such as networks of radar stations is difficult to find in the literature despite being a well-known problem in defence and security domains.

For example, Kulkarni et al. [6] identifies four challenges faced by WSNs: (1) the wireless ad hoc nature of the network, (2) mobility and changes in network topology, (3) energy limitations of nodes, and (4) physical node distribution. Of these, only the last is a common concern of fixed large-scale sensor networks. On the other hand, MANETs and WSNs share many challenges except that MANETs cover much larger areas. Consistent with the third WSN challenge, a common goal with MANETs is to minimize the energy consumed, often by minimizing the movement of mobile sensors or, in the case of small nodes, improving the data communications efficiency, in order to extend the lifetime of the network. This is not a significant concern for FSNs. Instead, FSNs generally look to maximize the coverage area while minimizing installation and operations costs [29, 30]. These are the same objectives generally faced by cellular networks. In situations where all sensors (or cellular base stations) are identical, the number of stations is used as a proxy for cost [31–34]. The general relationships between these different types of networks are summarized in Fig. 2.

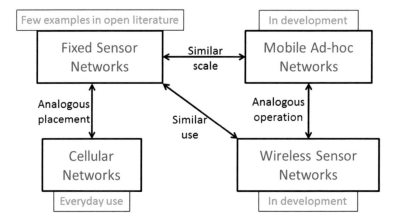

Fig. 2 Network relationships

Table 2 Methods and applications

Application	Method							
	Genetic algorithm	Greedy algorithm	Fuzzy logic	Potential field/virtual forces	Linear programming	Neural network	Swarm intelligence	Other
Intrusion detection	[5, 30, 31, 36]					[72]		[19, 57, 104]
Open area (volume) surveillance		[32]	[38, 86, 96]	[42]	[29]		[86]	[55, 86, 97, 98]
Confined area surveillance		[40]		[41]				[18]
Target of interest surveillance /tracking			[71]			[99]		[18, 42, 55, 57, 85, 103, 104]
Data analysis	[84]							[98, 101, 102]
Data aggregation								[11, 12]
Cell base station placement	[34, 35]	[33]						[37]
Other		[39, 100]	[82]					[54, 76]

Table 3 Methods and objectives

		Method							
		Genetic algorithm	Greedy algorithm	Fuzzy logic	Potential field/virtual forces	Linear programming	Neural network	Swarm intelligence	Other
Objective	Coverage/people served	[5, 30, 31, 34–36]	[32, 33, 39, 40]	[38, 82, 86, 96]	[41, 42]	[29, 39]		[86]	[37, 81, 86, 97]
	Cost/number of nodes	[30, 31, 34, 35]	[32, 33]		[41, 42]	[29]			[37, 56]
	Transmission energy	[5]		[86]				[81, 86]	[8, 9, 12, 28, 51, 52]
	Movement energy	[36]		[38]					
	Sensor energy ("on" time)								[19, 47, 51, 97]
	Other	[31]	[40, 100]	[71, 82, 96]			[72]	[81]	[11, 19, 54, 85]

Table 2 summarizes how various optimization methods have been applied to different sensor network applications in the literature, while Table 3 examines the optimization objectives that have been addressed with these methods. The numbers in these tables correspond to the references at the end of this chapter. Optimization methods include CI methods such as genetic algorithms (GA), multi-objective GAs, Swarm Intelligence (including Ant Colony Optimization and Particle Swarm Optimization), other heuristics such as tabu search as well as non-CI methods such as greedy algorithms or linear programming. In all cases where coverage is being used as an objective, it is blanket coverage (as defined in [14]) that is being measured.

There are several ways in which the competing objectives of maximum coverage and minimum cost are reconciled. Some studies use multiobjective optimization, resulting in a Pareto front of solutions [5, 30, 31, 35]. Others assign weights to create a combined objective function [34, 36, 37]. Several studies also treat one or the other as a constraint, either fixing the number of sensors (and thus the cost) and determining the maximum area coverage [30, 38–42], or fixing a minimum allowed coverage and determining the required number of sensors [29, 33].

Sensor locations are generally dealt with in one of three ways. The most restrictive is to allow sensors to be placed at predetermined locations, which may be

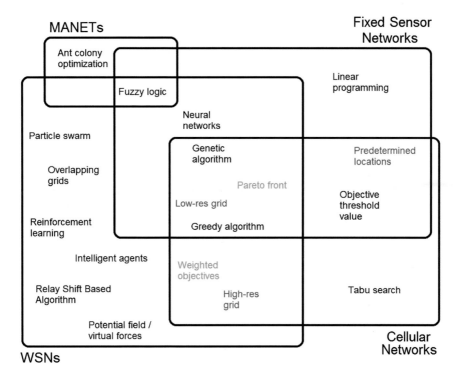

Fig. 3 Recurring themes across network types

appropriate when the sensors require some pre-existing infrastructure or specific terrain (e.g., FSNs and CNs). The most general is to allow sensors to appear at any location within the area of interest. When location is treated as a continuous variable, we refer to this case as a high-resolution grid. For the intermediate case, a low-resolution grid, sensors could be placed at the vertices of a grid with a finite number of points.

Various CI and data modelling methods are grouped in Fig. 3 based on the type of network they were used for. This diagram also identifies in red how sensor and base station locations were handled. In addition, many studies use methods to deal with conflicting objectives like multiobjective optimization (to create Pareto fronts of non-dominated solutions) and single-objective optimization with a weighted sum of several objectives. These two options are shown in green.

3   Wireless Sensor Networks

3.1   Background

The emergence of WSNs is a result of the development of small-size embedded microcomputer-based systems, which support a wide range of sensors. WSNs use a large number of small, inexpensive sensors instead of a smaller number of powerful sensors. As shown in Fig. 4, the main components of a wireless sensor node are: the sensor, embedded controller, memory unit, communication device and power supply. Sensing, actuating, communicating and processing capabilities of sensor nodes enable their capabilities to self-organize and communicate in the deployed areas. The low cost, miniaturized size and easy deployment, makes sensor nodes attractive for use in military applications with versatile requirements. Different sensor node architectures can be chosen based on the application requirements. Several comprehensive overviews of the research in the field have been written [1, 20, 21, 43, 44].

Some key points are that each sensor node, in addition to its sensing capability, has limited processing and data transmission capabilities. However, they are mainly deployed in large numbers, thus their computational load can be shared across all or a subset of nodes to save energy resources and extend the lifetime of the WSN. An example of effective use and conservation of the nodes' energy is to organize neighbouring sensor nodes into local clusters using a technique such as the Low-Energy Adaptive Clustering Hierarchy (LEACH) proposed by Heinzelman et al. [9], where each cluster is assigned to a cluster head. The cluster head gathers the sensed data from its cluster members and performs data processing and aggregation prior to transmission of the data to the sink node. Moreover, the cluster head role can be rotated between cluster nodes thus ensuring that the energy load is distributed evenly.

Fig. 4 Sensor node architecture

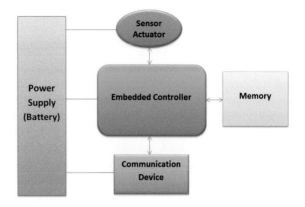

Individual sensor nodes can be considered expendable: the nature of wireless nodes requires them to be battery powered and when the battery dies, nodes are typically assumed to be irrecoverable [45]. The energy efficiency, which is closely related to the lifetime of the WSN, is one of the main constraints in the design of sensor nodes [46]. Thus, conserving battery life by minimizing the amount of work done by each node becomes a priority. Military applications may be data intensive and/or require WSNs to be deployed over large timeframes, thus making energy efficiency an important design characteristic. Energy-efficient topology control algorithms, data aggregation, routing, schedule-based protocols, sensor modes of operation (e.g., active, idle, sleep) can be all used to extend sensor network operation [47]. Furthermore, a WSN also has to be tolerant to the loss of individual sensors by exploiting redundant deployment of nodes, and/or use of a handoff mechanism, which enables the transfer of services to healthy neighbourhood sensor nodes to restore and maintain the connectivity of a failed link to a sink or destination node [48]. The Quality of Service (QoS) attributes of WSNs such as event detection, delay (latency of a sensor response), bandwidth (limited number of channels and data rate transmission capabilities typically in ranges of 250 kbit/s or less), etc. differ based on the choice of hardware/software platform for specific WSN applications [20, 49, 50]; however, they are important factors to be considered during the WSN design and deployment stages.

The scalability of WSN architectures and protocols based on the number of sensors deployed is another important aspect to be considered, especially for military applications given the necessity to deploy WSNs in settings from small villages to large battlefields. Based on WSN application requirements, the densities of sensors in specific deployed areas might be non-homogeneous, and the network should be able to adapt to such changes in configuration. Moreover, as WSN dynamics change due to the depletion of energy resources of individual sensor nodes or different assigned tasks, the network must still be able to self-configure, adapt and remain operational [51, 52].

3.2   Defence and Security Applications

Arampatzis et al. [23] provide a survey of WSN applications including a section on military applications where the areas of interest are not limited to information collection only, but also include enemy tracking, battlefield surveillance and target classification via networks consisting of sensor nodes equipped with seismic and acoustic sensing capabilities. He et al. [19] tackle an important aspect of WSN use in surveillance missions, where the sensors are deployed in large numbers with the ability to detect and track vehicles in a region of interest (1) in an energy-efficient manner, where only a subset of sensors nodes are active and monitoring at any one time, while the rest are in low power mode, and (2) in a stealthy manner, where the sensor network has a low probability of being detected given that sensors use minimal communications in the absence of events. Thus, by considering a trade-off between energy consumption and surveillance performance as a system design parameter, the sensor network is highly functional and long lasting while being adaptable to changes.

Đurišić et al. [53] examines some WSN military applications ranging in scale from sensors deployed across a large area such as a battlefield to detect infrared, chemical, or acoustic signatures, to multi-sensor systems used for perimeter protection to sensor networks worn by soldiers to monitor their vital functions. Liu et al. [54] test their Simulator for Wireless Ad hoc Networks against a scenario depicting chemical agent dispersal in an urban area. Although their chemical plume dispersion model has been simplified, it still illustrates the importance of networked chemical detection sensors.

Afolabi et al. [55] discuss viable options in combining different advanced technologies, such as UAVs and wireless sensing devices to enhance surveillance capabilities. The cooperation and integration of UAVs in a WSN improves the performance of surveillance missions by using an efficient deployment of sensor nodes; where the maximum coverage is attempted with the least possible number of nodes via equilateral triangulation (this type of grid has the smallest overlapping area as compared to grids based on squares or hexagons) [56]. Thus, the addition of UAVs may provide a relatively inexpensive surveillance solution when linked with deployed sensor nodes to cover a specific region of interest.

Song et al. [57] analyze the performance of Passive Infrared (PIR) sensors and their use as WSNs for surveillance systems. For example, these systems can be used for tracking intruders by detecting the movement of the temperature gradient between the warm person and their cooler surroundings. Processing data from PIR sensors is efficient with an output as simple as "nothing detected" or "movement detected" compared to a vision-based device, which would require a larger onboard memory and computational power due to more complex data processing required for image processing.

Finally, sensor networks can be used in conjunction with UAVs in applications such as collaborative surveillance missions (e.g., to aid military troops during combat operations) including the detection and tracking of enemy forces or the detection of hazardous biological, chemical, and/or explosive vapor [58]. Naturally, this merging of UAVs and WSNs leads to the necessity of studying how WSNs and MANETs (discussed in Sect. 4.3) could interact and the research challenges this would bring. This topic will be discussed in more detail in Sect. 4.5.

3.3 Review of Methods and Applications

A common goal for WSNs is to maximize the lifetime of the network, while meeting the application requirements. In particular, energy-aware design to ensure the prolonged life of surveillance missions should be of interest in WSN design [19]. While efficient network topology control [9] exploits the redundant deployment of sensor nodes, it restricts the set of nodes which are considered neighbors of a given node to overcome the energy limitations; hence, minimizing the number of retransmissions required to deliver data to the receiver (by only a few selected nodes). Similarly, sensor nodes communicate with a sink node (or base station) via multi-hop paths, thus in-network processing is also used to reduce the amount of data sent (thus reducing overhead) throughout the network [1, 51].

The reduced overhead is achieved by lowering the number of messages forwarded throughout the network by applying data aggregation principles within sensor nodes. Benefits of data aggregation depend on the sensor nodes' configuration. If the sensors are configured in a radial configuration as shown in Fig. 5a where all the sensor nodes are one hop away from the sink, data aggregation is not beneficial. However, in the case shown in Fig. 5b, where the sensor nodes are more than one hop away from the sink, data aggregation at intermediate nodes leads to lower message overhead.

For WSNs, the choice of sensor node configuration, i.e. flat versus hierarchical, depends on the application and the size of the deployed network. In flat networks all nodes are considered equal and the main emphasis of network topology is power usage control. However, the scalability of a network due to non-homogeneity remains a concern. In hierarchical networks, the emphasis is on the backbone or cluster connected topology, which takes advantage of heterogeneity and aids in constructing a self-organizing network. A large-scale WSN deployment, in the case

Fig. 5 Data aggregation: **a** radial configuration, **b** feasible configuration

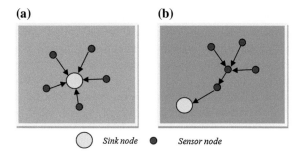

of battlefield and/or long-term missions would need to take into account the energy-awareness of a network, thus most likely utilizing hierarchical network topology.

The overall network coverage and energy efficiency depends on many factors including the existence of powerful mobile or fixed nodes (with lasting energy resources, a powerful processing core and transmission range) acting as intermediate nodes within a deployed sensor network. As an example, if energy constrained nodes transmit at longer distances frequently, the sensors' energy resources will be quickly depleted leading to node failures and sensor network lifetime reduction. However, if the role of transmission is taken by powerful nodes with considerable (or rechargeable) energy resources, prolonged operation of a network is possible. This is one of the reasons why node-based local clustering, data aggregation and in-network processing would be important for a viable and long lasting WSN [9, 59].

Jourdan and de Weck [5] aim to maximize the total sensor coverage, as well as the lifetime of the network. This was done by randomly deploying the available sensors to create individuals in a Multi-objective Genetic Algorithm (MOGA). This is one of the rare cases of truly optimizing across multiple objectives; in fact addressing this gap is stated as part of their motivation. To measure network life-time, they assumed that each sensor sends its data to a primary receiver once for each "sensing cycle." However due to multi-hoping, sensors may need to relay data other than their own. Each of these transmissions depletes some of the sensor's energy. The lifetime of the network is determined by the number of sensing cycles before any one sensor's energy is completely depleted. They concluded that a network of sensors whose communications range is more than double their sensing range is most efficient in a cluster configuration. A cluster allows multiple paths from any sensor in the cluster to the sink node. Otherwise, a hub and spoke configuration is more efficient.

Other studies [36, 38] aim to maximize sensor coverage while minimizing the movement of sensors (and thus the energy expenditure) after an initial random deployment. A weighting factor is used by Jiang et al. [36] to treat both objectives as one, while fuzzy logic [38] allows a move if the coverage state is improved, and sometimes allows a move if the coverage state is to remain the same. GA calculations are done at the nodes based on information exchanged with neighbours [36]. As processing consumes less energy than communications, this method is more energy efficient than having all nodes report their locations to a central processor which would then determine the new locations and send them back to the nodes.

Osmani et al. [38] also measure the resulting "message complexity," which is the number of messages exchanged; however, they don't treat it as an objective to be optimized by the algorithm. Minimizing the movement of wireless sensors is an important military objective given that battery power is at a premium. Deployed WSNs should thus try to adjust their position only when there is a higher likelihood of obtaining more information by moving than by staying in the current position (e.g., based on analysis of previously sensed data).

Liu et al. [54] provide a scalable framework for the simulation of sensor networks, and its use for studying the performance of routing algorithms. In this case, the authors were not attempting to optimize the network, but rather to demonstrate that their simulator for wireless ad hoc networks (SWAN) could be used to measure network capacity and performance of routing algorithms in sensor networks. Calculating the coverage of the network is outside of their scope; however, it should be a concern in the initial network layout. Their simulation environment allows proposed network configurations to be tested before being deployed. This way when a WSN is later implemented, it can use the most efficient configuration to route data to the sink.

Howard et al. [41] and Zou and Chakrabarty [42] simulate virtual forces (or potential fields) acting on the sensors, pushing them to spread out. Howard et al. [41] use a friction force to prevent the nodes from spreading out indefinitely, while Zou and Chakrabarty [42] apply a repelling force between nodes within some threshold distance and an attraction force between nodes outside some larger threshold distance.

The incremental development algorithm [40] addresses the issue of WSN coverage area; however, it assumes that the sensor nodes are deployed one at a time, which might not be a feasible solution in case of military applications (e.g., for large-scale deployment of thousands of nodes).

3.4 Research Challenges

WSNs place a premium on energy efficiency with transceiver and processor being the main energy consuming blocks. Energy scavenging utilizing solar cells, vibration and/or other alternative means to recharge sensor's battery needs to be considered since it may change network design. In certain applications, the number of nodes could be reduced given that fewer nodes might be assumed to have their batteries depleted. Minimizing the unnecessary transmission (and reception) of data and processing performed by sensor nodes is essential due to the limited energy resources.

Secure messaging is required due to the threat of cyber-attacks on military surveillance systems. This issue is not discussed often in the WSN literature. However, it needs to be addressed in particular where the security breach in the network might cause casualties of friendly military troops on the battlefield. Butun et al. [60] elaborates on Intrusion Detection Systems (IDS) initiatives in addressing future WSN security concerns such as jamming, flooding attacks, eavesdropping, etc. which might degrade and incapacitate WSNs. However, due to the limited energy resources and computational capabilities of WSN nodes, access control techniques used for traditional wired and/or wireless network security do not apply [61]. The use of existing or new CI techniques to detect security threats to WSNs would pose a challenge especially when considered in conjunction with scarce energy resources of sensor nodes. Thus, there should be more research in secure communications of devices with limited energy capacity and into techniques to help WSNs thwart cyber-attacks.

The design and deployment of WSNs has many challenges with respect to network fault tolerance, lifetime, self-organization, scalability, node hardware/ software considerations, feasible network architectures and between-node communication protocols to be adopted under different scenarios [62]. All of these characteristics are difficult to accommodate into a single optimal WSN solution. Thus, application-specific purpose-built WSNs should be studied. Furthermore, based on the overall trade-offs, selecting adequate design parameters of choice, which could provide an optimal solution with respect to cost and performance, poses another complex and interesting WSN design challenge, due to the dynamics and diverse requirements of military applications [2, 55, 58, 63].

The WSN design requirements could be different when considering the deployment of WSN for non-critical or peacetime missions, where the security of the network and its lifetime are not of prime importance. WSN challenges related to energy efficiency, sensor node battery life (energy scavenging), control topology, in-network processing and self-organization in order to prolong a lifetime of network, while at the same time conforming to the (required) guaranteed network connectivity and security aspect of networks might differ considerably. Consequently, comparing network architectures of nodes built for high threat environments versus low threat environments (i.e., commercial off-the-shelf sensor nodes) would provide interesting insights into military WSN design (e.g., would commercial off-the-shelf sensor nodes be good enough for a given peacetime application?).

Moreover, over-the-air firmware upgrade of sensor nodes under different circumstances (e.g., tactical military sensor network in remote large-scale areas) to accommodate different functionality of versatile sensor nodes, could be considered in future research.

WSNs may also be combined in various ways with LSNs in order to create more comprehensive SA. Currently, there is increasingly more research being done into the use of WSNs in conjunction with one or more UAVs or other assets. The UAVs in those cases might be the WSN information recipients and further relay the sensed data to base stations. How WSNs might increase the effectiveness of single or multiple UAVs (and even MANETs) could be of considerable interest. Furthermore, how WSNs would improve the SA of LSNs should also be studied since depending on the application, LSNs might not be able to gather all relevant data (e.g., from a battlefield).

4 Large Sensor Networks

4.1 Background

Large Sensor Networks include networks of sensors typically associated with defence and security, such as MANETs made up of airborne sensors or FSNs of large early warning radar systems. These are used for homeland security, rogue aircraft detection, drug smuggling detection, etc. LSNs also include many civilian sensors such as air traffic control, Automatic Identification System (AIS) [64], and satellite-based sensors [65, 66]. While the sensors are sophisticated, the networking aspect is not well studied. The fusion of data gathered from multiple sources should be a topic of interest ensuring that they may complement each other in the most efficient way possible. Like WSNs, these networks aim to provide the maximum coverage possible. Unlike WSNs, the replacement cost of a single sensor is a significant concern, while the energy expenditure is not.

While some of these large sensors are often used alone, the military should be interested in the ability of multiple sensors to provide a combined SA that is greater than the sum of its parts. Combining sensors with different sensing regimes, such as radar and optical sensors, allows the confirmation of detected objects between sensors and the detection of objects that might be visible in one medium but not others. Accordingly, if the coverage areas of multiple sensors overlap, then one sensor might be able to provide information that was missed by another. However, this overlap also represents a reduction in the total coverage area that could have been achieved by the same sensors if they were separated. Another approach to multi-sensor surveillance could be the use of a wide-area coarse resolution sensor to provide initial detections that are then followed up by a smaller area, higher resolution sensor.

A very large scale sensor network, the Distant Early Warning (DEW) Line created a radar barrier by placing sites along the "most northerly practicable part of North America" [67]. The NWS used many of the same sites [15]. Figure 6 shows how the main line of sites was not significantly altered during the transition from the DEW Line to the NWS.[1]

As part of LSNs, we also study methods used to optimize CN coverage. Although CNs are used for communication rather than surveillance, the base stations in each CN must be able to detect a cellular phone within their coverage area and, therefore, the placement of these base stations is analogous to the placement of sensors as both represent some type of coverage within a given radius of an installation. The goal of both CNs and FSNs is typically to maximize the amount of coverage provided by a network of installations (whether cell towers or radars) while minimizing the number of installations required.

4.2 Fixed Sensor Networks

Sakr and Wesolkowski adapt their MOGA to optimize across three objectives, while also accounting for multiple types of sensors [30]. In their implementation, each sensor type was characterized by a unique coverage radius and cost. Their objectives were to maximize the total coverage, minimize the total cost, and minimize the amount of coverage overlap. They assumed a fixed number of sensors,

---

[1]This figure was created in Google Maps [68, 69] using data from [15] and [70].

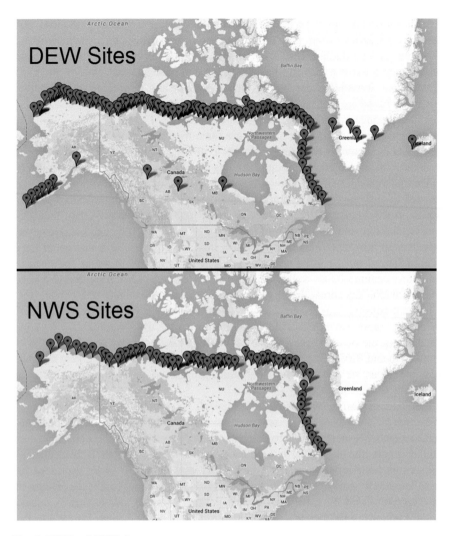

Fig. 6 DEW and NWS sites

which had the effect of reducing the search space. Although this work was framed
in the context of a WSN, the methodology has more in common with FSNs.
Specifically, it is limited to ten total sensors, and neither energy constraints nor the
ability of sensors to move are accounted for. While the sensors modelled could
indeed be wireless, this does not affect the methodology or results in this case. This
research further shows the usefulness of creating different network architectures
based on the emphasized objectives. The work could be extended more specifically
for WSNs by including an objective to examine energy consumption. Another
aspect of interest to FSNs would be to look at particular types of overlap coverage
(e.g., overlap by two or three sensors). From a defence perspective, it is significant

that this work accounts for a network of dissimilar sensors, although in this case the sensors are assumed to be redundant rather than complementary—hence the objective of minimizing, rather than maximizing, overlap. This work could be extended to seek to maximize the overlap of dissimilar sensors, while maintaining maximum coverage. In this case, increasing the limit on the total number of sensors would likely be required.

Oh et al. [31] examine the coverage by several sensors of different types. Each type of sensor is defined by a size and shape of its coverage area and each sensor can be placed anywhere on a grid; however, their algorithm does not allow the possibility of rotation of the coverage area about the sensor location. Their objectives are to maximize the coverage, to minimize the number of sensors used, to maximize a weight function based on a user-assigned sensor preference, and to minimize the distance of a randomly located target to the sensors. The objective of minimizing the number of sensors may be intended as a proxy for minimizing the total cost; however, another interpretation may be that a smaller number of sensors would be more manageable for the analyst receiving the data. The sensor preference function is unique. This could also be a proxy for cost although it is intended to be more situation dependant. Matching the right sensor to the intended target is an important consideration for defence surveillance, and this metric allows sensors to be ranked based on their appropriateness to the mission, while not exclusively considering the best sensor.

Church and ReVelle [39] set their objective as maximizing the number of people within a given service radius of any facility. They suggest solving this problem separately for a different number of facilities, essentially creating a Pareto front through a brute force approach. While this work was not presented in the context of surveillance, the number of people within a service radius could be substituted with the number of targets within a sensor range. This is a different perspective from which to look at the surveillance problem, and maximizing the number of targets in sensor range would be a preferable goal to maximizing the area coverage; however, it is also only measurable if target locations are known within some degree of certainty.

Miranda et al. [71] are not concerned with the implementation of an SN, but instead they address the problem of prioritising tasks assigned to available sensors using information provided by other sensors in the network. Their goal was to adjust priorities of radar tasks in such a way as to allow more effective scheduling; however, they have no metric for the effectiveness of the schedule. They specifically chose an example where their fuzzy logic algorithm performs differently from a hard logic version. The key point of this research is that sensors in a network can be used to inform the way in which other sensors belonging to the network can be used most efficiently. The priority of a sensing task was updated based on the current track quality on a target, the estimated hostility level of the target, the degree of threat, the appropriateness of the sensing platform's weapons systems, and the relative position of the target. All of this information is updated as awareness of the target improves.

Much of the work we have reviewed focuses on determining the ideal placement of sensors; however, CI methods can also be used in the data analysis that is required of a network of cooperating sensors. For example, Amato et al. [72] use neural networks to distinguish the movement of objects within a video from apparent movement due to the motion of the sensor itself. The United States Navy Cooperative Engagement Capability (CEC) does address the networking of multiple sensors but the network is not planned ahead of time, rather it combines the information provided by any available sensor in the same area [73]. The Brazilian system for vigilance of the Amazon region (SIVAM) similarly fuses data for environmental monitoring, air traffic control, and law enforcement [74]. These are large scale networks that combine elements of fixed and mobile sensors.

4.3 Mobile Ad hoc Networks

MANETs are flexible, dynamic, self-configuring (connected) mobile wireless multi-hop systems, which have become increasingly common for use in the areas where the deployment of a fixed wireless infrastructure is challenging. The applicability of MANETs is indispensable for use in network-centric warfare (NCW), which requires mission-critical systems to be highly robust and reliable. Hence, network design and analytical techniques are applied to design MANETs for use in NCW [75].

As a result of their wireless mobility, self-configuration and flexibility to be deployed in remote (or difficult to access) areas, MANETs are appropriate for numerous commercial and military applications such as natural disaster assistance, battlefield ISR, and surveillance and reconnaissance missions [24, 76]. A fixed wireless infrastructure is usually neither practical nor feasible in battlefield scenarios; as a result mobile wireless networks such as MANETs are essential for the rapid deployment and establishment of networks, consisting of adaptable, self-configurable mobile wireless nodes with real-time data, voice and video communications capabilities. The MANET system concept is instrumental in the development of vehicular ad hoc networks (VANET) and flying ad hoc networks (FANET), which are specialized MANETs. While in MANETs and VANETs, the focus is on moving nodes such as land vehicles, a FANET is a special form of MANET which addresses the concept of flying mobile nodes, i.e., multi-UAV systems [77].

The advantage of FANETs is in providing a more resilient and cost-effective solution compared to single UAV. Additionally, a FANET may extend the coverage area, survivability of a network, and speed of operation depending on the number of UAV systems included [78]. Nevertheless, due to high mobility of flying nodes and network dynamics (e.g., constantly changing node location), challenges exist with respect to multi-hop routing protocols. As an example, in airborne tactile networks, as speed increases, the successful delivery of the transmitted information (from all nodes to all nodes) drops [79]. Thus, the need for better interoperability of network layers is paramount such as for example leveraging link layer information for better cross-layer multi-hop routing decisions [80].

Sethi and Udagata [81] propose an efficient routing algorithm inspired by Ant Colony Optimization (ACO) techniques. The so called Ant-Efficient (Ant-E) algorithm improves the reliability of packet delivery by controlling the overhead and local transmission. The packets are divided into data and control packets, where data packets use information stored in the routing tables to reach the destination node. On the other hand, control packets, such as forward ant (FANT) and backward ant (BANT) are agents which are used to update the routing table and traffic information throughout the network. Lekova et al. [82] propose a delay tolerant event notification service utilizing fuzzy logic-based reasoning for sparse MANET networks in case of emergency or rescue situations, capable of capturing uncertainties in modeled data.

There are many similarities between MANETs and WSNs. For example, both network types do not need a fixed infrastructure and are self-configurable (adaptable to changes in network topology). They also rely on multi-hop routing for dissemination of data among network nodes. Power consumption is an important consideration in both MANETs and WSNs although of much more critical importance in WSNs. In comparison to WSNs, nodes in MANETs are typically equipped with more powerful and refuelable power systems.

On the other hand, some of the differences between MANETs and WSNs concern the number of nodes and their deployed densities. WSNs usually have many more nodes than MANETs; thus, scalability, while not a big concern in MANETs, can be an issue in WSNs. Moreover, while only a few nodes could be mobile in WSNs, usually all nodes are mobile in a MANET. Redundant deployment of nodes makes the use of data aggregation and in-network processing essential in WSNs, while it is mostly irrelevant in MANETs.

4.4 Cellular Networks

The objective in CN base station placement is to maximize the coverage area or the amount of cellular traffic served [5]. Studies also define a QoS level that must be achieved [35]. The trade-off is to minimize the required number of base stations while maximizing coverage. The CN coverage area is a similar objective to that used in surveillance networks, while traffic served is analogous to the number of targets detected. In contrast, the QoS calculation is not directly applicable to LSNs because in an LSN the communications infrastructure is separate from the sensors.

Meunier et al. [35] use three different types of base stations, distinguished by their antenna types: omnidirectional, small directive, and large directive. This is analogous to a sensor network that has access to omnidirectional, narrow FOV and wide FOV sensors. In addition, sites with directive base stations are allowed to have between one and three base stations. In addition to three objectives (minimize the number of sites, maximize the amount of traffic served, and minimize the interference from overlapping

M.G. Ball et al.

cells), they also consider two constraints: covering the entire area, and having a handover area between cells. The handover area is an area of cell coverage overlap which enables a moving cell phone user to be switched between the cell they are leaving and the cell they are entering. While the objectives would have to be adapted for use with sensor networks, the ability to account for multiple sensor types is important in defence applications.

Amaldi et al. [37] define an installation cost associated with each potential base station site, rather than with the base station itself. This accounts for a range of considerations such as pre-existing infrastructure or remote, difficult to access, locations. This same concept is important for sensor networks, where an ideal location from a coverage perspective may not be as important as taking advantage of an infrastructure left behind by an older network.

4.5 Current Research Challenges

CEC [73] and SIVAM [74] incorporate inputs from multiple sensors but there is no indication that SIVAM sensor locations were optimized for most efficient coverage, and CEC focusses on fusing data from onboard sensors from all ships in a group, whose locations will also not be based on optimum coverage.

The logic that is used to move sensors after an initial random deployment could be modified to determine optimal placement of sensors before deployment, simply not accounting for movement from initial positions. However this may be over-complicating the determination of ideal sensor locations.

It is difficult to measure how different systems should cooperate to provide the best overall SA. Using maritime surveillance as an example, suppose that satellites provide extensive coverage but no identification of vessels that aren't broadcasting legitimate AIS signals [64], while aircraft equipped with a visual sensor may be able to provide identification [83] if they know where to search. Some mix of both systems (or alternatives) is almost certainly the best approach, but while area coverage is easy to measure, the value added by covering the same area with more than one sensor depends on the targets being sought.

Just as the NWS replaced the DEW Line, the NWS will eventually need to be updated, replaced, or abandoned. If a replacement is considered, it may be useful to consider new locations for the radars. While global warming and technological advancement may make it possible to move the radar line farther north, advances in the radar technology may allow the radars to achieve the same capability while being positioned farther south. Well-defined objective functions should be used to capture the specific requirements of the mission. CI methods that were not available during the previous planning iterations could then be used to determine the best locations for the radars.

The convergence of MANETs and WSNs could be a unique dynamic system solution with "high resolution" sensing capabilities and mobility. The integration of such a system could pose a great challenge in itself. Taking into consideration that

VANETs and FANETs (UAVs) are also part of MANETs, the amount of available information, data dissemination and processing could prove to be very challenging. Finding effective CI techniques to be used and applied in the separation of "noisy data" from essential data for mission-critical scenarios should be of interest.

5 Future Directions

We have examined large fixed sensors, as well as mobile sensors with limited movement capability relative to their sensing range (WSNs) or large possible movements (MANETs). In a defence and security system of systems approach, both LSNs and WSNs should work together with patrolling sensors (e.g., foot patrols or aircraft-mounted sensors), all the way to polar-orbiting satellites. The fact that that these mobile sensors may move significantly compared to their sensing range introduces significant challenges in making a comparison to stationary sensors and thus finding a proper mix. Very few studies [30, 31, 35] allow different types of sensors or base stations to work together, and in these cases it is only the shapes and sizes of the coverage areas that are considered. Approaches from similar resource-based fields such as fleet mix computation [84] could also be adopted.

Optimal use of sensor networks continues to be a challenge. Handoff between sensors for the purposes of maintaining a track is discussed in [85]. The authors assume that all sensors are omnidirectional and the motivation behind handing off the tracking duty is that non-necessary sensors can sleep and conserve energy. How would this translate to large networks where sensors may be of different types, may be directional, and may have gaps in coverage where a target might temporarily disappear, but where energy conservation is not a driving concern?

Much of the work on WSNs focuses on the movement of sensors into position after an initial deployment, but to conserve energy sensors are rarely moved once they are in position [86]. LSNs also have varying levels of mobility; however, their movement tends to be an aspect of their use, rather than deployment. These range from movement while in use (such as satellite-based radar [65, 66] and Airborne Warning and Control Systems—AWACS [87, 88]), to movement between uses (moveable radars such as Russia's P-18 [26, 89] or Belarus' Vostok-D [90]), to no movement (permanent radar installations [16, 17]). Sensors that move after initial deployment introduce the complication that instantaneous coverage is not enough to measure their utility. It would also be advantageous for these various systems to be used together and so finding the correct sensor mix is another important challenge.

When coverage has been used as an objective, it usually refers to blanket coverage (as opposed to barrier or sweep [14]). Barrier coverage may be trivial as the required length is either covered, or not. Sweep coverage allows more area to be covered over time than would be possible if the sensors were stationary; however, measuring the effectiveness of this type of coverage remains a challenge. This is related to the difficulty of evaluating the performance of multiple types of sensors working together. Moving sensors such as satellites and AWACS provide sweep coverage while stationary sensors provide blanket coverage, adding an extra layer of complexity to their evaluation as parts of a system of systems.

Multiple conflicting objectives are dealt in one of three ways: the first method is to assign weights to each objective [34, 36, 37]; the second is to treat one or more objectives as constraints [29, 32, 33, 40, 86]; and the third is to perform a multi-objective optimization resulting in a Pareto front [5, 30, 31, 35]. The first two methods are related, as a constraint is equivalent to an objective with an arbitrarily large weight. No single method is ideal: weights attempt to rate the importance of each objective based on subjective individual preferences, and a Pareto front represents a large number of potential solutions from which to choose. A suggested course of action for future work is to first create the Pareto front and then examine the solutions using a multi-criteria decision tool [91].

Albeit currently at an infancy level, the convergence of MANETs and WSNs are instrumental in further development of new opportunities within Internet of Things (IoT) applications [92] where both technologies can be integrated for monitoring, public safety, surveillance and security applications. Ubiquitous sensing and the fast collection of data (supported by MANET and WSN) combined with computational intelligence could improve sensor network design.

Furthermore, as MANETs and WSNs merge with FSNs and IoT, many new challenges will arise including managing and processing this large amount of data. There will be big data analytics challenges where existing data processing methods do not apply. The shift towards big data in military ISR will require finding new methodologies capable of removing redundant information while extracting and processing essential data. The application of sophisticated new CI algorithms will be required.

Therefore, the synergy of IoT and big data technologies could offer an unparalleled opportunity towards using data driven discovery for military SA. The amount of data available from multiple sources of information could be used to predict and prevent natural disasters, potential dangers and threats, contributing to a safer future [93]. Therefore, leveraging CI methods within IoT and big data initiatives should be a focus for future military ISR applications [94, 95].

References

1. Akyildiz, I.F., Su, W., Sankarasubramaniam, Y., Cayirci, E.: A survey on sensor networks. IEEE Commun. Mag. **40**, 102–114 Aug 2002
2. Chong, C.Y., Kumar, S.P.: Sensor networks: evolution, opportunities, and challenges. Proc. IEEE **91**(8), 1247–1256 (2003)
3. Ramanathan, R., Redi, J.: A brief overview of ad hoc networks: challenges and directions. IEEE Commun. Mag. **40**(5), 20–22 (2002)
4. Wolfgang, K., Martin, M.: A survey on real-world implementations of mobile ad-hoc networks. In: Ad Hoc Netw. **5**(3), 324–339 (2007)

5. Jourdan, D.B., de Weck, O.L.: Layout optimization for a wireless sensor network using a multi-objective genetic algorithm. In: IEEE Semi Annular Vehicular Technology Conference (2004)
6. Kulkarni, R.V., Förster, A., Venayagamoorthy, G.K., Computational intelligence in WSN: a survey. IEEE Comm. Surv. Tutorials **13**(1) (2011)
7. Alirezaei, G., Mathar, R., Ghofrani, P.: Power optimization in sensor networks for passive radar applications. In: IEEE WiSEE, pp. 1–7, 7–9 Nov 2013
8. Lindsey, S., Raghavendra, C.S.: PEGASIS: Power-efficient gathering in sensor information systems. In: IEEE Aerospace Conference, vol. 3, pp. 3–1125, (2002)
9. Heinzelman, W.B., Chandrakasan, A.P., Balakrishnan, H.: An application-specific protocol architecture for wireless microsensor networks. IEEE Trans. Wireless Commun. **1**(4), 660–670 (2002)
10. Vodel, M., Hardt, W.: Data aggregation and data fusion techniques in WSN/SANET topologies—a critical discussion. In: IEEE TENCON, pp. 1–6, Nov 2012
11. Madden, S., Franklin, M.J., Hellerstein, J.M., Hong, W.: A tiny aggregation service for Ad-hoc sensor networks. In: Proceeding of 5th Symposium on Operating Systems Design and Implementation (OSDI), **32**, 131–146 (2002)
12. Dasgupta, K., Kalpakis, K., Namjoshi, P.: An efficient clustering-based heuristic for data gathering and aggregation in sensor networks. In: Wireless Communications and Networking, 2003, vol. 3, pp. 1948–1953 20–20 March 2003
13. Beyer, D.: Accomplishments of the DARPA SURAN Program. IEEE MILCOM **2**, 855–862 (1990)
14. Gage, D.W.: Command control for many-robot systems. In: Proceeding AUVS-92 (1992)
15. North Warning System. http://en.wikipedia.org/wiki/North_Warning_System. Accesssed 24 June 2015
16. AN/FPS-117. http://en.wikipedia.org/wiki/AN/FPS-117. Accesssed 24 June 2015
17. AN/FPS-124. http://en.wikipedia.org/wiki/AN/FPS-124. Accesssed 24 June 2015
18. Chu, M., Reich, J.E., Zhao, F.: Distributed attention for large video sensor networks. In: Intelligent Distributed Surveillance System 2004 seminar, London, UK (2004)
19. He, T., Krishnamurthy, S., Stankovic, J., Abdelzaher, T., Luo, L., Stoleru, R., Yan, T., Gu, L., Hui, J., Krogh, B.: Energy-efficient surveillance system using wireless sensor networks. In: ACM MobiSys '04, pp. 270–283. New York, NY, USA (2004)
20. Hogler, K., Willig, A.: Protocols and Architectures for Wireless Sensor Networks. Wiley, England (2007)
21. Harte, S., O'Flynn, B., Martinez-Catala, R.V., Popovici, E.M.: Design and implementation of a miniaturised, low power wireless sensor node. In: Proceeding of IEEE ECCTD, pp. 894–897, August 2007
22. AN/TPS-71 Relocatable Over-The-Horizon Radar (ROTHR). https://janes.ihs.com/CustomPages/Janes/DisplayPage.aspx?DocType=Reference&ItemId=++ +1498275&Pubabbrev=JC4IL. Accessed 24 June 2014
23. Arampatzis, T., Lygeros, J., Manesis, S.: A survey of applications of wireless sensors and wireless sensor networks. In: Proceeding of IEEE International Symposium on Mediterranean Conference on Control and Automation, pp. 719–724, 27–29 June 2005
24. Dali, W., Chan, H.A.: Analysis of the applications and characteristics of Ad Hoc networks. In: ICCT, pp. 1–4, Nov 2006
25. Jiejun, K., Jun-Hong, C., Dapeng, W., Gerla, M.: Building underwater ad-hoc networks and sensor networks for large scale real-time aquatic applications. IEEE MILCOM **3**, 1535–1541 (2005)
26. P-18 early-warning radar. https://janes.ihs.com/CustomPages/Janes/DisplayPage.aspx?DocType=Reference&ItemId=+++1498259&Pubabbrev=JC4IL. Accessed 24 June 2014
27. Doerry, A.W., Dickey, F.M.: Synthetic aperture radar. In: Optics and Photonics News, pp. 28–33, Nov 2004
28. Lindsey, S., Raghavendra, C.S.: PEGASIS: Power-efficient gathering in sensor information systems. In: IEEE Aerospace Conference, vol. 3, pp. 3–1130 (2002)

29. Chakrabarty, K., Iyengar, S.S., Qi, H., Cho, E.: Grid coverage for surveillance and target location in distributed sensor networks. IEEE Trans. Comput. **51**, 1448–1453 (2002)
30. Sakr, Z., Wesolkowski, S.: Sensor network management using multiobjective evolutionary optimization. In: IEEE CISDA, pp. 39–42 (2011)
31. Oh, S.C., Tan, C.H., Kong, F.W., Tan, Y.S., Ng, K.H., Ng, G.W., Tai, K.: Multiobjective optimization of sensor network deployment by a genetic algorithm. In: IEEE Congress on Evolutionary Computation, pp. 3917–3921 (2007)
32. Dhillon, S.S., Chakrabarty, K., Iyengar, S.S.: Sensor placement for grid coverage under imprecise detections. In: Conference on Information Fusion, vol. 2, pp. 1581–1587 (2002)
33. Bose, R.: A smart technique for determining base-station locations in an urban environment. IEEE Trans. Veh. Technol. **50**, 43–47 (2001)
34. Han, J.K., Park, B.S., Choi, Y.S., Park, H.K.: Genetic approach with a new representation for base station placement in mobile communications. In: Proceeding of IEEE Vehicular Technology Conference, vol. 4. pp. 2703–2707, Oct 2001
35. Meunier, H., Talbi, E., Reininger, P.: A multiobjective genetic algorithm for radio network optimization. In: Proceeding Congress on Evolutionary Computation, vol. 1. pp. 317–324 (2000)
36. Jiang, X., Chen, Y., Yu, T.: Localized distributed sensor deployment via coevolutionary computation. In: International Conference on Communication and Networking in China (2008)
37. Amaldi, E., Capone, A., Malucelli, F.. Signori, F.: UMTS radio planning: optimizing base station configuration. In: IEEE Vehicle Conference vol. 2, pp. 768–772, Sept 2002
38. Osmani, A., Dehghan, M., Pourakbar, H., Emdadi, P.: Fuzzy-based movement-assisted sensor deployment method in wireless sensor networks. In: IEEE Proceeding of the International Conference on Computational Intelligence, Communication System and Networks, India, (2009)
39. Church, R., ReVelle, C.: The maximal covering location problem. In: Papers of the Regional Science Association, **32**, 101–118 (1974)
40. Howard, A., Matarić, M.J., Sukhatme, G.S.: An incremental self-deployment algorithm for mobile sensor networks. Auton. Robots Special Issue Intell. Embedded Syst. **13**(2), 113–126 (2002)
41. Howard, A., Matarić, M.J., Sukhatme, G.S.: Mobile sensor network deployment using potential fields: a distributed, scalable solution to the area coverage problem. In: Proceeding International Conference on Distributed Autonomous Robotic Systems, pp. 299–308 (2002)
42. Zou, Y., Chakrabarty, K.: Sensor deployment and target localization based on virtual forces. Proc. IEEE INFOCOM **2**, 1293–1303 (2003)
43. Madden, S., Franklin, M.J., Hellerstein, J.M., Hong, W.: A tiny aggregation service for Ad-hoc sensor networks. In: Proceeding of 5th Symposium on Operating Systems Design and Implementation (OSDI), vol. 32, pp. 131–146 (2002)
44. El Kateeb A., Ramesh A., Azzawi, L.: Wireless sensor nodes processor architecture and design. In: Proceeding of IEEE CCECE, pp. 1031–1034, May 2008
45. Antolin, D., Medrano, N., Calvo, B.: Analysis of the operating life for battery-operated wireless sensor nodes. In: IEEE IECON, pp.3883–3886, Nov 2013
46. Qela, B., Wainer, G., Mouftah, H.: Simulation of large wireless sensor networks using Cell-DEVS. In: WinterSim Conference, pp. 3189–3200, 13–16 Dec 2009
47. Schurgers, C., Tsiatsis, V., Ganeriwal, S., Srivastava, M.: Optimizing sensor networks in the energy-latency-density design space. IEEE Trans. Mob. Comput. **1**(1), 70–80 (2002)
48. Geetha, D.D., Nalini, N., Biradar, R.C.: Active node based fault tolerance in wireless sensor network. In: IEEE INDICON, pp. 404–409, 7–9 Dec 2012
49. Kateeb, A.El., Ramesh, A., Azzawi, L.: Wireless sensor nodes processor architecture and design. In: Proceeding of IEEE CCECE, pp. 1031–1034, May 2008
50. Harte, S., O'Flynn, B., Martinez-Catala, R.V., Popovici, E.M.: Design and implementation of a miniaturised, low power wireless sensor node. In: Proceeding of IEEE 28th ECCTD, pp. 894–897, August 2007

51. Cerpa, A., Estrin, D.: ASCENT: adaptive self-configuring sensor networks topologies. IEEE INFOCOM **3**, 1278–1287 (2002)
52. Sousa, M.P., de Alencar, M.S., Kumar, A., Araujo Lopes, W.T.: Scalability in an adaptive cooperative system for wireless sensor networks. In: Ultra Modern Telecommunications and Workshops, ICUMT '09. pp. 1–6, 12–14 Oct 2009
53. Đurišić, M.P., Tafa, M.P., Dimić, G., Milutinović, V.: A survey of military applications of wireless sensor networks. In: Mediterranean Conference on Embedded Computing (2012)
54. Liu, J.X., Perrone, L.F., Nicol, D.M., Liljenstam, M., Elliott, C., Pearson, D.: Simulation modeling of large-scale ad-hoc sensor networks. In: European Simulation Interoperability Workshop 2001
55. Afolabi, D., Man, K.L., Liang, H.-N., Lim, E.G., Shen, Z., Lei, C.-U., Krilavicius, T., Yang, Y., Cheng, L., Hahanov, V., Yemelyanov, I.: A WSN approach to unmanned aerial surveillance of traffic anomalies: some challenges and potential solutions. In: Design and Test Symposium 2013, pp. 1–4, 27–30 Sept 2013
56. Akshay, N., Kumar, M.P., Harish, B., Dhanorkar, S.: An efficient approach for sensor deployments in wireless sensor network. In: Emerging Trends in Robotics and Communication Technologies INTERACT, pp. 350–355, 3–5 Dec 2010
57. Song, B., Choi, H., Lee, H.S.: Surveillance tracking system using passive infrared motion sensors in wireless sensor network. In: ICOIN 2008. International Conference on Information Networking, pp. 1–5, Jan 2008
58. Hussain, M.A., Khan, P., Sup, K.K.: WSN research activities for military application. In: Adv. Comm. Tech. 2009. ICACT 2009, pp. 271–274, Feb 2009
59. Heurtefeux, K., Valois, F.: Topology control algorithms: a qualitative study during the sensor networks life. MASS **2007**, 1–7 (2007)
60. Butun, I., Morgera, S.D., Sankar, R.: A Survey of intrusion detection systems in wireless sensor networks. IEEE Comm. Surv. Tutorials **16**(1), 266–282, First Quarter 2014
61. Butun, I., Sankar, R.: A brief survey of access control in wireless sensor networks. In: IEEE CCNC, pp. 1118–1119, Jan 2011
62. Jenkins, L.: Challenges in deployment of wireless sensor networks. In: Industrial and Information Systems (ICIIS), Dec 2014
63. Lee, S.H., Lee, S., Song, H., Lee, H.S.: Wireless sensor network design for tactical military applications. In: Remote large-scale environments. IEEE MILCOM, pp. 1–7, 18–21 Oct 2009
64. IMO—Automatic Identification System. http://www.imo.org/OurWork/Safety/Navigation/Pages/AIS.aspx. Accessed 24 June 2015
65. RADARSAT-2—Canadian Space Agency. http://www.asc-csa.gc.ca/eng/satellites/radarsat2/. Accessed 24 June 2015
66. SPOT 1 to 5. http://www.geo-airbusds.com/en/4388-spot-1-to-spot-5-satellite-images. Accessed 24 June 2015
67. Herd, A.W.G.: A Practicable Project: Canada, the United States, and the construction of the DEW line. In: Calgary Papers in Military and Strategic Studies Occasional Paper Nr 4, 2011 —Can. Arctic Sovereignty and Sec.: Hist. Perspectives, pp. 171–200 (2011)
68. North Warning System. https://maps.google.com/maps?q=http://tools.wmflabs.org/kmlexport/%3Farticle%3DNorth_Warning_System%26section%3DStations%26usecache%3D1&output=classic&dg=feature. Accessed 12 Feb 2015
69. List of DEW Line Sites. https://maps.google.com/maps?q=http://tools.wmflabs.org/kmlexport/%3Farticle%3DList_of_DEW_Line_Sites%26usecache%3D1&output=classic&dg=feature. Accessed 12 Feb 2015
70. List of DEW Line Sites. http://en.wikipedia.org/wiki/List_of_DEW_Line_Sites. Accessed 24 June 2015
71. Miranda, S.L.C., Baker, C.J., Woodbridge, K., Griffiths, H.D.: Fuzzy logic approach for prioritisation of radar tasks and sectors of surveillance in multifunction radar. IET Radar Sonar Navig. **1**(2), 131–141 (2007)

72. Amato, A., Di Lecce, V., Piuri, V.: Neural network based video surveillance system. In: IEEE International Conference on CI for Homeland Security and Personal Safety (2005)
73. The cooperative engagement capability. In: Johns Hopkins APL Tech. Dig. **16**(4) (1995)
74. Jensen, D.: SIVAM: Communication, navigation and surveillance for the Amazon. Avionics Mag. http://www.aviationtoday.com/av/military/SIVAM-Communication-Navigation-and-Surveillance-for-the-Amazon_12730.html. Accessed 04 June 2014
75. Kant, L., Young, K., Younis, O., Shallcross, D., Sinkar, K., Mcauley, A., Manousakis, K., Chang, K., Graff, C.: Network science based approaches to design and analyze MANETs for military applications. IEEE Commun. Mag. **46**(11), 55–61 (2008)
76. Jiejun, K., Jun-Hong, C., Dapeng, W., Gerla, M.: Building underwater ad-hoc networks and sensor networks for large scale real-time aquatic applications. IEEE MILCOM **3**, 1535–1541 (2005)
77. Ilker, B., Ozgur, K.S., Samil, T.: Flying Ad-Hoc networks (FANETs): a survey. Ad Hoc Netw. **11**(3), 1254–1270, ISSN 1570-8705 (2013)
78. Haiyang, C., Cao, Y., Chen, Y.Q.: Autopilots for small fixed-wing unmanned air vehicles: a survey. In: Mechatronics and Automation, 2007, ICMA (2007)
79. Bow-Nan, C., Moore, S.: A comparison of MANET routing protocols on airborne tactical networks. MILCOM **2012**, 1–6 (2012)
80. Bow-Nan, Ch., Wheeler, J., Veytser, L.: Radio-to-router interface technology and its applicability on the tactical edge. IEEE Commun. Mag. **50**(10), 70–77 (2012)
81. Sethi, S., Udagata, S.K.: The efficient ant routing protocol for MANET. Int. J. Comput. Sci. Eng. **2**(7), 2414–2420 (2010)
82. Lekova, A., Skjelsvik, K., Plagemann, T., Goebel, V.: Fuzzy logic-based event notification in sparse MANETs. AINAW'07 **2**, 296–301 (2007)
83. Global Discovery' Maritime Patrol Aircraft. https://janes.ihs.com/CustomPages/Janes/DisplayPage.aspx?DocType=Reference&ItemId=+++1598078&Pubabbrev=JC4IA. Accessed 24 June 2015
84. Wojtaszek, D., Wesolkowski, S.: Military fleet mix computation and analysis. IEEE Comput. Intell. Mag. **7**(3), 53–61 (2012)
85. Zhao, F., Shin, J., Reich, J.: Information-driven dynamic sensor collaboration for tracking applications. IEEE Sig. Process. Mag. **19**, 61–72 (2002)
86. Wu, X., Cho, J., d'Auriol, B.J., Lee, S.: Mobility-assisted relocation for self-deployment in wireless sensor networks. IEICE Trans. **90-B**(8), 2056–2069 (2007)
87. Boeing E-767 AWACS.: https://janes.ihs.com/CustomPages/Janes/DisplayPage.aspx?DocType=Reference&ItemId=+++1337360&Pubabbrev=JAU_. Accessed 24 June 2015
88. Boeing E-3 Sentry.: https://janes.ihs.com/CustomPages/Janes/DisplayPage.aspx?DocType=Reference&ItemId=+++1337336&Pubabbrev=JAU_. Accessed 24 June 2015
89. P-18.: "Spoon Rest D". http://www.radartutorial.eu/19.kartei/karte909.en.html. Accessed 24 June 2015
90. Vostok-D/E mobile surveillance radar. https://janes.ihs.com/CustomPages/Janes/DisplayPage.aspx?DocType=Reference&ItemId=+++1721035&Pubabbrev=JC4IL. Accessed 24 June 2015
91. Bonissone, P.P., Subbu, R., Lizzi, J.: Multicriteria decision making (MCDM): a framework for research and applications. IEEE CI Mag. **4**(3), 48–61 (2009)
92. Bellavista, P., Cardone, G., Corradi, A., Foschini, L.: Convergence of MANET and WSN in IoT urban scenarios. IEEE Sens. J. **13**(10), 3558–3567 (2013)
93. Gang-Hoon, K., Silvana, T., Ji-Hyong, Ch.: Big-data applications in the government sector. Commun. ACM **57**, 78–85 (2014)
94. Zhou, Zhi-Hua, Chawla, N.V., Jin, Yaochu, Williams, G.J.: Big data opportunities and challenges: discussions from data analytics perspectives. IEEE CI Mag. **9**(4), 62–74 (2014)
95. Choi, A.J.: Internet of things: evolution towards a hyper-connected society. In: IEEE Solid-State Circuits Conference (A-SSCC), pp. 5–8, 10–12 Nov 2014
96. Shu, H., Liang, Q., Gao, J.: Wireless sensor network lifetime analysis using interval type-2 fuzzy logic systems. IEEE Trans. Fuzzy Syst. **16**(2), 416–427 (2008)

97. Zhang, H., Hou, J.C.: Maintaining sensing coverage and connectivity in large sensor networks. Ad Hoc & Sens. Wireless Netw. **1**, 89–124 (2005)
98. Abielmona, R., Petriu, E.M., Harb, M., Wesolkowski, S.: Mission-driven robotic intelligent sensor agents for territorial security. IEEE CI Mag. **6**(1), 55–67 (2011)
99. Garcia-Rodriguez, J., Angelopoulou, A., Mora-Gimeno, F.J., Psarrou, A.: Building visual surveillance systems with neural networks. In: Computational Intelligence for Privacy and Security, pp. 181–198 (2012)
100. Bulusu, N., Heidemann, J., Estrin, D.: Adaptive beacon placement. In: Proceeding of International Conference on Distributed Computing System, pp. 489–498, April (2001)
101. Li, D., Wong, K.D., Hu, Y.H., Sayeed, A.M.: Detection, classification, and tracking of targets. IEEE Signal Process. Mag. **19**, 17–29 (2002)
102. Meesookho, C., Narayanan, S., Raghavendra, C.: Collaborative classification applications in sensor networks. In: Proceeding of IEEE Multichannel and Sensor Array Signal Processing Workshop, Arlington, VA (2002)
103. Sinopoli, B., Sharp, C., Schenato, L., Shaffert, S., Sastry, S.S.: Distributed control applications within sensor networks. Proc. IEEE **91**(8), 1235–1246 (2003)
104. Arora, A., Dutta, P., Bapat, S., Kulathumani, V., Zhang, H., Naik, V., Mittal, V., Cao, H., Demirbas, M., Gouda, M., Choi Y., Herman, T., Kulkarni, S., Arumugam, U., Nesterenko M., Vora A., Miyashita, M.: A line in the sand: a wireless sensor network for target detection, classification, and tracking, computer networks. Int. J. Comput. Telecom. Netw. **46**(5), 605–634, 5 Dec 2004

Sensor Resource Management: Intelligent Multi-objective Modularized Optimization Methodology and Models

Boris Kovalerchuk and Leonid Perlovsky

Abstract The importance of the optimal Sensor Resource Management (SRM) problem is growing. The number of Radar, EO/IR, Overhead Persistent InfraRed (OPIR), and other sensors with best capabilities, is limited in the stressing tasking environment relative to sensing needs. Sensor assets differ significantly in number, location, and capability over time. To determine on which object a sensor should collect measurements during the next observation period k, the known algorithms favor the object with the expected measurements that would result in the largest gain in relative information. We propose a new tasking paradigm OPTIMA for sensors that goes beyond information gain. It includes Sensor Resource Analyzer, and the Sensor Tasking Algorithm (Tasker). The Tasker maintains timing constraints, resolution, and geometric differences between sensors, relative to the tasking requirements on track quality and the measurements of object characterization quality. The Tasker does this using the computational intelligence approach of multi-objective optimization, which involves evolutionary methods.

Keywords Sensor resource management (SRM) · Multi-objective optimization · Adaptive models · Integer linear programming · Evolutionary computing · Dynamic logic · Optimization under uncertainty

1 Introduction

The configurations of sensor platforms can include dozens of global radars and EO/IR sensors (ship-based, sea-based, ground-based, space-based, and air-based) and thousands of local sensors with different bands and capabilities. The challenge

B. Kovalerchuk (✉)
Department of Computer Science, Central Washington University, Washington, USA
e-mail: borisk@cwu.edu

L. Perlovsky
LP Information Technology and Harvard University, Cambridge, USA

© Springer International Publishing Switzerland 2016
R. Abielmona et al. (eds.), *Recent Advances in Computational Intelligence in Defense and Security*, Studies in Computational Intelligence 621,
DOI 10.1007/978-3-319-26450-9_25

and opportunity of the Sensor Resource Management (SRM) is related to a large difference in resolutions, errors and uncertainties of the sensors. There are situations where none of the sensors individually can improve certainty/resolution of the object to the required level. However, it is possible with assigning a pair of sensors with optimal lines of sights to the object. For instance, the range resolution of radars could be about 1 cm, but angular resolution could be only 1 km, thus two orthogonal radars could increase resolution significantly.

This leads to the tasks to optimize sensor resource use *across assets* in real time under the operational *constraints* of each sensor type, and to use for *planning*. Among the common SRM goals are: maximizing available sensor resources for *search,* optimizing sensor resources for *tracking,* and, better *defending* the high priority assets in a raid environment.

The SRM goals often contradict each other. Consider typical goals: (i) decreasing the overall sensor resource utilization, (ii) increasing the probability that all threat objects in a raid are tracked, and (iii) decreasing potential overload of sensors at individual platforms/units. The chances that all these goals will not contradict each other and all will be satisfied by a particular solution (assignment of sensors) are low. This consideration leads to the necessity of **multi-objective optimization** approach, which is pursued in this work using the computational intelligence approach.

It is possible that the most resource utilization is a full 100 % load of N sensors at unit A without any room for handling more areas of interests and objects, while only 10 % of N sensors at the unit B are used. A more even use/allocation/tasking of sensors at units A and B, that decreases potential *overload* of sensors, may require more sensors, say 60 % at unit A and 60 % at unit B used at time t. In the first scenario (100:10) at some moment unit A may have not enough sensors to defend itself not only to track objects of interest, while in the second scenario (60:60) there is a plenty room for extra load, but more sensors are used. This is another reason for multi-objective optimization.

There is extensive literature on sensor resource management [3, 5, 8, 9, 11–13, 18–22, 25, 26, 32, 33, 35, 39].

Information gain is one of most actively used approaches in SRM. The relative information gain is a *scalar* measure between the *prior and posterior probability density functions* $p(x_k|z_{1:k-1})$ and $p(x_k|z_{1:k})$ based on the Renyi o-divergence [12]

$$D_\alpha(z_{1:k}||z_{1:k-1}) = \frac{1}{\alpha-1} log \int \left[\frac{p(x_k|z_{1:k})}{p(x_k|z_{1:k-1})}\right]^\alpha p(x_k|z_{1:k-1})dx_k$$

To determine on which object a sensor should collect measurements during the next observation period k the algorithm in [21] adopts the strategy from [12, 23] that favors the object whose expected measurements would result in the largest gain in relative information:

$$\mu_v = \arg max_v({}^a\widehat{\Delta}_k^{\mu v})$$

where

$${}^a\widehat{\Delta}_k^{\mu v} = \int p(z_k^{\mu v}|z_{1:k-1}^v)D_\alpha(z_k^{\mu v}, z_{1:k-1}^v||z_{1:k-1}^v)dz_k^{\mu v}$$

is the expected local information.

This formulation that selects the object with the largest gain in relative information has an important weakness of locality: it computes the increase of certainty of the measurement (at time k) of a given object A by applying itself to a single sensor S relative to the certainty of measurement at time k−1 recorded in the system track [18].

We may have a situation where none of the sensors individually will improve certainty/resolution of object to the required level as shown in Fig. 1a, b. However, reaching the required certainty is possible with assigning a pair of sensors (radars) with Lines of Sights (LOS) to the object that are close to be orthogonal as Fig. 1c shows. In Fig. 1 the blue ellipse shows the original uncertainty of the location of the object at time k-1. The ellipses of uncertainty of the radars R1 and R2 are narrow ellipses that have a long intersections with the blue ellipse of the original uncertainty of the object location as shown in Fig. 1a, b. Figure 1c shows that the area of uncertainty due to overlap of uncertainty areas for R1 and R2 is much smaller with dramatic information gain. Note that individually both radars R1 and R2 equally improve relative information gain used in [21].

The proposed approach goes beyond the state of the art described in [21] including the powerful idea of learning the parameter a in Renyi o-divergence. It also overcomes a potential conflict between assignments of the sensors to the objects in the independent assignment of "best" sensors for each object. In the independent sensor assignment a particular sensor can get conflicting "best" assignments to two or more objects that are at different locations that the sensor cannot cover at the same time.

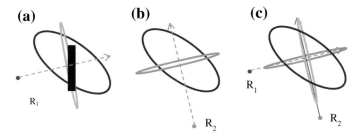

(a)   **(b)**   **(c)**

Fig. 1 Information gain with allocation of single radar versus allocation of two radars to the object

In the information gain SRM formulation [21], the state estimation and SRM are tightly coupled. This leads to the complex models and computational challenges. In addition it is difficult to change one of the components in a tightly coupled formulation. For instance, it is difficult to add more optimization criteria beyond the information gain and to incorporate the effect of a new interaction between the sensors. In general the progress in tracking, discrimination, fusion, and SRM technologies is not synchronous. Therefore the tightly coupled formulation makes it difficult to integrate these technologies into a more advanced system. Therefore we consider the state estimation itself as an *external,* but connected task to the SRM.

The proposed new sensor tasking system OPTIMA includes a *Sensor Resource Analyzer* and the *Sensor Tasking models and algorithms* (Tasker).

The **requirements** for the SRM solution are:

1. Minimize the number of sensors for a given coverage, and maximize the effectiveness of each sensor given its performance and resource constraints.
2. Provide dynamic tasking of sensors where multiple sensors cooperate in search, detection, tracking, and identification.
3. Maximize the probability of successfully covering all threat objects.

The fusion center can provide requirements for SRM: desired resolution, angle, band, level of decrease of uncertainties in object characterization, and others. The challenge is that requirements (1)–(3) likely contradict each other which requires using a multi-objective optimization approach. The types of questions that the Sensor Resource Analyzer of the OPTIMA system intends to answer are:

- Will particular configuration C of platforms provide the full coverage of some areas of interests **A** with the required capabilities **R** for tracking and discrimination?
- What is a *minimal configuration* **C** of platforms to provide the full coverage of some areas of interests **A** with the required capabilities **R** for tracking and discrimination?
- What part of the areas of interest **A** will not be covered at required capabilities **R** for tracking and discrimination if configuration **C** of platforms will be used?

Sensor coverage can be degraded due to multiple reasons at any time. Natural environments, engagement conditions, and high noise background can impact radars and EO/IR sensors. As a result, degradation can take multiple forms: inadequate signal to noise ratio, degraded specific range cells and azimuthal directions in the Field of View (FoV), too much energy on the focal plane of IR sensor, aspects of degradations compensated by the sensor itself.

Reconstructing the scene degradation from this varying information is a challenge. As a result inadequate input information can corrupt tracking and discrimination of objects. Information collected from widely distributed sensors can be used to determine areas where and how sensor coverage is degraded to allocate alternate resources to compensate. While individual sensors (EO/IR, Radar) can generally determine when a particular portion of the scene is degraded, the challenge is in an effective use of this information for efficient sensor tasking. We consider tasking sensors in the degraded environment as a *generalization* of the sensor resource management (SRM) task for a *degraded environment*. This means that input *data messages* to the SRM describe not only normal, but also degraded sensor capabilities and the degraded environment.

This paper is organized as follows. Section 2 presents the SRM optimization models. Section 3 presents translation of tracking and discrimination requirements into flags and solution of optimization models using Computational Intelligence techniques. The paper concludes with the description of the related and future work.

As this paper structure shows, it is focused on the conceptual development of methodology and models. The experimental studies are a subject of a separate work based on the domain specific extensive input sensor data. Typically these specific data are outside of scope of other domains. This paper is for a wider audience at general methodological level. There is also an extensive computational aspect of the methodology that we address as a separate work.

1.1 OPTIMA System Architecture with Computational Intelligence Solution

This work proposes a **new sensor tasking method** for both long-time planning and for real-time SRM based on Intelligent Multi-objective Modularized Optimization Model. The SRM system OPTIMA and its context are illustrated in Fig. 2. The OPTIMA maintains timing constraints, resolution and geometric differences between the sensors relative to the tasking requirements on track quality and the measurements of object characterization quality. The solution is based on the computational intelligence approach that involves evolutionary methods, dynamic logic, and multi-objective optimization. The system design allows a user to select the version of an objective function of the minimal configuration such as minimal number of platforms, minimal cost/value/capabilities of sensor platforms. In the version of the model presented below, it is assumed that all motions of sensor platforms are known, as well as the capabilities (possible degraded), and status of the sensors onboard the platforms.

The OPTIMA Model involves:

- multiple sensors of different types and with varying capabilities.
- sensor locations with respect to the object complex,
- timing constraints
- requirements for track quality, and
- requirements for object measurements characterization (discrimination) quality

Tracking and discrimination may require different sensors and these multi-objective requirements can change dynamically [1]. Thus the model is updated with such new input, and the new output is produced (see Fig. 2).

Fig. 2 Context of sensor
resource management

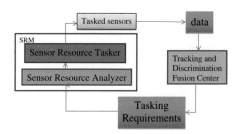

The uniqueness of this approach is in the use of intelligent multi-objective optimization of SRM Model with adaptable Integer Linear Programming (ILP) models, Cramer-Rao Bounds (CRBs) and algorithms accounting for the association part of tracking and fusion problem. These CRBs able to evaluate object characterization (classification features) and therefore object values. Another uniqueness of the approach is in using flags within SRM, which encompass all the information external to the main goals/task (such as information from tracking algorithms). These flags are readily computed from available information or information adaptively estimated in real time. These benefits surpass existing state of the art and permit more accurate overall sensor coordination.

1.2 Modularized Design

This work follows the *modularized design paradigm* where tracking, discrimination, fusion, and SRM as separate, but communicating modules to allow the SRM algorithm to work when tracking, discrimination and fusion algorithms are changed/upgraded.

To implement the modularized design, we build a set of integer linear programming (ILP) models described below with both continuous and binary variables that extensively use the concept of the *flags*. One of the flags is a binary flag, $f(a_i, s_j, t, r)$. If sensor s_j is *capable* of *covering/observing* area of interest a_i at the time interval t with the required resolution r, then $f(a_i, s_j, t, r) = 1$, else $f(a, s, t, r) = 0$. Another flag is a *stochastic flag,* which is a probability that sensor s_j is capable of covering/observing area of interest a_i at the time interval t with the required resolution r. Flags serve as a mechanism to *link* tracking, discrimination, fusion modules, models and algorithms with SRM optimization models and algorithms.

The advantage of this approach is that it allows separating: (i) rigorous formulation of the SRM optimization models (objective functions and constraints), (ii) multi-objective models that combine them, and (iii) feasible and fast computations for solving these models.

The OPTIMA system model tasking sensors in the both normal and degraded environment that is input data messages represent not only normal, but also degraded sensor capabilities and degraded performance, and environment

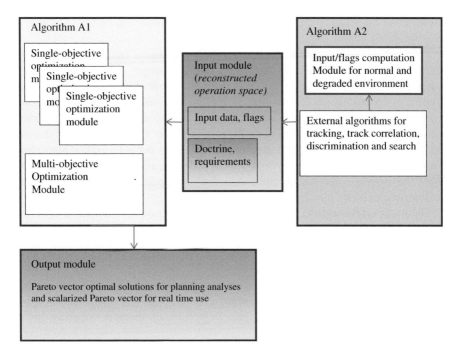

Fig. 3 Modularized architecture of the SRM OPTIMA system with operation space characterization

description. For instance, instead of entering to the SRM systems the normal resolution r of the sensor S another value r-Δ is entered. Such generalization means that a set of additional input module/modules with appropriate algorithms need to be developed to generate degradations Δ. Such a module is shown in Fig. 3 as a block with a red frame. Figure 3 shows also the whole modularized ***architecture*** of the OPTIMA system.

This architecture assumes two types of algorithms: (A1) optimal sensor assignment for reducing sensor resource utilization using estimated (A2) adaptive algorithm for working with the first algorithm to support input of sensor metadata including sensor "health" data, tracking, track correlation data, and data fusion data. The first algorithm exploits sensor geometry including a possibility that two sensors observing the same object (at approximately orthogonal geometry) could produce more accurate results in a shorter time.

The algorithm of the first type solves the problem of optimal sensor assignment. The algorithm of the second type provides the input data for the first one allowing the use of tracking and fusion algorithms that are at the level of the Cramer-Rao Bound (CRB), which sets up the best possible performance accounting for the associations between sensors, objects, and tracks [28]. These algorithms exploit a novel technique [27] developed for air-ground radars. In the case of the degraded environment the CRB is computed for the degraded sensor environment by using the appropriate sensor models in CRB computation.

This architecture combines the mathematical techniques (multi-objective optimization with the analysis of the Pareto border, Integer Liner Programming, Adaptive methods) and physical considerations (sensor phenomenology and geometry of locations relative to objects). It allows the addressing of contradictory requirements that cannot be addressed by the classical optimization methods without setting up a tradeoff between them in advance. However in the dynamic environment the tradeoff must be dynamic as well. Thus the optimization algorithm must adapt to such a dynamic environment in real time, and this is proposed in this paper via dynamic updates of input data, looping optimization cycles, and the flexible selection of objective functions.

The **main mathematical advantage** of this architecture in comparison with the state of the art is that it *decouples* tracking and track estimation algorithms/filters from the optimization and environment estimation. The other advantages of a new architecture are that it allows:

1. a *variety* of *external tracking and discrimination algorithms* by computing "flags" representing external algorithms,
2. *multiple optimization criteria* by selecting/changing modules adapting for a particular scenario,
3. *multiple tradeoffs between multiplicity of optimization criteria* in multi-objective setting providing a mathematically rigorous solutions.

These advantages make this modular architecture more adaptable and universal than other architectures, which is important for practical applications.

The important aspect of modularized design is that it computes flags (parameters of optimization models) outside of the optimization module and outside of tracking module as well. This is a *buffer* idea that has been very successful in many other areas including computer architecture design with cache memory as a buffer between the primary memory and CPU. Another example is separation/decoupling data from computational modules by database management systems. A user can change/fix the computational module without changing database (DB). Similar separation was made between a Knowledge Base (KB) and Computational modules (Reasoning Engines) in Artificial Intelligence.

The modular architecture separates the operation space characterization from both types of computations: sensor tasking optimization and object tracking and discrimination. Another advantage of this separation is clearer mathematical formulation of the SRM as an optimization task and use of the powerful techniques developed in this mathematical area for decades.

Many other popular SRM approaches are tightly integrated with Information Gain (IG) maximization that gives this single criterion dominance over the other criteria. Such *monocentric* approach has fundamental weaknesses. Maximization of the IG can contradict to minimization of the number of sensors and to the need to provide relatively even load to different sensors.

The new approach is free from tight integration while allowing using information gain approach too. Information gain criterion is very sensitive to the accuracy of estimates of probability distributions such as covariance matrixes, accuracy of tracking data association/correlation and so on. The proposed approach is free from such unrealistic assumption too.

1.3 Computational Intelligence Methodologies

Biologically-inspired computational intelligence approaches are promising for SRM modeling. One of the biological inspirations is coming from similarity between ultimate goals of SRM and the foraging/hunting models in mathematical ecology that have natural analogy with "hunting" targets. Foraging had already thousands of years evolution time that it much longer than SRM evolution.

Especially interesting here are the tradeoffs between time and gain observed in the natural foraging systems, and the respective models and objective functions that combine them. In one of the models [36, 38] the predator attempts to maximize E/ (h + s), where s is the search time involved. For a range of prey, the predators average intake rate is

$$E_{average}/(h_{average} + s_{average}),$$

where $E_{average}$ is the average energy of all prey items in the diet, $h_{average}$ is the average handling time and $s_{average}$ is the average search time.

In terms of SRM models the energy E is in line with a class of gain functions. The most interesting part of E is the tradeoff between handling time and search time. The weak part of many bio-inspired methods that they do not go further than shallow inspiration ending up with algorithms with little interpretation of tradeoffs borrowed from other fields like formula $E_{average}/(h_{average} + s_{average})$ above.

Beyond the weakness of *justification* of the optimization criteria, the search of the optimal solution is also *highly heuristic* in bio-inspired algorithms. Therefore we focus on the mathematically *rigorous* solution as a *benchmark* for the heuristic solutions that may have some benefits of faster computations and simplicity.

The key idea of the proposed approach is to combine computational intelligence techniques (multi-objective optimization based on the Pareto border, Integer Liner Programming under uncertainty, and Adaptive Learning methods) with physical considerations (sensor phenomenology and geometry of locations relative to targets).

The Computational Intelligence methodologies that are applicable to solve the SRM OPTIMA models are evolutionary computing methods including adaptive multi-objective optimization that exploit genetic algorithms, colony optimization, particle swarm optimization, interval, stochastic and fuzzy optimization, and adaptive dynamic logic of phenomena. A multi-objective evolutionary approach helps: (1) to speed up funding the Pareto border points, and (2) to explore found

Pareto points for identifying the "best" ones. Genetic Algorithms (GA) for (1) and (2) use different fitness functions or different parameters of the same function. These computational issues are a topic of the separate work.

Contradictory goals of the SRM represented by multiple objective functions cannot be reached by the classical optimization without setting up a tradeoff between them and/or constrains in advance. However in the dynamic environment the tradeoff must be dynamic as well. Thus the optimization algorithm must learn and adapt to such dynamic environment in real time. This is challenging task with the growing number of assets, closely spaced objects and real time constrains for the algorithm. The main idea of the approach is borrowed from **nature—the modularized optimization**. Such bio-inspired approach mimics a cooperative team hunting/foraging in nature with the abilities of dynamic learning, adapting, and self-tasking.

The existing models in mathematical ecology [24] such as Optimal Foraging Theory (OFT) and its Digestive Rate Model (DRM) of foraging that are now linked to the cumulative prospect theory (Tversky, Kahneman, Nobel Prize 2002) of human decision making under uncertainty are valuable sources of novel ideas to improve the SRM models under conflicting objective functions. In particular, OFT and DRM deal with optimizing the tradeoff between foraging times.

The OPTIMA architecture assumes two types of algorithms:

- Algorithms (A1) that provide optimal sensor assignment (including deliberately reduced sensor utilization to be able to sensor more objects later) using estimated performance including degraded one and
- Algorithms (A2) that support input of sensor data and metadata to A1 including sensor "health" data, tracking, track correlation data, and data fusion data.

The first class of algorithms exploits sensor geometry including a possibility that two sensors observing the same object (at approximately orthogonal geometry) can produce more accurate results in a shorter time. The algorithm of the first type solves the problem of optimal sensor assignment. The algorithm of the second type provides the input data for the first one, allowing the use of tracking and fusion algorithms that are at the level of the Cramer-Rao Bound (CRB) which sets up the best possible performance accounting for associations between sensors, objects, and tracks [28]. These algorithms exploit a novel computational intelligence technique [27] developed for air-ground radars, which have not been applied to the degraded scenarios.

The proposed SRM methodology for sensor planning includes:

- simulating different input data/scenarios (including different locations of assets and levels of degradation of sensors),
- analyzing values of all objective functions of interest for the best Pareto solutions.

Pareto solutions can be very different, e.g., with maximization of objective functions F_1 and F_2 the Pareto border may include pairs of their values (0.9, 03) and (08, 04). In summary the approach consists of: (1) Intelligent SRM models and algorithms, (2) Translation of tracking and discrimination requirements to flags for sensors in the SRM model, (3) the SRM model and algorithm for paired sensors.

1.4 Notation

Below we introduce the notations:

- T is a *track*.
- G is a *object*.
- U_T is a *track ambiguity descriptor* (*message*) at time t in the form of a model $U_T(t) = \langle A_T, \Omega_T \rangle$, where A_T is a set of *ambiguity characteristics of the track* T and Ω_T is a set of relations on A_T at time t.
- $U_G(t)$ is a *object ambiguity descriptor* (*message*) *of the object* G at time t in the form of a model $U_G = \langle A_G, \Omega_G \rangle$, where A_G is a set of *ambiguity characteristics of the object* G and Ω_G is a set of relations on A_G at time t.
- The triple $E(t) = \langle U_T(t), U_G(t), C(t) \rangle$ is called a *sensing environment* at time t, where C(t) is a sensor model (a set characteristics of the sensors C such as locations, orientations, FOV, resolution, and others).
- M(E(t)) is a *vector of measures of environment degradation* E(t).
- $V(C_K)$ is an **environment operator** (*algorithm*) that produces a new environment $E_K(t)$, $V(C_K) = E_K$.
- K(E(t) is a **tasking operator for sensors** (*algorithm*) at environment E(t), that is K assigns a new set of characteristics to the sensors C_K (sensor model). In these mathematical terms we need to design a tasking operator K to **decrease the environment degradation,** that is $M(E(t)) > M(E_K(t + 1))$. Below the design of such an operator is provided via a set of single-objective Integer Linear Programming (ILP) models and multi-objective optimization models on the top of ILP models as outlined in Fig. 1.
- $\{a\}_j$ is a set of *Areas of Interests* (AOI) that sensor s_j can observe at time interval t. Each area of interest may contain an *Object Complex* (TC). For a search/scanning sensor only the area can be known, for a tracking sensor an object complex can be known. $a_{ij} \in \{a\}_j$ is a *marked area of interest* in $\{a\}_j$. Only *one* area is marked for each sensor. This marking can have different interpretations depending on the version of sensor tasking. The examples of interpretations of marked areas are: (1) a marked area a_{1j} is the main focus area of sensor s_j that is assigned by the requirements and tasking objectives, (2) a marked area a_{2j} is an area that has the best viewing geometry from s_j, (3) a marked area a_{3j} is an area that has the best resolution from s_j at time interval t, (4) a marked area a_{4j} is an area that has the worst resolution from s_j at time interval t.
- r is a resolution of the sensor. Cramer-Rao Bound (CRB) is one of the ways to assign value to r, because CRB is the error of localization and tracking of an object with the given sensor. CRB gives the best possibly achievable resolution.

Also CRB is useful for assigning resolution to discriminate RV and decoy. CRB can include coordinates and velocities, and in addition, classification features related to the classification/discrimination probability.

- v is a Field of View (FOV) of a sensor at a particular time.
- $f(a_i,s_j,t,r)$ is a binary flag, $f(a_i,s_j,t,r) = 1$, if sensor s_j is *capable* of *covering/observing* area of interest a_i at the time interval t with the required resolution r, else $f(a,s,t,r) = 0$.
- $f_s(a_i,s_j,t,r)$ is a stochastic flag, $f(a_i,s_j,t,r) \in [0,1]$. It is a *probability* that sensor s_j is *capable* of *covering/observing* the area of interest a_i at the time interval t with the required resolution r.
- $p(s_j,t,v)$ is a *time to point* sensor s_j to get FOV v by time t. This requires changing its line of sight (LOS).
- $f^*(a_i,s_j,t,r,v)$ is a binary flag, $f^*(a_i,s_j,t,r,v) = 1$, if $f(a_i,s_j,t,r) = 1$ and a_i is a marked area a_{ij} of sensor s_j at the time interval t within FOV v, else $f^*(a_i,s_j,t,r,v) = 0$. In other words, If sensor s_j observes several AOIs within FOV v then only the marked AOI will get flag $f^*(a_i,s_j,t,r,v) = 1$, other AOI a_k within the same FOV v of sensor s_j will get $f^*(a_k,s_j,t,r,v) = 0$. This flag will be used in the optimization task formulation to minimize the number and cost of the required sensors.
- $f^*_s(a_i,s_j,t,r,v)$ is a *stochastic* flag (*probability*), $f^*_s(a_i,s_j,t,r,v) = f_s(a_i,s,t,r)$ if and a_i is a marked area a_{ij} of sensor s_j at the time interval t within FOV v, else $f^*(a_i,s_j,t, r,v) = 0$.
- $\{s\}_i$ is a set of sensors that can observe AOI a_i at time interval t.
- $s_{ij} \in \{s\}_i$ is a *marked sensor* in $\{s\}$. More than *one* sensor can be marked for each area. This sensor can have different interpretations depending on a version of the sensor tasking interest. The examples of interpretations of marked sensors are: (1) sensors s_j that are assigned to AOI a_i by the requirements and tasking objectives, (2) sensors that have the best viewing geometry to AOI a_i, (3) sensors that have the best resolution of AOI a_i at time interval t, (4) sensors that have the worst resolution from s_j at time interval t, (5) Aegis radars, and others.
- $\{g_l(s_j)\}$ is a set of binary flags, $g_k(s_j) = 1$ indicates that sensor s_j is of *type/category* g_k, else $g_k(s_j) = 0$. The tasking requirements may include specific types of sensors, e.g., $g_1(s_j) = 1$ can indicate that s_j is a staring sensor, $g_2(s_j) = 1$ indicates that s_j is a scanning sensor (or scanning mode of the sensor), and $g_3(s_j) = 1$ indicates that s_j is a sensor on a specific platform.
- $\{h_k(s_j,s_q)\}$ is a set of binary flags, $h_k(s_j,s_q) = 1$ indicates that sensors s_j and s_q have a specific relationship, e.g., $h_1(s_j,s_q) = 1$ can indicate that the angle between their LOSs at time t is in the interval $[75^0,105^0]$ that is close to orthogonality.
- x_{ijt} is a *binary variable*, $x_{ijt} = 1$ indicates that sensor s_j is *tasked* to observe area a_i for the time interval t, else $x_{ijt} = 0$. Finding values of x_{ijt} is the goal of optimal sensor tasking.
- $\{c_j\}$ is a set of *objective function coefficients*. The interpretation of these coefficients depends on the specification of the sensor tasking objectives. The examples of interpretation of $\{c_j\}$ are: (1) costs of the sensors $\{s_j\}$ and their platforms (2) pointing time, (4) errors of sensors (e.g., covariance matrixes of

LOS), (5) capability characteristics of sensors such as resolution or sensitivity of the sensors (6) information gain that the sensors add relative to the current knowledge of the object complexes, and others.

Coefficients of types (1)–(4) lead to minimization models (min of number of sensors, min of cost, min of pointing time, min of errors). Coefficients of types (5)–(6) lead to maximization models (max of capabilities, max of information gain). In essence we have two opposite categories of objective functions: *cost* (1–4) and *gain* (5–6) with wide interpretations of costs and gains. It is not required to interpret them literally. The multi-objective approach will allow seeing the optimal value of each objective function in the context of the values of other objective functions before a tradeoff between objective functions is made. In contrast the popular weighting approach combines such objective functions into their weighted sum "in the dark" without such analysis.

The **Pareto multi-objective SRM model** that combines models with these objective functions allows the analysis of the *optimal value* of each objective function, in the context of the values of other objective functions, *before a tradeoff* between objective functions is made. In contrast, the popular weighting approach combines objective functions into their weighted sum "*in the dark*" without such an analysis. The detailed elaboration of the multi-objective SRM model is a topic of a separate paper.

1.5 ILP Models for Time T

Consider optimization *objective functions for a fixed time*:

$$ext \sum_{i=1}^{n} \sum_{j=1}^{m} c_j \cdot f^* \left(a_i, s_j, t, r, v \right) \cdot x_{ijt}$$

where *ext* (extremum) stands for max or min with the constraints presented below.

Coverage constraints (all areas of interest $\{a_i\}$ must be covered at least by one sensor):

$$\sum_{j=1}^{m} f(a_i, s_j, t, r) \cdot x_{ijt} \geq 1, \quad i = 1, 2, \ldots, N$$

Constrains (all variables that assign sensors to areas at time t must be binary; sensor either assigned or not to the AOI):

$$x_{ijt} \in \{0, 1\}, \quad i = 1, 2, \ldots, N; j = 1, 2, \ldots, M$$

Sensor types constraints (sensors of all required types $g_1, g_2, \ldots, g_K$ must be used):

$$\sum_{j:\,g_k(s_j)=1} x_{ijt} \geq 1, i=1,2,\ldots,N, k=1,2,\ldots,K$$

Sensor relationship constraints (sensors with all required relations $h_1, h_2, \ldots, h_L$ must be used in required quantities H_{il} for each AOI a_i),

$$\sum_{j:\,h_l(s_j,s_q)=1} x_{ijt} \geq H_{il}, i=1,2,\ldots,N, l=1,2,\ldots,L$$

If $\{c_j\}$ are the *costs* of observing all the areas a_i with the object complexes, then we use the first objective function and minimize the total cost of the observation under constraints. This will provide the least expensive solution under constraints. If costs of all sensors are considered to be equal then in fact this objective function minimizes the *total number of sensors*. Similarly if $\{c_j\}$ are *capabilities* of the sensors, then we minimize total capabilities under constraints.

If $\{c_j\}$ are *information gains* then we use the second objective function and maximize the total information gain. Flags $f^*(a_i, s_j, t, r, v)$ in the objective function allow to optimize the use of sensors for the situations when some individual sensors observe several object complexes within a single FOV avoiding double counting such sensors.

The coverage constraints require that at least one sensor will be *tasked* to observe/*cover* each area a_i with the required resolution of observation. The sensor type constraints require that at least one sensor of each required type will be used. These sensor type constraints can be generalized by substituting 1 on the right side of the inequality with another required number of sensors of type g_k.

The relationship constraints allow the incorporation into the optimization model of multiple desired geometric relationships between the locations of sensors and objects, such as orthogonality relative to the object. If there are no sufficient resources to track each objects with two sensors, then we can choose which objects require two sensors, and which will be tracked with one sensor. The models also allow dedicating only a small amount of time of the second sensor for tracking the same object, which significantly increases the resolution in terms of Cramer-Rao Bounds (CRB) as a function of two parameters: time on object for a sensor from platform 1, and time on object for a sensor from platform 2. This optimization framework allows highly modular SRM where some set of modules is responsible for computing all flags and updating them.

1.6   ILP Model for Larger Time Interval

The optimization model described above assumes a *short time interval* where all flags do not change their values significantly. For the larger time intervals this assumption is not true, the values of flags change dynamically within the larger time intervals. This leads to the model modification with additional summation for all time moments from t = 1 to T:

Objective functions for a longer time interval:

$$ext \sum_{t=1}^{T} \sum_{i=1}^{n} \sum_{j=1}^{m} c_j \cdot f^*\left(a_i, s_j, t, r, v\right) \cdot x_{ijt}$$

with the constraints presented below. If $\{c_j\}$ are *costs* of observing all areas a_i with object complexes, then we use the first objective function and minimize the total cost of observation under constraints. This provides the least expensive solution under constraints. If costs of all sensors are considered to be equal, then in fact this objective function minimizes the *total number of sensors*. Similarly if $\{c_j\}$ are *capabilities* of the sensors, then we minimize the total capabilities under constraints.

If $\{c_j\}$ are *information gains* then we use the second objective function and maximize the total information gain. Flags $f^*(a_i, s_j, t, r, v)$ in the objective function allow optimizing the use of sensors for the situations when some individual sensors observe several object complexes within a single FOV avoiding double counting such sensors.

Coverage constraints (all areas of interest $\{a_i\}$ must be covered at least by one sensor):

$$\sum_{j=1}^{m} f\left(a_i, s_j, t, r\right) \cdot x_{ijt} \geq 1, \quad i = 1, 2, \ldots, N, t = 1.2, \ldots, T$$

Binary constrains (all variables that assign sensors to areas at time t must be binary; sensor either assigned or not to the AOI):

$$x_{ijt} \in \{0, 1\}, \quad i = 1, 2, \ldots, N; j = 1, 2, \ldots, M, \quad t = 1.2, \ldots, T$$

Sensor types constraints (sensors of all required types g_1, g_2, ..., g_K must be used):

$$\sum_{j:\, g_k(s_j) = 1} x_{ijt} \geq 1, i = 1, 2, \ldots, N, \ k = 1, 2, \ldots, K, \ t = 1.2, \ldots, T$$

Sensor relationship constraints (sensors with all required relations h_1, h_2, ..., h_L must be used in required quantities H_{il} for each AOI a_i),

$$\sum_{j:\, h_l(s_j, s_q) = 1} x_{ijt} \geq H_{il}, \ i = 1, 2, \ldots, N, \ l = 1, 2, \ldots, L, \ t = 1, \ldots, T$$

The coverage constraints require that at least one sensor will be *tasked* to observe/*cover* each area a_i, with the required resolution of observation. The sensor type constraints require that at least one sensor of each required type will be used.

These sensor type constraints can be generalized by substituting 1 on the right side of the inequality by another required number of sensors of type g_k.

The relationship constraints allow incorporating into the optimization model the multiple desired geometric relationships between the locations of the sensors and the objects, such as orthogonality. If there are no sufficient resources to track each of the objects with two radars, then the algorithm selects objects requiring two sensors, using attained accuracy and object classifications. The model also allows dedicating only a small amount of time of the second radar for tracking the same object, which will also significantly increase track resolution and object characterization. This optimization framework allows the highly modular SRM using modules that compute flags.

1.7   Stochastic ILP Model for Larger Time Interval

The optimization models described above assume deterministic flags in the objective functions. Below we present stochastic versions of the objective function that can explicitly capture *uncertainty* of the operation space situation. It requires changing deterministic flags f* to stochastic flags f*$_s$ in the objective functions and constraints:

$$ext \sum_{t=1}^{T} \sum_{i=1}^{N} \sum_{j=1}^{M} c_j \cdot f_s^* (a_i, s_j, t, r, v) \cdot x_{ijt}$$

with the constraints presented below.

Coverage constraints (all areas of interest {a_i} must be covered with confidence F_i or greater:

$$\sum_{j=1}^{m} f_s^* (a_i, s_j, t, r, v) \cdot x_{ijt} \geq F_i, \ i = 1, 2, \ldots, N, \ t = 1.2, \ldots, T$$

Binary constrains (all variables that assign sensors to areas at time t must be binary; sensor either assigned to the AOI or not):

$$x_{ijt} \in \{0, 1\}, i = 1, 2, \ldots, N; j = 1, 2, \ldots, M, t = 1.2, \ldots, T$$

Sensor relationship constraints (sensors with all required relations $h_1, h_2, \ldots, h_L$ must be used in required quantities H_{il} for each AOI a_i),

$$\sum_{j: h_l(s_j, s_q) = 1} x_{ijt} \geq H_{il},$$

$$i = 1, 2, \ldots, N, l = 1, 2, \ldots, L, t = 1, \ldots, T$$

Sensor relationship constraints (sensors with all required relations h_1, h_2, ..., h_L must be used):

$$\sum_{j:\, h_l\left(s_j,\, s_q\right)\,=\,1} x_{ijt} \geq 1, \quad i = 1, 2, \ldots, N,\ l = 1, 2, \ldots, L, t = 1, 2, \ldots, T$$

1.8 Sensor Message Constructs and Operation Space Reconstruction

The algorithm for sensor message constructs is as follows:

1. Select the sensor composition (e.g., two radars of one type, two radars of another type, and a constellation of 12 EO/IR sensors).
2. Select a representative operation scenario with the sensors listed in (1) above. Describe these sensors in terms of:

   a. U_T (*track ambiguity descriptor/message*) at time t in the form of a model $U_T(t) = \langle A_T, \Omega_T \rangle$, where A_T is a set of *ambiguity characteristics of the track* T and Ω_T is a set of relations on A_T at time t;

   b. $U_G(t)$ (*object ambiguity descriptor/message*) *of the object* G at time t in the form of a model $U_G = \langle A_G, \Omega_G \rangle$, where A_G is a set of *ambiguity characteristics of object* G and Ω_G is a set of relations on A_G at time t;

   c. $E(t) = \langle U_T(t), U_G(t), C(t) \rangle$ triple (*sensing environment* at time t), where C(t) is a sensor model (a set characteristics of the sensors C such as locations, orientations, FOV, resolution, "health", and others);

   d. $M(E(t))$ (*vector of measures of environment degradation $E(t)$*);

   e. $V(C_K)$ (**environment operator**/*algorithm*) that produces a new environment $E_K(t)$, $V(C_K) = E_K$.

3. Describe the items listed in step 2 in the form of messages with a specific format.

The example of the messages going to the *sensing environment* at time t, E (t) = $\langle U_T(t), U_G(t), C(t) \rangle$ is a binary or numeric flag f_{Energy}, which is "too much energy is on the focal plane". It is accompanied by additional flags that indicate the consequences of this degradation such as f_{SNR}, which indicates the decreased SNR, $f_{detection}$ which indicates decreased detection sensitivity, f_{range} which indicates that the range is uncertain, f_{angle} which indicates that the angle and pointing vector are is uncertain.

In the notation section several flags have been introduced. These flags are used to define sensor messages and are parts of the messages.

The binary flag $f(a_i, s_j, t, r)$ indicates whether sensor s_j is *capable* to *cover/observe* area a_i at the time interval t with the required resolution/capability r. This flag is computed directly for each value of its variables: the identified areas a_i, sensor s_j, and required resolution/capability r at time t. LOS and FOV with their errors for sensor s_j at time t, location of the AOI a_i are used to check if a required resolution/capability can be reached. If the AOI a_i contains a detected moving object then the known dynamic properties of the trajectory of the object are used to identify the next location of the AIO a_i and to compute flag f for time t + 1. These properties are derived from the external tracking, track correlation algorithms. Similarly each sensor resolution relative to objects are derived, other capabilities and their adequacy relative to required ones are computed.

The flag $f^*(a_i,s_j,t,r,v)$ is directly computed by the algorithm from $f(a_i,s_j,t,r)$, when an area a is marked for a sensor s_j. The marking of the area is identified from tasking requirements and tasking doctrine. Flags $\{g_k(s_j)\}$ that indicate types of sensors are computed from the database of specifications of sensors. Relations flags are computed based on the definitions of the relations. In some cases all flags are specified in advance. In other cases these flags are computed to achieve the required resolution of the sensor system. This includes the Monte-Carlo Simulation and the Cramer Rao Bounds (CRB). The CRBs are computed for each object and sensor, as well as for each object and a pair of good and degraded sensors as required. If required it is done by using tracking and fusion algorithms outlined below.

These flags can tell a story: 'we were unable to track and/or characterize the object without excess ambiguity'. This information is used to guide the SRM to find a sensor which can view the object complex, from a less impacted viewing angle, or in a wavelength that is less susceptible to the degrading agent in the SRM models.

1.9 SRM Model with Orthogonalization

Consider sensors $S_1, S_2,..., S_n$. Assume that for each sensor we know the sensor model M which includes its location, orientation, and capabilities (FOV, resolution, band, response time, and others). Thus we have models $M(S_1), M(S_2),...,M(S_n)$.

Assume that there is no single sensor that provides the required resolution for the area of interest a_i. In this situation we search for a pair of sensors (S_k, S_m) that will have the angle between their LOSs closer to $90°$ than any other pair of sensors when pointed to the center of a_i (See Fig. 4). In other words, we order all pairs of sensors according to the dot products of vectors of LOSs and search for the OPIR sensors with max of dot products,

$$\text{Arg max}_{i=1,n,\ k=1:n}(\text{LOS}(S_i) \cdot \text{LOS}(S_k))$$

After all these pairs of sensors are found for all areas of interests the algorithm checks their consistency that is no sensor assigned to two or more areas. If such inconsistency is found it is resolved by removing a conflicting sensor from the pair with the least value of dot product and assign the best sensor to this pair that is not used yet for any object. If the set of such sensors is empty then wait until the busy sensor will be released from the area ai, or use the round robin method.

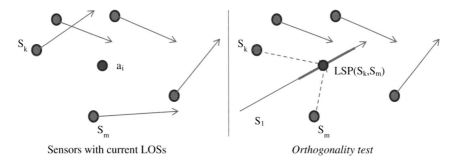

Sensors with current LOSs   *Orthogonality test*

Fig. 4 Orthogonality test for pairs of sensors

If the center of the area a_i is not known or this area is very large then the algorithm searches a pair of sensors (S_k, S_m), which has *"better"* angles between their *possible* LOSs relative to the several subareas within the area of a_i, than any other pair of sensors. The angles are "better" if they correspond to the *largest sets of points*, $LSP(S_k, S_m)$ in the subarea where the angle between $LOS(S_k)$ and $LOS (S_m)$ is closer to $90°$ than for any other pair of sensors (*"orthogonality" test*).

This base algorithm can be enhanced to deal with the motion of platforms and areas of interests. Specifically the subarea can be selected based on the tracking of the object in the area a_i as shown in Fig. 4. Tracking by nearly orthogonal sensors such as radars and EO/IR brings significant accuracy improvement.

1.10 Multi-objective SRM Optimization

To resolve the **contradictory goals** such as decreasing the overall sensor resource *utilization*, increasing the *probability* that all threat objects are tracked, and decreasing potential *overload* of sensors at individual platforms/units the **multi-objective optimization approach** is used.

The multi-objective optimization model is built on a set of *Single-Objective Tasks* (SOT) as shown on Fig. 1. Let F_1, F_2,...,F_n are objective functions of respective single-objective SRM tasks.

The optimal solution of SOT_1 provides the value f_1 of F_1. We also can compute values f_{i1} of F_2, F_3, ..., F_n for this solution for SOT_1. This produces a vector $(f_{11}, f_{21}, f_{31},...f_{n1})$. Similarly values f_i and associated values f_{ij} are produced for all other SOT_i with objective functions F_i. Each of these vectors constitutes a *vector solution*. The Pareto border is a set of all vector solutions that cannot be improved.

Consider an example with two vector solutions {(0.6; 0.2), (0.2; 0.8)} for two objective functions, F_1 and F_2 that are maximized. Here the best solution relative to objective function F_1 is 0.6, but it is very weak (0.2) relative to F_2. Similarly, the best solution for the F_2 is 0.8, but it is very weak (0.2) for F_1. A solution (0.5; 0.5) may exist and be a good tradeoff between these two solutions. These vectors are a part of the Pareto border.

The advantage of the Pareto approach is that we analyze a much wider set of possible solutions than a set of "optimal" solutions provided by scalar cost functions such as those based on information gain [18]. In essence, it is *better to introduce cost functions within the Pareto set than without it*. The same Pareto approach is used by us for discrimination, and for the combination of the tracking and discrimination.

An innovative approach based on the analysis of T-norms [16] allows coming to the optimal solution by combining several objective functions. To move from the Pareto boundary to a **trade-off solution**, multiple fusion (aggregation) operators have been proposed. The class of fusion operators used in fuzzy logic for the membership functions is known as T-norms. T-norms can distort the Pareto order property dissolving the important difference between the nodes [16]. For example, suppose that W = {(0.0; 0.5), (0.2; 0.8), (0.6; 0.2)} then the best points (Pareto points) are P = {(0.2; 0.8), (0.6; 0.2)}. First, we need to know that T-norms do not contradict the Pareto optimum. In fact, a T-norm such as the popular-in-fuzzy-logic min can add new 'best" points that do not belong to the Pareto optimum.

Consider W = {(0.0;0.5), (0.2;0.8), (0.6;0.2), (1.0;0.8), (0.9;0.8)}. Here Pareto optimum includes only (1.0; 0.8), but the T-norm as minimum gives also (0.9; 0.8) as a best point which is wrong. Now we see how the lack of interpretability is translated into a lower accuracy of the solution.

Setting up a trade-off preference relation H between alternative vectors (nodes) in the Pareto set must be consistent with a meaningful preference of assigning a sensor to objects. Unfortunately relation H rarely is known completely. Each T-norm serves as a compact approximation of H. However a T-norm can be far away from modeling H satisfactorily. It is desirable that T-norms preserve the strict order for all pairs (x,y) that is

$$(x, y) < (v, u) \Rightarrow T(x, y) < T(v, u).$$

However this is true only for some pairs, e.g.,

$$(0.3; 0.5) < (0.4; 0.7) \Rightarrow \min(0.3; 0.5) < \min(0.4; 0.7),$$

but it is not true for (0.3; 0.5) < (0.4; 0.7), where min(0.3; 0.5) = min(0.3; 0.7). Thus, T-norms can distort the order dissolving the important difference between the nodes.

We measure this distortion by introducing a *Pareto set distortion factor* k_2 and use the **least distorted** T-norms. Factor k_2 computes a ratio of two numbers m and h = r(r−1)/2: where m is the *number of unequal pairs* of nodes of the lattice that have equal T-norm values and (2) h = r(r−1)/2 is the *total number of different pairs of nodes* of lattice L that has r nodes. Next we use the *quantified Pareto set distortion factor* k_3 that is modified factor k_2 to address the requirement (3) of sufficient power of scale given by T-norm [16],

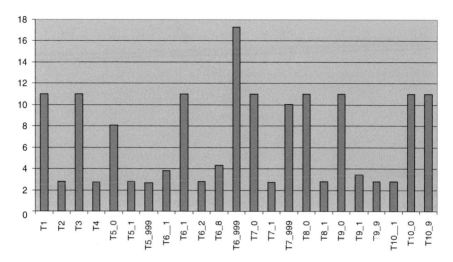

Fig. 5 Distortion factor k_1 for considered T-norms

$$k_2 = \frac{2m}{r(r-1)}, k_3 = \frac{\sum_{i=1}^{11} \binom{2}{m_i}}{\binom{2}{r}}$$

where r is the same as for k_2 and mi is the number of nodes in the ith subinterval of the lattice L. The ith component of the sum in k_3 gives the number of "glued" pair nodes of the lattice in the subinterval i. Figure 5 shows lattice distortion factor k_1 by different T-norms. For speeding up computations of the Pareto border, we use the theory of Monotone Boolean Functions [17]. The main idea of cutting the computation time is finding the attributes that are relatively independent and as such they can be processed relatively independently in parallel.

1.11 Degraded Sensors and Environment

Helping a degraded IR sensor with a single IR sensor. Figure 6 shows a case when the first EOIR sensor is degraded and the second sensor can be used instead of the first one. The second EOIR sensor has a better viewing geometry. However, the first degraded EOIR sensor S_1 cannot guide the second sensor S_2 how to change its FOV, because the first sensor has no range information to the object complex. It has only 2-D directional information of the line of sight (LOS).

Consider a scenario where at time t the sensor S_1 observes an Area of interests (AOI) a_i that contains an object complex. The 3-D center of the a_i is already identified. For instance, it can be done jointly by this sensor S_1 and another sensor

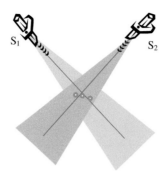

Fig. 6 Situation with a better viewing geometry for the second sensor and lack of guidance from the degraded first sensor on the location of the object complex

S_2 that was available at time t. At the next time t + 1 sensor S_1 is downgraded to the level that another sensor S_3 should substitute it to continue accurately tracking the object complex in a_j. The second sensor needs to scan the FOV of the first sensor to find the Area of Interest (AOI) that contains the objects.

Algorithm outline:

Step 1:   Search for a substitute sensor S_3 that has the same or very similar capabilities as s_1 and similar geometry relative to the AOI a_j. It is identified by the thresholds on differences in distances and FOVs,

$$\text{Capab}((s_1)) \approx \text{Capab}(s_3), \ \text{Dist}(s_1, s_3) < T_{\text{distance}}, \ \text{diffFOV}(s_1, s_3) < T_{\text{diffFOV}}$$

Step 2:   If Step 1 did not find a sensor S_3 that satisfy the requirements of Step 1 then the search is conducted under modified relaxed requirements. Sensor S_3 should satisfy only specific capabilities requirements such as resolution. Let R_1 be a resolution that sensor S_1 provides for area a_j. Another sensor can be a more powerful sensor located further from a_j than S_1, but it still can provide resolution R_1,

$$\text{Resolution}(S_3, a_j) \geq \text{Resolution}(S_1, a_j) = R_1.$$

Additional requirements can be imposed on SNR and on differences in FOV: $\text{diffFOV}(S_1, S_3) < T_{\text{diffFOV}}$. The dynamic adjusting requirements can be modeled in accordance with the Dynamic logic process [15].

Helping degraded IR sensor by selecting two IR sensors. Consider a degraded sensor S_1 which needs to be substituted by other sensors, and there is no simple solution; that is, there is no other sensor with very similar or better capabilities that can be quickly reoriented to the same area as S_1. The algorithm searches for a pair of sensors (S_k, S_m) that are able to substitute sensor S_1. The idea of the search algorithm is to find (S_k, S_m) with Lines of Sight (LOS) closest to LOS of S_1 within $90°$ limits.

$$\forall S_i \neq S_k,\ S_i \neq S_m\ 90° \geq ||LOS(S_1) - LOS(S_i)|| \geq ||LOS(S_1) - LOS(S_k)|| \quad \& $$
$$90° \geq ||LOS(S_1) - LOS(S_i)|| \geq ||LOS(S_1) - LOS(S_m)||$$

In other words we order all sensors according to the dot products of vectors of LOS and search for the sensors with max of dot products,

$$\text{Arg max}_{i=2,n}(LOS(S_1) \cdot LOS(S_i))$$

In addition a pair of sensors (S_k, S_m) should have "*better*" angles between their *possible* LOSs, relative to $LOS(S_1)$, than any other pair of sensors. The angles are "better" if they correspond to the *largest sets of points*, $LSP(S_k, S_m)$, on the LOS (S_1), where the angle between $LOS(S_k)$ and LOS (S_m) is closer to 90° than for any other pair of sensors ("**orthogonality**" **test**).

Now assume that sensor S_1 is *degraded partially*, that is some of its information is useful not only its LOS. We have already assumed that LOS of S_1 carries useful information, e.g., there are some object detections in its FOV with the given LOS. This means that it makes sense to continue to observe the environment in the area captured by FOV of S_1. Let the additional uncorrupted information from S_1 be the *Direction to the Cluster of Objects* (DCT), that is we have vector $DCT(S_1)$ in addition to $LOS(S_1)$. Now we can search for the pair of sensors (S_k, S_m) that have the *largest sets of points*, $LSP(S_k, S_m)$ on $DCT(S_1)$, not only on the $LOS(S_1)$. This increases the accuracy of the information that (S_k, S_m) provides (see Fig. 7).

If we have an uncorrupted $DCT(S_1)$ only in the part of the FOV of S_1, then a solution based on DCT is used in this part of the FOV and the solution based on the $LOS(S_1)$ is used in the corrupted part of FOV. In the case of multiple directions to the Cluster of Objects for sensor S_1 the optimal pairs of sensors are computed for each direction.

Helping degraded radar by selecting two IRs. Now consider a situation when S_1 is a degraded radar, or a radar that observes a too-dense scene and sensors S_2, S_3 ,..., S_n are EOIR sensors that can be used to help or substitute S_1, depending on the

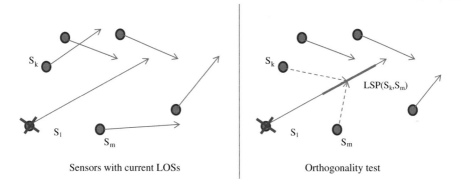

Sensors with current LOSs   Orthogonality test

Fig. 7 Orthogonality test in degraded environment

level of degradation or complexity of the scene. Assume that radar gives 3-D location L of the object with acceptable accuracy to point an EOIR sensor. The algorithm searches for the EOIR sensor with the LOS that is most close to 90° to the radar LOS, and closest to the location L to provide a better object resolution,

$$\text{Arg min}_{i=2,n}(\text{LOS}(S_1) \cdot \text{LOS}(S_i)), \text{Arg min}_{i=2,n}(L(T), L(S_i)).$$

For two IRs that we consider here the formulation is similar to Task 3.2 relative to orthogonality of IRs.

Selecting two Radars to help degraded IR. Of the two radars, choose the one the LOS of which is closest to 90^0 to the LOS of IR. If location of the object is too uncertain (from IR data), divide the IR-LOS into two pieces, one—best for the radar 1, and part two—best for the radar 2.

The worst case is when a degraded sensor S_1 does not provide any useful information. In this case LOS (S_1) is degraded too. This task is equivalent to a general task of tasking $n-1$ sensors and has very little specifics relative to the general sensor management task to be exploited efficiently. This task is out of the scope of this paper.

2   Computational Intelligence Solution for SRM Model

2.1   Exact and Heuristic Algorithms

The proposed above ILP SRM models require efficient algorithms to solve them. The classical Linear Programming models can be solved by the simplex method for relatively large N and M. For instance, 15 OPIR sensors (N = 15) and 15 areas of interests (M = 15) lead to 225 variables for each time interval t.

The proposed ILP SRM tasks are NP-hard problems that cannot be solved exactly when the number of parameters is large. For smaller number of parameters multiple computational methods solve it exactly [7]. Thus, depending of the size of the SRM model and time constraints to solve it (planning or real-time tasking) exact, approximate or heuristic methods are needed.

Multiple generic heuristics can solve the ILP SRM tasks: (1) Tabu search, (2) Hill climbing, (3) Simulated annealing, (4) Reactive search optimization, (5) Ant colony optimization, (6) Hopfield neural networks, (7) Genetic algorithms and others. The last three classes of methods are effective Computational Intelligence methods that have been successful in SRM [2, 10, 30, 31]

The development of a specialized algorithm is also advantageous for the SRM to get a better performance by taking into account the specifics of the IL SRM models. This includes using bio-inspired methodologies of Dynamic Logic [15, 27].

The exact algorithms to solve the SRM task include: (1) cutting plane methods (solving the LP relaxation and adding linear constraints that drive the solution towards being integer without excluding any integer feasible points), (2) variants of the branch and bound method (the branch and cut method combines both branch and bound and cutting plane methods). The solutions of the LP relaxations give a worst-case estimate of how far from optimality the returned solution is.

The relaxation method to solve this SRM task consists of converting this discrete Linear Programming (LP) task to the LP task with continuous variables by substituting the binary constrains to the constraints where x_{ijt} are non-negative numbers limited by 1 (discrete constraint relaxation),

$$0 \le x_{ijt} \le 1, \quad i = 1, 2, \ldots, N; j = 1, 2, \ldots, M$$

This classical LP task can be solved by the simplex method for large N and M. The next step is exploring the vicinities of vertices produced by the simplex method. This exploration includes finding the feasible binary points in the vicinity, computing the value of the objective function on them, and selecting the best ones. The size of the vicinities and the number of simplex vertices to be explored can be adjusted to minimize the computations.

A simple suboptimal version of this approach is to interpret non-integer component of the solution, x_{ijt} as a confidence measures that sensor sj should be assigned to the AOI a_i at time t. If such a confidence measure x_{ijt} is, say, above 0.8 then we can use rounding of x_{ijt} to 1 to get an integer solution. While such use of rounding is commonly criticized that it does not lead to the optimal solution, but its deviation from the optimal non-integer solution can be estimated and a suboptimal reasonable solution can be produced for large datasets using the classical LP techniques.

2.2 Generalization of ILP SRM Models to Uncertain Numbers

The natural generalization is coming from the fact that some coefficients and flags are uncertain in ILP SRM models. This uncertainty can be modeled by defining coefficients/flags as *uncertain numbers* given as: (1) intervals, (2) probability distributions, or (3) fuzzy sets. Respectively it leads to different classes of models and algorithms of optimization under uncertainty: *interval, stochastic,* or fuzzy optimization models and algorithms.

In the optimization under uncertainty the key issue is defining and justifying a way to sum up uncertain numbers (summands in the ILP formulation). The definition of the sum depends on assumptions and goals. Below we consider three categories them.

Case 1: (classical interval math): All points in [a, b] interval are equally belong to this interval. We are interested only in the low and upper limits of the sum:

$$[a, b] + [c, d] = [a + c, b + d].$$

Case 2: (discrete pdf math): All discrete points are distributed in the [a, b] interval, that is, for all x and y in [a, b], probabilities p(x) and p(y) are given and independent from the discrete pdf on [c, d]. We are interested in the distribution p(w) of sum points in the sum interval [a + c, b + d] including most likely sums:

$$p(w) = \sum_{x+y=w} p_{[a,b]}(x) p_{[c,d]}(y) \quad \text{for all } x + y = w$$

Case 3: (fuzzy math): All points in [a, b] are given with fuzzy logic membership function $m_{[a,b]}(x)$ in [a, b] interval and all points in [c, d] are given with fuzzy logic membership function $m_{[c,d]}(x)$. We are interested in getting a membership function of the sum points in [a + c, b + d] interval including a most possible sum.

Zadeh [40] asserts that case 3 must be solved by applying his *extension principle*, described in his 1965 paper, because *fuzzy math* is based on this principle:

$$m_{[a,b]+[c,d]}(w) = \min \, \max \left(m_{[a,b]}(x), \, m_{[c,d]}(y) \right)$$

for all x, y such that x + y = w, where x and y are from [a,b] and [c,d], respectively. The discussion at BISC in 2014 revealed disagreement on case 3 because the reference to the *extension principle* is not sufficient to justify the minmax formula above. This formula has a status of the hypothesis in general and in ILP SRM models with uncertain numbers given my membership functions in particular. Our approach for the case 3 is an adaptation of the case 2 [14].

2.3 Computation of Flags—Parameters of ILP Problem

Computing flags requires translating *requirements for tracking* in the form of track accuracy and the uncertainty into the sensor capabilities in terms of the flags. The similar translation is needed for the *discrimination requirements*. This *translation* allows a user flexibility to use both measurement information and tacit expert knowledge.

The binary flag $f(a_i, s_j, t, r)$ indicates whether sensor s_j is *capable* to *cover/observe* area a_i at the time interval t with required resolution/capability r. This flag is computed directly for each value of its variables: the identified areas a_i, sensor s_j, and required resolution/capability r at time t. LOS and FOV with their errors for sensor s_j at time t, location of the AOI a_i are used to check if a required resolution/capability can be reached. If the AOI a_i contains a detected moving object then the known dynamic properties of the trajectory of the object are used to identify the next location of the AIO a_i and to compute flag f for time $t + 1$. These properties are derived from the external tracking, track correlation algorithms as outlined in Fig. 1. Similarly each sensor resolution relative to objects is derived along other required for the model characteristics.

The flag $f^*(a_i,s_j,t,r,v)$ is directly computed from $f(a_i,s_j,t,r)$, when an area a is marked for a sensor s_j. The marking of the area is identified from tasking requirements and tasking doctrine. Flags $\{g_k(s_j)\}$ that indicate types of sensors are computed from the database of specifications of sensors. Relations flags are computed based on the definitions of the relations.

Flags can be assumed to be known or computed to achieve the required resolution of the sensor system. This includes the Monte-Carlo Simulation and the CRBs. The CRBs is computed for each object and sensor, as well as for each object and a pair of sensors as required. It can also be done by using tracking and fusion and algorithms outlined below.

2.4 Solutions with CRB and Dynamic Logic

Lambert and Sinno [21] discuss in detail that significant part of errors in tracking and fusion might originate from incorrect associations between sensors and objects; and therefore their results using CRBs that do not account for associations are only approximate. The proposed OPTIMA system uses CRBs with algorithms for tracking and fusion accounting for the association part of these problems obtained in [27, 28].

It is well known that the best algorithms currently used for tracking and fusion in difficult conditions cannot attain the best theoretically possible performance as specified by the Cramer-Rao Bounds for the given difficult conditions [28]. This deficiency is due to high *computational complexity* of current tracking and fusion algorithms [28, 29]. This limits sensor resource utilization. This fundamental difficulty of algorithms currently in use has been overcome by a computational intelligence *dynamic logic* approach [15, 27].

Dynamic logic starts not from the actual LP model M, but modifies both the objective function and the constraints of M to produce a model M_1, and then solves M_1 as a solution for M. The full dynamic logic methodology process has multiple stages that *generate dynamically a sequence of models* $M_1, M_2, \ldots, M_n$, where only model M_n is a solution of M. Models $M_1, M_2, \ldots, M_{n-1}$ provide only intermediate solutions.

Some algorithms for the ILP problem exploit the general idea, which is a core of the dynamic logic approach described above. The LP relaxation algorithm is in this category, when some constraints are removed and the objective function is modified by adding a penalty summand. LP decomposition methods also modify LP models [37].

The cutting plane algorithm [6] creates model M_2 by adding linear constraints (called cuts) to the relaxed model M_1 to drive the solution to be integer. The relaxed model M_1 removes constraints that variables are integers. The cut removes the current non-integer solution from a set of feasible solutions to the relaxation. This process of constructing new models M_i is repeated until an optimal integer solution is found in model M_n. However, these methods *do not change the dimensionality* of the search space and variables beyond converting discrete variables into continuous ones. The Dynamic Logic approach expands this by pointing out this underused opportunity. This is a fundamentally biologically inspired approach within the Computational Intelligence paradigm.

One of the drawbacks of the heuristic computational intelligence approach is in the difficulties of estimating how far its solution is from the optimal one. Another drawback is in the uncertainty of the situation, when this algorithm does not find any solution. In this case we do not know whether the optimal solution does not exist or just was not found. This is a motivation for developing specialized computational intelligence algorithms for the SRM integer linear programming models.

The dynamic logic idea for solving the SRM ILP model is finding some preliminary candidate solutions, then adding more constrains, and getting more accurate solutions. In this process we change an optimization criterion to find all feasible solutions that are under the specified constraints, and then solving the task again for the feasible solutions under the new constraints.

Within the dynamic logic methodology the original objective function F is not used at the beginning of the process as an objective function but as a source to construct a new objective function F_1. Similarly, the original constraints C are used to build new constraints C_1. For instance, we can solve the classical LP problem by removing a constraint that all variables are binary and then search for the binary solution using the classical solution as guidance.

Similarly the objective function F can be modeled much "rougher" by substituting coefficients by their interval estimates, e.g., a = 5 is substituted by the interval [4, 6] in function F_1. For instance, if F_1 is computed in way where positive coefficients are taken with their min value from their interval while negative coefficients are computed with their max value, we will get a lower estimate for both F and F_1. Similarly we can get upper estimates for these functions. In extreme cases if interval of the coefficient is [0, 10] then low limits means removing this coefficient from the constraint, which simplifies computations.

3   Applications, Related and Future Work

This paper had shown conceptual advantages of the proposed SRM methodology and the models relative to known approaches. Known approaches focus on: (1) tightly integrate SRM with tracking and (2) use a single scalar efficiency criterion such as Information Gain (IG).

Does it ensure that the new methodology and models will outperform known IG-based approaches and methods in applications? This question assumes that a *single performance criterion* exists that allows us to ask about *outperformance*. The main point of our methodology is that this assumption is an extreme simplification of the SRM situation. In many SRM tasks it is illusion that such single criterion exists.

On the other side when a single criterion such as IG is well justified not just postulated for an SRM task, the question of outperformance is an *incorrect* question too. How other methods can outperform IG if the performance criterion is IG? The criterion of the outperformance must differ from IG to be able to talk about the outperformance.

This is exactly the core concept of the proposed methodology—making selection of criteria a part of the methodology that first decouples SRM from tracking and discrimination. At the best of our knowledge this is the first systematic attempt of this kind in SRM. Therefore the direct comparison "apple to apple" with existing methodologies is not practical at this moment. When more methodologies of this kind emerge it will be a base for an "apple to apple" comparison.

There are several aspects of the SRM that can be incorporated into the proposed methodology. One of them is distributed network coordination. Often it leads to the tradeoff between the rigor of the mathematical formulation of the SRM task and lack of information for finding the optimum. In other words, it brings additional optimization criteria such as minimization of communications, and the increase in system robustness.

An idea of distributed network coordination is discussed in [32, 34] as a way to minimize communications and increase system robustness. The argument is that in the traditional approach sensors accept tasking orders from networked tracking elements that may have only *uncertain knowledge about sensor's capability*, limitations and other tasks such as self-defense. In particular, a real-time fuzzy control algorithm in [32, 33] running on each UAV gives the UAV limited autonomy allowing it to change course immediately without consulting with any other entity. In a similar development a resource manager based on fuzzy logic is optimized by evolutionary algorithms.

In [32, 33] a fuzzy logic resource allocation algorithm enables UAVs to automatically cooperate. The algorithm determines the trajectory and points each UAV for measurements. This fuzzy logic model takes into account the UAVs' risk, risk tolerance, reliability, mission priority for sampling, fuel limitations, mission cost, and other uncertainties. While the scope of this work differs from our task, expanding our optimization design to accommodate such factors as mission cost and related uncertainties is one of the topics of the future work. It is also an important direction in further development of fuzzy optimization.

The discrete optimization formulation for large-scale sensor selection in decentralized networks is proposed in [34]. It considers a situation without central fusion center. Each Fusion Center (FC) communicates only with the neighboring FCs. Our model can be expanded for this situation too.

The methodology and the OPTIMA class of the SRM optimization models are quite universal. Therefore, the development and application of multi-sensor fusion systems based on these models opens significant opportunities for detection and classification in a wide range of areas from bio-surveillance, monitoring, fault diagnostics, medical diagnosis, to cargo inspection, inspection of infrastructure, and others.

Optimization of the wireless sensors and phone communications is one of them to increase the efficiency of communications. Others include environment monitoring, and management of any business resources: mapping of fires, detecting and mapping pollutions, air-quality, water-quality by a network of distributed sensors. Recent accidents with high speed trains motivate development of the SRM in this area [8]. Maturing and integrating SRM models and algorithms will make solution of the above outlined problems more efficient.

The resource saving is turned into more accurate monitoring to save electricity, water, paper, etc. Networked, temperature sensors can automatically map insulation leaks in buildings and reduce energy waste. Temperature sensors, just like radars have coverage areas, sensitivity diagrams, etc. The ease of installation leads to the sensors that discover each other and communicate their measurements. This requires intelligent and adaptive algorithms such as OPTIMA. The same applies to irrigation, humidity, insect, soil chemical composition, etc. sensors in agriculture. Large number of low-cost, solar-powered, mesh-networked humidity and other sensors placed on a plantation can be optimized to help farmers save water, increase crop yields and lower cost.

4   Conclusion

Optimal SRM opens the opportunity: (1) to maximize the available sensor resources for search, (2) to optimize sensor resources for tracking, and, (3) to better defend the high priority assets. The models and algorithms proposed in this work allow the decreasing of the overall sensor resource usage, while increasing the probability that all threat objects in a raid are tracked. In addition, target characterization (discrimination) is optimized by using the same class of model but with different parameters (flags) that are specific for discrimination. Our unique approach is in multi-objective SRM optimization model and algorithms, as well as in the use of Cramer-Rao Bounds (CRBs), and the algorithms accounting for the association part of the tracking and fusion problem. These CRBs allow the evaluation of target characterization (classification features), and therefore target values. Another uniqueness of our approach is in using the flags within the SRM, which encompass all of the information external to the main goals of the program (such as the information from tracking algorithms). These flags are readily computed from the available information or information adaptively estimated in real time. These benefits surpass existing state of the art and permit efficient sensor coordination.

References

1. Abbass, H.A., Bender, A.: The Pareto operating curve for risk minimization. Artif. Life Robot. J. Springer **14**(4), 449–452 (2009)
2. Burgess, D., Levins, C.: Intelligent sensor resource management using evolutionary computing techniques. In: Integration of Knowledge Intensive Multi-Agent Systems/KIMAS, pp. 325–329 (2003)
3. Borndörer, R.M.: Designing telecommunication networks by integer programming (2012)
4. Bar-Shalom, Y., Li, X.R., Kirubarajan, T.: Estimation with Applications to Tracking and Navigation. Wiley, New York (2001)
5. Castanon, D.A.: Approximate dynamic programming for sensor management. In: Proceedings of the 36th IEEE Conference on Decision and Control (1997)
6. Cornuejols, G.: Valid inequalities for mixed integer linear programs. Math. Program. Ser. B 112, 3–44 (2008)
7. Junger, M., et al.(eds): 50 Years of Integer Programming 1958–2008. Springer, Berlin (2010)
8. Junji, K., Makoto, S., Akihiko, S., Kentarou, H., Yutaka, S., Koichi, N., Takashi, M., Tetsuo, K.: Development of 100 mbps-ethernet-based train communication network. In: 9th World Congress on Railway Research, (5), 1–12 (2011)
9. Hall, D., Llinas, J.: An introduction to multisensor resource management. Proc. IEEE **85**(1), 6–23 (1997)
10. Hanlon, N., Cohen, K., Kivelevitch, E.: Sensor resource management to support UAS integration into the national airspace system. In: AIAA (2015) http://dx.doi.org/10.2514/6.2015-0360
11. Hero, A., Cochran, D.: Sensor management: past. Present Future. IEEE Sensors J. **11**(12), 3064–3075 (2011)
12. Hero, A., Keucher, C., Blatt, D.: Information theoretic approaches to sensor management. In: Hero, A., Castanon, D., Cochran, D., Kastella, K. (eds.) Foundations and Applications of Sensor Management, New York: Springer (2008) 33–57
13. Hero, A., Castanon, D., Cochran, D., Kastella K., (eds.): Foundations and Applications of Sensor Management. Chapter 3, pp. 33–58. Springer (2007)
14. Kovalerchuk, B.: Summation of Linguistic Numbers, In: NAFIPS IEEE, (2015) (in print)
15. Kovalerchuk, B., Perlovsky, L., Wheeler, G.: Modeling of phenomena and dynamic logic of phenomena. J. Appl. Non-classical Logics **22**(1), 51–82 (2012)
16. Kovalerchuk, B.: Interpretable fuzzy systems: analysis of T-norm interpretability. In: 2010 IEEE World Congress on Computational Intelligence, Barcelona (2010) doi:10.1109/FUZZY.2010.5584837
17. Kovalerchuk, B., Triantaphyllou, E., Despande, A.S., Vityaev, E.: Interactive learning of monotone boolean functions. Inf. Sci. **94**(1–4), 87–118 (1996)
18. Kreucher, C., Hero, A., Kastella, K., Morelande, M.: An information-based approach to sensor management in large dynamic networks. Proc. IEEE **95**(5), 978–999 (2007)
19. Kreucher, C., Hero, A., Kastella, K.: A comparison of task driven and information driven sensor management for target tracking. In: 44th IEEE Conference on Decision and Control, pp. 4004–4009 (2005)
20. Kreucher, C., Hero, A., Kastella, K., Chang, D.: Efficient method of non-myopic sensor management for multi-target tracking. In: 43rd IEEE Conference on Decision and Control, pp. 722–727 (2004)
21. Lambert, H.C., Sinno, D.: Bioinspired Resource Management for Multi-Sensor object Tracking Systems. MIT Lincoln Laboratory Project Report MD-26 (2011)
22. Liggins, M., Hall, D., Llinas, J. (eds): Handbook of Multisensor Resource Management. CRC, Boca Raton (2008)
23. Manyika, L., Durrant-Whyte, H.: Data Fusion and Sensor Management: A Decentralized Information-Theoretic Approach. Ellis Ilorwood, New York (1994)

24. Pastor, J.: Mathematical Ecology of Populations and Ecosystems. Wiley-Blackwell, Chichester (2008)
25. Patsikas, D.: Track Score Processing of Multiple Dissimilar Sensors. NPS, Monterey CA (2007)
26. Perillo, M., Heinzelman, W.: Sensor Management. Wireless Sensor Networks, pp. 351–372. Springer, Berlin (2004)
27. Perlovsky, L., Deming, R.W.: Maximum likelihood joint tracking and association in a strong clutter without combinatorial complexity. Int. J. Adv. Robot. Syst. **9**, 1–9 (2013)
28. Perlovsky, L.: Cramer-Rao bound for tracking in clutter and tracking multiple objects. Pattern Recogn. Lett. **18**(3), 283–288 (1997)
29. Perlovsky, L.: Conundrum of combinatorial complexity. IEEE Trans. PAMI **20**(6), 666–670 (1998)
30. Severson, T., Paley, D.: Optimal sensor coordination for multiobject search and track assignment. IEEE Trans. Aerosp. Electron. Syst. **50**(3), 2313–2320 (2014)
31. Shea, P., Kirk, J., Welchons, D.: Adaptive sensor management for multiple missions. In: Proceeding of SPIE 7330, Sensors and Systems for Space Applications III, 73300M, doi:10. 1117/12.818892 (2009)
32. Smith III, J.F., Nguyen, T.V.H.: Distributed autonomous systems: resource management, planning, and control algorithms. Proc. SPIE vol. 5809, p. 65 (2005)
33. Smith III, J.F., Nguyen, T.V.H.: Fuzzy Logic Based UAV Allocation and Coordination, p. 11. DTIC, Naval Research Lab Washington, Washington (2006)
34. Tharmarasa, R., Kirubarajan, T., Sinha, A., Lang, T.: Decentralized sensor selection for large-scale multisensor-multiobject tracking. IEEE Trans. Aerosp. Electron. Syst. **47**(2), 1307–1324 (2011)
35. Tian, X., Bar-Shalom, Y.: Exact algorithms for four track-to-track resource management configurations: all you wanted to know but were afraid to ask. In: 12th International Conference on Information Fusion, pp. 537–554. Seattle, USA (2009)
36. Van Gils, J.: Digestive bottleneck affects foraging decisions in red knots Calidris canutus. I. prey choice. J. Anim. Ecol. **74**(1), 120–130 (2005)
37. Vanderbeck, F., Wolsey, L.: Reformulation and decomposition of integer programs. In: Junger, M. Liebling, T. et al. (eds) 50 Years of Integer Programming 1958–2008, pp. 431–504. Springer, Berlin (2010)
38. Verlinden, C., Wiley, R.H.: The constraints of digestive rate: An alternative model of diet selection. Evol. Ecol. **3**(3), 264–273 (1989)
39. Weir, B.S., Sokol, T.M.: Radar coordination and resource management in a distributed sensor network using emergent control. Proc. SPIE **7350**, 73500I (2009)
40. Zadeh, L.: Berkeley Initiative on Soft Computing/BISC group, http://mybisc.blogspot.com Accessed 04 July 2014

Bio-Inspired Topology Control Mechanism for Unmanned Underwater Vehicles

Jianmin Zou, Stephen Gundry, M. Umit Uyar, Janusz Kusyk
and Cem Safak Sahin

Abstract Unmanned underwater vehicles (UUVs) are increasingly used in maritime applications to acquire information in harsh and inaccessible underwater environments. UUVs can autonomously run intelligent topology control algorithms to adjust their positions such that they can achieve desired underwater wireless sensor network (UWSN) configurations. We present a topology control mechanism based on particle swarm optimization (PSO), called 3D-PSO, allowing UUVs to cooperatively protect valued assets in unknown 3D underwater spaces. 3D-PSO provides a user-defined level of protection density around an asset and fault tolerant connectivity within the UWSN by utilizing Yao-graph inspired metrics in fitness calculations. Using only a limited information collected from a UUV's neighborhood, 3D-PSO guides UUVs to make movement decisions over unknown 3D spaces. Three classes of applications for UWSN configurations are presented and analyzed. In 3D encapsulation class of applications, UUVs uniformly cover the underside of a maritime vessel. In planar distribution class of applications, UUVs form a plane to cover a given dimension in 3D space. The third class involves spherical distribution of UUVs such that they are uniformly distributed and maintain connectivity. Formal analysis and experimental results with respect to average protection space, total underwater movement, average network connectivity and fault tolerance demonstrate that 3D-PSO is an efficient tool to guide UUVs for these three classes of applications in UWSNs.

Keywords Particle swarm optimization · Self-organizing networks · Topology control · Node spreading · Underwater wireless sensor networks · Unmanned underwater vehicles

J. Zou (✉) · M. Umit Uyar
The City College of New York, New York, USA
e-mail: jzou@ccny.cuny.edu

S. Gundry · J. Kusyk
United States Patent and Trademark Office, Alexandria, VA, USA

C.S. Sahin
MIT Lincoln Laboratory, Lexington, MA, USA

© Springer International Publishing Switzerland 2016
R. Abielmona et al. (eds.), *Recent Advances in Computational Intelligence in Defense and Security*, Studies in Computational Intelligence 621,
DOI 10.1007/978-3-319-26450-9_26

1 Introduction

Recent developments in *unmanned underwater vehicles* (UUVs) have created a revolution in oceanographic science, commercial exploration, and military operations.
Improved battery technology and lower-power consuming electronics allow UUVs to
operate longer in underwater environments and explore larger spaces. Meanwhile,
the reduced cost and increased sophistication of unmanned vehicles have made it
possible for UUVs to cooperatively coordinate their efforts towards solving complex
tasks. Guided by robust network topology control algorithms, teams of unmanned
vehicles can efficiently use available resources and information to explore or protect
areas of interest during various civilian and military tasks. UUVs, which typically
are able to navigate freely within a three-dimensional underwater space, may be
equipped with sensors and communication equipment to monitor its surroundings.
Driven by sea-based economic development, the security of harbors, maritime vessels, and submarines have become important topics. Hostile underwater acts against
coastal facilities, ships, and submarines may be prohibitively difficult to continuously observe with human operators. Therefore, deploying UUVs to quickly detect
possible threats to maritime targets is a practical necessity. Once a UUV detects a
potential threat, it reports to a data collection point via a self-organized underwater wireless sensor network (UWSN) and allows security forces to take appropriate
countermeasures.

1.1 Challenge of Topology Control in Underwater Conditions

Due to dynamically changing underwater conditions, it may be infeasible to use a
central controller to guide the behavior of UUVs for underwater tasks. In realistic applications, each UUV needs to independently make intelligent choices about
its next movement location to protect critical assets. Autonomous topology control
algorithms allow UUVs to adapt their movements while basing their decisions only on
local information extracted from their surroundings. These algorithms do not require
a centralized infrastructure and, thus, eliminate reliance on unpredictable and under-
provisioned wireless networks. Another challenge for UUVs operating autonomously
in harsh environments is that there is often a high probability of becoming disabled
due to malfunction or attack. Therefore, it is an important goal of a UWSN is to provide a fault-tolerant topology control mechanism that guides UUVs to automatically
adjust their locations to provide sensor coverage for disabled vehicles.

1.2 Our Existing Research

In the earlier stages of our work, Sahin et al. [1] introduced a force-based genetic
algorithm (FGA) to make intelligent decisions for unmanned mobile vehicles operating in two dimensional spaces. Urrea et al. [2] implemented dynamical system model

for (FGA) demonstrating that an autonomous vehicle utilizing our FGA will move-ment decisions will improve uniformity among all nodes deploying themselves over an unknown two-dimensional area. This work was extended by using a homogeneous Markov chain model to systematically demonstrate that groups of autonomous vehi-cles guided by our FGA have a high probability of achieving desirable geometric con-figurations [3]. To avoid selfish and less efficient behavior of mobile nodes utilizing an elitist approach, a new algorithm was presented by combining game theory and FGA in [4, 5]. In [6], a GA-based topology control algorithm called 3D-GA was intro-duced to guide the mobile vehicles moving in three dimensional space. We show a three-dimensional particle swarm optimization (PSO) based topology control mech-anism, called 3D-PSO [7], to guide the UUVs operating in harsh 3D environments. We formally proved that 3D-PSO is able to guide UUVs toward a uniform distribution even when there are significant errors in location information from neighbors. Initial versions of asset protection algorithms for UWSNs appeared in [8].

1.3   Contribution of This Article

Building on our existing research, we introduce a new topology control mecha-nism that can provide a user-defined level of protection density and fault tolerant connectivity by integrating shield density and Yao graph [9] inspired metrics into our fitness calculations. This new topology control mechanism is then applied to a class of problems where network sensor protection of valuable water-based assets is required. We prove that our algorithm is highly robust and can be an effective solu-tion for 3D encapsulation, and planar and spherical protection classes of problems with civilian and military applications such as protection of maritime vessels, har-bors, and submarines. Despite the fact that the geometry of the classes of problems varies significantly, each UUV running our 3D-PSO can autonomously make mean-ingful next movement decisions based solely on the information obtained from other UUVs located within its communication range. We evaluate the performance of our 3D-PSO by measuring the percentage of a target surface or volume that UUVs cover, the total distance traveled by all vehicles, and the average connectivity of the UWSN. Formal analysis and simulation experiments demonstrate that our 3D-PSO algorithm can form a fault tolerant UWSN with a user-defined level of UUV protection around the assets.

1.4   Article Organization

The rest of this paper is organized as follows. Currently reported research on UUVs, area protection, and topology control for UWSNs is summarized in Sect. 2. Our 3D-PSO for UUVs moving in a three dimensional space to protect assets are given in Sect. 3. Section 4 presents the fitness function for 3D-PSO and a formal analysis of its coverage and connectivity properties. Performance metrics of our topology control mechanism are defined in Sect. 5. The results of our simulation experiments are presented in Sect. 6.

2 Related Work

Early UUV technology was very limited and often required line of site remote control for underwater vehicles. During the post World War II era, technologies related to UUVs advanced significantly. In 1957, Stan Murphy created the first autonomous underwater vehicle (AUV) [10]. For the following years, UUVs have continued to progress through various preliminary stages. Notably, during the late 1990s, Draper Laboratory created two testbeds for the U.S Navy that were significant in the development of many enabling technologies for underwater applications [11].

Although UUVs have been widely used in many fields, there are still some limitations hindering their widespread deployment. One of the most important challenges is to develop topology control algorithms that are able to efficiently guide the movement of UUVs. Underwater environmental conditions may change continuously, and therefore, UUVs need to adjust their positions using topology control algorithms that are able to accommodate for these changes. Specifically, topology control for UUVs in a UWSN which has been studied in different contexts. Rodoplu and Meng [12] introduce a GPS guided location based network topology control algorithm to reduce the energy consumption of UUVs. Li et al. [13] extend the findings of [12] and introduce a cone based distributed topology control algorithm, which only requires the directional information of nodes and can reduce the transmission power consumed by mobile nodes. In [14], a topology control mechanism based on the various aspects of graph theory is proposed to adjust transmission power levels and maintain a minimum k-connectivity which prevents network partitions in a 3D environment. Chen et al. propose a graph theoretic algorithm to find a balance between a sparse and dense k-connected graph called the Interference Prediction Based Topology Control algorithm (IPBTC) for 3D wireless sensor networks [15].

3 Our 3D-PSO

In this section, we present our particle swarm optimization based algorithm, called 3D-PSO, to guide the movement of UUVs operating in a UWSN. In our mobility model, each UUV can move freely and autonomously in 3D Cartesian space and is equipped with a sensor device that can monitor a limited area around a UUV. A set of UUVs can work together to sense the approach of hostile intruders that attempt to come near protected assets. Once a UUV detects an unknown object in its sensing area (e.g., a submersible explosive device, underwater surveillance vehicle, etc.), it will report the information back to a data collection point such as a surface vessel or terrestrial based data center. UUVs must be able to adapt to ever-changing underwater environmental conditions and maintain connectivity in a UWSN so that urgent sensing information can be returned to a data collection point in a timely manner.

Initially, 3D-PSO will randomly generate n_s candidate solutions, called particles, inside N_i's movement area R_{mov}. Here, n_s is the total number of candidate solutions in the swarm and R_{mov} is the maximum distance that a UUV can move in a single step. Each particle in the swarm represents a potential solution that contains the speed and direction of the next movement for N_i. Using sensors or communication devices, a UUV can determine the positions of its neighbors, which are located inside its communication area R_{com}. The fitness function, presented in Sect. 4 for 3D-PSO has been developed to evaluate the quality of potential movement locations. Solely based on information from the neighbors, this fitness function will evaluate the goodness of each particle and calculate the velocity of each particle for the next generation of movement decisions. Once the maximum number of generations ($iter_{max}$) is reached, node N_i will select the particle location with the best fitness as its next position to move (only if the fitness is better than the fitness of its current location).

Let $L_a(\tau)$ represent the position of particle a in the solution space at generation τ for $\tau \leq iter_{max}$. The position of particle a within the solution space is iteratively updated by adding a velocity vector v_a to the current position:

$$L_a(\tau + 1) = L_a(\tau) + v_a(\tau + 1) \tag{1}$$

The velocity vector for the next generation $v_a(\tau + 1)$ is calculated as:

$$v_a(\tau + 1) = \omega \times v_a(\tau) + C_1 \times \varphi_1 \times [pBest_a(\tau) - L_a(\tau)] + C_2 \times \varphi_2 \times [gBest(\tau) - L_a(\tau)] \tag{2}$$

where ω is the inertial weight, which linearly decreases over the course of all generations [16]. C_1 and C_2 are positive constants which represent the particle's confidence level to its own solution and the best solution found in the swarm, respectively. The variables φ_1 and φ_2 are uniformly distributed random numbers in the range of [0,1]. $pBest_a(\tau)$ is the solution with the best fitness that particle a has found as of generation τ. $gBest(\tau)$ is a solution with the best fitness for all particles found from first generation, as of generation τ. They are defined as:

$$F(gBest(\tau)) = min\{F(pBest_0(\tau)), \ldots, F(pBest_{n_s}(\tau))\} \tag{3}$$

where F is the fitness function to evaluate the goodness of a candidate solution and $gBest(\tau) \in \{pBest_0(\tau), \ldots, pBest_{n_s}(\tau)\}$.

4   Implementation of Fitness Function for 3D-PSO

Fitness functions (or, objective functions) are used in many heuristic computational techniques to evaluate candidate solutions. For our 3D-PSO, we use a force-based fitness function to evaluate each particle location for a UUV with respect to its distance to a protected object or area as well as the distance of a candidate location to its neighbors within R_{com}. The neighbors positioned closer to a node N_i will have larger virtual forces (i.e., less preferable) compared to the neighbors distanced farther away from N_i within R_{com}.

In this section, we first introduce two methods that our fault tolerance mechanisms that 3D-PSO uses to provide a user-defined level of asset protection. Next, we present three different types of fitness functions that guide UUVs to distribute themselves for different classes of applications such as 3D encapsulation, planar or spherical distributions.

4.1 Fault Tolerant Network Topologies

In general, the level of protection required for an asset is based on a tradeoff between the asset's level of importance and the likelihood of a fault. In the first fault tolerance method that 3D-PSO uses is based on *shield density* Ω, which is a user defined parameter indicating the preferred distance among UUVs in 3D encapsulation and planar distribution classes of applications. Figure 1 shows examples of UWSN topologies that have different shield densities, where For small values of Ω, UUVs will be closer to each other, whereas larger Ω values result in a less tightly packed node distributions.

The second method that we introduce for fault tolerance is based on Yao graphs [9] which is used 3D-PSO fitness function to control the number of neighbors for each UUV. Several geometrical structures using Yao graphs have been proposed for simple power adjustment problems in 2D and 3D sensor networks [17]. Figure 2a shows an example of a 2D Yao graph. Here, the circular region around a particle $a(\tau)$ within R_{com} at generation τ is divided into p equally spaced Yao partitions that do not inter-

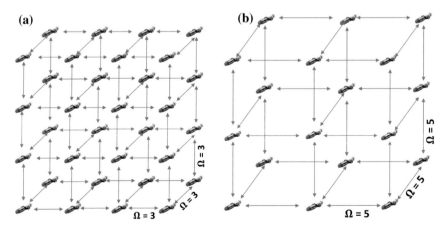

Fig. 1 Examples of shield densities for UUVs with **a** $\Omega = 3$, and **b** $\Omega = 5$

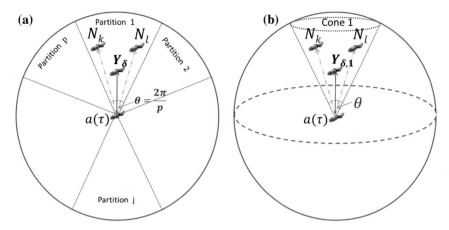

Fig. 2 Examples of Yao graph structure for **a** 2D, and **b** 3D spaces

sect with each other. The angle for each region is equal to $\frac{2\pi}{p}$. For 3D spherical protection, the fitness function inspired by 3D Yao graphs preserves a minimum desired
number of connections for a node with its neighbors. The initial 2-D version of this
approach has been studied in our previous work where we implemented a differential
evolution algorithm that utilized a Yao graph inspired fitness to uniformly distribute
mobile nodes over an unknown two dimensional area [18]. In our 3D-PSO, as shown
in Fig. 2b, the space for a particle $a(\tau)$ of node N_i within R_{com} has been separated into
several similarly shaped X Yao cones. The angle for each cone, θ, is a user-defined
parameter that can be adjusted in the range of $\left(0 < \theta \leq \frac{\pi}{2}\right]$. A cone x in which at
least one neighbor is located is called an *active cone*, $Y_{N_{i,x}}$ $(x = 1, \ldots, X)$. Let k be the
user-defined minimum number of active cones for a given UUV. If $Y_{\delta,x}$ is the closest neighbor for an active cone x of node N_i, let $d_{min,p}$ be the distance between the
closest neighbor and node N_i. In our 3D-PSO algorithm, we use only $Y_{\delta,x}$ neighbor
in each active cone x to calculate the fitness value of a candidate location. The line
between the closest detected neighbor, $Y_{\delta,x}$, and a candidate location, $a(\tau)$, is used
as the initial axis to create the Yao partition. Therefore, the orientation of the cones
for each candidate location is highly dynamic and changes in each generation of the
fitness calculation.

4.2 Fitness of 3D-PSO

The fitness F_i for each candidate solution for node N_i with σ_i neighbors is defined as
follows:

$$F_i = \begin{cases} F_{max} & \text{if } \sigma_i = 0 \\ min[F_{max}, \sum_j^{\sigma_i} f_{ij}] & otherwise \end{cases} \qquad (4)$$

where F_{max} is the maximum penalty applied to any location that causes a node N_i to disconnect itself from all its neighbors, and f_{ij} is the virtual force applied on N_i by its neighbor N_j. In this definition, smaller fitness values indicate fitter positions to move for a given node. Virtual force values applied to a node by its neighbors are calculated based on the class of problems as presented in the following sections.

In our earlier work [19], we proved that a UUV using 3D-PSO will only move to a location if this location improves its fitness:

Lemma 1 (Taken from [19]) *Using* 3D-PSO, *the fitness* F^{t+1} *of node* N_i *at time* $(t + 1)$ *is better than or at least as good as its fitness* F^t *at time* (t).

4.3 Fitness Function for 3D Encapsulation Class of Problems

The 3D encapsulation class of problems are defined as using unmanned vehicles to generate a parabolic surface of protection underneath a floating vessel to detect intrusions occurring below the vessel. Typical examples for this class of problem is protection for underside of ships, floating oil platforms, and other surface based assets. All UUVs running 3D-PSO will spread out the underside of an asset while maintaining a pre-defined distance Δ_ω from the asset surface throughout their deployment. The fitness f_{ij} for 3D encapsulation applications can be calculated as:

$$f_{ij} = \psi * F_{pen} + \begin{cases} F_{pen} & \text{if } \sigma_j = 1 \\ \frac{R_{com}}{d_{ij}} - \daleth & \text{if } 0 < d_{ij} < \Omega \\ \lambda * (d_{ij} - \Omega) * F_{pen} & \text{if } \Omega \le d_{ij} \le R_{com} \\ F_{pen} & \text{if } d_{ij} > R_{com} \end{cases} \tag{5}$$

where ψ is the difference between the pre-defined target distance to asset surface Δ_ω and the current distance between a candidate location and the underside of an asset. The variable d_{ij} is the Euclidean distance between the position of one of the candidate solutions and a neighboring node N_j. The total number of neighbors within the R_{com} for a neighboring node N_j is σ_j. The variable $\daleth$ is defined as $\frac{R_{com}}{\Omega}$ which incorporates the desired shield density into the fitness calculation. F_{pen} is a penalty fitness for a candidate location which would prevent a neighboring node N_j (with $\sigma_j = 1$) becoming disconnected from N_i or N_i moving too far from the asset surface. λ $(0 < \lambda < 1)$ is used to scale the fitness penalty for the shield density. Larger values of Δ_ω create a larger protection surface that allows for potential threats to be detected earlier, but require more UUVs to fully surround the asset.

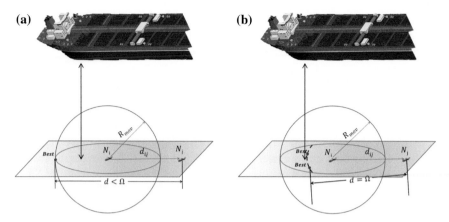

Fig. 3 Example of two UUVs separated such that **a** $d_{ij} + R_{mov} < R_{com}$, and **b** $d_{ij} + R_{mov} \geq R_{com}$ (grey shaded area represents the surface that is distanced Δ_ω from the protected surface)

Definition 1 In 3D-PSO the penalty fitness F_{pen} is bounded by F_{max} (i.e., $F_{pen} \leq F_{max}$).

Definition 2 In 3D-PSO, since two UUVs cannot occupy the same location, $d_{ij} \neq 0$, where $F_{pen} \gg \frac{R_{com}}{d_{ij}}$ for small d_{ij}.

The following lemma shows that a UUV at Δ_ω distance away from the surface of a target asset and with only one neighbor will not move further away from the asset surface.

Lemma 2 *If a node N_i with one neighbor N_j, autonomously running 3D-PSO for 3D encapsulation class of problems, reaches the user-defined distance of Δ_ω to the target surface at time (t), it will maintain this distance at time $(t + 1)$.*

Proof (Proof sketch) Typically, the desired distance between a node and a target surface (i.e., Δ_ω) is much larger than the movement distance for a single step of R_{mov}. For large target vessels, we can assume that the vessel surface is flat with respect to the size of a UUV. The surface inside of node N_i's movement space, which maintains a distance of Δ_ω units to the target surface, is shown in grey in Fig. 3. Based on Eq. (5), if the distance between N_i and its neighbor N_j is less than $(\Omega - R_{mov})$, the best movement choice for node N_i is the furthest point in the surface of movement which is depicted in Fig. 3a as *Best*. When $(d_{ij} + R_{mov}) > \Omega$, there exists an arc (marked as a dashed line in Fig. 3b) representing the next movement locations for node N_i which will result in zero virtual force between N_i and N_j. Our 3D-PSO will find the best location for N_i, which will be over the flat surface and Δ_ω units away from the target surface at time $(t + 1)$. Therefore, N_i will not move away from the asset surface.

Let us now consider the case where a node N_i is at a distance of Δ_ω units to a target surface and has multiple neighbors. We can prove that node N_i's movement away from the target surface is bounded:

Theorem 1 *Using* 3D-PSO, *once a node* N_i *with* σ_i *neighbors* ($\sigma_i > 0$) *is located* Δ_ω *units away from a target surface at a time* (t), *it will not move from the target surface more than* $\psi \leqslant \max[1, \lambda * (R_{com} - \Omega)]$ *units at time* ($t + 1$).

Proof When all candidate solutions for N_i at time ($t + 1$) are located Δ_ω units away from the target surface, N_i can select any of the candidates and still maintain the preferred distance to the target surface. However, if at least one of the candidate positions of N_i is located ($\Delta_\omega - \psi$) units away from the surface (as shown in Fig. 4), the neighbors of node N_i can be separated into two groups. The first group G_1 consists of all σ_{G_1} neighbors which are closer than Ω units (i.e., $N_j \in G_1$), whereas the second group G_2 includes σ_{G_2} neighbors that are further than Ω units away from N_i. Since N_i is located at Δ_ω units from the target surface at time (t), using Eq. (5), the fitness value for N_i, which is only based on its distance to its neighbors, can be calculated as:

$$
\begin{aligned}
F_i(t) &= \sum_{j \in G_1} f_{ij} + \sum_{k \in G_2} f_{ik} \\
&= \sum_{j \in G_1} \left(\frac{R_{com}}{d_{ij}} - 1 \right) + \lambda * F_{pen} \sum_{k \in G_2} (d_{ik} - \Omega) \\
&\leqslant \sigma_{G_1} * F_{pen} + \lambda * F_{pen} \sum_{k \in G_2} (d_{ik} - \Omega)
\end{aligned}
\tag{6}
$$

On the other hand, for a candidate location $L_a = \Delta_\omega - \psi$, the fitness is affected by both the virtual forces from neighbors and the distance to the target surface (Fig. 4). The fitness of L_a will determined as $F_{L_a}(t) = \sigma_i * \psi * F_{pen}$. Based on Lemma 1, it will only be possible for node N_i to choose the location L_a as its next location to move at time ($t + 1$), if $F_{L_a}(t) \leq F_i(t)$:

Fig. 4 Example where a candidate location for the next step is located between the node and the target surface (*grey shaded area represents the surface that is distanced* Δ_ω *from the protected surface*)

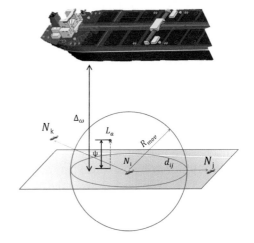

$$\Rightarrow \sigma_i * \psi * F_{pen} \leqslant \sigma_{G_1} * F_{pen} + \lambda * F_{pen} \sum_{k \in G_2} \left(d_{ik} - \Omega \right)$$

$$\Rightarrow \sigma_i * \psi \leqslant \sigma_{G_1} + \lambda * \left(\sum_{k \in G_2} d_{ik} - \sigma_{G_2} * \Omega \right)$$

$$\Rightarrow \sigma_i * \psi \leqslant \sigma_{G_1} + \lambda * \left(\sum_{k \in G_2} R_{com} - \sigma_{G_2} * \Omega \right)$$ (7)

$$\Rightarrow \sigma_i * \psi \leqslant \sigma_i - \sigma_{G_2} + \lambda * \left(\sum_{k \in G_2} R_{com} - \sigma_{G_2} * \Omega \right)$$

When σ_{G_1} or σ_{G_2} equals to zero, Eq. (7) has its maximum value:

$$\sigma_i * \psi \leqslant \begin{cases} \sigma_i & \text{if } \sigma_{G_2} = 0 \\ \lambda * \sigma_i (R_{com} - \Omega) & \text{if } \sigma_{G_1} = 0 \end{cases}$$ (8)

Therefore, we prove that at time $(t + 1)$, ψ is bounded by 1 or $\lambda * (R_{com} - \Omega)$, whichever is greater.

Theorem 1 shows that the maximum value of ψ is bounded by $\max[1, \lambda * (R_{com} - \Omega)]$. A UUV cannot move away from the target surface at any time once it reaches the surface Δ_ω units away from the target. Based on Theorem 1, we can state the following corollary for the 3D encapsulation class of problems:

Corollary 1 *For* 3D *encapsulation class of problems, once a node N_i running* 3D-PSO *is located at Δ_ω units away from the target surface (i.e., $\psi = 0$) at time (t), the distance between N_i and a target surface is bounded by $\Delta_\omega \pm \max[1, \lambda * (R_{com} - \Omega)]$ units at time $(t+1)$.*

4.4 Fitness for Planar Distribution Class of Problems

Planar distribution class of problems represents the applications where mobile nodes must be distributed along one Cartesian plane of a 3D space. This type of protection is used for situations where the boundary between hostile and safe spaces can be defined along a single plane. A typical example of this would be port or harbor protection applications, where maritime surface-based or submersible vessels would need to go through the entrance of a harbor, and afterwards the harbor entrance needs to be sealed by the UUVs to protect the harbor from hostile entities located outside. In this class of applications, UUVs utilizing our 3D-PSO will automatically adjust their locations over the planar surface at the entrance of the harbor to detect potential intruders. We show in this section that, our 3D-PSO is not confined to any predetermined axis and the protection plane can be oriented arbitrarily. The fitness f_{ij} for a candidate location for node N_i is based on its distance to the protection plane δ_p and the virtual forces inflicted by its local neighbor N_j, which is calculated as:

$$f_{ij} = \mathfrak{y}_i * F_{pen} + \begin{cases} F_{pen} & \text{if } \sigma_j = 1 \\ \frac{R_{com}}{d_{ij}} - \mathbf{1} & \text{if } 0 < d_{ij} < \Omega \\ \lambda * (d_{ij} - \Omega) * F_{pen} & \text{if } \Omega \le d_{ij} \le R_{com} \\ F_{pen} & \text{if } d_{ij} > R_{com} \end{cases} \qquad (9)$$

where $\mathfrak{y}_i$ is the norm distance of a candidate location to the δ_p plane.

Similar to Theorem 1, if a node N_i is located at the target plane (i.e., $\mathfrak{y}_i = 0$) at time (t), N_i will not move to a location further than $\max[1, \lambda * (R_{com} - \Omega)]$ units from the target plane at time $(t + 1)$:

Corollary 2 *For planar distribution class of problems, once a node N_i running 3D-PSO is located at target plane δ_p at time (t), its maximum movement away from δ_p is bounded by $\mathfrak{y}_i \le \max[1, \lambda * (R_{com} - \Omega)]$ units at time $(t + 1)$.*

In our pervious work [19], we prove that, using 3D-PSO, the mobile nodes will separate apart and that the sum of the distances among the nodes will only increase towards a uniform distribution or, if further improvement not possible, stay at the same locations as the node deployment progresses. This spreading will continue as long as the distance between neighboring nodes is less than the desired node density Ω, as stated by the following theorem:

Theorem 2 (Taken from [19]) *Let node N_i running 3D-PSO have σ_i neighbors $(\sigma_i > 0)$ and $F_i^{t+1} \le F_i^t$, then sum of distances between N_i and its neighboring nodes will not decrease at time $(t + 1)$ such that $\sum_{j=1}^{\sigma_i} d_{ij}^{t+1} \ge \sum_{j=1}^{\sigma_i} d_{ij}^t$ for $0 < d_{ij} \le \Omega$.*

For planar distribution class of problems, the following lemma shows that the virtual force inflicted on N_i by its neighbor N_j will be minimum when the distance between them is Ω units.

Lemma 3 *For a node N_i running 3D-PSO for planar distribution class of problems with σ_i neighbors $(\sigma_i > 0)$, the virtual force inflicted by a neighboring node N_j will be minimum if $d_{ij} = \Omega$.*

Proof If a position for N_i will not cause the degree of a neighbor node N_j to become zero (i.e., N_j will not be isolated by N_i's movement), using Eq. (9), the virtual force inflicted on N_i by its neighbor N_j is reduced as:

$$f_{ij} = \begin{cases} \frac{R_{com}}{d_{ij}} - \mathbf{1} & \text{if } 0 < d_{ij} < \Omega \\ \lambda * (d_{ij} - \Omega) * F_{pen} & \text{if } \Omega \le d_{ij} \le R_{com} \end{cases} \qquad (10)$$

where $(\frac{R_{com}}{d_{ij}} - \mathbf{1})$ is a monotonically decreasing function, while $(\lambda * (d_{ij} - \Omega) * F_{pen})$ is a monotonically increasing function. Therefore, when the distance to N_j is equal to Ω, N_i will receive a minimum virtual force from N_j.

The following theorem shows that UUVs autonomously running 3D-PSO for planar distribution class of problems will not move to locations farther than Ω units away from its neighbors.

Theorem 3 *Mobile nodes autonomously running* 3D-PSO *for planar distribution class of problems will not move to locations farther than Ω units away from its neighbors.*

Proof (Proof sketch) Based on Theorem 2, node N_i will move away from its neighbors to improve its fitness and, hence, the planar area coverage. Based on Lemma 3, the ideal distance between two neighbors is Ω units. The nodes will not move to the locations that do not improve their fitnesses as stated in Lemma 1. Therefore, N_i will move away from its neighbors until its distance to them is Ω units if the protection space is large enough.

4.5 Fitness for Spherical Distribution Class of Problems

For spherical distribution class of problems (e.g., fully submerged submarine protection applications), UUVs autonomously running our 3D-PSO surround the underwater asset and adjust their locations to maintain a spherical surface of protection with the asset at the center of the sphere. We implemented a 3D Yao graph inspired fitness function to provide fault tolerant connectivity and a user-defined level of protection density.

As discussed in Sect. 4.1, a Yao structure can be created by dividing the area around a mobile node into distinct regions. In 3D spaces, it is not possible to create Yao partitions that do not have overlapping space. For our 3D-PSO, we developed an algorithm, called 3D-YAO-PART, to construct conical Yao partitions using a similar approach given in [17] as presented in Alg. 1.

Algorithm 1 3D-YAO-PART: Creating conical Yao partitions for Node N_i

---
 1: Collect the neighbor positions for Node N_i
 2: **for** $i = 1$ to the total number of candidate locations n_s **do**
 3:     Sort neighbors N_j by the distance to the location of particle a such that $d_{L_a,N_j} \leq R_{mov}$ and $d_{L_a,N_j} \leq d_{L_a,N_{j+1}}$
 4:     For all σ_i neighbors, set Visited$[\sigma_i]$ = false
 5:     **for** $j = 1$ to σ_i **do**
 6:         **if** Visited$[N_j]$ = false **then**
 7:             Create a cone $\Upsilon_{N_{i,j}}$ using the line from a to N_j as the axis and θ as the angle of cone
 8:             Set Visited = true for all neighbors located in the cone $\Upsilon_{N_{i,j}}$
 9:         **end if**
10:     **end for**
11: **end for**

It is important to show that, even tough the cones overlap with each other, there is at least one unique UUV at each active cone. The following lemma shows that each active cone $Y_{N_i,x}$ of node N_i ($x = 1, \ldots, X$) will have a unique closest neighbor $Y_{\delta,x}$ to N_i.

Lemma 4 *Algorithm* 3D-YAO-PART *guarantees that, for a node N_i with σ_i neighbors, there are no two active cones $Y_{N_i,x}$ and $Y_{N_i,y}$ such that the respective closest neighbors $Y_{\delta,x}$ and $Y_{\delta,y}$ are the same.*

Proof (*Proof sketch*) In our 3D Yao partition algorithm 3D-YAO-PART, each sorted neighbor of N_i will be assigned into a cone, and all other neighbors located in the same cone will be associated with this cone. Suppose two active cones $Y_{N_i,x}$ and $Y_{N_i,y}$ have the same closest neighbor which is node N_j. In this case, both cones will have the same axis based on line 7 in Algorithm 1. Since the cone angle θ is the same in all cones, the cone $Y_{N_i,x}$ must be the same as $Y_{N_i,y}$ (i.e., two different active cones will always have different respective closest neighbors). Therefore, it is not possible to have two different active cones sharing the same closet neighbor to node N_i.

The total number of cones surrounding a submerged asset is bounded by the size of each cone's angle, as presented by the following lemma whose proof is presented in [17]:

Lemma 5 (Taken from [17]) *The angle β for any two cones must greater than $\theta/2$, and the maximum number of cones (i.e., X) is bounded by $2/[1 - \cos(\theta/4)]$.*

For fully submerged asset protection applications, our fitness function will guide the UUVs to create a spherical protection space surrounding the asset. Let $\mathfrak{R}$ be the radius of the protection sphere, and L_{asset} the location of the asset. The radius $\mathfrak{R}$ should be smaller than R_{com} so that UUVs can communicate with the protected asset and report information about potential intruders. The fitness function implemented in our 3D-PSO for spherical distribution applications contains two parts. The first part, called *distance fitness*, is determined based on a node's distance to the closest neighbor in each Yao cone. The second part of the fitness, called *critical fitness*, is used to maintain the minimum of k neighbors for each UUV. For node N_i with a neighbor N_j, if N_j has k or fewer neighbors, it will send a broadcast message to the nodes within R_{com}. Receiving this message, node N_i will assign a penalty fitness to all potential movement decisions that will result in N_j having less than k neighbors. This fitness function can be expressed as:

$$f_{ij} = \underbrace{\Lambda_{ij}}_{critical\,fitness} + \underbrace{\begin{cases} F_{pen} & \text{if } \sigma_j = 1 \text{ and } d_{min,p} > R_{com} \\ \frac{R_{com}}{d_{min,p}} - 1 + \varepsilon * | d_{i,\varsigma} - \mathfrak{R} | & \text{if } 0 < d_{min,p} < \Omega \\ \lambda * (R_{com} - d_{min,p}) * F_{pen} & \text{if } \Omega \le d_{min,p} \le R_{com} \end{cases}}_{distance\,fitness} \qquad (11)$$

$$\Lambda_{ij} = \begin{cases} 0 & \text{for } \sigma(i) \geq k \\ F_{pen} & \text{for } \sigma(i) < k \text{ or } d_{crit} > R_{com} \end{cases} \tag{12}$$

where the $d_{i,\varsigma}$ is the Euclidean distance between a candidate solution and the asset location L_{asset}, ε is a scale factor to guide vehicles toward the surface of the spherical protection space, Λ_{ij} is the critical fitness and d_{crit} is the distance between a candidate solution and the critical neighbor(s) ($\sigma(j) \leq k$).

The following theorem states that, using the fitness given in Eq. (11), UUVs running 3D-PSO will maintain at least k neighbors:

Theorem 4 *If a node N_i running* 3D-PSO *for fully submerged asset protection has k or more neighbors at time (t), N_i will maintain k-connectivity from time (t) until the end of the* UWSN *deployment.*

Proof (Proof sketch) From Eqs. (11) to (12), the critical fitness is assigned the maximum penalty if a potential movement location causes node N_i to have less than k neighbors. Also, the penalty fitness F_{pen} will be applied to potential locations that cause a node N_i to move away from its neighbor(s) N_j that rely on node N_i to maintain k connectivity. Lemma 1 guarantees that the mobile nodes will not move to locations which do not improve their fitnesses. Therefore, if node N_i can have k connectivity, it will maintain it throughout the deployment.

5   Our 3D-PSO Performance Metrics

In this section, we evaluate the performance of our 3D-PSO topology control mechanism with respect to: (i) *area protection coverage*, (ii) *total* UUV *movement*, and (iii) *average connectivity*.

5.1   Area Protection Coverage

Area protection coverage (APC) is the ratio of the area protected by UUVs to the total space to be protected [8]. It is important to note that the total protection space changes based on the application class. For example, in 3D encapsulation applications, the protection space can be any parabolic shape such as the underside of any given ship. For planar distribution, such as the harbor protection applications, the area of protection is the physical size of the port entrance whereas for fully submerged assets APC is a spherical surface that surrounds the entire target.

In order to quantify the coverage of the protected area by a UUV N_i, we create a projection from the coverage of N_i to the entire protection space and calculate the size of the protected area as P_i. This way, the overlapping coverage by multiple nodes are discounted. The percentage of the total space covered by all mobile nodes is given as:

$$\text{APC} = \frac{\bigcup_{i=1}^{\mathcal{N}} P_i}{S} \times 100 \tag{13}$$

where S is the total required protection space, and $\mathcal{N}$ is the total number of UUVs. When APC is 100 %, the space is fully protected.

5.2 Total UUV Movement

Since autonomous mobile nodes have limited energy resources, the total UUV movement (TUM) until achieving a uniform distribution is an important metric to quantify the performance of our algorithm. The location of N_i and its coordinates at time (t) can be defined as $L_i[t]$ and (x_i^t, y_i^t, z_i^t), respectively. Let $d(L_i[t-1], L_i[t])$ denote the distance travelled by N_i during a single step (i.e., from time $(t-1)$ to (t).) TUM(t) for $\mathcal{N}$ UUVs can be defined as:

$$\text{TUM}(t) = \sum_{i=1}^{\mathcal{N}} d(L_i[t-1], L_i[t]) \tag{14}$$

5.3 Average Connectivity

One of the fundamental problems in dynamic and harsh underwater environments is achieving and maintaining connectivity among the UUVs [18]. We use average connectivity for all UUVs to show the effectiveness of our algorithm. Average connectivity D_{avg} is defined as the mean node degree for all UUVs running 3D-PSO. This metric measures the amount of fault-tolerance existing for UUVs within a UWSN. The average connectivity at time (t) is given as:

$$D_{avg}(t) = \frac{\sum_{i=1}^{\mathcal{N}} D_i}{\mathcal{N}} \tag{15}$$

6 Simulation Experiments

We have developed a Java based simulation environment to evaluate the effectiveness of 3D-PSO algorithm. The MASON library [20] was used to visualize the experimental results generated by our software. In our experiments, the default size of the area of deployment is a cube with dimensions of (100, 100, 100). For all of the experiments, we defined the maximum movement range for a UUV as $R_{mov} = 5$. In order to reduce stochastic noise in the observed data, all experiments were repeated 30 times and the results were averaged.

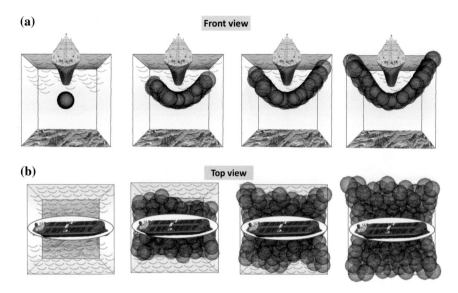

Fig. 5 Front and top views of encapsulation protection at steps 0, 10, 40, and final step of 200 with 120 UUVs and $R_{com} = 10$

Figure 5 depicts (from the front and top views) 120 mobile nodes which are providing the encapsulation type of protection, at steps 0, 10, 40, and finally at step 200 with $R_{com} = 10$. In Fig. 5, each shaded sphere represents the R_{com} of a UUV located at its center. Initially, all UUVs are deployed at an entry point under the ship. Without loss of generality, this step can be changed to an arbitrary exit point at the naval vessel from which all UUVs are initially dropped. Each UUV is guided by 3D-PSO to spread underneath the naval vessel. We can observe from Fig. 5 that the UUVs spread apart quickly during the initial steps, and that most of the space beneath the ship has been protected by the time 3D-PSO reaches step 40. After that, the UUVs continue to make small adjustments to find better locations and improve the coverage of the protection area.

In Fig. 6a, b, the protected coverage area and the total traveled distance are shown for R_{com} values of 8, 10 and 12. We can observe that UUVs almost cover the entire protection area for R_{com} values of 12 and 10 after running 3D-PSO for 80 steps. Approximately 90 % coverage is obtained with $R_{com} = 8$ at step 200. As can be seen from Fig. 6a, UUVs converge faster when R_{com} is larger. Note, however, that larger R_{com} implies more energy consumption by each node. For all cases, the most significant increase in APC occurs during the first 50 steps.

Figure 6b depicts that the total UUV movement decreases as each experiment progresses. At the beginning of an experiment, all nodes move apart quickly. As the experiment continues, UUVs move slower to make smaller adjustments to their positions. For example, TUM for each step decreases from 450 to 15 from step 0 to 200 for $R_{com} = 10$. Comparing the outcomes from three different R_{com} values, we see that

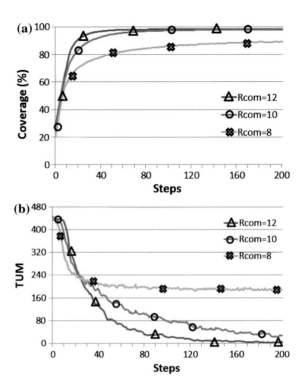

Fig. 6 APC (**a**) and TUM (**b**) for ship protection using 120 UUVs with $R_{mov} = 5$, and $R_{com} = 12$, 10, and 8

UUVs travel less when R_{com} is larger. The UUVs reach a stable distribution and stop moving at step 200 for $R_{com} = 12$. Similarly, the total UUV movement for all nodes in a single step is TUM = 190 for $R_{com} = 8$ after step 80, which does not diminish until the end of each experiment. This indicates that UUVs continue to move back and forth and try to achieve better locations. This result implies that each UUV guided by our algorithm considers a larger number of nodes in its neighborhood and makes more meaningful decisions when R_{com} is larger. However, UUVs need to spend more energy for larger R_{com} values. Considering that node movement is the most energy consuming operation, increasing the R_{com} is likely a better use of resources for many ship protection missions.

Figure 7a–c show the coverage, total UUV movement, and the average connectivity for different network density (Ω) values in 3D encapsulation class of protection problems. From the experiments, when Ω is larger, UUVs can cover larger spaces, and the coverage is 62, 80 and 99 % for $\Omega = 8$, 9 and 10 at step 200, respectively. The selected value of the shield density has a large effect on the protection coverage. For example, when the Ω value decreases from 10 to 9, the coverage is approximately 20 % smaller; however, the average connectivity D_{avg} for each UUV increases by almost 400 % (from 1 to 4) based on Fig. 7c. On the other hand, if we compare $\Omega = 9$ and 8, the difference in coverage is approximately 17 %, and the D_{avg} increases by almost 25 % (from 4 to 5). Figure 7b shows that UUVs with $\Omega = 8$ converges better

Fig. 7 APC (**a**), TUM
(**b**) and average connectivity
(**c**) for ship protection using
120 UUVs with $R_{mov} = 5$,
$R_{com} = 10$ and $\Omega = 8, 9,$
and 10

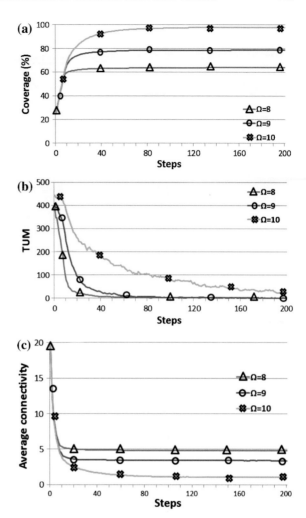

for the first 50 steps compared to the case of $\Omega = 9$. After step 50, the UUVs are almost immobile for $\Omega = 8$ and 9, which means that they have reached stable positions and do not find better positions to move. TUM also decreases as the experiment progresses, when the shield density reaches $\Omega = 10$, but the rate of decrease is lower than when Ω is 8 and 9. Larger Ω values require significantly more steps to reach stable configurations.

Based on Fig. 7a–c, the network reliability is the lowest for $\Omega = 10$ since the average connectivity $D_{avg} \approx 1$. For harsh underwater environments, there is a distinct possibility that an underwater vehicle may malfunction during a mission and create a vulnerability for the protected assets. In our experiments, reducing the shield density (e.g., from $\Omega = 10$ to 9) results in a more compact UWSN, reduced TUM and, hence, extending UWSN lifespan. Incrementing the shield density also improves the

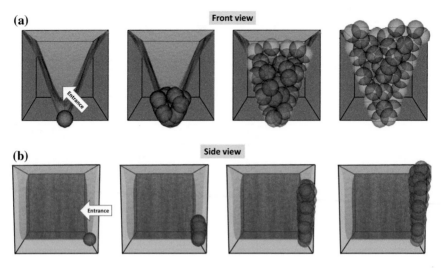

Fig. 8 Front and side views for harbor protection using 50 UUVs with $R_{com} = 10$ at steps 0, 10, 40, and the final step of (200)

average connectivity since more nodes can communicate with each other. However, the increases in shield density come at the cost of the reduced coverage (nodes are more concentrated in smaller areas). Also, we notice that while the value for the shield density Ω is reduced to provide a more robust network, there is a threshold after which increasing shield density does not improve TUM.

Figure 8 shows the simulation results for 50 UUVs with $R_{com} = 10$ spreading to protect the entrance of a harbor. To represent more realistic applications, all nodes are placed at the bottom of the sea floor at the port entrance. This corresponds to a situation where all UUVs are dropped from a maritime vessel at the entrance of a harbor, which begin to spread apart after reaching the sea floor. It should be noted that our 3D-PSO does not require a specific initial deployment location and the performance does not significantly vary for different types of initial deployment conditions. For example it is possible that the initial starting position for UUVs may at the surface of the water or at a location close to the shore.

In Fig. 9a, for planar distribution class of problems where an entry to a harbor to be sealed by the UUVs, we notice that the nodes cover approximately 80 % of the harbor entrance during the first 40 steps for $R_{com} = 10$. The harbor entry is fully protected after 100 steps for R_{com} values of 10 and 12, while APC = 88 % at step 200 for $R_{com} = 8$. For this class of problems, increasing communication range does improve the total coverage, however, larger R_{com} values significantly improve the rates that UUVs converge towards a full coverage. This result is expected since the UUVs can process information from a larger area and make better movement decisions. An additional benefit of the large R_{com} is the higher average connectivity achieved during the spreading process.

Fig. 9 APC (**a**) and TUM (**b**) for port protection using 50 UUVs with $R_{mov} = 5$, and $R_{com} = 12$, 10, and 8

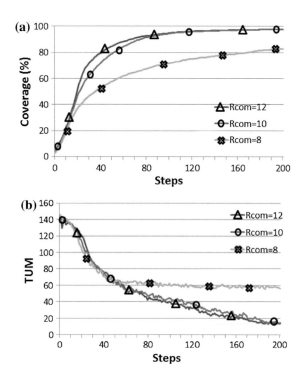

Figure 9b presents TUM values for 50 AUVs with different R_{com} values. Here, the total movement distance for both $R_{com} = 10$ and 12 are almost equal. However, both are better (i.e., smaller) compared to the case when $R_{com} = 8$. UUVs travel almost the same distance during the first 50 steps for all planar distribution experiments. However, when $R_{com} = 8$, the nodes cannot converge to a stable distribution which indicates that the communication range is insufficiently small for this type of mission. When the value of R_{com} is greater than 10, the distance traveled by all UUVs does not significantly change; therefore, increasing R_{com} above a certain value does not necessarily reduce TUM and can potentially decrease network lifespan (due to unnecessary movements) for the planar distribution missions.

Figure 10a shows the network coverage for planar distribution applications with different shield density values. At step 200, UUVs cover approximately 99, 65, and 35 % for Ω values of 10, 9, and 8, respectively. Changing the value of shield density Ω can significantly affect the network coverage for planar distribution type of problems. For example, when the density is reduced by 10 %, the coverage decreases by 30 %. Figure 10b shows that smaller values of Ω results in reduced UUV movement (i.e., the UUVs become stable quicker). In Fig. 10b, for 50 UUVs, to reach TUM = 20 requires 175, 80 and 20 steps for the Ω values of 8, 9, and 10, respectively.

Fig. 10 APC (**a**), TUM
(**b**) and average connectivity
(**c**) for port protection using
120 UUVs with $R_{mov} = 5$,
$R_{com} = 10$ and $\Omega = 8, 9,$
and 10

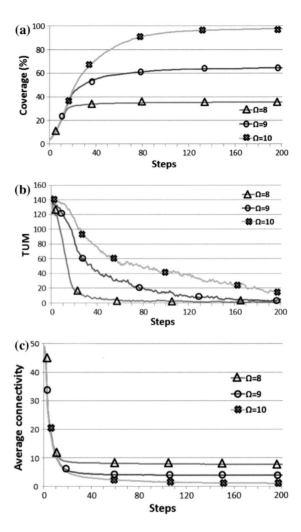

The average network connectivity for planar class of problems is presented in Fig. 10c. Shield density values of $\Omega = 8$, 9, and 10 result in the average network connectivity of $D_{avg} = 8$, 4 and 2, respectively. This result matches with the APC results from Fig. 10a since UUVs with smaller values of Ω cover less planar space. Also, since all UUVs are close to each other for small values of Ω, the connectivity becomes larger. Compared to 3D encapsulation class, we observe that planar distribution problems are more sensitive to changes in Ω values with respect to network coverage. However, reducing Ω in planar distribution problems does not significantly reduce TUM (i.e., the network lifespan) nor increase D_{avg}.

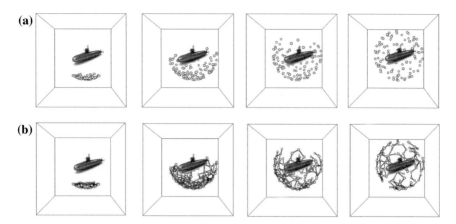

Fig. 11 Submarine protection using 85 UUVs with $R_{com} = 10$ at steps 10, 20, 60, and the final step of 200

Figure 11 depicts 85 UUVs, each autonomously running our 3D-PSO to surround a submarine for protection, where each line between two UUVs represents a local connection. Figure 11a shows that UUVs are initially deployed under the submarine. At step 60, almost 80 % of the area around the submerged asset has been covered by the UUVs. Figures 11b show that the network connectivity becomes more sparse as the experiment progresses, but all UUVs remain connected to each other, and no network partitions occur.

Figure 12 shows the experimental results for the user-defined minimum connectivity values of $k = 2$ and 3. In this experiment, we choose the angles for all cones θ in the Yao graph as 60°. In Fig. 12a, UUVs using our 3D-PSO with $k = 3$ cannot cover the entire protection space, since the larger value of k creates a more tightly bound network topology. However, as shown in Fig. 12b, UUVs travel less when they operate with a large minimum connectivity parameter (e.g., $k = 3$ in our experiment). As the experiment progresses, TUM decreases for both $k = 2$ and 3, which indicates that UUVs reach stable configuration. Figure 12c confirms that using 3D-PSO with a 3D Yao graph inspired fitness function generates an average network connectivity which is larger than the user defined k value.

7 Concluding Remarks

In this article, we present a particle swarm optimization (PSO) based topology control mechanism, called 3D-PSO, allowing UUVs in an underwater wireless sensor network to cooperatively protect valued assets in unknown 3D spaces. Each UUV running our 3D-PSO autonomously makes movement decisions using only local neighborhood information. Using our shield density parameter and 3D Yao graph inspired fitness function, our 3D-PSO can provide a user-defined level of protection for different

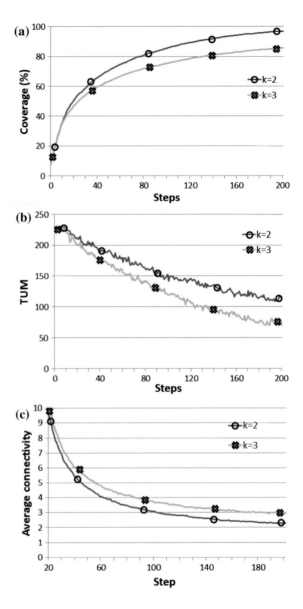

Fig. 12 APC (**a**), TUM (**b**) and average connectivity (**c**) for submarine protection using 85 UUVs with $R_{mov} = 5$, $R_{com} = 10$, $\theta = 60°$, $k = 2$ and 3

maritime vessel applications. Three classes of distribution application in UWSNs are presented and analyzed. In 3D encapsulation class, UUVs uniformly spread over the underside of maritime vessels to detect any hostile or unexpected underwater intrusions. Another application class is the planar distribution of UUVs to create 2D formation for any given dimension of a 3D protection space. In spherical distribution class of applications, the UUVs form a shpere around a given asset such as a fully submerged submarine.

Formal analysis and experimental results for average protection space, total underwater movement, and average network connectivity demonstrate that our 3D-PSO can quickly and efficiently spread UUVs apart uniformly in unknown 3D underwater spaces while providing fault tolerant connectivity within the UWSN.

References

1. Şahin, C.Ş., Uyar, M.Ü., Gundry, S., Urrea, E.: Self organization for area coverage maximization and energy conservation in mobile ad hoc networks. Trans. Comput. Sci. **15**, 49–73 (2012)
2. Urrea, E., Şahin, C.Ş., Uyar, M.U., Conner, M., Bertoli, G., Pizzo, C.: Estimating behavior of a ga-based topology control for self-spreading nodes in manets. In: Proceedings of the International Conference on Military Communications (MILCOM), pp. 1281–1286 (2010)
3. Gundry, S., Zou, J., Urrea, E., Şahin, C.Ş., Kusyk, J., Uyar, M.: Analysis of emergent behavior for ga-based topology control mechanism for self-spreading nodes in MANETs. In: Advances in Intelligent Modelling and Simulation, Vol. 422 of Studies in Computational Intelligence, pp. 155–183. Springer, Berlin, Heidelberg (2012)
4. Kusyk, J., Şahin, C.Ş., Zou, J., Gundry, S., Uyar, M., Urrea, E.: Game theoretic and bio-inspired optimization approach for autonomous movement of MANET nodes. In: Handbook of Optimization, Vol. 38 of Intelligent Systems Reference Library, pp. 129–155. Springer, Berlin, Heidelberg (2013)
5. Kusyk, J., Şahin, C.Ş., Uyar, M.U., Urrea, E., Gundry, S.: Self-organization of nodes in mobile Ad Hoc networks using evolutionary games and genetic algorithms. J. Adv. Res. **2**, 253–264 (2011)
6. Zou, J., Gundry, S., Kusyk, J., Uyar, M.U., Şahin, C.Ş.: 3D genetic algorithms for underwater sensor networks. Int. J. Ad Hoc Ubiquitous Comput. **13**(1), 10–22 (2013)
7. Zou, J., Gundry, S., Kusyk, J., Şahin, C.Ş., Uyar, M.: Particle swarm optimization based topology control mechanism for autonomous underwater vehicles operating in three-dimensional space. In: Special Issues in Marine Robotics Book, Springer (2013)
8. Zou, J., Gundry, S., Uyar, M., Kusyk, J., Şahin, C.Ş.: Bio-inspired topology control mechanism for autonomous underwater vehicles used in maritime surveillance. In: IEEE Conference on Homeland Security Technologies (HST 2013) (2013)
9. Yao, A.C.-C.: On constructing minimum spanning trees in k-dimensional spaces and related problems. SIAM J. Comput. **11**(4), 721–736 (1982)
10. Blidberg, D.R.: The development of autonomous underwater vehicles (AUVs); a brief summary. In: IEEE International Conference on Robotics and Automation, vol. 4. Seoul, South Korea (May 2001)
11. Wernli, R.L.: Low Cost uuv's for Military Applications: Is the Technology Ready? Technical Report, DTIC Document (2000)
12. Rodoplu, V., Meng, T.H.: Minimum energy mobile wireless networks. IEEE J. Sel. Areas Commun. **17**, 1333–1344 (1998)
13. Li, L., Halpern, J.Y., Bahl, P., Wang, Y.-M., Wattenhofer, R.: A cone-based distributed topology-control algorithm for wireless multi-hop networks. IEEE/ACM Trans. Netw. **13**(1), 147–159 (2005)
14. Wang, Y., Cao, L., Dahlberg, T.: Efficient fault tolerant topology control for three-dimensional wireless networks. In: Proceedings of 17th International Conference on Computer Communications and Networks, 2008, ICCCN '08, pp. 1–6 (2008)
15. Chen, B., li Wang, L., jin Ai, Y.: Link interference prediction-based topology control algorithm for 3d wireless sensor networks. In: 2nd International Conference on Information Science and Engineering (ICISE), pp. 2168–2171 (2010)

16. Shi, Y., Eberhart, R.: A modified particle swarm optimizer. In: Proceedings of IEEE World Congress Computational Intelligence. The 1998 IEEE International confrerence on Evolutionary Computation, pp. 69–73 (1998)
17. Wang, Y., Li, F., Dahlberg, T.A.: Energy-efficient topology control for three-dimensional sensor networks. Int. J. Sens. Netw. 68–78 (2008)
18. Gundry, S., Zou, J., Kusyk, J., Uyar, M.U., Şahin, C.Ş.: Fault tolerant bio-inspired topology control mechanism for autonomous mobile node distribution in manets. In: Proceedings of IEEE International Conference on Military Communications (MILCOM), pp. 1–6 (2012)
19. Zou, J., Gundry, S., Kusyk, J., Şahin, C.Ş., Uyar, M., Particle swarm optimization based topology control mechanism for autonomous underwater vehicles operating in three-dimensional space. In: Advanced in Marine Robotics, pp. 9–36. Lambert Academic Publishing (2013)
20. Luke, S., Cioffi-Revilla, C., Panait, L., Sullivan, K., Balan, G.: Mason: a multiagent simulation environment. Simulation **81**(7), 517–527 (2005)